# TEMPERATE FRUITS

## *Volume 2: Nuts and Berries*

## The Authors

**Professor (Dr.) Sisir Kumar Mitra** worked (until 28.02.2014) at the Bidhan Chandra Krishi Viswavidyalaya, Mohanpur, West Bengal, India. He served as Head, Department of Fruits and Orchard Management (1996-2000), Dean, Faculty of Horticulture (2007-2009) and Dean, Post Graduate Studies (2009-2012) of the university. He did Post-doctoral research (1985-1986) at the Horticulture Research International, East Malling, UK as Commonwealth Fellow and worked as Visiting Professor at the Institute Coltivazoni Arboree, Italy, University of Lavras, Brazil, and Kyoto University, Japan, Adjunct Professor, UBKV, Cooch Behar, West Bengal. He has been conferred with Honorary Fellowship by the International Society for Horticultural Science (ISHS), Confederation of Horticulture Associations of India (CHAI) and Fellow Horticulture Society of India (HSI).

Professor Mitra guided 24 Ph.D. and 32 M.Sc. (Horticulture) students and has published 252 scientific papers, 74 popular articles, 11 books, 18 book chapters, 1 IBPGR Descriptors and 28 technical reports.

Served as Member of the International Scientific Committee, Chaired, Technical Sessions and delivered keynote/invited lecture in 43 International events. He was the Chairman, International Society for Horticultural Science (ISHS) Section, Tropical and Subtropical Fruits, (2010-2018). Presently he is serving as Board Member responsible for publication of ISHS, Science Editor, ISHS and Editor-in-Chief, Fruits: The International Journal of Tropical and Subtropical Horticulture. He is also the Chairman, Editorial Advisory Board of Acta Horticulturae and Chronica Horticulturae, Chairman, ISHS Working Group on Avocado, and Jackfruit and other Moraceae. He has been Convener of seven International Symposia and Editor of eight volumes of Acta Horticulturae.

He has served as Technical Advisor of the Board of Trustee of International Tropical Fruits Network of FAO, and as FAO Consultant. He served as Member, in many Research Advisory Committees (RAC) of ICAR Institutes and also a member of QRT. He is Advisor, CAB International, UK, and Member of Editorial Board of many professional journals. He is in the Board of Directors, of the West Bengal State Food Processing and Horticulture Development Corporation Limited, and Technical Advisor in many academic and professional organizations.

# TEMPERATE FRUITS

## *Volume 2: Nuts and Berries*

**Temperate Fruits**

***Edited by: S.K. Mitra, D.S. Rathore and T.K. Bose (1990)***

***Revised and Enlarged by***

**S.K. Mitra**

2020

Daya Publishing House®

*A Division of*

Astral International Pvt. Ltd.

New Delhi – 110002

First Published 1990
Revised and enlarged Edition 2020
ISBN 9789390371228 (Int. Edition)

*Publisher's Note:*

*Every possible effort has been made to ensure that the information contained in this book is accurate at the time of going to press, and the publisher and author cannot accept responsibility for any errors or omissions, however caused. No responsibility for loss or damage occasioned to any person acting, or refraining from action, as a result of the material in this publication can be accepted by the editor, the publisher or the author. The Publisher is not associated with any product or vendor mentioned in the book. The contents of this work are intended to further general scientific research, understanding and discussion only. Readers should consult with a specialist where appropriate.*

*Every effort has been made to trace the owners of copyright material used in this book, if any. The author and the publisher will be grateful for any omission brought to their notice for acknowledgement in the future editions of the book.*



*Published by* : **Daya Publishing House®**
A Division of
**Astral International Pvt. Ltd.**
– ISO 9001:2015 Certified Company –
4736/23, Ansari Road, Darya Ganj
New Delhi-110 002
Ph. 011-43549197, 23278134
E-mail: info@astralint.com
Website: www.astralint.com

# Acknowledgement

Many people have, in one way or another, helped in the production of this book; most were mentioned in the 'preface' in the first edition and I hope they will forgive me if I do not mention them all again. I express my gratitude to all the contributors of the first edition of book from where I worked to publish this revised and enlarged edition.

I like to thank Professor Eddo Rugini, University of Tuscia, Viterbo, Italy, Professor Sadiye Gozlekci, Akdeniz University, Turkey, Late Professor Norman Looney, Pacific Agri-Food Research Centre, Agriculture and Agri-Food Canada, Dr. Damiano Avanzato, International Horticulture Consultant, Rome, Italy, Dr. K.K. Pramanick, IARI Regional Station, Shimla, Himachal Pradesh, India, Professor B. Ghosh, Dr. S. Debnath, Dr. D. Majhi and Sri. S. Sau, Bidhan Chandra Krishi Viswavidyalaya, Mohanpur, West Bengal, India for support, information, photographs and ideas.

I express my sincere gratitude to my respected teacher Professor T.K. Bose who was also the editor of the first edition. He is the one behind this revised edition, felt the needs of students, teachers and researchers and encouraged me to work for compiling with updated information.

My special thanks to Dr. P.K. Pathak, KVK, W.B. University of Animal and Fishery Sciences, West Bengal, India for his editorial assistance without which possibly I could not have finished.

*Professor S.K. Mitra*

# Preface to the Revised and Enlarged Edition

This book was first published in 1991 and was well received by students, teachers and researchers in many universities and institutes in India and abroad. In the past 28 years ( since its first publication), tremendous progress have been made in temperate fruit research, notably in developing varieties and rootstocks, canopy management, understanding the flowering physiology, storage technology, biological control of pests and diseases, resistance breeding, biotechnology, *etc.*, Revision of the book with updated information was considered necessary.

In the process of a comprehensive literature survey it was felt impossible to compile all the information in one volume. This enlarged edition on temperate fruits is, therefore, published in two volumes. The first volume consists of pome and stone fruits (apples, pears, peaches, plums, apricots and cherries) in more than 550 pages. The second volume deals with temperate nuts and berries (almonds, chestnuts, hazelnuts, pecans, pistachios, walnut, strawberries, currants, gooseberries, raspberries, blueberries and cranberries) in about 650 pages.

Apart from thorough revision with updated scientific information on various aspects of mentioned pome, stone, nuts and berries, substantial reorganization of the text has been made. Emphasis has been laid to include released cultivars and rootstocks, recent trends in propagation and orchard management, pruning and canopy management, nutrition and tissue analysis, water management, improved technology for harvesting, postharvest handling, storage and ripening. Special emphasis has been laid to include the development in breeding and improvement through biotechnological approaches like molecular markers, genetic transformation, genetic map and microbes for N and P input. Other information like organic production technology, protected cultivation and impact of climate change are included, where sufficient information is available.

Though the subject is vast, this two volumes book presents the concepts in a condensed, informative and lucid manner. It will be of immense help to the students, teachers and researchers of horticulture, plant science, plant physiology and plant protection.

***Professor S.K. Mitra***

# Preface of the First Edition

The present title has been planned and designed to meet the long-felt need for a book covering major aspects of important temperate fruits. The considerable large volume of research done on these fruits and the wealth of information accumulated over the years made the publication of such a volume imperative. It thus becomes a useful textbook for the students of horticulture all over the world as well as important reference book for the researchers on this subject.

As a well-prepared and comprehensive textbook, the title incorporates recent literature, updated scientific and general information and important tables that in many cases, have not compiled before. As such, the book will make the readers conversant with the current researches and development in this field.

I am grateful to all the contributors for their efforts and care with which they prepared the different chapters of the book. I express my gratitude Dr. S.K. Mitra and Dr. D.S. Rathore, the co-editors who helped me immensely in editing as well as during printing process. I wish to record my appreciation to my colleagues-Prof. N. Mukherjee, Dr. B. Ghosh and also to Dr. B. Chowdhury of Horticulture and Allied Publishers, Mr. K. Dasgupta, Mrs. K. Neyogi and Mr. S. Bhowmick for their help.

This publication has been financed by United Bank of India, Old Court House Street Branch, Calcutta and I am thankful to Mr. J.V. Shetti, Chairman and Managing Director of the Bank for his interest and cooperation.

**Calcutta** *T.K.Bose*

**August, 1991**

# Contents

## Volume 1
## Temperate Fruits: Pome and Stone Fruits

# Contributors of Temperate Fruits: Nuts and Berries Sections (1990)

**Bist, H.S.**

Department of Fruit Culture and Orchard Management, Dr. Y.S. Parmar University of Horticulture and Forestry, Solan 173230, Himachal Pradesh.

**Chauhan, J.S.**

Department of Fruit Culture and Orchard Management, Dr. Y.S. Parmar University of Horticulture and Forestry, Solan 173230, Himachal Pradesh.

**Jindal, K.K.**

Regional Horticulture Research Station, Dr. Y.S. Parmar University of Horticulture and Forestry, Mashobra, Shimla 171008, Himachal Pradesh.

**Karkar, B.K.**

Regional Horticulture Research Station, Dr. Y.S. Parmar University of Horticulture and Forestry, Mashobra, Shimla 171008, Himachal Pradesh.

**Mehta, K.**

Department of Fruit Culture and Orchard Management, Dr. Y.S. Parmar University of Horticulture and Forestry, Solan 173230, Himachal Pradesh.

**Mitra, S.K.**

Faculty of Agriculture, Bidhan Chandra Krishi Viswavidyalaya, Mohanpur, West Bengal 741 235,

**Monastra, F.**

Instituto Sperimentale per la Fruitticoltura, MAF, Rome, Italy

**Pathak, P.**

Faculty of Agriculture, Bidhan Chandra Krishi Viswavidyalaya, Mohanpur, West Bengal 741 235,

**Rathore, D.S.**

National Bureau of Plant Genetic Resources, Regional Station, Phagli, Shimla, Himachal Pradesh 171 004

**Rugini, E.**

Universita degli studi della Tuscia, Instituto di ortofloroarboricultura, 01100 Viterbo, Italy.

**Sharma, S.D.**

Regional Horticulture Research Station, Dr. Y.S. Parmar University of Horticulture and Forestry, Sharbo, Rekong Peo, Kinnaur, Himachal Pradesh.

**Thakur, B.S.**

Regional Horticulture Research Station, Dr. Y.S. Parmar University of Horticulture and Forestry, Kandaghat, Solan 172215, Himachal Pradesh.

**Thakur, K.S.**

Department of Postharvest Technology, Dr. Y.S. Parmar University of Horticulture and Forestry, Solan 173230, Himachal Pradesh.

## Chapter 7

# Almonds

The almond is one of the major and oldest, tree-nut crops in the world. World production of almond reached 1.09 million metric tonnes. In many countries, almond is cultivated in marginal lands under non-irrigated and non-intensive culture conditions, by using cultivars selected in different environments or by employing low quality ecotypes, with consequent low yield. Nevertheless, when almond is cultivated in fertile soils and under good irrigation conditions, very high yields are recorded. The USA and Spain produce more than 50 per cent of the world production, followed by Italy, Iran and Syria. Both USA and Spain produce highly marketable nuts, mostly from young orchards, combining uniformity, low moisture content and good size.

## Composition and Uses

Almonds are a nutrient-dense source with respect to lipids, protein, dietary fibre and minerals (Yada *et al.*, 2011). Protein, ranging from 16 per cent to 22 per cent of kernel weight between cultivars, with the major type globulin (about 74 per cent) and albumin (about 21 per cent). Amino acids, mainly found in almond protein-amandin, with low content of free amino acid (> 200mg/100g). Lipids, accounting for more than 50 per cent of kernel weight, with the majority of unsaturated fatty acids. The contents of oleic acid are about 70 per cent and that of linoleic acid is about 20 per cent. Vitamins, lipid-soluble (lipophilic) E type excessively in rich, with the major homologue α-tocopherol, and other manor homologues γ, β- tocopherols, α-tocotrienol; and water soluble (hydrophilic) B type in small amount, (thiamine $B_1$, riboflavin $B_2$, niacin $B_3$, pantothenic acid $B_5$, pyridoxine $B_6$, biotin $B_7$, folate $B_9$). Dietary fibres, making up 10.8 per cent -13.5 per cent of the total kernel, including pectin, cellulose and xyloglucans. The starch and sugar are, however, very low. The kernel, contents a range of minerals such as calcium, iron, magnesium, phosphorus, potassium, zinc, manganese and selenium of which high in potassium (705mg/100g) and phosphorus (484mg/100g).

The high proportion of monounsaturated fatty acid (MUFA) (oleic acid C18:2) and polyunsaturated fatty acid (PUFA) (linoleic acid C18:2) are attributed to almond's cardiovascular health benefits (Wein *et al.*, 2010, Jaceldo-Siegl *et al.*, 2011). The abundance of vitamin E in almond lipids play an outstanding role portraying almond's putative health profile.

The nuts are usually eaten when they are ripe. Kernels obtained after blanching, roasting, frying and salting are highly delicious and in great demand. Almond oil is used in confectionery and also for pharmaceutical and cosmetic preparations. Almond butter is free from starch which has good demand for diabetic food. However, green nuts can also be consumed. Green nut are picked before the outer husk begins to lose its colour and harden. The nuts are then opened and the kernels eaten in the milky stage with lightly sugared cream cheese.

A precise evaluation of the quantity of almonds used in the confectionery industry is impossible but it is probably about 75-80 per cent, while 20-25 per cent are used for direct consumption (Anon., 1981) including use in cookery (Woodroof, 1967).

Bitter almond kernels are used in perfume, pharmaceuticals and a small amount is used for making pastries. They contain amygdalin, a cyanogenic glucoside which differs from all the other known cyanogenic glucosides in having gentiobiose as its sugar unit (Harborne, 1973). The by-products of almonds are used in chemical industry.

## Origin and Distribution

Cultivated almond *Prunus dulcis* (Miller) D.A. Webb. (syn. *Prutrus amygdalus* Batsch.) belongs to the family Rosaceae, it is native to central Asian mountain areas (India, Iran and Pakistan), where it is still possible to find many wild species. In ancient times the species spread as far as the Mediterranean basin. Almond introduction in the Mediterranean regions, occurred through the Phoenicians, and around 450BC almond culture spread from Greece to all Mediterranean coast, with several important centres appearing in the Iberian Peninsula, France, Italy, Morocco, Tunisia and Turkey (Kester and Ross, 1996).

According to Russian scientists (Popov, 1929; Vavilov, 1930) cultivated almond is a hybrid between two species: *Prunus fenzliana* and *P. bucharica*, because both are closely related to the cultivated almond. In fact, it is possible to find natural hybrids of *P. bucharica* with a sweet kernel. Probably other wild species such as *Prunus webbii* spread in many countries (Alberghina, 1978, Godini, 1979), have contributed genes to the cultivated species, including the rare characteristic of self-compatibility found in some cultivars in Italy (Crescimanno, 1960; Godini, 1979) and in Spain (Felipe and Socias I Company, 1977).

Until the end of the 1800s, propagation by seeds, derived almost exclusively from cross-pollination, favoured the appearance of a high number of geographic groups 'ecotypes' making it difficult to identify varieties from genuine hybrid strains (Grasselly and Ciossa-Raynaud, 1980). These ecotypes continue to increase in number even now, due to the continuing use of this propagation method in some countries. In many other countries, however, the use of multiplication method by grafting has not succeeded efficiently in reducing variability among commercial almond kernels (different shell consistency, size and taste), because ecotypes without an efficient selection have been used.

Now, in more agronomically advanced countries, the tendency is to replace old orchards with new ones by using a few cultivars, grafted onto clonal rootstocks. This practice leads to the loss of invaluable germplasm and, therefore, it is necessary to increase the number of germplasm collections or to enlarge the existing ones (Egea *et al.*, 1984).

Almond is cultivated mainly in countries between the latitudes 36° and 45° N, although under certain mild climatic conditions, cultivation may extend further north. It is also found in South Africa, South America, Australia and India. The limiting factor in almond cultivation is spring frost, particularly during full bloom, when it is often responsible for an irregular yield, even if growers try to grow several cultivars with different flowering dates in the same orchard.

In the past, and even now in some Mediterranean countries, the almond has been confined to marginal and non-irrigated lands, being considered a robust species. In these countries, therefore, the almond production has remained constant or has fallen. The situation in the USA is completely different. Apart from an initial period from the early 1830s, when the plant was introduced from Europe, to the 1930s, when cultivation was still confined to marginal lands, American growers began to plant almond in more fertile and irrigated areas, using the most advanced agronomic methods. In addition, in order to adapt ideal varieties to their environmental conditions, they rapidly selected them from the offspring of crosses between European ecotypes which showed a low adaptability in the USA. As a result, the USA has become the main world producer of shelled almonds. The US production derives from soft and paper-shell types, mostly from a small number of cultivars such as: Nonpareil (55 per cent), Merced (7.5 per cent) and Ne Plus Ultra (6.5 per cent).

## Area and Production

The USA is the dominant producer of almond in the world, producing 898,167 MT annually. California is the primary producing region, about 70 per cent of the almond is exported in shelled form and the rest processed or sold unshelled.

Spain is the second largest producer, after the USA. Spain produces 229,959 MT annually. It grows in the region of Valencia, Catalonia, Andalusia, Murcia and Aragon. The new orchards are with cvs. Ferragnes, Ferradual and Cristomorto.

Until the early 1960s, Italy was the main producer of almonds in the world, but gradually more profitable crops such as citrus, peaches and grapes have replaced almond orchards; nevertheless the south of Italy is very suitable for almond cultivation and still produces a good amount of the world almond. Italian production comes mainly from the old orchards containing a mixture of hard-

and semi-hard-shell cultivars, such as, Tuono, Genco, Filippo Ceo, Avola and Fascionello. There are few old specialized orchards, most of them are consociated with olive and herbaceous plants. Only a few new orchards have been planted in the recent years and Italy does not produce enough for its own use, although a list of recommended cultivars and ideal areas has been suggested from long time (Monastra and Fideghelli, 1977; Monastra *et al.*, 1980). The other important almond producing countries are Iran, Syria, Morocco, Tunisia, Turkey, Australia and Greece (Table 7.1).

**Table 7.1: Top Ten Almond Producing Countries**

| *Countries* | *Annual Production (MT)* |
|---|---|
| USA | 898,167 |
| Spain | 229,959 |
| Italy | 100,664 |
| Iran | 99,551 |
| Syria | 84,485 |
| Morocco | 72,360 |
| Tunisia | 49,605 |
| Turkey | 49,048 |
| Australia | 46,381 |
| Greece | 45,150 |

## Taxonomy

There is a good deal of diversity in the views of almond classification; in fact, it has been classified in the genera *Prunus* and also in the genera *Amygdalus*. According to Rehder (1921), the genera *Prunus* is subdivided into the subgenera *Amygdalus*, to which belong *P. persica*, *P. armeniaca* and others, including the cultivated almond, *Prunus amygdalus* or *P. dulcis* to distinguish from variety *amera* which has bitter taste due to high content of glucoside cyanogenetic amygdalin. In the last century many wild almonds have been classified as species. Since classification is ardous, due mainly to common self-sterility of almonds which cause their offspring to be very heterogeneous, most of them, very probably, are only varieties.

The cultivated almond is a tree reaching heights exceeding 20m and with a few exceptions, of upright bearing. The branches are glabrous and when they are one-year-old are pale green to reddish brown in colour. In most Mediterranean varieties the leaves have petioles and are lanceolate or ovate with an acute apex. The tree can survive for 100 years or more.

## Cultivars

**California Paper shell:** Tree upright, bear's flowers on both spurs and on long shoots. Regular bearer, matures about 150-155 days of full bloom.

**Cristomorto**: Tree upright and vigorous, flowers borne mostly on spurs, self-incompatible and highly productive. Main pollinators are Rachele and Texas.

**Almond Cultivar Ne Plus Ultra**

**Almond Cultivar Nonpareil**

**Verticillium Wilt**

**Almond Flower**

**Almond tree in bearing**

**Desmayo Larqueta**: Spreading tree, moderately vigorous, flowers born on spurs or on one-year old shoots, self-incompatible and moderate yielder. The pollinator cultivars are Marcona, Ne Plus Ultra and Nonpareil.

**Fascionello:** Tree upright and moderately vigorous, flowers borne mostly on spurs, self-incompatible and medium yielder. The cvs. Pizzuta Aovala and Romana are good pollinators.

**Ferragnes:** Tree upright, moderately vigorous, flowers borne on spurs and on 1-year-old shoots, self-incompatible and high yielder. The cvs. Cristomoto, Ferraduel, Filippo Ceo, Texas and Tuono are good pollinators.

**Filippo Ceo**: Tree upright, moderately vigorous, flowers borne only on spurs, self-compatible and high yielder. The main pollinators are Genco and Tuono.

**Fra Giulio Grande:** Tree spreading, moderately vigorous, flowers borne only on spurs, self-compatible and high yielder. The cultivars Ferruduel, Ferragnes and Ferrdstar are good pollinators.

**Genco:** Tree spreading and vigorous, flowers borne mostly on spurs, self-compatible and highly productive. The main pollinators used are Ferrdstar, Filippo Ceo, Texas and Tuono.

**Makhdoom:** Tree growth habit is spreading or drooping. It bears flowers on spurs and also one-year-old shoot. Regular bearer, medium yielder, fruit matures after 140-145 days of full bloom.

**Marcona:** Tree drooping and vigorous, flowers borne on spurs or on one-year-old shoots, self-compatible and medium yielder. The cultivars Ne Plus Ultra, Nonpareil and Texas are good pollinators.

**Ne Plus Ultra:** Tree spreading vigorous, flowers borne on spurs or one-year-old shoots, self-incompatible and medium yielder. The cultivars Desmayo L, Marcona and Nonpareil are the main pollinators.

**Nonpareil:** Tree spreading, moderately vigorous, flowers borne on spurs or on one-year-old shoots, self-incompatible and medium yielder. The main polinator are Desmayo L, Drake, Marcona, Ne Plus Ultra. In India it is a regular and early maturing cultivar, mature after 140 days of full bloom.

**Peerless:** Tree spreading and vigorous, flowers borne mostly on spurs, self-incompatible and medium yielder. Main pollinators are Ne Plus Ultra and Nonpareil.

**Pizzuta d' Avola:** Tree spreading and vigorous, flowers borne mostly on one-year-old shoots, self-incompatible and medium yielder. The cultivars Fascionello and Romana are good pollinators.

**Shalimar:** Tree spreading or drooping. It bears flowers on long shoots and spurs. A regular bearer cultivars, ready to harvest after 142-145 days of full bloom.

**Tardy Nonpareil:** Tree spreading and vigorous, flowers borne on spurs or one-year-old shoots, self-incompatible and medium yielder. The cultivars Al and Ferradstar are good pollinators.

**Texas:** Tree upright and vigorous, flowers borne on spurs or one-year-old shoots, self-incompatible and medium yielder. Good pollinator cvs. are Cristomoto, Tuono, Ferrudual and Ferragnes.

**Tuono:** Tree spreading and vigorous, flowers borne only on spurs, self-incompatible and high yielder. The main pollinators are Ferragnes, Fillippo Ceo, Fra Giulio G, Genco and Texas.

**Vesta:** Tree spreading and vigorous, flowers borne on spurs or one-year-old shoots, self-incompatible and medium yielder. The cvs. Drake, Marcona and Nonpareil are good pollinators.

**Waris**: Tree upright, suitable for high density planting, regular bearer, bears flower on long shoots and spurs.

## New Cultivars

### Developed from United States

**Durango:** Tree medium, upright-spreading, vigorous growth habit. Peerless-type kernel; paper shell, low proportion of double kernels.

**Kochi:** Tree medium, upright-spreading, vigorous growth habit, flowers born on spurs and terminal shoots. Nonpareil-type kernel; low proportion of double seed coat.

**Sweetheart:** Tree upright-spreading, moderate vigorous, flowers born on spur and on terminal shoots. Nuts are well-sealed paper shell; readily blanched and low proportion of double kernel. It is partially self-compatible and cross-compatible with major cultivars.

**Winters:** Tree upright and vigorous, flowers borne on spurs and on terminal shoots, cross-compatible with Nonpareil. Nuts similar to Nonpareil in shape and moderately sealed paper shell.

### Developed from Spain (CITA de Aragon czaragoza)

**Blanquerna:** (Genco-open pollinated), self-compatible, large kernel, excellent quality.

**Cambra:** (Ferragnes x Tuono), self-compatible, hard shell.

**Felisia:** (Titan x Tuono), self-compatible, small kernel.

**Belona:** (Blanquerna x Belle d' Aurons), self-compatible, large kernel.

**Mardia:** (Felisia x Bertina), self-compatible, disease tolerant, very high content of oleic acid and low linoleic acid (Socias-I-Company *et al.*, 2008).

**Marinada:** (Lauranne x Glorieta), self-compatible, good yielder.

**Vairo:** (Primorskij x Cristomorto) x Lauanne, self-compatible, high vigour, medium branching.

**Marta:** (Ferragnes x Tuono), self-compatible, upright tree, high vigour.

**Penta:** (S5133 x Lauranne), self-compatible, hard shell.

**Table 7.2: Industrial Potential of Blanched Kernels of each Almond Cultivar, Sorted Alphabetically (Average from 2005 to 2008 of cultivars grown in three locations, three trees per cultivar and place)***

| *Cultivar* | *Hard Turron* | *Soft Turron* | *Marzipan* | *Toasting Colour* | *Chocolate Bars* | *Slices* | *Sticks Length* | *Flour* | *Light Foods* | *Shelf Life* |
|---|---|---|---|---|---|---|---|---|---|---|
| Constanti | medium | medium | high | whitish | high | low | short | high | low | very long |
| D'Largueta | medium | medium | medium | brown | low | high | long | high | low | long |
| Francoli | medium | medium | medium | brown | high | medium | medium | high | low | very long |
| Glorieta | high | high | high | yellow | medium | low | long | high | low | very long |
| Guara | medium | high | high | whitish | high | low | medium | high | low | long |
| Lauranne | medium | high | high | brown | high | high | medium | medium | low | long |
| Marcona | high | high | high | whitish | high | low | short | high | low | medium |
| Marinada | low | low | low | brown | high | medium | medium | high | medium | low |
| Marta | medium | high | very high | - | medium | high | long | medium | low | - |
| Masbovera | medium | high | high | yellow | medium | low | long | high | low | very long |
| Nonpareil | low | low | very low | brown | high | high | medium | medium | medium | very low |
| Tarraco | medium | high | high | yellow | low | medium | long | high | low | medium |
| Vairo | medium | medium | medium | yellow | high | high | medium | high | low | long |
| 12-350 | high | high | high | whitish | medium | low | medium | high | low | very long |
| 13-531 | high | high | high | yellow | high | high | medium | medium | low | long |
| 24-53 | low | low | low | whitish | high | low | long | high | medium | low |
| 23-173 | low | low | low | brown | medium | medium | long | high | low | long |

[x] Colour of blanched and toasted kernels.

* Romero *et al.* (2011).

**Table 7.3: Nut and Kernel Characteristics of the some Almond Cultivars***

| *Cultivar* | *Shape* | *No. Nuts (Kernels) kg* | *Long (mm)* | *Large (mm)* | *Thickness* | *Shell Softness* | *Shelling Yield (per cent)* | *Double Kernel (per cent)* | *Kernel Colour Intensity* | *Kernel Shrivelling* |
|---|---|---|---|---|---|---|---|---|---|---|
| Cristomorto | cordate (elliptic) | 222 (625) | 35 (23) | 23 (14) | 18 (8) | hard | 36 | 20-25 | intermediate | intermediate |
| Desmayo L. | cordate (oblong) | 218 (768) | 39 (28) | 24 (14) | 15 (8) | hard | 29 | 0 | dark | intermediate |
| Fascionello | cordate (elliptic) | 141 (589) | 42 (31) | 25 (15) | 18 (8) | stone | 24 | 2-10 | light | intermediate |
| Ferragnes | cordate (oblong) | 286 (664) | 36 (29) | 21 (13) | 16 (8) | hard | 41 | 0 | light | intermediate |
| Filippo Ceo | cordate (elliptic) | 237 (622) | 32 (23) | 22 (13) | 19 (10) | hard | 37 | 40-45 | light | slightly wrinkled |
| Fra Giulio G. | cordate (oblong) | 169 (255) | 39 (27) | 22 (13) | 16 (7) | hard | 30 | 5-20 | light | intermediate |
| Genco | oblong ovate (ovate) | 292 (831) | 27 (20) | 21 (14) | 17 (8) | hard | 35 | 0-3 | intermediate | intermediate |
| Marcona | round (round) | 178 (664) | 31 (21) | 25 (16) | 19(9) | hard | 27 | 0-3 | dark | intermediate |
| Ne Plus Ultra | cordate (oblong) | 398 (664) | 42 (27) | 20 (12) | 14 (8) | paper | 61 | 29 | intermediate | intermediate |
| Nonpareil | cordate (oblong) | 475 (712) | 33 (25) | 19 (13) | 12 (8) | paper | 65-70 | 0-2 | light | slightly wrinkled |
| Peerless | cordate (elliptic) | 286 (712) | 35 (24) | 23 (13) | 17 (12) | semi-hard | 40 | 11 | intermediate | intermediate |
| Pizzuta d'Avola | cordate (oblong) | 154 (712) | 45 (29) | 24 (16) | 13 (6) | stone | 21 | 3 | dark | intermediate |
| Tardy Nonpareil | cordate (oblong) | 499 (831) | 34 (24) | 19 (12) | 13 (7) | paper | 61 | 0 | light | slightly wrinkled |
| Texas | cordate oblong (elliptic) | 330 (664) | 29 (22) | 20 (13) | 17 (9) | semi-hard | 50 | 39 | dark | intermediate |
| Tuono | cordate round (oblong- elliptic) | 261 (622) | 35 (24) | 24 (14) | 18 (8) | hard | 41 | 30 | light | intermediate |
| Vesta | cordate (oblong-elliptic) | 415 (664) | 21 (13) | 21 (13) | 15 (9) | paper | 62 | 0 | light | slightly wrinkled |

* Monastra *et al.* (1982).

Romero *et al.* (2011) tested industrial potential of new almond cultivars released from Institut de Recerca i Tecnologia Agroalimentaries (IRTA), Spain. The studied cultivars were Constanti, Francoli, Glorieta, Marinada, Masbovera, Tarraco, Vario, and the advanced new selections 12-350, 13-531, 23-173, 24-53 and 29-148. Reference cultivars were also tested to compare for four years (2005-2008). Raw, blanched, and roasted nuts were analyzed. There were significant differences in potential industrial applications of different cultivars. (Table 7.2) Nuts and Kernels characters of some almond cultivars were described by Monastra *et al.* (1982) in Table 7.3. Some of the new cultivars have better nutritional value and shelf life. The physic-chemical composition of some the new cultivars bred from the IRTA's (Vargas *et al.*, 2011) are presented in Tables 7.4 and 7.5.

**Table 7.4: Nut Characteristics of New and Reference Almond Cultivars: Kernel weight (g), kernel yield (shelling percentage, per cent), double kernels (per cent), and kernel appearance (scale 1 to 9, with 9 the highest mark). Mean values of samples analyzed during 7 to 30 years.***

| *Cultivar* | *N*[1] | *Kernel Weight (g)* | *Kernel Yield (per cent)* | *Double Kernel (per cent)* | *Kernel Appearance (scale 1-9)* |
|---|---|---|---|---|---|
| New: | | | | | |
| Vairo | 37 | 1.20 | 29 | 0.1 | 7.0 |
| Constanti | 42 | 1.19 | 27 | 1.1 | 6.2 |
| Marinada | 32 | 1.32 | 31 | 0.3 | 6.8 |
| Tarraco | 27 | 1.67 | 32 | 0.1 | 6.9 |
| Reference : | | | | | |
| D'Largueta | 85 | 1.34 | 27 | 1.4 | 6.7 |
| Marcona | 209 | 1.33 | 26 | 2.7 | 6.5 |
| Masbovera | 180 | 1.35 | 28 | 0.4 | 6.4 |
| Ferragnes | 262 | 1.49 | 34 | 0.1 | 6.4 |
| Guara | 89 | 1.33 | 35 | 11.4 | 6.3 |

[1]Number of samples analyzed.

* Vargas *et al.* (2011).

### Developed from France

**Stelette:** (Ferranges x Tuono), self-compatible, semi-hard shell, medium vigour.

**Mandaline:** (Ferragens x Tuono), self-compatible, hard shell, tall and upright tree.

## Climate

Favourable environmental conditions are essential for success in the cultivation of almond.

**Table 7.5: Chemical Composition of Blanched Kernels from New and Reference Almond Cultivars, Mean Values of Three Locations and Four Years***

| *Cultivar* | *Oil (per cent)* | *Protein (per cent)* | *Soluble Sugars (per cent)* | *Total Fiber (per cent)* |
|---|---|---|---|---|
| *New:* | | | | |
| Constanti | 53.2 | 23.9 | 2.6 | 9.3 |
| Marinada | 47.9 | 24.2 | 4.1 | 11.2 |
| Tarraco | 53.4 | 24.6 | 2.9 | 9.3 |
| Vairo | 52.7 | 24.5 | 3.0 | 9.0 |
| *Reference :* | | | | |
| Marcona | 54.2 | 25.7 | 2.2 | 8.8 |
| Nonpareil | 45.5 | 25.3 | 2.1 | 9.7 |
| Masbovera | 54.9 | 23.0 | 2.7 | 7.8 |

[1]Romero *et al.*, 2009.

## Temperature

Temperature is the most important climatic factor and almond requires a mean winter cold for a regular bud sprouting in spring; has a moderate tolerance to low temperature during winter and possesses a low tolerance to early spring frost because of its early blossom.

The large selection of ecotypes, carried out by man in the course of centuries, has allowed them to adapt to local climate. However, in very hot areas, *i.e.*, in North Africa, a delayed opening of buds and a fall of flower buds may occur after a too mild winter (Grigorian, 1975). In Spain, Tabuenca *et al.* (1972) have put 38 cultivars of almond into 5 groups according to their requirements for winter chilling and warm temperature in early spring. The chilling requirement for normal bud opening depends on the cultivars and varies from 100 to 700 hours at below 7.2°C (Tabuenca, 1979). Bud dormancy has been related to high concentration of total phenols in the buds (Fadl *et al.*, 1977).

As regards the tolerance to low winter temperature, almond can tolerate temperature as low as –20°C. A fall, but not a sudden fall, to this temperature calls for a greater resistance. Resistance to cold is closely related to bud dry matter content and to high levels of oligosaccharides.

Analysis of air temperatures in the Ebro Central Valley (Zaragoza, Spain) for 24 years shows drastic changes in temperature during winter and spring and high variability among years (Alonso *et al.*, 2011). Under these conditions, there is an excess of chill and most almond cultivars satisfy their chilling requirements very early, during the first two weeks of December, when endodormancy finishes. As a consequence, bloom dates are ruled mainly by the heat requirements. Early-blooming cultivars showed a higher variation in dates of full bloom because they are more dependent on warm temperatures during January and February, where heat accumulation is slower. Late-blooming cultivars showed more stable blooming dates because their greater heat requirements are quickly satisfied by the higher March temperatures.

The limiting factor in almond cultivation is the spring frost, especially during full bloom or immediately after fruit set. Following dormant period, with the completion of meiosis in the pollen sacs (Denisov, 1968) in the flower buds results in a decrease of oligosaccharides (Jablonskij and Markovich, 1970) and consequently in a decrease of resistance to cold resulting in a damage to the ovaries and to the stamen base (Milella, 1963). The most unfavourable conditions, however, are a rapid fall of temperature after a warm and humid period. The tolerance to low temperature during full bloom is dependent on the cultivar; the cvs. Jordanolo, Nonpareil and Ne Plus Ultra (Grasselly and Crossa-Raynaud, 1980), Etnea, Nuciddara and Desmajo rojo (Alberghina and Continella, 1981) have been found to be the most resistant. South facing slopes are considered the best for almond growing, but also flat land may be suitable, if well breezy. For those areas where spring frost is common during spring time, late blossoming almond cultivars are obviously the most suitable.

## Light

Although the almond is cultivated in regions where light intensity does not represent a limiting factor, it is not suitable for high-density planting. In particular, certain varieties, observed by Grasselly and Crossa-Raynaud (1980), are 'photosensitive' if grown in a low light intensity, have poorly lignified branches and a harder endocarp.

## Rainfall

Almond when grafted on almond seedlings, can tolerate high summer temperatures and very low humidity better than any other temperate fruit tree. With its deep root system up to 3 m and leathery leaves resistant to the drying up, almond can withstand long periods of summer dry weather, typical of the Mediterranean climate. In these areas almond trees are not usually irrigated because the water is used for other more profitable crops. So, in areas having a mean rainfall below 250 mm a year, almond trees are planted at a wide spacing (50-60 trees per hectare). The planting density increases with the increase in annual rainfall. Although some varieties differ, almond behaves as a 'xerophyte'; however, in order to obtain a reasonable almond yield under semi-arid conditions, irrigation is required. With adequate irrigation it is possible to increase the yield 3-4 times, either because the grower can plant the tree at a higher density or because the tree is not subject to water stress, which results in early leaf senescence and reduction in growth (Fereres *et al.*, 1981). It is important for almond to irrigate at the right moment, particularly when it is grafted on peach rootstock. In fact, if not irrigated in the drought season, almond falls in the quiescent phase which preserves it to the water stress. On the contrary if the phase is interrupted by a single irrigation almond shows an adverse effect and may also die. Severe water stress together with high temperatures during summer can reduce hull split and consequently promote the occurrence of hull-tight nuts with higher percentage of empty and wrinkled kernels (Palasciano *et al.*, 2014). However, in some cultivars, it has been found an inverse relationship between the quality of irrigation water supplied and the oil content (Romojaro and Berenguer, 1984).

## Soil

Almonds are often grown in infertile soil. Therefore, a soil analysis should be carried out to correct any nutrient deficiencies before planting. Leaf analysis is also useful for the diagnosis of mineral nutrition, even though leaf composition depends on many other factors, such as, soil, root distribution, irrigation, stage of development, probably not closely dependent on cultivars (Averna *et al.*, 1984). Furthermore, leaf P could be superior in mycorhized plants (Roldan-Fajardo *et al.*, 1982).

## Rootstock

### Almond Seedling

Almond seedling rootstock from the bitter or sweet cultivars are used because of their tolerance to drought, lime soils and iron chlorosis, and their longevity, but their roots are susceptible to several diseases especially in irrigated soils. In winter climates and irrigated areas, peach seedlings are used. They have more uniform vigour and survival during transplanting is higher. (Grassely and Crossa-Raynaud, 1980). The seeds of the cv. Nemaguard, which have a thinner shell and is resistant to nematodes; and cv. Lovell which produces uniform seedlings, are particularly used in the USA. In France GF 305 is used, although it is susceptible to *Agrobacterium tumefaciens* and the root-knot nematodes (Bernhard and Grasselly, 1969). In general, peach seedlings are more susceptible to viruses which reduce the percentage of successful grafts or the life of plants, and they are less lime-tolerant, so may not live very long.

### Selected Rootstocks

Peach x almond hybrids (GF 677; GF 557); plum (Myrobalan 2032; Marianna 2624; Marianna GF 8/1); and hybrids of myrobalan x peach or myrobalan x almond.

**Hybrids of peach x almond**: GF 677 hybrid is almost exclusively used in intensive almond orchards (Monastra, 1972, 1976), on well-drained soils containing 5 per cent to 12 per cent active lime (Grasselly, 1973). It induces rapid growth and precocity (Monastra and Pennone, 1980). This group includes Hansen, Nickels, Bright's, Cornerstone, Titan, *etc.* As a group, these rootstocks are very vigorous, have good anchorage, are relatively drought tolerant and are highly resistant to root knot nematodes.

It can be propagated by leafy softwood cuttings, collected from a hedge of mother plants in June or July, by mist propagation system with two longitudinal basal woundings and treated with 1000-2000 mg/l IBA plus a fungicide. Rooting percentage usually is not very high. However, Rodriguez Navarro (1981) reported high rooting percentage (99 per cent) by using woody cuttings chilled for a total of 560 h (40 h below 7°C in the field + 92 h stratificition below 7°C in sand + 428 h at 4°C and 75 per cent RH) prior to dipping in IBA (1000 ppm at the basal end and 200 ppm at the distal end) and placed in perlite with 20°C basal heating.

Loreri *et al.* (1985) after studying the seasonal rooting ability of GF 677 under mist, found that the natural rooting potential was very high in October when leaf drop was around 50 per cent. They found, however, that hardwood cuttings, propagated with bottom heat, gave the best result.

At present, however, it is propagated almost exclusively *in vitro*.

## Plums

Of the potential rootstocks only Marianna 2624 and Marianna GF 8/1 are used for intensive orchards. Both are used for poorly-drained soils subject to water logging (Bernhard and Grasselly, 1969); both show good resistance to nematodes and to *Armillaria mellea* and are easily propagated asexually. They reduce vigour resulting in better productivity. A disadvantage is their limited compatibility with almond (Grasselly, 1969; Monastra and Pennone, 1980) which usually causes early leaf fall. Some cultivars, however, are more compatible: Marianna 2624 with Marcona, Texas, Ne Plus Ultra and Ai (Bernhard and Grasselly, 1969; Grasselly, 1969); Marianna B/1 with Ferragnes, Ai, Peerless, Texas, Ne Plus Ultra, Fascionello and Pizzuta d'Avola (Bernhard and Grasselly, 1969; Monastra, 1979). Both rootstocks are propagated by 25 cm long hardwood cuttings. They are taken in winter and the basal part is treated with 2000 ppm IBA and fungicide and then planted directly in the nursery. It is better to give bottom heat to promote callusing for 3-4 weeks at 18-20°C with the top portion of the cuttings kept at cool outdoor temperatures before planting in the nursery.

## Other Hybrids

Hybrids between *Prunus cerasifera* x *P. persicri* and *P. cerasifera* x *P. dulcis* are difficult to obtain and they show a high degree of incompatibility with almond. Until now they have not been used commercially.

Nemaguard has been the main rootstock of the California almond industry for many decades. Nemaguard is vigorous, resistant to root knot nematode, adopted to well-drained non-calcareous loam and sandy loam soil, doesn't sucker much and has good anchorage. However, it is susceptible to salt toxicity (sodium, chloride and boron), "wet feet", phytophthora root-rot, oak root fungus, crown gall and 'heart rot'.

## New Rootstocks

**Felinem:** A clonal almond x peach (Garfi almond x Nemared peach) hybrid rootstock, developed in Spain, adapted to calcareous soil; resistance to lime-induced chlorosis, resistant to broad range of root-knot nematodes, however sensitive to waterlogging.

**Garnem:** A clonal almond x peach (Garfi almond x Nemared peach) hybrid, developed in Spain. It is also adopted to calcareous soil, resistant to lime-induced chlorosis and a broad range of root-knot nematodes.

**Adafuel:** Natural hybrid selection (probably 'Marcona' seedling) developed at Spain, very vigorous adopted to calcareous soil.

**Atlas:** Interspecific hybrid of peach, almond, apricot and plum, developed at USA, very good anchorage, somewhat resistant to root-knot nematode, susceptible to bacterial canker and verticillum wilt, tolerant to alkalinity.

**Hansen 536:** A almond x peach hybrid, developed at USA, good anchorage, resistant to drought but susceptible to ring nematode, bacterial canker, oak root fungus, crown gall and verticillum wilt.

**Nickels:** Almond x peach hybrid from USA, vigorous rootstock, resistant to drought and soil fungi.

**Mariana M40**: *P. cerasifera x P. munsoniana* hybrid, developed at USA, good anchorage, susceptible to bacterial cranker.

**Mariana 2624:** Good anchorage, resistant to root-knot nematode and crown gall, developed at USA.

**Viking:** Interspecies cross to *P. blirena*, developed at USA, excellent anchorage, resistant to ring nematode, root knot nematode; susceptible to verticillum wilt and tolerant to alkaline as well as wet soils.

**Sirio:** Selection made at Italy from open pollinated seedlings of INRAGF 557, good anchorage.

## Propagation

Almond may be propagated by seed; by grafting on seedlings of almond, peach or plum; by grafting on selected rootstocks; by micropropagation. Vegetative propagation of clones either for rootstocks or to produce own-rooted plants is very difficult and it is not used. Kester and Sartori (1966) were unable to obtain any roots from cultivars and from their off springs; however, Damavandy Kozakonane (1970) succeeded in obtaining roots by hardwood cuttings from a clone of the cv. Marcona, and the percentage of rooting was around 40. All other cultivars tested showed a rooting ranging from 0 to 15 per cent.

### Seed Propagation

Until the end of the 1800s, propagation by seeds was the only technique used and is still used today in countries like Iran, Afghanistan, Greece and Morocco. This method is useful for finding new ecotypes, but it is not suitable for propagation, because almond is highly heterozygous and consequently does not breed true.

In semi-arid and less fertile regions sowing is made directly in the field or in a seed bed, where the seedlings will be grafted. When sown in the field, usually in late autumn, 3 nuts, which have been soaked in water for 24 hours, are put in each 8-10 cm deep hole. The best plant will be grafted in the following autumn by budding or in the winter by a scion.

A rational method is the use of stratified seeds from the current year, because almond seeds lose viability in 4-5 months and peach seeds in 5-6 months. So the seeds are placed in polythene bags with sterile sand or vermiculite and kept at 2-3°C in a well-ventilated room for 2-3 months for almond, and 5-6 months for peach. The seeds can be sown in a seed bed in the spring; the plants are usually well grown and can be grafted in June and can be transplanted in the autumn. Smaller plants or any failed grafts can be grafted in autumn or in the next spring. When seeds are sown in autumn directly in the seed bed moist-chilling occurs naturally. Usually they will be ready for grafting in the autumn and any re-grafting can be done in the spring with the remaining small plants.

## Vegetative Propagation

The usual method of grafting almond is budding.

## Budding

Budding is carried out in autumn, spring and summer at a height of 20 cm from the ground in the nursery and 30-40 cm in the field. Autumn budding is done from the end of July to the end of September using buds taken from the middle or from the base of freshly cut shoots. Irrigating a few days before budding will facilitate bark separation. Rootstocks are cut back above the bud as soon as growth starts in spring and the plants are ready for transplanting in the autumn. Spring budding is carried out when autumn buds fail. Buds from shoots collected in the winter and kept in polythene bags at 2-3°C are used. The rootstocks are cut back after about two weeks and can be transplanted in the autumn. Because almond grows quickly, June budding is also possible in regions with a longer growing season, carried out on seedlings germinated in the previous spring. So tree production has been done in less than one year.

## Micropropagation

Almond rootstock, GF 677, was one of the first rootstocks to be micropropagated (Zuccherelli, 1979) and is still largely propagated by this method with an annual production of 7-8 millions of plants in Europe. Almond cultivars may also be micropropagated. The cultivars Ferragnes, Tuono, and Fascionello have given satisfactory results in rooting and have been successively passed through the acclimatization phase (Rugini and Verma, 1982, 1983; Rugini, 1986). In those areas where seedlings or seedling almond rootstocks are used, even the *in vitro* own-rooted almond may be advantageous, although, there is a no demand for own-rooted plants, due to the lack of data on their field-performance.

Almond cultivars have some difficulty in rooting; however, that may be overcome by taking into account two essential conditions: (i) *in vitro* rejuvenation by means of several subcultures before initiating rooting; (ii) to start the rooting phase in the dark for 7-10 days (Rugini and Verma, 1983; Rugini, 1986). Since the dark period causes a decreased survival in pots, a new technique was used, giving some good results (Rugini *et al.*, 1987a). The following suggestions are useful to achieve success:

1. Use shoot tips or nodes in dormant or growing season;
2. Disinfect the surface with 7-9 per cent of Ca or Na hypochloride for 20-25 min; however, a mixture of 12 per cent Domestos and 350 mg/l $HgCl_2$,

under vacuum for 3-4 min and rinse twice in sterile water under vacuum for 20-40 min is recommended (Rugini, 1986);

3. Aseptically collect small explants (0.4-1 mm long) or nodes and place in half strength Murashige and Skoog (1962) medium plus 2 per cent sucrose, 0.7 mg BAP and 0.7-0.8 per cent agar. Shoot tips are found to be more suitable material than nodes because they present fewer problems for bacterial contamination which can appear in the culture even after many months. After 2-3 subcultures, use full strength MS medium with 3 per cent sucrose and leave the other conditions the same. Use 23-24°C in 16/8 h photoperiod. If shoot vitrification appears for unknown reasons, subculture the shoots in a medium solidified with 0.4 per cent agar plus 0.5-0.7 per cent pectin from grape of must (M-gel from Migross, Switzerland) (Rugini *et al.*, 1987b). The newly emerging shoots will have more expanded leaves and be of normal appearance. However, pectin of grape increases apical dominance and bacterial contamination if the cultures are not perfectly clean. Bacteria may be controlled by using 200 mg/l of vancomycin (Ely Lilly) which is active against many plant bacteria without any secondary effect on plant growth (Rugini, unpublished). Subculture should be carried out not more than 18-20 days to avoid the decline of the culture;
4. Shoot elongation, before rooting phase, in a medium containing less BAP (0.3-0.4 m/l) plus $GA_3$ (10 mg/l), is not really necessary;
5. For the rooting phase, use Bourgin and Nitsch (1967) medium with macroelements reduced to half concentration plus NAA or IBA (1-2 mg/l), or dipping the explants in 100 mg/l NAA or IBA for 10 seconds and place them in a hormone-free medium, in darkness for about 7-10 days, then place the explants in a 16/8 h photoperiod. An initial period of darkness is essential for root induction in most cultivars tested.

   A simple method, developed by (Rugini *et al.*, 1981a) showed good result. The technique does not take into account the initial period of continuous darkness which causes shoot senescence with consequent deleterious effects on plant survival. The base of the shoots are kept in continuous darkness by painting the outside of the vessels up to polycarbonate (3-4 mm in ) after agar solidification. Usually rooting occurs at a higher percentage, with more roots per explants and higher survival after transplanting,
6. Transfer the plantlets to peat moss and perlite (1:1) for the hardening phase at 18-20°C in high light intensity using 16/8 h photoperiod under high humidity and well-aerated conditions. If plantlets do not grow and form a 'rosette', spray $GA_3$ or $GA_{4+7}$ at 200 mg/l once per week for some weeks.

GF 677 is a vigorous rootstock, used for both peach and almond worldwide. Nowadays, it is propagated mainly through micro-propagation. Widening and undesirable growth of leaves as well as poor rooting are the major problems during its *in vitro* culture. Sepahvand *et al.* (2011) reported that myoinositol at 10mg/l could prevent undesirable widening of leaves as well as resetting of shoots. Thiamine at 1.6mg/l was found highly effective in rooting shoots.

Iskalan *et al.* (2008) developed protocol for *in vitro* propagation of cv, Nonpareil. Zygotic embryos isolated from mature seeds was placed in MS medium containing 30g sucrose/l, 0.5 and 1.0mg N6-BA/l and 7g agar/l which resulted multiple shoot initiation at the rate of 11±1.32 and 14.7±2.12 shoots/explants, respectively, in 28 days of culture. New shoot production was obtained from MS medium supplemented with 1.0mg BA/l. Rooting was achieved in half-strength MS medium supplemented with 8.0mg IAA/l. The *in vitro* raised plants were successfully transplanted in the field after acclimatization. Parada-Ponce and Vilegas Monter (2009) generated a protocol for propagation of almond x peach hybrid H1 from explants obtained from adult plants grown under field conditions. The best establishment *in vitro* (44 per cent) was achieved with stem nodal sections growing on the wood plant medium (WPM). For the *in vitro* multiplication, the best treatment reported was the WPM medium added with 2mM of ammonium nitrate and 2.5μm of BA. *In vitro* rooting was achieved by adding IBA. The rooting and acclimatization can be performed simultaneously with Radix 10000 at 20 per cent (p/p) if it is applied during transplanting.

Apart from micropropagation, other *in vitro* techniques have been studied. Plants have been regenerated from seedling callus (Mehra and Mehra, 1974), from seedling roots, from which a highly morphogenetic line has been selected and plants have been established in the field (Rugini, 1986). Cell suspensions have been established and maintained for long periods. Casein hydrolysate has been found to be essential (Rugini and Verma, 1982); however, it can be replaced by an amino acid mixture (Rugini, 1986).

A protocol to obtain protoplasts from cell suspensions and to regenerate callus from those, has been successful (Wu and Kuniyuki, 1985; Rugini, 1986).

Culture of immature embryos has been carried out by Zdruikovskaya Rikhter (1980). Mature embryos germinated rapidly after harvesting on Bourgin and Nitsch (1967) medium supplemented with (0.25 mg/l) of $CoCl_2 \cdot 6H_2O$ (Rugini, 1986).

The in *vitro* almond culture programme aims at the genetic improvement and cryopreservation of germplasm (Rugini, 1986; Kester *et al.*, 1986). Cell and tissue culture biotechnoiogy and molecular genetics incorporated into a conventional almond breeding programme could complement and shorten the screening and selection process and substitute for screening and testing of whole plants for many traits (Kester *et al.*, 1986). Selection for nematodes, *Fusicoccum*, *Armillaria*, herbicide resistance, salt tolerance, early screening for graft compatibility with plum, introduction of the easy-to-root ability into species via *Agrobacterium rhizogenes*, regenerating plants from transformed roots (Rugini, 1984, 1986), induction of polyploids and homozygous plant from eggs or pollen, or induced mutation by physical and chemical methods *in vitro*, could be an example of potential application in almond (Rugini, 1986; Kester *et al.*, 1986).

## Planting

Seedlings or grafted plants of one-or more-year-old with well-formed wood, healthy roots, and in leafless condition are planted in well-prepared soil fertilized with organic fertilizer.

The best period for planting is between November and December. The distance between trees varies according to the variety, rootstocks, soil fetility, and irrigation system. Five x 5 m seems to be the best distance for cultivars grafted on peach seedlings or on peach x almond hybrids.

### Windbreak

Effective shelter must be well established before almond trees are planted. Wind-breaks are made with trees or with artificial shelter belts. The latter are expensive but are effective immediately while living shelter belts become effective only after 2 years. *Cupressus arizonica* and *Cupressus macrocarpa* are best among the windbreaks used for almond.

## Training and Pruning

Pruning is carried out at three different stages in the life of the almond tree: (i) during early growing, (ii) when cropping, and (iii) to reinvigorate old plants.

1. The early pruning is practised to shape the tree. It is carried out immediately after planting by cutting at 1 metre from the ground, when the commonest 'vase' shape is required. The primary scaffold limbs (3 or 4) developing within 10-15 cm from the top will form the tree crotch. Summer pruning is carried out 2-3 times in order to retain shoots for main branches and to eliminate unwanted branches and suckers. Although summer pruning reduces tree growth, it is advisable, while pruning in winter, to remove only suckers. Summer pruning should be carried out in the second year to form the secondary framework scaffold at 50-60 cm from the crotch.

   Further, winter pruning will be required in the 3rd and 4th years when the tree should have 3-4 main branches each with 5-7 smaller branches. To allow light penetration the centre of the tree should be kept free of shoots.

2. When the tree is cropping, regular pruning should be carried out every year to ensure high yield. Almond flowers are usually located on spurs which live about 5 years and in order to renew about one-fifth of fruiting wood every year it is a practice to stimulate new shoots by pruning. Amount of pruning depends on the previous crops, efficiency of pollination, kind of fertilization, cultivar, tree vigour and climate.

3. Reinvigorating pruning is carried out only when trees are old and unproductive but healthy. The main branches can be cut back and many new shoots will grow out to replace them in the spring.

## Nutrition

Nitrogen deficiency in almond is characterized by small, pale leaves with little vigour, drying of small twigs, reduced fruit set and yield (Lilleland, 1951). Application of nitrogen in the form of urea markedly improved the yield by increasing blossom receptivity and percentage of fruit set (Weinbaum *et al.*, 1980a). In K-deficient soil, application of $K_2SO_4$ at 1 to 2.5 tonnes per acre markedly increased yield (Lilleland, 1951). Fatali-Zade and Semochkina (1972) suggested application of 180 kg N and 360 kg $P_2O_5$ per hectare.

The deficiency of zinc causes little leaf in almond. The leaves show chlorosis between veins. The deficiency symptoms respond to dormant spraying of zinc sulphate at 15 to 25 pounds per 100 gallon water (Proebsting and Serr, 1966).

The plants showed poor growth, rough bark condition and shrivelling of kernel due to copper deficiency (Kester *et al.*, 1956). Foliar spraying of copper EDTA was found effective for correcting the deficiency.

The tip burn in leaves usually appeared due to boron deficiency (Hass, 1945). Hansen *et al.* (1962) corrected boron deficiency by broadcasting borax in soil at 50-100 lb/acre which was found effective for 4 years.

Soria *et al.* (1982) obtained highest yield when the leaf contains 2.7 per cent N, 1.35 per cent P, 1.25 per cent K, 3.15 per cent Ca, 0.95 per cent Mg, 100 ppm Fe, 57 ppm Mn and 36 ppm Zn. Annual variations of the mineral elements in the leaves were recorded (Monstra, 1975; Agabbio *et al.*, 1977; Averna *et al.*, 1984). The total amount of N, P and K in flower buds and in the flowers and fruits was found to increase progressively during the growing season but as the percentage of dry weight increased, these elements decreased during fruit development. Leaf N and P decreased from spring flush to leaf fall, whereas shoot N and P increased in late summer and then decreased from early autumn to spring flush, flowering and fruit development (Agabbio *et al.*, 1971).

It should be a routine practice to apply 200 kg per hectare of both P and K before planting. Subsequently, 100-120 kg N, 50-60 kg each of P and K should be applied annually to maintain the fertility in bearing orchards. P and K should be applied in a single dressing in December or January and N in three split doses, in February, March and April (Monastra, 1979).

## Irrigation

Irrigation is the most important practice to improve the growth and yield of almond trees, because summer rainfall is insufficient in the areas where they are grown. In the USA summer irrigation is practised in almost all the orchards. The amount of water to apply and the frequency of irrigation treatments are related to water retention capacity, the depth of the soil and rainfall. However, about 1500-2000 $m^3$/ha of water should be applied from spring up to 10 days before harvesting.

The quality of irrigation water is certainly important; in some almond growing areas (*i.e.*, North Africa) the water often contains sodium chloride (NaCl) which, above a certain concentration, is very dangerous for almonds. Water with a high salt concentration should not be used for irrigation, particularly to shallow clay soil; in any case, a winter irrigation contributes to desalinize the soil. Overhead irrigation should not be used because almond leaves are very sensitive to salts.

Surface (flood) and sprinkler irrigation and recently also drip irrigation are the usual systems. The choice of the system depends on the quality of water available. Drip irrigation is usually used when the land is very undulated.

Almond is regarded as tolerant to water deficit conditions. However, irrigation scheduling has considerable effects on improving the quality of nuts (Nanos *et al.,* 2002). Sensitivity to almond orchards towards water stress is higher during postharvest period compared to the kernel filling stage. In fact, applying deficit irrigation at this period limits growth in the following season, which consequently reduces the differentiation of reproductive buds (Esparza *et al.,* 2001). Reduction of irrigation during kernel-filling through to harvest is suggested. The Reduction of irrigation water by 40-60 per cent or even 80 per cent during this phase-but avoiding water deficit during rapid vegetative and first growth (Martin and Kester, 1978, Giorna and Marsal, 1995) has been found generally productive (Romero, 2002). Alcon *et al.* (2013) studied and assess the long-term economic viability of deficits irrigation (DI) strategies in almond tree cv. Marta grown in a semiarid area (southeast Spain). They observed that regulated deficit irrigation (RDI) and mild-to-moderate sustained deficit irrigation (SDI) were the most economically feasible treatments over the 6-year period.

## Soil Management

Soil management is particularly important in almond orchards because the nuts have to be picked up from the soil surface. However, tillage is required to eliminate perennial weeds, permit water penetration and for good soil aeration and mineral nutrient incorporation. The ground should be cultivated to a depth of 10-15 cm during autumn and winter and repeated in spring and summer according to conditions and extent of weed growth. The chemical control of weeds should be carried out, one month after petal fall, by using oxyfluorfen plus either glyphosate or paraquat for controlling summer annual weeds (West *et al.,* 1983).

Although almond, in many countries, is consociated with herbaceous crops like potato, beans, *etc.,* it should be avoided because they leave in the soil several harmful micro-organisms for almond. It can also be damaged by the weedicides -glyphosate, diuron, linuron, simazine used to control weeds of the intercrops.

Monthly application of steam or flame or application of d-limonene (every 5-6 wk) adequately suppress weeds in organic almond orchards (Shrestha *et al.,* 2012). Soil herbicides; Stomp 33EC at 4 or 6 l/ha and Trevissimo (Diurom + Glyphosate) at 10 l/ha, applied all at once or in two instalments over 45 days effectively controlled weeds in row strip and showed no negative effect on vegetative or reproductive growth of trees (Djeneva and Rankova, 2011).

## Flowering

Plants from seedlings show juvenile characteristics: small leaves, bushy habit sometimes with thorns. Even shoots from buds collected from mature trees and grafted on seedlings have a juvenile period but it is shorter than that of seedlings. The length of the juvenile period. however, depends on the volume of the canopy and the method of pruning (Grasselly and Crossa-Raynaud, 1980).

Cultivar, such as, Tuono, Marcona, Ai, Ferragnes, Texas, Nonpareil, Peerless, Ne Plus Ultra start cropping in 3-4 years; other cultivars, such as, Bartre, Davey, Kapareil take more than 5 years (Grasselly and Crossa-Raynaund, 1980).

Flower induction occurs in the year preceding blooming, usually from May to June. Agabbio and Ortu (1974) demonstrated by removing the leaves that the period of induction depends on the cultivar; the more floriferous cultivars tending to differentiate earlier, around the middle of May.

Heavy crop may reduce flower bud differentiation, although biennial bearing depends on the cultivar, even in good growing conditions. Nitrogen fertilization increases the number of flowers/tree rather than by increasing blossom receptivity and percentage of fruit set (Weinbaum *et al.*, 1980b).

Almond is often considered to be a moderately alternate bearing species but historical yield data typically do not exhibit clean patterns of alternate bearing at the orchard level. While research has indicated that spurs (the main fruit bearing unit in almond trees) rarely produce fruit in two subsequent years. Tombesi *et al.* (2011) analyze the bearing behaviour of almond trees at both the orchard and the individual spur level over multiple years to explain this apparent paradox. The 10-year yield patterns of three almond cultivars grown at three different sites within California were studied. At the orchard level one cultivar (Nonpareil) showed a tendency for alternate bearing at one site but other cultivars and sites did not. The orchard and the individual trees in which the spur population was conducted showed tendencies for alternate bearing but the total tagged spur population bore fruit in any given year and therefore while individual fruiting spurs exhibited a high level of non-bearing after fruiting the previous year the spurs that did produce fruit in any year did not constitute enough of the total spur population to exhibit alternate bearing at the whole population level. The results suggested that annual bearing fluctuations in almond are probably mainly due to year-to-year variations of parameters affecting fruit set and that high rates of fruit set in a given year may involve a larger-than-normal percentage of a spur population in fruit bearing. This would limit the size of the spur population available for flowering in the subsequent year and could cause alternate year bearing.

According to Grasselly and Crossa-Raynaund (1980); Ferragnes, Nonpareil and Peerless have a slight tendency to biennial bearing, while cultivars Marcona and Tuono have a more tendency and Rachele has a strong tendency.

Morphological differentiation takes place in August, and at the beginning of autumn flower primordia appear in the buds. Depending on the cultivar, pollen grain formation (microsporogenesis) occurs between December and January, while the ovary maturation (macrosporogenesis) occurs just before flowering, in contrast to other *Prunus* species (Pimienta and Polito, 1983).

The flower buds develop on three types of shoot: (i) around the tip of short spurs, 3-10 cm long with less than 8 nodes, which have 1-5 or more flower buds, while the terminal bud is always vegetative, which maintains the elongation of the spur for subsequent years; (ii) on longer spurs or short shoots with 8-25 nodes; (iii) on long shoots, with more than 25 nodes, mainly at the distal end, where each pointed vegetative bud is subtended by swollen, rounded flower buds which normally contain only one flower, exceptionally two.

Flowers are hermaphrodite with white or pink petals, 5 sepals, a single and unicarpel pistil which usually contains 2 ovules in orthotropic position, and 20-40 stamens. Flower anomalies in different cultivars are described and illustrated by several authors (Serafimov 1977; Dhatt and Dhillon 1981; Dhatt *et al.*, 1982). In some cultivars ovary abortion occurs in flowers on the long shoots, rarely this occurs in spurs, Ovule abortion was studied by Pimienta and Polito (1982) in Nonpareil almond, the earliest indication is the deposit of callose, 2 days after pollination, starting from chalazal region occupying progressively the nucellus, blocking translocation of metabolite supply, but whether this was the cause or consequence of ovule abortion was uncertain.

Flowering precedes leaf emergence and takes place in very early season, occurring between January and April depending on the cultivar. Date of blossom depends on the climatic conditions and it may change considerably from one year to another. The time of flowering is also variable, from 10 to 30 days, depending on the cultivars and climate; in continental areas blossom is almost simultaneous for all the cultivars, on the contrary in Mediterranean-like climate it is quite long.

Almond is the earliest temperate tree nut to bloom in winter/spring, showing wide adaptability to different environments. Barros *et al.* (2012) of Genomics of Plant Stress Laboratory, Portugal investigated the role of two almond *CBFs* over the natural autumn/winter dormancy period in adult dormant trees selected for their early blooming. The expression analysis and GA metabolism was also studied. Gene expression analysis performed in shoots and flower buds showed that *PdCBF1*, *PdCBF2* and *PdDHN1* were active during the cold acclimation process, reducing expression after bud. In flower buds, this response occurred in parallel with the induction in the GA biosynthesis pathway. Barros *et al.* (2012) observed that *PdCBF2* showed a progressive increase in transcript abundance during the autumn in close correlation with cold acclimation, while high transcript levels of *PdCBF1* and *PdDHN1* genes were found to sharply reduce transcriptition in flower buds and internodes, suggesting damping of CBF-mediated cold signalling during growth resumption, in correlation with cold hardening decline. Flower bud break was also found followed by a decrease in the expression of *PdGA20OX*, a candidate gene involved in gibberellins biosynthesis, and an increase in the expression of two homeotic genes related to floral organ development, *PdMADS1* and -3. *PdMADS1* transcript accumulation was only detected after bud break, mainly from the green-tip stage. This study provides the first molecular evidence for the role of CBF- mediated low-temperature signalling in fruit tree flower buds, also indicating putive markers to study the complex network of events occurring prior to anthesis.

## Pollination and Fertilization

### Pollination

Although some cultivars, such as, Tuono, Filippo Ceo, Genco (Grasselly and Olivier, 1976); Falsobarese, Ferrante, Palatina, Sannicandro, Scorsa Verde (Godini, 1977) and Mollisona (Crescimanno, 1960) are self-fertile; almost all others are self-incompatible. Only under very favourable environmental conditions the pollen tube can occasionally overcome this incompatibility (Ducom, 1968). So, almond should be cross-pollinated (Milella, 1959). Usually two pollenizerss possibly of commercial importance are used; one flowering slightly before and the other just after the main cultivar (Micke and Kester, 1981). Good pollination can be obtained by single row of the main cultivar and the pollinator. The percentage of pollenizers (15-20 per cent), normally used for many other crops, is not enough for almond. Liu Xiaoltua (2008) suggested pollenizers for 4 leading almond selections of China. The best pollenizers for Xin 1 were suggested as Xin 2 and Xin 9; for Xin 2 were Xin 1 and Xin 18; for Xin 9 were Xin 18 and Xin 1 and the best pollenizar for Xin 18 were Xin 2 and Xin 9.

Almond are fully pollinator-dependent crop. In California, pollination occurs in late Febraury through to early March. Honey bees (*Apis mellifera*) are the main pollinator and many producers depend on renting hives from beekeepers to achieve full crop yield. Usually one or two colonies per hectare, but four or five colonies will ensure good pollination even during less favourable seasons with lower temperatures, more competition from herbaceous plants or wind (Thorp, 1981).

The development of the sac, was studied by Pimienta and Polito (1983) in the cultivar Nonpareil. They observed that it takes place before anthesis, and it is stimulated by the presence of compatible pollen tubes in the style; cross-pollination promotes its final elongation growth. They also found irregularities in megagametophyte development, including delayed differentiation of the megaspore mother cell, embryo sac abortion and lack of polar nuclei fusion and embryo sac elongation were frequently noted in ovules of self-and non-pollinated flowers.

### Fertilization

Pimienta *et al.* (1983) observed that in the self-pollinated incompatible cultivar male gametophyte was rejected on the stigma and also in the style. It is characterized by lower pollen grain retention on the stigma, a reduction and delayed germination of the pollen and when the pollens were able to germinate, a lower frequency of pollen tubes grew through the style. The pollen tube growth slowed down greatly after the tubes entered the ovarian locule before they reached the ovule.

A temperature of 15-23°C (lower than that for other deciduous fruits), although it depends on the cultivar, is necessary for pollen germination and pollen-tube growth, which reaches the embryo sac in 3 days. Loreti and Viti (1984) observed that the pollen tubes of the early flowering Kapareil germinated at lower temperature (2-6°C) compared with others. Garcia Garcia *et al.* (1980) studied the influence of temperature on pollen germination of several almond varieties. According to Garcia Garcia (1977), pollen-tubes grown *in vitro*, elongated about 3 times more at 15°C than at 10°C, although there was a marked difference between cultivars. In self-compatible cultivars the pollen-tube growth is slower and the best temperature is around 25°C instead of 15°C for frozen pollen (Socias I Company, 1976; Godini, 1981). Stigma receptivity declines rapidly within 3-5 days. At petal fall stage a temperature lower than l°C kill the styles and ovules (Brewer, 1981). Usually one egg is fertilized, depending on the cultivars and climatic conditions. Fertilization of both eggs gives rise to double kernel. Gulgan (1975) reported formation of small seeds probably arising from parthenocarpic development of the egg.

## Flower and Fruit Drop

A high fruit set is desirable because smaller nut size is not a real disadvantage; however, it should not be so heavy as to affect flower differentiation in the following year.

Apart from the flowers which drop early due to the defective, unfertilized and degenerated ovaries (Garcia Garcia *et al.*, 1980), after fertilization almond nuts drop at 3 distinct stages: (i) immediately after fertilization. (ii) within 30 days after flowering. (iii) after 4-5 days from flowering (in April). Although flower and fruit drop always occur, improvement in management of orchards reduces the drop at different stages. Usually the percentage of harvested nuts is 15-35 per cent of initial number of flowers.

## Fruit Growth and Maturity

The almond fruit is an egg-shaped drupe with a sigmoid growth pattern. Fruit development stops at 70-75 days after blooming. Fruit maturity is not related to the date of flowering. The pericarp (hull) is almost always thin (5-15 mm) and may be pubescent; in July it begins to split open freeing the endocarp. The endocarp-shape, surface appearance and consistency are characteristics of the cultivar, but it can change with environmental conditions (Grasselly and Crossa-Raynaud, 1980). The endocarp contains one or two kernels varying in size, weighing from 0.5 to 1.5 g, with brown integuments at maturity. The embryo grows very slowly at the beginning, 50 days after fertilization it measures only 100-200 μm (Grigorian, 1972). When the fruit reaches full size the embryo grows very fast and uses all the endosperm which remained in a semi-solid consistency. At maturity the embryo, with its big cotyledons, fills the whole seed. Fruit thinning normally, is not required.

The first sign of nut ripening is the dehiscence of the hull along the ventral suture and consequently the reduction of the removal force. Usually growers start to harvest one month and a half after the first sign of hull dehiscence. Nuts growing at the periphery of the tree usually mature earlier than those located in the centre. Even treatment with ethylene promoters cannot synchronize ripening; furthermore, if nuts dry out before their hull dehisces the result is sticktight, which is increased by water or nutritional stress. The advantage of harvesting before ripening is a reduction of fruit losses due to navel orange worms, but the plant may be damaged by excessive shaking required for harvesting, the kernel remains smaller, there is a loss of flavour and drying costs will be higher.

Tombesi *et al.* (2010) working with 28 almond cultivars in Central Valleys of California observed that the length of the period between full bloom to hull split was negatively correlated with the accumulation of degree-days between full bloom and 90 days after full bloom (mean $R^2 = 0.51 \pm 0.13$) with generally poor correlations with degree-days to 30 or 50 days after full bloom (mean $R^2 = 0.31 \pm 0.02$) and 0.36 ±0.3, respectively). This result suggested that temperatures in the first 90 days after full bloom are the primary factor influencing the time of nut maturity in almond cultivars in California.

## Harvesting and Yield

Almonds can be harvested green or dry. Nuts of paper variety are also harvested at the green stage for direct consumption, but it is limited to some Mediterranean countries (only 2 per cent of production). Dry harvest is made from August to October. In, non-irrigated orchards the nuts ripe earlier than in the irrigated ones.

In almond, the date of harvest is not always closely-related to the date of fruit maturity. Harvesting can be delayed until the fruits are dry enough to harvest. The nuts are shaken and picked-up by machines from the orchard floor. A timely harvest is essential to maintain nut quality and to minimize post-harvest microbial contamination (Danyluk *et al.*, 2007). Nuts from each cultivar must be harvested separately from other cultivars in order to optimise hulling, shelling and marketing. Hence the maturity dates of individual almond cultivars in the same orchard must be sufficiently different to prevent undesirable mixing of the nuts (Tombesi *et al.*, 2010).

The orchards are often irrigated, two weeks before harvesting for better detachment of fruits. The soil should be as smooth as possible and free from weeds. Nuts are mechanically shaked, possibly when the hull is still attached to shell and swept into the rows from where they are picked up by machine and then conveyed to carts or trailers for hulling. Ethylene promoters associated with the mechanical shaking have shown to be of some advantages (Vitagliano and Viti, 1983). Yield per hectare varies from 10 to 20 quintals of shelled almonds depending on cultural practices including irrigation and cultivar.

## Storage, Grading and Kernel Use

After harvesting, the nuts are removed from the hull by machine. The machines for hard-shell almonds are not suitable for soft-shell or paper-shell, and adjustments and modification are necessary. Then nuts are dried in the sun for 3-4 days until moisture content is reduced to 8.0-8.5 per cent, assuming that there will be a further 1 per cent loss during storage. Storage of soft-shell nuts may become a problem due to the growth of mould (Lichev and K'osev, 1983). The classical way of storing almonds nuts is to keep them in the shell after natural drying until their consumption or use in industry (Garcia-Pascual *et al.*, 2003). Unshelled almonds stored at ambient temperature do not show significant chemical and biochemical changes for one year (Sensi *et al.*, 1996). In order to prevent the off-flavour of almonds due to rancidity, different packaging materials or modified atmosphere can be used. Peeled kernels could be stored up to 9 months without a serious loss of quality when packaged in high barrier packaging at 4°C or ambient (Senesi *et al.*, 1991).

Almonds for table consumption are bleached with sulphur dioxide to improve their appearance. The remainder are shelled and calibrated mechanically, and the kernels are sold by cultivars (*e.g.*, Marcona, Ne Plus Ultra) size or according to town or region of origin.

Shelling yield (kernel weight/kernel + shell weight) x 100, of cultivars, varies according to shell consistency and is generally classified as follows:

1. Stone or very hard-shell (shelling yield= 20-25 per cent)
2. Hard-shell (shelling yield= 25-35 per cent)
3. Semi-hard-shell (shelling yield = 35-45 per cent)
4. Soft-shell (shelling yield = 45-55 per cent)
5. Paper-shell (shelling yield = 55-65 per cent)

Grades of kernels for the international trade are:

*Unselected kernels*: consist of broken and double kernels containing impurities which, after chopping or grinding, are used in chocolate and pastry industry;

*Selected kernels*: selected according to the size and unit weight and after mechanical peeling, with water at 98°C for 2 min, are utilized for the food industry (confectionery, ice-cream, salted, sugar-coated or toasted almonds);

*Broken, double, misshaped and bitter kernels*: ground or in pieces are used in pastry, confectionery or for extracting oil used for cosmetics and in dermatology for its anti-inflammatory properties.

Loreti and Xiloyannis (1978), and Loreti *et al.* (1979) have put the commercial cultivars into two groups according to their use in confectionery as follows:

1. *Almonds to be used whole*: Pizzuta d'Avola, Nonpareil, Fasciuneddu, Davey, Ferraduel, Ferragnes, Peerless, Thompson, XX, Ibba, Provvista,
2. *Almonds to be used in pieces*: Filippo Ceo, Yaltano, Picantili, Ne Plus Ultra, Bocchino, Casu, Rachele, Romana, Occhio Rosso. Other well-known cultivars, such as, Marcona and Dasmayo Larqueta may be included in the first group, while Texas may be included in the second one.

In the market, almonds compete with other nuts such as hazelnuts, walnuts, cashews, pecans or other products used in confectionery, so the demand depends on their price.

## Pests

A large number of pests including birds, insects, mites and nematodes are found to attack almond.

### Naval Orange Worm (*Amyelois transitella*)

This is the main insect pest of almonds in California. It is a pyralid producing

3-4 generations and overwinters as larvae in mummified nuts left on the trees after harvest. In almond it feeds on the kernel before and after hull split only in soft-shelled cultivars. Neonates tunnel into the nut, consume the nutmeat, and cause direct damage by producing frass and webbing, and feed internally until pupation (Curtis and Barnes, 1977). This damage facilitates infection by *Aspergillus* fungi, which produce aflatoxins (Palumbo *et al.*, 2014). They are controlled by removing mummified nuts from the trees and early harvesting. Chemical control only gives 50 per cent control of larvae (Engle and Barnes, 1983). Continued use of the broad-spectrum pyrethroids is in question; biofenthrin resistance has already been reported (Demkovich *et al.*, 2015). Reduced risk pesticides, including spinosyns, such as spinosand and spinetoram, have been examined as alternatives to the broad-spectrum pyrethroids. Spinosyns are produced as a natural fermentation product from soil actinomycetes and thus are approved for use in organic agriculture (Cleveland *et al.*, 2002).

## Peach Twig Borer (*Anarsia lineatella*)

It is a small butterfly which occurs almost worldwide. Overwintering as larvae under bark, it produces 2 generations and attacks buds, flowers, shoots and fruits. The larvae of the first generation attack shoots making them wither; the second one attacks fruitlets which then drop off or ripen prematurely. It can be a serious pest in the nursery. It is controlled by removing affected buds in spring or spraying the tree with carbamate (sevin) or organophosphate in summer.

## *Malacosoma neustria*

It is a butterfly with a wingspan of 30-40 mm, bright ochre in colour with two darker transverse zones. The adults are nocturnal. They appear in July and lay their eggs (about 300) in a characteristic ring shape around the shoots of the tree, where they overwinter. As larvae live in whitish web containing the leaves on which they feed. The infestations do not occur every year. They are controlled by destroying eggs in winter and by sprayings carbamate or organophosphate in spring.

Other lepidoptera which damage the almond occasionally are: *Argyroplace variegana*, *Zeuzera pyrina*, *Cossus cossus*, *Hyponomeuta padaellus*, *Aglaope infausta* and *Lymantria dispar*.

## *Monosteira unicostata*

The insects are more prevalent in the Mediterranean area. The adults overwinter in group under the bark or in other crevices. Eggs are laid in groups of 5 or 6 in rows along the midrib on the under side of the leaf where the larvae feed on the sap, causing the characteristic chlorosis and early leaf fall. If necessary they can be controlled with organophosphate or carbamate in June or July or in late summer before harvesting.

## Black Peach Aphid (*Brachycaudus persicae*)

It attacks both the roots, shoots and leaves of the trees. Wingless forms which overwinter on the roots and in early spring migrate at bud swell to early summer. Feeds on the newly developing flowers, leaves and shoots. Some fruit distortion may occur. Natural enemies, such as ladybirds, lacewings and parasitic wasps, can be very important in controlling aphids.

## San Jose Scale (*Diaspidiotus perniciosus*)

It is widespread and is a very harmful pest in California. The female is yellow with a 0.2 mm circular protective covering; it overwinters as young larva, and also as adult in areas with mild winter. The scale feed on bark, foliage and fruit, sucking plant juices and injecting a toxin, resulting in loss of tree vigour, growth and productivity. The presence of gum on wood is a common sign of scale attack. San Jose Scale can be managed with natural enemies, including ladybirds, parasitic wasps and lacewing larvae. At low level of infestation, one application of pyriproxyfen at 0.07 per cent, pheroxcarb at 0.05 per cent, or mineral oil at 1 per cent in spring reduced the number of nymphs fixed per twig, at a level similar to that of a traditional treatment of chlopyrifos at 0.08 per cent (Sazo *et al.*, 2008).

## Ulmi Comma Scale (*Mytilococcus ulmi*)

It is similar to the species which lives on citrus, yellow and comma-shaped; it overwinters as eggs, produces 1-3 generations and attacks shoots and other branches.

## Almond Dark Grey Scale (*Pelomphala inopinata*)

It is distributed widely in the Mediterranean basin and it also lives on pear and pistachio (Roberti and Salerno, 1982). It overwinters as an adult, viviparous and the nymphs appear in July.

## Olive Parlatoria (*Parlatoria oleae*)

Another scale insect which causes occasional damage to almond. Scale insects suck the juice from the tree which decline in vigour and productivity, They are controlled with 4-5 per cent, heavy mineral oil in winter for a light infestation, or light mineral oil plus organophosphate to kill the larvae. Every 2-3 years the trees should be sprayed with polysulphures to remove the oil residue. The seasonal flights of the males, particularly of San Jose scale can be monitored with pheromone traps or stiky tapes (Barnes *et al.*, 1985).

## *Anthonomous amygdali*

The pest is widespread in Mediterranean countries. It is reddish brown in colour with white spots, 3-4 mm long. The legless larvae are yellow with a brown head and are 5-6 cm long. It overwinters as adult, feeding on flower buds where it lays its eggs, during the winter the larvae feed on softer parts where they hatch. They are controlled by spraying carbaryl in winter.

## *Capnodis tenebrionis*

It has a two-year life cycle. Adults appear in June and feed on the bark of young shoots, buds and petiole. Eggs are laid at the base of the tree and the larvae bore into the bark of the trunk and roots. They are controlled with carbaryl applied around the base of the trunk in July and August.

## *Scolytus amygdali*

The adult is black and 2.0-2.5 mm long. It produces 2-3 generations and overwinters as larvae which are whitish and legless. The adults appear in April and bore holes at base of petioles and shoots, where females lay eggs in the galleries. The larvae bore more galleries causing shoots to die. They are controlled by removing and destroying infested shoots and carbamate sprays in the spring.

Other insects which are less common on almond are *Scolytus rugulosus; Melolontha melolontha,* the larvae of which damage the roots and the adults feed on leaves; *Tropinota hirta* and *Tropinota squallida* and *Oxytirea funesta* damage the flowers; *Anisandrus* (*Xyleborus*) *dispar* mines the branches.

## *Cimbex quadrimaculata*

In spring the female lays eggs on the leaves and the larvae feed on the margins of the leaves. It overwinters in the soil as pupae and produces only 1 generation. It is controlled with organophosphate.

## *Caliroa limacina*

The female lays eggs on the leaves, and the sticky yellow-green larvae eat the parenchyma tissue leaving the veins and epidermis. It produces 2-3 generations and is controlled with organophosphate.

According to Rice (1981) there are six species of mites that may cause severe damages to almond trees. Mites suck juice from the leaves, reduce fruit size, increase the number of mummified nuts and can reduce yield after 1 or 2 years of heavy infestation.

## Brown Mite (*Bryobia rubrioculus*)

It is brownish red in colour, lives on the woody parts of the tree and feeds on leaves at night. The mites pierce the leaf epidermal cells and suck out the contents. Infested leaves develops whitish grey spots on the upper surface and a dull appearance overall. It overwinters as eggs under the bark. If infection occurs during the growing season, dicofol application is recommended.

## European Red Mite (*Panonychus ulmi*)

The adult is dark red with 6-8 white spots in its back. It overwinters in the egg stage and produces 5-10 generations per year. Eggs are laid on young shoots. Biological control by using brown lacewing (*Hemerabius* sp.) was found very effective.

Other mites which attack almond are *Panonychus citri* (citrus red mite); *Tetranychus pacificus* (Pacific mite); *Tetranychus urticae* (two spotted mites); *Tetranychus turkestani* (strawberry spider mite) (Zalom *et al.*, 1984). Chemical control with miticides such as propargite (omite) or cyhexatin (plictran) is sometimes necessary, but usually their predators are sufficient to control them, if they are not killed by insecticides (Hoy *et al.*, 1984).Three spraying Pyrida, Ethion or Agrifol was recommended to control *T. urticae* (Khan *et al.*, 2009)

## Nematodes

The damage due to nematode infestation is more severe in sandy soils or in soils with a sandy subsoil. They attack the root system causing a reduction in growth and yield. The major nematodes are:

### Root-knot Nematodes

They produce galls on the roots; *Meloidogyne incognita* the commonest and the most widespread; *M. hapla*, occurs in more northerly areas than the normal cultivated ones; *M. javanica* and *M. arenaria*, more common in warmer areas.

### Root-lesion Nematode (*Partylenchus vulnus*)

It causes lesions on the surface and inside the roots.

### Ring and Dagger Nematodes (*Mesocriconema xenoplax* and *Xiphenema americanum*)

Both live on the surface roots and transmit viruses. Control measures include the use of resistant rootstocks, healthy plant material and soil fumigation (with, 1,3-dichloropropene) before planting, if necessary.

## Diseases

Almonds are susceptible to several diseases and very little is known about the susceptibility of the new cultivars. Although it is difficult to control the pathogens, it is advisable to take precautions with good soil preparation, watering regime, hygiene and sprays to control diseases.

## Crown Gall (*Agrobacterium tumefaciens*)

It occurs worldwide and is the main bacterium to attack almond. Nursery stocks suffer severe damage. The gall size varies from a few millimetres to 30 centimetres. Young plants appear weak with small leaves and can die. Control in the field is almost impossible. Besides a good hygiene in the nursery, young plants can be protected by dipping their roots in a suspension of *Agrobacterium radiobacter* for 30 second before transplanting in the orchard. However, the treatment will not eradicate the existing galls. The bacterium requires an entry point, usually a wound, and therefore minimisation of injuries during planting, training and pruning is important in the management of this disease.

## Bacterial Canker (*Pseudomonas syringae*)

The most commonly seen symptoms in young almonds are bud death and blackened blossoms (blossom blast), and the sour-smelling, sunken, elliptical, rough, dark cankers from which watery exudate ooze.

Blast is a sporadic disease, causing the death of buds, flowers, young leaves and shoots. This bacterium, furthermore, is responsible to form ice crystals at higher temperatures than normal, which damage the plants. It can be controlled by reducing the ring nematodes which may infect the trees with this disease (Bostock *et al.*, 1985). Vigorous trees are less susceptible to bacterial canker, while young trees,

2-8 years old, are more affected. Good practices that promote tree health and vigour may help deter bacterial canker. Pre-planting fumigation, frequent irrigation and post-plant nematicides in each fall for the first eight years of orchard life are some of the recommendations to control.

## Verticillium Wilt (*Verticillum dahliae*)

It represents a serious problem in almond orchards when they are intercropped with susceptible crops (herbaceous or woody plants). Fungi enter to the young roots, block the xylem vessels and give rise to a xylem discolouration. The roots soon stop functioning causing the leaves to turn yellow and die although remaining on the tree. Symptoms appear on the side of the plant where the root system has been infected. Cool spring weather and soil moisture favour fungal development, while hot summer weather kills the fungus. Young trees are more susceptible to the disease which usually does not kill the plants but damages them severely inhibiting growth and reducing yield. All rootstocks use for almond are susceptible except Marianna 2624.

## Node Canker (*Fusicoccum amygdali*)

Buds and young shoots are mainly attacked. The infected buds produce short shoot with yellow leaves. The typical symptoms appearing in summer are necrotic dark spots in a shield shape with gumming. Buds and flowers are also attacked. The spread of the disease is favoured by high humidity and also due to the susceptibility of some cultivars (Piglionica, 1967). To control the disease application of 2 per cent Bordeaux mixture in autum and captafol, benomyl or thiophanate-methyl in spring has been recommended (Piglionica and Ferrara, 1979).

## Brown Blossom Rot and Twig Blight (*Monilia laxa* or *M. fructicola*)

It attacks leaves, flowers, fruitlets and the base of shoots carrying the flowers, forming gum. Cushion-like conidial sporodochia are visible on shoots in early spring. From blossom infection the fungus grows into spurs causing twig dieback and branch cankers. High humidity and a temperature around 20°C increase susceptibility.

The infected shoots are removed in summer, sprayed with benomyl plus light oils in autumn to destroy conidial sporodochia, and with benomyl after flowering for controlling the disease. Control of brown rot depends on protecting blossoms from infection from popcorn stage through bloom. Application should begin in pink bud and to be continued on at full bloom and petal fall. Azoxystrobin, Captan, Cyprodinil and Thiophanate-methyl (Topsin-M) are effective to control brown rot.

## Shot Hole or Gummy Parasite (*Wilsonomyces carpophilus* syn. *Stigmina carpophila*)

It causes shot hole spots or lesions on young shoots, flowers, fruits and leaves, and small purplish areas surrounded by a yellow halo appear. Subsequently these areas turn light brown and the dead tissue may fall out, or the whole leaf drop off. Shot hole also attacks young fruits and can caused premature fruit drop. To control, spray with Thiram or Ziram, Captan and Captafol after flowering, and in autumn with 2 per cent Bordeaux mixture. The same kind of holes in the leaves may be caused by *Cercospora circumscissa*. Azoxystrobin, Trifloxystrobin and Improdione are also effective fungicides for managing shot hole.

## Stone Fruit Rust (*Tranzschelia pruni-spinosae*)

It is widespread all over the world, usually it is not harmful because it mainfests itself late in the season. It induces an early leaf drop. Chlorotic spots, containing red-brownish pustules, appear on leaves mainly on the underside. It usually occurs after a heavy spring rainfall. Control is carried out with water-wettable sulphur and Zineb.

Other parasites which may also cause some damage to the leaves are: *Cladosporium carpophilum; Spheroteca pannosa; Taphrina deformans; Polistigma ochraceum; Ascochyta chlorospora; Sphaceloma siculum; Pleospora herbarum*. Furthermore, *Fomes fulvus* also causes the wood rot.

## Phytophthora Crown and Red Rot (*Phytophthora* spp.)

At least 11 species infect almond rootstocks in cool and moist soils. It can invade roots, crown, trunks and branches. The symptoms include trunk and branch canker, root rot and decline of the tree. The decline may be fast or slow and death occurs with the advent of warm weather. The bark and tissues turn reddish brown in colour; mycelial growth, rhizomorphs or fruiting bodies are not observed. A proper water management and resistant rootstocks help to control it.

## Strain of Tomato Ring Spot Virus (TomRSV)

It causes Yellow Bud Mosaic. It is widespread and is the virus which can damage seriously to the almond plants. The symptoms are several and include rosette growth, growth which is crinkly and distorted, and chlorotic spots; affected leaves standing out from those which are normal. In summer the symptoms can disappear. Some Italian cultivars are susceptible (Scaramuzzi, 1957) but in California only the cv. Mission seems to be seriously damaged (Bostock *et a1.*, 1985). The plant usually does not die but is susceptible to other parasites.

## Strain of Prunus Necrotic Ring Spot Virus (PNRSV)

It induces Almond Calico. This strain belongs to a group of related viruses of *Prunus Ring Spot virus*. It seems to be transmitted by pollen. On the same tree there may be several strains interacting with each other. Different cultivars show a different reactions to the same strain. The leaves of a tree affected by calico develop characteristic chlorotic rings or bands which become creamy white particularly in cool spring. The virus does not kill the tree, the tree affected by calico often develops bud failure (BF) (Rice, 1981).

Another etiologically complex viral disease is the so-called mosaic due to several viruses, such as: prunus necrotic ring virus (PNRV), prunus dwarfing virus (PDV), apple mosaic virus (ApMV) and apple chlorotic spot virus (ACI-SV). In addition another kind of disease concerned with a stem pitting condition, described by Quacquarelli and Martelli (1972), is associated with ApMV, ACLSV and PDV.

Chemical control of virus has not given satisfactory results and is not used in practice. Only nematode-resistant rootstocks should be used, because viruses are often transmitted by them. It should be mentioned that the almond can easily be freed from virus by thermotherapy (Kester *et al.*, 1986). A 4- to 5- week exposure at 38°C can inactivate the virus in the buds, while a treatment lasting 8-10 weeks can completely recover the whole plant (Quacquarelli and Martelli, 1972).

A virus-like disease, *rickettsia*, causes leaf scorch in California. It mainfests itself first with necrosis of the edge of leaves of isolated shoots and afterwards the whole plant.

### Infectious-Bud Failure

The disease results in vegetative and flower bud failure to open, the flowers opened are normal while the fruits are deformed with thick hull. In plants affected by this disease a strain of PNRV has been found. (Mircetich *et al.*, 1976). The disease probably spreads by vegetative materials.

### Non-Infectious-Bud Failure

This is also known as crazytop and muletail. It is classified as a genetic disorder because it is not transferred from an infected plant to a healthy plant, but it is transmitted by vegetative propagation and through seeds by both male and female gametes. This disorder is the result of the failure of the vegetative buds, which appear necrotic, to grow in spring. Flower buds are not affected once they have formed. All buds on the middle of shoots can be affected and many at the tips, the remainders growing at a wide angle and often change growth direction by giving the branch a crooked, tangled appearance (Bostock *et al.*, 1985). This disorder occurs in regions with high summer temperatures (Hellali and Kester, 1979). Rough bark on shoots which have grown from the affected buds, is sometimes associated with this disorder.

Weinbaum *et al.* (1980b) found a higher stomatal resistance in BF-sensitive subclones compared with leaves of normal subclones of the same cultivar. Marked differences were observed at temperature between 26° and 38°C.

### Non-Productive Syndrome

Vigorous trees with non-productive syndrome, which are sometimes seen in orchards, have malformed nuts and reproductive organs. This is probably due to the use of buds with affected cells when budding (Bostock *et al.*, 1985). In addition, adverse weather, nutritional disorders and soil conditions may produce other symptoms and a consequent reduction in crop.

## Breeding and Crop Improvement

Several new cultivars have already been obtained from controlled breeding programmes, while others are still at the evaluation stage (Crossa-Raynaud, 1977;

Spiegel-Roy *et al.*, 1982; 1984). All American and French cultivars have been obtained by this procedure by using Mediterranean ecotypes.

In an all out attempt to initiate a proper breeding programme, several research workers from the Mediterranean Basin countries formed GREMPA (Groupe de Reserches et d'Etudes Mediterraneens pour l'Amandier) to promote and coordinate research to obtain better varieties with the following characteristics: 1) late blossoming; 2) cold-tolerant flowers and fruitlets; 3) early bearing in young trees; 4) regular and heavy crops; 5) resistance to pests and diseases; 6) absence of double kernels; 7) good fruit quality; 8) a shelling yield ranging from 35-45 per cent (semi-hard-shell cultivar); 9) well-formed trees; 10) early ripening; 11) self-fertile; 12) tolerance to winter cold.

The group began by choosing a list of cultivars with exceptional characteristics (Crossa-Raynaud, 1977) to be grown in each country for evaluation and those with the desirable characteristics would then be used in breeding programmes. Almond breeding programmes started at the end of 19[th] century, but only recently the heritability of certain traits has been identified and reported as follows:

## Heritability

### Time of Blossoming

Time of blossoming is influenced by winter chilling requirement and heat requirement in spring. It is inherited as a quantitative trait and depends on multiple factors (Kester, 1965; Grasselly and Gall, 1967). Crosses between early-blooming cultivars result in early-blooming offspring. Similarly, crosses between late-blooming cultivars give late-blooming hybrids, while crosses between early and late-blooming cultivars result in offspring following a normal distribution. An exception to the rule is Tardy Nonpareil (a late-blooming mutant of Nonpareil) whose offspring had a bimodal distribution in the F1 (Kester, 1965; Grasselly, 1977), to which Kester (1965) attributed to the mutation of dominant major gene controlling blossom time. Late flowering in the offspring is associated with lower cropping and vigour, while Grasselly (1977) observed a correlation between late blossom and absence of double kernels. Grasselly and Olivier (1983) obtained an $F_2$ population showing a good bloom density with a normal distribution by crossing the most fertile individuals of the $F_1$ hybrids, having very late blossom time.

### Blossoming Density

Another quantitative inherited trait is bloom density. Grasselly and Crossa-Raynaud (1980) observed that both the very fertile cultivar Marcona and the less fertile cultivar Bartre transmit their own characteristic, respectively, to their offspring.

### Self-fertility

Almond is considered a self-sterile species, except for a few cultivars, due to gametic incompatibility. Modifiers of gene expression are probably responsible for the variation in the degree of self-fertility. Although no detailed information is available on the inheritance of this trait, it would seem that the self-fertility system (*sf*) is dominant over the self-incompatibility system (*si*). According to Grasselly and Olivier (1983) self-incompatible parents play an important role in the inheritance of this trait, they found that from 66 per cent to 100 per cent of offsprings may be self-compatible depending on which self-incompatible parent is used. Self-fertility is subject to environmental influences (Kesrer and Asay, 1975), furthermore, the offspring appears to have low vigour (Grasselly and Crossa-Raynaud, 1980). Self-fertility has been introduced also into almond by crosses with *Prunus persica, P. davidiana* and *P. mira* (Kester and Asay, 1975).

Development of self-compatibile cultivar is the main breeding objective of most Mediterranean programmes. The first self-compatible cultivars releases from Spain were 'Guara', 'Ayles' and 'Moncayo'. The 'Guara' has covered more than 50 per cent of new plantings in Spain. Subsequently several other self-compatible cultivars, including 'Blanquerna', 'Cambra' and 'Felisia' (Socias i Company and Filipe, 2007) were developed.

## Self-incompatibility

Almond cultivars are mostly self-incompatible and in them the S-RNaces and the F-box protein (SFB) have been identified as components of gametophytic self-incompatibility (GSI). Sexual reduction in almond is regulated by GSI, a mechanism governed by a multigene complex with at least two linked genes located within the *S*-locus that all expressed in female and male organs (Gomez *et al.*, 2014). In the pistil, GSI is expressed as stylar glycoproteins of around 30kDa with ribonuclease activity (*S*-RNases) that show an allele-specific polymorphism in the extracellular matrix of the transmitting tissue, where they enter pollen tubes acting as cytotoxins by specifically inhibiting the growth of incompatible pollen (McClure *et al.*, 1989). Recent studies are concentrated in identifying the cause(s) of self-compatibility in a few individuals with identical *S*-haplotype. Thus, although this confirmed findings of SI studies in almond indicate that the degradation model could explain the functioning of GSI, it could be speculated a more complex system with the existence of other factors occurring (Gomez *et al.*, 2014).

## Tree Characteristics

The insertion angle of the branches, apical dominance, vigour and the hardness of the wood determine the characteristics of the plant. Almonds possess great variability; it has compact cultivars (Tuono), many-branched cultivars (Ai) and upright bearing and unbranched cultivars (Bartre). The above characteristics, although not connected with a large number of genes, seem to be additively inherited. Grasselly and Crossa-Raynaud (1980) reported an inbreeding depression as regards vigour. Crosses between closely related ecotypes or from self-pollinated plants gave rise to an $F_1$ offspring characterized by very low vigour and a long juvenile period: crosses between distantly related ecotypes, on the other hand, show different characters in the offspring.

## Foliage and Flower Types

Almond cultivars show variation in the density, size and colour of leaves. Certain cultivars such as Harriot and Jordanolo have very dense foliage which is transmitted to their offspring (Kester and Asay, 1975). The leaves of the offspring originating from a free-pollinated cultivar resemble the mother cultivar in colour and flexibility. Petal size and the number of stamens seem to be transmitted quantitatively. Furthermore, Grasselly and Crossa-Raynaud (1980) reported that in genetic crosses large number of stamens tend to predominate.

### Pericarp (Hull)

Time of ripening (hull dehisces) shows a large genotype by year interaction. Inheritance appears to be polygenic with a high heritability (Kester *et al.*, 1977). Pubescent epicarps are dominant over glabrous ones (Grasselly and Crossa-Raynaud, 1980).

### Endocarp (Shell)

Semi-hard-shell cultivars are used in breeding programmes, avoiding both very-hard-shell and susceptibility to parasites, respectively. Hardness is regulated by two major alleles: D: very hard-shell (dominant) and d = soft-shell (recessive). However, in the case of semi-hard-shell cultivars (Ai, Ferragnes, Tuono) it seems that other genes are associate with the major gene. Grasselly and Crossa-Raynaud (1980) observed that certain cultivars such as Cristomorto, Desmayo Larqueta, Ferraduel and others are homozygous (*DD*) for hard-shell characteristic; other cultivars such as Bartre and Marcona are heterozygous (*Dd*), while others are homozygous recessive (*dd*) with soft-shell, *e.g.*, Nonpareil, Texas, Ne Plus Ultra, Greek, *etc.*

### Kernel Size and Shape

Kernel size and shape are interrelated characteristics. Size, although influenced by the quantity of fruit produced, characterizes the cultivars showing high heritability. The Cultivar Bartre's offspring inherits very large kernels and also low fertility, both characteristics of the mother plant. Grasselly (1972) found that cultivars Texas and Ardechoise have small kernels and this character is transmitted to their offspring.

Endocarp shape, determining also the shape of kernels, is quantitatvely inherited. Crosses between cultivars having pointed endocarp (Ne Plus Ultra, Cristomorto) produce offspring with an identical endocarp shape, while cultivars with rounded endocarp (Ai, Marcona) give rise to offspring with rounded endocarp. Crosses between the two types produce offspring of intermediate shape.

### Double Kernels and Twin Embryos

The inheritance of double kernel characteristic is complicated and unpredictable. According to Spiegel-Roy and Kochba (1974, 1977) double kernel appearance is favoured by low temperature at bloom time. Grasselly and Gall (1967) observed that the percentage of double kernels increases with the increase of the field. Double kernel characteristic has high heritability and seems in most cases to be quantitatively inherited, although cultivars, such as, Texas, Ne Plus Ultra and Cristomorto transmit it to a higher degree than that they themselves possess (Grasselly and Crossa-Raymond, 1980).

Twin embryo is an uncommon characteristic found in some cultivars more than in others with an incidence as high as 10 per cent. This characteristic may be a potential source of haploids but no haploid has been identified (Kester and Asay, 1975).

### Bitterness

Sweet, semi-sweet and bitter are the three main characteristics of almond kernels. According to Heppener (1926) bitterness is due to a single recessive gene (*s*). Cultivars such as, Ferragnes, Ai, Jordan, Greek and Bartre are homozygous for the sweet characteristic (*SS*), while cultivars, such as, Nonpareil, Ne Plus Ultra, Peerless, Texas, Marcona, Tuono and Cristomorto are heterogeneous (*Ss*). In certain almond wild species and cultivars, the characteristic of bitterness seems to be the dominant character. Grasselly and Crossa-Raymond (1980) reported that the cultivar Texas (*Ss*) x *Amygdilus fenziliana*, gave rise to individuals with bitter kernels.

Breeding studies crossing sweet and bitter genotypes showed that bitterness is a monogenic recessive trait. Its locus, Sweet kernel (SK) locus, was localized in linkage group five (G5) in an almond genetic linkage map, although its biochemical functions remain unknown. Molecular marker studies identified four microsatellites as linked to and flanking the SK/sk gene, that can be very useful for the selection of this character in almond breeding with an efficiency rate of 97 per cent (Sanchez-Perez, 2014). Genome studies performed in two sweet and two bitter almond genotypes have shown more than 200 single nucleotide polymorphisms in a 3Mb region in which sk locus is included within G5. The elucidation of the polymorphisms defining bitterness will be decisive to identify the sk gene and the development of a molecular marker for bitterness in almond (Sanchez-Perez, 2014).

## Pathogen and other Disorder Resistance

A large variation in cultivar susceptibility to pathogenic agents was observed by Grasselly (1972) who stated that resistance is usually controlled by quantitative genes. By crossing-almond (susceptible) with peach (resistant), Kochba and Spiegel-Roy (1912) obtained $F_1$ hybrids showing almost complete dominance to the nematode *Meloidogyne javanica*.

The disease non-infectious bud failure (BF) is heritable at different levels in somatic tissue. Offspring produce evidence of BF after they have been grown and propagated for a number of years (Kester, 1968). According to Kester (1968) this disorder may be explained on the basis of the concept that paramutation involves the progressive regression of a dominant factor *N*, essential for the normal bud development, to an altered form $N^1$. This altered $N^1$ allele (or segment of the chromosome) would be recessive and BF expression would result only when $N^l N^l$ alleles are combined. It is, therefore, obviously advisable to exclude from breeding programmes the cultivars affected by this disease.

Wild almond species, showing resistance to major diseases, early ripening and very late bloom, have been used by Russian research workers to introduce these characteristics into cultivated almond.

## Rootability

By crossing almond, which normally has an inconsistent rooting capacity, with peach, which shows high capacity to root, Kester and Sartori (1966) obtained in the $F_1$ a heterogeneous hybrid population, ranging from individuals having difficult-to-root by cuttings to those which rooted better than peach cuttings.

### Mutagenesis

Mutagenesis is still employed together with cross-breeding for crop improvement in almond. The exposure of buds of the cultivars Fascionello to radiaton with $^{60}Co$ at a dose of 3 Krad during vegetative dormancy resulted in a branch which bloomed 15 days later (Monastra and Fideghelli, 1977).

*Agrobacterium rhizogenes* has been used to get transformed roots (Rugini, 1984, 1986). An attempt to regenerate plants from them and so to introduce easy-to-root ability in almond species is in progress.

Seeds of the cvs. Ne Plus Ultra, Drake and Texas irradiated with gamma rays at 1000, 2000 and 4000 rad, respectively, caused the plants, after 7 years, to produce higher seed oil content, while the most susceptible gamma irradiated Krymskii offspring showed low seed oil content (Veliev *et al.*, 1979).

## QTL Mapping

Almond breeding is increasingly taking into account kernel quality as a breeding objective. In different parameters considered in evaluating almond quality includes nut and kernel physical parameters, such as width, thickness, size, geometric diameter, spherical index *etc.*, as well as chemical components such as protein, oil content, fatty acid tocopherol content. SSR markers have become a very useful tool for constructing linkage maps and locating genes controlling phenotypic variability. The study of almond cross 'Vivot' x 'Blanquerna' (V x B) has allowed a linkage map of this progeny to be constructed (Fernandez i Marti *et al.*, 2011). Fernandez i Marti *et al.* (2014) constructed a map with 56 SSR markers for an almond population showing a wide range of variability for the chemical compositions and for the nut and kernel physical parameters in almond. A total of 26 putative quantitative trait loci (QTL) controlling these chemical and physical traits have been detected in this analysis, corresponding to seven genomic regions of the eight almond linkage groups (LG). Some QTLs were clustered in the same region or shared the same molecular markers, as supported by the correlations between the traits. The limit of detection values for any given trait ranged from 2.06 to 5.17, explaining between 13.0 and 44.0 per cent of the phenotypic variance of the trait. Regarding chemical traits, a total of 12 putative QTL have been detected (Font i Forcada *et al.*, 2012), corresponding to seven genomic regions of the eight almond linkage groups. The logarithm of the odds (LOD) values for any given trait ranged from 2.12 to 4.87, explaining from 11.0 to 33.1 per cent of the phenotypic variance of the trait. This the first complete genetic framework map for physical and chemical components of almond nut and kernel, associated with a large number of parameters controlling physical traits in almond (Fernandez i Marti *et al.*, 2014).

## REFERENCES

Agabbio, M. and Ortu, S. (1974). Riv. Ortoflorofrutt. Ital., 58:315-19.

Agabbio, M., Frau, A and Ortu, S. (1977). Studi Sassaresi III, 24: 43-55.

Alberghina, O. (1978). Tecnica Agricola, 30: 385-93.

Alberghina, O. and Contineila, G. (1981). Options Medirerreneennes, 1: 93-108.

Alcon, F., Egea, G. and Notes, P.A. (2013). Irri. Sci., 31: 931-41.

Alonso, J.M., Anson, J.M., Espiau, M.T. and Socias I Company, R. (2011). Acta Hort., 912: 337-42.

Anonymous (1981). IRVAM Bull. December, 1981. Rome.

Anonymous (1984-1985). California Fruit and Nuts Statistics.

Averna, V., Pisa, A. DI, Crescimanno, F.G., Marco L. DI and Inglese, P. (1984). Tecnica Agricola, 36:49-79.

Barnes, M.M., Barnett, W.W., Bentley, W.J., Curtis, R.K., Hoy, M.A., Rice, R.E., Weakley, C.V. and Zalom, F.G. (1985). Integrated Pest Management for Almonds, Univ. of California, publ., 3308:48-91.

Barros, P.M., Goncalves, N., Saibo, N.J.M. and Oliveira, M.M. (2012). Jr. Exp. Bot., 63: 4585-96.

Bernhard, R. and Grasselly, C. (1969). Bulletin Technique d'Information, 241.

Bostock, R.M. and English, W.H., Mircetich, J.M., Nyland, G., Ogawa, J.M. and Teviotdale, B.L. (1985). Integrated Pest Management for Almonds, Univ. of California, Publ., 3308:92-126.

Bourgin, J.P. and Nitsch, J.P. (1967). Ann. Physiol. Veg., 9:377-82.

Brewer, R.F. (1981). Almond orchard Management, (eds W.C. Micke and D.E. Kester), Univ. of California, Publ., 4092:62-67.

Cleveland, C.B., Mayyes, M.A., and Cryer, S.A. (2002). Pest Management Sci., 58:70-84.

Crescimanno, F.G. (1960). Riv. Ortoflorofrutt. Ital., l-2: 13-23.

Crossa-Raynaud, P. (1977). Ann. Amelior. Plant, 27: 491-507.

Curtis, R.K. and Barnes, M.M. (1977). Jr. Econ. Entomol., 70:395-98.

Damavaniy Kozakonane, H. (1970). Thesis, Universite de Bordeaux.

Danyluk, M.D., Nozawa-Inoue, M., Hristova, K.R., Scow, K.M., Lampinen, B. and Harris, L.J. (2007). Jr. Appl. Microbiol., 104: 1391-99.

Demkovich, M., Siegel, J.P., Higbee, B.S. and Brenbaum, M.R. (2015). Environ. Entomol., 44:855-63.

Denisov, V. (1968). Dokl. Vresop. Akad. Nauki. Lenina, SSR., 9.

Dhatt, A.S. and Dhillon, D.S. (1981). Sci. and Cult., 47: 270-71.

Dhatt, A.S., Dhillon, D.S. and Jawanda, J.S. (1982). Punjab Hort. J., 22:181-84.

Djeneva, A. and Rankova, Z. (2011). Acta Hort., 912: 679-83.

Ducom, P. (1968). La Pomologie Francaise. l0:159-74.

Egea, L., Garcia, J.E., Egea, J. and Berenguer, T. (1984). Options Mediterraneennes, 2:13-26.

Engle, C.E. and Barnes, M.M. (1983). Calif. Agric., 37:19.

Esparza, G., Dejong, T.M., Weinbaum, S. A. and Klein, I. (2001). Tree Physiol., 21: 1073-79.

Fadl, M.S., El-Barkouki, M. and EI-Deen, S.A.S. (1977). Egyptian J. Physiol. Sci., 4:157-59.

Fatali-Zade, F.A. and Semochkina, L.G. (1972). Subtrop. Kultury, 2:128-1131.

Felipe, A. and Socias i Compnay, R. (1977). Atti 3 Riunione GREMPA, Valenzano, Bari., pp. 145-48.

Fereres, E., Aldrich, T.U., Schulbach, H. and Martinich, D.A. (1981). Calif. Agric., 35:11-12.

Fernandez i Marti, A., Font i Forcada, C. and Socias I Company, R. (2012). BMC Genet., 13:47.

Fernandez i Marti, A., Howad, W., Tao, R., Alonso, J.M., Arus, P. and Socias I Company, R. (2011). Tree Genet Genomes, 7: 629-39.

Font i Forcada, C., Fernandez i Marti, A. and Socias I Company, R. (2012). BMC Genet., 13: 47.

Garcia Garcia, J.E. (1977). Thesis Univ. of Valencia, Spain, p. 300.

Garcia Garcia, J.E. and Egea Ibanez, L. (1979). Anales de Edafologia y Agrobiologia, 38: 2181-93.

Garcia Garcia, J.E., Egeu, J. and Berenguer, T. (1980). ITEA (Informacion Tecnica Economica Agraria), 11:15-22.

Garcia-Pascual, P., Nateos, M., Carbonell, V. and Salazar, D.M. (2003). Biosystem Eng., 84: 201-209.

Giorna, J. and Marshal, J. (1995). Rie Deficitario controlado Fundamentos 4 aplicaciones, Madrid. Pp. 99-118.

Godini, A. (1977). Atti 3 Riunione GREMPA, Valenzano, Bari, pp. 150-59.

Godini, A. (1979). Scienza e Tecnica Agraria, 19: 1-10.

Godini, A. (1981). Riv. Ortoflorofrutt. Ital., 65:135-42.

Gomez, E.M., Dicenta, F. and Ortega, E. (2014). Acta Hort., 1028: 127-32.

Grasselly, C. (1969). Ann. Amel. Plant., 19:3.

Grasselly, C. (1972). Ph. D. Thesis, Univ. de Bordeaux, n. 79.

Grasselly, C. (1973) Bulletin technique d'information, n. 279.

Grasselly, C. (1977). Pomologie Francaise, 5:77-84.

Grasselly, C. and Crossa-Ralnaud, P. (1980). G-P. Maisonneuve et Larose, Paris.

Grasselly, C. and Gall, H. (1967). Ann. Amel. Pl., 17: 83-91.

Grasselly, C. and Olivier, G. (1976). Ann. Amel. Pl., 26:107-13.

Grasselly, C. and Olivier, G. (1983). Options Mediterreneennes, 2:133-36.

Grigorian, V. (1972). Thesis, Univ. de Bordeaux, n. 164.

Grigorian, V. (1975). Ann. Amel. Pl., 25: 371-80.

Gulgan, R. (1975). Proceeding 2 Colloque GREMPA

Haas, A.R.C. (1945). Proc. Amer. Soc. Hort. Sci., 6:69-77.

Hansen, C.J., Kester, D.E. and Uriu, K. (1962). Calif. Agric., 16: 6-7.

Harborne, J.B. (1973). Chapman and Hall, London.

Hellali, R. and Kester, D.E. (1979). J. Amer. Soc. Hort. Sci., 104:375-78.

Heppener, M.J. (1926). Genetics, 8:390-92.

Hoy, M.A., Barnett, W.W., Hendricks, L.C., Castro, D., Cahn, D. and Bentley, W. J. (1984). Calif. Agric., 38: 18-20.

Iskaln, C., Akbas, F.A., Naml, S., Tilkat, E. and Basaran, D. (2008). African J. Biotech., 7:1875-80.

Jablonskij, E.A. and Markovich, Z.Y. (1970). Trud. Gosudar. Nikits. Bota. Sada., 46:63-83.

Jaceldo-Siegl, K., Sabate, J., Batech, M. and Frascer, G.E. (2011). Nutrition, Metabolism and Cardivascular Disease, 21: 7-13.

Kesrer, D.E. and Asay, R. (1975). Advance in Fruit Breeding (eds. J. Janick and J. N. Moore), pp. 387-419. Purdue. Univ. Press, Lafayette, India.

Kester, D.E, Liu, L., Fenton, C.A.L. and Durzan, D.J. (1986). Biotechnology in Agriculture and Forestry, Vol. 1 Trees (ed. Y.P.S Bajaj) Springer-Verlag, Berlin,

Kester, D.E. (1965). Proc. Amer. Soc. Hort. Sci., 87:214-21.

Kester, D.E. (1968). Proc. Amer. Soc. Hort. Sci., 92: 16-28.

Kester, D.E. and Sartori, E. (l966). Proc. Amer. Soc. Hort. Sci., 88:219-23.

Kester, D.E., Brown, J.G. and Aldrich, T. (1956). Calif. Agric., l0:13.

Kester, D.E., Hansche, P.E., Beres, V. and Asay, R.N. (1977). J. Amer. Soc. Hort. Sci., 102: 264-23.

Khan, I., Khalil, S.K., Wahad, F.I., Khattak, S.U.K. and Saeed, A. (2009). Pak. J. Zool., 41: 445-53.

Kochba, J. and Spiegel-Roy, P. (1972). HortSci., 7: 503-05.

Lichev, P. and K'osev, Kh. (1983). Gradinarska i Lozarska Nauka, 20: 14-20.

Lilleland, O. (1951). In: Fruit Nutrition, (ed. N.F. Childers) Rutgers State Univ., New Brunswick, N J., p 264.

Liu, XiaoHua, (2008). China Fruits, 1: 36-37.

Loreti, F. and Viti, R. (1984). Options Mdditerraneennes, 2: 177-83.

Loreti, F. and Xiloyannis, C. (1978). Riv. Ortoflorofrutt. Ital., 62:147-54.

Loreti, F., Morini, S. and Grilli, A. (1985). Acta Hort., 173:261-69.

Loreti, F., Viti, R. and Xiloyannis, C. (1979). Atti Migloramento della Coltura di Mondorlo e Nocciolo. Aspetti genetici e Tecnici. Catania, pp. 371-18.

Martin, G.C. and Kester, D. (1978). In: Almond Water Management (eds. W.C. Micke and D. Kester). Univ. California, Barkley, pp. 46-51.

McClure, B., Haring, V., Ebert, P.R., Anderson, M.A. Simpson, R.J., Sakiyama, F. and Clarke, A.E. (1989). Nature, 342: 955-57.

Mehra, A. and Mehra, P.N. (1974). Bot. Gaz., 132: 6l-73.

Micke, W.C. and Kester, D.E. (1981). Almond Orchard Management, University of California, Publ., 4092:52-56.

Milella, A. (1959). Riv. Ortoflorofrutt. Ital., 5: 322-32.

Milelta, A. (1963). Studi Sassaresi, 3:13-22.

Mircetich, S.M., Lowe, S.K., Moller, W.J. and Nyland, G. (1976). Phytopath., 66:17-24.

Monastra, F. (1972). Rev. Ortoflorofrutt. Ital., 6:325.

Monastra, F. (1975). Colloque GREMPA. Montpellier.

Monastra, F. (1976). Atti Congreso International de Almendra Y Avelldna, Reus (SPAIN)

Monastra, F. (1979). Edagricole, Bologna, p. 78.

Monastra, F. and Fideghelli, C. (1977). Atti 3° fuunione GREMPA, pp. 55-64, Valenzano (Bari).

Monastra, F. and Pennone, F. (1980). Atti AI Convegno Pomologico Caserta.

Monastra, F., Crisafulli, A., Marchese, F., Ondradu, G., Pavia, R. and Rrvalta, L. (1982). Ministero Agricoltura e Foreste, ROMA.

Monastra, F., Della Strada, G., Fideghelli, C. and Quarta, R. (1980). 4 Colloque du GREMPA, Smirne.

Murashige, T. and Skoog, F. (1962). Physiol. Plant., 15: 473-97.

Nanos, G.D., Kazantzis, L., Kefalas, P., Petrakis, C. and Stavroulakis, G.G. (2002). Scientia Hort., 96: 249-56.

Palasciano, M., Logoluso, V. and Vllegas-Monter, A. (2009). Revista Fitotecnia Mexicana, 32: 103-109.

Palumbo, J.D., Mahoney, N.L., Light, D.M., Siegel, J.P., Puckett, R.D. and Michailides, T.J. (2014). Pl. Disease, 98: 1194-99.

Parada-Ponce, D.M. and Villegas-Monter, A. (2009). Revista Fitolecnia Mexicana, 32: 103-109.

Phillips, D.J., Purcelt, S.L. and Stanley, G.L. (1980). Agr. Rev. and Manuals US Dept. of Agric., No. FARM-W-20, pp 12.

Piglionica, V. (1967). Phytopath. Medit., 6: 138-43.

Piglionica, V. and Ferrara, G. (1979). Atti 3° Riunione CREMPA, p. 321-30, Valenzano (Bari.).

Pimienta, E. and Polito, V.S. (1982). Amer. J. Bot., 69:913-20.

Pimienta, E. and Polito, V.S. (1983). Ann. Bot., 51: 469-79.

Pimienta, E., Polito, V.S. and Kester, D.E. (1983). J. Amer. Soc. Hort. Sci., 108:643-47.

Popov, M.G. (1929). Bull. Appl. Bot., 22:241-83.

Proebsting, E.L. and Serr, E.F. (1966). In: Fruit Nutrition. (ed. N.F. Childers). Rutgers State Univ., New Brunswiek, N.J., p. 264.

Quacquarelli, A. and Martelli, G.P. (1972). Actas III Congr. Un. Med. Oeiras, 22-28 October.

Rehder, I. (1921). J. Arnold Arboretum, 3: 24.

Rice, R.E. (1981). Almond Orchard Management, Univ. of California, Publ., 4092:117-24.

Rodriguez Navarro, J. (1981). Ann. del Ist. Nacional de Invest. Agrarias Agricola, 16:31-43.

Roldan-Fajardo, B.E., Barea, J.M., Ocampo, J.A. and Azcon-Aguilar, C. (1982). Plant Soil, 68:361-67.

Romero, A., Vargas, F.J., Tous, J., Ninot, A. and Miarnau, X. (2011). Acta Hort., 912: 399-404.

Romero, P. (2002). Doctoral thesis. Univ. Murcia, Spain. pp. 307.

Romojaro, F. and Berenguer, T. (1984). Options Mdditerrannennes, 2:185-88.

Rugini, E. (1984). Proc. 41st Conf. in Plant Tissue Cult. and its Agriculture Applications. Nottingham 17-21, September (Abstract).

Rugini, E. (1986). Handbook of Plant Cell Culture. (eds. D.A. Evans, R.W. Sharp and P.V. Ammirato) Vol 4, pp. 515-612 Macmillan, New York.

Rugini, E. and Verma, D.C. (1982). Proc. 5th Intern. Congr. Plant Tissue and Cell Culture (ed. F. Fujiware), pp. 141-742, Tokyo.

Rugini, E. and Verma, D.C. (1983). Plant Sci. Let., 28:273-81.

Sanchez-Perez, R. (2014). Acta Hort., 1028: 73-78.

Savino, V. (1985). L'Italia Agricola, 4-6:115-24.

Sazo, L., Araya, J.E. and Esparza, S. (2008). Chilean J. Agric. Res., 68: 284-89.

Scaramuzzi, G. (1957). Atti Ist. Bot. Lab. Critt. Univ. Pavia, Sez. V., 14:329-38.

Sensi, E., Rizzolo, A. and Sarlo, S. (1991). Italian J. Fd. Sci., 3: 209-18.

Sensi, E., Rizzolo, A., Colombo, C. and Testoni, A. (1996). Italian J. Fd. Sci., 8: 115-25.

Sepahvand, S., Ebadi, A., Kamali, K. and Ghaemmaghami, S.A. (2012). Jr. Agric. Sci. Toronto, 4: 275-80.

Serafimov, S.S. (1977). Gradinarska i Lozarska Nauka, 14:15-19.

Shrestha, A., Moretti, M. and Mourad, N. (2012). Weed Tech., 26: 110-16.

Socias I Company, R. and Felipe, A.J. (2007). HortSci., 42: 704-06.

Socias I Company, R., Kodad, O., Alonso, J.M. and Felipe, A.J. (2008). HortSci., 43: 2240-42.

Socias, I Company, R. (1976). Thesis Mast. Sci., Univ. California, Davis.

Soria, J., Palacios, S.J. and Velasco, E.E. (1982). Anales de Edafologia Y Agrobiologia,4l: 1003-25.

Spiegel-Roy, P. and Kochba, J. (1974). Z. Pflanzenzucht, 7l: 319-29.

Spiegel-Roy, P. and Kochba, J. (1977). Atti 3° Riunione GREMPA, pp. 163-79, Valenzano, (Bari.)

Spiegel-Roy, P., Kochba, J. and Iris, R. (1982). HortSci., 17: 271.

Spiegel-Roy, P., Striem, M. and Baron, I. (1984). Alon Hanotea, 39:259-531.

Tabuenca, M.C. (1979). Ann. Est. Exp. de Aula Dei., 14:519-31.

Tabuenca, M.C., Mut, M. and Herrero, S. (1972). Ann. Est. Exp. de Aula Dei., 3-4:378-95.

Thorp, W.R. (1981). Almond Orchard Management, Univ. of California, Publ. 4092:57-64.

Tombesi, S., Lampinen, B.D., Metcalf, S. and Dejong, T.M. (2011). Tree Physiol., 31: 1413-21.

Tombesi, S., Sealia, R., Connell, J., Lampinen, B. and Dejong, T.M. (2010). J. Hort. Sci. Biotech., 85: 317-22.

Vavilov, N.I. (1930). Intern. Hort. Congr., London.

Veliev, I.M., Khudadatov, A.I. and Cuseinova, F. (1979). Izvestiya Akad. Nauk Azerbaidzhanskoi SSR, Biologicheskikh Nauka, 4: 39-43.

Vitagliano, C. and Viti, R. (1983). Riv. Ortoflorofruticol. Ital., 67: 139- 1145.

Wein, M.A., Bleich, D., Raghuwanshi, M., Gould-Forgerite, S., Gomes, J., Monahan-Couch, L. and Oda, K. (2010). Jr. Amer. Clinic. Nutr., 29: 189-97.

Weinbaum, S.A., Uriu, K., Micke, W.C. and Meith, H.C. (1980a). HortSci., 15:78-79.

Weinbaum, S.A., Even-Chen, Z. and Kester, D.E. (1980b). HortSci., 15: 583-85.

West, L.D., Hildreth, R.C. and Schlieselman, J. (1983). Proc. Western Society Weed Sci., 36:189-90.

Woodroof, J.G. (1967). Tree Nuts, Vol. 1, pp. l3z-182. The AVI publ., Westport. Connecticut.

Wu, S.C. and Kuniyuki, A.H. (1985). Plant Sci., Ireland, 4l:55-60.

Yada, S., Lapsley, K. and Huang, G.W. (2011). Jr. Food Chem. Anal., 24: 469-80.

Zalom, F.G., Wilson, L.T., Hoy, M.A., Barnet, W.W. and Smilanick, J.M. (1984). Calif. Agric., 38: 17-19.

Zdruikovskaya-Rikhter, A. (1980). Byull. Gl. Bot. Sada, 115: 85-90.

Zuccherelli, G. (1979). Frutticoltura, 41:15-20.

# Chapter 8

# Chestnuts

The chestnuts (*Castanea* sp.) have been an important item of human diet in Northern Hemisphere since time immemorial. The remains of carbonized plants recovered from the sites destroyed in 79 AD furnish evidence of many staple food plants used then and chestnut is one of them (Meyer, 1980). It has been grown in China and Japan for more than 4000 years for food, fuel, timber and beautification. The popularity of chestnut is attributed to its productivity, ease of harvesting; simplicity of preparation for eating, low fat content and diversity of uses for high and low income people (Woodroof, 1979).

## Composition and Uses

The composition of Chinese chestnuts as estimated by Watt and Merrill (1963) is presented in Table 8.1. Among all the major nuts, only chestnuts have a very low fat content and consequently, can be processed by boiling, drying and stored at moderate temperatures without the problem of rancidity. For the same reason, roasted and boiled nuts are more easily digestible (Westwood, 1987).

**Table 8.1: Composition of Chestnuts (100 g edible portion)***

| *Ingredient* | *Fresh* | *Dried* | *Flour* |
|---|---|---|---|
| Water (per cent) | 52.5 | 8.4 | 11.4 |
| Protein (g) | 2.9 | 6.7 | 6.1 |
| Fat (g) | 1.5 | 4.1 | 3.7 |
| Carbohydrates (g) | 42.1 | 78.6 | 76.2 |
| Fibre (g) | 1.1 | 2.5 | 2.0 |
| Ash (g) | 1.0 | 2.2 | 2.6 |
| Calcium (mg) | 27 | 52 | 50 |

| *Ingredient* | *Fresh* | *Dried* | *Flour* |
|---|---|---|---|
| Phosphorus (mg) | 88 | 162 | 164 |
| Iron (mg) | 1.7 | 3.3 | 3.2 |
| Sodium (mg) | 6 | 12 | 11 |
| Potassium (mg) | 454 | 875 | 847 |
| Magnesium (mg) | 41 | - | - |
| Vitamin A (IU) | - | - | - |
| Thiamine (mg) | 0.22 | 0.32 | 0.23 |
| Riboflavin (mg) | 0.22 | 0.38 | 0.37 |
| Niacin (mg) | 0.6 | 1.2 | 1.0 |
| Ascorbic acid (mg) | - | - | - |
| Energy (cal) | 194 | 377 | 362 |

* Watt and Merrill (1963).

Chestnut fruits are very rich in starch, free sugars and fibre. Two forms of starch-amylose (33 per cent of starch) and amylopectin (67 per cent of starch) was reported (Pereira-Lorenzo *et al.*, 2006). Both mono-and di-saccharides (glucose, fructose, sucrose and maltose) readily absorbed sucrose are present in the nuts. Chestnut is low in fat. The major unsaturated fatty acids presents are oleic, linoleic and α-linolenic constituting more than 85 per cent of the total fatty acid content (Borges *et al.*, 2007). The major minerals of chestnut fruit (*C. sativa*) was reported as Zn (5099.4 mg/kg), Mn (3031.9 mg/kg), Na 1058.6 mg/kg) and Ca (308.5 mg/kg) (Yldz *et al.*, 2009). Dried chestnuts contain a modest content of sodium (17 mg/100g), but a very high content of potassium (738 mg/100g average). Two important vitamins: riboflavin ($B_2$) and nicotinic acid content are present in good amount in chestnuts. Other vitamins include $B_1$ (thiamine, 0.1 mg/100g), C (23 mg/100g) and pantothenic acid (0.9 mg/100g) (Kunsch *et al.*, 1999). European chestnuts contain less starch and more dextrose and other water soluble carbohydrates, whereas wild small American chestnuts are richer in lipids but contain less starch (Woodroof, 1979). McCarthy and Meredith (1988) have compared the chemical composition of American, Chinese and European chestnuts (Table 8.2). American and Chinese

**Table 8.2: Proximate Composition of Raw Chestnut Kernels (100 g edible portion)***

| *Proximates* | *American* | | *Chinese* | | *European* | |
|---|---|---|---|---|---|---|
| | *Mean (g)* | *SE* | *Mean (g)* | *SE* | *Mean (g)* | *SE* |
| Water | 43.70 | - | 43.95 | - | 54.88 | - |
| Protein (N x 5.30) | 4.83 | 0.06 | 4.20 | 0.01 | 1.98 | 0.00 |
| Fat | 1.32 | 0.04 | 1.11 | 0.05 | 1.63 | 0.03 |
| Total carbohydrate | 48.57 | - | 49.07 | - | 40.28 | - |
| Crude fibre | 1.91 | 0.01 | 1.64 | 0.04 | 1.36 | 0.04 |
| Ash | 1.58 | 0.02 | 1.67 | 0.00 | 1.23 | 0.00 |

*McCarthy and Meredith (1988).

chestnuts are high in protein, crude fibre, ash, iron, magnesium and phosphorus than European chestnuts. However, European chestnuts have higher fat values than American and Chinese chestnuts.

The chestnut fruit processing generates large amounts of residue as pericarp (outer shell; 8.9-13.5 per cent) and integument (inner shell; 6.3-10.1 per cent). These materials clearly have the potential as source of valuable co-products. The analysis of the pericarp and integument of four Portuguese chestnut cultivars (Judia, Longal, Martaimha and Lada) reveald significant content of total phenolics, low molecular weight phenolics (gallic and ellagic acid), condensed tannins and ellagitannins including castalagin, vescalagin, acutissimin A, and acutissimin B. The integument tissue had the highest levels of total phenolics and condensed tannins. These materials could be used for the extraction of valuable phenolics (Vasconcelos *et al.*, 2010).

The chestnut with best organoleptic characteristics are sold fresh, after conditioning (sterilization with or without curing) to guarantee the safety of the products. The wide range of processed itmes like dried chestnuts, flour, chestnut creams, candied chestnut, chestnut flakes, vacuum packed peeled chestnut are known and appreciated since centuries. The seeds are used in paediatric to treat gastroenteritis and are also suitable component in gluten-free diets for people suffering from coeliac disease. Chestnut honey is characterized by a good percentage of isomalt, a high fructose/glucose ratio, a low glucose/water ratio and in terms of minerals it presents a significant larger amount of potassium than other honeys (Bounous, 2009).

Chastnut wood is used for timber, paper, or in fuel production. The timber is straight grained, and closely resembles oak in colour and texture. Chestnut trees have ornamental value in landscape planting and also provide shade. To develop timber trees for reforestation, hybrids between Chinese chestnut (*Castanea mollissima*) and American chestnut (*Castanea dentata*), for uniform progeny, are being raised in the USA (Diller *et al.*, 1964).

## Origin and Distribution

The different species of chestnuts are native to western Asia, Europe and northern America (Anon., 1959; Vavilov, 1951). Chinese chestnut (*Castanea mollissima*) is native to China, Japanese chestnut (*C. crenata*) to Japan and South Korea, and European chestnut (*C. sativa*) is native to Mediterranean countries of southern Europe and Asia Minor (McKay and Jaynes, 1969). Western America is the native place for American chestnut (*C. dentate*). Beside chestnuts, chinkapins also belong to this genus and *C. pumila*, *C. ozarkensas* and *C. alnifolia* (trailling chestnut) are native to southeastern USA, Henry chestnut (*C. henryi*) and Seguin chestnut, (*C. seguinii*) are shrubs, native to China. Jaynes (1969) has described the common name, chromosome number and range of different species of chestnut and chinkapins (Table 8.3).

## Production

Chestnut forests are found in China, USA, Europe, Australia, New Zealand,

India, Turkey and many other countries. China is the largest producer of chestnut producing 1,620,000 tones annually followed by South Korea (82,200 tons), Turkey, Bolivia, Italy, Japan, Portugal, Greece, North Korea and France (Table 8.4). In India, the production figures are not known as it grows scattered in Jammu and Kashmir, Himachal Pradesh and North Eastern Hill Region.

**Table 8.3: Common and Botanical Name, Chromosome Number, and Range of Chestnut (*Castanea*) Species***

| *Common Name* | *Botanical Name* | *Chromosome Number* | *Range* |
|---|---|---|---|
| Chinese chestnut | *C. mollissima* | 24 | China |
| Japanese chestnut | *C. crenata* | 24 | Japan, Korea |
| European chestnut | *C. sativa* | 24 | South Europe, Asia Minor |
| American chestnut | *C. dentate* | 24 | East USA-Maine to Alabama |
| Seguin chestnut* | *C. seguinii* | 24 | China |
| - | *C. davidii* | | China |
| | **Balanocastanon Section** | | |
| Allegany chinkapin | *C. pumila* | 24 | Arkansas, Missouri |
| Ozark chinkapin | *C. ozarkensis* | 24 | North Carolina to Florida |
| Ashe chinkapin | *C. ashei* | 24 | Gergia and Florida to Texas |
| Florida chinkapin | *C. floridana* | | Southeastern United States |
| | *C. pauscispina* | | |
| | **Hypecastanon Section** | | |
| Henry chinkpin | *C. henryi* | 24 | Southwestern China |

*It is a chinkapin more closely related to chestnut; hence included in the chestnut section.

*Jaynes (1969).

**Table 8.4: Major Chestnut Producing Countries of the World**

| *Countries* | *Production (tones)* |
|---|---|
| China | 1,620,000 |
| Korea (South) | 82,200 |
| Turkey | 59,171 |
| Bolivia | 53,577 |
| Italy | 42,700 |
| Japan | 23,500 |
| Portugal | 22,400 |
| Greece | 20,900 |
| Korea (North) | 10,700 |
| France | 9,530 |

FAOSTAT (2010).

Chestnut Blight

Chinese Chestnut

Chestnut Orchard

Male Flower Stock

## Taxonomy

The genus *Castanea* belongs to the sub-family Castaneoideae and family Fagaceae. The chestnut tree is geographically distributed in three main areas. Asia with *Castanea crenata* Sieb. and Zucc. (Japan) and *C. mollissima* Bl. (China and Korea), North America with *C. dentate* Borkh. and Europe with *C. sativa* (Bounous, 2005). The tree is deciduous species with hypogeal germination. Chestnut is a monoecious species and has two different types of inflorescence: the unisexual staminate catkins in the lower parts of the shoot and the bisexual catkins in the terminal end of the shoots (Valdiviesso *et al.*, 1993). Nuts are large, brown with a pale scar at the base. Generally, 1-3 nuts per involucre or bur are present but rarely 5 or 7 nuts are also seen (Bailey, 1949, Westwood, 1978). Important species of *Castanea* are described below:

## Chestnut

### *C. mollissima* (Chinese chestnut)

The name *mollissima* means soft hair and this species is recognized by dense hair on young leaves, and in winter, downy yellow terminal parts of the branchlets. Leaf blade is thicker and, in general, mature leaves are broader in relation to length than other species (Woodroof, 1979). The species is the most promising from the nut production viewpoint. The nuts have small scar or hilum. The pellicle or thin membranous skin on the nuts is thin and peels readily from the kernel. The trees are spreading type and long-lived with a round top. During dormancy, trees can withstand cold temperature up to -28.8°C. It is resistant to chestnut blight fungus and most of the cultivars belong to this species. Chinese chestnut has been introduced into many countries for its adaptability to different pedoclimates. It grows from 50 to 2800 msl. in a wide range of climatic conditions.

### *C. crenata* (Japanese chestnut)

Narrow leaves with glandular scales on the under surface are the characteristics of this species. The nuts are large among all the species, but poor in quality. The kernel is less sweet as compared with Chinese chestnut and the pellicle is fibrous and adheres with the kernel. The scar on the nuts is large. The trees are more susceptible to chestnut blight and are less winter-hardy. It spread from Japan to Korea and to northeast China, and it was naturalized in South Korea and in Taiwan (Tanaka *et al.*, 2005).

### *C. sativa* (European chestnut)

It is also known as Italian or Spanish chestnut. Leaves are long with coarse-toothed margins and hairy. Tall tree of majestic appearance; it is vigorous and can exceed 30 m height and 400 years of age. Nuts are dark brown in colour with light-coloured stripes. Pellicle is more fibrous and frequently folded into the folds of kernel and thus is removed with difficulty. The species is highly susceptible to the dreadful disease, chestnut blight.

### *C. dentate* (American chestnut)

Trees are tall and straight and often reaching a height of 30 m in forests with a stem diameter of 1-3 m. Leaves oblong-lanceolate, acute, long-pointed at the apex, coarsely serrated except towards wedge-shaped base, and hair is absent on both the surfaces. The nuts are broader and small (75-150 per pound) and have a tuft of hair at the apical end. The nuts are sweet with easily separating pellicles and probably the best of all the nuts. The trees are highly susceptible to blight and this caused a complete annihilation of entire plantations in the United States during the beginning of 20$^{th}$ century (McKay and Jaynes, 1969).

## Chinkapin

Chinkapins unlike chestnuts have only one nut in each bur and this is the characteristic difference between chestnuts and chinkapins (Jaynes, 1969).

### *C. pumila* (Allegany chinkapin)

Trees of the most commonly occurring chinkapin rarely attain the height of 3 metres. The leaves have tuft of hair on the underside. The burs are round (2.5 cm diameter) with soft spines. The nuts are small, sweet and flavoured. The trees of this species are grown for ornamental purpose and for plantation in wildlife sanctuaries for food. The species can also be used as dwarf rootstock, but abundant suckering is a drawback.

### *C. ozarkensis* (Ozark chinkapin)

The species can be grown into small trees and often produces multi-stemmed shrub. It grows wild in Arkansas and Missouri states of USA and serves as wild-life food.

### *C. alnifolia* (Trailing chinkapin)

This species produces small shrubs, one metre in height and is naturally propagated by stolons in sandy barrens. Leaves are large, oval-lanceolate, generally obtuse and slightly hairy on the underside (Bailey, 1949).

### *C. henryi* (Henry chinkapin)

It is a forest tree growing in wild forms in southwestern China. The leaves are toothed and narrow. The species is susceptible to blight.

### *C. seguinii* (Seguin chestnut)

The small trees or shrubs are native to China. Plants are dwarf in nature and bears heavily at young age and these attributes are important for breeding programme. The burs have three nuts unlike American chinkapins.

## Cultivars

Generally the seedling selections of different species are grown in the regions of their origin.

### Abundance

The cultivar was selected in the USA from the Chinese chestnut seedlings. The nuts are attractive in appearance with excellent flavour and the pellicle separates easily from the kernel (Reed and Davidson, 1954). Fully grown trees produce undersized nuts making 'Abundance' unsuitable as future cultivar.

### Nanking

Nanking is a popular Chinese chestnut cultivar originated in Georgia. The nuts are dark brown, bearing heavy crops regularly. It is a mid-season cultivar (Brooks and Olmo, 1972).

### Meiling

The cultivar is a selection from the seedlings raised from the nuts introduced from China and has been released by the USDA. The nuts resemble so much to Nanking that they are almost indistinguishable. The trees are vigorous, upright and bear early heavy crop.

### Orrin

The seedling tested as Good C-7 was named Orrin by the USDA. The nuts are dark mahogany, almost black in colour with a slightly pubescent tip (Woodroof, 1979). 'Orrin' flowers later and matures earlier than other blight-resistant cultivars (McKay, 1963). The nuts are superior in flavour and keeping quality than most of the other cultivars. The trees are prolific bearers and fairly tolerant to cold.

### Crane

It originated from the seeds imported from China and later on tested by grafting on seedling rootstocks and grown with Nanking at different locations in the USA. The nuts are dark, cherry-coloured and nearly completely glabrous. Well-cured nuts are excellent in flavour and keeping quality is better than all other cultivars. The grafted trees start bearing in the second year. This mid-season cultivar should be planted with Nanking for cross-pollination and better fruit set (Brooks and Olmo, 1972).

### Sleeping Giant

It is a cross between Chinese chestnut and hybrid of Japanese chestnut x American chestnut. Large nuts are shiny and slightly conical. The trees are blight resistant; grow large with green, glossy leaves being suitable for ornamental purpose.

### Hemming

The nuts are of good quality and mahogany colour but variable in size. The scion of this cultivar makes good union with the seedlings of Chinese chestnut.

### Willoka

It originated in British Columbia as an open-pollinated seedling of *C. mollissima*. The nuts are large with yellow kernels and pleasant flavour and the tree produces good crop (Brooks and Olmo, 1972).

### Jersey Gem

It is a hybrid between Orrin and Nanking. The kernel is sweet, crunchy with yellow meat and pleasant flavour. It is harvested in the first week of October. The tree is vigorous, round and productive. Nuts mature uniformly and drop in a few days (Brooks and Olmo, 1983).

### Marron de Saint-Vincent

This French cultivar is also known as d'Olargues. It fruits in 4-5 years after planting. It is tolerant to spring frost. The quality is good but fruit is rather small (80-100 dry fruits/kg). The cultivar is, however, susceptible to ink disease and blight (Casini *et al.*, 1987).

## Chinese Cultivars

### Bayue Hong

Chinese chestnut cultivar, selected from seedlings of 'Luotain' chestnut (JunYong *et al.*, 2011). High yielder, big nut size, good quality and barrenness tolerance. The nut is dark red.

### Daiye Zaofeng

Chinese chestnut cultivar, selected from seedlings of 'Mountain Tai'. Early ripening, good quality and high yielder. Nut is glossy, medium to large in size (GuangNing *et al.*, 2011).

### Yunliang

A selection of *C. mollissima* developed in 1987 (Bin *et al.*, 1989). Average yield 3.5kg/tree, nuts are rich in protein (7.23 per cent), average nut weight 11.0g and good quality.

### Yunzhen

It is also a selection of *C. mollissima* with similar yield and nut weight of Yunliang. Nuts have protein content of 8.35 per cent and starch content of 43.78 per cent (Bin *et al.*, 1989).

### Heishanzhai 7

This Chinese cultivar with shorter catkins developed from the 'Yanshan' variety of *C. mollissima* (WuGang *et al.*, 2009). Lengths of most catkins are only 0.3 cm. Average nut weight 8.7g, suitable to grow in drought mountainous region.

## Japanese Cultivars

### Tskuba (Ganne x Hayadama)

Tree upright and vigorous, average nut weight more than 25g, reddish brown, good kernel quality and suitable for processing.

### Tanzawa (Otomune x Taisyowase)

Tree vigorous and spreading. Nuts are light brown, average weight 23g, good kernel quality and matures in early September.

### Ginyose

A chance seedling, located at Toyono, Osaka, nuts are compressed round, average weight more than 25g, dark brown with good kernel quality.

### Kunimi (Tanzawa x Taisyowase)

Tree spreading, small canopy, nuts are round to roundish. It is resistant to gall wasp.

### Ishizuchi (Ganne x Kasaharawase)

Tree moderately vigorous, nuts are round, reddish brown and average weight 25g. It is resistant to chestnut gall wasp.

## Important Cultivars of California (USA)

### Colossal (European x Japanese)

Produces uniform sized, very large, sweet nuts and average 40 per kg. Productive trees, however moderately susceptible to blight.

### Silverleaf (European)

Large, very sweet nut that peels easily, but the shell has a tendency to crack, used as pollenizer of 'Colossal.

**Fowler (European)**

Nuts are of excellent quality and size with a 'Marrone' type nut.

**Montesol (European)**

Highly productive, produces very large nuts, with good quality.

**Neveda (European x Japanese)**

Pollenizer for 'Colossal', very productive, nuts are smaller than 'Colossal'.

**Douglas Hybrid (Chinese x American)**

Good sized nuts, late flowering, blight resistant.

**Eurobella (European x Japanese)**

Pollenizer for 'Colossal', nuts are smaller.

Chestnut cultivars growing in different countries is presented in Table 8.5.

**Table 8.5: Chest Nut Cultivars Growing in Different Countries**

| | | |
|---|---|---|
| Australia | : | Buffalo Queen, Red Spanish, Purton's Pride, Bouche de Betizac, Marrone di Chiusa Pesio. |
| Brazil | : | Okuniwase, Isumo, Taisho-Wase, Tamatsukuri, KM2, Tiodawase, Kinshu. |
| China | : | Chu shu hong, Jiu jia zhong, Duan zha, Qin zha, Jiao zha, Jian ding yu li. |
| France | : | Marigoule, Bouche de Betizae, Goujounac (P), Belle Epine (P), Precoce Migoule. |
| India | : | H6-30, H-6, H5-29, H-29, H-27, H-12, HN-15. |
| Italy | : | Chiusa Pesio, Villar Focchiardo, Luserna, S.Giorio, Bruzola, Perosa Argentina, Roncegno, Drena Combai, Monfenera, Siriato, Pulfero, Zocca. |
| Japan | : | Tanzawa, Tsukuba, Ishizuchi, Ginyose, Kunmi. |
| Korea (South) | : | Okkwang, Kwangjujoyul, Sandae, Sunseong, Daab. |
| New Zealand | : | 1002, 1005, 1015. |
| Portugal | : | Longal, Lada, Aveleira, Boa Ventura, Cota, Lamela, Negra, Trigueira, Amarelal, Benfeita, Bebim, Judia, Baria, Viana, Colarinha. |
| Romania | : | Mara, Iza, Gureni, Hobita, Polovragi, Tismana, |
| Spain | : | Negral, Garrida, Loura, Amarelante, Inxerta, Famosa, Ventura, Longal, Pelona, Parede, Chamberga. |
| Switzerland | : | Luina, Verdesa, Marrone-di Cuneo, Pinca, Berogna. |
| Turkey | : | Albayrak, Altinay, Unal, Salipazari, Erfelk, Marigoule, Isiklar. |
| USA | : | Amy, Bouche de Betizac, Colossal, Gideon, Maraval, Marki, Meiling, Nanking, Paragon, Primato, Qing, Okei, Revival, Shing, Silverleaf, Smith, Tsukuba, Yolo Grande, Naveda, Benton Harbor. |

## Climate

Chestnut grows in temperate zone lying between 30° and 50° latitudes (Westwood, 1978). The chestnut species do well in a wide range of climatic conditions but respond differently to winter temperatures. Chinese chestnut trees are more cold-hardy than Japanese chestnut (Yasunobu, 1968). Cold-hardiness of the twigs increases rapidly after late November and lasts until early March, being most marked in mid-December when they could withstand –10°C (Yasunobu, 1966). Chinese chestnut can withstand temperature up to –32°C without injury when fully dormant (Crane, 1960). The chestnut trees blossom late in the spring, but young flowers are formed soon after the buds begin to swell and are subjected to killing due to frost. Therefore, the trees should be located on high sloping grounds for adequate air drainage to prevent damage from late spring frost. Bencat and Bolvansky (1984) studied the influence of climatic factors on Spanish chestnut at different locations and suggested that temperature had a dominant influence and rainfall a secondary influence on fruit weight. The annual rainfall in chestnut producing areas is 75 cm and precipitation of 2.5 cm is received during the dry period of August. Adequate moisture, especially in summer months and sufficient winter chill received at high altitude are the critical factors for nut production in Turkey. The chilling requirement of most cultivars of *C. sativa* is low. Some cultivars of *C. dentata* and its hybrids with other species seem to have a chilling requirement as that of Santa Barbara (Chandler, 1957).*

The European chestnut influenced by oceanic climate which is characterized by annual sunlight of 2400-2600 h, rainfall ranging between 600 and 1500 mm, mean annual temperature between 9-13°, 27°C being the mean of maximum temperature (Heiniger and Conedera, 1992; Gomes-Laranjo *et al.*, 2008). Chestnut growing regions must have 1900-2200 degree-days between May and October (Dinis *et al.*, 2011). The impact of climate change on the phenology of Chinese chestnut (*C. mollissima*) for 45 years (1963-2008) was studied by Liang *et al.* (2013). They reported that flowering dates in Beijing advanced by 1.6 days per decade over the length of the record. The growing season expanded by 4.3 days per decade. Warming during the period of chill accumulation delayed rather than advanced spring phenology.

## Soil

Generally, chestnuts are grown on wide range of soils, but the well-drained, deep and fertile soils are considered the best. Chinese chestnut trees grow tall and spread well in fertile soils (McKay and Jaynes, 1969). Soil should be slightly acidic, with pH 5.6-6.5 (Crane, 1960). Chestnut growth in relation to pH was studied in solution culture experiments (Ishitsuka *et al.*, 1979). The trees grew better in nutrient solutions of pH 5 rather than at 6 or 7. Adequate soil moisture ensures proper growth and production of quality nuts, but well-established trees can withstand drought conditions. Near Oklairoma City one plantation produces crop regularly even under frequent drought conditions (Hirschi, 1946). Chestnut does not thrive in waterlogged soils; in adverse conditions, growth, yield and quality are affected. Sweet chestnut (*C. sativa*) grows well in sedimentary or siliceous soil. Their roots tend to decay in poorly drained soils, which explain why they grow well in hills and mountainsides (Pereira *et al.*, 2011).

## Rootstocks

The seedlings of the same species should be used as rootstock in chestnuts. Chinese chestnut seedlings are used as rootstocks for most cultivars due to the fact that these are resistant to chestnut blight and cold injury and most of the cultivars belong to Chinese chestnut. However, the chestnut species hybridize freely when planted near each other and these mixed seedlings may lead to incompatibility (Macdaniels, 1955). Torikata and Higuchi (1962) studied the compatibility between Japanese chestnut (*C. crenata*) and Chinese chestnut (*C. mollissina*) using the cultivars Gin-Yose (Japanese) and Boji (Chinese) as scions, and seedlings of Gin-Yose and Shibaguri (both Japanese) and Miyugawa No. 20 (Chinese) as rootstocks. They found that Boji on Japanese stocks did not make satisfactory growth, whereas all other combinations were successful. Analysis showed that crude ash, K and Ca contents were higher and Mg and Mn contents were lower in Boji scions than in Gin-Yose scions on either Japanese or Chinese stocks. The Japanese stocks tended to accumulate more crude ash, K and Ca, and less Mg and Mn, than Chinese stocks.

Chinese (*C. mollissima*) and Japanese (*C. crenata*) species are resistant to ink disease but have poor peeling. These species were used to develop hybrid rootstocks resistant to ink disease. Rootstock developed from France (Ferosacre CA90, Maraval CA74, Marigoule CA15, Marlhac CA118 and Marsol CA07-) and Spain (CHR-162 (7521), CHR-151 (HS), CHR-168 (110) and CHR-161 (100) showed different degree of resistant to ink disease (Breisch, 1995; Pereira-Lorenzo and Fernandez-Lopez, 1997). 'Marsol' and 'Maraval' were developed by crossing *C. sativa* with *C. crenata*. 'Marsol CA07' imparted superior vigour and has good productivity. 'Maraval CA74' showed broad graft compatibility with French cultivars and slightly low vigour, which allows for higher planting density (Breisch and Hennion, 2004). In California, nuts of the same cultivar are used to raise seedling rootstocks. In south west France, chestnut growers opted for planting new orchards with 'Marigoule' and 'Marsol' (Hennion, 2010). Fuxing (2003) selected 'Jinpingaichuili' an extra-dwarfing rootstock of Chinese chestnut (*C. mollissima*). It has weeping growth habit. The height of 5-year-old trees grafted into it was only around 2.5m.

## Propagation

### Seed Propagation

Chestnuts are mostly propagated by seeds, but the established cultivars are raised by vegetative method. To ensure uniform germination, the seed requires stratification. The nuts before stratification should be given hot water treatment at 50-52°C for 30 minutes, to kill weevil eggs and larvae of *Curculio caryotrypes* (Ackerman and Jaynes, 1980). The nuts are kept in moist condition at 0°C or 2°C for 2 months (Woodroof, 1979). The temperatures much below freezing point or above 4°C should be avoided during stratification. Too wet and too dry conditions are also undesirable. Seeds of Chinese chestnut usually germinate in 16±2 days from sowing (Shepard *et al.*, 1989). The sprouting nuts are usually planted in the nursery beds 5-7 cm deep and 15 cm apart in rows at a distance of 90 cm. If the seedlings are to be kept in the nursery for 2 years, the distance between plants in row is increased to 30 cm. Taproot, 20-25 cm below the ground level is cut at least a year before transplanting to avoid mortality. One to three nuts can be planted *in situ* retaining the best seedling thereafter. Srivastava *et al.* (2008) recorded significantly higher seed germination (48.67 per cent), seedling height and diameter by using extra large seeds of *C. sativa*. Large seeds stored in sand or aluminium foil resulted good germination and seedling growth. The best media for seed germination was found as soil + sand. Large seeds showed maximum germination when sown in the month of November.

## Vegetative Propagation

### Budding and Grafting

Propagation by budding chestnut trees has been attempted for many years with limited and sporadic success (Hartmann and Kester, 1968, Jaynes, 1972, Davie, 1977). Becker (1961) studied various methods of budding and suggested that greenwood budding was the best method when the trees were still in active growth.

A technique of budding on the epicotyls of sprouted chestnut seeds was described by Ackerman and Jaynes (1980). The germinated seedlings were transferred to 7.5 cm pots after 6 days. After another 6 days modified chip budding with scions (dormant) of *Castanea pumila* (chinkapin) was done at 1 cm above the cotyledon attachment. The epicotyl at this time was only 4-6 cm high and 2.8-3.6 cm in diameter at this position of budding, bud shields of 1.5-2 cm were used. The union was tied with rubber rings. Topping of upper 3-4 cm portion was done at budding and the rest was cut off 5 days later. Bud started sprouting after 16 days. Ackerman and Jaynes (1980) conducted trials on epicotyl budding with three cultivars and 62-100 per cent success was obtained.

However, budding is not widely practised, because the wood is fluted or grooved and cambium of scion and stock does not join properly (Woodroof, 1979). Moreover, several other grafting techniques have proved more reliable (Hartmann and Kester, 1968, Wilmoth, 1977).

Splice and whip grafting using scion and stock of the same diameter are the most successful methods (Woodroof, 1979). For a better success in grafting, the rootstock should be in active growth stage whereas the scion-wood should be in dormant condition. If the grafting is done in early spring, the scion-buds may be killed by spring frost. The best time of grafting is when rootstock produces 3-4 leaves during May and survival rate was 60 per cent (Kayo, 1967). The highest percentage of plants that sprouted and survived (85-90 per cent) were obtained with tongue grafting (Pathak *et al.*, 1979). For improving the winter hardiness of chestnuts (*C. sativa*), approach grafting has been recommended at a height of 1 metre on *Quercus borealis* and *Q. polttstris* (Makhmet and Katsalap, 1979).

Lliev *et al.* (2013) recorded maximum success of splice grafting (86.7 to 93.3 per cent) during spring (end of Feb.) using 'dormant' and 'active' rootstock compared with 69.33 to 77.33 per cent in side veneer grafting and 40.00 to 81.33 per cent in cleft grafting. Regardless of the techniques used, the success in often low, which is believed to be due to scion/rootstock incompatibility with restricted flow of nutrients through the union, low winter temperature that injured the developing tissue in the graft union (MacDaniels, 1955), or using the chestnut blight infected scions (McKay and Jaynes, 1969). Secondary phloem fibres in chestnut rootstock may also interfere with graft union formation. Warmund *et al.* (2012) observed that

Chinese chestnut seedlings had 2/5 phyllotaxy with five alternate buds present in two revolutions around the stem. Vascular cambium in 1-year-old rootstock was arranged in a regular five-lobed cylinder in cross-sections. Primary phloem fibres were present in a discontinuous ring and secondary phloem fibres were usually clustered outside the cambium between the lobes. Grafts were successful when scion buds were placed on the rootstock disrupting secondary phloem fibres with cambia of the budded tissue aligned. Warmud *et al.* (2012) also reported that containerized rootstock at 56 per cent and 62 per cent growing medium moisture content before grafting had 25 per cent graft union success rates, whereas non-irrigated plants at 48 per cent moisture had a 75 per cent success rate.

Solar *et al.* (2010) studied the grafting performance of 'Marsol' grafted onto different rootstocks. They reported 100 per cent graft success in the combination of 'Marsol'/'Marsol' and 65 per cent in 'Kozjak'/'Marsol'. In the following year, the growth of central shoots was recorded maximum when 'Marsol' was grafted onto 'Bouche de Betizac', 'Marsol' and 'Precoce Migoule'.

**Nut Grafting**

This method, also known as nurse grafting, has the following advantages :

a) Stock/seedlings are not necessary.
b) Grafting is done indoor.
c) Time of grafting is not critical.
d) Less thick scion is used.
e) Incompatibility of stock-scion at initial stages can be determined in about 5 weeks.
f) Juvenile period can be reduced considerably.

In this method, a dormant scion is grafted into a nut, after removing its germinating shoot and root. The method was initially used for inducing rooting in chestnut cuttings (Moore, 1963) and later on modified (Jaynes, 1964). The grafted nuts are kept in heated frames at 23.5°C for about 4 weeks for hardening. After the root initiation, liquid fertilizers are applied at every 3 weeks until mid-August for good growth. Success with this method was reported to be 0-90 per cent depending on the skill and experience of the propagator and the compatibility between stock and scion.

Hypocotyl grafting in chestnut clones, *viz.*, Nana (*C. sativa*), clone 431 (*C. crenata*) and HS (*C. sativa* x *C. crenata*) was studied by Vieitez and Vieitez (1981). The percentage of successful grafts was found highest with HS on *C. crenata* (100 per cent) followed by clone 431 on either *C. crenata* or *C. sativa* or HS, the success being 98.6, 94.0 and 89.3 per cent, respectively.

Duman and Serdar (2006) studied the effect of inverted radical and hypocotyls grafting methods on potted nursery tree production. Newly germinated chestnut seeds and young seedlings of SA5-1 genotype were used as rootstocks and sprouts with dormant buds of the same genotypes were used as scions. They reported that potted chestnut nursery trees could be produced in one year using either of the grafting methods. Inverted radical grafting however, showed 83.6 to 96.7 per cent success.

## Stooling

Propagation by stooling was described by many workers (Ichii, 1962, Jaynes, 1969). Initially, the clones to be rooted are grafted very low on stock plants. After one year, these are cut down in early spring, leaving 2 or 3 buds above the union. Two or three shoots will develop and with the increase in age up to five years, the number increases up to 25. In June, a wire is twisted at the base of each shoot and soil is mounded to 1-5 cm. Pathak *et al.* (1979) obtained 90 per cent rooted shoots by treatment with IBA (2500 ppm).

## Cutting

Softwood cuttings taken in June-July from field-grown stock plants showed excellent rooting under mist by treatment with auxin (Jaynes. 1976). Cuttings (20-30 cm long) from current season's growth rooted better than from the apex of the shoots (Pellati *et al.*, 1977). Etiolation of shoots prior to taking cuttings increased rooting (Rinallo *et al.*, 1987). Etiolation in mid-May by wrapping the apical 6-12 cm in foil and taking the cuttings in late October showed 100 per cent rooting by reatment with auxin (Vieitez and Ballester, 1988). Hunter and Norton (1985) reported 35.5, 5.0 and 1.6 per cent rooting in cuttings of cultivars Au-Leader, Au-Homestead and Au-Cropper, respectively by treating the cuttings for 5 seconds in an aqueous solution of IBA at 3g/l. Iglesias *et al.* (1988) recorded highest rooting (30 per cent) in the month of January by treatment with IBA at 3 g/l in a greenhouse with 25°C bottom heat under intermittent mist.

American chestnut are difficult-to-root by cuttings. Galic *et al.* (2014) developed a two-step protocol to root American chestnut cuttings. In the first year, a shoot from a selected tree is grafted onto a rootstock. In the second year, it is moved into a greenhouse in February and new shoots taken and rooted in April and May. In the second experiment, 46 per cent of cuttings rooted and all cuttings leafed out and 15 per cent overwintered successfully. They also suggested using 1-2 per cent IBA as rooting hormone.

## Top Working

The chestnuts are top worked either to induce early fruiting in a seedling or to increase productivity of old plants. Caraffini (1986) described the method of bringing a 100-year-old sweet chestnut plantation back into commercial production. Between 5 and 15 April, 3- to 4-year-old coppice shoots 4-6 cm in diameter, 1.6 m above the ground were grafted. The scions prepared from 1-year-old wood, 2-3 cm in diameter with sound bark and good buds were used. The grafting success rate was 65-70 per cent. Nut production began 2-4 years later depending on the cultivar grafted.

## Micropropagation

Keys and Cech (1987) found considerable variation among the *Castanea* species (*C. dentata*, *C. mollissima*, *C. sativa* and *C. crenata*) in the requirements for *in vitro* propagation. They developed a nutrient medium to provide uniform growth of callus cultures of all 4 *Castanea* spp, using a modified Knop's medium. The medium consists of Greshoff and Doy major salts, Murashige and Skoog minor salts and organic constituents with IAA and kinetin. They suggested that cultures should be grown in l6-h photoperiod with 27°/20°C day/night temperature. Jacquiot (1966) obtained active proliferation in the cambium tissue of *C. sativa*. However, organ differentiation was first induced by Vieitez *et al.* (1978). Proliferatron was induced by adding 2.4-D (1 mg/l) and coconut milk with the basal medium instead of a cytokinin. Root regeneration was obtained with both IBA and NAA (10 mg/l) supplemented with kinetin or BA (0.5 mg/l).

Micropropagation of mature chestnut was successful by using stump shoots collected in winter and grown on a MS half-nutrient media (Vieitez *et al.*, 1983). BA at 0.2 mg/l was optimum for multiplication of shoots while shoot cuttings were dipped in IBA solution (1 mg/l) for 2 minutes for rooting. Chauvin and Salesses (1988) also reported successful plantlet production in presence of BA, IBA and $GA_3$ from axillary buds of stools and greenhouse layers taken in mid-May.

Vieitez and Vieitez (1980) reported that BAP caused the most satisfactory effect on promoting the proliferation of axillary shoots, whereas zeatin slightly inhibited the development of axillary shoots but increased the induction rate and caused more vigorous shoots. Tetsumura and Yamashita (2004) achieved similar results with the addition of zeatin caused the highest proliferation rate. Tafazoli *et al.* (2013) developed protocol for regeneration of European chestnut (*C. sativa*) through indirect organogenesis. They reported that nodal segment was the best for callus induction (69.4 per cent) when cultured on MS medium with 0.2 mg/l thidiazuron (TDZ) and MS was the best medium for callus formation (74.6 per cent). The highest shoot multiplication (66.9 per cent) was observed on MS medium with 0.2 mg/l TDZ. Regenerated shoots were rooted *in vitro* on MS containing 1.5 mg/l IBA. Plantlets with well developed root and shoot systems were acclimatized inside the green house and 80 per cent of the plants survived on transfer to garden soil.

## Planting

Pits of one cubic metre in size are prepared and filled with mixture of soil and farmyard manure well in advance, so that the soil in pits settles down before planting. The best time for planting the chestnut trees is the fall, *i.e.*, November and December. The chestnut plants have a taproot system requiring extra care during transplanting. The top portion is pruned so that water requirement of the plant is met with the root system. The planting distance is generally kept 12 m but the Chinese chestnut seedlings may require a distance of about 20 m. In high density plantations a distance of 7.5 m is sufficient and the tree size is controlled by pruning. Initially, spacing of 7.5 m is kept, but when the plants grow and touch the crowns, half of the plants are removed to give a spacing of 15 m. The grafted plants require less spacing. For the vigorous Euro-Japanese hybrids (*C. crenata* x *C. sativa*) the distance ranged between 7m x 8m (178 plants/ha) and 8m x 10m (125 plants/ha). Spacing for European chestnut are 10m x 10m (100 plants/ha) and more according to climate and soil fertility.

Bencat (1968) suggested planting of 500 plants per hectare under high density system. In Japan, tree thinning is recommended. The main cultivars Tenzara, Tbuki or Tsukubawase (dwarf type) or Yamatowase (upright type) were planted at a distance of 3.5 m. The thinning of pollenizers is done after 8 years (Yasunobu *et al.*, 1972). In mountainous terrain of Shandong, China, grafted plants are planted at a distance of 2.5 m only (Wang, 1988). Pollenizer trees should be planted, for every 8-10 trees of the main cultivar, one pollenizer cultivar is required. If the nut size, shape, colour, maturity date of the pollenizer is different than the main cultivar, it is suggested to plant pollenizer in separate row to facilitate harvesting and sorting of nuts.

## Pruning

Initially the tree should be trained on single stem up to a height of 1.5 m retaining the scaffold branches at proper places. The Chinese chestnut tree forms a low-headed tree, if left unpruned. The unpruned trees come into bearing early. An ideal practice is to allow the trees to come into bearing before any pruning is done, then to remove only a few of the lowest branches each year until the tree is properly headed. Observations made at Beltsville by McKay and Crane (1951) indicate that early pruning of young Chinese chestnut trees causes severe dwarfing and consequent delay in the formation of catkins and bearing of nuts. It was suggested that all pruning operations should be delayed until the trees reach bearing age, and from then on low limbs may be removed gradually year by year until the trees are trained to the required height. Bencat and Tokar (1977) in an experiment with sweet chestnut, pruned two-year-old seedlings in late April before planting by shortening the leader (l-year-old wood) or by cutting back the 2-year-old wood of the trunk. The trees were found to tolerate pruning and had a good regenerating capacity. However, cutting of 1-year-old wood produced more shoots and cutting of 2-year-old wood produced taller plants.

The European, Japanese and Korean orchards are trained to form 'open center' trees which allow sunlight into the vase-shaped middle, as well as the tops and sides. A smaller tree may be more desirable for each of harvesting.

## Soil Management

Clean cultivation and straw mulch have been found better for the growth of young chestnut plants than uncultivated sward (Yasunobu, 1970). The trees in sward were mostly injured by cold, losing 50 per cent, of their buds at 2°C. In a comparison of different soil management systems, maximum growth and yield of chestnut trees were obtained on straw-mulched plots (Itakura and Shiraki, 1961). They further reported higher chestnut leaf P contents under orchard grass sod and mulch than under clean cultivation. The lower N content of the leaves found in plots under sod than in clean cultivated plots, should be increased to optimum level by the application of additional N. Woodroof (1979) recommended sod mulch system for hilly regions to control soil erosion particularly for established trees. Young chestnut orchards can be intercropped with row crops, such as perennial peanut, clover, hay or a small fruit crop until chestnut harvest begins in about five years of planting (Hochmuth *et al.*, 2009).

# Nutrition

## Role of Nutrients

### Nitrogen

Nitrogen is required for proper growth of chestnut trees. Deficiency of this element is common in nuts. Plants deficient in nitrogen have yellow foliage and poor growth. The nuts may be poorly filled due to the deficiency of N (Shear and Faust, 1980). Lack of vigour, reduced fruit set and nut yield with small and poorly filled nuts were the symptoms reported by Proebsting and Serr (1966). Ichii (1962) reported that nut size was positively correlated with the fruit and leaf nitrogen. For good nut size (20-25 g) the optimum nitrogen levels in the nuts and leaves were found to be l. 14 to I.33 per cent and 1.98 to 2.34 per cent, respectively.

Good growth was observed in the Japanese chestnut plants grown in granite soils, rich in N and the absorption coefficient of N was found to be 98 (Isoda. 1963). Hirabe and Watanabe (1965) studied the effect of N fertilization on growth of Japanese chestnut seedlings. N was applied at the rate of 2 g per pot and the fertilized seedlings showed better growth and top/root ratio than the unfertilized seedlings.

Young trees frequently failed to respond because the need of N is influenced by many factors such as initial fertility of soil, depth, moisture conditions, soil management and climatic conditions. Harvest losses of N may be considerable and according to one estimate 453 kg chestnut remove 4.5 kg of N (Stephens and Jaynes, 1969). Highly productive orchard may respond to the application of N at 90 kg per acre per year. Well-developed branches and good nutrient status of leaves was observed in the 7-year-old Japanese chestnut seedlings grown on volcanic soils rich in N (Tsumita and Watanabe, 1971).

In water culture experiments, N deficiency between 2 June and 28 July adversely affected the tree weight, shoot elongation and bur growth. The uptake of other nutrients was also influenced by the time of application of N (Ishitsuka *et al.*, 1974).

### Phosphorus

Nut species respond to P application rarely. Phosphorus deficiency may cause slender shoot growth and small dark green leaves with purple or reddish colour along the veins followed by premature leaf fall (Proebsting and Serr, 1966, Shear and Faust, 1980). Isoda (1963) observed poor growth of Japanese chestnut seedlings grown in granite soils without P application.

### Potassium

Potassium application did not affect the growth of chestnut trees (Isoda, 1963). Stephens and Jaynes (1969) described the deficiency symptoms as yellowish and mottled foliage with tip and marginal burring.

### Boron

Deficiency of boron affects both fruit setting and yield in chestnut (Portela *et al.*, 2011). Portela *et al.* (2015) applied 100g boron per tree as sodium tetraborate beneath the canopy of 15-year-old 'Judia' tree growing in B-deficient soil. Application of B increased fruit setting (32 per cent in control to 81 per cent with boron) and yield (9-16 kg/tree). The leaf B content increased from 6-10 mg/kg to 26-105 mg/kg due to B application. Results also show that adequate boron nutrition may improve the crude fat content of the kernel. Leaves sampled when flowers were in bloom, were the most efficient tissue for the early diagnosis of B-deficiency (Portela and Louzada, 2012).

### Magnesium

Magnesium deficiency symptoms showed interveinal yellowing in older leaves which started in the centre of the leaf, and in the most severe stages the whole leaf become yellow, with golden-brown spots near the main vein. Normal trees (symptoms free trees) have a mean Mg concentration of 2.7 g/kg while the deficient trees have 1.6 g/kg. The N/Mg ration of 23 was suggested as the threshold value above which chlorotic and non-chlorotic trees separate (Portela *et al.*, 2003).

## Fertilizer Recommendations

Chestnut plantations in depleted soils require regular fertilizer application and as a rule 500 g of 5-10-5 fertilizer per year age of the tree should be given (McKay and Jaynes, 1969). During planting, fertilizer should not be mixed with the soil to avoid burning of roots.

N, P and K at 4.6 kg ammonium sulphate, 3.1 kg superphosphate and 0.8 kg muriate of potash per hectare when applied to 3 cultivars of chestnut in mid-October or early March resulted in good growth and higher yield. Leaf N was 2.5 per cent, whereas P, K, Ca and Mg remained unaffected with the application of fertilizers (Tsumita and Watanabe, 1971). Growth of trees receiving high rates of farmyard manure was superior to that of control; at 12.5 tonnes per hectare, fewer but large-sized nuts were produced and the total yield was more than control (Sugimoto *et al.*, 1974).

Breisch (1995) suggested application of 50g N/plant in the first year, increased to 250 g/plant in the 5$^{th}$ year with increasing potassium, starting at 80 g/plant. From the 6$^{th}$ year, fertilizer should be broadcasted at N (60-80 kg/ha), P (9-13 kg/ha) and K (66-100 kg/ha).

Toprak and Seferoglu (2013) studied the seasonal changes of nutrient elements in leaves of 'Karaasi' chestnut at Turkey. They observed increased N, P and Mg content of leaf up to August, then they decreased. K and Cu contents gradually increased during the vegetation period to the end of harvest. Ca and Mn contents continuously increased during vegetation period while Fe content decreased up to August and it increase thereafter and decreased at harvest. Zn content was stable until September, then it increased up to harvest. They suggested sampling leaves between 15 July and 15 August for estimation of leaf nutrients.

## Irrigation

Supplemental watering helps promote tree and to reduce stress, especially in the initial years after planting. Once established, trees are relatively drought tolerant; however, irrigation will help bearing trees to produce larger nuts and improve nut fill (Kaiser and Ernst, 2016). Drip irrigation is recommended to conserve water and effectively provide water to the trees only, rather than weeds in non-crop areas between rows (Hochmuth *et al.*, 2009).

## Flowering, Pollination and Fruit Development

### Flowering

Chestnuts are monoecious, having pistillate and staminate flowers at different locations on the same tree. Staminate flowers appear on cylindrical catkins with l0-20 stamens and 6-parted calyx. Two kinds of catkins exist, one is a pure staminate group, and the other one contains mixed catkins that have both staminate and pistillate flowers (Abbe, 1974). European researchers have divided the chestnut into four different groups on the basis of the structure of staminate flowers. These are astamine (having no stamen), brachistamine (having short stamens within the perigon), mesostamine (having stamens as long as the perigon), and longistamine (heaving stemens longer than the perigon) (Breviglieri, 1951; Soligant, 1973). Soylu and Ayfer (1981) stated that anther structure and pollen production varied significantly in the male sterile genotypes and longistamine types are suitable pollenizers with respect to their pollen production and viability. Mert and Soylu (2006) reported that stamens of male-fertile cultivars (Firdola, Karamehmet, Sariaslama and Haciomer) showed more distinct flowering habit than male-sterile cultivars (Osmanoglu and Vakit Kestanesi), which failed to show visible flowering due to short stamens. Flower clusters on the catkins had six to seven flowers while the mean number of stamens per flower varied from 9.50 to 11.95 among the cultivars, except of 'Osmanoglu' with 2.60 stamens per flower. Some 'Osmanoglu' flower had no stamens or abnormal stamens (*e.g.* lacking anthers) or abortive anthers. The surface structure of male-fertile cultivars consisted of polygonal epidermis cells, which had a thread-like surface appearance. Male-sterile 'Vakit Kestansi' had epidermis cells with hollow middles and swollen edges. In male-sterile cultivars the length of the anthers was significantly less than that of the male-fertile cultivars. Male sterility in chestnut is conditioned by two (Soylu, 1992) or four genes (Bolvansky and Mendel, 1999).

Cytoplasmic male sterility (CMS) is present in progeny of interspecific crosses between American and Asian castanea species. $F_1$ progeny of *C. dentate* as female x *C. mollissima* as male are almost always male sterile, whereas $F_1$ progeny of reciprocal cross are male fertile. There are however, exceptions to this general rule. These exceptional *C. dentate* x *C. mollissima* trees are male fertile rather than male sterile. Analysis of the trnV-ndhC region of chloroplast DNA (cpDNA) of both male sterile and male fertile $F_1$ hybrids of *C. dentate x c. mollissima* demonstrated that male sterility was correlated with American chestnut D chloplast haplotype while fertile hybrids of the same genotype had either the P or M chloroplast haplotype. This variation in male fertility in hybrids of *C. dentate* x *c. mollissima* may be caused by the interaction of nuclear genes from *C. mollissima* with genes encoded in the mitochondria inherited from *C. dentate*, because in other species CMS has been found associated with variations in mtDNA (Sisco *et al.*, 2014).

Vilkomerson (1960) observed that the sequence of bloom of the staminate and prstillate flowers is not always the same in all the trees. Staminate flowers shed pollens before the pistillate flowers open. The female flowers are usually ready to receive the pollens when the male flowers of the bisexual catkins shed pollens. A distinct flowering pattern is observed in each cultivar which remains constant in each season (Zimmerman, 1972). The Chinese chestnut is one of the latest to bloom in the season, the flowers appearing early in the month of June. Secondary flowering, unusual flower arrangements in the inflorescence and vivipary, have also been reported (Bencat, 1964). No correlation between catkin length and stamen length was observed in European chestnut cultivars grown in different locations (Bencat and Bolvansky, 1983).

Flower bud differentiation in all types of chestnut cultivars, early, medium and late, occurs in the month of July, the most active floral bud developmental stage being in August (Rhee, 1978). Thereafter, floral bud remains inactive up to the month of February. Phenological observations recorded by Bencat and Tokar (1978) revealed that the buds grew actively from mid-March to mid-April, leafing out started in mid-April and anthesis took place in early June. Sex expression of Chinese chestnut has been reported to be influenced by ethephon, $GA_3$ and urea treatments (Young *et al.*, 1985). Ethephon (100-500 mg/l) inhibited the development of female flowers and promoted that of male flowers. $GA_3$ (50-500 mg/l) painted on the young leaves and buds decreased the number of male flowers but increased the number of female flowers. Soil application of urea decreased the number of male flowers and increased the number of female flowers. Spraying the trees with $GA_3$ or urea or a mixture of both at the onset of floral primordial differentiation increased the female to male flower ratio.

## Pollination

Since chestnuts are self-sterile, more than one cultivar or seedlings are planted together to ensure cross pollination. Fruit setting is low on young plants because of insufficient pollens for cross-pollinated fruit crop (Clapper, 1954). But McKay and Jaynes (1969) reported that it is both wind as well as insect pollinated. Some cultivars are considered to be good pollen producers, *viz.*, Catagno della Costa, Morethina and Morte Grossa (Bergamini, 1975). In California chestnut orchards 'Belle Epine', Eurobella' (for 'Colossal'), 'Marigoul', 'Neveda' (for 'Colossal'), 'Okei', 'Silverleaf' and 'Tsukuba' are considered as good pollenizers. In New Zealand, variety '1005' is self-fertile but assisted by '1002'. '1015' is a good pollenizer for '1002' and '1005' for '1015'.

## Fruit Development

The kernel consists of two cotyledons joined with a growing point and enclosed in a pellicle which is a membranous covering, originating as integument of the ovule. Fruits ripen in early September in some cultivars and the latest ripening period is late October or November. Seildov (1988) reported the total growth period of 153-189 days from leaf emergence to fruit ripeness. Fruit ripened earliest in cv. Farash and forms KKF5, ZTF 15 and ZTF 20 and was latest in cv. Barguava and forms ZADLF 30, KZF 36, KKF 90 and KKF 120.

Chinese chestnut flowers in mid-summer, with flowers typically reaching anthesis from late June through mid-July in Ohio, USA. About eight weeks after pollination, the kernel was found still very small and most of the shell was filled with the integument, appearing fuzzy and white. By 10 weeks, the yellow kernel expanded significantly, but not yet filled the shell. Approximately 12 weeks after pollination, the yellow kernel was fully expanded to fill the inside of the shell, and the integument has become a skin-like seed coat, brown and white in colour (the pellicle). The ripe kernel was found composed of two cotyledons and an embryonic axis with a radicle and plumule (Miller *et al.*, 2013). Harvesting occurs approximately 12 to 14 weeks after pollination, typically ranging from mid-September to mid-October in Ohio, USA. Bolavansky and Uzik (2014) observed differential effect of both air temperature and the amount of rainfall during the period June- September on fruit weight. In trees with larger fruit, fruit weight was found positively correlated with mean air temperature and negatively correlated with the amount of rainfall during the period of fruit development.

## Maturity

Chinese chestnut plants raised from seedlings usually start bearing after 5-8 years from planting, whereas grafted plants after the second year. Chinkapins fruit at an age of 3-4 years. Fruit ripening begins in early September and completes in early November, *i.e.*, it lasts from 10 to 50 days depending on genetic and ecological factors (Bencat and Tokar, 1978). Fruits mature individually on the trees; the bur dehisces forming quadrants similar to those of pecans. However, in contrast to pecans, the dehiscence zone of chestnut is not well defined. In some cases, it appears more as a tear than as a distinct zone of separating cells. When about 5 per cent nuts have dropped, the rest are matured for mechanical harvesting (Peterson and Monroe, 1980). The splitting and opening of the burs is a sign that the nuts inside are mature (usually from mid-September to mid-October in California). Chinese chestnut from seedlings do not have uniform ripening period and as a consequence the harvest is often a major production expense. Natural nut drop period is usually 10-15 days but it may range up to 30 days in some years, depending on environmental conditions. Fallen nuts should be piked up at least every 2 days. Ethephon has been applied for uniform maturity. Soylu *et al.* (1987) reported that spraying of ethephon (1000 ppm) enhanced uniform maturity, improved earliness by 5-7 days, increased number of burs opening by 40-50 per cent and reduced fruit removal force. Slight chlorosis and drop of leaves caused by ethephon can be minimized by using calcium.

## Harvesting

Since chestnuts are highly perishable, the nuts fallen to the ground are gathered every alternate day and the gathering period lasts from 16 to 30 days (Woodroof, 1979). On small orchards and for trees, hand-harvesting is the most viable option.

It is estimated that one person could harvest a 4 ha planting consisting of 1000 trees by hand. Hand-harvest also assures the highest quality since nutshells are not scratched (Vossen, 2000). In view of the efficiency and increasing cost of labour, the use of shaker, catching frame and burr removal device is desirable (Woodroof, 1979). Mechanical harvesting system has been reported by Peterson and Monroe (1980). An inertia shaker and half-roll out catching frame effectively harvest the chestnuts and a device removes and separates the burrs from the nuts. Domokos (1979) found that a Schaumann type vibrator with a 35 cm grip, 1200 revolutions/ minute frequency and 4-5 mm amplitude was suitable for use in mature orchards.

## Yield

The nut yield fluctuates greatly in chestnut seedlings. Kim *et al.* (1971) recorded more yield from superior grafted trees than from naturally regenerated ones. An average annual yield of 15.62 kg/tree has been obtained from a superior line 3115 (Anon., 1987). McKay and Jaynes (1969) reported 350.2 to 1089.2 kg average yield per hectare from a private chestnut orchard and an average yield of 6.36 kg from an 15-year-old tree. According to Hemming (1944), a 16-year-old tree produces about 18.1 kg of nuts. In Kashmir valley, India, 80-100 years-old chestnut tree produced 50-60kg nut yield per tree (Pandit *et al.*, 2009). Ludders (2004) stated that European chestnut produced 100-150 kg/tree, but large old trees can yield up to 300 kg.

## Storage

Compared to most other nuts, chestnuts have a higher respiration rate and due to this water loss and starch conversion to sugar is also high (Wills *et al.*, 2007). Postharvest conversion of starch enhance sweetness and flavour (Harte *et al.*, 2003). The chestnuts contain about 50 per cent, moisture at harvest which is lost rapidly at room temperature making the kernels hard and inedible. To prevent this, the moisture level of the nuts is brought to 10 per cent by drying at 4.4°C with well-circulated air of 70 per cent relative humidity. Drying or curing is not required if nuts are stored immediately after harvest. Curing is, however, desirable for improving the eating quality which happens on the conversion of starch into sugar during curing. Toryan *et al.* (1978) compared the effects of the thickness of polyethylene bags and storage temperatures and obtained best results with 30-35 µm thick ployethylene bags at 0°C. In a study of controlled atmosphere storage of Japanese chestnut, Lee *et al.* (1983) reported the best result on sealing the nuts in 0.1 mm polyethylene bags (30 cm x 40 cm) with 9-11 pin holes. The total weight loss was less than 20 per cent after 8 months, storage and $CO_2$ and $O_2$ concentrations in the bags reached 5-7 per cent and 7-9 per cent, respectively. Extra large nuts stored better than large nuts. A high $CO_2$ level in the storage environment decreased the metabolic activity of the chestnut fruits. Peano *et al.* (2014) successfully stored fresh nuts of cvs. Bouche de Betizac and Garrone Nero for 90 days in $CO_2$-enriched pallet bags ($CO_2$ 80 per cent) at 1±1°C and 90-95 per cent RH. Both cultivars exhibited a low water content (per cent) under MA storage condition. The shelf life of extra large nuts was 3 months at 1°C, 2 months at 5°C and 1 month at 10°C. Sucrose content of nuts stored at 1°, 5° and 10°C increased rapidly during the first month while starch content decreased (Kawano *et al.*, 1984). Vossen (2000) recommended storing of nuts at –2°C to 0.5°C which can extend the shelf life of nuts for three months. Inceday *et al.* (2014) reported that browning of peeled chestnut (cv. Saraslama) in storage (in plastic film wrapped trays at 4°C) can be inhibited by dipping in sodium metabisulphate or citric acid solution at 2-4 µm. Early harvesting, fungicide treatment and moisture retention with plastic wrapping during cool storage at 0°C are important steps in the successful storage of Spanish chestnuts (Dooley and Brohier, 1986).

Storage losses mostly occur due to mould which can be controlled by destroying the mould organism, lowering the moisture content of nutmeats to 10 per cent and relative humidity of atmosphere to 70 per cent or lower, and maintaining the storage temperature at 0°C or lower (Woodroof, 1979). Fungi involve in chestnut mold and rot include *Phomopsis endogena, P. viterbenis camici, Diplodia castanea, Alternaria* sp., *Aspergillus* sp., *Penicillum* sp. *Botrytis cinerea, Gloesporium* sp., *Fusarium* sp., *Dothiorella* sp., *Cytodiplospora castanea, Pestalotia* sp., *Acrosperira mirabilis, etc.* Most of the natural infection occurs in field, therefore, control measures are needed to restrict disease development in field. Partial control of the disease can be achieved by low temperature and CA storage following an initial period in a controlled carbon monoxide atmosphere (Washington *et al.*, 1998). Gonzalez *et al.* (2010) reported good control of chestnut mold and kernel decay by treatment with 2700ppm hydrogen peroxide + 200ppm peracetic acid and 0.92 ppm ozone solution, respectively when combined with good agricultural practices and postharvest management.

## Pests

### Chestnut Weevils (*Curculio auriger, C. propinquus, C. proboscideus, C. sayi, C. elephas, C. sikkimensis, Balanius rectus*)

Weevils are the major pests in chestnut production. Many species of weevils infest the nuts, long proboscis of the weevil pierces the nut and eggs are laid inside. Larvae, called grubs damage the nuts. *Curculio proboscideus* takes 1 or 2 years to complete its life cycle and *C. auriger* completes the life cycle in 2 or 3 years. To eradicate these weevils, the nuts are collected when these fall on the ground after ripening and stored in isolation so that the larvae can not escape on emergence. Newly emerged larvae are destroyed and thus weevil population can be checked to a limit (Johnson, 1969). The diapuse intensity in *C. sikkimesis* changes reversibly in response to changes in ambient tremperature (Higaki and Toyama, 2012). Payne *et al.* (1976) observed in field trial on Chinese chestnut trees that carbaryl (2 lb/100 gallon) sprayed 6 times at 7 days' interval, starting from August, was the most effective measure to control the pest. Immersion of nuts in hot water (52°C) for 30 minutes killed weevil larvae without affecting the nut quality (Payne and Wells, 1978).

Speranza and Paparatti (2010) reported that the two carpophagous weevils (*C. elephas* and *C. propinquus*) can cause yield loss of 20-90 per cent in different years at central Italy. Thiacloprid gave good control of weevils.

### Gall Wasp (*Dryocosmus kuriphilus*)

Female lay eggs in early summer inside buds where the larvae overwinter and remain immature until the next spring. At bud burst in spring, larval feeding induces the formation of green or rose-coloured galls (Otake, 1989). It caused production loss up to 80 per cent and is considered one of the major pests of *C. mollissima* and *C. crenata*.

Battisti *et al.* (2014) reported severe infestation of chestnut gall on *C. sativa* at Italy. The nut yield was negatively related with gall wasp infestation level, with losses as high as 80 per cent being reported when the number of current year galls was above six galls per 50 cm twig.

Classical biocontrol using parasite *Torymus sinensis* could offer an opportunity to regulate this pest (Murakami *et al.*, 1987; Rieske, 2007; Quacchia *et al.*, 2011).

### Chestnut Tortricid Moths

Tortricid moth species are constantly very noxious because their larvae feed on developing nuts, thus causing important harvest losses. Early chestnut moth (*Pammene fasciana*) caused early drop of fruits at the beginning of the development, more serious damages are caused by the intermediate (*Cydia fagiglandana*) and late (*C. splendana*) chestnut moths. Larvae penetrate the nuts and develop into them, digging a tunnel and eating the endosperm. Usually damaged fruits fall during the summer, but in some cases they may complete the ripening and can be harvested as part of the crop (Pedrazzoli *et al.*, 2012).

Tortricid sex attractant showed a potential control of both the Cydia species. Pherobank pheromone was found more efficient ( 75 per cent ) in capturing the chestnut adults moths (Santos *et al.*, 2012).

### Mites (*Oligonychus bicolor, Tetranychus urticae*)

*Oligonychus bicolor* is the most serious mite of chestnut. It feeds on the upper surface of the leaves and occasionally on the bark of young twigs, if present in large numbers. Eggs are laid along the mid-rib and veins and the infested leaves are discoloured along the veins. To control this mite, sprays of kelthane, parathion, *etc.*, are recommended and only one spray may be effective.

## Diseases

### Chestnut Blight (*Cryphonectria parasitica*, Syn. *Endothia parasitica*)

This destructive disease of chestnut has completely wiped out the forests of American chestnut trees (*Castanea dentate*) during the first quarter of $20^{th}$ century (Carter, 1975). Chinese chestnut is resistant to this blight and, therefore, used for commercial plantations.

The symptoms appear as yellowing and wilting of the foliage on American chestnut. The leaves eventually become brown and shrivelled. A rusty-brown, stem-encircling canker is the cause of these foliar symptoms, and it is readily apparent on young trees or branches that are smooth barked (Griffin, 1986). The disease is also widespread in all the regions where sweet chestnut (*C. sativa*) is a major component of the landscape.

Natural biological control of blight due to cytoplasmic hypovirulence apparently occurs in most of the European countries, and the United States (Elliston, 1981; Griffin *et al.*, 1983; Anagnostakis, 1987).

Chinese and Japanese chestnut have the greast level of blight resistance and the progeny of Chinese x American crosses had a level of resistance intermediate between the two species (Griffen, 1986). Growth of the fungus could be checked by spraying of carbendazin, benomyl and thiophamate methyl in rotation (Delen, 1980). Inoculating large branches and trees with dsRNA containing *C. parasitica* may produce effective biological control (Baines *et al.*, 2014).

## Oak Wilt (*Ceratocystis fagacearum*)

The Chinese chestnut is quite susceptible to this fungus (Bretz and Long, 1950). First symptom of the disease is drooping of leaves, and water-soaked areas appear later. These areas turn brown along the margin and tip and the leaves fall. On tree trunks, swollen vertical cracks develop on the bark under which the fungal mycellium is present. In case of severe infection the tree dies. Major reservoir of the disease is white oak and insects, specially belonging to the group of nitidulid beetles, are vectors; these insects feed on sap and move from tree to tree.

To control, the infected trees should be cut and burnt and stumps should also be removed. Fresh plantations may be made as the disease does not reoccur.

## Blossom End Rot (*Glomerella cingulata*)

The fungus was first noticed on Chinese chestnut in Georgia, USA. Small brown spots appear on the green burs 2-3 weeks after the nuts have attained full size and these spots are confined to stylar opening. Subsequently, burs turn brown. Mostly, the nuts start decaying at the blossom end. Sometimes, black spot appears near the stylar end on the shell and afterwards, the whole kernel decays. The disease is frequently seen in over-crowded orchards and, therefore, thinning of trees is suggested. When the nuts are developing, spraying with captan or zineb is recommended.

## Ink Disease or Root Rot (*Phytophthora cinnamomi, P. cambivora, P. cryptogea, P. citricola, P. pseudosyriangae*)

It causes root and collar rot of adult trees and of seedlings in nurseries, plantations and forests. Symptoms of the disease on adult tree include chlorotic leaves reduced in size, thinning of the crown, and immature husks remaining on the tree after leaf-fall. Flame shape dark necrosis is evident on the collar of the tree after debarking. Infected large roots produce black exudates that stain the surrounding soil (Vannini and Vettraino, 2001).

The disease develops rapidly when the soil remains excessively wet (Erwin and Ribeiro, 1996). Vuono and Vannini (2010) suggested Integrated Control Protocol (ICP), including management of superficial water flows, copper sulphate and metalaxyl application, and potassium phosphate (trunk injection of potassium phosphate). Trunk injection of potassium phosphate water solution at 0.77g $H_3PO_3$

per cm of trunk diameter applied once a year in June-July for 3-5 consecutive years effectively reduced disease symptoms (Gentile *et al.*, 2010). There is also interest in using antagonistic organisms to control root rot. Organism such as *Trichoderma*, *Gliocladium*, *Bacillus*, *Pseudomonas* and *Streptomyces* may suppress disease in nurseries where there is a well-drained and aerated potting mix.

### Verticillium Wilt (*Verticillium albo-atrum*)

In poorly managed orchards, this disease becomes a problem. The leaves of the infected plants become yellow with brown margins; the wood develops brown streaks and the yellow leaves turn brown prematurely. Removal of wilted branches, good sanitation, and application of fertilizer and irrigation to encourage good growth will check the disease.

## Breeding and Improvement

Both in the eastern and western endemic regions of chestnut, the seedling selections have been made for centuries, on the basis of nut size and productivity of the trees. The written records of such selections are mostly not available from China, but more than 145 cultivars in Japan and 200 clonal selections each in France and Italy have been reported (Camus, 1929, Motte, 1935). The clonal selection is still prevalent in China, Japan, Korea and European countries and many improved types have been released for cultivation (Anon., 1987, Hardovic *et al.*, 1983). Through strict screening, more than 10 superior lines of chestnut, including 3113 and Ho 20, have been selected in China. The line 3113 was found to start bearing in the following year after grafting with an average yield of 15.62 kg/tree. The nuts are of high quality. The line Ho 20 was found suitable for close planting and tolerant to poor infertile soils (Anon., 1987). Later, due to the prevalence of dreadful diseases and pests, more emphasis has been laid on evolving the resistant lines. In the United States, breeding for blight resistance was started by A.H. Graves. He first crossed the American chestnut with the Japanese chestnut and the resultant $F_1$ with the Chinese chestnut (Jaynes, 1975). The gall wasp is a very serious problem of chestnut cultivation in Japan and Korea which necessitated the development of gall wasp resistant clones in these countries (Kim *et al.*, 1965, Shimura *et al.*, 1971).

In the Mediterranean countries where chestnut has been a major crop, its culture has been threatened by two fungal diseases, *viz.*, the ink disease and chestnut blight. These diseases are mainly responsible for declining production of chestnut in Europe. Therefore, efforts are underway to evolve clones resistant to these diseases, particularly to ink disease, because chestnut blight is not as devastating in Europe as in America.

Research on Chinese, European, American and Asian chestnut with numerous chestnut populations across the world revealed that most populations are of natural origin, composed of a few hundreds or thousands of trees, and separated by geographical natural obstacles. Genetic structure of such population have formed and influenced by natural selection and have not changed significantly because the trees can be reproductive for a long time-a hundred and more years (Solar *et al.*, 2005). Great phenotypic diversity among the chestnut trees with in populations on morphological traits, productivity and fruit quality have been reported by many researchers (Pereira *et al.*, 1996; Serdar, 1999; Solar *et al.*, 2005). This research forms basis for the selection of the best types from natural population where domestic chestnut has been produced for centuries. The traditional assessment of chestnut populations which is based on morphological and agronomical trait evaluations has been improved with the use of biochemical and molecular markers to explain genetic structure of populations (Solar *et al.*, 2005).

The goal of the American Chestnut Foundation (TACF) is the restoration of the American chestnut as a viable species to its native range. Toward that end, TACF is using conventional backcross breeding to introgress blight resistance from Chinese into American chestnut. The identification of species-specific molecular genetic markers could potentially accelerate recovery of the American type by permitting selection against Chinese chestnut genes not associated with disease resistance, and identifying pairings of hybrid trees that can produce progeny with a higher proportion of American genetic background than expected by chance. Genotypes of forty-six Chinese and forty-nine American chestnut trees were examined using a Golden Gate Bead Array (Illumina) of American chestnut single-nucleotide polymorphisms (SNPs), and identified eighteen SNPs with strongly divergent allele frequencies in the two species. The web-based Allele-Specific Primer (WASP) designing tool successfully designed primers for fifteen of these eighteen SNPs. The corresponding sequences from American and Chinese chestnut were aligned to each other to identify potential problems in primer sites and aligned to peach or apple genomic sequence to identify likely intron positions that might interfere with amplification. The fifteen WASP primer sets were tested on a panel of two American, one interspecific $F_1$, and three Chinese chestnut trees. Twelve primer sets amplified an allele from American and hybrid but not from Chinese chestnut trees (allele potentially specific to American chestnut). Of these twelve, three also amplified an allele from Chinese and hybrid but not American chestnut (allele potentially specific to Chinese chestnut). For general use, species specificity of these alleles will need to be validated in a larger panel of both American and Chinese chestnut trees. However, for immediate use in the TACF breeding program, these alleles only need to differentiate the Chinese (resistance sources) and American chestnut parents (recurrent parents) used in the current backcross generations (Georgi *et al.*, 2014).

## Breeding Objectives for Scion Cultivars

As mentioned above, the breeding objectives for chestnut differ from country to country based on local and regional problems of cultivation. In, general, the objectives can be summarized as under:

### Endodormancy

From the start of autumn and throughout winter dormancy, starch stored in the stems and buds of woody plants is broken down into oligosaccharides, mainly raffinose family oligosaccharides (RFOs). RFO synthesis protects plants against winter cold; the enzyme galactinol synthase (catalyzing the first step in RFO synthesis) is thought to play an important role in this process. Ibanez *et al.*

(2013) characterized galactinol synthase gene (CsGols1) in chestnut and examined its expression in relation to tree dormancy. In chestnut branches, CsGols1 mRNA levels paralleled the induction and course of winter dormancy, starting the rise at the beginning of autumn and elevated levels, persisting until the onset of spring. It was also reported that, exposure of chestnut plants to 4°C induced the expression of CsGols1 gene in the stem. However, the permanent induction of stem CsGols1 during winter months, rather than being a direct response to the cold was found to be associated with the endodormancy state. This observation points to transcriptional mechanisms controlling the maintenance of endodormancy.

**Nut Size and Quality**

Breeding for large nut size and higher sugar content should continue because such nuts store well and sold at a premium price. The Chinese chestnut contains more soluble sugars than the European and Japanese species (Jaynes, 1975). Xibing *et al.* (2013) studied genetic variation analysis for nut phenotype and nutrient content of ninety-seven Chinese chestnut cultivars from Tend province of China. They observed (1) extensive genetic variation among cultivars for nut phenotype and nutrient. Coefficient of genetic variation of traits were large, and the largest reached 0.378, indicated that the genetic potential were large, and the repeatability of characters were above 0.928, which showed that these were controlled by strong genetic factor, (2) significant positive correlation among phenotype characters and also positive correlation among nutrient indexes, but there were negative correlation between the phenotype traits and nutrient indexes. Thus, phenotype and nutrient were believed to be two comprehensive factors which were mutually independent and also has certain associations and (3) forty three cultivars were selected through principal component analysis using eight indexes of nut phenotype and nutrient, have single or multiple advantages such as bigger nuts, or rich nutrient, or both, which can be used for future breeding programme.

**Yield**

Basically, bearing capacity is a genetical character as some seedlings yield more than others. A seedling tree at 12 years yield 29.54 kg nuts which was very high in comparison with other seedlings (McKay and Jaynes, 1969).

**Timber**

In the USA, emphasis is also laid on developing timber types, which means trees with stems large in length and diameter.

**Disease Resistance**

Russell (1987) estimated that about 4 billion trees were killed by chestnut blight disease in America. American species are generally susceptible to chestnut blight; however some cultivars of *C. mollissima* show resistance (Huang *et al.*, 1996). European chestnut (*C. sativa*) suffered severe damage in 1930's due to infection of blight fungus. Hypervirulent strains of *Cryphonectria parasitica* is widely used in Europe to reduce mortality of European chestnut, however, it shows no success on infected American chestnut (Milgroom and Cortesi, 2004).

Another important disease known as 'Ink disease', (*P. cambivora*) first reported from Portugal in 1830 which caused the death of thousands of European chestnut trees (Elena Corredoira *et al.*, 2017). Asian chestnuts (*C. mollisima* and *C. crenata*) are however, resistant to both blight and ink disease.

Charles Burnham hypothesized in 1981 that the blight resistance of *Castanea mollissima* could be backcrossed to *C. dentate* to produce a tree that would perform in the forest like the formerly dominant American chestnut. The American Chestnut Foundation (TACF) was formed in 1983 to test this hypothesis using a traditional plant breeding programme. Two B1 chestnut trees, 'Clapper' and 'Graves' produced by former breeding programmes of the U.S. Department of Agriculture and the Connecticut Agricultural Experiment Station, were used to jump-start the programme. These former breeding programs identified Chinese chestnut as the most promising source of blight resistance and developed most of the methods used in the current programme. 'Clapper' and 'Graves' each were backcrossed into approximately thirty American chestnut from the vicinity of Sugar Grove, Virginia, backcrossed again then intercrossed twice. Prior to each cross, trees were selected for blight resistance and recurrent morphological traits. The resultant $B_3F_3$ seedlings were planted into the forest for testing beginning in 2009. Tests of the parents of these $B_3F_3$ suggested that their blight resistance will fall somewhere between an, intermediate and a high level. Hebard *et al.* (2014) working in this project are now in the midway through selection in $B_3F_3$ seed orchards and resistance among $B_3F_3$ progeny, they believed to increase as more trees are culled. The ultimate blight resistance of these trees in the forest was expected to be influenced by their adaptation and also may be affected by hypoviruses and other factors; which may take decades for a clear picture to emerge. In addition to 'Clapper', and 'Graves', 'Nanking' Chinese chestnut has also been backcrossed into twenty American backgrounds. More than twenty other sources of blight resistance from Chinese chestnut have been backcrossed, to varying degrees, into American chestnut but into only a few backgrounds; by themselves they do not constitute a viable breeding population. Other *Castanea* species, *C. crenata, C. henryi* and *C. sequinii* have been used to a lesser extent, as well as large, surviving American chestnut trees. This breeding program has been extended throughout the range of the American chestnut in TACF's network of seventeen state chapters. Through the efforts of those chapters, the effective size of the core population (Ne) was expected to exceed five hundred. The chapters also plan to develop viable breeding populations from the additional sources of blight resistance (Hebard *et al.*, 2014).

## Pest Resistance

In Japan and Korea, gall wasp resistance is major breeding objective. Japanese chestnut breeding programme has successfully develop chestnut gall wasp resistant cultivars 'Tanzawa', 'Ibuki', 'Tsukuba', 'Ishizuchi', 'Kunimi' and 'Shiho'. Euro-Japanese hybrid 'Bouche de Betizac' showed a hypersensitive response to infestation causing death of larvae at budburst.

## Breeding Objectives for Rootstocks

Development of clonal rootstocks possessing ink disease resistance is the main objective of chestnut rootstock breeding. Sometimes, stionic incompatibility is experienced due to the rootstock raised from interspecific crosses which can be overcome with the provision of vegetatively propagated clonal rootstocks. *Castanea crenata, C. mollisima* and their hybrids are more resistant to ink disease than *C. sativa* but they cannot be used as rootstocks because of stock-scion incompatibility. Therefore, it is necessary to incorporate the resistant genes in the cultivated species. Hybrid rootstock resistant to ink disease has been developed in France (Ferosacre CA90; Maraval CA74; Marigoule CA15; Marlhac CA118; Marsol CA07) and Spain (CHR-162 (7521); CHR-151 (HS); CHR-168(110) and CHR-161 (100).

# REFERENCES

Abbe, A.C. (1974). Bot. Rev., 40: 159-261.

Ackerman, W.L. and Jaynes, H.T. (1980). HortSci., 16 : 186-87.

Anagnostakis, S.L. (1987). Mycologia, 79:23-27.

Anonymous (1959). The Wealth of India, Vol. I, CSIR, New Delhi

Anonymous (1987). Forest Sci.and Tech., 2:16-18.

Bailey, L.H. (1949). The Standard Cyclopedia of Horticulture, The Macmillan Co., New York, Vol. I, pp. 681-82.

Baines, A.D., Fulbright, D.W. and jarosz, A.M. (2014). Acta Hort., 1019: 23-29.

Barbina, A. (1964). Agric. Venezie, 18: 32-48.

Battisti, A., Benrengnu, I., Colombari, F. and Haack, R.A. (2014). Agric. Forest Entomol., 16: 75-79.

Becker, G. (1961). 52nd A.R. Northern Nut Grs' Assoc., p. 87.

Bencat, F. (1964). Biologia Bratislava, 19: 912-19.

Bencat, F. (1968). Pol'nohospodarstvo, 14:427-36.

Bencat, F. and Bolvansky, M. (1983). Folia Dendrologica, 10: 27-51.

Bencat, F. and Bolvansky, M. (1984). Folia Dendrologica, 11: 213-31.

Bencat, F. and Tokar, F. (1977). Pol'nohospodarstvo, 24: 448-58.

Bencat, F. and Tokar, F. (1978). Folia Dendrologica, 3: 49-89.

Bergamini, A. (1975). Rivista della ortoflorofrutticoltura Italiana, 53: 103-06.

Bin, Lu, ZeXia, S., WeMin, Y., RuChang, H and Delu, N. (2000). China Fruits, 3: 10-11.

Bolvansky, M. and Mendel, L. (1999). Acta Hort., 494: 339-44.

Bolvansky, M. and Uzik, M. (2014). Acta Hort., 1019: 31-37.

Borges, O.P., Carvalho, J.S., Correia, P.R. and Silva, A.P. (2007). Jr. Food Comp. Anal., 20: 80-89.

Bounous, G. (2005). Acta Hort., 693: 33-40.

Bounous, G. (2009) Scripta Hort., 9: 72-84.

Breisch, H. (1995). Chataignes et marrons. CTIFL, pp. 239.

Breisch, H. and Hennion, B. (2004). Infos CTIFL, 198: 35-38.

Bretz, T.W. and Long, W.G. (1950). Pl. Dis. Reptr., 34: 291.

Breviglievi, N. (1951). Nel torritoriodi Vvallombrosa Centro di Studio Sul Castagno., 1: 15-49.

Brooks, R.M. and Olmo, H.P. (1972). Register of New Fruit and Varieties, Univ. California Press, Berkley, Los Angels, pp. 205-11.

Brooks, R.M. and Olmo, H.P. (1983). HortSci., l8: 155-61.

Camus, A. (1929). The Chestnuts: Monograph of *Castanea* and *Castanopsis*. Paul Lechevalier, Paris.

Caraffini, B. (1986). Informatore Agrario, 42: 42-45.

Carter, J.C. (1975). J. Arboriculrure. l: 141-44.

Casini, E., Magini, M. and Alessendri, S. (1987). Marron d' Olargues Arboiculture Fruitire, 34: 62.

Chandler, W.H. (1957). Deciduous Orchards. Lea and Febiger, Philadephia, p. 482.

Chauvin, J.F. and Salesses, G. (1988). Acta Hort., 227:340-45.

Clapper, R.B. (1954). J. Hered., 45: 107-14.

Crane, H.L. (1960). 51st A.R. Northern Nut Grs' Assoc., pp. 22-29.

Davie, W.L. (1977). 68th A.R. Northern Nut Grs' Assoc., pp. 108-10.

Delen, N. (1980). J. Turkish Phytopath., 9: 27-47.

Diller, J.D., Clapper, R.B. and Jaynes, R.A. (1964). U.S. Forest Serv. Res. Note N.E., pp.25-28.

Dinis, L., Peixoto, F., Pinto, T., Costa, R., Bennett, R.N. and Cardose, J.V. (2008). Photosynthetica, 46: 63-72.

Domokos, A. (1979). Kertgazdasag, 11: 9-13.

Dooley, L.B. and Brohier, L. (1986). Abst. 22nd. Int. Hort. Congr., California. No. 1482.

Duman, E. and Serdar, U. (2006). Hort. Sci. (Prague), 33: 16-22.

Elena Corredoira, M., Teresa Martinez, M., Jose, C. and Caremon San Jose, M. (2017). Forests, 8: 394; doi: 10.3390/f8100394 (www.mdpi.com/journal/forests).

Elliston, J.E. (1981). Amer. Chestnut Cooperators' Meet. Tech. Rep. NE-64.

Erwin, D.C. and Ribeiro. O.K. (1996). Phytophthora disease world wide. St. Paul., pp. 562.

Fuxing, Qiu (2003). South China Fruits, 32(6):68.

Galic, d., Date, A. andf Alward, M. (2014). Acta Hort., 1019: 99-103.

Gentile, S., Valentino, D. and Tamietti, G. (2010). Acta Hort., 866: 417-24.

Georgi, L.L., Hebard, F.V., Nelson, C.D., Staton, M.E., Olukolu, B.A. and Abbott, A.G. (2014). Acta Hort., 1019: 105-12.

Gomes-Laranjo, J., Coutinho, J.P., Galhano, V. and Ferreira-Cardoso, J.V. (2008). Photosynthetica, 46: 63-72.

Gonzalez, I.R.D., Fulbright, D.W., Ryser, E.T. and Guyer, D. (2010). Acta Hort., 886: 563-70.

Griffin, C.J. (1986). Hort. Review, 8: 291-338.

Griffin, C.J., Hebard, F.V., Wendt, R. and Elkins, J.R. (1983). Phytopathology, 73: 1084-92.

GuangNing, S., GuiDong, M., ShouLe, T., Lin,Xu, and Xu, Liu, (2011). Acta Hort. Sinica, 38:1407-08.

Hardovic, H.F., Kester, Ferhtovic, S. and Hlisc, T. (1983). Jugoslovensko Vocarstvo, 17:119-23.

Harte, J.B., Dolan, K., Cameron, A. and Fulbright, D.W. (2003). NNGA, 94: 22-31.

Hartmann, H.T. and Kestar, D.E. (1968). Plant Propagation Principles and Practices (2nd ed.) Prentice Hall Inc. Englewood Cliffs, NJ, pp. 423-25.

Hebard, F.V., Fitzsimmons, S.F. Gurney, K.M. and Saielli, T.M. (2014). Acta Hort., 1019: 135-39.

Heiniger, U and Conedra, M. (1992). Proc. Int. Chestnut. Conf. West Virginia, pp. 175-78.

Hemming, E. Sam. (1944). A.R. Northern Nuts Grs' Assoc., 35:22-24.

Hennion, B. (2010). Acta Hort., 866:493-97.

Higaki, M. and Toyama, M. (2012). Jr. Insect Physiol., 58: 56-60.

Hirabe, M. and Watanabe, T. (1965). Bull. Kanagawa Hort. Exp., 13:45-50.

Hirschi, A.G. (1946). A.R. Northern. Nut Grs' Assoc., 37:32-33.

Hochmuth, R.C., Wallace, R.D., Van Blokland, P.J. and Williamson, J.G. (2009). UP/ IFAS Dept. Food Resource Economics, Staff report. #354.

Huang, H., Dane, F. and Norton, J.D. (1994). Theor. Appl. Genet., 88: 981-85.

Hunter, A.G. and Norton, J.D. (1985). Scientia Hort., 26:42-45.

Ibanez, C., Collada, C., Casado, R., Gonzalez-Melendi, P., Aragoncillo, C. and Allona, I. (2013). Trees: Structure and Function, 27: 1039-16.

Ichii, T. (1962). J. Jap. Soc. Hort. Sci., 31: 257-56.

Iglesias, L., Dias, T. and Gonzalez, E. (1988). Acta Hort., 221:173-75.

Inceday, B., Yldz, G. and Uylaser, V. (2014). Acta Hort., 1019: 147-52.

Ishitsuka, Y., Nagumo, M. and Komatsu, E. (1974). Bull. Ibaraki Ken Hort. Exp. Stn., Japan, 5: 23-28.

Ishitsuka, Y., Nagumo, M. and Komatsu, E. (1979). Bull. Ibaraki Ken Hort. Exp. Stn., Japan.

Isoda, R. (1963). J. Jap. Soc. Hort. Sci., 32: 34-41.

Itakura, T. and Shiraki, Y. (1961). Bull. Nat. Inst. Agric. Sci. Hiratsuka Ser. F, 9:215-52.

Jacquiot, C. (1966). Comptes Rendus Hebdomadaires des Seances de L'Academic des Sciences, D. (Sciences Naturalles), 243:510-12.

Jaynes, R.A. (1964). A.R. Northern Nut Grs' Assoc., 55:16-20.

Jaynes, R.A. (1969). In: Handbook of North American Nut Trees, (ed. R.A. Jaynes). The W.F. Humphrey Press Inc., New York, pp. 391-394.

Jaynes, R.A. (1972). US Dept. Agric. Forest Sanv. Res. Rep. p. 12.

Jaynes, R.A. (1975). In: Advances in Fruit Breeding. (eds. J. Janick and J.N. Moore) Purdue Univ. Press, pp. 490-503.

Jaynes, R.A. (1976). 65th A.R. Northern Nut Grs' Assoc., pp. 128-31.

Johnson, W.T. (1969). In: Handbook of North American Nut Trees (ed. R.A. Jaynes). The W.F. Humphrey Press Inc., New York, pp, 67-156.

JunYong, C., XiHua, Z., ChunYong, X., YongJie, X., AiHua, Li, ZhiJian, L., ZinZhu, Li. and ShanShan, X. (2011). Acta Hort. Sinica, 38:2415-16.

Kaiser, C. and Ernst. M. (2016). Center for crop Diversification, Agriculture and Natural Resources, University of Kentucky, USA.

Kawano, S., Onodesa, T., Hagakawa, A., Iwamata, M., Ohta, H. and Sugarwara, W. (1984). J. Jap. Hort. Sci., 53: 194-201.

Kayo, S.P. (1967). Res. Rep. Inst, Genet., 5: 75-84.

Keys, R.N. and Cech, F.C. (1987). West Virginia Forestry Notes, 13:19-24.

Kim, K.S., Park, M.K. and Ahn, W.Y. (1971). Bull. Seol Nat. Univ. Fores., 8:51-74.

Kim, K.S., Park, M.K. and Lee, M.D. (1965). Res. Rep. Rural Dev., Suwan, 8: 49-62.

Kunch, U., Scharer, H., Patrian, B., Hurter, J., Conedra, M. and Jelmini, J. (1999). Acta Hort., 494: 119-27.

Lee, B.Y., Kim, Y.B. and Han. P.J. (1983). Rep. Rural Dev. S. Korea, Soil Fert. Crop Prot. Mycology Farm Products Uti., 25: 170-81.

Liang, Guo, JunHu, Dai., Kar, S.R., JianChu, Xu and Luedeling, E. (2013). Agric. Forest meteor., 180: 164-72.

Lliev, N., Varbeva, L. and Milev, M. (2013). Forestry Ideas, 19: 41-48.

Lozovol, A.D. and Chernyshov, M.P. (1979). Ratital, mye Resursy, 15: 536-40.

Ludders, P. (2004). Erwerbsobstbau, 46(1): 7-12.

MacDaniels, L.H. (1955). Ann. Rept. Northern Nut Growers, USA. 46: 92.

MacDaniels, L.H. (1955). Proc. North. Nut. Gros. Assoc., 46: 92-97.

Makhmet, B.M. and Katsalap, V.I. (1979). Bull. Glavnogo Botanicheskoyo Sada, 114:14-17.

Matsui, S. and Torikata, H. (1972). J. Jap. Soc. Hort. Sci., 41: 344-52.

McCarthy, M.A. and Meredith, F.I. (1988). Econ. Bot., 42: 29-36.

McKay, J.W. (1963). Fruit Var. Hort. Dig., 17: 73-74.

McKay, J.W. and Crane, H.L. (1951). Rep. 42nd Ann. meeting Northern Nut Grs' Assoc.,

McKay, J.W. and Jaynes, R.A. (1969). In: Handbook of North American Nut Trees (ed. R.A. Jaynes). The W.F. Humphrey Press Inc., New York, pp. 264-86.

Mert, C. and Soylu, A. (2006). J. Amer. Soc. Hort. Sci., 131: 752-59.

Meyer, F.G. (1980). Econ. Bot., 34: 401- 37.

Milgroom, M.G. and Cortesi, P. (2004). Ann. Rev. Phytopathol., 42: 311-38.

Miller, A., Miller, D.D. and Pijut, P.M. (2013). 103rd Ann. Rept.Northern Nut Growers Assoc. USA, pp. 27-33.

Moore, J.C. (1963). A.R. Northern Nut Grs' Assoc., pp. 88-89.

Motte, J. (1935). Bull. Maison France-Japanaise, Tokyo.

Murakami, Y., Uemura, M. and Gyoutoku, Y. (1987). Proc. Assoc. Plant Protection, Kyushu, 33:195- 98.

Otake, A. (1989). Appl. Entomol. Zool., 37: 172-74.

Pandit, A.H., Kour, A., Wani, M.S. and Mir, M.A. (2009). Acta Hort., 815: 51-56.

Paparatti, B. and Speranza, S. (2003). Informatore Agrario, 59(38): 75.

Pathak, R.K., Singh, B.B. and Om, H. (1979). Punjab Hort. J., 19: 76-79.

Payne, J.A. and Wells, J.M. (l978). J. Econ. Entom., 71: 894-95.

Payne, J.A., Polles, G. and Lowman, H. (1976). 65th Ann. Rep. Northern Nut Grs' Assoc. Inc. Ontario, Canada, pp. 43-44.

Peano, C., Baudino, C., Giuggioli, N.R. and Girgenti, V. (2014). Italian J. Fd. Sci., 26: 74-80.

Pedrazzoli, F., Salvadori, C., Cristofaro, A. de., Santo, P. di., Endrizzi, E., Peverieri, G.S., Roversi, P.F., Ziccardi, A. and Angeli, G. (2012). IOBC/WPRS-Bull., 74: 117-23.

Pellati. E. De., Garsa, V. and Dalla, C. (1971). Informatore Agrario, 33: 331(55):52-53.

Pereira, M.G., Caramelo, L., Gouveia, C., Gomes-Laranjo, J. and Magalhaes, M. (2011). Nat. hazards Earth Syst. Sci., 11: 2729-39.

Pereira-Lorenzo, S., Fernaudez-Lopez, J. and Gonzales, N. (1996). J. Amer. Soc. Hort. Sci., 12: 183-89.

Pereira-Lorenzo, S., Ramos-Cabrer, A.M., Diaz-Hernandez, M.B., Ciordia-Ara, M. and Rios-Mesa, D. (2006). Scientia Hort., 107: 306-14.

Perira-Lorenzo, S. and Fernandez-Lopez, J. (1997). Jr. Hort. Sci., 72: 731-39.

Peterson, D.L. and Monroe, G.E. (1980). Mechanical Harvesting System for Chinese chestnuts, Fruit and Tree Res. Lab., Byron, Georgia, USA, pp. 19-24.

Polansky, M. and Murphy, E.W. (1966). J. Amer. Diet Assoc., 48: 109-11.

Portella, E. and Louzada, J. (2012). Jr. Pl. Nutr., 35: 304-10.

Portella, E., Roboredo, M. and Louzada, J. (2003). Jr. Pl. Nutr., 26: 503-23.

Potella, E., Ferreir-Cardoso, J. and Louzada, J. (2011). Jr. Pl. Nutr., 34: 1245-53.

Potella, E., Ferreir-Cardoso, J., Louzada, J. and Gomes-Laranjo, J. (2015). Jr. Pl. Nutr., 38: 973-87.

Proebsting, E.L. and Serr, E.F. (1966). In: Nutrition of Fruit Crops, (ed. N.F. Childers), Hort. Pub., New Burnswick, NJ. pp 262-16.

Quacchia, A., Piazza, E., Pavia, G. and Alma, A. (2011). Sherwood Forest ed. Alberi Oggi., 177: 13-16.

Reed, C.A. and Davidson, J. (1954). The Improved Nut Trees of North America. Devin Adair, New York.

Rhee, Y.S. (1978). J Korean Soc. Hort. Sci., 19:41-47.

Rieske, L.K. (2007). EPPO. Bull., 37: 172-74.

Rinallo, C., Gellini, R. and Fabbri, A. (1987). Adv. Hort. Sci., 1: 27-33.

Russell, E.W.B. (1987). Torrey Bot. Club, 114: 183-90.

Santos, A., Azevedo, J., Pimentel, R., Ventura, L., Mexia, A. and Lopes, D. (2012). Revista de Ciencias Agrarias Portugal, 35: 292-97.

Seildov, A.K. (1988). Subtrop. Kul'tury. 1: 118-27.

Serdar, U. (1999). Acta Hort., 494: 327-32.

Shear, C.B. and Faust, M. (1980). Hort. Rev., 2: 142-63.

Shepard, E., Miller, D.D. and Miller, G. (1989). HortSci., 24: 516.

Shimura, I. (1912). Bull. Hort. Res. St. Ser A., 11:1-13.

Shimura, I. (1971). Bull. Hort. Res. St. Ser. A., No. 10.

Sisco, P.H., Neel, T.C., Hebard, F.V., Craddock, J.H. and Shaw, J. (2014). Acta Hort., 1019: 215-22.

Solar, A., Podjavorsek, A. and Stampar, F. (2005). Genet Resour. Crop. Evol., 52: 381-94.

Solar, A., Slatnar, A. and Stampar, F. (2010). Acta Hort., 866: 309-14.

Soligant, G. (1973). Centre de Recherches Agronomiques de Bordwaux, Bull.d' Infor., 280: 425-50.

Soylu, A. (1992). Acta Hort., 317:181-85.

Soylu, A. and Ayfer, M. (1981). Bahce, 10: 45-65.

Soylu, A., Eris, A. and Sermenli, T. (1987). Pub. Uludag Universitesi, 7: 8-141.

Speranza, S. and Paparatti, B. (2010). Acta Hort., 866: 411-15.

Srivastava, K.K., Sharma, M.K., Sounduri, A.S., Wani, M.S. and Singh, S.R. (2008). India J. Pl. Physiol., 13: 166-71.

Stephens, G.R. and Jaynes, R.A. (1969). In: Handbook of North American Nut Trees (ed. R.A. Jaynes). The W.F. Humphrey Press Inc., New York. pp. 13-18.

Sugimoto, M., Koto, M. and Omori, S. (1974). Bull Kanagawa Hort. Exp. Sta., 22:42-53.

Tanaka, T., Yamamoto, T. and Suzuki, M. (2005). Breed. Sci., 55: 271-77.

Tazazoli, M., Nasr, S.M.H., Jalivand, H. and Bayat, D. (2013). African J. Biotechnol., 12: 7063-69.

Tetsumura, T. and Yamashita, K. (2004). HortSci., 1684-87.

Toprak, S. and Seferoglu, S. (2013). Amer. J. Res. Commu., 1(5): 1-8.

Torikata, H. and Hi-euchi, H. (1962). J. Jap. Soc. Hort. Sci., 31: 115-22.

Toryan, A.V., Alent'ev, P.N., and Zadorazhnyi, I.M. (1978). Sadovodstvo. 12: 28.

Tsumita, M. and Watanabe, T. (1971). Bull Kanagawa Hort. Exp. Stn., 19: 38-44.

Valdiviesso, T., Madeira, C. and Abreu, C.P. (1993). Proc. Int. Cong. Chestnut, Univ. Perugia Press, Italy. pp. 95-97.

Vannini, A. and Vettraino, A.M. (2001). For. Snow. Landsc. Res., 76: 345-50.

Vasconcelos, M.do.C.B.M.de, Bennett, R.N., Quideau, S., Jacquet, R., Rosa, E.A.S. and Ferreira-Cardoso, J.V. (2010). Indus. Crop. Produts., 31: 301-11.

Vavilov, N.L. (1951). The Origin, Variation, Immunity and Breeding of Cultivated Plants, Ronald Press, New York.

Vieitez, A.M., Ballester, A., Vieitez, M. and Vieitez, E. (1983). J. Hort. Sci., 58:457-63.

Vieitez, F.J. and Ballester, A. (1988). Phyton., 18: 13-19.

Vieitez, M.L. and Vieitez, A.M. (1981). An. Edafol. Agrobiol., 40: 317-26.

Vieitez, R.M., Grozalez, M.L. and Vieitez, E. (1978). Scientia Hort., 8: 243-47.

Vietez, A.M. and Vieitez, M.L. (1980). Jr. Hort. Sci., 55: 83-84.

Vilkomerson, H. (1960). Proc. Northern Nut Grs' Assoc., 31:114-16.

Vossen, P. (2000). Chestnut culture in California. Pub. 8010, 16 Calif. United States Div. Agric. Nat. Resou.

Vossen, P. (2000). Univ. California. Pub. 8010.

Vuono, G. and Vannini, A. (2010). Acta Hort., 866: 425-30.

Wang, T.L. (1988). Forest Science and Technology, 1: 17-19.

Warmund, M.R., Cumbie, B.G. and Coggeshall, M.V. (2012). HortSci., 47: 893-95.

Washington, W.S., Hood, V., Goubran, F., Hepworth, G., and Stewart-Wade, S. (1998). Australian J. Exp. Agric. 38: 295-303.

Watt, B.K. and Merrill, A.L. (1963). Agric. Res. Serv., US Dep. Agric. Handbook, p. 8.

WeGang, H., ZhiJun, Z., Li;I, C., ShenFan, C. and XiShan, He. (2009). Scintia Silva Sinicae, 45(6):177.

Westwood, M.N. (1978). Temperate Zone Pomology, W.H. Freeman and Co., San Fransisco, USA, pp. 15, 65 and 427.

Wills, R., McGlasson, B., Graham, D. and Joyce, D. (2007). Postharvest: An introduction to the physiology and handling of fruit, vegetable and ornamentals (5th Edition ed.) CABI, UK.

Wilmoth, L.H. (1977). 68th A.R. Northern Nut Grs' Assoc., pp. 98-99.

Woodroof, J.G. (1979). Tree Nuts. AVI Pub. Co. Inc., Westport, Connecticut, pp.232-64.

XiBing, J., BangChu, G., Dan, T., QingZhong, L., Xin, C., KaiYun, Wu, and Quanen, D. (2013). Acta Botanica Boreli Occidentalia Sinica, 33: 2216-24.

Yasunobu, Y., Watanabe, T., Kobayashi, K., Mizuno, N. and Nagayama, L. (1972). Bull. Kanagawa Hort. Expt. Stat., 20: 20-27.

Yasunobu, Y. (1966). Bull. Kanagawa Hort. Expt. Stat., 14: 25-32.

Yasunobu, Y. (1968). Bull. Kanagawa Hort. Expt. Stat., 16: 29-33.

Yasunobu, Y. (1970). Bull. Kanagawa Hort. Expt. Stat., 18: 75-81.

Yldz, M.U., Ozcan, M.M., Calsr, S., Demir, F. and Er, F. (2009). World Appl. Sci. J., 6: 365-72.

Young, Q., Ren, L. and Du, C. (1985). Scientia Hort., 26: 209-15.

Zimmerman, R.H (1972). HortSci., 7: 447-55.

## Chapter 9

# Hazelnuts

The hazelnuts belong to the family Corylaceae. The species are found growing in wild forms in temperate zone extending from Japan, China, Turkey, India, Europe to North America. The nuts are small-sized and have hard shell. *Corylus avellana* is the most important species growing in Europe, whereas another species known as *Corylus maxima* is native to southeastern Europe and western part of Asia. As a result of natural selection and hybridization, these cultivated hazels known as hazelnuts and cobnuts have appeared. Some of the writers prefer to consider the cultivars that are called cobnuts to be from *Corylus avellana*, whereas hazelnuts to those cultivars which are from *Corylus maxima*. There is not much distinction between the two but cobnut is often described as a short round nut which is not covered entirely by the husk called involucre. The hazelnut is entirely covered with the husk.

## Composition and Uses

Hazalnuts are rich source of proteins, fats and minerals. One hundred grams of hazelnuts provides 600-650 kcl energy. It contains 10-22 per cent carbohydrates, 1-3 per cent of celluloses and pectins (Baysal, 1996), 10-24 per cent protein (Pala *et al.*, 1996). Hazelnut kernels are good source of fat (50-73 per cent) and contain unsaturated fatty acids (Garcia *et al.*, 1994). Koksal *et al.* (2006) studied nutrient composition of 117 hazelnut cultivars of Turkey. They reported the main fatty acids in hazelnut were oleic (79.4 per cent), linoleic (13.0 per cent) and palmitic acid (5.4 per cent). The ratios of polyunsaturated/saturated and unsaturated/saturated fatty acids of hazelnut cultivars were found to be between 1.23 and 2.87, and 11.1 and 16.4, respectively. The average content of niacin (1.45mg/100g), vitamin $B_1$ (0.28mg/100g), vitamin $B_2$ (0.05mg/100g), vitamin $B_6$ (0.5mg/100g), ascorbic acid (2.45mg/100g), folic acid (0.043mg/100g), retinol (3.25mg/100g) and tocopherol (26.9mg/100g) content were reported in 17 cultivars. The amount of essential amino acids, mostly as arginine (2003mg/100g) and leucine (1150mg/100g), and the non-essential amino acids mostly as glutamic acid (2714mg/100g) and aspartic acid (1493mg/100g) were also determined in the hazelnut cultivars studied. Hazelnut is rich in minerals and contains (per 100g): K-863mg, Mn-186mg, Mg-173mg, Ca-5.6mg, Fe-4.2mg, Zn-2.9mg, Na-2.6mg and Cu-2.3mg (Koksal *et al.*, 2006).

Hazelnuts are energy dense foods, the distinctive composition of hazelnuts is critical for their health effects. There are consistent evidences from epidemiology and clinical studies of the beneficial effects of hazelnut consumption on risk of coronary heart disease and emerging cardiovascular risk factors. Hazalnuts target multiple cardiovascular risk factors and mechanisms, which help to diminish the risk for coronary heart disease without undue weight gains, or even with reduced adiposity (Pourfazad and Raouf Mehrpour, 2017). Hazelnut-enriched diets provide improvements on plasma lipid and lipoprotein levels in subjects affects by hypercholesterolemia (Santi *et al.*, 2017).

Hazelnuts are consumed as natural, blanched and roasted or their products, such as sliced, chopped, flour and hazelnut butter, in order to provide flavour in dairy, bakery, confectionery, candy and chocolate products (Pourfazed and Raouf Mehrpour, 2017).

## Origin and Distribution

The terms hazelnut and filbert are used commonly for the plants of genus *Corylus*. The term hazelnut has been used by Groeschke (1887) in his monograph on *Corylus*. Similarly, Bailey (1949) and Rehder (1956) have also used the term hazelnut. However, Howes (1948) has made a distinction between these two growing in Great Britain. The hazelnuts have longer husk than the nut, whereas reverse is true for cobnut. In the USA, the nuts of European species are called filbert and those obtained from American species are called hazelnut (Slite, 1930). The history of hazelnut culture dates back 5000 years, and its origin spans from Caspian Sea shores to China, which includes northern Anatolia, Turkey (Anil *et al.*, 2017). The Romans and the Greeks were using hazelnuts for medicinal purposes and knew its food value from very early times. It is because of continuous selection, many present-day cultivars have appeared and their cultivation have extended beyond their natural habitat where climatic conditions were congenial. The name of hazelnut has probably its origin in 'full-beard' meaning the nut is entirely covered with the husk (Bunyard, 1920).

As the views expressed by Bunyard (1920), Theophrastus recognized only two types, the one producing round nuts and another having oblong nuts. These nuts were known as Heracieatic nuts, introduced from Heraclea on the Black Sea. The species *avellana* gets its name from Abellana in Asia from where the nuts were introduced as observed by Pliny. The name avellana might have also originated after the name of Avellino, an important town in Italy where hazelnuts are found to grow for centuries. The Italian name of Avellana was adopted in Britain and eventually adopted by Linnaeus as the specific name of the indigenous species (Fuller, 1896). *Corylus americana* and *C. cornuta* are native of North America.

Bacterial Blight of Hazelnut

Catkins

Hazelnut Orchard

**Hazelnut Cultivar Barcelona**

From the very beginning, both long and short husked nuts were generally called filberts and hazels, respectively. However, this does not hold true because both *Corylus maxima* and *C. avellana*, cross readily and cultivars that we see today is the mixture. Due to the extensive cultivation of these two species together have led to the continuous variation in husk length and by no means it is now valid to identify these two on the basis of husk length (Trotter, 1947).

The German types called as Lambertsnuss is thought to have come from Langbart or long-beard' meaning long husk. Another thought given to its name was that Lambertsnuss had originated from Longobards or Lombards of Italy.

The distribution of the European hazelnut (*Corylus avellana*) is wide extending from Ireland to Norway to Ural Mountains, through the Caucasus to northwestern Iran, to Lebanon, and across the Mediterranean Sea to the Iberian peninsula (Mehlenbacher, 2009). Hazelnut production in 2016-17 estimated at 29,000MT. Northeast China (Liaoning, Heilongjiang and Jilin provinces) remains a tradition hazelnut production region and the planting area has expanded to other provinces including Shanxi and Xinjiang. The cultivated varieties of hazelnut (*C. avellana*) were introduced to the United States from Europe by Felix Gillet, a nursery manager in Nevada City, California. Between 1885 and 1905, he introduced several varieties and some of these are still grown in the Pacific Northwest today. Hazelnut is a new introduction to Australia. Potential area for production has been identified at the river valleys in north-eastern Victoria, the Central Tablelands of the New South Wales and parts of Tasmania.

## Area and Production

Turkey is the worlds leading producer of hazelnut producing about 70 percent (549,000 tons) of the total production in 2014. The other major producing countries are Italy (112,643 tons), USA (40,500tons), Georgia (39,700 tons), Azerbaijan (31,202tons) and China (23,000tons). Iran, Spain, France and Poland also produce considerable quantities of Hazelnut.

In Turkey, hazelnut production is found in 33 provinces, however 95 per cent of the total production comes from Sakarya, Giresum, Duzce, Samsun and Trabzon province. Around 4 million people are directly or indirectly involved in hazelnut production, and about 400,000 families solely depend on it at Turkey (Anil *et al.*, 2017).

In Italy, the chief centres of hazelnut production are Sicily, Compania, Lagio, Podmant and Venice provinces.

The four major production areas are influenced to some extent by large bodies of water, Turkey by Black Sea, Italy by the Mediterranean and Oregon by the Pacific Ocean.

## Taxonomy

The term 'hazelnut' is used for all species of the genus *Corylus* (Fideghelli and DeSalvador, 2009), which belongs to the Betulaceae. There are nine widely recognized species: Five of those are shrubs: *C. avellana, C. americana, C. cornuta, C. heterophylla*, and *C. sieboldiana*, and four are trees: *C. colurna, C. jacquemontii, C. chinensis* and *C. ferox* (Thompson *et al.*, 1996).

The studies of Erdogan and Mehlenbacher (2000a,b, 2002) based on morphology and phenology, the success of controlled pollinations, and comparisons of DNA-ITS sequences, placed *Corylus* species into three main groups: the leafy-husked shrubs (*C. avellana, C. americana, C. heterophylla*) the bristle-husked shrubs (*C. cornuta, C. californcia, C. sieboldiana*) and the tree hazels (*C. colurna, C. jacquemontii* and *C. chinensis*). Hybridization with in a group was generally easy, but crosses between species in different groups were generally unsuccessful. Two additional species are known: the paperbark hazel (known as *C. papyraceae* or *C. fargesii*) and *C. ferox* whose huks resemble the burs of a chestnut (Mehlenbacher, 2009). The clonal rootstocks "Newberg' and "Dundee' which produce few suckers, are considered to be hybrids of *C. colurna* x *C. avellana* (Lagerstedt, 1990).

There is a variation in the plant characteristics of different species. The plant having central leader are common in *Corylus colurna* and *C. chinensis*. All species of hazelnuts are deciduous but in some species the leaves are retained on the branches during winters in areas experiencing mild climate. The trees are not very tall as other trees are, but under fertile conditions plants may be comparatively taller and may attain a height between 10 to 15 metres.

The plants of this genus are monoecious, bear unisexual flowers. The conspicuous, bloom of the hazelnut is associated with the extention growth of the staminate catkins on leafless twigs with subsequent release of large amount of pollen.

There is variation in appearance of the husk which covers the nut. This variation is usually helpful in identifying the species and cultivars. The husks of *C. avellana* and *C. americana* are short and flared up which allow the nut to fall readily. This is called as 'free husking' and is useful for early harvesting. However, the husks of *C. colurna* are much thicker and forked deeply into lance-like lobes. After drying, the lobes curl back and the cluster of husks look like a spiny cactus. The husks of *C. chinensis* have got a constriction covering the nut and deeply lobed bracts recurve to resemble antlers. The husks of some species like *C. ferox* and *C. tibetica* are spiny and *C. maxima* has long husk, whereas the husk of *C. cornuta* is considerably constricted above the nut which make the removal of nut very difficult (Kasapligil, 1963).

### C. *maxima*

It is a native to southern Europe. Due to overlap of ranges and lack of genetic blocks between *C. avellena* and *C. maxima*, a number of interspecific hybrids have resulted.

### C. *avellana*

This is the commercially cultivated species and is known as European hazelnut. It crosses with many species of the genus and has resulted in large variation in the present-day cultivars. The trees are large (3-10m). The species is less hardy.

### C. *tibetica*

This Tibetan hazelnut is native to western China (Shun-Ching, 1935). It is shrubby, 3 to 7 m in height (Rehder, 1940).

### C. *americana*

This American hazelnut species is found growing in Florida, Oklahoma, Missouri and Minnesota (Rehder, 1940). Fruits are medium sized, enclosed with husk that are usually twice the length of nut. Trees are medium-small that is sucker free. The species is cold hardy and resistant to eastern fire blight.

### C. *cornuta*

It grows in the east coast of the North America, from Quebec to Georgia through Saskatchewan to Missouri. This is a beaked hazelnut and is very winter-hardy. It has undesirable traits : long husk, small and thick-shelled nut.

### C. *chinensis*

This species is found growing in Yunnan Hupei and Szechwan provinces of China (Shun-Ching, 1935). The trees are tall, grows up to 40 m and produce nuts that have husk strongly contracted above the nut.

### C. *heterophylla*

The species is well distributed in Japan and also found in China and Korea. It is shrubby in nature and grows about 2-3 m in height.

### C. *colurna*

This Turkish tree hazel is found in southeastern part of Europe through Turkey, Caucasia, northern part of Iran and in the Himalayan hills to China (Kasapligil, 1964). It grows well on steep stony slopes, attaining a height of 15-20 m or more, and bears irregularly (Anadolliev and Penev, 1975).

### *C. sieboldiana*

The species is found in Japan, Korea and China and also known as Manchurian and the Japanese hazelnut. It is shrubby, attaining a height of 3-5 m. The husks constrict into a tube above the nut.

### *C. jacquemontii*

This so-called Himalayan hazelnut is considered a variety of *C. colurna* (Kasapligil, 1963). The tree produces small, thick-shelled nuts.

### *C. ferox*

It is indigenous to northern India, Sikkim and Nepal, and grows successfully at an altitude of more than 3000 m.

The leaves of hazelnuts are flat, elliptical in shape with hairs on both surface and a double serrated margins (Snare, 2008). The female inflorescences, known as burrs, are born in tight clusters (Dale *et al.*, 2012). The male inflorescences, known as catkins which have 130-290 flowers.

## Cultivars

### Barcelona

In France it is known as Coutard hazelnut. It is fairly susceptible to nut gall mite (*Phytoptus avellanae*) and bacterial blight (*Xanthomonas campestris*). It is pollinated by Segorbe at the beginning of flowering, Butler at full flowering and Merveilte de Bollwiller at the end of flowering. It cannot be pollinated by Ennis or Tonde di Giffoni. It is also good pollenizer for Segorbe at the beginning of flowering. This cultivar is fairly late, nuts ripen in mid-September in the Bordeaux region and is susceptible to frost. Yield is generally good, sometimes irregular and the empty nut percentage may be high if there is a cold weather during the second half of June.

### Butler

This cultivar was selected as a seedling of unknown parentage in an Oregon orchard. It is biennial in bearing but very productive and bears viable pollens, which are compatible with Barcelona, Ennis and several other cultivars. It is winter-hardy and less susceptible to big bud mite (*Phytoptus avellanae*).

### Ennis

The cultivar of unknown source and parentage was brought to Oregon from Washington in about 1940 and has shown considerable advantage over Barcelona cultivar. It is high yielding; less prone to biennial bearing than Barcelona and the nuts are large. The weight of 100 nuts ranges between 400 g and 430 g. The light brown nuts are borne singly or in pairs in very short husks. Catkins are small, produce large amount of defective pollens and are not compatible with Barcelona. It is very vigorous and cold resistant.

### Istariki Dagutyasti

This cultivar seems to be better adapted in respect of nut characteristics to the inland site than to the coastal area. It shows variations in shell thickness and in kernel weight to shell weight ratio of the nut. The prospects of clonal selection are better with regard to whole fruit weight and kernel weight.

### Northamton

In Sarajioo, this cultivar flowers moderately early, the male catkins between 6 February and 21 March and the females between 21 February and 31 March. The flowers are mainly dichogamous and protogyny and protandry occur in equal proportions. The cultivar is almost self-sterile. Pollination with Princess Royal and White Lambert resulted in 53.95 and 52.00 per cent fruit set, respectively.

### Ana

This cultivar of *C. avellana* has weak to moderate growth and a spherical crown. It reaches a height of 2.6-2.8 m and a width of 1.6-1.8 m at 5 years of age. Kernel percentage is 58.6-59.4 and the yield is high. The short and small crown of this cultivar made it suitable for intensive commercial cultivation.

### Self Husker (Freehusker)

Originated in Newberg, Oregon by A.B. Scherf, from seedling selection. Nut falls free from husks. shell very thin, resembles DuChilly. The tree is moderately vigorous, upright, bears annually, produces 1-2 nuts per cluster.

### Skinner

Introduced in 1967, it is a cross between Dropmore x seedling of Italian Red. The nut is large, little wider than long; shell is medium in thickness, chestnut colour; kernel plump; flavour good, resembles Barcelona. It ripens in the last week of September. The tree is very hardy, tolerant to blight.

### Tonda Gentile Tomanu

It is a high quality hazelnut and the yield of kernel is about 44-48 per cent of the nuts. The tree is cold-resistant and early fruiting. Harvesting begins from mid-August.

### Arutela (Hall's Giant x Tonda Gentile Gentile delle Langhe)

Developed at Romania and introduced in 2005. Early maturing cultivars, round, small (2.2g) kernel, medium-brown, husk 70 per cent longer than nut, cold hardy, incompatibility alleles $S_2S_7$.

### Cozia (Hall's Giant x pollen mixture)

Developed at Romania and introduced in 1987. Large (4.0g), round nut, easy pellicle removal, early flowering cultivars.

### Feriale (Imperiale de Trebizonde x Butler)

Developed by INRA Bordeaux, France, introduced in 2004. Very large nut (4.1g), pellicle has little fiber, free husking. Precocious, high-yielding; resistant to bud mites.

### Ferwiller (Hall's Giant x Tonda Romana)

Developed by INRA Bordeaux, France, introduced in 2004. Medium sized (2.5g) nut. Pellicle has little fibre, free husking, very few blank nuts, precocious, resistant to bud mites.

### Natval (Valcea 22 x Ennis)

Developed at Romania, introduced in 2010. Medium sized (2.4g), round nut, easy pellicle removal, sime-spreading tree.

### Primval (Hall's Giant x pollen mixture).

Developed at Romania, introduced in 2010. Medium sized (2.5g), round nut, easy pellicle removal, semi-spreading tree, incompatibility alleles $S_{10}S_2$.

### Red Dragon (OSU 487.005 x OSU 367.039)

Developed at Oregon State University, Corvallis, USA, introduced in 2008. Small, slightly long and compressed nut. Late maturing, free husking, resistant to eastern fire blight, intermediate susceptibility to bud mites, incompatibility alleles $S_6S_{26}$.

### Romavel (Hall's Giant x pollen mixture)

Developed at Romania, introduced in 1991. Medium sized (2.9g), round shaped nut, easy pellicle removal, semi-erect, medium vigor tree, late flowering, incompatibility alleles $S_{10}S_2$.

### Tonda Pacifica (Tonda Gentile delle Langhe x Barcelma x Extra Ghiaghli)

Developed at Oregon State University, Corvallis, USA, introduced in 2010. Small (2.2g) nut, pellicle has little fibre, free-husking, excellent texture and flavour, susceptible to eastern fire blight, moderately susceptible to big bud mites, incompatibility alleles $S_1S_2$ (Mehlenbacher *et al.*, 2011a).

### Uroise de Valcea (Ennis x Purple Filbert)

Developed at Romania, introduced in 2002, very large (4.8g), ovoid nut, tree low to medium vigor, upright, incompatibility alleles $S_{11}S_{20}$.

### Valcea 22 (clone of Imperiale de Trebizonde)

Developed at Romania, introduced in 1980. Large sized (3.9g), oblate shaped nut, pellicle has little fibres, tree low vigour, spreading, precocious and high yielding, resistant to bud mites, early flowering, incompatibility alleles $S_2S_{10}$.

**Yamhill (OSU 296.082 x VR 8-32)**

Developed by Oregon State University, Corvallis, USA, Introduced in 2008, small (2.3g) nut, pellicle has little fibre, free husking few blank nuts, high nut yield, early flowering, highly resistant to bud mites, resistant to eastern filbert blight, incompatibility alleles $S_8S_{26}$ (Mehlenbacher *et al.*, 2009).

**McDonald (Tonda Pacific x Santiam)**

Released by the Oregon Agricultural Experiment Station, USA in January, 2014. (Mehlenbacher *et al.*, 2016). Trees are moderately vigorous, globose in shape, high nut yield, small to medium nut size, high kernel percentage (52 per cent), early nut maturity, excellent kernel quality and have high level of resistance to eastern filbert blight.

**Wepster (Tonda Pacific x OSU 440.005)**

Released by Oregon Agricultural Experimental Station, USA in January, 2013. The pedigree includes, germplasm from Italy ('Tonda Gentile delle Langhe', 'Montebello'), Spain ('Barcelona'). England ('Daviana'), Turkey ('Tombul Ghiaghli','Extra Ghiaghli') and the Pacific Northwestern United States ('Gasaway'). 'Wepster' is an early maturing high yielder and vigorous cultivar, small nut size, very good kernel quality and resistant to eastern filbert blight.

**Jefferson (OSU 252.146 x OSU 414.062)**

Released by the Oregon Agricultural Experiment Station in January, 2009. The nuts and kernels are similar in size to 'Barcelona'. It has been released for the in-shell market, however its kernel quality is suitable for many end uses, including chocolate products and baked goods. Trees are small and high-yielder, resistant to eastern filbert blight (Mehlenbacher *et al.*, 2011b).

**Lewis (OSU 17.028 x Willamette)**

It was released by the Oregon Agricultural Experiment Station, USA in January 1999 for the kernel market (Mehlenbacher *et al.*, 2000). 'Lewis' has a higher yield efficiency, smaller nuts (2.9g), higher percent kernel, earlier maturity, and greater quantitative resistance to eastern filbert blight.

**Sacajawea (OSU 43.091 x Sant Pere)**

Released by the Oregon State University, USA for Kernel market (Mehlenbacher *et al.*, 2008). Trees are smaller than 'Barcelona', higher nut yield efficiency, early nut maturity, higher percentage of kernel and better pellicle removal ratings. The excellent kernel quality is comparable to the best Italian cultivars.

**Dorris (OSU 309.074 x Delta)**

Developed at Oregon Agricultural Experiment Station, USA and released in February, 2011 (Mehlenbacher *et al.*, 2013). 'Dorris' is suitable to use as blanched kernel and in-shell markets. It combines high resistance to eastern filbert blight, high nut yield, medium nut size, attractive nuts, with excellent texture and flavour.

## Roverd (Red Lambert x Dariana)

Developed at Romania (Vicol *et al.*, 2009). Trees are medium to high in vigour, semi-spreading, nut small (1.4g), good kernel ratio (55.0 per cent), fruit ripen in the end of August.

## Santiam (OSU 249.159 x VR 17.5)

Developed by Oregon Agricultural Experiment Station, USA and released in February, 2005 (Mehlenbacher *et al.*, 2007). 'Santiam' has complete resistance to eastern filbert blight. Trees are smaller, higher yield efficiency, higher kernel percentage (51 per cent), and smaller nuts and earlier nut maturity. The frequency of blanks and poorly filled nuts is less than 'Barcelona'.

## Clark (Tombul Ghiaghli X Willamette)

Developed by Oregon Agricultural Experiment Station and released in 1999 (Menlenbacher *et al.*, 2001). 'Clark' has higher yield efficiency, smaller nuts, higher percentage of kernel, better suitability for the blanched kernel market, early maturity, and greater quantitative resistance to eastern filbert blight.

## Turkish Hazelnut Cultivars

### Tombul

Most important cultivar of Turkey. Nuts are rounded and tidy shape. Nut weight 1.8g, kernel weight 1.0g, nut/kernel ratio 54.4 per cent.

### Sivri

Highly productive cultivar, however sensitive to droght, high pre-harvest drop in dry season. Nut weight 2.0g, kernel weight 1.0g, nut/kernel ratio 49.9 per cent.

### Cakildak

Tolerant to late spring frost because it has very late bud burst. Nut weight 2.1g, kernel weight 1.2g, nut/kernel ration 55.8 per cent.

### Fosa

Nut weight 1.7g, kernel weight 0.8g, nut/kernel ratio 50-53 per cent.

### Palaz

Nut weight 2.1g, kernel weight 1.1g, nut/kernel ratio ratio 51.4 per cent.

### Okay 28 (Tombul x Kargalak)

Bud burst is one week later than 'Tombul' which is important for late spring frosts. Nut weight 2.9g, kernel weight 1.5g, nut/kernel ratio 54.6 per cent.

### Giresun Melezi (Tombul x Kargalak)

Nut weight 2.4g, kernel 1.2g, nut/kernel ratio 51.7 per cent.

### Allahverdi

Seedling selection, bud burst is 15 days later than 'Tombul'. Compared with 'Tombul' it produces more male flowers, pollination last one week longer. Nut weight 1.8g, kernel weight 0.8g, nut/kernel ratio 49.3 per cent.

The important hazelnut cultivars of the world are presented in Table 9.1.

## Climate

Climate of a particular place plays an important role in the successful cultivation of fruit. The hazelnuts require cool summer and moderately cold winter. Long cool summer is more favourable for the development of the nuts than long warm summer, and a rather long exposure to moderate cold is necessary to break the rest period. In humid areas, the trees are susceptible to foliage diseases. Hazelnuts are adapted to dry summer of the Pacific coast than to the climate in the other regions.

### Temperature

The hazelnuts are resistant to cold. The catkins and pistils are highly resistant to cold as compared with the open flowers. A temperature of –9.5°C is said to cause damage to the open pistils of the flower. In some areas, where large water bodies exist, the danger of cold injury is reduced to a considerable extent. In England, parts of Germany and Scandinavia, where hazelnuts are grown, the water bodies equalize the temperature, delaying the warm weather and opening of the pistillate flowers until danger of such cold nights have passed.

**Table 9.1: Important Hazelnut Cultivars of the World**

| | | |
|---|---|---|
| Romania | : | Valcea 22, Cozia, Uriase de Valcea, Romavel, Arutela, Valverd, Primval, Natval, Rovered. |
| Iran | : | Fertile de Coutard, Negeret, Rond du Pimount, Rasmi Pashmineh, Paeizeh, Shastac 2, Gerche, Gherdooii. |
| Turkey | : | Tombul, Sivri, Cakildak, Fosa, Palaz, Okay 28. |
| Australia | : | Barcelona, Tonda di Giffoni, Segorbe, Tonda Romana, TBC (Tokolyi/Brown field Cosford), Ennis. |
| United States | : | Barcelona, Butler, Daviana, DuChilly, Ennis, Hali's Giant, Montebello, Delta Gamma, Jefferson, Lewis, Tonda di Giffoni, Santiam, Clark and Yamhill. |
| Spain | : | Alcover, Casina, Closca Molla, Culpia, Negret |
| Italy | : | Camponica, Daria, Ghirara, Mortarella, Nocchione, Tonda Gentile delle Langhe, Tonda di Giffoni. |
| England | : | Cosford, DuChilly |
| Germany | : | Gunslebert |

A significant correlation was found between minimum and maximum air temperature during growth period and the flowering dates of male and female inflorescence. An increase in the sum of active temperatures retarded the appearance of the stigmas in the early flowering cultivars which require, from flower differentiation to flowering, a minimal active temperature sum of 2700-2800°C, whereas for medium and late cultivars the requirement is between 2900-3000°C. It was also observed that the production was accelerated when the mean daily temperature rose by 1.2-4.2°C but retarded when temperature increased further (Kavardzhikov, 1980, 1982). Hazelnut trees have poor tolerance to wind, heat and water stress. The trees achieve optimum photosynthesis at 25°C to 28°C (Schulze and Kuppers, 1979). Temperature above 30°-35°C compared with persistently low humidity may cause heat stress (Kerslake, 2012). *Corylus avellana* can resist, without damage, temperatures down to -15/-18°C (Kripina *et al.*, 1992). The young shoots can tolerate a temperature of -1°C (Fideghelli and DeSalvador, 2009).

Hummer *et al.* (1986) reported that the general order of tissue hardiness from least to most was pith, xylem, cambium and cortex. The vegetative buds in mid-winter were hardier than pistillate flowers in October but most pistillate flowers were hardier than the staminate ones only in January. Several hazelnuts cultivars had fully blooming pistillate flowers that were uninjured at –30°C in December and at –40°C in January.

The ideal annual rainfall is 800-1000mm well distributed during the year and the chilling required varies among cultivars and the kind of buds (male flowers: 100-1200 h, female inflorescence: 500-1200 h, vegetative buds: 600-1200 h). a breeze during blooming is the best condition for good pollination (Fideghelli and DeSalvador, 2009).

The opening of pistillate flowers is delayed in southern California and along the coast when chilling is inadequate more than it delays shedding of pollen, but fruiting seems better after the warmest winter in those areas as after the coolest. But when the buds of hazelnuts open after delay following a warm winter, they open more nearly altogether than apple or peach buds. The chilling requirement of 'Barcelona' leaf buds is considerably less than those of other cultivars.

The best site for hazelnut cultivation should have good air movement and protection from drying winds. Areas, where moderate rainfall occurs which is well-distributed through-out the year, are considered ideal for the successful cultivation of the hazelnuts/filberts (Chandler, 1987).

## Soils

The exact type of soil which is best suited to hazelnuts varies in different areas where hazelnuts are grown, but in all cases it should be deep, fertile and well-drained. Deep soils naturally afford a large supply of water and plant nutrients during the dry season. The hazelnut roots readily grow to a depth of 2.5- 4 m. Root growth is restricted by impervious hard pan, water table, sand or gravel and lack of aeration in the soil. The trees on medium soil fertility do not give good yields with increasing age and size of the trees and extreme variation in crop yield occurs. A good crop in one year may be followed by several years of low production. In a year of heavy production, new shoot growth is very short and unproductive and it may require two or more years for the tree to develop shoot growth prior to heavy bearing.

The soil should be well-aerated up to a depth of about 4 m. Hazelnuts seem to be more tolerant of rather compact soils than walnuts or pecans. The small trees or bushes will fruit well on poor hill-soils but the yield in comparison with the trees growing on deep fertile soils is poor. Neutral to slightly acidic soils having about pH 6 are favourable for hazelnut culture.

## Rootstocks

Rootstocks play an important role in respect of general performance and increase yield potential of the scion cultivar. Grafting on a rootstock increases hazelnut yield by 25-50 per cent compared with own-rooted trees and also facilitates training to produce trees with hollow crown on short stems (Mauhrer, 1975). Rival (1979) recommended *C. colurna* as a rootstock for *C. avellana* because of its vigour, extensive root system and non-suckering habit. Cultivars Geante de Halle, Imperial de Trebizonde, Fertile de Coutard and Bolwiller performed well on *C. colurna* rootstock. Grafting is possible at soil level, at 0.80-1 m above ground or even at 2 m or more for different purposes. Anadolliev and Bozhinova (1982) recorded that *C. colurna* selections were suitable as rootstocks for highly intensive plantations of single stem hazelnuts. The trees did not sucker or show symptoms of early incompatibility with scion cultivars of different vigour. For medium vigour, selections of *C. colurna*, such as, Perunika 4, Perunika 20, Nesebr 2 and Verna 2, whereas for dwarf growth, Sadovo 3, Perunika 15, and Sadovo 2 are recommended.

*Corylus colurna* has been found to improve drought and frost resistance (Ninic-Todorovic *et al.*, 2008), improves the resistance towards pests and disease (Ninic-Todorovic *et al.*, 2000) and improve fruit and kernel quality (Miletic *et al.*, 2009).

The primary factor determining hazelnut culture in the US is Eastern Filbert Blight (EFB) which has boh cultural and genetic implications for rootstock development. Historically, there have been challenges with grafting hazelnuts, which were overcome to a certain extent by Lagerstedt's development in the early 1980s of the heat-callusing pipe (Thompson *et al.*, 1996). In addition, Lagerstedt worked to develop non-suckering rootstocks, releasing two *C. colurna x c. avellana* hybrids ('Newberg' and 'Dundee') (Lagerstedt, 1993). Unfortunately, they were not used due to susceptibility to EFB (Grauke and Thomson, 2003). The effort to combat EFB led to the identification of 'Gazaway' as heterozygous for the single dominant resistant gene (Mehlenbacher *et al.*, 1991).

Rovira *et al.* (2014) evaluated the performance of 'Negret-N-9' selection hazelnut onto four clonal rootstocks ('Dundee' and 'Newberg' two open pollinated *C. colurna* seeds, low-suckering cultivar 'Tonda Bianca' and IRTA-MB-69 selection) in Spain. 'Dundee' and 'Tonda Bianca' rootstocks showed the highest vegetative growth, whereas, 'Dundee', 'Newberg' and IRTA-MB-69' showed less suckering. The highest yield was recorded by using 'Dundee' as rootstock.

The yields of 8 cultivars grafted on *C. colurna* rootstocks ranged from 2.2 kg/ tree in 'Daviana' to 4.5 k/tree in 'Lambert' (Meiss, 1985). Bauckmann (1979) also obtained higher yields from trees grafted on *C. colurna* than from cultivars on their own roots.

## Propagation

Most of the hazelnuts trees are of seedling origin. Growers sometimes obtain new trees by making mounds of soil around the trunks of trees in the orchard. The soil heap is removed and the suckers that have formed roots at their bases are cut and planted in the nursery or directly in the orchard.

### Seed Propagation

Blomme and Degeyter (1978) reported that the presence of the pericarp in hazelnuts greatly reduced germination percentage immediately after harvest. However, removal of the testa as well as pericarp increased germination to 81 per cent. Dormancy could be largely eliminated by a 6-month stratification period. With late harvested (19 September) nuts having hard pericarp, a warm stratificaiton period at 20°C was necessary, followed by cold period at 4°C; with early harvested (21 August) nuts, the endosperm and embryo were often insufficiently developed for good germination. Arias *et al.* (1976) reported that dormant seeds, when subjected to chilling for six weeks at 5°C and subsequently transferred to 20°C, resulted in the accumulation of gibberellin followed by germination.

Germination of more than 85 per cent was achieved when newly harvested hazelnuts were treated with 100 ppm $GA_3$, whether kept in darkness or light. With nuts dry stored for 4 weeks, 87 per cent germination occurred when intact seed was treated with $GA_3$ and germinated in darkness and 83 per cent germination occurred when embryos from nuts dry stored for 4 weeks were chilled at 5°C and then germinated in darkness (Bradbeer *et al.*, 1978). Shannon *et al.* (1983) found that intact seeds of cv. Kent Cobs did not germinate in the dark or when illuminated with white light for 30 minutes on 3 consecutive days. Removal of the testa stimulated germination only in the light. Excision of one whole cotyledon (half seed) was more effective in stimulating germination in the light treatment than only removing the testa. Red radicle irradiation promoted germination in seeds in which only the radicle end of the embryo axis was uncovered. When the axis was uncovered, the presence of the testa stimulated germination under white or red light but without uncovering of the embryo axis, the testa inhibited germination.

Gosling and Ross (1980) reported that the activities of glucose-6-phosphate dehydrogenase and 6-phosphogluconic acid dehydrogenase remained relatively constant in seeds stratified at 5°C. The major increase coincided with the break of dormancy. Bradbeer and Cotman (1967) observed that the dormancy of hazel seeds was broken by less than 30 days chilling at 5°C. Kowalski and Kawecki (1982) found the best germination when the freshly harvested and immediately stratified nuts were kept at 18°C for 3 weeks followed by stratification at 5°C. Storage of nuts decreased germination at both temperatures. The phospholipid content of the cotyledons and embryonic axes increased during germination, Early increase in the relative amount of phosphatidyl glycerol and decrease in the relative amount of phosphatidyl ethanolamine occurred in both cotyledons and axes. The later stages of germination were accompanied by further changes in the phospholipid of the axes. Increases in the degree of unsaturation of the $C_{18}$ fatty acids occurred in both tissues (Shewry *et al.*, 1974).

Pretreatment of intact dormant seed of hazel cultivar Kent Cob with either ethanol or mercuric chloride alone was without effect in stimulating subsequent germination in the light. However, when treated with ethanol and then soaked in mercuric chloride solution, intact seeds germinated when subsequently sown in the light. Treatment with ethanol and mercuric chloride did not replace the light requirement of the intact seeds. The effect of such treatment was similar to that of removal of the testa and may interfere with the movement of germination inhibitors from the testa, most likely ABA. Ethanol may be replaced by diethylether, chloroform or n-propanol (Jeavons and Jarvis, 1984).

## Vegetative Propagation

### Cutting

Propagation through cuttings has been found successful. Vanicheva (1984) reported that etiolation advanced rooting and ensured a stable and high rooting percentage. The best result was obtained with total etiolation followed by localized etiolation and the best time for taking cuttings is during rapid shoot growth. Thomson (1984) obtained 80-85 per cent success when treated with 6000 ppm IBA but mortality was high. He also recorded better success when branches were girdled in early October, 2-3 cm above the base of the previous season's growth and blanched by wrapping with black tape from 2 cm below the girdle to 8 cm above it. Cuttings 1 cm above the girdle were removed six weeks later, dipped in IBA solution and struck under mist. The success was 85 and 70 per cent in cvs. Victoria and Macrocarpa, respectively. Leaf retention was necessary for successful rooting. Genetic variations among the different clones in root regeneration from cuttings has been reported (Anon.,1978). Higher concentration of IBA (1000 ppm) caused thick and brittle roots in cuttings.

Roversi (1969) obtained better results by using gentle bottom heat in a peat/s and media with an air temperature of 13-15°C. A basal temperature of 25°C was better than 20°C. Cuttings from one-year-old shoots were better than those made from suckers. The cuttings taken from basal parts rooted better than those of distal ends. Rooting was stimulated by using IAA + IBA + NAA (2000 + 2000 + 1000 ppm). Under optimum conditions, 74 per cent cuttings produced roots within 65 days.

Paglietta (1973) in Italy was successful in two systems of propagating woody cuttings of hazelnut cv. Barcelona. In the first, by blanching shoots up to 50 cm in length which were taped from the base to a height of about 30 cm at the end of May, then cut off below the tape after 60 days, stripped of tape and rooted under mist. In the second, the cuttings were held in aerated water which permitted root inspection without damage for transplanting at the right stage. Pellati and De (1982) reported that woody cuttings taken in early March developed faster, were stronger and survived better after planting out than semi-woody cuttings taken in May and June-August. Piskornik *et al.* (1982) reported that juvenile cuttings rooted better than hardwood when treated with IBA at 50 ppm.

Rodriguez *et al.* (1986) examined the shoots of different age groups and found that 4-week-old shoots contain discontinuous sclerenchymatic ring made up of fibre bundles without scleroids, 1-year-old shoots possessed a practically continuous sclerenchymatic ring with fibre bundles containing cells with moderately lignified secondary cell walls and some scleroids scattered among them and 2-year-old shoots possessed a strongly lignified sclerenchymatic ring mainly consisting of scleroids which in some cases appeared to be fused and forming lignin blocks. The observations suggest that 1-year-old shoots may be used as cutting in place of 4-week-old shoots which are small in diameter and do not survive well.

### Layering

The success in air-layering with the application of IBA has been reported by Fontanazza (1973), with trench-layering by Bergamini and Cristoferi (1968) and with mound-layering by Fontanazza and Salleo (1968), Lagerstedt (1978), Pathak *et al.* (1978), Kazi-zade and Bozhko (1981) and Kavardzhikov and Krinkov (1981). In air-layering, complete success in rooting was obtained when the base of shoot was treated with lanolin containing 3000 ppm IBA, and the rooting media containing a 3:1 mixture of perlite and soil was wrapped with black polythene. High percentage of rooting was also recorded without a rooting medium or with perlite alone. However, when no medium was used fewer roots were produced and these were thick, little branched and generally inferior (Fontanazza, 1973). Tombest (1983) obtained good rooting of hazel layers by using peat as media and covering with black plastic. Girdling below the branch increased rooting.

Roversi (2015) developed a modified layering method to get a good number of rooted hazelnut plants from no-suckering hazelnut clones of 'Tonda Gentile delle Langhe'. Before the sprouting the branches of the plants were cut at 35 cm from the ground. The water sprouts that developed are usually very vigorous were ringed, for mound layering. The ringed water sprouts then covered with a container containing substrates. The water sprouts developed roots and suitable for planting directly in the orchard.

### Grafting and Budding

Me *et al.* (1981) obtained 90 per cent success when leafy shoots of cv. Negret were grafted on the seedlings of Tonda Gentile della in May. Pathak *et al.* (1978) reported success between 80 and 100 per cent with tongue or cleft grafting on 1-year-old seedling rootstock in March. Rahimi *et al.* (2016) developed hypocotyls cleft grafting method with coating the scion with a thin layer of paraffin (wax). 'Carmela' (*C. avellana*) hazelnut was grafted on etiolated hypocotyls of *C. americana.* Uncoated scions had only 9 per cent success rate, while the success rate increased to 85 per cent when the scion was coated with a thin layer of paraffin just after grafting. Anadolliev (1976) reported success between 75 and 95 per cent with machine grafting.

The callus formation increased by raising the temperature at grafting union up to 80°F (Yarris and Lagerstedt 1981; Lagerstedt, 1984). Manzo (1972) obtained 79.2 per cent success with budding in cv. Tonda di Giffoni and 67.3 per cent in Mortorella.

### Micropropagation

The tissue culture of hazelnut might be accomplished by using the nodal segments and shoot tips, however, as a recalcitrant species, using axillary buds and embryo culture was commonly used in micropropagation of hazelnut.

The success in micropropagation of different cultivars of *Corylus avellana* has been reported by Kai *et al.* (1984). They used axillary buds as explants, cultured in MS medium supplemented with Zuccherelli vitamins, NAA at 0.01 mg/1, $GA_3$ at 0.1 mg/l and BA at 5 mg/1. The percentage success of shoot formation was 75, 70 and 50 for cultivars Segorbe, Fertile de Coutard and Merveille de Bollwiller, respectively. The chlorosis was controlled by Sequestrene 138 Fe but not by FeEDTA. Treatment with IBA at 0.1 mg/l increased root production and growth was normal. The success in micropropagation of hazelnuts has also been reported by Perez *et al.* (1983, 1985, 1986), Anderson (1984), Rodriquez and Fernandez (1982) and Radojevic *et al.*, 1975).

Driver and Kuniyki Walnut Medium (DKW) (1984), which was originally developed for walnut, was more often used in the micropropagation of hazelnut. Damiano *et al.* (2005) developed an establishment stage basal medium, and multiplication medium. They found a medium that contains macro and minor nutrients from the formulation of Perez-Tornero *et al.* (1987) in combination with DKW organics was the best for establishment.

Microbial contamination and toxic effects of over disinfections are problems with achieving viable explants (Hand, 2013). Ethanol can be applied in the disinfection process by dipping the material in it before using the bleach (Nas, 2004). As a response of plants to excision, explants browning are another common problem in culture. There are methods that can be applied to reduce the browning rate such as storing the stock plants under low light/in the dark, soaking explants in water after excision and sub-culturing the explants frequently (Bonga and Aderkas, 1992; Compton and Preece, 1986).

## Planting

The land preparation and laying out systems are common to all nut fruits. The slopy land should be terraced before planting so that the plants could be planted at proper distance. This is essential because the basin is made to facilitate the application of manures and fertilizers and irrigation to the hazelnut plants. Care should be taken that while planting, there should not be any rock which may restrict the root growth and will ultimately affect the health and production of the tree.

In Oregon, where hazelnuts are grown commercially, planting is done in late winter or spring. Planting in early winter encourage more root development before development in spring (Pollack, 1998) while planting in early spring when the trees are dormant helps to settle the roots in the soil and starts growing when bud break (Leuty *et al.*, 2012). Hazelnuts are usually planted on mound after digging holes and the roots are extended outwards and downwards. Top soil is firmed over the roots and the area around the tree mulched with saw-dust or loose soil making sure that the tree is well supplied with water. Shallow planting is done to reduce the number of suckers and to make their removal easier (Sander, 1963). Newly planted plants are headed back at 50-75 cm and the trunks protected from sun scald by covering the trunks with white paint or wrapping with white paper.

The application of exterior quality white latex paint proved useful for protecting young hazelnuts from sun scald (Lagerstedt, 1978).

## Planting Distance

The tree density per hectare varies from a minimum of about 300 trees (or bushes) per hectare through to a maximum of about of 1700 (Fideghelli and Salvador, 2009). The trees which are not allowed to grow as bushes or trained in hedgerows, should be planted about 6 m apart. Trees have been found to be more productive than bushes. The central part of a bush is weak or even killed by the shade of the outer part. Planting distance depends on the intention to grow intercrops or plant fillers. In the hazelnut producing areas of Turkey, trees are planted in clumps with centres 5-7 m apart and little pruning is done (Serr, 1964). Tkhagushev and Merzhanjan (1975) reported that soil moisture reserves were always greater under densely planted bushes due to reduced evaporation and this created a favourable microclimate when compared between 5x5 m and l0xl0 m spacing.

In an experiment hazelnut trees were spaced at 4x4 m and trained with open centres or inrows at 4x1 m with free crown formation. After 9 years, half the trees were removed from the close rows. The yield over six years was 1480 kg/ha from widely spaced trees and 1630 kg/ha from the rows. The need for tree thinning and greater harvesting difficulties with the rows make this system less convenient (Damigella and Alberghina, 1978). The seedlings of *Corylus heterophylla* producc fruits three years after planting and, when these were planted up to 200 trees per acre with 10 suckers per tree, showed high yield (Kim *et al.*, 1981). Largerstedt (1973) reported that trees after 12 years, spaced at 25 feet (8 m) apart, began to yield more nuts per tree than those spaced at 15-20 feet (5-6.5 m).

Training systems and plant density of hazelnut orchards in Italy are based on bushes with 3-5 stems at a density of 400-500 plants per hectare. However, in many countries, as well as in Italy, new orchards are based on single trunk plants using higher-density plantation at 3 X 4.5m (Cristofori *et al.*, 2017).

**Table 9.2: Training System and Planting Density of the Hazelnut Orchards**

| *Training System* | *Distance* | | *Trees per Hectare* | *Reference* |
|---|---|---|---|---|
| | *Between Rows* | *Between Trees on the Row* | | |
| Multistem Bush (MB) | 4-6 | 4-6 | 277-625 | Romsiondo *et al.* (1983); Tous *et al.* (1994). |
| Vase | 4-6 | 3-5 | 416-833 | Romsiondo *et al.* (1983); Tous *et al.* (1994). |
| Vase or MB[1] (Dynamic density) | 4-6 | 2.5-3 (5-6) | 555-1000 (277-500) | Romsiondo *et al.* (1983); Bignami *et al.* (2005); Azarenko *et al.* (2005). |
| Hedgerow (V-shape-hedge) | 5 | 2.0+2.4 | 1666 | Romsiondo *et al.* (1983); Tous *et al.* (1994); Me *et al.* (2001). |

[1]50 per cent of the trees or bushes will be removed after 10-15 years from planting

Adopted from (Fideghelli and Salvador, 2009).

## Training and Pruning

The most common training systems are the Vase and the Multistem Bush. Another training system with certain importance is the Hedgerow (V-shape-hedge) (Romisondo *et al.*, 1983; Tous *et al.*, 1994; Me *et al.*, 2001; Bignami *et al.*, 2005). A limited practical extension has the Y system (Tous *et al.*, 1994) and the Monocone or Vertical axis (Lespinasse and Delort, 1986). The Multistem Bush is still very popular in traditional areas where mechanisation is not widely applied, while the Vase is more suitable and is carried out to the present time in mechanized orchards (Fideghelli and Salvador, 2009).

Regular annual pruning (by hand and/or mechanical) improves the light penetration into the canopy, the flower bud formation, fruit setting and nut quality. For production pruning, it is suggested to prune one-fifth of the orchard each year. Heading back trees into 3- and 4-year old wood is particularly helpful in stimulating new growth and maintains production in old crowded trees. However, heading cuts affect branch architecture by stimulating growth of vigorous shoots that are less prone to flower and produce nuts in the next year. Branch thinning is preferred over heading to maintain a branch architecture that are more favourable for flowering while increasing light penetration in the canopy (Tombesi and Farinelli, 2017).

Pavlenko and Slyusarchuik (1983) obtained higher yields on bushes thinned to 10-12 shoots. Manusev (1978) reported that plants of 5-7 years of age, trained as bushes with 6 stems, produced highest yield of 1054 kg/ha. He also suggested that multi-stem (4, 6 or 8) bushes gave better results than training on a single short trunk as a pyramid or as a goblet. Fregoni and Zioni (1964) reported that by combining the best features of goblet and bush training into U angle system, the yield of hazelnuts was markedly increased. The rows planted from North to South were more productive than planted from East to West, the average difference being 20 to 23 per cent.

However, Lagerstedt (1973) recorded higher yield from hazelnuts trained as trees than those trained as bushes, and trees were more suited to mechanical management practices. In hedgerow trials with 'Barcelona' the growth rate of plants spaced at 6 or 8 feet began to fall after a few years as compared with those planted 10 or 12 feet apart.

The importance of a regular pruning is directly correlated to the tree density. The pruning is done in December-January when trees are dormant.

## Nutrition

In general, hazelnut need less mineral nutrients compared to other deciduous fruit trees, bud quality and yield are correlated to mineral nutrition (Sentis *et al.*, 2004). In studies on a podzolic medium clayey chernozem soil, seedlings receiving N, $P_2O_5$ and $K_2O$ at 60:60:60: kg/ha removed 118 kg N, 48 kg $P_2O_5$ and 63 kg $K_2O$ from the soil, whereas non-fertilized seedlings removed 97 kg N, 40 kg $P_2O_5$ and 52 kg $K_2O$ (Saban, 1977).

In Italian province, the recommendations for autumn manuring are FYM 300 quintals and superphosphate 6 quintals/ha while for spring application, ammonium sulphate 4-5 quintals or calcium cyanide 3-4 quintals or ammonium nitrate 1.5 to 3 quintals/ha (De Rosa, 1961).

Chaplin and Dixon (1979) applied N as urea to *Corylus avellana* at 0.68, 1.36 or 2.72 kg N/tree annually from 1971-77. Increasing N significantly increased leaf N, yield and tree size. Leaf P level was reduced by raising N rates in all years. Soil pH after 7 years of N application significantly declined as the N rate increased, Leaf Mn level was increased by N in all years probably due to the decrease in soil pH. The standard ranges for leaf N in hazelnut were determined as follows: deficiency in which visible symptoms present had 2.0 per cent, below normal 2.0-2.2 per cent, normal 2.2-2.4 per cent and above normal more than 2.4 per cent.

Quickly available fertilizers like nitrate of soda are of little value when applied in the fall as they leach out so quickly that plants obtain only a small part of the nutrient. Application in late winter or early spring proved more effective. Ammoniated phosphate (16-20-0) and ammonium sulphate as sources of nitrogen applied either in the exact materials used, or amounts have been so adjusted that approximately 100 lb nitrogen have been applied per acre with 135 lb of phosphorus as $P_2O_5$, 50 lb sulphur and 100 lb potassium as $K_2O$ - varying a few pounds according to the materials used.

Boron sprays induced more fruit set over controls. Leaves contained more B in the treated trees than under controls throughout the season, and the amount of boron in young fruits increased two-fold with B sprays. The differences between treated and control fruits disappeared by mid-summer. Boron content of fruit in the month of May from unsprayed trees was found low for optimum nut development, indicating that annual B sprays are required. Since foliar sprays in April damaged young leaves and shoot tips, sprays untill second week of May is not recommended (Shrestha *et al.*, 1987). In similar studies by Shrestha *et al.* (1986) with 'Barcelona' hazelnut, boron spray was found to increase fruit set and decrease dry matter accumulation in the entire fruiting twig. In unsprayed twigs, prior to 5 June, leaves constituted 80 per cent of the total dry weight of fruiting twig. Later in the season on 5 August, the per cent dry weight was maximum in the nuts (66 per cent) followed by in leaves (11 per cent) and husks (13 per cent). Leaves contained 87 per cent of the total B in fruiting twigs early in the season indicating that the leaf was the major sink for B at that time. Ozenc and Ozenc (2015) reported that application of magnesium at 150 kg/ha increased yield, oil content, kernel ratio, shelled and kernel nut weight and percentage of healthy nuts. The treatment also decreased the percentage of empty and wrinkle nuts. Application of 0.8 to 1.6 kg Zn/ha was suggested by Ozenc *et al.* (2015) which improves mineral composition of nuts.

Foliar application of Coryl- Dry (7 per cent $CH_4N_2O$, 2 per cent N organic, 0.5 per cent $H_3BO_3$, 0.5 per cent Zn EDTA and 0.05 per cent Fe EDTA), 3-4 times from fifth leaf stage reduce the incidence of blank nut at harvest, improve qualitative characters of nuts at harvest, mostly kernel dry weight. The fat percentage was increased and also qualitatively improved by an increased percentage of unsaturated fatty acids (Pannico, 2014).

## Leaf Analysis

The hazel leaf N, P, K, Ca, Mg, Zn and Mn concentrations were found relatively stable from mid-August to mid-September. This is a suitable sampling time for diagnostic purposes. Boron was not stable during this time. The foliar concentrations of 2.2 per cent N, 0.8 per cent K, 1.44 per cent Ca, 0.21 per cent Mg, 0.14 per cent S, 8.8 ppm Cu, 19.5 ppm Zn and 204 ppm Fe appeared to be optimum for growth. Soil analysis were less sensitive for determining the need for nutrients but were regarded as useful additional information (Kowalenko, 1981, 1984). Crescimanno *et al.* (1981) sampled leaves of fruiting shoots of four hazelnut cultivars from 23 April to 8 November. During these months N levels fell from 1.87-2.38 per cent to 1.40-1.60 per cent, $P_2O_5$ from 0.26 -0.42 per cent to 0.13-0.14 per cent and $K_2O$ from 0.41-0.63 per cent to 0.22-0.39 per cent. Calcium levels increased from 1.04-1.23 per cent to 2.21-2.37 per cent and Mg level remained constant at about 0.25-0.29 per cent.

Canali *et al.* (2004) suggested analysis of leaf B and Ca in April, N. P and Ca in June and K and Mg in October. They suggested the following standards:

| *Nutrient* | *Sample time* | *g/kg* | |
|---|---|---|---|
| | | *Minimum* | *Maximum* |
| N | June | 18.44 | 21.29 |
| P | June | 1.12 | 1.68 |
| K | October | 6.16 | 8.42 |
| Ca | April | 6.49 | 8.09 |
| Ca | June | 9.60 | 10.81 |
| Mg | October | 1.06 | 1.39 |
| B* | April | 2.43 | 4.57 |

* mg/kg

Ozturk and Tarakcioglu (2016) stated that N, P, K, Ca, Mg, Fe, Cu, Mn and B content of leaves were almost stable in July and August, except for Zn which showed fluctuations during vegetative period.

## Irrigation

The work on the water requirements of hazelnut is limited. When the plantation is developed on fertile soils with good water holding capacity, the plants are rarely irrigated.

De Rosa (1961) reported that irrigation, at the rate of 2500-3500 cu m water per hectare, when applied from May to mid-August in trees growing in the Italian province of Avellino, proved to be profitable. Lagerstedt (1973) found marked increase in growth in hedgerow trees in response to irrigation but the first nut yield was not affected appreciably by irrigation or spacing.

Gispert *et al.* (2015) recorded higher yield and quality nuts of 'Pauetet' hazelnut in Spain by appliying irrigation through drip at 33 per cent wet soil volume. Ozmen (2016) stated that rainfed hazelnut trees may require supplement irrigation to avoid summer stress. It was concluded that hazelnut trees in Turkey can grow well and give normal yield for the areas which have about 950mm precipitations. Irrigation reduces the number and weight of empty and defects nuts, which were higher under rainfed condition (Ksenija *et al.*, 2016). Nut production is strongly affected by moisture stress, particularly when it occurs 10-15 days after fruit setting (Mingeau *et al.*, 1994). Bignami *et al.* (2008) found that an irrigation level equal to 75 and 100 per cent of crop evapotranspiration (Etc) guarantees the highest yield.

## Removal of Suckers

Removal of suckers from the base of the trunk is a common practice. The hazelnut plant is basically a shrub or bush and if left as such will form a thick mass of shoots which is difficult to manage. The suckers can be reduced by root pruning at the time of planting.

In order to remove the suckers effectively, it is necessary to remove the suckers 3-4 times in a season up to several years after planting. The buds from which suckers arise are formed on the underground base of the nursery tree. It is, therefore, essential to remove the new suckers as close as possible to their point of origin. Weed killers like 2, 4-D amine and several other compounds have been used to control suckering with some success (Parker *et al.*, 1955).

The suckering of *C. avellana* depends on cultivar (Radicati *et al.*, 1994). The meristem excision technique developed in Oregon applied on basal buds of young rooted suckers is very effective (Smith and Erdogan, 2001) but is not used in most of the producing countries (Fideghelli and Salvador, 2009). Desukering through heat is more efficient on young shoots and less costly using GPL flame than steam (Tomasone *et al.*, 2008).

## Weed Control

*Artemisia vulgaris*, an important weed in Turkish hazelnut plantations, could be controlled by fosamine at the rate of 700 ml/ha all at one or as split of 200 + 500 ml/ha (Guncan, 1985). Legerstedt (1977a) found that MBR-12-325 (mefluidide), when applied at the rate of 0.28, 0.56 and 1.12kg/ha in summer prior to mowing between the rows of non-tilled orchards, reduced weed growth for 4 weeks. Higher concentration was effective and application after mowing was not satisfactory. When applied at the end of August, prior to last mowing before harvest, weed growth was suppressed until all the nuts have been collected in mid-October. Kaya-Altop *et al.* (2016) evaluated string trimming alone or combined with herbicides in hazelnut orchards to control mugwort (*Artemisia vulgaris*), purple nutsedge (*Cyperus rotundus*) and burning nettle (*Urtica urens*) on a sandy clay soil at Turkey. They reported best control by a combination treatment of glyphosate plus carfentrazone-ethyl application 15 days after string trimming. The treatment is also cost effective.

## Flowering

The flowering in hazelnut starts after 3-4 years but trees usually take 5 to 6 years in the orchard before they come into bearing. The productive age is between 10 and 20 years. The branches between 16 and 25 cm long had the highest reproductive capacity (Kavardzhikov, 1984). The male inflorescences, termed catkins, are composed of 130 to 290 flowers. Each flower includes one bract and two bracteols protecting four bifurcated stamens, each one terminating in two anthers (Dimoulas, 1978). The female inflorescences, known as glomerules, are, in fact, compound buds (Dimoulas, 1978).

Koval (1972) studied seven phases of development of female hazelnut flowers and five phases in the development of the male flowers. The female flowers were initiated in the middle of the summer and completed development in the spring. The male flower of all varieties completed development in the year of initiation and showed low winter hardiness.

Incipient and developing buds of six cultivars were examined by stereomicroscopy at 7-day interval from their first appearance to flowering. Differentiation started with the catkins in mid-April and with female flowers in mid-May in hazelnuts (Kavardzhikov, 1984).

Hazelnuts are monoecious, wind pollinated and self-incompatible. Reproduction is resistricted by a sporophytic self-incompatibility system, which is controlled by a single locus with various *S*-alleles determining compatibility (Olsen *et al.*, 2000). Dominance or codominance of the alleles is expressed in the pollen, whereas all known *S*-alleles are codominant in the pistil (Mehlenbacher, 1997). Plants are typically either protandrous or protogynous depending on their genetic background and the climate of the region they are grown in regions with mild climates; protandry seems to be more common, whereas in regions with long, cold winters, protogynous or homogamous flowering typically occurs (Germain, 1994; Olsen *et al.*, 2000).

The male catkins become evident during the growing season as early as June. Pistillate flowers are borne in tight clusters and do not become visible until early winter under mild climate. The conspicuous bloom of the hazelnut is most commonly associated with the extension growth of the staminate catkins on leafless twigs with subsequent release of clouds of pollen. This release signifies the end of a development period that was initiated early in the previous year.

The pistillate flowers which develop into the nuts are usually borne in lateral and terminal buds on the previous season's growth. The staminate flowers which produce pollen are borne in naked catkins on short stalks either laterally on the previous season's growth or terminally on shoot spurs which may bear leaf or fruit bud. The studies conducted by Trotter (1948) in cultivar San Giovanni in Italy have shown that the pistillate flower begins to develop within a bud in a leaf axil during the end of June or beginning of July. The flowers arise as a simple meristematic mass or mammilla, which becomes trilobal. During development, the middle lobe disappears, the side lobes grow toward the exterior, forming two rudimentary carpels differentiated at the top into stigmatic style-like appendages.

At the bottom they are surrounded by a minute perigynous ring. Each flower is joined to the axil of a small bract connected to the receptacle. Several such flowers make up the compound flower or flower clusters. Fontanazza *et al.* (1971) observed that in both 10- and 25-year-old hazelnut cv. Tonda Romana, the number of female inflorescence increased with the length of the year-old shoot. In younger plants, most of the female inflorescence were in isolated positions, whereas in older plants they were more numerous and occurred mainly at the base of the male catkins. In short shoot the isolated female inflorescence occurred towards the apex, whereas in longer shoots they arose between the middle and apex. Giulivo and Pisani (1973) stated that isolated inflorescence arising laterally on the shoot away from the catkins were the most likely to set nuts, while inflorescence arising at the base of catkins or on spurs showed varying abilities depending on cultivar and season.

The heavy production in one year is usually followed by a light crop in the following year. Frost, rain and other factors may cause low yield or crop failure. It is thought that heavy renewal pruning may correct or modify the natural alternate bearing tendency (Painter, 1962).

## Pollination

Self-incompatibility in hazelnut cultivars is common which makes interplanting a necessity in order to provide cross-pollination. The blooming season under normal condition extends over a period of three months, beginning from pollen shedding by early cultivars to full bloom of late cultivars. Capik and Molnar (2014) studied the flowering phenology of eastern filbert-resistant hazelnut accessions in New Jersey. Results showed that the different accessions followed a similar progression of bloom each year (both staminate and pistillate flowers), however, the date of bloom and duration of bloom, especially for pollen shed, differed each year, largely corresponding to average air temperature trends. Generally about 12 per cent trees of pollenizer cultivar are sufficient to carry out pollination of main cultivar under favourable conditions. The use of about 20 per cent pollenizer trees is also not uncommon in areas having unfavourable conditions during flowering. Recommendations in Oregon include planting at least three different pollenizers that shed pollen at different times during the period that female flowers of the main crop cultivar are receptive to ensure consistent orchard pollination (Mehlenbacher *et al.*, 2009).

The blooming in hazelnut is dichogamous and, therefore, appearance of staminate and pistillate flowers may not be synchronized. Protandry, protogyny and homogamy occur depending on cultivar, growing conditions and season. Bodecs (1987) reported that almost all hazelnuts are partially self-sterile and crop successfully in mixed plantings. In open-pollinated orchards of 'Barcelona' hazelnuts supplied with several different pollenizers, the percentages of fruit set was 55 to 65 although over 90 per cent individual flowers were pollinated. A large proportion of the ovaries failed to mature up to the time of fertilization and only 15 per cent developed into mature nuts. These studies indicate that poor pollination is not a casual factor to have such a low fruit set but it is a varietal weakness (Thompson, 1968).

Romisondo *et al.* (1978) reported that ideal pollenizer should be intercompatible with the fruiting cultivar and should produce acceptable nuts. The hybrids 10 GE and 101 (Tonda Gentile della Langhe x Cosford) were the most efficient pollenizers for Tonda Gentile della Langhe. The cvs. Maria di Gesu, Carrello and Tonda Romana showed varying degrees of self-compatibility but were, in general, benefited by open-pollination. In Maria di Gesu, a significant increase in fertilization was observed when Favigna Tonda, Favigna Lunga and Carrello were used as pollenizers (Geraci, 1974).

Normal flowering in hazelnut was observed at temperatures between 0.4° and 14.6°C. Fruit set was highest under artificial pollination than self-pollination. 'Nemasa' was found as best pollenizer for cvs. Cherkessi-2, Zorinkii, Khachapura, Futkurami and Dedoplis-Titi (Makhno, 1981). Daviana, Lansing, and Butler were the most effective pollenizer for 'Barcelona'. Pollens of 'Woodford' and 'Montabello' reduced a high percentage of blank nuts. 'Barcelona' was nearly self-incompatible. The combination of *Corylus colurna* x *C. avellana* cv. Barcelona was completely incompatible (Zielincki and Thompson, 1967). Wide variations were observed in the time of receptivity of the pistillate flowers of different cultivars, whereas the male flowers shed their pollen over a short period. In general, American cultivars had receptive pistils before the pollens were shed but in European cultivars except 'Barcelona' and 'DuChilly', the reverse was true (Talbott, 1959).

Temperature is an important factor for pollination. Hosseinpour *et al.* (2015) studied the pollen compatibility of hazelnut cultivars in Iran. They reported the best pollinezers for the cultivars Pashmine, Tabestane, Shastak, Barcelona, Segorbe, Daviana and Marveille were Segrobe, Barcelona, Segrobe, Shastak, Shastak, Pashmine and Pashmine, respectively. When temperature in late winter were relatively high, protandry occurred but in cooler atmosphere protogyny was common. In Contiollei conditions the critical temperature for the development of protandry was found to be 9°C (Stritzke, 1962).

Pollination starts at 4°C to 5°C (Lukasiewicz, 1994). Temperature below –8°C or above 20°C can damage pollen and stigmas at the time of pollination (Baldwin, 2004). Piskornik *et al.* (1975) studied the pollen germination of *C. avellana* and 4 cultivars of *C. pontica* stored at 2°, 12° or 20°C and 60-70 per cent RH for up to 150 days. The optimum storage temperature was 2°C and satisfactory pollen germination was obtained after 30-40 days of storage, it declined thereafter. The highest germination (65.6 per cent) after 30 days was observed in cvs. *C. avellana* and *C. pontica,* while in cultivars Strzepiastookrywowy and Truchessa of *C. pontica,* the percentage of pollen germination were 54 and 33, respectively. Callon and Thompson (1986) suggested that pre-treatment of stigmas with 1 M or 2 M sugar solutions before the application of incompatible pollen prevented pollen rejection by the stigma surface. A 1:1 mixture of dry sugar and incompatible pollen overcame the stigmatic barrier.

The suspected causes of black nuts are cytomixis during megasporogenesis, genetic incompatibility between pollenizer and main crop cultivars, self-pollination, aneuploidy, lack of double fertilization, embryo abortion, seasonal variation and cultural practices (Lagerstedt, 1977b). All fertilized flowers of the hazelnut do not produce fruits. Different stages between fertilization and harvest, the growth of the ovule and the embryo may stop. When fertilization does not occur or when the kernel does not grow enough after fertilization, blank fruits may appear (Beyhan and Marangoz, 1999).

## Fruit Growth and Development

After pollination, the pollen tube grows to the basal region of the stigma and the nucleus enters a resting stage and remains so up to 4 or 5 months when the pollen tube resumes growth and the ovules are fertilized. A large number of stigmas are coming out from a bud of the pistillate flowers and the fruit is borne in clusters of several nuts, each in a leaf-like involucre. The shell is from oval wall, the wall of the ovary, whereas kernel is formed from embryo. There is no outward evidence of fruit development after several months of pollination. Thompson (1979) studied the growth and development of pistillate flower and fruit of 'Bracelona' from anthesis in January to maturity in September. He found that the ovary growth was very slow for about 4-5 months, proceeded very rapidly for six weeks. During this time about 90 per cent of increase in diameter occurred and then abruptly ceased. Fertilization was necessary for the formation of normal nut embryos. Full size nuts may be empty at maturity and it resulted from pollination without fertilization or from embryo abortion.

The mechanisms how phytohormones regulate ovule abortion in hazelnut are largely unknown. YonQuing *et al.* (2017) recently studied the phytohormone biosynthesis and signal transduction pathways in developing and abortive hazelnut ovules. RNA-seq identified 1209 differentially expressed genes with Kyoto Encyclopedia of Genes and Genomes (KEGG) pathway annotation between developing and abortive ovules in hazelnut. Hormone signal transduction (Ko04075), and several phytohormone biosynthesis pathway, were found to be significantly enriched after pathway enrichment analysis. Compared with developing ovules, the levels of IAA, GA, cytokinin (CTK) and salicylic acid (SA) were found significantly lower in abortive ovules, while levels of abscisic acid (ABA), ethylene (ETH), and jasmonic acid (JA) were found significantly higher. Rate-limiting enzyme involved in the biosynthesis of phytohormons, including 9-*cis*-epoxycarotenoid dioxygenase, ACC synthase, ACC synthase and phospholipase A1 were significantly up-regulated in abortive ovules. Some key genes in the ETH, JA and ABA signal transduction pathways were highly unregulated, including ethylene-responsive transcription factor 1 (ERF1/2) and the serine/threonine-protein kinase CTR1 in the ETH signal pathway, jasmonate ZIM domain-containing protein (JAZ) and the transcription factor MYC2 in the JA signal pathway, and ABA responsive element binding factor (ABF) in the JA signal pathway. Thus; ABA, ETH and JA signalling were found more active in abortive ovules than in developing ovules. Expression levels of *NCED, ABF, ACS, ERF1/2, PLA1* and *MYC2* revealed by q-RT-PCR were consistent with the results obtained by differentially expressed genes (DEG) analysis. The results suggested that multiple phytohormone signalling pathways are involved in the development of hazelnut ovules. Inhibition of IAA, GA, CTK and SA biosynthesis and signalling pathways, and activation of ABA, ETH and JA biosynthesis and signalling pathways may contribute to the formation of abortive ovules in hazelnut.

The increase in dry matter content and dimensions of the pericarp followed a 3-phase sigmoid curve with phases, 1 and 3 representing slow growth and phase 2 rapid growth. Lipid content increased fairly regularly throughout ripening and was accompanied by water loss (Roversi, 1973). The time period from fertilization to harvest in cv. Tombut was noted 89 and 96 days in two years of observation and for cv. Palaz it was 84 and 86 days in two years (Beyhan and Marangoz, 2007).

The nuts borne singly are larger in weight and fills better as compared with the nuts borne in clusters. The amount of polyunsaturated fatty acids (PUFA) decreased (from 30.09 to 10.28g/100g of oil) from early developed nuts to completely matured 'Katalonski' hazelnuts on the other hand, the amount of monounsaturated fatty acids (MUFA) increased from early to harvest stage (from 22.63 to 79.17g/100g oil). The tocopherol content increased from early to harvest stage and the amount of α-tocopherol was observed to be the highest among tocols (Ciemniewska-Zytkiewich *et al.*, 2015). The total sterol content reached the maximum values at the early stage and a significant reduction in the total sterol content occurred at the harvest stage (Ilyasoglu, 2015). Cristofori *et al.* (2015) working with three Italian cultivars ('Tonda Gentile Romana', 'Tonda di Giffoni' and 'Nocchione') observed that oil content increased continuously during the development of the kernel from 30-35 g/100g dry weight (DW) at the beginning of July to 61-68g/100g DW in the first week of September. Nitrogen content in the kernel decreased continuously during development, from 42g/100g DW to 2.9g/100g DW. Soluble sugars, mainly sucrose showed a wide fluctuation until late July, and the content observed in the mature seed was about 5g/100g DW. The organic acids, mainly represented by malic acid, showed a large decrease during the early stages (July), and reached on average a content of about 03-0.4g/100g DW at the mature stage. Starch content decreased gradually, with a peak in early August, and then decreased rapidly.

Pannico (2014) reported large variability occurring in seed growth and composition within hazelnut canopies. Dry matter partitioning to kernels increased from the bottom to the top of the canopy and this correlated with a similar increasing pattern in light availability. As a result kernels located in the top canopy layers are larger and contain larger total amounts of fats, carbohydrates, proteins and ashes compared to seeds located in lower layer of the canopy. Dry matter composition was also significantly affected by fruit position within the canopy. However, fat concentration decreased, while protein and ash concentration increased, from the bottom to the top layers of the canopy. The level of unsaturation of fatty acids decreased from the bottom to the top of canopy.

## Harvesting

Fully mature fruits of hazelnuts drop on the ground. Ripe nuts collected from the ground were superior to those picked by hand slightly before maturity, in yield of kernels per unit weight of nuts, ease of blanching and resistance to rancidity. During ripening vitamin E and palmitic and oleic acids increased, while stearic and liolenic acid decreased (Fregoni and Zioni, 1963).

The nuts on the ground should be picked up soon to avoid discolouration by wet soil. Nuts should be gathered 2 or 3 times during the season. Hazelnuts lying on the moist or damp ground for two weeks gradually darken. Rainy weather enhances the discolouration of shells. A person can harvest an average of 70-75kg cluster hazelnuts per day. This amounts to an average of 23-25kg of dry unshelled hazelnuts.

Different levels of mechanization of harvesting operations were made for harvesting mechanically. They vary from simple blowers like backpack dry blowers to tractor-mounted equipments and self-propelled sweepers and blowers. The mechanical harvesting of hazelnut needs the graden's surface to be completely cleaned by special tools before the harvesting operation starts because it is mechanically collected by sweeping and sucking from the ground. Hazelnut harvesting machine can collect the amount that 6.74 collectors can collect traditionally (Yildiz, 2016).

Two types of machines are used to gather hazelnuts from the ground. The first type gathers fruit by suction and dehusks them and the second type uses large brushes (Manzo and De, 1979). A prototype nut picker is driven by a two-wheeled tractor with a separate small motor to drive tire vibrator arm. The nuts drop onto a catcher and are transferred to boxes. The capacity of the prototype is 500 kg in 7 hours with two men (Burgaard, 1974). The harvesting, by vibration was practical and promising. A new tractor mounted high frequency vibrator with 10-14 shafts to capture all stems and equipped with roller grips was developed and successfully tested (Slepokurov *et al.*, 1984).

Ethephon at 750-1000 ppm accelerated nut drop by 2-3 weeks without causing excessive premature leaf-fall, reduction in nut quality or plant injury (Lagerstedt, 1972). Treatment at 2000 ppm resulted in an average of 95 per cent nut drop after 19 days as compared with 34 per cent drop in control plants (Table 9.3). Baldini *et al.* (1973) reported that 85 per cent of the nuts were harvested by applying CEPA at 200 ppm on 1 August compared with 60 per cent in the untreated controls. Spraying San Giovanni bushes on 1 August and Tonda Gentile Romana bushes on 10 August with 200, 400 and 600 ppm CEPA induced early nut drop which was proportional to CEPA concentrations. When the trees were harvested 20 days after spraying using portable pneumatic shakers, about 96 per cent of the nuts were recovered as compared with 70 per cent from the untreated control (Rotundo *et al.*, 1984).

**Table 9.3: Percentage of Hazelnut Nut-Drop due to Ethephon Application***

| *Ethephon Concentration Applied on September 4* | *Days after treatment* | | |
|---|---|---|---|
| | *12* | *19* | *27* |
| Control | 12 | 34 | 88 |
| 500 ppm | 43 | 76 | 100 |
| 1000 ppm | 41 | 84 | 100 |
| 2000 ppm | 71 | 95 | 100 |

* Lagersted (1972).

## Yield

The yield varies greatly according to the age and vigour of trees, planting distance, soil cultural practices and many other factors. Italian hazelnuts yield 1.5 t/ha and hazelnuts grown in the USA (mostly Oregon) yield 2.5 t/ha (USAID, 2011). Good growers harvest 1 t/ha by year 6 and 2.5 t/ha by year 10. Average yields in countries like Turkey with less intensive orchard management, fall below 1.5 t/ha (Redpath, 2016).

## Storage

Nuts are exposed to high levels of airborn inoculumn in the field, and also during harvest and/or processing. The nuts (usually in shell) are stored in warehouses from one season to next. Since the hazelnut kernel has a low respiration rate it might be expected to be a relatively stable commodity. However, due to large amounts of unsaturated fatty acids (mainly oleic acid), oxidative rancidity is a potential threat to the stability of the kernel. Mold and rancidity are the two factors which can strongly affect the market acceptability of hazelnuts.

The damp or undried hazelnuts should not be stored in sacks or large containers for more than two weeks, because kernels become mouldy or off-flavoured even if such condition is not evident from the appearance of shells. It is suggested to estimate, the probable storage period at harvest time, then grade and store the lots under conditions which would ensure adequate storage at minimum cost. After the hazelnuts are dried to an in-shell moisture content of 7-8 per cent or a shelled moisture content of 3-4 per cent the nut should be graded, fumigated and sealed in plastic lined boxes, bags or bins. The purpose of using plastic is to-stabilize the moisture and prevent the absorption of flavour from other products. They should not be allowed to change weight by substantial gain or loss of moisture. The storage life depends largely upon the temperature. At 70°F (20°C) or lower the hazelnuts will keep good for 14 months. At 32 -35°F (0-1.6°C) with 60-65 per cent relative humidity, hazelnuts will keep for two years without insect damage or mould growth, also the boxes or bags need not be sealed. At temperature between 25-27°F (–3.9 to –2.8°C) with 60-65 per cent relative humidity, hazelnuts may be stored for 4 years.

Martin *et al.* (2001) studied the modified atmosphere storage on 'Negret' hazelnut. The shelled and unshelled hazelnuts were stored with different $O_2$ concentrations (1 per cent, 5 per cent, 10 per cent and 20 per cent) and two different temperatures (7° and 25°C). They observed that after one year storage period, none of the storage modes had given rise to significant rancidity. It was also confirmed that the shell protected the hazelnut from oxidative deterioration; storage in atmosphere with less than 10 per cent $O_2$ levels significantly reduced auto oxidation and at low temperature lip rancidity was delayed. Ghirardello *et al.* (2014) reported that after one year storage of 'Tonda Gentile delle Langhe' hazelnut at ambient temperature, the acidity was higher (0.47 per cent oleic acid) than the value considered the acceptable limit after storage (0.40 per cent oleic acid), while storage at low temperature (4°C and 55 per cent RH) maintain low level of acidity and lip oxidation. However, in modified atmosphere (1 per cent $O_2$ and 99 per cent

$N_2$), showed the best storage condition with 0.13 per cent of oleic acid and 0.11 mEq $O_2$/kg lipid oxidation.

Nuts packed in low density polythene (LDPE) could be stored longer both at ambient temperature and in refrigeration (Guine *et al.*, 2015).

The shelled hazelnuts were stored in silos under different conditions. In cool (3.6°C) storage with about 60 per cent humidity and in warm (18-25°C) atmosphere with 55 per cent humidity, storage in 99.5 per cent nitrogen stabilized the water content of the nuts whereas storage in air caused a slow loss (Keme *et al.*, 1981). Sun drying of hazelnuts is considered best as artificial heat causes rancidity. Unshelled nuts can be stored for a year and shelled kernels for 4-5 months in winter but only for one month in warm conditions (De Rosa, 1979). The warm humid storage conditions increased deterioration by enzymic reaction and moisture uptake. The shell provided some protection and the moisture content should be kept low by initial drying (Hadorn, 1977).

## Pests

A large number of pests are found to damage the hazelnuts but those causing serious damage are given below:

### *Eoletranychus coryli*

This pest was not recorded earlier but it causes extensive premature leaf drop. However, in the past, the main pest of hazel had been curculinoid *Balaninus nucum*, which can be controlled by using endosulfan and carbaryl at the rate of 100 g active ingredient/100 l water (Nicita, 1986).

### *Curculio nucum*

It is a serious pest of hazelnut and frequently damages *C. glendium*. Three applications of ultracid (methidathion) 40 WP at 0.2 per cent proved most effective in reducing damage below 10 per cent. Sevin 85 WP at 0.25 per cent was also effective (Burges and Czencz, 1976). The pests were controlled effectively when a pyrethroid was applied at the beginning of adult activity and a compound containing phosphoric acid ester at the end (Toth, 1984).

### Beech Borer (*Agrilus viridis*)

It has become a serious pest of hazelnut, tunnelling deeply in the trunk and branches. This causes swelling and sometimes withering and death of infected branches, especially young planted trees (Petlegrino and Mozzone, 1985). A parasite of *Ooencyrtru* species has been found to destroy 45-60 per cent of eggs. Two application of lindane on 30 June and 15 July gave good control and killed about 90 per cent larvae (Ciampolini and Ugolini, 1975).

### Gall Mite (*Phytoptus avellanae*)

The gall mite is also a major pest of hazelnuts in Turkey, Italy and Spain destroying 30-40 per cent of the buds in some regions (Viggiani, 1972). Aliniazee and Krantz (1978) found best control with endosulfan at 1.2 g/l during mite migration.

It might, however, affect the biological equilibrium as it is toxic to entomophagous insects (Arzone, 1973, 1975). The migration of mites begins when the new hazel shoots have 2-3 leaves and continues for 40-50 days. The critical control period, however, appears to last for 30 days only. Two application of a zinphos-ethyl or endosulfan reduced bud infestation from 40 per cent to fewer than 10 per cent (Viggiani and Bianco, 1974). The observations carried out in Poland on infestation of buds with this mite indicate that there is variation with regard to susceptibility. The cvs. Starowlski, Warszawski, Czerwong, Krakows, Ki and breeding stocks 13/6 and 7/5 were very susceptible while Cud Z, Bolviller, Katalonski, Obrzyni, Halle and Kulistry Wolski were infested to a lesser extent. Mites were also found in buds showing no signs of infestation (Maziarz, 1983). The progeny of 'Barcelona' x 'Hall's Giants', 'Kruse' x 'Lansing' and 'Monlebello' x 'Comoton' showed high resistance to mite (Thompson, 1977). Spraying of entomopathogenic fungus *Beauveria bassiana* (Bb-024) suspension at 5 x $10^7$ conidia/ml caused significant control of big bud mite (Abramishvili *et al.*, 2016)

## Oyster Scale (*Quadraspidiotus ostreaeformis*)

Severe infestation of hazel by oyster scale was found in several areas of commercial cultivation in Campania region of Italy. Attacks by *Q. ostreaeformis* frequently occurred on trees infected by the fungus *Cytospora corylicola*. The scale has 1 generation a year, the 2nd-instar nymph overwintered and crawlers were found from the end of May to the beginning of July. Its most active natural enemies are coccinellis predator (*Chilocorus bipustulatus*) and aphelinid parasitoid (*Archenomus longiclava*). Insecticides were reported to cause harm to the natural enemies. It is, therefore, recommended that the insecticides should be applied only if really needed. Direct control of *Q. ostreaeformis* should be confined to application of yellow mineral oil (DNOC in oil) or barium polysulfide just before the resumption of tree growth in early spring (Viggiani, 1985).

## Filbertworm (*Cydia latiferreana*) and Filbert Weevil (*Curculio occidentalis*)

Filbertworm and filbert weevil larvae feed on nut kernels and can cause between 20 and 50 percent damage if untreated. Both species cause similar damage. Infested nuts have holes in the shell, and the kernels are damaged by feeding and contaminated with feceo (Walton *et al.*, 2007).

## Green Shield Bug (*Palomena prasina*)

It feeds on hazelnut fruit and causes premature nut dropping in early season and later kernel damage during nut development. Entomopathogenic fungi *Lecanicillium muscarium* and *Beauveria bassiana* were found effective for controlling *P. prasina* (Erper *et al.*, 2016).

## Nematode

Kepenekci (2002) reported several nematodes associated with hazelnuts: *Filenchus afghanicus, Hemicycliophora punensis, Pratylenchoides hispaniensis, Pratylenchus pratensiobrinus, Helicotylenchus crenacauda, Hemicycliophora sturhani,*

*Merlinius* (= *Scutylenchus*) *lenorus, Tylenchorynchus cylindricus* as plant parasitic species and *Ditylenchus anchilosposomus* as fungivorous species in the west part of Blacksea region of Turkey. *Criconemella, Meloidogyne, Paratylenchus* and *Pratylenchus* were found as important genera by means of the damage potential on hazelnut as plant parasitic group (Akyazi *et al.*, 2014).

## Diseases

### Bacterial Blight (*Xanthomonas arboricola* pv. *corylina*)

The bacteria attack mainly nursery and young orchards causing significant plant mortality (Pistta *et al.*, 2016). It enters the plant through the stomata, lenticels and small wounds caused by insects and attacks the parenchymatous tissue. Any organ except the roots are affected. The primary infection of buds, stems and branches occurs in the autumn and the young shoots, leaves and involucre of the fruit are infected in the spring. In France, the bacterium, in layer nurseries and orchards, suppressed buds on young branches and caused necrosis of shoots and branches and also cankers. Blotches appeared on leaves, cupules and nut (Garden and Devaux, 1987). The disease is controlled by removing all infected organs and then applying sprays of 1.5 per cent Bordeaux mixture in autumn when first rains occur; in spring, before bud burst, a further spray of 1 per cent Bordeaux mixture should be given. Good results have been obtained with copper oxychloride (50 per cent a.i.) used at a rate of 500 g/100 l of water in spring (De Rosa, 1969). Pistiia *et al.* (2016) suggested hot water treatment (42°C for 45 min.) to propagative materials to control and spread of the disease.

### Eastern Filbert Blight (*Anisogramma anomala*)

Eastern Filbert Blight (EFB) is a destructive disease of European hazelnut (*C. avellana*), while the wild north-American hazelnut, *C. americana,* only experiences minor symptoms. It causes perennial cankers that girdle stems and branch dieback, leading to eventual death of plants (Johnson and Pinkerton, 2002). Trees become susceptible to infection by *A. anomala* only after vegetative buds have initiated active growth in spring (budbreak), and are most susceptible during the periods of leaf emergence and initial shoot elongation. Infection occurs over a relative broad range of temperatures (8 to 25°C) and facilitated by a 24 to 72 hour period of high humidity (Stone *et al.*, 1992). None of the control methods, however, has been 100 per cent effective. Fungicides including chlorothalonil, copper hydroxide, fenarimol and propiconazol, have protected European hazelnut from infection of EFB when applied during March-April (Pinkerton, 1996). Disease scouting and therapeutic pruning of diseased branches were suggested during the winter, when leaves do not visually obstruct canker locations. A number of hybrids resistance to EFB has been developed by Oregon state University since 2009 (details are available under 'cultivars' and 'breeding').

### Brown Rot or Anthracnose

The disease is caused by *Gloeosporium coryli,* and results in 20-22 per cent leaf and 18-20 per cent fruit infection in a year. The effective control was obtained with

Bordeaux mixture or thiophanate when applied twice and reduced leaf and fruit infection to 3 and 4 per cent, respectively (Beradze and Dzimistarishvili, 1981).

### Grey Mould

It is a common disease in some areas where hazelnuts are grown and has been controlled effectively by Bordeaux mixture followed by Zineb and Benomyl. The first spray should be made in mid-May and another in June when the fruits appear (Beradze and Dzimistarishvili, 1985).

### Viral Disease

A mosaic disease of hazel trees was found transmitted by grafting. The symptoms on severely affected young plants were eliminated by heat treatment at 38°C for 110 days (Marenaud and Germain, 1975).

## Breeding and Improvement

The hazelnuts breeding programme, has not received due importance like other nut crops. Many workers have crossed *Corylus avellana* with other species to incorporate specific characters found in these species. There are some interested or observant growers who have planted seedling orchards and seedling nurseries in search of desirable selections. Due to the keen interest of the growers and nurserymen, the numbers of seedling selections have originated from open-pollination of available seedling population. The important seedling selections that have been named are 'Nonpareil', 'Nooksack', 'Royal', 'Woodford', 'Gem', 'Fitzgerald' and 'Brixnut'. However, these selections could not compete with 'Barcelona' and 'Daviana'. Several seedlings from open-pollinated 'Nonpareil' which have desirable characteristics have been obtained by Olgar Jemtegaard in Oregon. His Selection 5 was multiplied and tested for suitability to grow with 'Barcelona'. It is of similar size, shape, appearance and is a good pollenizer for 'Barcelona'. It is also easy to graft and is a fairly vigorous growing tree.

Gellatly (1956, 1966) found that most of the crosses between the European and Turkish hazelnuts resulted in seedlings of intermediate characters. A range of variation has been found for many characters, such as, yield, suckering habit, upright habit, hardiness, nut size, shell thickness and husk type. When the European hazelnut was used as the pollen parent, the resulting trees produced nuts that were thick-shelled but larger than those of Turkish hazelnut. They had tree-type growth, intermediate in size and exhibited the bark characteristics of European parent.

Olgar Jemtegaard selected a tree with striking intermediate characters from a *C. avellana* seedling population. The tree was named 'Filcorn' because its nuts resemble that of acorn. The 'Filcorn' has originated from a cross between *C. avellana* and *C. colurna* and again back-crossed with *C. avellana*. Several of its seedlings have shown variation in suckering habit but many of them have been selected for little or no suckering and are being multiplied for further evaluation as rootstocks (Lagerstedt, 1973). Besides 'Filcorn' seedlings, other chance seedling of *Corylus* species are being collected to study their suitability for rootstocks. This approach will be the solution to produce non-suckering habit which may be obtained by back-crossing 'Filicorn' to *C. colurna*.

'Barcelona' is one of the most important cultivars grown in the USA but it has some drawbacks, such as, moderate productivity, poor kernel percentage (40 per cent) high susceptibility to brown stain, unattractive kernel, lack of early maturity, high blank production, moderate tree vigour, moderate nut flavour.

Nearly 20 countries of the world have programme to conserve local ecotypes and cultivars (Cristofori, 2006) and use of the diversity in breeding.

The world largest producer Turkey has 16 traditional commercial hazelnut cultivars. In addition to these old cultivars obtained from natural hybridization, the Hazelnut Research Institute has developed three new cultivars and four candidate cultivars through intensive breeding programmes using selection from a very rich biodiversity and subsequent hybridization (Anil *et al.*, 2017).

The Oregon State University (OSU) hazelnut breeding programme initiated in 1969, continues to develop new cultivars that combine suitability to the blanched kernel market with resistance to Eastern Filbert Blight (EFB) caused by *Anisogramma anomala,* about 5000 seedlings have been planted each year during the past two decade, and 12 cultivars and 10 pollinizers have been released. Corvallis collections contain ~ 900 accessions and include representative of all 11 *Coryllus* species and ~ 500 clones of *C. avellana*. All *Coryllus* species are diploid have small genome (0.48 pg/1c, 380MB), and an estimated 35,000 genes. Morphology, phenology and all marker types investigated to date have shown *C. avellana* to be highly polymorphic.

Complete resistance of EFB was first discovered in the obsolete pollinizer 'Gasaway' (Cameron, 1976). Genetic studies showed that complete resistance is conferred by a single dominant gene (Mehlenbacher *et al.*, 1991). 'Gasaway' has been the major source of resistance used in the OSU breeding programme. 'Gasaway' has low yields and undesirable nut and kernel characteristics, thus requiring considerable effort to combine the resistance gene with many attributes required of a commercial acceptable cultivar (Chen *et al.*, 2007). OSU's hazelnut breeding programme resulted development of several EFB resistant cultivars by Dr. Shawn Mehlenbacher and his group, the first of which 'Santiam' was released in 2005, followed by 'Yamhill' in 2008, 'Jefferson' in 2009 and Dorris in 2012. These new cultivars have reinvigorated the hazelnut industry in Oregon. Two late-shedding pollenizers for 'Jefferson' were released, 'Eta' and 'Theta' in 2009 (Mehlenbacher, 2009).

A worldwide genetic resource collection has been made with complete resistance to EFB in *C. avellana* discovered in the following genotypes. 'Gasaway' from Washington, 'Zimmerman' from Oregon, 'Weschcke seedling' from Minnesota, 'Ratoli' from Spain, Georgian 759010 from the Republic of Georgia, OSU495.072 from Russia, COR157 from Finland 'Uebov' from Cacak, Serbia, 'Crvejne 3/96' from Cacak, Serbia, 'Culpla' from Spain and Moscow No1, No2 from Moscow, Russia (Vidyasagar *et al.*, 2010). These genotypes are being integrated into the OSU hazelnut breeding programme to produce new cultivars expressing a wide diversity of genes for resistance to EFB (Honglin *et al.*, 2007). 'Gasaway' and some of its offspring have been infected by EFB isolates from New Jersey, Minnesota and Michigan. Efforts to create new cultivars with durable EFB resistance include identifying and studying new resistance sources. Bhattarai *et al.* (2017a) studied resistance accessions *C. americana* 'Rush' and interspecific hybrid selection 'Yoder#5' which were crossed with susceptible *C. avellana* selections and the resulting segregating seedling populations were inoculated both in pot grown plants and in field. Resistance from both sources segregated in a 1:1 ratio, indicating control by a single locus and a dominant allele for resistance. DNA extracted from the seedlings was amplified with previously mapped microsatellite marker. Resistance from both *C. americana* 'Rush' and 'Yoder#5' was placed on linkage group 7 in the same position as resistance from *C. avellana* 'Ratoli'.

Segregation for EFB resistance from 'Uebov' a new source from Serbia was studied (Bhattarai *et al.*, 2017b) in three progenies. The frequency of resistant seedlings following structure exposure was about 20 per cent in all three progenies. The ratio fail to fit the expected 1:1 ratio but did fit a ratio of 1 resistant: 3 susceptible, which would be expected if resistance were conferred by dominant alleles at two independent loci. Seedlings from a cross of susceptible selection OSU 741.105 and 'Uebov' were used to study correlation of disease response and presence of alleles at microsatellite markers loci. Resistance was found highly correlated with the presence of alleles at marker loci on linkage group 6 (LG6), and these markers also showed segregation distortion. The study concluded that EFB resistance from 'Uebov' maps to a single locus on LG6 in the same region as resistance from 'Gasaway' although only about 20 per cent of the seedlings are resistant because of segregation distortion.

New source of resistance to EFB are being searching from seedling populations from Estonia, Lativia, Lithuania, Uzbekistan, Poland, Ukraine, Russia and Kyrgyzstan and integrated them into advanced selection (Olsen, 2011). The breeding programme has also selected for other traits beneficial to commercial production such as uniformly early nut maturation, larger kernel size expressed as a percentage of the whole nut weight and improved kernel quality (Olsen, 2011).

Hazelnut breeding at the University of Torino, Italy was started in 1960's which resulted in the release of Selection 104E ('Tonda Gentile delle Langhe' x 'Cosford') as 'Daria'- clonal selection 'Tonda Roman' was made at University Perugia (Tombesi *et al.*, 1994), clonal selection in 'Tonda di Giffoni' at the instituo Sperimentale per la Frutticoltura in Caserta and clonal selection in 'Tonda Gentile delle Langhe' at Torino (Valentini *et al.*, 2001). Hazelnut breeding at INRA, France led to the release of 'Corabel' an open pollinated seedling of 'Barcelona'. Superior clonal selections 'Negret' and 'Gironell' have been identified from Spain (Rovira *et al.*, 1997). Romanian hazelnut breeding programme resulted hybrids 'Uraise de Valcea', 'Valcea-22' and 'Rovred'.

Work with isozymes shared great variability among *Corylus avellana* cultivars. RAPD and AFLP markers have similarly shown a high frequency of polymorphism. Research on *C. avellana* has generated a linkage map comprised of several hundred molecular markers that can be applied to work with other *Corylus* species (Mehlenbacher, 2012, Mehlbacher *et al.*, 2014). A real-time PCR assay that is pathogen-dependent provides opportunity to screen new resistance sources prior to identifying resistance alleles (Molnar *et al.*, 2013). The real-time PCR assay serves as an early generation phenotypic selection tool by detecting *A. anomala* in its symptomless latent phase (Molnar *et al.*, 2013). Colburn *et al.* (2017) utilized the 'Jefferson' transcriptome sequence to develop microsatellite markers for hazelnut. Microsatellites were mined from transcriptome sequence and aligned with the corresponding genomic sequence using a BLASTN search, and primers were designed from the genomic sequence. From a total of 1432 identified microsatellites, 382 primer pairs were designed for repeats with motifs of three base pairs or longer and at least five repeats, and a few di-repeat sequences. The primers were initially screened for polymorphism with a set of 24C *avellana* accessions, and those that were polymorphic were characterized using set of 63 accessions. One hundred nine polymorphic microsatellite markers that generated one or two fragments per accession were characterized with AAG, AGC and ACT being the common motifs. Two additional SSR primer pairs generated three PCR products in a few accessions. Of the 109 loci, 53 were placed in the linkage map of the mapping population. OSU 252.146 X OSU 414.062 and 22 were assigned to linkage groups using alternate segregation population. A dendrogram constructed from the SSR fingerprints at the 109 new marker loci mostly showed clustering of accessions from similar geographic origins and confirmed the tremendous amount of genetic diversity present within *C. avellana.*

Akin *et al.* (2016) screened eight cultivars to assess polymorphism allele size range and ease of scoring. Fourteen polymorphic, easy-to-score SSRs were selected and amplified in a single multiplex. The 14-SSR fingerprinting set was used to genotype 102 hazelnut accessions from different origins. The fingerprinting set distinguished unique accessions mainly according to parentage and in some cases based on geographic origin. They identified each of the cultivars released from the Oregon State University breeding programme and confirmed the parentage of six cultivars. This robust and cost-effective fingerprinting set for hazelnuts using microsatellite markers developed by Akin *et al.* (2016) could be useful to identify and parentage confirmation.

The IRTA de Mas Bover (Spain) investigating promising low suckering rootstocks ('Dundee' and 'Newberg' and two open-pollinated *C. colurna* seeds, 'Tonda Bianca' and 'Negret' clone MB-69) to solve the problem of high suckering (Rovira *et al.*, 2014). Seedlings of *C. colurna* were evaluated as a rootstocks compared with four own rooted cultivars ('Rimski', 'Istarki Dugi', 'Tonda Gentile Romana' and 'Cosford'). Results showed large nuts and higher fruit weight in trees grafted on *C. colurna* and seedlings and these trees were more vigorous and more productive (Miletic *et al.*, 2009).

## REFERENCES

Abramishvili, T., Burjanadze, M., Broufas, G., Knapp, M., Clereq, P. de., Walzer, A., Zemek, R. and Palevsky, E. (2016). 10BC/WPRS Bull., 120: 1-4.

Akin, M., Nyberg, A., Postman, J.A., Mehlenbacher, S.A. and Bassil, N.V. (2016). European J. Hort. Sci., 81: 327-38.

Akyazi, F., Yildiz, S. and Felek, A.F. (2014). Jr. Entomol. and Nematol., 6(8): 112-21.

Aliniazee, M.T. (1977). Ann. Entom. Soc. America, 69 : 820-24.

Aliniazee, M.T. and Krantz, G.W. (1978). Anzeiger Fur. Schadlingskunda Pflanzanschutz Ummettschutz, 51 : 37-39.

Anadolliev, G. (1976). Dvoshcharstvo, 65: 39-42.

Anadolliev, G. and Bozhinova, K. (1982). Gradinarska i Lozarska Nauka, 91:25-33.

Anadolliev, G. and Penev, S. (1975). Gradinarska i Lozarska Nauka, 12: 54-58.

Anderson, J.C. (1984). Proc. Int. Pl. Propagators Soc., 33:132-37.

Anil, S., Kurt, H., Akar, A. and Kose, C.B. (2017). Chronica Hort., 56(4): 30-36.

Anonymous (1965). Ore. State Univ. Coop. Exten. Circ. No. 645.

Anonymous (1978). J. Newzealand Tree Crops Assoc., 3:21-22.

Arias, I., Willams, P.M. and Bradbeer, J.W. (1976). Planta, l31: 135-39.

Arzone, A. (l973, 1975). Annali della Facolta di Scienze Agrarie della Universita degli Studi di Torino, 9: 37l-38.

Azarenko, A.N., McCluskey, R.L., Chambers, W.C. and Orchards, W.J. (2005). Acta Hort., 686: 237-42.

Bailey, L.H. (1949). Mannual of Cultivated Plants. Rev. Edition. Macmillan Co. New York.

Baldini, E., Costa, G., Pisani, P. and Kamina, A. (1973). Rivista della Ortoflorofrutticoltura Italiana, 57: 106-11.

Baldwin, B. (2004). In Nuts. RIRDC, Australia, pp. 393-401.

Bauckmann, M. (1979). Erwerbsobstbau, 21: 242-45.

Baysal, A. (1996). Beslenme, Hatipoglu Yayinevi. Ankara, p. 494.

Beradze, L.A. and Dzimistarishivili, N.B. (1987). Subtrop. Kul'tury, 1:111-13.

Beradze, L.A. and Dzimistarishvili, N.B. (1985). Subtrop. Kul'tury, 2:156-57.

Bergamini, A. and Cristoferi, G. (1968). Frutticoltura, 30 : 93-95.

Beyhan, N. and Marangoz, D. (1999). Utusal Bahee Bitkileri Kongresi, Ankara, pp. 585-89.

Beyhan, N. and Marangoz, D. (2007). Scientia Hort., 113: 208-15.

Bhattarai, G., Mehlenbacher, S.A. and Smith, D.C. (2017a). Tree Genetics and Genomes, 13(2): 45.

Bhattarai, G., Mehlenbacher, S.A. and Smith, D.C. (2017b). Jr. Amer. Soc. Hort. Sci., 142: 289-97.

Bignami, C., Bertazza, G., Bizzarri, S., Bruziches, A., Cammelli, C. and Cristopri, V. (2005). Acta Hort., 686: 263-70.

Bignami, C., Cristofori, V., Ghini, P. and Rugini, E. (2008). 7$^{th}$ Int. Congress on Hazelnut, Viterbo, Italy, June 23-27, 2008. Book of Abstract: 47.

Blomme, R. and Degeyter, L. (1978). Verbondsnieuws voor de Belgische Sierteelt, 22 : 173-75.

Bodecs, L. (1987). Kerteszeti es Szoleszet, 36: 7.

Bonga, J.M. and Aderkas, P.V. (1992). In Vitro Culture of Trees (eds. J.M. Bonga and P.V. Aderkas). Kluwer Acad. Publs., pp. 51-68.

Bradbeer, J.W. and Colman, B. (1967). New Phytol., 66: 5-15.

Bradbeer, J.W., Arias, *I.E.* and Nirmala, I.S. (1978). Pesticide Sci., 9: 184-86.

Bunyard, F.A. (1920). J. Royal Hort. Soc., 45: 224-32.

Burgaard, E. (1974). Fruglavleren, 3: 113-15.

Burges, G. and Czencz, K. (1976). Kerteszeti es Szoleszet, 25 : 4.

Callon N.W. and Thompson, M.M. (1986). Theoretical and Applied Genetics, 7l: 657-61.

Canali, S., Nardi, P., Neri, U. and Gentili, A. (2004). Acta Hort., 686: 291-94.

Capik, J.M. and Monlar. T.J. (2014). HortTech., 24(2): 196-208.

Carden, L. and Devaux. M. (1987). Eppo Bull., 17:241-54.

Casella, D. (1958). Ann. Fac. Scr. Agrar Univ. Napoli, 23: 1-22.

Chandler, W.A. (1957). Deciduous orchards, Lea and Febiger, Philadelphia, pp. 448-55.

Chaplin, M.H. and Dixon, A.R. (1979). J. Amer. Soc. Hort. Sci., 104:710-12.

Chen, H., Mehlenbacher, S.A. and Smith, D.C. (2007). HortSci., 42: 466-69.

Ciampolini, M. and Ugolini, A. (1975). Informatore Fitopatologica, 25:21-27.

Ciemniewska-Zytkiewiczs, H., Pasini, F., Verardo, V., Brys, J., Koczon, P. and Caboni, M.F. (2015). European J. Lipid Sci. Technol., 117: 710-17.

Colburn, B.C., Mehlenbacher, S.A. and Sathuvalli, V.R. (2017). Mol. Breed., 37(2): 16.

Compton, M.E. and Preece, J.E. (1986). Newslett. Intern. Assoc. Pl. Tissue Cult., 50: 9-18.

Crescimanno, F.G., Sottile, I., Averna, V. and Bazen, E. (1981). Tecnica Agricola, 33: 355-74.

Cristifori, V., Blasi, E., pancino, B., Stelliferi, R. and Lazzari, M. (2017). Acta Hort., 1160: 165-71.

Cristofori, V., Bertazza, G. and Bignami, C. (2015). Fruits (Paris), 70: 311-22.

Dale, A., Galic, D., Leuty, T., Filotas, M. and Currie, E. (2012). Ministry of Agriculture, Food and Rural Affairs, Ontario, Canada.

Damiano, C., Catenaro, E., Giovinazzi, J. and Frattarelli, A. (2005). Acta Hort., 686: 221-25.

Damigella, P. and Alberghina, D. (1978). Tecnica Agricola, 30: 243-55.

De Rosa, M. (1961). Progr. Agric. Bologna, 7: 740-46.

De Rosa, M. (1969). Inform Agrar. Verona, 25: 1404.

De Rosa, M. (1979). Frutticoltura, 41 : 28-30.

Driver, J.A. and Kuniyuki, A.H. (1984). HortSci., 19: 507-09.

Durusoy, Y.Z. (1968). Agric., 6: 7-8.

Engler, H.F. (1953). Proc. Nut Growers Soc. Ore. Wash., 39:182-85.

Erper, I., Sarutian, I., Akea, I., Aksoy, H.M. and Tuncer, C. (2016). Egyptian J. Biological Pest Control, 26: 573-78.

Ferslake, F. (2012). Wealth from Water. Factsheet.

Fideghelli, C. and DeSalvador, F.R. (2009). Acta Hort., 845: 39-52.

Fontanazza, G., Salleo, S. and Ballatori, E. (1971). Annali della Facolta di Agraria, 26: 323-47.

Fontanrzza, G. (1973). Rivista della ortoflorofruitticoltura Italiana, 57:38-47.

Fontanszza, G. and Salleo, S. (1968). Ital Agric., 105: 277-89.

Fregoni, M. and Zioni, E. (1963). Frutticoltura, 25 : 651-56.

Fregoni, M. and Zioni, E. (1964). Frutticoltura, 26: 521-23.

Fuller, A.S. (1896). The Nut Culturist. New York Orange.

Garcia, J.M., Agar, I.T. and Streif, J. (1994). Turkish J. Agric. Forestry, 18: 199-202.

Gellatly, J.U. (1956). Ann. Rept. Nut Growers Assoc., 47: 110-12.

Gellatly, J.U. (1966). Ann. Rept. Nut Growers Assoc, 57: 98-101.

Geraci, G. (1974). Tecnica Agricola, 26: 1047-63.

Germain, E. (1994). Acta Hort., 351: 195-210.

Ghirardello, D., Zeppa, G., Rollel, L., Gerbi, V., Contessa, C., Valentini, N. Botta, R. and Griseri, G. (2014). Acta Hort., 1052: 315-18.

Gispert, J.R., Ramirez, de Cartagena, F., Villar, J.M. and Rovira, O.M. (2015). ITEA, 111(2): 109-26.

Giulivo. C. and Pisani, P.L. (1973). Informatore Agrario, 29: 14009-13.

Gosling, P.G. and Ross, J.D. (1980). Planta, 148: 362-66.

Grauke, L.J. and Thompson, T. (2003). Acta Hort., 622: 553-66.

Groeschke, F. (1887). Die Haselnuss, Thu, Arten Und Three Kultuee, 1099, Berlin

Guine, R.P., Almeida, C.F.F. and Correia, P.M.R. (2015). Jr. Food Measurement and Characterization, 9: 11-19.

Guncan, A. (1985). Doga Bilim Dergisi, 9:177-82.

Hadorn, H. (1977). Rev. Choc. Confect. Bakery, 2′ 16: 39.

Hand, C.R. (2013). M.S. Thesis, Oregon State University, USA.

Honglin, C., Mehlenbacher, S.A. and Smith, D.C. (2007). HortSci., 42: 466-69.

Hosseinpour, A., Seifi, E., Javadi, D. and Ramezanpour, S.S. (2015).Adv. in Hort. Sci., 29: 13-16.

Howes, F.N. (1948). Nuts-their production and everyday uses, Faber and Faber, London, pp. 1-264.

Hummer, K., Lagerstedt, H.B. and Kim, S.K. (1986). J. Amer. Soc. Hort. Sci., 111: 474.

Ilyasoglu, H. (2015). Int. J. Food Properties, 18: 456-63.

Jeavons, R.A. and Jarvis, B.C. (1984). New Phytol., 96: 551-664.

Johnson, K.B. and Pinkerton, J.N. (2002). Compendium of nut crop disease in temperate zones (eds. T.E. Teviotdale, J. Michalides and J.B. Pscheidt), APS Press, Amer. Phytopathol. Soc. St. Paul, M.N. pp. 44-46.

Kabir, H., Leklem, J. and Miller, L.T. (1983). J. Fd. Sci., 48: 1472-25.

Kai, H.A.L., Salesses, G. and Mouras, A. (984). Agronomie, 4: 399-402.

Kasaphgil, B. (1963). Cal. Hort. Soc. J., 24:95-104.

Kasaphgil, B. (1964). Adansonia, 4: 43-90.

Kavardzhikov, L. (1980). Gradinarska i Lozarska Nauka, 17 : 9-10.

Kavardzhikov, L. (1982). Gradinarska i Lozarska Nauka, 19: 11-15.

Kavardzhikov, L. (1983). Gradinarska i Lozarska Nauka, 20: 10-15.

Kavardzhikov, L. (1984). Gradinarska, i Lozarska Nauka, 21 : 7-15.

Kavardzhikov, L. and Krinkov, K.H. (1981). Gradinarska i Lozarska Nauka, 16: 33-40.

Kaya-Altop, E., Haghnama, K., Sariaslan, D., Phillippo, C.J., Mennan, H., and Zandstra, B.H. (2016). Crop. Protec., 80: 7-14.

Kazi-Zade, F.N. and Bozhko, N.V. (1981). Sadovodstvo, 11: 25.

Keme, T., Messerli, M., Shejbal, J. and Vitali, F. (1981). Zucker and Susswarenwitscheft., 34:360-64.

Kentsurashvili, S.E. (1987). Subtrop. Kul'tury, 1:31-33.

Kepenekci, I. (2002). Nematropica, 32: 83-85.

Kim, Y.H., Cheon, J.Y., Hyeon, M.C. and Kang, S.J. (1981). Res. Rept. Office of Rural Dev., S. Korea. Hort. and Sericulture, 23: 44-48.

Kirpina, I., Curlje, M. and Vujevi, P. (1992). Acta Hort., 351: 329-33.

Koksal, A.I., Artik, N., Simsek, A. and Gunes, N. (2006). Food Chem., 99: 509-15.

Koval, G.K. (1972). Sel'skokhozyaistvennaya Biologiya, 7 : 295-91.

Kovalenko, N.V. and Makhno, V.G. (1976). Subtrop, Kul'tury, 3 and 4: 129-32.

Kowalenko, C.G. (1981). Res. Rev. Res. Stn. Agassiz B.C., 1981, Nov. 8-9.

Kowalenko, C.G. (1984). Canad. J. Soil Sci., 64: 147-49.

Kowalenko, C.G. and Maas, E.F. (1982). Canad. J. Soil Sci., 62:209-11.

Kowalski, R.M. and Kawecki, Z. (1982). Roczniki Nauk Rolniczych A., 105 : 177-90.

Ksenija, M., Pejic, B., Belic, M., Jankovic, D. and Pavlovic, L. (2016). Res. J. Agric. Sci. Romania, 48: 75-81.

Lagerstedl, H.B. (1978). HortSci., 13: 533-34.

Lagerstedt, H. (1993). Proc. Nut Growers Assoc. of Oregon, 78: 89-101.

Lagerstedt, H.B. (1972). J. Amer. Soc. Hort. Sci., 97:738-40.

Lagerstedt, H.B. (1973). Western Fruit Grower, 27:6,8, 11, 13.

Lagerstedt, H.B. (1976). Proc. 67th A.R. of Northern Nut Growers Assoc., pp.56-60.

Lagerstedt, H.B. (1977a). Proc. Western Soc. Weed Sci., 18/30: 70.

Lagerstedt, H.B. (1977b). Econ. Bot., 3: 153-59.

Lagerstedt, H.B. (1984). Amer. Nurserymen, 160: 113-17.

Lagerstedt, H.B. (1990). Ann. Rep. Northern Nut Grower's Assoc., 81: 60-63.

Lespinasse, J.M. and Delort, J.F. (1986). Acta Hort., 160:139-55.

Leuty, T., Galic, D., Bailey, P., Dale, A., Currie, E. and Filotas, M. (2012). Ministry of Agriculture, Food and Rural Affairs, Ontario, Canada.

Lukasiewicz, J. (1994). New Crops News, 4:1.

Makhno, V.G. (1981). Subtrop. Kul'tury, 5: 69-74.

Manusev, B. (1978). Jugosolvensko Vocarstvo, 12: 31-37.

Manzo, P. (1972). Annali dell Instituto Spc. per la Frutticolture, 3: 411-26.

Manzo, P. and De, R. (1979). Informatore Agrario, 35 : 8417-19.

Marenaud, C. and Germain, E. (1975). Annals de phytopathologie, 7: 133-45.

Martin, S., Fernandez-Garcia, M.B., Romero, T. and Lopez, A. (2001). Jr. Fd. Processing and Preservation, 25: 309-21.

Mauhrer, K.J. (1975). Baumschulpraxis, 5: 111-13.

Maziarz, B. (1983). Polskie Pismo Entomologiczne, 53: 649-55.

Me, C.D., Brozold, L.R. and Caransano, F.C. (1981). Frutticoltura, 43: 31-35.

Me, G., Valentini, N. and Miaja, M.L. (2001). Acta Hort., 556: 321-26.

Mehlenbacher, S.A. (1997). Theor. Appl. Genet., 94: 360-66.

Mehlenbacher, S.A. (2009). Acta Hort., 845: 33-38.

Mehlenbacher, S.A. (2009). Acta Hort., 845: 33-38.

Mehlenbacher, S.A. (2009). Proc. Nut Grower's Soc. Oregon, Washington and British Colombia, pp. 57-63.

Mehlenbacher, S.A., Azarenko, A.N., Smith, D.C. and McCluskey, R.L. (2007). HortSci., 42: 715-17.

Mehlenbacher, S.A., Azarenko, A.N., Smith, D.C. and McCluskey, R.L. (2000). HortSci., 35: 314-15.

Mehlenbacher, S.A., Azarenko, A.N., Smith, D.C. and McCluskey, R.L. (2001). HortSci., 36: 995-96.

Mehlenbacher, S.A., Smith, D.C. and McCluskey, R.L. (2008). HortSci., 43: 255-57.

Mehlenbacher, S.A., Smith, D.C. and McCluskey, R.L. (2009). HortSci., 44: 845-47.

Mehlenbacher, S.A., Smith, D.C. and McCluskey, R.L. (2011b). HortSci., 46: 662-64.

Mehlenbacher, S.A., Smith, D.C. and McCluskey, R.L. (2013). HortSci., 48: 796-99.

Mehlenbacher, S.A., Smith, D.C. and McCluskey, R.L. (2014). HortSci., 49: 346-49.

Mehlenbacher, S.A., Smith, D.C. and McCluskey, R.L. (2016). HortSci., 51: 757-60.

Mehlenbacher, S.A., Smith, D.C., McCluskey, R.L. and Thompson, M.M. (2011a). HortSci., 46: 505-08.

Mehlenbacher, S.A., Thompson, M.M. and Cameron, H.R. (1991). HortSci., 26: 410-11.

Mehlenbacher, S.A., Thompson, M.M. and Cameron, H.R. (1991). HortSci., 26: 410-11.

Meiss, W, (1985). Deutsche Baumschule, 37: 42.

Miletic, R., Mitrovic, M. and Rakicevic, M. (2009). Acta Hort., 845: 283-85.

Miletic, R., Mitrovic, M. and Rakicevic, M. (2009). Acta Hort., 845: 283-86.

Mingeau, M., Ameglio, T., Pons, B. and Rousseau, P. (1994). Acta Hort., 351: 305-14.

Molnar, T.J., Walsh, E., Capik, M., Sathuvalli, V., Mehlenbacher, S.A., Rossman, A.Y. and Zhang, N. (2013). Pl. Dis., 97: 813-18.

Nas, M.N. (2004). Turkish J. Agric. Forestry, 28: 189-94.

Nicita, G. (1986). Informatore Agrario, 42: 42-69.

Ninic-Todorovic, J. (2000). FAO-CIHEAM-Nucis-Newsletter, No. 9.

Ninic-Todorovic, J., Cerovic, S., Ognjanov, V., Golosin, B., Bijelic, S., Jacimovic, G. and Kurjakov, A. (2008). Paper presented at 7$^{th}$ Int. Congrs. On Hazelnut, Vitorbo, Italy, Abst. No. 64.

Olsen, J. (2011). HortSci., 46: 343-44.

Olsen, J., Mehlenbacher, S.A. and Azarenko, A.N. (2000). HortTech., 10: 113-15.

Ozenc, N., Ozenc, D.B. (2015). Academic Journal, Lagos, Nigeria, 10: 465-70.

Ozenc, N., Ozenc, D.B. and Blackwell, W. (2015). Jr. Sci. Food. Agric., 95: 1956-62.

Ozmen, S. (2016). Erwerbsobstbau, 58: 173-80.

Ozturk, Y. and Tarakcioglu, C. (2016). Akademik Ziraat Dergisi, 5(2): 87-96.

Paglieua, R. (1973). Pl. Propagator, 19: 20.

Painter, J.H. (1962). Proc. Nut Growers Soc. Ore. Wash., 48:20-23.

Pala, M., Ackurt, F., Loker, M., Yildiz, M. and Omeroglu, S. (1996). Turkish J. Agric. Forestry, 20: 43-48.

Pallegrino, S. and Mozzone, G. (1985). Informatore Agrario, 41: 68-70.

Pannico, A. (2014). Ph.D. thesis, Universita Degli Studi di Napoli, Federico II, Italy.

Parker, W.B. Wesseler, P. and Roberts, W. (1955). Proc. Ore. Sta. Hort. Soc., 47: 1223-24.

Parmia, P., Onea, I. and Stanciu, N. (1979). Lucrari Sttinti. Inst. de. de. Cu. Prod. Pen-Pomicultura Pitesti, 8 :193-99.

Pathak, R.K., Sinha, M.M. and Pandey, V.S. (1978). Indian J. Agric. Sci., 48: 556-58.

Pavlenko, F.A. and Slyusarchuik, V.E. (1983). Sadovodstvo, 11: 26.

Pellati, E. and De, V. (1982). Informatore Agrario, 38: 22893-22.

Perez, C., Fernandez, B. and Rodriquez, R. (1983). Plant Cell Rept., 2: 226-28.

Perez, C., Rodriguez, R. and Tames, R.S. (1985). Plant Cell Rept., 4:137-39.

Perez, C., Rodriquez, R. and Tames, R.S. (1986). Plant Science Irish Republic, 45: 59-64.

Perez-Tornero, O., Lopez, J.M., Egea, J. Rodriguez, R. and Sanchez-Tames, R. (1987). Acta Hort. 221: 505-10.

Pinkerton, J. N. (1996). Pl. Disease, 80: 1308-16.

Pisetta, M., Albertin, I., Petriccione, M. and Scorthihini, M. (2016). Scietia Hort., 211: 187-93.

Piskornik, Z., Piskornik, M. and Goc, F. (1975). Zeszyty Naukowe Akademii Rolniczej w Krakowie. Ogrodnictwo, 104: 183-96.

Piskornik, Z., Piskornik, M. and Goc, F. (1982). Acta Agrobotanica, 35: 169-74.

Pollack, S. (1998). Economic Research Service, USDA.

Pourfarzad, A. and Raouf Mehrpou, G. (2017). EC Nutrion, 83: 101-05.

Radicati, L., Martino, I. and Vergano, G. (1994). Acta Hort., 351: 489-94.

Radojevic, L., Vujicic, R. and Neskovic. M. (1975). Zeit. fur. Planzenphysiologie, 77: 33-41.

Rahemi, A.R., Taghavi, T. and Dale, A. (2016). Jr. of Nuts, 7(2): 83-87.

Redpath, M. (2016). Bulltin 1: An introduction to growing hazelnuts in New Zealand, Hazelnut Growers Assoc. NZ. pp. 5.

Reed. C.A. and Davidson. J. (1958). The improved nut trees of North America and how to grow them. Devin Adair. New York.

Rehder, A. (1940). Mannual of Cultivated Trees and Shrubs. Macmillan, New York.

Rivala, P. (1979). Rerue. Horticole Suisse., 52 : 67-68.

Rodriguez, A., Albuerne, M. and Tames, R.S. (1986). Phyton Argentina., 46: 27-31.

Rodriquez, R. and Fernandez, B. (1982). Anales, de Edafologia Y. Agrobiologia. 41: 2295-301.

Romisondo, P., Me, G., Limongelli, F. and Radicati, L. (1978). Rivista della Ortoflorofruitticoltura Italiana, 62: 162-68.

Romisondo, P., Me, G., Manzo, P. and Tombesi, A. (1983). Convegno Naz. Sul Nocciolo, Avellino 22-24 settembre, 61-78.

Rotundo, A., Pasquarella, C. and Forlani, M. (1984). Rivista della Ortoflorofrutticoltura Italiana, 68 :225-35.

Roversi, A. (1969). Inform. agra. Varona., 25: 1729-30.

Roversi, A. (1973). Annal della Facolta di Agraria Univ. Cattolia del Sacro. Curore Milano., 13:318-38.

Roversi, A. (2015). Bulgarian J. Agric. Sci., 21: 355-57.

Rovira, M., Cristofori, V., and Silvestri, C., Celli, T., Hermoso, J.F., Tous, J. and Romero, A. (2014). Acta Hort., 1052: 145-50.

Rovira, M., Cristofori, V., Silvestri, C., Celli, T. Hermoso, J.F., Tous, J, and Romare, A. (2014). Acta Hort., 1052: 145-50.

Rovira, M., Romero, M. and Clave, J. (1997). Acta Hort., 445: 145-50.

Saban, B.A. (1977). Nauchnye Trudy. Lvov. S.kh. Inta., 73: 140-43.

Sander. E. (1963). Proc. Nut Growers Soc. Ore. Wash., 49: 43-44.

Santi, C., Giorni, A., Terenzi, C.T., Altavista, P. and Bacchetta, L. (2017). Food and Nutrition Sci., 8: 633-46.

Schreider, W.K. (1947). Proc. Nut. Growers Soc. Ore. Wash., 33: 101-108.

Schulze, E.D. and Kuppers, M. (1979). Planta, 146: 319-26.

Sentis, X., Ferran, J. Tous, J. and Romero, A. (2004). Acta Hort., 686: 281-84.

Shannon, P.R.M., Jeavons, R.A. and Jarvis, B.C. (1983). Plant and cell Physiol., 24:933-36.

Shewry, P.R., Pinfield, N. J. and Stobart, A.K. (1974). J. Exp. Bot., 24: 1100-105.

Shrestha, G.K., Thompson, M.M. and Righetti, T.L. (1986). Commun. Soil Sci. Plant Analysis. 17 : 1033-40.

Shrestha, G.K., Thompson, M.M. Righctt, T.L. (1987). J. Amer. Soc. Hort. Sci., 112:412-16.

Shun-Ching (1935). Forest Botany of China. Commercial Press.

Slate, G.L. (1930). Hazelnuts. N.Y.S. Agr. Expt. Sta. Bull. Geneva. No.558.

Slepokurov, V.L., Gavrilov, N.P, and Makhno, V.C. (1984). Sadovodstvo. 11:27-28.

Smith, D.C. and Erdogan, V. (2001). Acta Hort., 556: 263-67.

Snare, L. (2008). Primefacts. Profitable and Sustainable Primary Industry, 765: 1-8.

Stritzke, S. (1962). Arch. Gartenb. 10 : 573-608.

Talbott, J.E. (1959). 50th A.R. North Nut Growers. Assoc., pp.76-78.

Thompson, G.E. (1984). Australian Hort., 82: 50-62.

Thompson, M.M. (1968). Bett. Fruit. Belt. Veg., 62: 6. 14.

Thompson, M.M. (1969). Proc. Nut. Grow. Soc. Ore. Wash., 55: 80-84.

Thompson, M.M. (1972). Proc. Nut Grow Soc. Ore. Wash., 57: 43-46.

Thompson, M.M. (1977). J. Amer. Soc. Hort. Sci., 102:39-42

Thompson, M.M. (1979). J. Amer. Soc. Hort. Sci., 104: 427-32.

Thompson, M.M., Lagerstedt, H.B. and Mehlenbacher, S.A. (1996). In: Fruit Breeding, Willey, New York, 3: 125-84.

Tkhagushev, N.A. and Merzhanjan, Yu, A. (1975). Trudy Kubauskoge. S.kh. Instituta. 139: 66-70.

Tomasone, R., Colorio, G., Cedola, C. and Pagano, M. (2008). 7th Int. Congress on Hazelnut, Viterbo, Italy, June 23-27, 2008. Book of Abstract: 137.

Tombesi, A., Preziosi, P. and Boco, M. (1994). Acta Hort., 351: 119-22.

Tombesi, S. and Farinelli, D. (2017). Acta Hort., 1160: 141-44.

Tombest, A. (1983). Annali della Facolta. di Agrario Univ. deg. Stu. de Per., 37: 86-96.

Toth, L. (1984). Novenyvedelen, 20: 366-68.

Tous, J., Girona, J. and Tasias, J. (1994). Acta Hort., 351: 395-418.

Trotter, A. (1947). Rend. Acad. Lincri. Sec., 2: 745-49.

Trotter, A. (1948). Rend. Acad. Lincri Sec., 4: 659-66.

Valentini, N., Marinoni, D., Me, G. and Botta, R. (2011). Acta Hort., 556: 209-18.

Vanicheva, S.G. (1981). Int. Spo. Vy. Pos. Mat. Sad. Kul. Moscow. USSR. pp. 111-17.

Vicol, A., Botu, I. and Botu, M. (2009). Ingineria Mediului, 14: 43-46.

Vidyasagar, R.S., Mehlenbacher, S.A. and Smith, D.C. (2010). HortSci., 45: 1116-19.

Viggiani, G. (1972). Annali della Facolta di Scienze Agrarie della Universita degli studi de Napoli, Portici, 6 :160- 67.

Viggiani, G. (1985). Bollettino del Laboratorio di Entomologia Agraria Filippo Silverstti, 42:3-9.

Viggiani, G. and Bianco, M. (1974). Bollettino del Ldboratorio di Entomologia Agraria Filippo Silverstri, 31 :30-53.

Watson, V.M., Chambers, U., Dreves, A., Bruck, D.J. and Olseh, J. (2007). Oregon State University. Extension Service, 8946.

Woodworth, R.H. (1929). Bot. Caz., 88: 383-99.

Yarris, L. and Lagerstedt, H.B. (1981). Agricultural Research. USA. 30: 12-13.

Yildiz, T. (2016). Turkish J. Agric. and Forestry, 40: 301-10.

YunQing, C., YiXin, Z., JianFeng, L., BoWen, Y. and YueTong, M. (2017). Pl. Growth Regu., 81: 147-57.

Zielinski, Q. B. and Thompson, M.M. (1967). Proc. Amer. Soc. Hort. Sci., 91: 187-91.

# Chapter 10

# Pecans

Pecan nut is a valuable horticultural gift of North America to the world occupying fifth rank among leading tree nuts because of its excellent nutty flavour and superb tree characteristics. In the USA, pecan is considered the 'queen of nuts' because of its value both as a wild and as a cultivated nut and acceptable quality of meats (Woodroof, 1979). As compared with other nut fruits which have been grown since time immemorial, pecan has been under cultivation only for about one hundred years but has gained enormous popularity (Wetzstein and Sparks, 1986). Initially, American Indians carried and planted the nut, which they called 'pecan', along the water courses. The common North American crow did play a significant role in the spread of the nut. Besides, the Indians also brought about considerable improvement in the nut quality by selecting for large size and thin shell (Madden and Malstrom, 1975).

## Composition and Uses

Pecan nut is a rich source of fat, protein, carbohydrates and minerals. The proximate composition of 100g kernel has been determined as fat, 71.2g; protein, 9.29; carbohydrates, 14.69; fibre, 2.3g water, 3.4g; calcium, 73 mg; phosphorus, 289 mg; potassium, 603 mg; magnesium, 142 mg; iron, 2.4 mg; vitamin A, 130 I.U.; thiamine, 0.86 mg; riboflavin, 0.13 mg; niacin, 0.9 mg; ascorbic acid, 2 mg and food value, 687 calories (Watt and Merrill, 1963). Lipids are the most abundant components of pecan, which was estimated 71.97 and 74.27 g/100g for raw and roasted pecans, respectively. (Anon., 2012). Carbohydrates, calculated by difference, were found to be 13.8g/100g nutmeat (raw) with total dietary fibre making up 69 per cent of the carbohydrate portion. Proteins were found to be the next largest constituent with a content of 9.17g/100g nutmeat (raw). The lipid content, however, can vary from 65 to 75 per cent, depending on method of production, age of the trees, cultivars, and the previous year productivity (Santerre, 1994). The minor lipid constituents of the nut are more complex lipids including other acylglycerols, tocopherol and sterols (Senter *et al.*, 1980). Kernels contain six main phenolic acids: gallic, gentisic, vanillic, protocatechuic, *p*-hydroxybenzoic acid, and *p*-hydroxyphenylacetic acid as well as coumaric and syringic acid in small quantities (Senter *et al.*, 1980). The presence of phenolic compounds with high antioxidant capacity indicates that pecan kernels should be added to the list of important dietary sources of antioxidants (Lombardini *et al.*, 2007). The oil composition of the cultivar Stuart was reported as: oleic acid 68.8 per cent; linoleic acid 21 per cent; palmitic acid 6.6 per cent; stearic acid, 2.2 per cent; linolenic acid 1.1 per cent and arachidonic acid 0.2 per cent (French, 1962). In different pecan cultivars, the oleic acid ranges from 51-77 per cent. Jacobs (1951) has reported the properties of pecan oil as specific gravity at 15°C, 0.9184; refractive index at 20°C, 1.4708; iodine number, 106.0 and saponification number, 198.0. The predominant amino acids recorded in the matured pecan nut are arginine, phenylalanine, valine, glutamic acid, histidine, cystine and lysine in decreasing order (Wood and Reilly, 1984). It has acceptable levels of sulphur amino acids except the lysine content which is low (Elmore and Polles, 1980).

Traditionally, pecan is considered a highly nutritive and energy-giving nut fruit which is usually taken roasted or salted to supplement normal diet. It has reported that consumption of pecans lowers the risk of heart disease (Yahia, 2010). Almost 90 per cent of the nuts are sold shelled and rest in the shell. The nuts are commonly used to add aroma, flavour, crispness, mealiness, tenderness, or to garnish a large number of dishes. The greatest use of nut is, however, in baking products, such as, fruit cakes, custard pies, cookies, nut bread and cake fillings. Another important item for consuming the nutmeat is ice-cream. Pecan ice-cream is a top selling one after the vanilla and chocolate (Woodroof, 1979). Salted nuts, pecan butter, candies, *etc.*, are the other outlets of the nut use. In fact, thousands of recipes are available for serving pecan nuts. Above all, these are best gifts on auspicious occasions.

The important uses of pecan shell, a by-product, are manufacture of tannin, charcoal and abrasives in hand soap; flour as a filler for plastic wood and veneer wood; fillers for feeds, insecticides and fertilizers; and fuel for heating. The shell flour of various sizes is also used as soft grit in non-skid paints, adhesives, dynamite and polishing metals.

Pecan tree is an exceedingly beautiful plant which can be used for shade and ornamental purposes. The tree has a surpassing value on timber market due to its strength and hardness ranking third only after black walnut and wild cherry, respectively (Woodroof, 1979). There is a great demand for its veneer and lumber for use in decorative panelling, fine furniture, flooring and in pallet manufacture. Hickory is considered as the best burning wood for imparting the flavour to the smoking of meats (Madden and Malstrom, 1975).

## Origin and Distribution

Pecan is a native of North America but the opinions differ whether it originated further north and was water borne to the south or appeared simultaneously over a wide region. Romberg (1968) opined that it originated during Cretaceous period in

Canada and moved south in recent geological times. On receding of the last ice caps, the North American crow and Indians played a significant role in its spread over a wide region. The early Spanish explorers on their arrival found the pecan growing along the Mississippi river and its tributaries. The native habitat of pecan ranges from Texas in the southwest to Alabama in the southeast extending up to southern Illinois (Hanna, 1987). Its cultural range, however, overlapped the native range, extending beyond into New Mexico, Alabama, Georgia, Florida, South Carolina and to a lesser degree into other fringe states (Madden *et al.*, 1969). About 93 per cent of the world's annual pecan production is produced in 15 US southern states and Mexico. In 2016, the pecan production in the world exceeds 250 million kg of which 47 per cent produced by Mexico and 46 per cent by USA. The other producing countries are South Africa (4 per cent), Australia (1 per cent) and China (1 per cent). Peru, Chile and Israel together produced 1 per cent of the world's production.

Besides the USA, the worldwide distribution of pecan is confined to only Australia, Canada, Egypt, India, Israel, Mexico, Morocco, Peru, Turkey and South Africa probably because the crop is rather new and have a longer gestation period than other fruits.

## Taxonomy

The accepted botanical name of pecan is *Carya illinoinensis* (Wang.) K. Koch, which is a member of genus *Carya*, an ancient Greek name of walnut, belonging to the *Juglandaceae* family and comprises 20 species. Various old names of pecan such as *Juglans pecan, Juglans illinoinensis, Hicoria pecan* and *Carya pecan* have been discarded. All the species of the genus *Carya* are together known as 'hickories' and natural hybrids between pecan and other hickories are called 'hicans' (McKay,1961). Natural hybridization among hickories is the main source of confusion in the characterization of *Carya* species which are twenty in number and except for all other are native to the Americas (Table 10.1). On account of its excellent flavour, pecan is the only commercial hickory while shagbark (*C. ovata*), shellbark (*C. laciniosi*) and mokcrnut (*C. tomentosa*) also have sweet kernels but poor in cracking quality. Some characters of important hickories are given in Table 10.2.

## Cultivars

The native seedlings of pecan are preponderant in North America accounting for 59 per cent of the total acreage and 39 per cent of the total production (Thompson, 1984). The seedling groves are gradually receding with a concomitant expansion of the plantings of improved cultivars. The choice of pecan cultivars should be based on the climate, incidence of diseases, market needs and modern cultural trends. The chilling and total number of days required by a cultivar to mature a crop are important climatic considerations. Accordingly, American cultivars have been grouped into the northern, the southern and the western groups (Woodroof. 1979). The cultivars originating in the north need more chilling but mature crop in a shorter growing season (165 days). In contrast, southern cultivars have a long maturation period (200 days), and are more susceptible to fungal diseases. Western cultivars are suited to relatively dry climate. In the USA, 90 per cent, of pecan is

**Table 10.1: Common and Latin Names, Chromosome Number, and Range of Hickory (Carya) Species***

| *Latin Name* | *Somatic Chromosome Number* | *Common Name* | *Range* |
|---|---|---|---|
| APOCARYA SECTION | | | |
| *C. illinoinensis* | 32 | Pecan | Illinois and Iowa, South to Alabama and Mexico |
| *C. myristicaeformis* | 32 | Nutmeg H. | S. Carolina to Arkansas and Mexico |
| *C. carythayensis* | 32 | Cathay H. | East China |
| *C. aquatic* | 32 | Water H. | Virginia to Illinois South to Florida and Texas |
| *C. cordiformis* | 32 | Bitternut H. | Quebec to Minnesota South to Florida and Louisiana |
| *C. palmeri* | | | Mexico |
| *C. tonkinensis* | | | South East Asia |
| *C. poilanei* | | | South East Asia |
| CARYA SECTION | | | |
| *C. glabra* | 64 | Pignut H. | Maine to Ontario South to Florida and Mississippi |
| *C. ovalis* | 64 | Red H. | Massachusetts to Wisconsin South to Georgia and Mississippi |
| *C. pallida* | | Sand H. | Alabama and Tennessee |
| *C. tomentosa* (*C. alba*) | 64 | Mockernut H. | Massachusetts to Ontario and Nebraska South to Florida and Texas |
| *C. laciniosa* | 32 | Shellbark H. | New York to Iowa, South to Alabama and Oklahoma |
| *C. ovate* | 32 | Shagbark H. | Quebec to Minnesota South to Florida and Texas |
| *C. carolinae-septentrionalis* | 32 | Caroline H. | North Carolina to Georgia, Tennessee and Alabama |
| *C. florodana* | 64 | Scrub H. | Peninsular Florida |
| *C. texana* | 64 | Black H. | Texas to Oklahoma and Arkansas |
| *C. leidermis* | | Swamp H. | South East United States |
| RHAMPHOCARYA SECTION | | | |
| *C. sinensis* | | Chinese H. | South East Asia |

* Jaynes (1969).

sold as shelled which means the nut size, shell thickness and cracking qualities are not very important, but these qualities have a premium in other countries where pecan is marketed in-shell. Recent cultural trend in fruit growing is toward high density orcharding. The cultivars should, therefore, be precocious, prolific and respond to pruning and good orchard management practices. Nearly half of the improved trees in USA consisted of three cultivars: 'Stuart' (22 per cent); 'Western Schley' (14.6 per cent) and Desirable (10.9 per cent). All these cultivars

**Table 10.2: Characteristics of Important Hickories (*Carya* sp)**

| *Hickories* | *Plant Height (m)* | *Bark Characters* | *No. of Leaflets* | *Shape of Leaflets* | *Length of Leaflets* | *Nut Shape* | *Nut Size (cm)* | *Nut Characters* | *Nut Flavours* |
|---|---|---|---|---|---|---|---|---|---|
| *C. illinoinensis* | 50 | Furred greyish brown | 11-12 | Oblong lanceolate | 10-18 | Ovoid oblong | 2.5-7 | Smart, light brown | Excellent |
| *C. ovate* | 40 | Shaggy | 5-7 | Oblong | 10-15 | Angular | 3.5-6 | Hardshell | Excellent |
| *C. laciniosa* | 40 | Shaggy | 5-7 | Oblong | 10-15 | Four angled | 2-5 | Hardshell | Sweetest |
| *C. cordiformis* | 30 | - | 5-9 | Oblong | - | Roundish | 2-4 | Thinshell | Bitter |
| *C. glabra* | 30 | Shaggy | 5-7 | - | - | Roundish | 2-3 | Thinshell | Sweet |
| *C. tomentosa* | 30 | Finely grooved | 7-9 | Oblong | 8-18 | Pear shaped | 3.5-5 | Sharply ridged | Sweet |
| *C. myristicaeformis* | 30 | Yellow brown scales | 5-11 | - | 7.5-13 | - | 2-2.5 | Reddish brown | Sweet |
| *C. carythayensis* | 20 | - | 5-7 | - | 10-14 | Ovoid | 2.5-3.5 | Four angled wrinkled | Sweet |
| *C. aquatica* | 25 | Light brown | 7-13 | Oblong | 8-12 | Ovoid | 2-3 | Four angled, wrinkled | Bitter |
| *C. pallida* | - | Rough pale bark | 7-9 | - | 8-12 | Ovoid | | Flat skinshell | - |

were developed in late nineteenth or early twentieth century. The other cultivars growing in USA are Wichita, Schley, Cheyenne, Success, Cape Fear, Moneymaker, Mohawk, SanSaba Imp. Mahan, Moore, Choctaw, *etc*.The salient features of some cultivars are described below.

### Stuart

This is a popular old cultivar originated as Alabama seedling (Thompson, 1984). The tree gives uniform yield of high quality nuts and is moderately resistant to scab disease. The large size nuts yield 44-50 per cent kernel. It is also recommended for homestead planting. It, however, slow to come into bearing and being protogynous should be planted with a protandrous cultivar like Desirable (Thompson, 1988).

### Western

Western is another important cultivar of USA having originated as a cross between Seedling and Texas (Thompson, 1984). Medium-sized, thin-shelled nuts have high kernel percentage (54-59 per cent) and good kernel quality. The cultivar is recommended for high density planting because it is precocious, very prolific and self-pollinated. It is, however, unsuitable for humid regions due to scab susceptibility.

### Desirable

Desirable is probably a hybrid of Success x Jewett or Russell and originated at Ocean Springs. Mississippi (Brooks and Olmo, 1972). The nut is larger than Stuart. 52 per cent kernel. The tree comes into bearing early, very prolific and regular bearer and is also resistant to scab. Desirable is recommended for both home and high density planting (Woodroof, 1979).

### Wichita

A hybrid between Halbert x Mahan. 'Wichita' is precocious and prolific bearer (Thompson, 1984). The nuts are of medium size, attractive appearance, and high quality and have high kernel percentage (58-63 per cent). The tree is susceptible to scab and has high zinc requirement. The cultivar is protogynous and should be grown with Western or Cheyenne cultivar for pollination.

### Cheyenne

This is a cross of Clark x Odom and released from Brownwood, Texas. Tree is very precocious, very productive, scab resistant and being recommended for high density purposes as its size can be controlled by pruning. It is a protandrous type and needs to be grown with protogynous cultivars, such as, Wichita. Medium-sized nuts are slightly pointed at the apical end and have 57-61 per cent kernel which is slightly wrinkled, loose in shell but has excellent flavour (Brooks and Olmo, 1972).

### Mohawk

It originated as a cross of Success x Mahan and was introduced in 1965. The tree is precocious, prolific, vigorous and fairly resistant to diseases. In spite of large

Downy Spot Symptom

Pecan Cultivar Desirable

Pecan Cultivar Western

Pecan Orchard

size, the nut matures exceptionally early (Madden *et al.*, 1969). Elongated nuts are thin-shelled, very attractive, having kernel percentage of 55-60. The kernels are quite smooth, easily shelled and of high quality.

## Mahan

Brooks and Olmo (1972) report its origin as a Schley seedling in Mississippi which was introduced by F.A. Mahan. The tree is vigorous, precocious, prolific bearer with good foliage. The leaves fall late in the season. The nuts are extra large resembling Mohawk, long, shell thin and tend to poor filling on older trees. It has given the highest yield among the various cultivars in India and Zimbabwe (Rathore, 1983; Robinson, 1977).

## Pawnee

This is a selection from Mohawk x Starking Hardy Giant crosses, which is early, prolific and fast growing in young age (Thompson and Hunter, 1985). It is one of the popular cultivar being propagated worldwide. The yield from 11-year-old trees was 3027 kg/ha. Nut quality is excellent and the size, appearance and kernel percentage are all good. Nut matures one or two weeks before Mohawk. The plant is tolerant to scab disease.

## Moreland

Most probably, it is a Schley seedling which has been released for cultivation under Louisiana conditions which is better in quality, yield and disease resistance to majority of the currently grown cultivars (O'Barr *et al.*, 1989). Nuts are easily shelled and the shelling percentage is 55.7. In size and shape, nuts are similar to Eastern Schley nuts. Moreland is protogynous requiring protandrous cultivars (Caddo, Desirable) for cross-pollination.

## Ukulinga

This has originated as a chance seedling in South Africa showing high promise for humid tropical regions (Allan, 1987). Ukulinga is not only precocious but has given higher yield than the cultivars Moore and Mahan (Haulik and Holtzhausen, 1987). The nuts are large, shell medium thick, kernel percentage 52-54 per cent, and kernel quality is fair (Haulik and Holtzhausen, 1988). It is comparatively less prone to irregular bearing.

**Apalachee (Moore x Schley):** Developed from USDA-ARS Pecan Breeding programme, scab disease resistant, 185 nuts/kg, avegare yield 1793kg/ha, mature early to mid-season, medium susceptibility to yellow aphid complex and black aphids (Thompson and Grauke, 2010a).

**Byrd (Wichita x Pawnee)**: Developed from university of Georgia, USA. High nut quality, early nut maturity, high yielder, moderately susceptible to pecan scab.

**INTA Delta I**: Selection made at Agrentina. Vigorous, densely limbed pecan cultivar with large, attractive nuts, 143 nuts/kg, 49 per cent kernel.

**INTA Delta II**: Another selection from Argentina. Low vegetative vigour and spars to medium branching; suitable to use as dwarf rootstock. Nuts are oval elliptic, 187 nuts/kg, 59 per cent kernel.

**Mandan: (BW-1 x Osage)**: developed from USDA-ARS Pecan Breeding Programme. Early maturing pecan with excellent yield potential high nut quality and good scab resistant (Thompson and Grauke, 2010b).

**Lipan (Cheyenne x Pawnee):** Developed from USD-ARS. High nut quality, high yield potential, medium-early nut maturity, resistant to pecan scab (*Fusicladium effusum*). It is well adopted to many pecan-growing areas (Thompson and Grauke, 2012).

**Moshan 1**: A seedling selection developed at China (GuangQin *et al.*, 2011). High yielder, hardy and disease resistant. Average nut weight 8.6-8.9g with 47.56 to 48.67 per cent kernel. Male and female flowers bloom almost in the same period.

**Hopi (Schley X McCulley)**: It was released by USDA-ARS (Thompson and Grauke, 2000). Medium and uniform yields with outstanding nut quality. It is resistant to vein spot (*Gnomonia nerviseda*).

**Nacono**: Released by USDA-ARS and the Texas Agricultural Experiment Station. 'Nacono' tested as selection 74-5-55 is a progeny from the cross between Cheyenne x Sioux. It has superior nut quality (nut weight 10.5g, kernel 57.1 per cent), moderate resistant to scab (Thompson and Grauke, 2001).

Badyal and Upadhayay (2004) evaluated seven pecan cultivars (Mahan, Nellis, Burkett, Western Schley, Success, Halbert and Calaro) under subtropical climate of Palampur, HP, India. the highest yield was recorded in Western Schley and lowest in Mahan. The weight and size of nut was noted highest in Halbert. The cv. Halbert also showed the highest Kernel percentage and fat content (Table 10.3).

## Climate

### Temperature

The climate is probably the main constraint in the expansion of pecan cultivation. It needs warm temperate climate meaning thereby that the extremes of both temperate and subtropical climates are undesirable. Characterizing the ecology of pecan trees, Wolstenholme (1979) suggested the following requirements for an ideal pecan environment: (i) growing season of 240-280 days, (ii) mean temperature above 26.7°C, (iii) heat unit accumulation (base 10°C) of 5000 degree days for the 7-month nut-growing period, (iv) mean temperature for the 3 coldest months between 7.2° and 12.8°C with at least 400 h of chilling temperature at or below 7.2°C, (v) medium to low rainfall, (vi) high levels of solar radiation during the growing season, and (vii) deep fertile soil with pH between 6.7 and 7.0.

Pecan is influenced by minimum temperature and number of chilling hours during winter, frost during spring and growing season temperatures. Kuden *et al.* (2013) studied the chilling requirement of pecan cultivars at Turkey. The male flowers buds of Kiowa, Schley, and Texhan were found to be the lowest chilling requiring cultivars suitable to warm climatic conditions. The female flowers of

**Table 10.3: Physico-chemical Characteristics of Various Pecan Cultivars***

| *Cultivar* | *Nut Yield (Kg/tree)* | *No. of Nuts per kg* | *Nut Weight (g)* | *Nut Size (cm)* | | *Kernel Weight (g)* | *Kernel Percentage* | *Protein (Per cent)* | *Fat (Per cent)* |
|---|---|---|---|---|---|---|---|---|---|
| | | | | *Length* | *Breadth* | | | | |
| Mahan | 11.8 | 190-195 | 5.26 | 5.40 | 2.86 | 3.57 | 56.8 | 10.20 | 64.06 |
| Nellis | 19.6 | 260-265 | 3.84 | 4.28 | 2.73 | 2.12 | 55.2 | 11.35 | 63.20 |
| Burkett | 18.4 | 230-235 | 4.34 | 3.60 | 3.06 | 2.70 | 62.2 | 9.60 | 66.70 |
| W. Schley | 21.2 | 220-225 | 4.54 | 4.08 | 2.36 | 2.56 | 67.8 | 10.25 | 63.90 |
| Success | 15.3 | 225-230 | 4.43 | 3.68 | 2.70 | 2.80 | 63.2 | 9.75 | 68.40 |
| Halbert | 13.5 | 175-180 | 5.71 | 4.26 | 3.40 | 3.90 | 68.3 | 10.30 | 69.30 |
| Calaro | 16.7 | 280-285 | 3.59 | 3.16 | 2.32 | 2.10 | 58.4 | 9.95 | 67.10 |
| CD (P=0.05) | 4.01 | | 0.10 | 0.19 | 0.16 | 0.19 | 8.70 | 0.55 | NS |

*Badyal and Upadhyay (2004).

cvs. Harris, Super, Mahan, Moneymaker and Wichita requires low chilling while cvs. Desirable, Schley and Woodard needs more chilling. Heavily fruited trees of alternate bearing cultivars which had lower tissue levels of starch, sugar and potassium, experienced more mortality due to severe winter cold than the trees of moderately bearing cultivars (Wood, 1986), Rajashekar and Reid (1989) observed super cooling in stems and apical buds of all pecan cultivars which was maximal in early January and least in early spring. In early April, stem samples of all cultivars were killed at or below –20.1°C, whereas apical buds were killed at –16° to –10°C. Wells (2008) studied the effect of late-spring freeze on cvs. Desirable and Kiowa. It was reported that cv. Desirable produced pistillate flowers from secondary buds after the freeze; however, many of these flowers were abnormal in appearance. Freeze-damage 'Desirable' trees exhibited shorter shoots, reduced flower and fruit retention, a lower chlorophyll index and decreased leaf N compared with non-damaged trees. Freeze-damaged 'Kiowa' trees had longer shoots and failed to produce a crop of pistillate flowers from secondary buds on most shoots. Freeze damage led to the appearance of mouse-care leaf symptoms and reduced leaf chlorophyll index, N and Mg concentrations of leaf. Smith and Cheary (2010) evaluated freeze damage bud injury of different pecan cultivars at Oklahoma. They reported higher bud death due to freezing temperatures (on 7 April, 2009) in cvs. Pawnee, OK642 and Mohawk. The cultivars with the best bud survival were Giles, kanza and Mount. Smith and Couch (1984) found that the frost damage varied with the cultivar and suggested various remedial measures, *viz.* (i) rootstock selection, (ii) budding at 30 cm height, (iii) selection of resistant cultivars, (iv) proper fertilization, and (v) maintenance of foliage in good condition to assist maximum carbohydrate production. Smith (2002) recorded maximum autumn freeze damage in bearing trees of cvs. Gratex, Shoshoni and Squerrel's Delight.

Like other temperate fruits, pecan also needs adequate chilling hours at a specific threshold temperature to foliate and fruit normally, insufficient winter chilling results in delayed foliation and uneven development of fruit buds. McEachern *et al.* (1978) determined a chilling requirement of 500 h below 7.2°C for the cultivars Desirable and Mahan, and 600 h for Stuart. Trees native to northern region of United States are more cold-tolerant than those native to and grown in the southern United States (Volk *et al.*, 2009). Trees overly stressed by crop load and premature defoliation, either died or were severely damaged at –3°C in mid-November. Orchard damage was found associated with death of tree roots during the dormant season (Wood and Reilly, 2001). Various chemicals (DONC, $KNO_3$, thiourea and DNOSBP) failed to supplement chilling on bud burst of pecan (Lagarda, 1987). Mean monthly temperature range from 24° to 29.5°C with little diurnal variation, during growing season, is necessary for best growth and fruiting (Madden *et al.*, 1969). The largest nuts are produced when heat unit accumulations from April through October total well over 3315 degree days at 10°C base (Wolstenhlome, 1979). Southern cultivars do not mature and fill properly on growing near the northern limit of pecan in the absence of sufficient heat units.

### Humidity and Rainfall

High humidity affects adversely the growth, fruiting and regularity of bearing in pecan by preventing pollination, increasing the incidence of diseases on leaves and nuts, premature sprouting of nuts in husks and delaying husk splitting (Madden *et al.*, 1969). An annual mean rainfall of 50 to 100 cm is recorded in the pecan growing tract of the USA and higher rainfall is generally associated with more fungal diseases. Hailstorms, floods and winds of high velocity are harmful to the trees.

## Soil

Good soil is an essential factor for the successful cultivation of pecan (Madden *et al.*, 1969). Although it is grown in varied soil conditions, sandy loam, deep, fertile soil, rich in organic matter, well-aerated and well-drained but moisture retentive soil is the best. Since pecan has a long taproot, a good soil depth of 6 m is desirable (Hanna, 1987). Symbiotic mycorrhiza and upper fibrous root system permeates the top one metre of the soil. It grows satisfactorily in the soil pH range from 5 to 8 but zinc deficiency is more severe on saline soils, thus soil pH of 6.4 is the best (Woodroof, 1979). Soil salinity and sodicity have been recognized as the major concerns where fields are being flood irrigated. Reclamation of these salt-affected lands is difficult due to low permeability. Native calcium sources can be effectively used to counter sodicity, can improve soil permeability and reduce amelioration costs. Application of polyacrylamide (PAM) at10 mg/l during the first irrigation improved water movement throughout the effective root zone that resulted in Na leaching. PAM application reduced soil electrical conductivity (ECe) in the top 60cm and sodium adsorption ratio (SAR) from 13.1 to 5.7 mmol 1/2 L-1/2 in the top 75 cm depths. Improved soil conditions increased pecan nut yield by 34 per cent in PAM-treated fields (Ganjegunte *et al.*, 2011).

Soil aeration is another important factor in pecan growing which has been studied in detail in the seedlings by Smith *et al.* (1989). They reported that restricted soil aeration induced partial stomatal closure and decreased leaf number, leaf area, and leaf, trunk and root dry weight. Restricted aeration increased N and P concentrations in the roots by decreasing their translocation to the leaves. Leaf concentrations of Ca, Mg and Mn were lower whereas that of Zn and Fe in roots higher due to lower aeration.

## Rootstocks

In absence of any clonal rootstock, the seedlings of various cultivars are presently used as rootstocks (Thompson, 1984). Nurserymen prefer the Riverside seedlings in Texas, while the Elliot and Curtis seedlings serve as rootstocks in southern states (Hanna, 1987). The rootstock effect on scion is very well known in fruit trees and the marked variation found in pecan cultivars may be ascribed to it because all seedlings rootstocks have great heterozygosity. Meilke (1981) reported the production of earliest and largest crops of pecan trees on Barton and Mohawk roots. Though statistically insignificant, Riverside rootstock increased the trunk diameter by 12 per cent (Hanna, 1987). The influence of rootstocks on the yield, yield efficiency and alternate bearing of Stuart and Schley scion cultivars was demonstrated by Wood (1989a). He also reported that, the yield was closely related to root starch concentration in January.

Since hickories and hicans are generally compatible with pecan, these have been experimentally tried as rootstock and interstock to test their adoptability and dwarfening effect. The information so far available is inconclusive. *Carya tomentosa* (mockernut) rootstock reduces both growth and productivity of pecan scion. The bitter pecan (*C. aquatica*) is adapted to poor drainage, flooding condition and low pH but it also decreases nut production (McKay, 1964). Several dwarfing rootstocks are under evaluation in the USA (Hanna, 1972; Jones, 1980). Wood (1989b) has identified the high yielding and alternate bearing reducing rootstocks. 'Riverside', 'Burkett' and 'Apache' are widely used as salt-resistant rootstock for orchards of west of central Texas. 'Riverside', 'Burkett', and 'Apache' are widely used as salt-resistant rootstocks for orchards of west of Central Texas. The following rootstocks are used in different US states (Table 10.4) (Thompson, 1990). Smith *et al.* (2001) observed that trees on 'Peruque', 'Giles', 'Colby', 'Apache' and native seedlings from Stillwater, Oklahoma and Brunswick, Missouri, USA were least injured by an exceptionally early autumn freeze on 8 and 9 October, 2000.

**Table 10.4: Rootstocks Used in different US States***

| *State* | *Cultivar* |
|---|---|
| Alabama | Elliott, Curtis, plus others |
| Arizona | Riverside and many others |
| Arkansas | Mainly natives |
| California | Riverside, Apache, VC1-68, plus others |
| Florida | Elliott, Curtis, Waukeenah, plus others |
| Georgia | Elliott, Curtis, plus others |
| Kansas | Giles, plus natives |
| Kentucky | Natives |
| Louisiana | Stuart, Moore, Elliott, Desirable, Candy, natives, plus others |
| Mississippi | Owens, Big Dan, Moore, water hickory |
| Missouri | Mainly natives |
| New Mexico | Riverside, Burkett |
| North Carolina | Cape Fear, plus others |
| Oklahoma | Riverside, Apache, Giles, plus others |
| South Carolina | Curtis, Stuart, Elliott |
| Tennessee | Gerardi, plus natives |
| Texas | Riverside, Apache, plus many others |

* Thompson (1990).

## Propagation

Pecan is commercially propagated by budding or grafting onto seedling rootstocks.

## Seed Propagation

Pecan seeds are recalcitrant in nature and the seeds showed higher germination, if shown within four days after harvest (Dalkle, 2013).The opinions differ about the stratification need for pecan nuts. According to Staden *et al.* (1976), shell act as a mechanical barrier to germination which can be overcome by germinating the nuts between 30° and 35°C. Caminada (1979) got 9.6 per cent germination on incubating the nuts at 30°C whereas 86 per cent germination resulted after stratification at 4°C for 60 to 90 days. Sometimes the nuts germinate while still on the tree. The germination is definitely higher and quicker at higher temperatures irrespective of stratification. Cold stratification of the nuts for 45 days followed by soaking in $GA_3$ (500-1000 ppm) for 48 h has given the best germination and subsequent seedling growth (Khalil *et al.*, 1982; El-Nabawy *et al.*, 1980; Knox and Smith, 1981). During the stratification, inhibitors gradually decreased and GA-like substance increased, reaching a maximum at the end of the 6-week stratification (El-Nabawy *et al.*, 1980). The larger nuts produce taller and thicker seedlings than smaller nuts (Herrera and Martinez, 1982). The seedling growth is dependent for the cotyledons for the first three weeks (Wetzstein *et al.*, 1983).

In order to facilitate transplanting and good seedling growth, deep, fertile sandy soils are preferred for nursery purpose. Good drainage is essential for soil aeration. Pecans do not stand wet-feet. The soil should be well prepared by cultivation. In late fall or early spring, the nuts are sown in rows, the spacing being determined by the interculture implements. The nuts are covered by about 10 cm thick soil layer. Since the seedlings are very succulent, enough care should be taken in carrying out interculture operations, *i.e.*, application of fertilizers and spraying of pesticides and fungicides. The blight disease is to be controlled by spraying Bordeaux mixture and the budworm by applying a suitable insecticide. Inferior small seedlings are removed during the first year of growth. Application of $GA_3$ (5000 ppm) in lanolin paste increases the stem girth, length and fresh and dry weight of plants (Khalil *et al.*, 1983). The plants remain in the nursery for 2 to 3 years in order to attain proper size for budding and grafting and subsequent growth but they are transplanted after one year by shortening the taproot to 20-25 cm length which enhances vigorous fibrous root system and high survival rate (Pichkhaya, 1987).

## Vegetative Propagation

### Budding and Grafting

Many budding and grafting methods have been tried for pecan propagation but patch budding and whip grafting are the most successful. Both methods are equally popular. The rootstocks of more than 2 cm diameter are usually patch budded and thinner rootstocks are whip-grafted but if rootstocks are more than 4 cm in diameter, inlay-bark grafting is adopted. The selection of a good scion-wood appears imperative for good success in all the methods. It should be straight, well developed, more than 2 cm in diameter with 5 to 8 cm space between the buds. The scion-stick should be approximately of the same diameter as the stock. Joolka and Rindhe (2000) evaluated methods and time of budding in Himachal Pradesh, India. Two-year-old seedlings of pecan cv. Burkett was used as rootstock and cv. Mahan as scion. Annular, forkert and chip budding were compared in different times. The highest bud sprouting was recorded in annular budding performed on 30 June, followed by in 20 July. Annular budding also showed highest survival percentage. Patch budding performed on 30 June also resulted high bud take success (Chandel and Ananda, 2002).

Maximos *et al.* (1979) investigated patch budding and reported 90-100 per cent bud take in August, whereas Bhardwaj (1983) got 75 per cent bud take in May. The operation should be accomplished by a double blade budding knife when bark peels easily in spring or summer season. The buds of preceding season are used for spring budding while well-developed mature buds of the current season are patch-budded in late autumn season. The bud is inserted more than 20 cm high on the stock from the ground and secured tightly with polyethylene or waxed cloth strips.

Whip grafting gives about 90 per cent success in late dormant season, the scion is grafted either on underground root of the stock in the field or on a root piece as bench grafting (Bolt, 1983; Gustafson and Storey, 1982; Sampaio and Barbin, 1980, 1981). Similar success has been reported by this method in softwood grafting on stem in August (Misra, 1985). If scions are smaller than the stock, matching of the cambium layers of stock and scion on one side of the graft is essential. Tongue grafting (cv. Mahan on cv. Burkett) performed on 3 March gave the highest bud sprouting compared with cleft and crown grafting. However, crown grafting gave the earliest bud sprouting (25.33 days) while survival rate was recorded higher in tongue grafting (Joolka *et al.*, 2001). In USA, whip grafting, bark grafting and four-flap grafting are the frequently used techniques for pecan grafting. Dormant scions, collected and stored for 60-120 days at 2°C before spring grafting season (Nesbitt *et al.*, 2002). JunNan *et al.* (2008) working with American pecan cultivars in China suggested that for good grafting, the rootstocks must be vigorous and healthy, and have a diameter at ground level of 0.8-1.5cm. The scions should be collected in winter, cut to a length of 20-25cm and cut ends should be treated with liquid wax at 90-100°C. Scion bundles should be buried in sand at 60 per cent humidity and stored at 5°C. The best grafting period was suggested as early April when the air temperature is 25-29°C for 20-25 days. The best position for side grafting was suggested as 3.0-3.5cm above the ground and the length of scion cut surface should be around 4.5-5.5cm.

## Cutting

Attempts to propagate pecan by hardwood cuttings have not been encouraging because the cuttings fail to establish and survive even after rooting (Hanna, 1987). In fact roots emerged from cuttings are thick and long instead of being fibrous. Allan *et al.* (1980) obtained 90 per cent rooting and 70-80 per cent survival when the shoots of topworked stumps were treated with IBA (1 per cent) and planted in a mist bed with bottom heat. The cuttings made from etiolated shoots gave 96 per cent rooting in mist after 5000 ppm IBA treatment (Rathore, 1985). Asraf (1977) reported 100 per cent rooting under mist in dormant cuttings after 3000 ppm IBA treatment when he inserted the cuttings between the cotyledons of germinating nuts from which the radicle and epicotyl had been removed. JianHong *et al.* (2014) noted higher rooting by quick dip treatment with IBA at 600mg/kg and NAA 900mg/kg. Annual (cuttings from current season's growth) rooted better than old cuttings.

Root cuttings (200 mm long and 10-15 mm in diameter) of 2-year Barton seedling when taken in dormant season and planted under 50 per cent shade gave 100 per cent rooting and 250 mm long shoot growth after one year (Bolt, 1983).

Air-layering in many cultivars using 3 per cent IBA was attempted by Grauke and Gel (1983) but they could obtain only 50 per cent rooting and 33.8 per cent survival.

### Stooling

Clonal propagation of pecans by stooling has been quite successful (Badiyala and Badyal, 1986; Sharma and Rosanghura, 1987; Wood, 1989c). Girdling and IBA application (3000 to 6000 ppm) are necessary to induce good rooting in stools. The mother stools should, however, be established from the cuttings or air-layers of the commercial cultivars before resorting to propagation by this method.

### Topworking

Pecan seedlings and inferior cultivars are commonly converted to improved cultivars as many orchards are established by grafting *in situ* and long life of the tree makes the venture more fruitful than planting young trees. The plants are beheaded in dormant season by cutting the limbs which are less than 10 cm in diameter but not more than two-third of the top should be removed at a time. The inlay bark graft covering the graft wound with aluminium foil and plastic bags has been found very satisfactory for topworking (Madden *et at.*, 1969). The early growing season is the proper grafting time. Precut dormant scions can be used as it has many advantages. If grafts fail, the new sprouts on the stubs can be patch budded. Likewise, native sprouts emerged on the thicker branches (more than 10 cm diameter) after cutting back should be budded in the summer or spring of the following year. A good care of the grafts is required after they start growing. All the sprouts on the stock below the scion should be removed.

## Micropropagation

Hansman and Novoa (1986) reviewed the literature on micropropagation of temperate nut trees. Smith (1977) initiated the work on micropropagation of pecan but he could only obtain callus growth. Significant progress in this regard was, however, made by Knox and Smith (1981) when they produced shoots and roots from the spring flush of the cultivar Riverside.

The role and doses of various plant regulators (BA, isopentenylaminopurine, IBA, IAA and GA) in *in vitro* propagation have been evaluated by Wood (1982). Successful micropropagation from greenhouse grown seedlings of the cultivar Desirable has been achieved by Hansman and Lazarte (1984) and Lazarte (1983, 1984). Shoot multiplication was obtained with nodal cuttings placed on a WP medium enriched with 2 per cent glucose and 3 mg IBA/l. The excised shoots were placed for rooting in test tubes containing WP medium with 2 per cent glucose and 3 mg IBA/*l*. The rooted shoots were then transferred to peat pellets covered with plastic caps. Later, these plantlets grew vigorously and got acclimatized to greenhouse conditions. Xiasyun *et al.* (2013) reported that nodal stem segments excised from the newly-grown shoots of the seedlings when cultured on modified MS media supplemented with 6-BA at 4.0mg/l +IBA at 0.01mg/l induced adventitious buds, with maximum induction up to 87 per cent. Modified ½ MS media with 6-BA at 2.0mg/l + IBA at 0.1mg/l was found suitable for tube shoot multiplication. Hassanen and Gabr (2013) in Egypt developed an efficient method for *in-vitro*-rooting. Stem segments of pecan were cultured on McCow's Woody Plant Medium (WPM) containing 3mg/l BAP for multiplication. After four weeks from culture, elongated shoots were excised. Shoots were splitted before culture, then shoots were cultured on ¼ and ½ strength WPM medium supplemented with different concentrations (0,1,2 and 3mg/l) of IBA and 1.5mg/l silver nitrate ($AgNO_3$). The best rooting (75 per cent) was obtained on half strength WPM medium supplemented with 3.0mg/l IBA + 1.5mg/l $AgNO_3$. The rooted plantlets were transferred to peat moss and sand (1:2 v/v). Eighty per cent plantlets were successfully acclimatized in greenhouse. Thus, the commercial micropropagation of pecan appears feasible. Somatic embryogenesis of pecan *in vitro* has been reported by Wetzstein *et al.* (1988).

## Planting

Since pecan root is long, deep ploughing besides the levelling and removal of extraneous material is required. The trees planted on soil chiselled (sub-soiled) 50-55 cm deep made 25 per cent, more growth than trees transplanted on non-chiselled soil (Madden, 1982). Late winter or early spring just before bud swelling is the best planting period. Because of slow development of fibrous roots, a considerable mortality is experienced on planting. Smith and Johnson (1981) obtained 100 per cent survival and good growth of transplanted trees on pruning the tops by 50 per cent and keeping the taproot 50 cm long. Laiche *et al.* (1983) did not find any significant difference between field-grown and container-grown plants five years after transplanting.

Thompson *et al.* (1982) compared the four methods of orchard establishment, *viz.* (i) planting pecan seeds in place, (ii) transplanting 1-year-old seedlings, (iii) transplanting 3-year-old seedling, and (iv) transplanting budded trees. Trees in treatments 1,2 and 3 were budded in the field. During the first 10 years, the yield was more in treatment 4 than any of the other treatments which showed little variation in yield and tree development. Treatment differences in yield and trunk measurements decreased over 10 years.

## Plant Density

Tree spacing depends on the cultivar, climate, soil and cultural practices to be followed. Pecan is a long-lived huge tree occupying large space on maturity. It is generally planted at a distance of 16-20 m where the soil is very fertile and rainfall is more than 75 cm. The distance can be reduced to 12 m if the soil fertility and rainfall are low. When the conventional planting distance of 16-20 m is adopted, temporary rows between the permanent rows at half-a-distance can be provided. As the permanent trees occupy the space the crowding begins, then the temporary trees should be removed in phases. Bearing pecan trees can be transplanted with a large tree spade when the trees and dormant. Transplants can survive with little or no-pruning (Wood *et al.*, 1990).

The response of pecan to high density planting has been found encouraging as the increased density of the trees give higher yield and return per unit of land and time (Witt *et al.*, 1989). In a comparison of high density orchard all establishment costs were recovered in 8 years in the former while it took 24 years in the latter for the recovery (Law *et al.*, 1980). According to Overcash *et al.* (1983), the closely spaced trees of precocious cultivars give more yield per hectare but less per tree than further spaced trees. The requirements of high density planting in pecan are planting of precocious and prolific bearing cultivars, annual pruning and improved orchard management including fertilizer, irrigation and control of pests and diseases. Cultivars such as Cape Fear, Summer and Choctaw which have the early bearing characteristics and bear nuts on internal branches instead of terminal branches are suitable to plant at closer spacing (10m x 10m). These trees should be pruned every year to confine them to their original spacing. Economic feasibility of increased density in pecan has been analyzed by Witt *et al.* (1989). The income-maximizing spacing's are 10.7x10.7 m and 13.7x13.7 m for the precocious cultivar Desirable and non-precocious cultivar Stuart, respectively, and annual pruning after the canopy closed gave the highest income for both the cultivars (Table 10.5).

Paclobutrazol has been found effective in controlling the growth of both young and large pecan trees (Wood, 1988a, b). When applied as a soil drench to 3-year-old trees of the cultivar Cheyenne, paclobutrazol (0.05 to 2.6 g/cm$^2$ trunk cross sectional area) promoted lateral branching and decreased tree growth, shoot extension and trunk and limb growth (Anderson and Aldrich, 1987). A single soil application of paclobutrazol (PBZ), uniconazole (UCZ) and flurprimidol (FPD) each at 132, 264 and 528 µmol/cm$^2$ trunk cross sectional area suppressed shoot elongation of young trees by 50-90 per cent for a period of three years (Wood, 1988c) but the yield was adversely affected if growth is reduced more than 60 per cent because it decreased the leaf area and increased leaf shedding due to shoot compactness. All the three chemicals have the commercial potential for controlling tree size.

In valleys on levelled land, the orchard is laid out by the square method, but contour planting is advisable on rolling lands. The holes should be made big enough to accommodate the entire root system after pruning. The roots should be set in the hole followed by the firm placement of soil. A basin needs to be prepared around the plant for watering purpose, which should be applied immediately after planting and thereafter frequently during summer season depending on the occurrence of rainfall. Fertilizer application is not required at planting time but it can be done later in the summer season if the plants are not making satisfactory growth. The stem should be whitewashed with white latex in water (50:50) to prevent sun scalding (Madden, 1982).

## Training and Pruning

As the pecan trees advance in age, they become larger and get crowded making pruning, spraying and harvesting operations difficult. Pruning in pecan is, therefore, desirable but is often neglected. Training begins with the planting when about one-

**Table 10.5: Per Hectare Net Present Value ($/ha) of Returns after Pruning from 'Desirable' and 'Stuart' Pecan Orchards under Various Pruning Practices***

| | | *Per cent of Crown Diameter Removed* | | | | | | | | |
|---|---|---|---|---|---|---|---|---|---|---|
| *Cultivars*** | *Discount Rate (Per cent)* | *30* | | *35* | | *40* | | | *20* | *10* |
| | | *Prune Once* | *Prune Twice* | *Prune Once* | *Prune Twice* | *Prune Once* | *Prune Twice* | *None* | *Biennial Pruning* | *Annual Pruning* |
| Desirable | 3.85 | 11,510 | 13,536 | 12,117 | 12,830 | 11,213 | 12,557 | 7157 | 13,747 | 14,981 |
| | 6.0 | 4,747 | 6,016 | 5,104 | 5,557 | 4,513 | 5,367 | 2063 | 6,188 | 6,980 |
| Stuart | 8.85 | 4,926 | 6,030 | 4,516 | 5,701 | 4,291 | 5,546 | 1808 | 6,139 | 6,551 |
| | 6.0 | -284 | 337 | -535 | 124 | -672 | 17 | -2186 | 394 | 625 |

**Desirable' is spaced 10.7x 10.7m and 'Stuart' is spaced 13.7x l3.7m.

* Witt *et al.* (1989).

third to one-half of the top is removed. From the second year, the plant is trained to central leader system and branches are allowed to emerge as high as 1.5 to 2 m from the ground on the trunk. No narrow angled crotch should form between two main links on the tree.

Smith and Hinriches (1980) described the training of high density plantings of the cultivar Western. Initially, the trees were planted at 5.83 x 8.83 m spacing and trained to central leader system which after 8 years pruned to a hedgerow 2m wide and topped at varying heights from 3.23 to 6.66 m. Nut yield increased with tree height to a maximum of 6 m and declined thereafter at 6.66 m.

The experiments have proved that pecan pruning does not adversely affect the production or nut quality (Malstrom *et al.*, 1982; Witt *et al.*, 1989). On the contrary, judicious annual pruning increases production besides controlling the tree size. Since pecan is a terminal bearer, it requires only light to moderate pruning. Heavy pruning results in extremely vigorous growth with no production on compensatory growth for several years (Hanna, 1987). Dried and broken branches should be removed and overcrowding branches be thinned.

## Nutrition

Some aspects of pecan nutrition should clearly be understood before taking decisions about fertilization programmes. Pecan responds very slowly to fertilizers, *i.e.*, the effects of nutrients applied in any one year will be reflected in growth and yield of the trees after two or three years. For the same reason, tissue analysis is not very helpful in depicting the nutritional condition of the tree. The yield does not decrease even at leaf N levels lower than the established threshold levels (Malstrom *et al.*, 1982). Filling of the nuts is negatively correlated with the number of nuts on a tree. A fertilizer programme which increases the number of nuts will adversely affect their filling and thus quality. The nutrients should not only be available for nut set but throughout the period of their development as well (Woodroof, 1979). Hence pecan nutrition should be managed for a number of years rather than only for a few years.

## Role of Nutrients

### Nitrogen

N deficiency symptoms in pecan have been described by Alben *et al.* (1942). Light green to yellowish green foliage, weak shoot growth and earlier fall defoliation are the main symptoms of N deficiency. In a 4-year trial of N fertilizer application on sandy loam and clayey soils, the deficiency symptoms appeared on unfertilized sandy soil when tissue N level was lower than 2 per cent (Malstrom *et al.*, 1983).

Smith *et al.* (1985) applied 0, 56, 112 and 224 kg N/ha for six consecutive years to the mature trees of Western cultivar and got a rapid response. The yield was increased by 56 kg N $ha^{-1}$, but not by 112 to 224 kg significantly, above 56 kg N $ha^{-1}$ (Table 10.6). They, however, justified the higher N rates on the basis of better tree appearance, shoot growth and production of new shoots than 56 kg N rate. Leaf N concentration was not a good indication of the total N absorption because the difference was relatively small between no N and highest N rate (2.13 per cent to 2.35 per cent). Studies with $^{15}N$ revealed that early spring growth, flowering, and embryo development used fertilizer-N applied in the previous year, as well as that applied during the current year (Kraimer *et al.*, 2001). Kraimer *et al.* (2004) further reported that late-season application of N-fertilizer during the kernel filling stage was started in perennial tissues (root and wood) for use in the following year; very little was used for current seasons growth of annual tissues (leaf), shuck and nut. Increased accumulation of perennial storage N caused during a heavy cropping on-year and may moderate the alternate-bearing trend in pecan by providing a greater reservoir of N in the following year. Smith *et al.* (2004) did not found any major differences in the percentage of fruiting shoots or cluster size associated with N rate (75 or 100kg/ha) or applying N as single (March) or split application (March and June).

**Table 10.6: The Influence of N Rate on Yield of 'Western' Pecan***

| *N Rate (kg/ha)* | *1979* | *1980* | *Yield (kg/ha) 1981* | *1983* | *Cumulative Yield* |
|---|---|---|---|---|---|
| 0 | 460 | 531 | 1333 | 577 | 2901 |
| 56 | 569 | 689 | 1841 | 862 | 3961 |
| 112 | 598 | 657 | 1792 | 856 | 3903 |
| 224 | 445 | 823 | 1861 | 820 | 3949 |
| Linear | NS | ** | ** | NS | ** |
| Quadratic | ** | NS | * | ** | ** |

Significant at 5 per cent (*), 1 per cent (**), and non-significant (NS).

* Smith *et al.* (1985).

Leaf tissue analysis showed that, N concentration of young leaves was as high as 2.5 per cent and in older leaves had dropped to 1.9 per cent (Driver *et al.*, 1984). Walworth *et al.* (2006) reported that leaf N concentrations decreased relatively constantly at a rate of 0.003 per cent per day, from flowering to fruit maturity and the overall change was from 2.97 to 1.80 per cent. Smith *et al.* (1985) recorded yield increase upon addition of the first 56kg N/ha, but leaf N concentration continued to increase as applications increased to 224 kg/ha. The maximum leaf N concentration reached in this study was 2.39 per cent, and the authors suggested threshold limit of 2.25 per cent.

Elmore and Polles (1980) studied the effect of N fertilization on amino acid composition of nutmeat and found that ammonium nitrate had increased the glutamate and proline content more than nitroform (slow release).

**Phosphorus**

Alben *et al.* (1942) described the symptoms of P deficiency as unthrifty growth and change of colour from dull green to yellowish green in severe cases. The shoot growth is short and slender and older leaflets become necrotic as the deficiency accentuates. But P deficiency symptoms have been rarely observed under field conditions. Sparks (1986a) compared the effect of root-applied P with foliar-applied P under greenhouse conditions and found that root application was better than foliar application in preventing P deficiency symptoms and increasing its concentration in the leaf, trunk and root. P sprays also caused leaf scorching. Sparks (1988b) obtained increased nut growth and decreased leaf scorch and premature defoliation when 2.2kg $H_3PO_4$/tree was applied in the field. Leaf P value for maximum nut growth and minimum leaf scorch were greater than 0.14 and 0.16 per cent, respectively.

Soil band application of P alleviated leaf deficiency symptoms, increased leaf P concentration, and improved flowering (Smith and Cheary, 2013).

**Potassium**

Deficiency symptoms of this element begin with the loss of green colour from the margins towards the mid-vein of the leaves which in acute cases appear as marginal chlorosis or leaf scorch (Alben *et al.*, 1942). Trees low in K, even without showing any, symptom, get severely injured or killed by low winter temperatures (–7.2°C) and the yield from the damaged trees also declines (Sharpe *et al.*, 1952). Sparks (1977a) found that heavy fruiting induced leaf scorch and defoliation.

Pecan trees are very inefficient in absorbing soil-applied K. Surface-applied K usually increases residual soil K by moving up to a depth of 70 cm (Smith *et al.*, 1985., Worley *et al.*, 1975). Smith *et al.* (1987) tried foliar application of K but without success.

Smith *et al.* (1985) found that K application at 0, 93 or 186 kg/ha, significantly increased the yield of mature Western cultivar trees but not in all years (Table 10.7). K application also increased significantly leaf K concentration but no relationship existed between this concentration and yield.

**Table 10.7: The Influence of K Rate on Yield of 'Western' Pecan***

| *K Rate (kg/ha)* | *Yield (kg/ha)* | | | | *Cumulative Yield* |
|---|---|---|---|---|---|
| | *1979* | *1980* | *1981* | *1983* | |
| 0 | 453 | 699 | 1698 | 724 | 3474 |
| 93 | 521 | 661 | 1661 | 769 | 3612 |
| 186 | 582 | 673 | 1779 | 852 | 3886 |
| Linear | ** | NS | NS | NS | * |

Significant at 5 per cent (*), at 1 per cent (**), and non-significant (NS).

* Smith *et al.* (1985).

Wood *et al.* (2010) studied the influence of elevating tree potassium on fruit drop and yield of pecan cv. Desirable in two locations in which generally accepted foliar analysis standards indicate satisfactory tree K nutritional status (*i.e.*, 0.75 per cent or greater to 2.5 per cent K/dry weight). Multiyear field studies of two such orchards found that elevating leaf and fruit K concentrations through soil banding of potash over drip irrigation emitters: (i) increased fruit retention by reducing stage II (June drop) fruit drop; (ii) increased in shell-nut yield; and (iii) increased nut quality by increasing kernel percentage. Potash applied through soil banding elevated foliar and fruit K concentration by ~ 0.1 per cent to 0.4 per cent units within a few months post-application depending on the amount applied; however, the beneficial effects of a single potash soil band application was found to diminished after the first year. A comparison of the K concentration of retained fruit versus abscised fruit during the stage II fruit drop window found that retained fruit possessed endogenous K concentration of 1.2 per cent to 1.7 per cent (dry weight basis) in one orchard and 1.5 per cent to 1.9 per cent in another orchard, whereas aborted fruit possessed K at 0.65 per cent to 1.2 per cent and 0.75 per cent to 1.2 per cent, respectively, these establishing ~1.25 per cent K as a "drop threshold". Smith and Cheary (2013) stated that soil band application of K require 3 or more years before a positive response could be observed. Higher rates of K-application increased the percentage of number-1 kernels.

### Calcium

Deficiency of Ca in pot culture causes yellowing, bronzing and reduction in leaflet size. As the deficiency persists, the leaflets come closer and overlap due to shortening of rachis (Alben *et al.*, 1942). Cole and Hunter (1962) reported that leaves from Ca-deficient trees were abnormally high in Zn, and the soil around the trees was acidic (pH 4.7 to 4.9). Liming improved the growth of the trees and reduced Zn content of trees.

Lime induces the leaching of P and K by increasing the soil pH which is conducive to plant growth and fixation of nitrogen by leguminous cover crops. Lime is generally surface applied to raise the soil pH and it along with NPK fertilizer increases top and root growth. White (1982) studied the effect of liming top loamy soil and subsoil sandy clay on root growth. Subsoil liming was more effective than topsoil liming. On raising the pH of the subsoil from 5.1 to 6.5, growth of taproots increased by 64 per cent. Subsoil application of lime at 2 and 3 meq Ca/100g soil (pH 6.0 and 6.5) accentuated the yield of small feeder roots 3- and 6-fold, respectively, over the control (subsoil pH 5.1) (Martin *et al.*, 1980).

### Magnesium

Marginal yellowing, interveinal chlorosis and necrosis are the characteristic symptoms of Mg deficiency which has often been observed on sandy soils and is attributed to leaching (Alben *et al.*, 1942).

Application of magnesium sulphate at the rate of 34 kg/ha annually for 5 years or as a single dose of 224 kg/ha increased leaf Mg level after 5 years of initial application (Worley *et al.*, 1975). Dolomite raised soil pH and soil Mg but not leaf Mg level. Single foliar spray of $MgSO_4$ or $Mg(NO_3)_2$ also did not increase leaf Mg. Sparks (1986b) reported that foliar applied $MgSO_4$ (0.24 per cent to 0.96 per cent) accentuated leaf concentration of Mg, and suppressed but could not prevent the deficiency symptoms. Root-supplied Mg enhanced the leaf Mg and growth more than the foliar sprays but the latter alleviated the nutritional imbalances induced by Mg deficiency.

**Zinc**

Alben *et al.* (1932) suggested Zn deficiency as the cause of 'rosette' which is a major problem in pecan production on sandy soils. Zn deficiency symptoms appear on the affected plants in the form of bronzing and crinkling of the leaflets, reduction in the leaflets size and shortening of internodes which give the appearance of a rosette of leaves on the shoot terminal. Kim *et al.* (2002) observed distorted leaf growth and severe rosetting of shoots under Zn deficiency. Interveinal mottling followed by interveinal chlorosis, interveinal necrosis and marginal curling. As the deficiency accentuates, shoots and branches die back, tree growth declines, nuts become small and production ceases, and ultimately the tree gets killed. Some limbs on a tree are more affected than the other. Zn deficiency reduces carbonic anhydrase activity which can serve as a test for determining its deficiency in pecan leaves (Snir, 1983). Kim *et al.* (2002) stated that in pecan, there are genotypic differences in Zn efficiency. The cv. Stuart shows Zn deficiency more frequently than cv. Curtis. Sparks and Payne (1982) tested the reliability of the satisfactory range of 50-100 ppm Zn in the leaflets and found the occurrence of the deficiency symptoms only when Zn levels in the leaflets had fallen below 40 ppm. Leaves of high yielding pecans of a data bank of 3000 observations from fertility trials with different cultivars (> 58kg nuts/tree) showed an average Zn content of 126 µg/g in leaves, with a cv. of 60 per cent (Beverly and Worley, 1992). Pond *et al.* (2006) reported that high-yielding pecan cv. Western Schley trees contained 174 µg/g of Zn with a cv. of 65 per cent.

Zn deficiency has been mostly corrected either by applying Zn to the soil or its spray on the foliage. It is commonly applied as zinc sulphate which gives quicker response than other compounds. Zinc oxide can be used in soil but its availability is rather slow. Zinc chelates and zinc-coated nails have also been used to correct the disorder. Upadhyay and Badyal (2008) suggested spraying of 0.5 per cent $ZnSO_4$ to increase nut yield and improving nut quality. Medina and Chavez (1999) observed that Zn-defecincy is common when grown on alkaline and Calcareous soil. The Zn concentration in foliage should be maintained above 50mg/kg in such situation by foliar spraying of Zn. They suggested more than two sprays of Zn for cv. Western and more than three sprays for cv. Wichita.

Since zinc is mobile, it is applied directly to the soil surface singly or combined with regular fertilizer programme. Soil application of Zn is helpful only on acidic soils and a higher dose is required on heavier than on light soils. Payne and Sparks (1982) compared the broadcast and band application methods and found more rapid tree response from the former than the latter. Leaf Zn level of 50-year-old Stuart and Schley trees rose to sufficiency level with broadcast of 40 kg $ZnSO_4$/ha but 3.2 kg $ZnSO_4$/tree was required in band application to create the same effect. In alkaline soil, soil banding of Zn-EDTA at 19kg Zn/ha can very effectively increase Zn uptake if placed in close proximity to the tree roots (Nunez-Moreno *et al.*, 2009). As soon as leaf zinc reaches the optimum level, its application should be discontinued to avoid toxicity.

Zn application on calcareous soil is ineffective and on other soils it is available slowly being unable to cure severe deficiency problems causing permanent injury to the framework of the tree. Foliar Zn application has, therefore, been recommended to alleviate the severe rosette promptly with $ZnSO_4$ or $Zn(NO_3)_2$ at the rate of 86g/100 l water. But Smith and Storey (1979) suggested a reduction in the rate as they got increased yield with $Zn(NO_3)_2$ at 10.89/100 l in comparison with the recommended rate. Zinc can also be sprayed along with the Bordeaux mixture (Sharpe and Gammon, 1958). Alben (1962) obtained best result with the application of Zn-chelate at 3 lb/100 gallon water followed by zinc sulphate at 2lb/100 gallon. According to Worley *et al.* (1980), the most efficient method of correcting rosette disorder is by pressure trunk injection using 8 lb solution per tree at 1g $ZnSO_4$/2.5 cm of trunk circumference.

**Manganese**

Mn deficiency has been characterized as 'mouse ear' or 'little leaf' which usually occurs at high soil pH in coastal areas and dooryard plantings (Gammon and Sharpe, 1956). The shortening of mid-vein of the leaflets makes them round, wrinkled, turning into a cup (Shear and Faust, 1980). The deficiency usually occurs when the leaf Mn level falls below 100 ppm in comparison with the normal range of over 300 ppm. To correct this condition, manganese sulphate (1 to 4 kg/tree) and ammonium sulphate (1 to 7 kg/tree) can be applied to the soil by evenly broadcasting under the tree canopy. The quantity of these compounds will vary with tree size and severity of symptoms (Woodroof, 1979). In the northern region of Coahulia, Mexico where soil is alkaline, limits to availability of Mn, Cu and Zn. Medina *et al.* (1999) suggested spraying of Mn as $MnSO_4$ at 2000 mg/l (for young trees) and 4000mg/l (for trees in production). They suggested a single spray on 18 April or two split sprays on 18 April and 9 May to increase foliar Mn levels. Smith *et al.* (2001) suggested three foliar application of $MnSO_4$ at 2.15kg/ha plus a surfactant, beginning at bud break, then twice at 3-week interval to correct Mn deficiency.

**Boron**

In case of B deficiency, the leaves become small and uniformly chlorotic and the shoots short, slender and die eventually (Alben *et al.*, 1942). The symptoms of B deficiency resemble to the rosette of pecan. Generally, B deficiency is not observed in the orchard. Water-stage fruit split (WsFs) is a relatively common and often major problem of certaion pecan cultivars. Wells and Wood (2008) advocated spraying of B to reduce WsFs. Nickel (Ni) sprays also found to reduce WsFs in Cv. Summer Fruit. The normal leaf B content ranges from 10 to 90 ppm. Sharpe and Gammon (1958) suggested application of 10 to 15 kg borax/ha to the soil along with the fertilizer to encounter the deficiency of B.

**Nickel**

Mouse-ear (ME) is a potentially severe anomalous growth disorder affecting pecan trees. It was found in second generation sites throughout much of the Gulf Coast Plain of the south eastern USA. The cvs. Desirable and Elliot treated either Cu or Ni indicated that foliar application of Ni corrects ME. A single foliar spray of nickel sulphate in mid-October prevented ME symptoms. These studies indicated that ME in pecan is due to Ni deficiency at bud break (Wood *et al.*, 2004).

## Fertilizer Recommendations

Worley *et al.* (1974) applied NPK fertilizer (0-1344 kg/ha) for a period of 10 years to the mature 'Stuart' pecan trees growing on loamy sandy soil but without any appreciable response. The highest yield was, however, recorded with 448 kg/ha per year (Table 10.8). Increasing cost of synthetic fertilizer dramatically reduces the profit margin of pecan producers of USA. Wells (2012) evaluated clover (*Trifolium incarnatum*) and poultry litter as alternative nitrogen sources. Results revealed that both poultry litter and clover provided adequate nutrition for pecan. This was reflected pecan leaf concentration, yield and alternate bearing. Poultry litter and clover may also provide additional benefits for pecan nutrition, primarily through the addition of organic matter which enhance the orchard soil environment (Wells, 2011). Poultry litter with and without clover can also provide many additional nutrients required by pecan trees, including P, K, Fe, Cu, and Zn. Where clover is used in the orchard, additional K application may be necessary (Walls, 2012).

In a field trial comparing the effects of calcium ammonium nitrate at 0, 2.4 or 4.8 kg, superphosphate at 0, 3.0 or 6.0 kg and potassium chloride at 0, 0.75 or 1.50 kg per tree per year in all possible combinations, Koen *et al.* (1978) obtained the highest yield per tree with N at the highest level and P and K at intermediate levels for which the corresponding leaf N,P and K levels were 2.59, 0.13 and 1.21 per cent, respectively. Woodroof (1979) recommended the NPK ratio of 8-1-3 for pecan.

## Tissue Analysis

In conventional sampling, 100 pairs of leaflets are picked from the centre of the middle leaves of the current season's growth for leaf analysis. O'Barr *et al.* (1984) compared the conventional sampling technique with another in which all leaflets on a leaf were used but could not find any significant difference between the two sampling techniques. The levels of various nutrients at different nutritional status in leaf of pecan trees are given in Table 10.9 (Westwood, 1978). One of the most important techniques in nutritional diagnosis in plant tissue analysis is the Diagnosis and Recommendation Integrated System (DRIS) is used as an alternative for its interpretation. DRIS norms were developed for pecan tree cv. Western with 68 observations from leaf analysis and yield per tree (Medina-Morales, 2004). The data bank information was taken from 670 trees from 66 orchards in 3 regions of northern Mexico, Comarca Lagunera, southern Chihuahua and northern Coahuila. The DRIS norms were tested in 2 experiments in which yield and foliar nutrient concentrations were evaluated when fertilizers were applied. These norms were the mean foliar concentrations and the coefficient of variation of the 45 relationship among 10 nutrients (N, P, K, Ca, Mg, Fe, Mn, Zn, Cu and B) of a high yield population (more than 60 kg/tree). The DRIS norms effectively predicted the results obtained in the field in the 2 Experiments.

The seasonal analysis of pecan leaves from fertilized trees revealed that N level dropped from about 2.6 per cent in June to about 2.1 per cent in September (Hammer and Hunter, 1949). Because of diminishing return effect of N uptake to supply, leaf N range associated with cardinal growth points are relatively narrow (Sparks and Baker, 1975). Leaf K content also declined steadily from about 1 per

**Table 10.8: Annual Yield of 'Stuart' Pecans as Affected by Rates of 10:4.4:8.3 Fertilizer-Kennedy Grove. Tifton, Ga.***

| *Fertilizer Rate (kg/ha)* | *Yield (kg/tree)*x* | | | | | | | | | | *Cumulative Yield Y* |
|---|---|---|---|---|---|---|---|---|---|---|---|
| | *1962* | *1963* | *1964* | *1965* | *1966* | *1967* | *1968* | *1969* | *1970* | *1971* | |
| 0 | 0 | 66a | 27a | 38a | 52a | 21a | 12a | 55a | 10a | 78a | 359ab |
| 448 biennially | 0 | 70a | 23a | 37a | 57a | 18a | 20a | 54a | 11a | 99b | 389ab |
| 448 annually | 0 | 79a | 29a | 40a | 44a | 15a | 13a | 79b | 64a | 105b | 468b |
| 896 annually | 0 | 67a | 22a | 38a | 47a | 18a | 15a | 65ab | 8a | 94ab | 374a |
| 1344 annually | 0 | 61a | 24a | 52a | 53a | 16a | 22a | 66ab | 16a | 98ab | 408b |

x-Yield means with the same letter within a column are not significantly different at 5 per cent.

y-Cumulative yield is not always sum of year yield due to rounding and omission of some trees from cumulative totals because of nut loss due to squirrels.

* Worley *et al.* (1974).

**Table 10.9: Nutrient Levels in Pecan Leaves Indicating the Range from Below Normal to Excess***

| *Nutrient Level* | *Per cent Dry Weight* | | | | | *ppm Dry Weight* | | | | |
|---|---|---|---|---|---|---|---|---|---|---|
| | *N* | *K* | *P* | *Ca* | *Mg* | *Mn* | *Fe* | *Cu* | *B* | *Zn* |
| BN | 1.6 | 0.4 | 0.08 | 0.2 | 0.18 | 140 | 40 | 1 | 40 | 10 |
| N | 2.3 | 1.0 | 0.12 | 0.7 | 0.30 | 200 | 75 | 4 | 60 | 18 |
| AN | 2.8 | 1.5 | 0.30 | 1.5 | 1.0 | 500 | 150 | 50 | 100 | 100 |
| EX | 3.0 | 2.0 | 0.70 | 3.0 | 2.0 | 1000 | 300 | 100 | 600 | 200 |

BN: Below Normal; N: Normal; AN: Above Normal; Ex=Excess.

* Westwood (1978).

cent, in June to about 0.35 per cent in September (Hammer and Hunter 1948). On the basis of sand culture experiments, Upson and Sparks (1969) considers leaf K level as a better indicator of its availability than the levels in roots or stems, and if leaf K level is lower than 1 per cent its deficiency will occur. The minimum acceptable leaf K level is 0.75 per cent. The annual variation in leaf K level is inversely related to annual rainfall patterns. Peer and Kessler (1984) proposed analysis of the shucks as an indication of the K status of the tree. Westwood (1978) suggested the normal, excess and below and above normal nutrients content of pecan leaves (Table 10.9).

Diver and Smith (1984) investigated the influence of fruit development on seasonal elemental concentrations and distribution in fruit and leaves of pecan. Leaf concentrations of Ca, Mg, Zn, Fe and Mn were more on fruiting than vegetative shoots sampled from 15 May to 15 October. N was higher on the fruiting shoots for 10 weeks and thereafter, higher on vegetative shoots. Leaf K content was lower on fruiting than on vegetative shoots after 2-5 June. Fruiting had no influence on leaf P values. Fruit N, P, K and Zn increased slowly up to 10 weeks from full bloom and then fell until maturation.

Kim and Wetzstein (2005) studies the changes in carbohydrates and nutrients in pecan related to development changes during a growing season. They reported that during leaf expansion, foliar levels of fructose and glucose declined from 12mg/g dry weight (DW) to 5mg/g DW and from 7mg/g DW to 3mg/g SW, respectively. After fruit maturation, stems exhibited an increase in sucrose, from 15mg/g DW to 37mg/g DW and a marked decrease in starch, from 59mg/g DW to 8mg/g DW. Seasonal variations in elemental content indicated developmental stage-related nutrient movement. During leaf development, all nutrients accumulated in leaves, with more than a 70 per cent increase in N, K, Ca, Mg, S, B and Fe content. During fruit development, the foliar content of many nutrients decreased, suggesting fruit are a strong sink for essential elements, as well as carbohydrates. However, nutrient changes was found to differ with the stage of fruit development. Foliar content of Mg and Zn decreased by 23 per cent and 24 per cent, respectively during cotyledon formation; while P, K, S, B, Cu and Mn levels decreased during reserve accumulation by 4, 15, 7, 12, 8 and 10 per cent, respectively. During the period leading-up to leaf abscission, leaf nutrient resorption was observed with decreased content of N (6 per cent), P (14 per cent), K (7 per cent), Ca (7 per cent), Mg (10 per cent) and S (3 per cent).

## Irrigation

Water requirement of pecans is higher than other nut fruits. Almost all growth and productivity parameters in pecan have benefited from irrigation. Supplemental water treatments significantly increased kernel yield, nut weight and diameter, kernel specific gravity, oil content and appearance (Heaton *et al.*, 1982). The yields were 25-30 per cent higher than the control plot (Daniell, 1982). Stein *et al.* (1989) reported increased yield, nut size, trunk diameter and kernel percentage due to summer and fall irrigation. Interestingly, they reported the need of water even before shuck opening as late season irrigation reduced sticktights and viviparous nuts. Godoy-Avila and Lopez-Montoya (2000) suggested 3 or 4 irrigations at

14 days intervals during kernel growth. This resulted high kernel weight and quality. It also promotes normal shuck opening. The fruit water potential was suggested to monitor for scheduling irrigation and then leaf water potential (Godoy-Avila *et al.*, 2005a). Sparks (1989a) found that drought stress induced fruit abortion particularly during third fruit drop in pecan. These experimental evidences suggest that pecan needs the maintenance of proper soil moisture all over the year. Besides the arid and semi-arid regions where irrigation is essential, it is also beneficial in humid regions in increasing the nut yield and quality (Daniell, 1979).

The common methods of applying water are flooding, furrows, sprinklers and drip irrigation. When water is available in plenty and the land is fairly levelled, the flood and furrow irrigation systems are used which are quite popular all over the world. Since high water losses occur in these systems due to evaporation and deep percolation and irrigation water has become a scarce commodity, sprinkler and trickle irrigation systems are gaining popularity day by day. All types of soils and more area can be irrigated by these systems because of higher irrigation efficiency. In pecan orchards, low angle nozzles of sprinkler irrigation system operate under foliage without wetting it. Pecan trees growing in course soil with a deep water table (>30m) and irrigated at 50KPa have an effective root zone of about 140-150cm (Sorensen and Jones, 1999).

Subsurface irrigation with six laterals required 30.6 per cent less water than flood irrigation and application of subsurface irrigation during nut filling stage increase 24 per cent nut production compared to flood irrigation. The critical period of high sensitivity to water stress occurred during the first 18 days of the nut filling stage. (Godoy-Avila *et al.*, 2005b).

The trickle or drip irrigation has many advantages over the other systems. It makes possible irrigation of twenty to forty per cent more land with the same amount of water, does not interfere in cultural operations and harvesting, reduces N and K application rates without significantly reducing growth or leaf nutrient levels and helps the use of systematic insecticides to pecan trees for controlling aphids and mites (Aitken and Gorsuch, 1985; Daniell, 1979). In this system, only about 25 per cent of the root system is provided with adequate soil moisture to prevent moisture stress of the trees. To determine the quantity of water required in trickle irrigation, Daniell (1979) suggested the use of class A pan and derived an equation for weekly water requirements to irrigate pecan trees. Using weather, soil and crop data, Mizamoto (1984) developed a computer programme adaptable to microcomputers for predicting soil water storage and irrigation needs.

Kallestad *et al.* (2008) developed a simple, practical irrigation scheduling tool specifically for flood irrigation pecan production in Mesilla valley, New Mexico. The information presented in the tool was derived using 14 years of archived climate data and model simulated consumptive water use. Using this device, farmers can estimate the time interval between their previous and the next irrigation for any date in the growing season.

## Soil Management

Pecan has a long growing season, meaning thereby that soil moisture and nutrients must be available for a considerably long period to ensure good plant growth and productivity. Adequate soil moisture is, particularly, essential during the rapid enlargement and filling of nuts. Since cover crops compete with fruit plants for moisture and nutrients and hamper cultural operations, the clean cultivation has been found better in pecan than the other soil management systems (Woodroof, 1979). Kennedy (1981) compared three management systems, *viz.*, disking, mowing and untreated control on clay soil and reported that periodic disking significantly increased tree height, diameter and survival by making available more moisture. Disking also resulted in significantly higher Na and Ca and lower P and Mg concentrations in the foliage than the other treatments. Likewise, Allan (1987) reports the best yield and growth (trunk circumference) of the trees of five cultivars under clean cultivation followed by cover crop and grass cover systems in decreasing order, respectively. A practice of continuously long cultivation, however, reduces the water holding capacity of the soil by decomposition of the organic matter. Furthermore, the leguminous cover crop enriches the soil by nitrogen addition. The clean cultivation conserves the soil moisture by weed control which can also be achieved by the use of herbicides. Consequently, the soil management programme in pecan orchards assumes great significance and should be adopted with due care.

## Cover Crops

The cover crops are grown during summer and winter seasons, especially on rolling topography to check soil erosion and leaching of nutrients. Summer cover crops are taken only in young orchards in which bearing has not commenced. Many crops can be sown as cover crops but legumes are more beneficial. The crops demanding more water and nutrients and occupying more space, such as, cotton and corn, are unsuitable as cover crops. The common summer cover crops are soybeans, velvet beans, cowpeas and *Crotalaria* spp., whereas arrow leaf and clovers are winter cover crops. Before sowing these crops, it must be ensured that soil has enough calcium and magnesium with a pH around 6.5.

The cultivation of winter cover crops is more essential because they utilize excess moisture and nutrients in winter season without competing with the fruit trees and releasing the nutrients later after decomposition. Apart from a single crop, combinations of crops, such as, vetch and peas, vetch and rye or oats, and peas and oats, have been found more satisfactory.

When planted for the first time, the seed of winter legumes is required to be inoculated with the correct strain of nodule bacteria for the best result (Woodroof, 1979).

## Weed Control

With the increasing cost of cultivation, the use of weedicides is gaining popularity. Pre-emergence and contact herbicides are generally used for controlling the weeds. Diuron and Simazine (2 to 5 kg/ha) are widely adopted to control grasses and broadleaf weeds as pre-emergence weedicides but should not be used before the pecan plants have established for three years. When a single herbicide is used, perennial weeds can flourish which need to be checked by the use of contact herbicides, such as, paraquat or glyphosate. McEachern and Storey (1984) got satisfactory control of pecan weeds with glyphosate (roundup) in a 10-year trial. Paraquat (1 kg/ha) has long been a reliable weedicide of pecan orchards which should not be allowed to come in direct contact with green stem and foliage. Smith *et al.* (2005) also suggested to use glyphosphate. Limited contact of glyphosate with the lowest 5-8cm of the trunk of the young pecan trees, which usually occurs during conventional orchard weed management, is unlikely to result in adverse effects on young pecan trees (Foshee *et al.*, 2008). Palmer amaranth (*Amaranthus palmeri*), however, showed resistant to glyphosate in New Mexico. Trifluralin and metolachlor showed control of glyphosate resistant weeds (Mohseni-Moghadam *et al.*, 2013). Application of a single herbicide may lead to a build up of tolerant weeds. It is, therefore, advisable to use a mixture of herbicides and similarly two herbicides can be used in rotation. Semi-parasitic evergreen mistletoe (*Phoradendron flavescens*) is an increasingly serious weed causing loss of nut yield and tree vigour in pecan orchards of the southeastern USA. Spot treatment of mistletoe clusters with 2,4-D at 1.2 to 2.4g/l (plus 2 per cent cropoil), 2-3 weeks before bud break, gave effective long-term control of mistletoe (Wood and Reilly, 2004).

In view of many limitations of cultivation, the sod-herbicide system has been suggested giving the advantage of both cultivation and sod culture (Daniell, 1979). The weeds are controlled by using herbicides 2 m each side of the row while sod system is maintained between the rows. The weeds competing with the trees are eliminated by the weedicides without mowing, thus the mechanical injury to trunks and feeder roots does not occur. The weed-free strip facilitates the mechanical harvesting and the sod reduces soil erosion.

## Flowering and Pollination

### Flowering

The pecan trees begin fruit production in a period of 2 to 6 years depending on the cultivar and vegetative growth. The staminate and pistillate flowers are borne separately on the same tree and organized into catkin and spikes, respectively. The pecan has compound buds enclosing floral and mixed buds in separate bud scales but within a common outer scale covering (Sparks, 1986b). On developing, floral buds produce catkins but mixed buds grow either into a vegetative shoot or in a single pistillate inflorescence. The catkins are borne on current seasons shoot and not on 1-year-old wood (Wetzstein and Sparks, 1984, 1986).

Initiation and differentiation of staminate flower occur about one year in advance of anthesis when shoot growth begins in spring and expanding leaves are 2-to 3-week-old. Thus, the inflorescence differentiates in early mid-May and flower primordia by early June. Thereafter, the development of bracted and imbricated floral primordia is different in protandrous and protogynous cultivars (Wetzstein and Sparks, 1984). In the former, the floral apex, bracteoles and anther primordia differentiate in June and July, whereas in the latter this organogenesis occurs late, just before anthesis in March and April. The bilobed anthers develop in March in protandrous cultivars but in April in protogynous cultivars. On the basis of light and scanning microscopy, pecan pollen morphology and germination have been described by Wetzstein and Sparks (1985). They further reported that the pollen remains viable *in vitro* only for a few days, *i.e.*, less than 1 per cent germination after storage for 5 days at room temperature. This is in contrast to earlier report by Romberg and Smith (1950) who could achieve pollination by the pollen stored for one year at 0.6°C and 25 per cent relative humidity.

According to Amling and Amling (1983), exposure to low temperature is essential for pistillate flower induction which takes place in late July or early August. The pistillate flower development occurs few weeks before anthesis on the resumption of current season's growth in spring. The sequence of development has been categorized into 9 stages (Wetzetein and Sparks, 1983). The number of female flowers on a spike varies from 2 to 10 determined by the cultivar and vigour of the shoot. The phenomenon of occurring both protandrous and protogynous individuals in pecan is known as heterodichogamy. Adriance (1931) and Dodge (1939) grouped the cultivars as protogynous, protandrous and fluctuating. But Smith and Romberg (1940) have repudiated the fluctuating group confirming only two groups, protandrous and protogynous, which remain stable in their behaviour every year though the time of pollen shedding or stigma receptivity may vary from season to season.

Catkins from protandrous genotypes are short, thick and compact compared with those of protogynous genotypes. Thus, catkins from protogynous types are long and narrow relative to those from protandrous type (Sparks, 2000). Fayek *et al.* (2008) working with pecan cultivars Cherokee, Desirable, Choctaw, Cape Fear and Graking at Egypt stated that pecan buds passed through 11 growing stages and the two types of flowers (male and female) did not mature at the same time (dichogamy). Treatment with potassium nitrate (5 per cent) and hydrogen cyanamide (Dormex, 3 per cent) to one-year-old shoots resulted in the highest percentage of flowers; both treatments induced the earlier flowering of female flowers (by 2 weeks). They concluded that application of Dormex or potassium nitrate synchronize the time of full bloom of male and female flowers. The treatments also increased the initial and final fruit set percentage. Spraying of Dormex (2 per cent) at the end of January and ethrel (750ppm) at the first week of September increased bud opening percentage and advanced flowering and harvesting time as compared with control in pecan cvs. Wichita and Western Schley. The treatment also increased fruit weight, kernel percentage, fruit dimensions, oil content and yield (El-Sharkawy and Osman, 2009).

## Pollination

Because of heterodichogamy, the pecan is mostly cross-pollinated, and self-pollination is very rare. The pollen is light, dry and pale yellow in colour. Wind is a pollinating agent which can carry pollen for about 900 m. planting of 5 per cent pollenizers has been recommended to ensure adequate pollination (Romberg, 1931; Woodroof, 1930). There is increasing evidence of substantial pollination related crop loss by pecan orchards. These most likely occur in block-type orchards consisting of only one or two cultivars, but can also occur at locations with a great number of different genotypes nearby. Main crop cultivars should generally be within about 2 rows of pollenizers to ensure cross-pollination. Thus, block widths exceeding about 4 rows between pollenizers are especially likely to exhibit serious pollination problems (Wood, 2000). Tree age/size and spring temperatures influence the characteristics of flower maturity windows and are probably primary factors contributing to pollination-related fruit-set losses in many block type orchards. Flower maturity tends to be earlier in older/larger trees while warmer springs accelerate catkin development relative to that of pistillate flowers. Often, within a genotype, there is some overlap of pollen shedding and stigma receptivity in which case dichogamy is incomplete. This contrasts with complete dichogamy where there is no overlap of pollen shedding and stigma receptivity. By far, imcomplete dichogamy is more common than complete dichogamy (Sparks, 2000).

The duration of both pollen dehiscence and stigma receptivity ranges from about 8 to 16 days depending on the seasons and cultivars (Sparks, 1986b). Other facters sometime causing pollination related crop losses are either abnormally wet weather, or strong dry winds during the pollination period or abnormally warm or cool springs. Pollination problems can be visually detected by noting premature non-insect related post pollination fruit drop (Wood, 2000). The pecan stigma is of dry type. The pollen tube emergence is visible within 3 hours of pollination (Wetzstein, 1989).

Sparks and Yeats (2002) reported that viability of pollen stored in liquid nitrogen was not diminished in comparison to fresh pollen even after 13 years. Thus, liquid nitrogen may often a means of haploid preservation of pecan.

## Alternate Bearing

Alternate bearing occurs in almost all nut fruits but is more pronounced in pecan. Sparks (1986b) prefers to call it irregular bearing because the sequence of 'on' and 'off' year does not necessarily follow an 'alternate' or 'biennial' pattern. Late spring freeze can damage the entire crop in an 'on' year resulting in a heavy crop in the next year which in turn will bear a poor crop in the third year. Defoliation due to prevailing drought or to the incidence of pests and diseases and likewise unfavourable climate induce the irregular bearing pattern. The alternate bearing in pecan occurs due either to inadequate pistillate flower production or to poor fruit set in 'off' year, the latter condition is an ultimate indicator of the phenomenon. Lockwood and Sparks (1978) have proved that initial spring growth of the pecan shoot and pistillate inflorescence is dependent on the substrates accumulated in the previous season. Better flower production and fruit set are obtained on longer shoots (Malstrom and McMeans, 1982). Thus, irregular bearing in pecan has been considered mainly related to carbohydrate accumulation.

High intensity of irregular bearing in pecan has been explained on the basis of its long fruit maturation period, delayed dry weight accumulation and high fat content of the kernel which altogether impede the annual bearing (Sparks, 1986b). Consequently, the degree of return fruiting is positively associated with early ripening cultivars (Smith *et al.*, 1986).

The flowering and fruit set in pecan are positively correlated with reserve carbohydrates which get accumulated in storage tissues during the previous growing season (Sparks, 1974, 1983: Worley, 1979). Shoots arising on 1-year-old vegetative wood produced 26 per cent more flowers and 32 per cent more nuts than those growing on formerly fruited shoots (Malstrom and McMeans, 1982). They also reported that low production was due to a failure to produce pistillate flowers which is in contrast to the opinion of Sparks (1986a) that failure of weak pistillate flowers to set fruits is the major cause of alternate bearing. According to Smith *et al.* (1986), two mechanisms appear to control flower production in pecan, *i.e.*, carbohydrate concentration in the trees and phytohormonal or growth regulator balances. On the basis of his findings, Wood (1982) suggested the possibility of a key regulatory role of IAA, possibly through its influence on regulation of bud cytokinin and gibberellin levels via their metabolism in roots.

Horticultural manipulation of pistillate flowering and mitigation of alternate bearing in commercial pecan orchards currently targets minimization of tree stress with orchard management strategies directly or indirectly targeting key exogenous biotic and abiotic stressors (Wood *et al.*, 2003). Wood (2011) examines the influence of key bioregulators *i.e.*, auxin (as NAA), cytokinin (6-BA), ethephon, and an auxin transport inhibitor, [TIBA (2,3,5-triiodobenzoic acid)] on subsequent season pistillate flower. It was reported that $GA_3$ and NAA inhibited whereas prohexadione-calcium (P-Ca, Calcium 3-oxido-5oxo-4-propionylcyclohex-3-enecarboxylate), ethephon, and BA+TIBA promoted floral initiation when topically applied to canopies before the kernel filling stage of seed development. These bioregulators exhibit potential for integrated into a bioregulator-based strategy to mitigate alternate bearing in pecan by selective and timely use in 'Off' and 'On' cycle years, depending on the bioregulators. Field studies provide evidence that a 'cytokinin-gibberellin balance' with partial modulation by auxin and ethylene, acts in the endogenous primordial environment of floral meristems as a 'second-level signal' regulating a key step in three-step process for initiation of pistillate flowers in pecan.

A good orchard management involving better nutrition, water relation, and pest and disease control has considerably reduced the magnitude of alternate bearing. Smith *et al.* (1986) stated that the data support the use of early ripening cultivars and management practices that maintain healthy foliage. Such practices should reduce alternate bearing by increasing carbohydrate production and storage and maintaining source of flower-promoting substances. Pruning should help in reducing the shading of branches and also increase the leaf-to-fruit ratio.

## Fruit Drop

The fruit set stage in pecan is not well defined because fruit drop occurs at all stages from anthesis to harvest. Abortion of pistillate flowers 'and fruits is governed by nutrition, soil moisture, cultivar and 'on' and 'off' year of the tree. It occurs at 4 stages during the growth cycle of the fruit (Sparks and Madden, 1985). The first drop consisting of weak or undeveloped flowers which take place immediately after full bloom and varies inversely with shoot vigour (Sparks and Heath, 1972). The intensity of this drop is high in the 'off' year of the alternate bearing cycle. The second drop coincides with abortion of non-pollinated flowers as is evident from self-pollination studies (Sparks and Madden, 1985). Failure of the first division of the zygote causes the third drop. Following the splitting of the fruit at water stage and discolouration, the fourth drop occurs and is related to high soil moisture and embryo abortion (Sparks and Madden, 1985). Besides the internal physiological conditions, pecan scab, hickory shuck worm and drought during the water demanding stage can also accentuate the fruit drop. For a good crop, a fruit set of more than 50 per cent is necessary (Smith and Romberg, 1940).

Comparative anatomical studies (Yates and Sparks, 1995) of aborting and non-aborting fruits indicated that second drop (12 to 47 day interval after pollination) is the absence of a zygote, third drop (55 to 98 days after pollination) results from a deficiency in cellular endosperm, and fourth drop (110-130 days after pollination) is due to delay in embryo development (Sparks, 2000).

Although, stage II drop (12-47-days after pollination, June drop) can be beneficial in excessively heavy crop-load years, however it is undesirable when crop-load is low to moderate. Wood *et al.* (2009) suggested post pollination spray of ReTain® (formulated from aminoethoxyvinylglycine by Valent BioSciences) at 32mg/l a.i. (11.7 oz/acre) to canopies increased fruit retention of pecan cv. Desirable and increase crop yield by 16 per cent to 38 per cent in trees carrying a 'moderate to heavy' crop. ReTain® did not detectably increase fruit retention on trees carrying a 'light' cropload. Retain® appears to offer commercial potential as a crop-load management tool for 'Desirable' through regulation of Stage II drop (*i.e.*, June drop).

## Fruit Thinning

Alternate bearing in pecan could be managed by adopting several strategies including fruit thinning and pruning. Mechanical fruit thinning increases nut size, kernel percentage and kernel grade and also return bloom (Smith and Gallot, 1990; Reid *et al.*, 1993; Wells *et al.*, 2009). Marquard (1987) studied the influence of leaf to fruit ratio on nut quality, shoot carbohydrates and photosynthesis. To fill one kernel, in the cultivars Sioux and Western, two leaves were sufficient but four leaves were required in Mohawk. Several chemicals have been tried but none has attained the commercial status for pecan fruit thinning. Harris and Smith (1957) obtained 10 per cent to 60 per cent pistillate flower thinning with maleic hydrazide (0.04 to 0.08 per cent), CIPC (200 ppm) and 2,4,5-T (10 to 25 ppm). Higher concentrations of the chemicals caused overthinning and slight twig injury. Hinriches *et al.* (1971) found ethephon (30 to 100 ppm) very successful in fruit thinning without adverse effect on foliage but both timing and concentration were critical. Ethephon (25 to 200 ppm) caused selective thinning ranging from 25 per cent to 75 per cent depending on the concentration and fruit age (Wood, 1983). Khalil *et al.* (2016) working with 27-year-old tree of pecan cv. Wichita at Las Cruces, USA, applied ethephon at different concentration (200,400 and 600 ppm) twice at 3 and 21 August for thinning. Foliar spray of ethephon at 600 ppm produced good quality nuts by reducing number of nuts in the excessive fruit loading year, which had resulted in greater fruit set in the following year.

## Fruit Growth and Development

After pollination, fruit does not grow for about 45 days and slowly for the next 15 days and thereafter enlarges rapidly for about 30 days reaching almost its final volume. From then onwards after 90 days from pollination untill maturity, embryo growth and nut adopts a subtle double sigmoid curve. Dry weight mostly accumulates during the last half of the growth period of the fruit (Dozier and Amling, 1974). Wood and McMeans (1982) and Wood and Reilly (1984) estimated the carbohydrates, fatty acids and proteins in developing fruit and kernel. During endosperm expansion, fructose and glucose accumulated rapidly and fatty acids and proteins increased and reducing sugars and inositol declined while sucrose increased. During the last 70 days of fruit growth, more than 80 per cent of mineral constituents of the fruit accumulate, *i.e.*, potassium increases about eightfold, copper sixfold, nitrogen, phosphrous, iron and zinc about fivefold (Sparks, 1977a; Thor and Smith, 1935).

The total lipid content increased during ripening. Saturated and polyunsaturated fatty acids decreased significantly, whereas, monounsaturated fatty acids increased during the ripening of nut. Maximum levels of total tocopherol (279.53 mg/kg oil) and xanthophyll (6.18mg/kg oil) in Tunisian grown pecan nuts were detected on 20$^{th}$ weeks after the flowering date. These amounts decreased gradually as ripening advances. The early stages of pecan ripening seem to have nutritional and pharmaceutical interests (Bouali *et al.*, 2013).

The relationship between the growth promoting and inhibiting substances and fruit growth could not be established (Lipe *et al.*, 1969). Likewise, $GA_3$ and BA when applied at low rates had no effect on vegetative growth, flowering and yield of pecan (Worley, 1989). Low ethylene production rate increases rapidly about a week prior to shuck dehiscence and again resumes the normal rate after the shuck dehiscence (Lipe and Morgan, 1970). Exogenous ethylene application increases shuck dehiscence indicating that the dehiscence is mediated through ethylene (Finch, 1937). Ethylene is, however, produced endogenously by the kernel because shucks fail to dehisce in absence of kernels (Lipe and Morgan, 1973). Ethephon spray alone or in combination with NAA for inducing shuck dehiscence is not practicable because of its concurrent induction of leaf abscission which suppresses return bloom and fruit set (Hinrichs, 1970; Kays *et al.*, 1975; Wood, 1989b). Due to the limited success of ethephon from foliar application, Stein *et al.* (1987) tried the method of tree injection at the critical stage of shuck separation from nut. Ethephon (100ppm) significantly advanced the harvest date over the non-treated trees by two weeks without affecting the crop in the following year.

## Maturity

Maturity of the pecan nuts can be defined according to whether the nuts are harvested early or late (Sparks, 1989a). Early harvest and subsequent artificial drying have become more common because of the high profitability associated with access to early markets (Sparks, 1989b, Herrera, 1994). Nuts from the late harvest are left on the tree until they dry naturally. Early harvest can impair the nut quality but not the late harvest. Beginning from October, the pecan harvesting continues until January. The development of colour markings as dots and streaks on the shell is an indication of nut maturation. When shuck separates from shell, some cells from both these tissues get deposited on the shell creating the markings

(Woodroof and Woodroof, 1927). Initially, the faint markings begin from blossom end and proceed gradually towards stem end as the shuck dehiscence increases. The faint markings intensify later except in case of poor filling. With the onset of shuck dehiscence along the sutures, the shell colour changes from white to brown (Kays and Wilson, 1977). Shuck dehiscence, colour markings on shell and clean separation of packing tissues from kernel indicates that the fruit is ready for picking. In case of delay or absence of shuck dehiscence in some geographical regions, the nuts are harvested and shuck is removed mechanically. Testa colour darkened and opalescence decrease (in cvs. Wichita and Western Schley) as the nuts matured. Bitterness of Western Schley pecans decreased with maturity. Both nut-in-shell and kernel moisture decreased with maturity whereas oil and sucrose contents increased (Wansri *et al.*, 2001). Singanusong *et al.* (2003) stated that Western Schley pecans grown in Australia should be harvested after the shuck has opened and it is either green or brown in colour to maximize total lipid content and quality. Pecan fruits of cvs. Wichita and Western Schley can be harvested earlier (about 2 weeks) when the trees are sprayed with Domex (2 per cent) at the end of January followed by ethrel (750ppm) on first of September (El-Sharkawy and Osman, 2009). Sparks (1989b) developed an equation for predicting nut maturity with high precision from heat units (base 18.5°C) accumulated in March and April. Maturity date is influenced more by temperature variation in April than in March.

## Harvesting

Generally, the nuts are thrashed down from the trees with bamboo poles or allowed to fall naturally on the ground wherefrom these are gathered manually. Hand shaking and gathering are possible only when plantings are small and labour is cheap and easily available but the method is slow, tedious and expensive.

For large-scale plantations, mechanical harvesting is adopted which involves tree shaking, picking and cleaning of nuts. Woodroof (1979) has described many types of machines for pecan harvest. Sarig *et al.* (1982) have designed a pecan harvester for heavy and wet soils. The harvester accomplishes nut pick-up, separation of nuts from trash and conveyance of nuts to a handling system. Nut pick up is achieved by a new mechanism utilizing an inclined endless chain elevator and a rubber fingered reel. The harvester is compact, requiring the use of only a small orchard tractor, and performs well.

Harvesting should be done in dry weather conditions. The land needs to be cleaned, levelled and firmed up and sod be close-cut for harvest (Gammon, 1976).

Verma *et al.* (1985) have developed a new dehusking method using two high friction belts to remove green shucks from the nuts.

## Yield

The commercial production in pecan begins by 7th to 10th year of planting depending on the precocity of a cultivar and growth of the tree (Madden *et al.*, 1969). Furthermore, yield varies greatly with the cultivar, age of the trees, 'off' and 'on' year and planting density. The yield ranging from 20 to 100 kg per tree is expected from fully grown trees. From South Africa, Haulik and Holtzhausen

(1987) reported an average annual yield of 25.1 kg per tree over a 22-year period and the extrapolated average yield of 2642 kg/ha for 10 years which compares favourably with the yields reported from the USA. Based on the collected data on nut count (NC), trunk circumference (TC), competition factor (CF), days from bud break (DAY), and high or low crop year (YR), Wright *et al.* (1990) developed an yield prediction model. The model developed predicted the natural logarithm of the total nuts on the tree which is

[In (NUTS)=2.122+[0.634 X In (NC)+ (0.00119 X TC) –(0.0701 X In(CF)) + (0.00639 X DAY) + (0.728 X YR)]

However, the model is not sufficiently accurate to predict individual tree yields well, but additional data show an ability to accurately predict yield.

## Vivipary

Crop loss to vivipary (*i.e.* germination of nuts on the tree) is potentially a major profit-limiting problem affecting yield and quality of nutmeats for certain pecan cultivars in certain locations. Improved orchard management tools and strategies are needed to avert major economic loss. Field research on 'Cheyenne' pecan revealed that incidence of vivipary on irrigated was 25-fold greater that on non-irrigated trees, and incidence on trees receiving excessive N fertilization was 21-fold greater than that of trees not receiving N. There was a strong irrigation x n interaction, with trees receiving irrigation + N exhibiting 41-fold more vivipary than did 'non-irrigated-N trees. The incidence of vivipary in orchards therefore, appears amenable to management via an appropriate irrigation and N-fertilization strategy. Treatment of developing 'Oconee' fruit with (±) ABA[(2-cis,4-trans)-5-(1-hydroxy-2,6,6-trimethyl-4-oxo-2-cyclohexen-1-yl)-3-methyl-2,4-pentadie-noic acid] or fluridone [(1-methyl-3-phenyl)-5-3-(trifluoromethyl) phenyl-4 (1H)-pyridinone)], an inhibiter of ABA biosynthesis, reduced and increased the incidence of vivipary, respectively. Additionally it was also observed that treatment with developing 'Summer fruit' with fluridone reduces the germination time of harvested pecan nuts. This indicates the influence of supplemental irrigation and N on incidence of vivipary is likely via alteration in ABA metabolism within developing fruit (Wood, 2015).

## Drying and Storage

### Drying

When pecans lie on the wet soil after harvest for few days, they discolour and get infected with mould and meat quality deteriorates. In order to prevent moulding, discolouration and rancidity, pecan nuts are dried soon after harvest to bring the moisture level of kernels to around 4.5 per cent. The drying also imparts satisfactory colour, texture, flavour and stability to the nuts which occurs naturally under field conditions if the weather and soil remain dry at harvest. The nuts should be dried at room temperature by constant circulation of low moisture air for about three days. Artificial heat drying of nuts is undesirable.

Woodroof (1979) recommended drying of nuts in rapidly circulated dry air (65 per cent RH) at refrigerated temperatures. Ghate and Chhinnan (1983) reported that direct contact and dehumidified air gave the best result in drying with silica gel. The direct contact decreased kernel moisture from 14.3 per cent to 4.3 per cent within 15 hours with 1:1 pecan-gel mixture or 21 hours with a 3:1 pecan-gel mixture. The kernel moisture level was brought down to about 4 per cent within 24 hours by dehumidified air at 30°C with a velocity of 18 to 31 metres/minute. They (1984) further achieved the desired drying by placing freshly harvested nuts in direct contact with bentonites. Chhinnan (1984) evaluated four mathematical models and found the page equation to be the most suitable for describing thin-layer drying of in-shell pecans.

## Storage

Pecan is consumed over a long period after harvest necessitating proper storage because of its semi-perishable nature. If nuts are not stored properly, they become inedible due to mould, insects and rancidity along with discolouration and staleness. In this context, drying, proper packing and storage temperature are the most important factors. Only high quality pecans, *i.e.*, having good size, colour, flavour and oil content should be stored to fetch high price in the off-season market compensating the extra cost involved in refrigeration.

The kernel moisture content and storage temperature are two important factors affecting the storage life of the nuts. Of these two factors, high moisture in nuts is even more responsible for deteriorating the quality which arises from delayed harvesting on wet soil, excessive moisture immediately before or after cracking and high relative humidity under storage (Woodroof, 1979). The relative humidity requirements of unshelled and shelled pecans for storage for more than 3 months are 75-80 per cent and 65-70 per cent, respectively.

The storage life of pecans can be extended in direct proportion of lowering the temperature, beginning from few months at room temperature to many years at –7.6°C. The refrigerated storage preserves natural flavour, colour and texture as long as eight years, retards rancidity and staleness and prevents moulding and insect damage (Woodroof, 1979). Freezing does not damage the nuts due to its low moisture content. The critical storage temperature has not been specified to preserve the nuts for a particular period. Unshelled and shelled pecans can be kept at room temperature (21.1°C) for 6 and 3 months, respectively, but the chances of discolouration, rancidity and moulding are more. El Wahab *et al.* (1984) found the cold storage of unshelled nuts for 20 months at 5°C to be the best which delayed decay and rancidity. At –6.7° to –3.3°C unshelled nuts can be preserved for 30 months and shelled halves for 18 to 24 months without deteriorating the quality (Woodroof, 1979). He further stated that shelled and unshelled nuts may be held for 6 to 10 years at -17.8°C. Nuts can be stored for about 120 days by storing in nylon-polyethylene plastic film under vacuum or in polypropylene plastic at ambient temperature (Oro *et al.*, 2008). Pecan in shell can be stored for 12 months at 0°C, 9 months at 10°C and 6 months at 20°C (Maness, 2004, Ortloff, 2009). The nutmeat can be stored for 18 months at –10°C, 10 months at 0°C and 3 months at 20°C (Maness, 2004). If pecans are refrigerated or frozen, they should be placed in air-tight container. Pecan meats readily absorb odors from other foods, resulting in off flavours. If pecans are to be stored at room temperature for an extended period of time, they should be held in containers that are adequately ventilated.

A number of insects and fungi become a major problem in storage at temperatures above 7ºC. The insects are controlled by fumigation but there is no standard practice for that, and fungal growth is prevented by maintaining the low moisture level of kernel (3 per cent to 4 per cent). Wells and Payne (1980) controlled in-shell weevils and reduced the mycoflora in pecan by storing in high $CO_2$ atmosphere at 7ºC and 65 per cent relative humidity. After 5 months in atmospheres of 21 per cent $O_2$ and 30 per cent $CO_2$ *Curculio caryae* mortality was 100 per cent, total mycoflora was significantly reduced and off-flavours had not developed in kernels. After 6 months, objectionable off-flavours were, however, observed in high $CO_2$ atmosphere, and *Alternaria, Penicilluim* and *Fusariurn* spp. were most frequently isolated.

## Pests and Diseases

A large number of pests and diseases attack the pecan making it difficult to control each of them individually. Gentry and Smith (1982) therefore, tried successfully a multi-pest management programme in a pecan orchard which included black-light trap suppression of *Cydia caryana, Gretchena bolliana,* and other lepidopterous species, an annual systematic insecticide application to control leaf feeding pests, a single soil fumigation to control *Curculio caryae,* and a reduced fungicide programme using 4 applications at double strength. This programme gave a higher net return per hectare than a conventionally managed orchard. Dutcher *et al.* (1983) suggested a pressurized trunk injection system to control foliar insect pests, zinc deficiency and foliar diseases.

## Pests

### Hickory Shuckworm (*Cydia caryana*)

This moth is a serious pest of pecan and other hickories which lays eggs on foliage and green nuts. Upon hatching, the cream coloured small caterpillars attack the immature nuts which prematurely fall to the ground. When the shell becomes hard, the caterpillers tunnel in the shuck hampering the kernel development. Overwintering of fully grown caterpillers occur in shucks on the ground. All the shucks after harvest should be burnt after collection. It is difficult to control the pest by insecticidal sprays because of overlapping generations and reinfestation from adjoining unsprayed trees. First application of guthion should be made at the beginning of shell hardening followed by another two applications at 2-week interval (Woodroof, 1979). Orgenophosphates, chlorpyrifos, imadin and pyrethroids are effective to control hickory shuck worm.

### Pecan Nut Case Bearer (*Acrobasis nuxvorella*)

The pest causes great damage to pecan nuts as a single larva can destroy the entire cluster. The moths overwinter as larvae in small cocoons located near the buds on the stem. As the buds begin growth in spring, the larvae activate and feed on them and later bore into new shoots and pupate there. The moth emergence in April-May coincides with the nut set and moths lay greenish white eggs on the blossom end of developing nuts. Thereafter, first generation eggs hatch and larvae feed on growing nuts destroying the crop in May and June. Later generations damage the shucks and foliage. Insecticidal sprays of parathion and malathion are quite effective in controlling this pest which should be made when about 10 per cent moths have emerged after pupation or tips of small nuts have become brown (Johnson, 1969). Pherocon II trap was found effective to catch months in the second and third flight periods (Grantham *et al.*, 2002).

## Pecan Weevil (*Curculio caryae*)

The weevil damages the nuts both in the orchards and storage. The type of damage includes premature nut drop, black spots on the kernel, shrivelled kernel and destruction of kernel by larvae (Hall *et al.*, 1981). After pupating in the soil, the adults move to the tree and damage the nuts before shell hardening during August and September and lay eggs after shell hardening. As soon as grubs hatch, they start feeding on kernels. Nuts infested with larvae result in complete destructions of the kernel (Colcote, 1975).

After emergence, pecan weevils enter the tree by either crawling up the bole or flying directly to the canopy or bole (Raney and Eikenbary, 1968). Cottrell and Wood (2008) found that newly-emerged weevils predominantly crawl up the bole to enter the canopy and search for viable fruit.

Early maturing cultivars are more attacked than late maturing ones. Five foliar applications of Carbaryl at 2.7 kg a.i./ha at interval of 7 to 10 days show significantly less damage by the weevil than 8 to 9 soil applications of Carbaryl (Payne *et al.*, 1985). The first spray should be applied as soon as the nuts begin to harden in the early cultivars. The entomopathogenic fungi *Beauveria bassiana* (Balsamo) Vuillemin and *Metarhizium anisopliae* (Metschnikoff) Sorokin are pathogenic to and are being developed as microbial agents for pecan weevil (Shapiro-Ilan *et al.*, 2009). Trunk sprays of *B. bassiana* suppress adult pecan weevil (Shapiro-Ilan *et al.*, 2009). Further studies (Shapiro-Ilan *et al.*, 2013) indicated that *B. bassiana* can be applied in soil and fertilizer treatment did not impact *B. bassiana* persistence. Circle trap (two traps encircling the tree) has been widely adopted on pecan, provided greater confidence in detecting and accurately timing treatments for pecan weevil, particularly as the pecan weevil emergence season progressed (Mulder and Grantham, 2007).

## Aphids

Many aphids cause injury to pecans. Black pecan aphids (*Monellia caryaefoliae*) feed on both sides of leaves and appearance of large yellow angular blotches is the first sign of its injury. The blotches later turn brown and further damage causes premature leaf fall adversely affecting the current as well as subsequent crops.

Three species of aphids- *Monellia caryella, Monelliopsis pecanis* and *Melanocallis caryaefoliae* often occur as outbreaks during the spring and late summer (Tedder, 1978). Black pecan aphids typically occur as outbreaks in the late summer and fall in the southeastern U.S. The spring outbreaks were found mixed populations of yellow pecan and blackmargined aphids (Dutcher *et al.*, 2012). Tedders and Wood (1985) estimated the damage of mixed aphid populations (*Monellia caryella, M. pecansis, M. nigropuncta*) on mature 'Stuart' pecan trees. Early season feeding by aphids had a major adverse effect on tree productivity, growth and energy reserves. Nut volume and weight, shoot starch and reducing sugars were reduced during early fruit development. Likewise, nut volume and weight, kernel weight and lipid content, shoot starch and reducing sugars and nut yield were reduced at fruit maturity.

A single application of aldicarb (28 kg a.i/ha) by placement near the drip emitters during the spring resulted in rapid control of aphids (Alverson, 1985). No aphid was found 6 weeks after application and the efficacy was maintained for the rest of the season. Aldicarb also controlled mites for the entire season (Dutcher and Harrison, 1984). Pesticide residues in the nuts were very low (Villiers and Viljohn, 1987). A complex of natural and introduced aphidophagous insects spray on the aphids throughout the season (Edelson and Estes, 1987) and biological control with these predators is an effective and sustainable method for maintaining relatively low aphid population. Aphidophagons bettle, *Harmonoia axyridis* effectively control foliar pecan aphids in a four week period (Abbas *et al.*, 2012). Dutch *et al.* (2012) suggested application of an aqueous solution of fermenting molasses to the foliage during an aphid outbreak which significantly increase the abundance of ladybeetles and lacewings and significantly reduced the abundance of yellow pecan, blackmargined and black pecan aphids.

## Armoured Scales

Three scales, *viz.*, obscure scale (*Melaspis obscura*), scarfy scale (*Chionaspis furfura*) and San Jose scale (*Aspidiotus perniciosus*) are most common which feed on hickories, chestnut, walnut, filbert and almond. They suck the plant sap and devitalize it. Dead bark and twigs are the symptoms of their attack. It is very difficult to identify these insects because of their resemblance to the bark. A single spray of any systematic insecticide (malathion, metasystox) in early spring is sufficient to kill these pests.

## Pecan Leaf Case Bearer (*Acrobasis juglandis*)

As the name indicates, the caterpillar keeps itself closed in a leaf case. The caterpillers emerging from overwintering cases in early spring feed on buds, leaves and flowers and later enter in the shoot. The broken shoots are symptoms of their attack. The moth lays eggs on the underside of a leaf along mid-vein. It has only one generation in a year (Johnson, 1969). The pest should be controlled by pre-pollination spray of an insecticide (parathion or malathion). If the insecticidal sprays are carried out to control other pestes, a separate application is not required.

## Fall Web Worm (*Hyphantria cunea*)

The caterpillars attack limbs and can cause defoliation, occasionally. During summer, a loose silken web is formed covering leaves, twigs and small branches. The insect damages a number of fruits but has a preference for pecan, hickory and persimmon. The egg masses are deposited on leaves and hatching takes place in about one week. Depending on the region, the moth has one or two generations in a year. It passes the winter in pupal stage in cocoons on the ground. Insecticidal sprays to control shuckworm are enough to eradicate this as well.

## Hickory Shoot Curculios

Four species of curculios commonly found attacking pecan are *Conotrachelus affinis*, *C. aratus*, *C. hicoriae* and *C. anaglyptius*. *C. affinis*; deposits eggs in immature nuts of various hickories and larvae feed inside the nut causing its premature fall. *C. hicoriae* damages young green nuts of pecan and may attack other hickories. *C. aratus* is smaller and more widely distributed than other species and feeds on young shoots and unfolding buds. Life history and feeding and oviposition damages of *C. schoofi* have been described and compared with other species by Tedders and Payne (1986). Lindane and endosulfan provide good control to all the species when sprayed early in the season or on the appearance of adults.

## Leaf Scorch Mite (*Eotetranychus hicoriae*)

A heavy infestation of this mite may cause serious injury to foliage. They feed mainly on the lower side of leaflets along the midrib. Initially, a brown discolouration appears at the midrib which subsequently covers the entire leaflets. The insecticides applied for the control of aphids will also check the mites. A voluntary phase-out of the manufacture of dicofol, the pricciple miticide used for chemical of leaf scorch mite, and the discovery and evaluation of western predatory mite, *Galendromus occidentales* as, a biological control of pecan leaf scorch mite have led to the registration of the selective miticide, bifenazate, as a possible replacement of dicofol for control of pecan leaf scorch mite in pecan orchards of US (Dutcher *et al.*, 2009). Bifenazate was reported as an effective miticide and had the additional benefit over dicofol of conserving phytoseiid mites. The lowest effective concentration as a foliar spray application was suggested as 0.3g actual bifenazate/l water (Dutcher *et al.*, 2009).

## Serpentine Leafminer (*Stigmella juglandifoliella*)

The leaf-miners feed between lower and upper surfaces of leaflets creating characteristic designs and the heavy infestation results in defoliation. The nepticulid has a single developmental stage dominant in commercial orchard at any time. A single application of the dimethoate, diflubenzuron or fenvalerate coinciding with peak adult emergence is sufficient to control the pest (Heyerdahl and Dutcher, 1985). The adult emergence period can be forecast by larvae sampling.

## Nematodes

The pecan root-knot nematodes, *Meloidogyne partiyla* was found infecting pecan in South Africa, USA and Mexico. The other *Meloidogyne* spp. infecting pecan includes *M. incognita, M. arenaria* and *M. javanica*. Affected trees exhibited dead branches in the upper canopy and stunted growth (Nyczepir *et al.*, 2002). Thomas *et al.* (2001) noted chlorosis of foliage and substantial dieback of branches in the upper canopy in trees infested with *M. partiyla*. Use of clean planting materials free from nematode infestation and sites free from nematodes should be selected for developing pecan orchard. Sanitation practices should be maintained to avoid the spread of the nematode within and between nursery stock producing sites and pecan orchards.

### Birds

Crows and bluejays are well known to cause great losses to pecans. Crows are particularly a menace to pecan as they start damaging the nuts soon after kernel filling either on the tree or by carrying them away. The damage is more serious to small, well-filled and thin-shelled nuts. Crows attack the nuts mostly early in the morning. After killing a crow, it should be hanged high on the top of a pecan tree to scare away the crows. Automatic (scare-away machine) explosion devices can also be used as they are more economical than employing watchmen for bird scaring.

## Diseases

### Pecan Scab (*Fusicladium effusum*)

Among all the diseases of pecan, scab is the most serious and occurs wherever the plant is grown. The cultivars differ in their susceptibility to the disease and northern cultivars are more tolerant than southern cultivars of pecan. The characteristic symptoms of scab are irregular, olive-green to black spots on leaves or young shoots and small dark brown to black circular lesions on the nuts (Johnson, 1969). The fungus attacks primarily the rapidly growing leaves, shoots and fruits and the damage is further accentuated if it coincides with rains during spring and early summer. The mature tissues are, however, immune to the disease. Gottwald and Bertrand (1988) demonstrated that fungicide sprays on fully grown nuts had no significant effect on crop yield or nut quality. Therefore, they suggested an abbreviated scab disease control programme depending on the cultivars. Protective measures such as burning of all diseased leaves, twigs and shucks are as important as the fungicidal sprays of Bordeaux mixture, zineb, cyprex and benlate which should commence about the middle of April and continued at 3-week intervals. Sterol-inhibitors (DMIs) such as propiconazole and fenbuconazole provide good control, especially early in the season for leaf scab. The strobilurins (azoxystrobin, kresoxim-methyl, pyraclostrobin) are very effective for scab control, both on leaves and nuts. Fentin hydroxide showed excellent nut scab control and most widely used fungicide for scab control. Management of the disease in tall trees is challenging due to the technical difficulties of getting sufficient spray coverage to the upper parts of the canopy. The use of trunk application of phosphite, which is systemically transported in the trees, however, not effective. Different methods of trunk application, for example injection of phosphite, might be effective (Bock *et al.*, 2013). Trees in pecan orchards are often >10-15m tall and older trees can reach > 30m, so to achieve adequate scab control additional aerial application of fungicide may be required (Bock *et al.*, 2015).

### Bacterial Leaf Scorch (*Xylella fastidiosa*)

A symptom of bacterial leaf scorch is necrosis of leaflets beginning at the tips and margins; necrotic leaflet tissue usually has tan to light brown colouration. The necrosis quickly progresses in an unbroken pattern toward the base and midrib and leaflets generally abscise soon after symptoms appear. Symptoms on the compound leaves frequently begin on the older leaflets and advance towards the leaf apex. Leaflet defoliation follows this same pattern so that rachis may be nearly bare with only the youngest leaflets still attached. Symptoms can develop throughout a tree or be continued to individual limbs (Sanderlin and Heyderich-Alger, 2000). The weight of the nuts from terminals with leaf scorch was lower than the uninfected trees (Sanderlin and Heyderich-Alger, 2003).

Pecan cultivar Cape Fear is particularly susceptible to this disease. The disease is highly transmissible through graft unions from infected rootstock into new growth developing from scions (Sanderlin and Melanson, 2006). Submersion of scion wood in 46°C for 30 min. greatly reduced transmission of *X fastidiosa* following grafting. The hot-water treatment did not affect grafting success (Sanderlin and Melanson, 2008).

## Downy Spot (*Mycospharella caryigena*)

The disease is only serious on neglected, declining trees and nursery plants and the latter are usually the source of infection. The symptoms include the appearance of olive-green, velvety tufts on the under surface in early summer and yellow spots on the upper surface of leaflets in late season. The black pimple like structures of perfect stage give a black shiny blotched look to the leaflets in the mid summer. The leaves may fall on severe infection. The scab control measures are also applicable to this disease. Fentin hydroxide is effective to control downy spot.

## Vein Spot (*Gnomonia nerviseda*)

Vein spot is a foliage disease which is more serious on several cultivars in some regions. The fungus infects the vascular tissue at the junction of petiole to the rachis and base of rachis causing premature leaf fall (Sanderlin *et al.*, 1983). In May and June, the infection is initiated by ascospores released after rainfall. Benomyl has been found to be the most effective fungicide against the vein spots as the benomyl-treated trees have significantly fewer lesions and less defoliation than untreated trees.

## Liver Spot (*Gnomonia caryae* var. *pecanae*)

Dark brown circular spots appear on the lower side of the leaflets adjoining the midvein in late spring. Later, the spot changes to cinnamon-brown in September and October. The spores from fallen leaves are the sources of further infection. Schley has been reported as a resistant cultivar to liver spot (Johnson, 1969). Initial three fungicidal sprays for scab will also control this disease.

## Anthracnose (*Glomerella cingulata*)

A secondary disease of pecan. Causes irregular lesions on leaves and dark, sunken lesions on nut shucks. Spraying of propiconazole and fenuconazole provide good control. A combination of propiconazol and fentin hydroxide is effective for fungicide resistance management.

## Crown Gall (*Agrobacterium tumefaciens*)

The crown gall is a bacterial disease which invades many plants and is worldwide in distribution. Tumours or wort-like growth from few centimetres to 30 cm in diameter develop on the collar and adjoining roots of the tree often protruding out of the soil. Besides killing the nursery plants, it may debilitate and eventually kill the mature trees. No satisfactory control measures have been developed to control crown gall. All diseased small plants may be destroyed. The injury to plants by cultivation should be avoided. After removing the gall, the affected tissue may be painted with the Elgetol-wood alcohol rnixture (Johnson, 1969). Pre-plant treatment with biopesticide GallTrol (*Agrobacterium radiobacter*, strain K-84) is suggested.

## Bunch Disease

It is caused by a virus which is supposed to be transmitted by sucking insects. The characteristic symptom of bunch disease is the bushy growth of slender willowy shoots. In spring and early summer, bunch growth begins earlier and is more prolific than healthy shoots. The branches of a severely affected plant get clustered with thick bloomy sucker growth (Johnson, 1969). The affected shoots should be pruned and burnt. The diseased wild plant growing in the vicinity of the orchard should also be removed. Highly susceptible Cultivars should not be grown in the bunch disease prone regions. The cultivar Stuart is resistant to this virus (Romberg, 1967).

# Breeding and Improvement

Natural selection has played a prominent role in the pecan improvement favouring comparatively vigorous and disease-free plants (Madden and Malstrom, 1975). Pro-longed juvenile phase of pecan enables it to develop extensive root system and large tree canopy to compete with other wild plants. In this struggle for existence, dwarf and diseased plants get eliminated. The genetically dwarf plants are not available in pecan. The Native American Indians played a significant role in the improvement of pecans as they selected the largest and thinnest shelled nuts and spread during their nomad life in pre-historic times. The seed propagation was another important factor which enabled the selection of good quality disease free types from large seedling populations. As a result, a number of cultivars were released and notable ones are Stuart, Schley, Moneymaker, Mahan, Curtis, Moore and Western Schley. Successful vegetative propagation of pecan developed in the later part of l9th century made a significant contribution in releasing and distribution of the selected seedlings as cultivars.

Pecan hybridization was started by Forkert (1914) who evolved 'Admirable', 'Dependable' and 'Desirable' cultivars. The 'Desirable' is still very popular in the southeast USA because its nuts are of high quality and the tree is disease tolerant. Reed developed the cultivar 'Caddo' from a cross of 'Brooks' x 'Alley' which was released in 1968 (Madden and Malstrom, 1975). L.D. Romberg has been the most successful pecan breeder in Texas. As many as 12 cultivars have been released out of his crosses (Madden and Malstrom, 1975). Of these, 'Wichita', 'Sioux', 'Cheyenne', 'Cherokee', 'Chiksaw' and 'Shoshoni' are very popular because of their precocity, prolificacy and compact growth habit which are required for high density culture. Thompson and Young (1985) listed more than thousand pecan cultivars, many of which are now extinct, but some are still popular on and in cultivation. 'Stuart', 'Western Schley' and 'Desirable' which were developed in late nineteenth or early twentieth century are still very popular in USA.

## Objectives for Breeding of Cultivars

In the past, the breeding objectives were to identify genotypes which were highly productive, disease resistant and bearing attractive in-shell nuts of high quality. More emphasis was laid on yield and quality of nuts. But the current trends of fruit culture, marketing, processing and mechanization have dramatically altered the situation and modern objectives of pecan breeding are given below (Madden and Malstrom, 1975):

### Precocity

In view of the high cost of land and planting, a pecan grower does not want to wait for about 10 years for a commercial crop and he demands precocious cultivars to give him the return on his investment in about five years. A cultivar must make sufficient vegetative growth in early age and be able to pass on to the reproductive phase as quickly as possible.

### Regular Bearing

Irregular bearing is a serious problem in pecan production and the phenomenon to a large extent is controlled by genetic factors. Regular bearing is, therefore, a very important objective of pecan breeding.

### Early Maturity

Early maturity is desirable on account for three reasons. Firstly, the early maturing cultivars have a tendency towards regular bearing; secondly, it helps in better utilization of labour and machinery by extending harvest period and thirdly, the nuts fetch a premium price in early market. 'Pawne' which is an early harvesting cultivar with large and high quality nut is being extensively used in breeding programme to incorporate early harvest date and good quality of nuts in new cultivars.

### Adaptability to High Density

High density cultivation is a modern trend of fruit growing. For this purpose, a cultivar must respond to high fertilizer, irrigation, and pruning practices by giving higher production.

## Disease and Pest Resistance

A number of disease, pest and nematodes attack pecan. Host plant resistance to scab has been observed in many improved cultivars. Cultivars have been developed for resistance to aphid. 'Pawnee' has shown high level of resistance to the yellow pecan aphid (Thompson and Grauke, 1998), while 'Cape Fear' is resistance to black and yellow pecan aphid complex (Kaakeh and Dutcher, 1994). The old Florida cultivar 'Elliot' is used to incorporate scab resistance into new cultivar. 'Byrd' ('Wichita x 'Pawnec'), an early maturing cultivar with high kernel quality has been released from UGA-Athens Campus.

## Mechanization

Unavailability and high cost of labour have given way to mechanisation, particularly in pruning and harvesting operations. The plant should not be very tall and be able to withstand the harvest damage. The size and appearance of nuts and thinness of shell are no longer important because pecan is mostly marketed as shelled kernels. Better-filled, medium-sized nuts having thick shell and strong sutures suitable to mechanical shelling are preferable.

### Inheritance of Characters

Polygenic inheritance and ecological influence make the genetical manipulations very difficult in fruit breeding. The characters like nut size, shape and markings, relative shell and kernel weights and kernel appearance are influenced by environment while characters, such as, type of catkins, form and colour of stigma, dichogamy type, shuck thickness and types of suture, lenticel type and distribution and type of branching are unaffected by environmental fluctuation (Romberg and Smith, 1950). Dichogamy and stigma colour are governed by single gene, protogyny and red stigma characters being dominant. The donor parents for precocity are 'Evers', 'Candy', 'Brooks', 'Shoshoni', 'Cherokee', 'Mahan', 'Moore' and 'Western', and those for early ripening are 'Candy', 'Barton', 'Moore' and 'Evers' (Madden, 1969). 'Mahan' and 'Schley' the two most productive parents have ben used in USDA breeding programme. A number of hybrids such as 'Choctaw' ('Success' x 'Mahan'), 'Wichita' ('Halbert' x 'Mahan'), 'Apache' (('Burkett' x 'Sachley'), 'Sioux' ('Schley' x 'Carmichael'), 'Mohawk' ('Success' x 'Mahan'), 'Shawne' ('Schley' x 'Barton'), 'Cherokee' ('Schley' x 'Evers'), 'Tejas' ('Mahan' x 'Risien 1'), 'Kiowa' ('Mahan' x 'Desirable'), 'Oconee' ('Schley' x 'Barton'), 'Hopi' ('Schley' x 'McCully'), 'Lakata' ('Mahan' x 'Major'), 'Apalachee' ('Moore' x 'Schley'), have been developed using 'Mahon' or 'Schley' as one of the parent.

## Genetic Diversity

Isozymes systems had been developed and used for genetic diversity of pecan populations and identifying pecan cultivars (Conner and Wood, 2001; Garke *et al.*, 2003; Dong Jia *et al.*, 2011). Conner and Wood (2001) have demonstrated the value of RAPD markers in determining genetic relationship and fingerprinting pecan cultivars. Although the technique was clearly informative the dominantly inherited (RAPD) fragments are not capable of the resolution in heritability studies that co-dominant procedures would offer (Grauke *et al.*, 2003). SSRs have been found to be highly variable and are inherited in co-dominant fashion, making them very informative in population genetic analysis. Grauke *et al.* (2003) developed a microsatellite-enriched library from 'Halbert' a native pecan selection from Coleman County, Texas. A genomic library enriched for simple sequence repeats (SSR) containing 6144 clones was archived in 384 well plates for screening. In total, 439 clones were identified after Southern hybridization di-and tri-nucleotide repeats as probes. In total, 125 positive clones were sequenced and primers were designed for 24 repeats. The SSR markers chosen for analysis include di-(CT and GA) and tri-nucleotide repeats (CCT, GAA and GAT). Of the 24 primer pairs tested, 19 successfully amplified microsatellites from 'Halbert'. DNA was isolated from 48 pecan and hickory accessions. The 19 SSR primers that produced good amplification products in 'Halbert' were used to evaluate the collection, with 11 revealing polymorphism. The number of fragments amplified with different primer combinations ranged from 4 to 32 in the 48 genotypes tested. Don Jia *et al.* (2011) studied 77 samples grown in China including 14 introduced cultivars, 12 domestic seedling breeding cultivars, and 49 fine pecan plants with no cultivar data, together with *Carya cathayensis* and *Juglans nigra*. They reported that Chinese-grown pecan cultivars and fine plants had significant diversity at the DNA level. 'Zhohgshan 25', 'Huangshan 1', Gan 2' and Nanjing 148' cultivars always grouped with 'Farley', showing their close relationship with 'Farley and also suggested possible breeding source. IISR and SSR techniques were found suitable for genetic diversity analysis and the identification of pecan resources.

Interspecific variation of pecan in Central Mexico was studied by Cerna-Cortes *et al.* (2003) using AFLP markers. The genetic differences in the pecan populations evaluated ranged from 3 to 17 per cent. In addition, a certain association among populations according to geographic origin was observed. AFLP markers associated with dichogamy were identified, constituting an original finding which could serve as a model for the prediction of a particular characteristic of a perennial plant from seedling stage. Beedanagari *et al.* (2005) reported the first genetic linkage maps of pecan, using RAPD and AFLP markers. Independent maps were constructed for the cultivars 'Pawnee' and 'Elliot' using the double pseudo-testcross mapping strategy and markers, including 217 RAPD, 258 AFLP, and two morphological markers were used in linkage analysis. The 'Pawnee' linkage map has 218 markers, comprising 176 testcross and 42 intercross markers, placed in 16 major and 13 minor (doublets and triplets) linkage groups. The 'Pawnee' linkage map covered 2,227cM with an average map distance of 12.7cM between adjacent markers. The 'Elliot' linkage map has 174 markers comprising 150 testcross and 22 intercross markers placed in 17 major and nine minor linkage groups. The 'Elliot' map covered 1,698 cM with an average map distance of 11.2 cM between adjacent markers. Segregation ratios for dichogamy type and stigma colour were not significantly different from 1:1, suggesting that both traits are controlled by single loci with protogyny and green stigmas dominant to protandry and red stigmas. These loci were found tightly linked (1.9cM) and were placed in 'Elliot' linkage group 16.

## Genome

Grauke *et al.* (2001) determined the genomic size of pecan leaves (cvs. Bryzozski-Vlasek and CSV 14-12), the distal 5mm of root tips (Little Jewel-seedling rootstock) and embryos from mature seeds (selections 74-20-16 and CSC 7-1) by flow cytometry. They reported that samples from pecan root and embryo produced distinct and consistent peaks, while leaf tissue did not. Genome size (2c) estimated from 2c root (1.64pg) and embryo nuclei (1.66, 1.68 and 1.69 pg; mean 1.68pg). The mean 2c genome size of pecan, therefore, is ≈1.7pg (1c=0.85pg or 816 Mbp). Jenkins *et al.*

(2015) generated low-cost, short read sequences for two pecan cultivars ('Pawnee' and '87M3-2.11') and SNP calls are available at http://www.hags.org/pecan.

The conservative fragment sequence of MADS-box genes from pecan was cloned and there phytogeny was assayed in order to provide references for cloning pecan new MADS-box family genes related to flower development and exploring molecular mechanism of pecan flower development (ZhengHai *et al.*, 2013). The total RNA was extracted from male flower of cv. Mahan and cDNA was synthesized using inverse transcription. The conservative fragment of MADS-box genes was cloned using RT-PCR, and the phylogenetic analysis was conducted using deduced amino acid sequences with known Arabidopsis MADS-box gene subfamily. Twenty eight cDNA fragments of MADS-box gene were obtained by RT-PCR, and the length of these fragment was 137 bp including the initiation of codon genes. The homology of these nucleotide sequence was between 65.7 per cent and 98.5 per cent. The 11 deduced amino acid sequences were different. According to phylogenetic tree analysis, all these gene fragments involving different types of gene in ABCDE model were divided into Arabidopsis MADS-box gene subfamily. It was concluded that there were many MADS-box genes in pecan. These gene fragments included all types of floral development genes in ABCDE model.

## Objectives for Breeding of Rootstocks

Research work on pecan rootstocks is still in its infancy and no tangible result has been achieved (Hanna, 1987). In developing clonal rootstocks, the following objectives should be taken into consideration. The USDA rootstock breeding programme is looking for identifying parental materials which can uptake more zinc and low sodium and chlorine ion.

### Dwarfness

All the pecan cultivars being very vigorous require a dwarfing rootstock to control the tree size for pruning, spraying and harvesting operations. A dwarfing stock is an urgent need of high density culture. The use of various related species of hickories as rootstock and interstock needs to be investigated thoroughly.

### Adaptability

Soil problems such as water logging, zinc deficiency and chloride toxicity can be tackled if rootstocks are bred to counteract these problems. Hanna and Storey (1972) have reported that the seedling rootstocks differ in their tolerance to zinc deficiency and chloride toxicity. Again, the related hickory species can provide the rootstocks for different soil conditions.

### Productivity

The effect of rootstocks on productivity has been reported by Wood (1989b,c). The characterization and clonal propagation of productive rootstocks are necessary for pecan culture.

## REFERENCES

Abbas, G., Karar, H. and Dutcher, J.D. (2012). J. Entomol. Sci., 47:116-20.

Ablen, A.O. (1962). Proc. Amer. Soc. Hort. Sci., 80: 312-14.

Adriance, G.W. (1931). Bot. Gaz., 91: 144-66.

Aitken, J.B. and Gorsuch, C.S. (1985). In: Drip/trickle irrigation in action, vol. II St Joseph, Michigan, USA, ASAE, pp. 495-99.

Alben, A.O., Cole, J.R. and Lewis, R.D. (1932). Phytopathology, 22: 979-81.

Alben, A.O., Hammer, H.E. and Sitton, B.G. (1942). Proc. Amer. Soc. Hort. Sci., 41:53-60.

Allan, P., Brursch, M.O., Smith, *I.E.* and Wolstenholme, B.N. (1980). Pecan Quart., 14:7,9, 11-14.

Altan, P. (1987). Citrus and Subropical Fruit J., 638:17-19.

Alverson, D.R. (1985). Special Publ., Georgia, Agri Exp Stn., 38: 31-42.

Amling, H.J. and Amling, K.A. (1983). J. Amer. Soc. Hort. Sci., 108:195-98.

Anderson, P.C. and Aldrich, J.H. (1987). HortSci., 22: 79-82.

Anonymous (2012). U.S.D.A. National Nutrient Database for Standard Reference, release 25. http://www.ars.usda.gov/nutrientdata.

Asraf, R. (1977). Fruits, 32: 309-19.

Badiyala, S.D. and Badyal, J.M. (1986). Punjab Hort. J., 26: 102-04.

Badyal, J.M. and Upadhayay, S.K. (2004). Acta Hort., 662: 167-69.

Ball, J.C. (1982). Environ. Entom., 11: 1001-04.

Beedanagari, S.R., Dove, S.K., Wood, B.W. and Conner, P. (2005). Theoretical and Applied Genet., 110: 1127-37.

Beverly, R.B. and Worley, R.E. (1992). HortSci., 27: 271-72.

Bhardwaj, K.N. (1983). Thesis Abstracts. Haryana Agri. Univ., 9: 6-5.

Bock, C.H., Brenneman, T.B., Hotchkiss, M.W. and Wood, B.W. (2013) Crop Protection, 54: 213-20.

Bock, C.H., Hotchkiss, M.W., Cottrell, T.E. and Wood, B.W. (2015). Acta Hort., 1070:171-81.

Bolt, L.C. (1983). Pecan Quart., 17: 20-21.

Bouali, I., Trabelsi, H., Abdallah, I.B., Albouchi, A., Martine, L., Gregoire, S., Bouzaien, G., Gandour, M., Boukhchina, S. and Berdeaux, O. (2013). Jr. Amer. Oil Chemists, Soc., 90: 1869-76.

Brooks, R.M. and Olmo, H.P. (1972). Register of New Fruit and Nut Varieties, Univ. of Calif. Press, Berkeley, pp. 467-81.

Caleote, V.R. (1975). Jr. Econ. Entomol., 68: 4-6.

Caminada, P. (1979). Hortus, 26:42-43.

Cerna-Cortes, J.F., Simpson, J., Martinez, O. and Martinez-Peniche, R.A. (2003). Jr. Fd. Agric. Environ., 1(3/4): 168-73.

Chandel, J.S. and Ananda, S.A. (2000). Jr. Appl. Hort., 4: 45-46.

Chhinnan, M.S. (1984). Trans. ASAE, 27: 610-15.

Cole, J.R. and Hunter, J.H. (1962). Plant Dis. Reptr., 46:894-95.

Conner, P. and Wood, P. (2001). Jr. Amer. Soc. Hort. Sci., 126: 474-80.

Cottrell, T.E. and Wood, B.W. (2008). Agric. For. Entomol., 10: 363-73.

Cottwald, T.R. and Bertrand, P.F. (1988). Pl. Disease. 722: 27-32.

Crane, H.L., Hardy, M.B., Loomis, N.H. and Dodge, F.N. (1934). Proc. Amer. Soc. Hort. Sci., 32:29-32.

Dalklc, Z. (2013). Jr. Fd. Agric. Environ., 11: 879-82.

Daniell, J.W. (1979). In: Tree Nuts (ed. J.G. Woodroof) AVI Publishing Co. Inc., Westport, Connecticut, pp. 366-67.

Daniell, J.W. (1982). Res. Bull. college of Agri. Exp. Stn., Georgia Univ., 289:13.

Diver, S.G. and Smith, M.W. (1984). Comm. Soil Sci. and Pl. Anal., 15:619-37.

Dodge, F.N. (1939). Proc. Amer. Soc. Hort. Sci., 27: 503-08.

Dong Jia, X., Wang, T., Zhai, M., Li, Y.R. and Guo, Z.R. (2011). Molecules, 16: 10078-92.

Dozier, W.A., Jr. and Amling, H.J. (1974). Ala. Agric. Exp. Stn. Bull., 463: 1-34.

Driver, S.G., Smith, M.W. and McNew, R.W. (1984). Comm. Soil. Sci. Pl. Anal., 15:619-39.

Dutcher, J.D. and Harrision, K. (1984). J. Econ. Entom., 77: 1037-40.

Dutcher, J.D., Fonsah, G.E. and Hudson, W.G. (2009). Jr. Entomol. Sci., 44: 98-110.

Dutcher, J.D., Karav, H. and Abbas, G. (2012). Insects, 3: 1257-70.

Dutcher, J.D., Worley, R.E. and Littrelt, R.H. (1983). Res. Bull. College of Agri. Exp. Stn., Univ. of Georgia, 296:12.

Edelson, J.V. and Estern, P.M. (1987). Entomol. Sci., 22: 336-347.

Elmore, C.D. and Potles, S.C. (1980). J. Agri. Fd. Chem., 28: 902-04.

El-Nabawy, S., Rawash, M.A., Desovky, I., Et-Hamady, A. and Khalil, F. (1980). Ann. Agri. Sci. Ain Shams Univ., 25: 323-38.

El-Sharkawy, S.M.M. and Osman, I.M.S. (2009). Egyptian J. Hort., 36: 29-46.

El-Wahab, F.K.A., El-Hamady, A.M., El-Nabawy, Rawash, M.A. and Hogagg, L.F. (1984). Gartenbauwissenschaft, 49: 67-64.

Fayek, M.A., Fayed, T.A., El-Said, E.H. and El-Hamed, E.E.A. (2008). Res. Jr. Agric. Biological Sci., 4: 310-20.

Finch, A.H. (1937). Proc. Amer. Soc. Hort. Sci., 34: 74-77.

Forkert, C. (1914). Proc. Natl. Nut Growers Assoc., 13: 26-30.

Foshee, W.G. III., Blythe, B.K., Goff, W.D., Faireloth, W.H. and Patternson, M.G. (2008). HortSci., 43: 399-402.

French, R.B. (1962). J. Amer. Oil Chem., 39: 176-78.

Gammon, N. Jr. (1976). Pecan South, 5:214-16.

Gammon, N. Jr. and Sharpe, R.H. (1956). Proc. Amer. Soc. Hort. Sci., 68:.195-200.

Ganjegunte, G.K., Sheng, Z.P. and Braun, R.J. (2011). Jr. Environ. Quality, 40: 1314-21.

Gentry, C.R. and Smith, J.S. (1982). Protection Ecol., 6: 339-51.

Ghate, S.R. and Chhinnan, M.S. (1983). Energy in Agric., 2:11-20.

Ghate, S.R. and Chhinnan, M.S. (1984). Trans. ASAE. 27:635-39.

Godoy-Avila, C. and Lopez-Montoya, I. (2000). Terra, 18: 305-11.

Godoy-Avila, C. Reyes, J.I. and Torres, E.C.A. (2005b). Agriculture Technica in Mexico, 31: 203-12.

Godoy-Avila, C., Xopiyaxtle, J.Z., Reyes, J.I. and Torres, E.C.A. (2005a) Terra, 23: 505-13.

Grantham, R.A., Mulder, P.G. Jr. and Cuperus, G.W. (2002). Southwestern Entomolgist, 27: 21-30.

Grauke, L.J., Iqbal, M., Reddy, A. and Thompson, T. (2003). Jr. Amer. Soc. Hort. Sci., 128: 374-80.

Grauke, L.J. and Gel, M. (1983). Louisiana Agric., 272: 969-70.

Grauke, L.J., Price, H. and Jonhston, J.S. (2001). HortSci., 36(4): 814.

GuangQin, L., Haizun, Z., Xu, Z., Jingya, S. and Beibei, Z. (2011). Jr. Fruit Sci., 28: 1132-33.

Gustafson, W.A. and Storey, J.B. (1982). HortSci., 17: 967-70.

Hall, M.J., Hedger, G.H., Eikenbary, R.D. and McNew, R.W. (1981). Environ. Entom., 19:668-72.

Hammer, H.E. and Hunter, J.E. (1948). Proc. S.E Pecan Growers Assoc., 411: 36-43.

Hammer, H.E. and Hunter, J.E. (1949). Plant Physiol., 24: 16-30.

Hanna, J.D. (1972). Ph.D. dissertation. Texas A and M University.

Hanna, J.D. (1987). In: Rootstocks for Fruit Crops, (eds. R.C. Rom and R.F. Carlson). John Wiley and Sons, New York, pp. 401-10.

Hansen, J.D. and Storey, J.B. (1972). Pecan Quart., 6: 17.

Hansman, D. and Novoa, C.O.YD. (1986). CAB. Int. Hort. Abstr., 56:403-16.

Hansman, K.C. and Lazarte, J.E. (1984). HortSci., 19: 237-39.

Harris, O.W. and Smith, C.L. (1957). Proc. Amer. Soc. Hort. Sci., 70:204-08.

Hassanein, S.A. and Gabr, M.F. (2013). World Applied Sci. J. 21: 315-19.

Haulik, T.K. and Holtzhausen, L.C. (1987). South African J. Pl. and Soil, 4:70-74.

Haulik, T.K. and Holtzhausen, L.C. (1988). South African J. Pl. and Soil, 5: 1-4.

Heaton, E.K., Daniell, J.W. and Moon, L.C. (1982). J. Food Sci., 47: 1271-75.

Herrera, E. and Martinez, D. (1982). Pecan Quart., 17: 18-19.

Heyerdahl, R. and Dutcher, J.D. (1985). J. Econ. Entom.,78:1121-24.

Hinrichs, H.A. (1970). Proc. Okla. Pecan Growers Assoc., 40: 29-33.

Hinrichs, H.A., Ramming, D.W. and Hopfer, D.A. (1971). Ann. Rep. North Nut Growers Assoc., 62: 105-08.

Jacobs, M.B. (1951). The Chemistry and Technology of Food and Food products, Vol.-II, Interscience Publishers, New York, p. 970.

Jaynes, R.A. (1969). In: Handbook of North American Nut Trees (ed. R.A. Jaynes). The Northern Nut Growers Assoc. Tennessea, pp. 385-87.

JianHong, Z., JuanJuan, S., GuoHua, X., YouJun, H., Jianain, H. and ZhenhJia, W. (2014) Jr. Anhui. Agric. Univ., 41: 203-208.

Jenkins, J., Wilson, B., Grimewood, J., Schmutz, J. and Grauke, L.J. (2015). Acta Hort., 1070: 101-108.

Johnson, W.T. (1969). In: Handbook of North American Nut Trees. (ed. R.A. Jaynes)., The Northern Nut Growers Assoc., Tennessee, pp- 67-156.

Jones, R.W. (1980). Proc. S.E. Pecan Growers Assoc., 73: 179.

Joolka, N.K., Rindhe, A.B. (2000). Indian J. Hort., 57: 51-53.

Joolka, N.K., Rindhe, A.B. and Sharma, M.K. (2001). Indian J. Hort., 58: 212-14.

JunNan, L., XinWu, X., XueLiang, X. and SuJong, Y. (2008). South China Fruits, 6:65-66.

Kaakeh, W. and Dutcher, J.D. (1994). J. Econ. Ent., 87: 951-56.

Kallestand, J.C., Mexal, J.G., Samis, T.W. and Heerema, R. (2008). Hort. Tech., 18: 714-25.

Kays, S.J. and Wilson, D.M. (1977). Sci.Hort., 6:213-22.

Kays, S.J., Crocker, T.E. and Worley, R.E. (1975). J. Agric Fd. Chem., 23: 1116-18.

Kennedy, H.E., Jr. (1981). Plant and Soil, 63: 307-16.

Khalil, F.A., Hamouda, A.M., El-Din, F.S. and El-Said, M.E. (1982). Agric. Res. Rev., 60:111-24.

Khalil, F.A., Hamouda, A.M., Shenab, LS.A. and Noaman, V.F. (1983). Ann. Agric. Sci., Ain Shams Univ., 28:1689-98.

Khalil, S.K., Mexal, J.G., Khalil, I.H., Wahab, S., Rchman, A., Hussain, Z., Khan, A.Z. and Khattak, M.K. (2016). The Pharmaceutical and Chemical J., 3(4): 150-56.

Kim, T. and Wetzstein, H.Y. (2005). Jr. Hort. Sci. Biotechnol., 80: 681-86.

Kim, T., Mills, H.A. and Wetzstein, H.Y. (2002). Jr. Pl. Nutrition, 25:1987-2000.

Knox, C.A. (1980). Ph. D. dissertation, Texas A and M University.

Knox, C.A. and Smith, R.A. (1981). Pecan Quart., 15: 23-24.

Koen, T.J., Langenegger, W. and Smart, G. (1978). Citrus and Subtrop. Fr. J., 540:10-12.

Kraimer, R.A., Lindemann, W.C. and Herrera, H.A. (2001). HortSci., 36:308-12.

Kraimer, R.A., Lindemann, W.C. and Herrera, H.A. (2004). HortSci., 39:256-60.

Kuden, A.B., Tuzeu, O., Bayazit, S., Yildirim, B. and Imrak, B. (2013). African J. Agric. Res., 8: 3159-65.

Lagarda, A. (1987). Acta Hort., 199:145-49.

Laiche, A.J. Jr., Kilby, W.W. and Overcash, J.P. (1983). HortSci., 18: 328-29.

Law, J.M., Ator, D.W. and Young, W.A. (1980). Pecan South, 2: 36-41.

Lazarte, J.E. (1983). Comb. Proc. Intern, Pl. Propagator's Soc., 33: 614-18.

Lazarte, J.E. (1984). Amer. Nurserymen, 159: 79-80, 82.

Lipe, J.A. and Morgan, P.W. (1970). HortSci., 5: 266-67.

Lipe, J.A. and Morgan, P.W. and Storey, J B. (1969). J. Amer. Soc. Hort. Sci., 94: 668-71.

Lockwood, D.W. and Sparks, D. (1978). J. Amer. Soc. Hort. Sci., 103:38-45.

Lombardini, L., Villarreal-Lozoya, J.E. and Cisneros-Zevallos, L. (2009). Acta Hort., 841: 91-96.

Madden, G.B. (1969). Proc. S.E. Pecan Growers Assoc., 61: 27-52.

Madden, G.D. (1982). Pecan Quart., 16: 23-24.

Madden, G.D. and Malstrom, H.L. (1975). In: Advances in Fruit Breeding, (eds., J. Janick, and J.N. Moore)., Purdue Univ. Press, West Lafayette, Indiana, pp. 420-38.

Madden, G.D., Brison, F.R. and McDaniel, J.C. (1969). In: Hand-Book of North American Nut Trees, The Northern Nut Growers Association, Tennessee, pp. 163-89.

Malstrom, H.L. and McMeans, J.L. (1982). HortSci., 17:.970-72.

Malstrom, H.L., Fenn., L.B. and Riley, T.D. (1983). Pecan Quart., 17:5-13.

Malstrom, H.L., Riley, T.D. and Jones, J.R. (1982). Pecan Quart., 16:6-15.

Maness, N. (2004). USDA Agricultural Handbook, 66.

Marquard. R.D. (1987). HortSci., 22: 256-57.

Martin, G.C., Campbell, R.C. and Carlson, R.M. (1980). Proc. Amer. Soc. Hort. Sci., 105: 34-37.

Maximos, S.E., Rawash, M.A., Dessokey, Antoun, N. and Youssef, N. (1979). Egyptian J. Hort., 6:157-67.

McEachern, G.R. and Storey, J.B. (1984). Pecan Quart., 18:27-30.

McEachern, G.R., Wolstenhotme, B.N. and Storey, J.B. (1978). HortSci., 13:694.

McKay, J.W. (1947). J. Agric. Res., 74: 263-83.

McKay, J.W. (1961). Proc. Texas Pecan Growers Assoc., 40: 66-76,

McKay, J.W. (1964). Proc. Texas Pecan Growers Assoc., 43: 67-70.

Medina, M. del-C., Medina-Moreno, E-de.J., Aguilar-Perez, J.H. and Garcia-Garza, S.J. (1999). Terra, 17: 317-23.

Medina, M.del-C (2004). Terra, 22: 445-50.

Medina, M.del-C and Chavez, G.J.F.J. (1999). Terra, 17: 293-98.

Mielke, E.A. (1981). Pecan Quart., 15: 9-10, 12, 14-18.

Misra, R.S. (1985). Progressive Hort., 17: 80-82.

Mizamoto, S. (1984). Trans, ASAE, 27: 456-63.

Mohsenl-Moghadam, M., Schroeder, J., Heerema, R. and Ashigh, J. (2013). Weed Technol., 27: 85-91.

Mudler, P.G. and Grantham, R.A. (2007). Oklahoma Coop. Ext. Serv. Fact. Sheet. No. EPP-7079.

Nesbitt, M.L., Goff, W.D. and Stein, L.A. (2002). Hort. Technol., 12: 257-60.

Nunez-Moreno, H., Walworth, J.L., Pond, A.P. and Kilby, M.W. (2009). HortSci., 44: 1736-40.

Nyczepir, A.P., Reily, C.C., Wood, B.W. and Thomas, S.H. (2002). Pl. Dis., 86: 441.

O'Barr, R.D., Kowalzuk, J., Koonce, K., Rachel, S. and Gel, M. (1984). Louisiana Agric., 28:22-24.

O'Barr, R.D., Sherman, W., Young, W.A., Meadows, W.A., Calcote, V. and Kenknight, G.E. (1989). Louisiana Agri. Exp. Stn., 129:18.

Oro, T., Ogliari, P.J., Amboni, R.D.de.M.C., Barrera-Arellano, D. and Block, J.M. (2008). Grasas y Aceites Sevilla, 59: 132-38.

Ortloff, S. (2009). American Peanut Council, Alexandria, VA.

Osborn, M.R., Pierce, W.C., Phillips, A.M., Cole, J.R. and Kenknight, G.E. (1966). USDA Handbook, 24: 55.

Overcash, J.P., Laiche, A,J.J. and Kilby, W.W. (1983). Pecan South,3:22-31.

Payne, J.A., Dutcher, J.D. and Ellis, H.C, (1985). J. Agric. Entom., 2: 1-5.

Pee'r, S. and Kessler, B. (1984). Scientia Hort., 24: 323-29.

Pichkhaya, N. Sh. (1987). Subtrop. Kul'tury, 1:106-10.

Pond, A., Walworth, J., Kilby, W.M., Gibson, R., Call, R. and Nunez, H. (2006). HortSci., 41:1339-41.

Rajashekar, C.B. and Reid, W. (1989). HortSci., 24:348-50.

Raney, H.G. and Eikenbary, R.D. 91968). Cand. Entomol., 100: 1091-95.

Rathoie, D.S. (1983). Progressive Hort., 15: 8-10.

Rathore, D.S. (1985). Indian J. Hort., 42: 61-62.

Reid, W., Huslig, S.M., Smith, M.W. and Maness, N.O. (1993). HortSci., 28: 800-802.

Robinson, J.C. (1977). Hortus, 24: 45-47.

Romberg, L. D. (1967). Proc. Texas Pecan Crowers Assoc., 46:72-75.

Romberg, L.D. (1931). Proc. Texas Pecan Growers Assoc., 11: 56-61.

Romberg, L.D. (1968). Proc. S.E. Pecan Growers Assoc., 61: 56-65.

Romberg, L.D. and Smith, C.L. (1950). Proc. Texas Pecan Growers Assoc., 29:12-21.

Sampaio, V.R. and Barbin, D. (1980). Anais de Escola Superior de Agricultura Zuig de Querroz, 33: 329-34.

Sampaio, V.R. and Barbin, D. (1981). Revista de Agricultura, Piracicoba, 56:279-85.

Sanderlin, R.S. and Heyderich-Alger, K.I. (2000). Pl. Dis., 84: 1282-86.

Sanderlin, R.S. and Heyderich-Alger, K.I. (2003). Pl. Dis., 87: 259-62.

Sanderlin, R.S. and Melanson, R.A. (2006). HortSci., 41: 1455-56.

Sanderlin, R.S. and Melanson, R.A. (2008). Pl. Dis., 92: 1124-26.

Sanderlin, R.S., Hunt, A.S. and Babcock, D.K. (1983). Pl. Dis. 67:1209-13.

Santerre, C.R. (1994). Pecan composition, In: Pecan Technology. (ed. C.R. Santerre) Chapman and Hall, New Yark, pp. 98-100.

Sarig, Y., Nahir, D., Emek, M., Grorz, F., Beres, H. and Rasis, A. (1982). Trans. ASAE, 25:586-89.

Senter, S.D., Horvat, R.J. and Forbus, W.R. (1980). Jr. Fd. Sci., 45: 1380-82,92.

Sharma, S.D. and Rosanghura, S. (1987). Fr. Science Rep., 14:119-23.

Sharpe, R.H. and Gammon, N.Jr. (1958). Fla. Agr. Exp. Sta. Bull., pp. 601-67.

Sharpe, R.H., Blackman, G,H. and Gammon, N.Jr. (1952). Proc. S.E. Pecan Growers Assoc.,45:81-82.

Shear, C.B. and Faust, M. (1980). Hort. Rev., 2: 142-63.

Singanusong, R., Mason, R.L., Arcy, B.R. and Nottigham, S.M. (2003). Jr. Agric. Fd. Chem., 51: 406-12.

Smith, M.W. (1977). Ph. D. dissertation, Texas A and M University.

Smith, M.W. (2002) HortSci., 37: 398-401.

Smith, M.W. and Cheary, B.S. (2005). HortSci., 48: 1411-15.

Smith, M.W. and Cheary, B.S. (2010). Jr. Amer. Pom. Soc., 64: 92-100.

Smith, M.W. and Couch, G. (1984). Pecan South, l8: 11-14.

Smith, M.W. and Gallot, J.C. (1990). HortSci., 25: 414-16.

Smith, M.W. and Hinrichs, H.A. (1980). Pecan Quart, 14:4-5,6.

Smith, M.W. and Johnson, J.L. (1981). Pecan Quart, l5: 20-22.

Smith, M.W., McNew, R.W., Ager, P.L, and Cotten, B.C. (1986). J. Amer. Soc. Hort. Sci., 111: 558-61.

Smith, M.W. and Romberg, L.D. (1940). J. Agric. Res., 60: 551-664.

Smith, M.W. and Storey, J.B. (1979). J. Amer. Soc. Hort. Sci., 104: 474-77.

Smith, M.W., Ager, P.L. and Endicott, D.S.W. (1985). J. Amer. Soc. Hort. Sci., 110:446-50.

Smith, M.W., Cheary, B.S. and Carroll, B.L. (2001). Jr. Amer. Pomol. Soc., 55: 124-28.

Smith, M.W., Cheary, B.S. and Carroll, B.L. (2004). HortSci., 39: 1412-15.

Smith, M.W., Cheary, B.S. and Carroll, B.L. (2005). HortSci., 40: 1723-25.

Smith, M.W., Cotton, B.S. and Ager, P.L. (1987). HortSci., 22: 82-84.

Smith, M.W., Wazir, F.K. and Akers, A.W. (1989). Comm. Soil Sci. and Pl. Anal., 20: 335-44.

Snir, I. (1983). Plant and Soil, 74: 287-89.

Sorenson, R.B. and Jones, T.L. (1999). Hort. Tech., 9: 402-08.

Spapiro-Ilan, D.I., Cottrell, T.E., Gardner, W.A., Behle, R.W., Ree, B. and Harris, M.K. (2009). Southwestern Entomologist, 34(2): 111-20.

Spapiro-Ilan, D.I., Gardner, W.A., Wells, L., Cottrell, T.E., Behle, R.W. and Wood, B.W. (2013). Environ. Entomol., 42(2): 253-61.

Sparks, D. (1974). Ann. Rep. North Nut Growers Assoc., 65: 145-58.

Sparks, D. (1977). HortSci., 12: 497-98.

Sparks, D. (1983). Ann. Rep. North Nut Growers Assoc., 74: 197-230.

Sparks, D. (1986a). HortSci., 2l:108-09.

Sparks, D. (1986b). HortSci., 2l: 451-53.

Sparks, D. (1988a). HortSci., 23: 570-57.

Sparks, D. (1988b). J. Amer. Soc. Hort. Sci., 113: 850-59.

Sparks, D. (1989a). HortSci., 24:78-79.

Sparks, D. (1989b). HortSci., 24: 454-55.

Sparks, D. (2002). Acta Hort., 527: 35-48.

Sparks, D. and Heath, J.L. (1972). HortSci., 7: 402-04.

Sparks, D. and Madden, G.D. (1985). J. Amer. Soc. Hort. Sci., 102:219-23.

Sparks, D. and Payne, J.A. (1982). HortSci., l7: 670-71.

Sparks, D. and Yates, D. (2002). HortSci., 37: 176-77.

Sparks,D. and Baker, D.H. (1975). J.Amer. Soc. Hort. Sci., 100:392-99.

Staden, J. van., *et al.* (1976). HortSci., 11: 261-62.

Stein, L.A., McEachern, G.R. and Storey, J.B. (1987). Acta Hort., 201: 175-19.

Stein, L.A., McEachern, G.R. and Storey, J.B. (1989). HortSci., 24:607-11.

Tedders, W.L. (1978). USDA Tech. Bull., 1579,29.

Tedders, W.L. and Payne, J.A. (1986). J. Econ. Entom., 79:490-96.

Tedders, W.L. and Wood, B.W. (1985). J. Econ. Entom., 78:642-46.

Thomas, S.H., Fuchs, J.M. and Handoo, Z.A. (2001). Pl. Dis., 85: 1030.

Thompson, T.E. (1984). Fruit Var. J., 38: 103-12.

Thompson, T.E. (1988). Fruit Var. J., 42: 118-19.

Thompson, T.E. and Grauke, L.J. (1998). Jr. Amer. Soc. Hort. Sci., 123: 85-90.

Thompson, T.E. and Grauke, L.J. (2000). HortSci., 35: 308-09.

Thompson, T.E. and Grauke, L.J. (2001). HortSci., 36: 1162-63.

Thompson, T.E. and Grauke, L.J. (2010a). HortSci., 45:1386-87.

Thompson, T.E. and Grauke, L.J. (2010b). HortSci., 45: 1392-93.

Thompson, T.E. and Grauke, L.J. (2012). HortSci., 47: 121-23.

Thompson, T.E. and Hunter, R.E. (1985). HortSci., 20: 776

Thompson, T.E., and Young, E.F.Jr. (1985). Rept. Grower's Assn. College. Station.

Thompson, T.E., Hunter, R.E., Madden, G.D. and Brown, E.J. (1982). HortSci., 17:972-73.

Thor, C.J.B. and Smith, C.L. (1935). J. Agric. Res., 50:97-121.

Upadhyay, S.K. and Badyal, J. (2008). Haryana J. Hort. Sci., 37: 6-8.

Upson, J.R. Jr. and Sparks, D. (1969). Proc. Amer. Soc. Hort. Sci., 94: 125-27.

Verma, B.P., Heaton, E.K. and Zaltzman, A. (1985). Paper, ASAE, 85,30, 41 p 9.

Vitliers, E.A. De and Viljoen, H.M. (1987). Inf. Bull., Citrus and Subtrop. Fr. Res. Inst., South Africa, 175: 16-20.

Volk, G.M., Waddell, J. and Towill, L. (2009). HortSci., 44: 317-21.

Walworth, J. Pond, A., Sower, G.J. and Kilby, W.M. (2006). HortSci., 41: 275-76.

Wansri, R., Mason, R.L., Wakeling, L.T. and Nottingham, S.M. (2001). Food Australia, 53: 562-67.

Watt, B.K. and Merrill, A.L. (1963). Agri. Res. Serv., USDA, Handbook-8.

Wells, J.M. and Payne, J.A. (1980). Pl. Disease, 64: 991-99.

Wells, M.L. (2008). Hort. Tech. 18: 455-59.

Wells, M.L. (2011). HortSci., 46: 306-10.

Wells, M.L. (2012). HortSci., 47: 927-31.

Wells, M.L. and Wood, B.W. (2008). HortSci., 43: 1437-40.

Wells, M.L., Carlson, D.S. and Edwards, R.P. (2009). Hort. Tech., 19: 518-20.

Westwood, M.N. (1978). Temperate-Zone Pomology, W.H. Freeman and Co. San Francisco, p. 153.

Wetzstein, H.Y. (1989). Acta Hort., 240: 193-96.

Wetzstein, H.Y. and Sparks, D. (1983). J. Amer. Soc. Hort. Sci., 108:997-1003.

Wetzstein, H.Y. and Sparks, D. (1984). J. Amer. Soc. Hort. Sci., 109: 245-52.

Wetzstein, H.Y. and Sparks, D. (1985). J. Amer. Soc, Hort. Sci., 110:778-81.

Wetzstein, H.Y. and Sparks, D. (1986). Hort. Rev., 8: 211-55.

Wetzstein, H.Y., Merkle, S.A., Ault, J.R., and Sommer, H.E. (1988). SAAS Bull. Bioch. Biotech., 1:64-68.

Wetzstein, H.Y., Sparks, D. and Long, G.A. (1983). HortSci., 18:331-33.

White, A.W. Jr. (1982). HortSci., 17: 380-81.

Witt, H.J., Allison, J.R. and Daniell, J.W. (1989). J. Amer. Soc. Hort. Sci., 114:61-64.

Wolstenholme, B.N. (1979). Pecan Quart., 13: 14-19.

Wood, B.W. (1982). HortSci., 17: 890.

Wood, B.W. (1983). HortSci., 18: 53-56.

Wood, B.W. (1986). J. Amer. Soc. Hort. Sci. 111: 533-37.

Wood, B.W. (1988a). HortSci., 23: 341-43.

Wood, B.W. (1988b). HortSci., 23: 1026-28.

Wood, B.W. (1988c). J. Amer. Soc. Hort. Sci., 113:374-77.

Wood, B.W. (1989a). HortSci., 24: 260-62.

Wood, B.W. (1989c). J. Amer. Soc. Hort. Sci., 114:279-82.

Wood, B.W. (2000). Hort. Tech., 10: 120-26.

Wood, B.W. (2011). HortSci., 46: 870-77.

Wood, B.W. (2015). Acta Hort., 1070: 33-42.

Wood, B.W. and McMeans, J.L. (1982). J. Amer. Soc. Hort. Sci., 107: 47-50.

Wood, B.W. and Reilly, C.C. (1984). HortSci., 19: 661-63.

Wood, B.W. and Reilly, C.C. (2001). HortSci., 36: 298-301.

Wood, B.W. and Reilly, C.C. (2004). HortSci., 39: 110-14.

Wood, B.W. (1989b). J. Amer. Soc. Hort. Sci., 114: 223-28.

Wood, B.W., Canner, P.J. and Worley, R.E. (2003). HortSci., 38: 361-66.

Wood, B.W., Lombardini, L. and Heerema, R.J. (2009). HortSci., 44: 1884-89.

Wood, B.W., Payne, J.A. and Jones, O. (1990). HortSci., 25: 916-18.

Wood, B.W., Reilly, C.C. and Nyczepir, A.P. (2004). HortSci., 39: 1238-42.

Wood, B.W., Wells, L. and Funderburke, F. (2010). HortSci., 45: 911-17.

Woodroof, J.G. (1930). J. Agric. Res., 40: 1059-104.

Woodroof, J.G. (1979). Tree Nuts. AVI Publishing co. Inc., Westport. Connecticut.

Woodroof, J.G. and Woodroof, N.C. (1927). J. Agric. Res., 3: 1049-63.

Worley, R.E. (1974). J. Amer. Soc. Hort. Sci., 99: 49-57.

Worley, R.E. (1979). HortSci., 14: 51-52.

Worley, R.E. (1989). HortSci., 24: 151.

Worley, R.E., Carter, C.L. and Johnson, A.W. (1975). J. Amer. Soc. Hort. Sci., 100:487-90.

Worley, R.E., Harmon, S.A. and Carter, R.L. (1974). J. Amer. Soc. Hort. Sci., 99:57-62.

Worley, R.E., Litterell, R.L. and Dutcher, J.B. (1980). Arboriculture. 6:253-57.

Wright, G.C., Benton Storey, J., Harris, M.K. and Spinz, P.T. (1990). HortSci. 25: 698-700.

Xiaoyan, D., Zeping, J., Chun, J. and YoyungRong, L. (2013). Jr. Jiangou Forestry Sci. Technol., 40(3): 10-14.

Yahia, E.M. (2010). In: Phytochemicals: Chemistry, Nutritional and stability. Wiley-Blackwell, Chapter 1, pp. 3-51.

Yates, *I.E.*, and Sparks, D. (1995). Jr. Amer. Soc. Hort. Sci., 120: 446-53.

ZhengHai, Mo., JiYu, Z., Min, Z., XiaoDong, J., JiPing, X., YongRong, L. and ZhongRen, G. (2013). Jr. Southern Agric., 44(5): 730-34.

# Chapter 11

# Pistachios

Pistachio nut (*Pistacia vera* L.) is an exceedingly delectable and valuable nut is in great demand and its acreage and production have increased manifold recently in many growing countries. But its rather exacting climatic requirements are the main impediments in expanding its cultivation to new regions. Pistachio can, however, withstand drought and adverse soil conditions comparatively well and can be grown in areas having shallow soils.

## Composition and Uses

Pistachio nut is relished very much as dessert after roasting and salting but it is mainly used as an ingredient of sweetmeats, confectionery, bakery goods, meat dishes and ice-creams. The nut is known for its excellent nutty flavour and digestive, sedative and tonic values. Marmalade is also made from the fruit husks. The kernel oil finds its use in confectionery and pharmacy.

According to Gopalan *et al.* (1987), the nut contains per 100 g: water 5.6 g; protein 19.8 g; fat 53.5 g; minerals 2.8 g; fibre 2.1 g; carbohydrates 16.2 g; calcium 140 mg; phosphorus 430 mg; iron 7.7 mg; carotene 144 µg; thiamine 0.67 mg; riboflavin 0.28 mg; niacin 2.3 mg and energy 626 k cal. In addition, Watt and Merrill (1963) have reported potassium 972 mg; magnesium 150 mg and ascorbic acid 30 mg. On an average, the pistachio kernels yield about 56 per cent oil which may go up to 68 per cent (Clarke *et al.*, 1976; Jacob, 1951; Ermakov *et al.*, 1982) The percentage fatty acid composition of the oil is: myristic 0.6; palmitic 8.2; stearic 1.6; oleic 69.6; and linoleic 19.8 (Anon., 1969). Depending on the extraction methods, the cold pressed oil is golden yellow while solvent extracted is dark green having aromatic odour. The oil has saponification value 191-195; iodine value 84-94; R.M. value 0.5; and acid value 0.5-4.0.

Compared to almonds, peanuts, pecan and walnuts, pistachios have fewer calories and less fat per serving. The unique biochemical composition of pistachios yields beneficial effects on cardiovascular health by impacting lipid status, glycemic control, blood pressure and measures of adiposity (Kris-Etherton *et al.*, 2008). Pistachios are rich in unsaturated fatty acids, phytosterols, lutein, dietry fibre, protein, magnesium, caretenoids, gamma tocopheral, phenolic compounds and potassium (U.S. Department of Agriculture, 2015) (Table 11.1).

**Table 11.1: Average Nutrient Composition of Pistachio Nuts per 100g**

| | |
|---|---|
| Energy (kcl) | 562 |
| Total Fat (g) | 45:39 |
| Saturated Fatty Acid (g) | 5.556 |
| Monounsaturated Fatty Acid (g) | 23.82 |
| Polyunsaturated Fatty Acid (g) | 13.74 |
| Total Protein (g) | 20.27 |
| Plant Sterols (mg) | 214 |
| Fibre (g) | 10.3 |
| Calcium (mg) | 105 |
| Magnesium (mg) | 121 |
| Potassium (mg) | 1025 |
| Lutein and Zeaxanthin (μg) | 1405 |
| Beta carotene (μg) | 249 |
| Gamma tocopherol (μg) | 22.6 |

Adopted from USDA (2015).

Pistachios contain an abudance of both water-and fat-soluble antioxidants including carotenoids, gamma-tocopherol and phenolic compounds and have been listed among the top fifty foods highest in antioxidants (Halvorsen *et al.*, 2006). Ozrenk *et al.* (2012) working with 14 well-adopted pistachio genotypes of Eastern Turkey reported 8.16-9.33 per cent palmitic acid, 0.54-0.68 per cent palmitoleic acid, 2.35-4.21 per cent stearic acid, 67.81-76.82 per cent oleic acid, 9.42-18.32 per cent linoleic acid, 0.27-0.38 per cent linolenic acid and 0.19-0.33 per cent arachidic acid. The range of selenium, alpha-tocopherol, gamma-tocopherol, delta-tocopherol, alpha-tocotrienoid, gamma-tocotrienoid and total carotenoid of these promising genotypes were between 11.44 and 190.71ng/g, 1.36 and 26.93, 36.17 and 170, 0.45 and 2.61, 0.96 and 3.76, 2.33 and 37.72 and 1.01 and 4.93mg/kg, respectively. The oil fraction of pistachio is a good source of essential fatty acids and phytosterols, which play a primary role in preventing cardiovascular disease (Ballistreri *et al.*, 2011).

Shokraii and Esen (1988) indicated that 66 per cent of the total protein was globulin (NaCl-soluble) while albumins (water soluble), glutelins (sodium borate-soluble) and prolamins (tert-butyl alcohol-soluble) contributed 25 per cent, 7 per cent and 2 per cent, respectively, of the total protein. Nine essential amino acids are present and lysine is predominant (Clarke *et al.*, 1976). The main sugar found in the kernel is sucrose which is followed by glucose, fructose and inositol (Crane and Al-Shalam, 1974). This is in contrast to the view of the predominance of glucose (Clarke *et al.*, 1976). A good dessert cultivar should have 58-60 per cent oil, more than 4 per cent sugars but less than 13.5 per cent protein (Chernova *et al.*, 1986). The main anthocyanin pigment in pistachio nut skin has been identified as cyanidin-3-0-β-galactoside (Miniati, 1981).

## Origin and Distribution

The place of origin of pistachio is though not known with certainty, yet central Asia is considered to be its home due to the large stands of wild trees found there (Whitehouse, 1957; Zohary, 1952). It is believed to have originated in Central Asia 80 million years ago (Kafkas and Perl-Treves, 2001). Zohary (1952) stated that *Pistacia vera* is native to the barren, dry foothills in the elevated deserts of Afghanistan, Iran and Turkey, and is thought to be the ancestral species of the modern-day pistachio. Its first commercial cultivation began in Iran. Turkey, Syria and other countries in the vicinity of wild pistachio forests and the fruit was probably introduced into Mediterranean Europe at the commencement of first century (Whitehouse, 1957). Smolsky and Smirnov (1931) considered the Badkhyz pistachio stands as the largest in central Asia which are located at Kushka in the south of Turkmenistan near the Afghanistan border. A large number of wild pistachio trees are found growing in northern Afghanistan beyond the river Amu Daria in low mountains of the Pamir-Alai.

## Production

The principal pistachio growing countries are Iran, USA, Turkey, China, Syria, Italy, Greece and Afghanistan, whereas on a small-scale it is cultivated in Tunisia, Spain, Jordan and Cyprus. The production of pistachio in different countries are presented in Table 11.2.

**Table 11.2: Pistachio Production in different Countries**

| *Country* | *Production in Tons 2014* |
|---|---|
| Iran | 415531 |
| USA | 233146 |
| Turkey | 80000 |
| China | 76943 |
| Syria | 28786 |
| Greece | 5700 |
| Italy | 3555 |
| Afghanistan | 2761 |
| Tunisia | 2500 |
| Spain | 2468 |

FAOSTAT (2015).

## Taxonomy

The genus *Pistacia* to which pistachio belongs is a member of the *Anacardiaceae* family. Zohary (1952) has recognized eleven species in the genus on the basis of leaf characteristics because floral characters are simple and uniform. The taxonomic relationships among its species of genus *Pistacia* is not well understood due to the fact they have no genetic barriers. Al-Saghir and Porter (2012) revised the taxonomy of this genus in details. It includes the following taxa: *Pistacia atlantica* Desf., *P. chinensis* Bunge subsp, *falcate* (Bess ex Martinelli) Rech. f., *P. chinensis* supsp. *integerrima* (J.L. Stew, ex Brandis) Rech. f., *P. eurycarpa* Yalt., *P. khinjuk* Stocks, *P. lentiscus* L. susp. *lentiscus*, *P. lentiscus* subsp. *emarginata* (Engl.) AL-Saghir, *P. mexicana* Humb., Bonpl., and Kunth, P. X saportae Burnat, *P. terebinthus* L., *P. vera* L., and *P. weinmannifolia* poiss. Ex Franch. The genus is divided into two sections: section Pistacia and section Lentiscella. A key to the 14 taxa that have been recognized by this study is included. The new combination *P. lentiscus* subsp. *emarginata* (Engl.) AL-Saghir is made, and the names *P. chinensis* subsp. *integerrima* (J.L. Stew. Ex Brandis) Rech. F., *P. terebinthus* L., and *P. vera* L. are lectotypified. The diploid chromosome numbers of *Pistacia vera* are 30 (Zohary, 1952). Among the various *Pistacia* species, only *P. vera* produces the commercially acceptable edible nuts while other species serve as rootstocks or ornamental trees. *Pistacia atlantica* and *P. terebinthus* bear small nuts being eaten locally and used for oil extraction. A high grade resin, mastic, is obtained from *P. lentiscus* (Anon., 1969).

Pistachio is a deciduous tree with broad-spreading branches and imparipinnate leaves, most often two paired (Zohary, 1952). Each leaf has one to five pairs of thick wide, leathery pinnate leaflets. A leaf at each node subtends a single axillary bud which mostly differentiates into floral bud during April and grows to the ultimate size by late June (Takeda *et al.*, 1979). Generally, one or two axillary buds located distally on the new growth are vegetative. Pistachio bears laterally on wood produced in the previous season (Crane and Nelson, 1971). It is a dioecious plant and both staminate and pistillate inflorescence consist of hundred to several hundreds individual flowers. Both types of flowers are apetalous and wind is the pollinating agent. The fruit is a semi-dry drupe. The nuts have a tendency to split open at one side without discharging the kernel. Each nut is enclosed within a thin lightly adhering, wrinkled and reddish skin. The kernel is single solid piece with pleasing mild flavour.

A brief description of various *Pistacia* species which have been tried as rootstocks of pistachio is given in Table 11.3.

## Cultivars

In the market pistachio nuts are known by the country of their origin, *i.e.*, Afghan, Iranian, Sicilian, Syrian and Turkish. The price is determined by the green colour and flavour of the nut. Large, light yellow and dry Iranian nuts are considered inferior to the Sicilian, Syrian and Turkish nuts, the latter having rich oily nut-like flavour and green colour (Whitehouse, 1951). The dark green colour of Italian nuts fetches them a premium price. Likewise, the smaller Afghan nuts are preferred by the ice-cream and pastry industries for the deep green colour of the kernels.

**Table 11.3: Brief Description of *Pistacia* Species Used as Rootstocks**

| *Name of the Species* | *Origin and Distribution* | *Plant Height (m)* | *No. of Leaflets* | *Description of Leaflet* | *Description of Fruit* |
|---|---|---|---|---|---|
| *P. vera* | Central Asia Mediterranean region | 9 | 2-10 | Thick, wide, leathery, pinnate leaflets, lanceolate | Fruit a semi-dry drupe |
| *P. atlantica* | Mountainous parts of Mediterranean region | 12-15 | 7-11 | Alternative lanceolate, obtuse, glabrous and sessile with narrow, winged petioles | — |
| *P. chinensis* | China | 18 | 10-14 | Short petiole, lanceolate, long-pointed | Fruit is an obovoid rotundate drupe compressed about 1cm long |
| *P. integerrima* | — | 12 | 8-10 | Opposite, lanceolate from an oblique base with short petioles, long pointed | Fruit a drupe broader than long |
| *P. khinjuk* | Western Asia | — | 2-4 | Oblique, broadly ovate or oblong | Small, obovate, compressed |
| *P. lentiscus* | Southern Europe, northern Africa and Algeria | 4-5 | 4-10 | Leathery, elliptic oblong or ovovate, obtuse | Small, orbicular, slightly annulate |
| *P. mutica* | Native to Asia Minor, Iran and Afganistan | 11 | 4-8 | Oblong or oblong-ovate, obtuse, petioles marginate or slightly winged | Fruit a drupe. Long, obovate, flattened obliquely apiculate |
| *P. terebinthus* | Distributed throughout southern Europe, northern Africa | 5 | 9-13 | Mucronate, petioles slightly winged | Small orbicular, slightly flattened dark purple and wrinkled |

The seedling selection have been made in various important pistachio growing countries on the basis of bearing and fruit quality and these are planted for the expansion of its cultivation. The best cultivars of Iran are Momtaz, Owhadi, Agah, Safeed and Kaleghoochi, which have been selected at Rafsinjan, the world-famous pistachio growing area in Iran. 'Momtaz' produced light coloured nuts in large clusters while 'Owhadi' nuts are smaller but better in flavour than 'Momtaz'. 'Agah' nuts have good qualities of both 'Momtaz' and 'Owhadi' cultivars (Whitehouse, 1957).

'Uzun' and 'Kirmizi' are the two popular cultivars of Turkey. The nuts of the former are long and plump whereas the nuts of the latter are medium in size, red-hulled but green kerneled. The important Syrian pistachio cultivars are Alemi, Obiad, Mirhavy Achoury, Ayimi, El Bataury and Ashouri (Whitehouse, 1957).

After introducing a large number of selections and cultivars from the pistachio growing regions of the world, the following cultivars have been released in the USA for cultivation after selection.

### Kerman

It is the main cultivar of California pistachio industry producing large and high quality nuts (Crane, 1984). It was selected in 1929 at California, from seed imported from Rafsanjan, Iran and released for trial in 1977. The tree is vigorous, upright, late bloomer and high yielder. It has, however, several drawbacks such as low natural nut splitting, blank nuts and tendency towards biennial bearing. Above all, epicarp lesion is the most serious defect which may cause crop losses up to 30 per cent in some years (Crane, 1984).

### Joley

It originated as an open-pollinated seedling and released in 1980 by the University of California. Both blooming and nut maturity are earlier than 'Kerman'. In comparison to 'Kerman', it produces nuts of smaller size but shell splitting is superior and blank production is less (Crane, 1984).

### Sfax

It originated in Tunisia. It is the earlier maturing than 'Kerman' bur the per cent of split shells is higher and nut quality is better. Its chilling requirement is also lower than 'Kerman' (Crane, 1984).

### Lassen

It originated at the US Plant Introduction Station, Chico, California, as a seedling selection from the seeds introduced by Whitehouse from Rafsanjan, Iran. The nut size is above average; shell splits well and is easily opened by hand. Roasted quality is good and yield is above average. It blooms frequently late in some years to escape frost (Brooks and Olmo, 1972).

### Sirora

Sirora originated from an open-pollinated seedling of Red Aleppo. The plant is

Gummosis in Pistachio

Male Inflorescence of Pistachio

of spreading type and slightly more vigorous than 'Kerman'. The nut is intermediate in shape and slightly flattened at the suture. It bears comparatively regularly, showing uniform bud break (Brooks and Olmo, 1972).

## Pontikis

It originated as a chance seedling of the cultivar 'Aegenes'. The tree is of medium size (4.5 to 5.0 m tall and 5-6 m in spread) and semi-erect. Flowering (15-20 April)

**Pistachio Tree**

**Pistachio cultivar Kerman**

and ripening (10-15 September) occur 10-15 days later than in 'Aegenes'. 'Pontikis' can be pollinated by *P. vera*. The fruits are large (25 x 16 mm) and the kernels are good in flavour and crisp with deep green cotyledons. The cultivar is mostly grown in Greece (Brooks and Olmo, 1972).

## Golden Hills

It was developed by university of California and released in 2006. It is a high-yielding, high-percent split, early maturing female *Pistacia vera* cultivar for harvest before 'Kerman'. Shell size and shape are similar to 'Kerman'.

## Kalehghouchi

It was developed in Iran. It is a high-yielding, female *P. vera* cultivar with large, high quality nuts which is more difficult to shake harvest than other cultivars. In-shell size and shape much larger than 'Kerman', harvest date and yield similar to 'Kerman'.

## Lost Hills

Developed by the university of California and released in 2007. It is an early maturing, low-alternate bearing, large nut, femal *P. vera* cultivars for harvest before 'Kerman'.

## Ruchle

Developed at USA. Female *P. vera* cultivar with potentially better nut quality than 'Kerman'. Nuts are similar to 'Kerman' but larger kernel, less susceptible to epicarp lesion than 'Kerman'.

## Gumdrop

It is a new female pistachio cultivar for California (Kallsen and Parfitt, 2017). It matures earlier than all commercial cultivars with equivalent yield and nut quality to 'Kerman'. 'Gumdrop' has very good yield, nut quality and processed nut appearance similar to 'Golden Hills' and 'Kerman'. The early nut maturity of 'Gumdrop' makes it less susceptible to insect damage from navel orange worm, a major pest of pistachio implicated in the occurrence of aflatoxin contamination.

Important cultivars of pistachio growing in different countries are presented in Table 11.4.

**Table 11.4: Pistachio Cultivars Growing in different Countries**

| | | |
|---|---|---|
| Australia | : | Sirora |
| China | : | Zaoshu, Duanguo, Changguo, Kerman, Lassen, Sirona, Notarolo. |
| Cyprus | : | Lamarka |
| Greece | : | Aegina, Thynal, Pontikis, Nichati, Fountoukati. |
| India | : | Kerman, Peter, Red Aleppo, Joley |
| Iran | : | Momtaz, Ouhadi, Kaletighouchi, Gherneza, Tbeahimi, Ogah, Wahadi, Akbari, Rezaei, Chrook, Rezaei-e-Zodras, Badami-e-Zodras. |
| Iraq | : | Ashoury |
| Israel | : | Kastel, Rashti, Kerman, Larnaka, Sfax, Aegina, Aleppo, Shufra, Arie. |
| Italy (Sicily) | : | Napoletana, Silvana, Gloria, Agostana, Girasola, Notaloro, Cappuccia, Femminello, Trabonella, Bronte. |
| Lebanon | : | Red Aleppo |
| Mexico | : | Kerman, Peters, Sfax, Chico. |
| Morocco | : | Achouri, Mateur. |
| Pakistan | : | Kerman, Ibrahmin, Owhadi, Safeed, Shasti, Wahedi. |
| South Africa | : | Ariyeh, Shufra, Sirora. |
| Spain | : | Aegina, Ashoury, Avdat, Avidon, Kerman, Larnaka, Mateur, Napoletana, Peters. |
| Syria | : | Ashouri Al Grahy, (Red Aleppo), Red Oleimy, White Batoury, Nab Al Dajaml, Ajamy, Baidy, Marawhy, Zaroory. |
| Tunisia | : | Sfax, Mateur, El Guettar, Nouri, Meknassy, Thynal. |
| Turkey | : | Uzun, Kirmizi, Red Aleppo, Alemi, El Bataury, Obiad, Ayimi, Siirt, Tekin, Halebi, Krmz, Ohadi. |
| USA (California) | : | Kerman, Joley, Lassen, Golden Hills, Lost Hills, Red Aleppo, Bronte, Trabonella, Kastel, Rashti Ibrahmim, Ohadi, Safeed, Lassen, Shasti, Buenzle. |

## Pollenizers

### Peters

It was selected and named by A.B. Peters in Fresno, California. The tree produces pollen in abundance and the blooming period synchronizes with that of 'Kerman' as well as some of the earlier blooming pistillate cultivars. It is considered as a universal pollenizer (Crane, 1984).

### Chico

It originated as a seedling selection of *P. vera* in Chico, California. The seeds were introduced from Aleppo, Syria. The tree is a prolific pollen producer, blooms early and is a suitable pollenizer for 'Bronte', 'Red Aleppo' and 'Trabonella' but blooms too early for 'Kerman'. Both 'Chico' and 'Peters' together are effective complementary pollenizers for most of the pistachio cultivars which are either early or late blooming (Joley, 1969).

### Randy

Developed by the University of California in 2007. Male *P. Vera* pollenizer for 'Golden Hills' and 'Lost Hills'. It has long flowering period, more and larger inflorescences than 'Peters' male, more viable pollen than 'Peters'.

## Climate

### Temperature

Pistachio is a deciduous plant which can stand a very wide range of temperature from –30°C in winter and 42.2°C in summer without being damaged (Joley, 1969, Whitehouse, 1957). In fact, there is no report about cold injury to the plant due to low temperature. The optimum temperature for growth and development of pistachio has been reported between 25 and 35°C (Ferguson, 2003). For proper bud opening and subsequent growth, pistachio trees, however, require 700 to 1000 chilling hours at or below 7°C (Crane, 1984). Which depends on the tree age, location and even cultivars (Beede *et al.*, 2005). Esmaeilizadeh *et al.* (2006) studied the chilling requirements of three Iranian pistachio cultivars and reported the chilling requirements of cvs. Ahmadaghaei, Fandoghi-Ghafouri, and Chorouk cultivars as 1000, 1200 and 1400 hours, respectively. Inadequate chilling results in delayed and irregular blooming and foliation along with conspicuous changes in leaf form and bearing behaviour (Crane and Takeda, 1979). Instead of normal five leaflets, the leaves had only three leaflets and even turned into simple form. Likewise, terminal bearing on current shoots has been observed due to mild winter (Crane, 1984).

On an average, a good crop of pistachio was harvested only once in a period of five years due to lack of chilling at Sacation, Arizona (Crane, 1984). Mineral oil, dinitro-orthocresol and hydrogen cynamide sprays were found effective in overcoming delayed blooming due to lack of chilling (Procopious, 1973; Pontikis, 1989). Arreola-Avila *et al.* (2011) reported that hydrogen cyanamide applied at 50 per cent at around 30 days before expected bud break can increase budbreak and substitute for chilling when pistachio is grown in environments with mild winters.

## Rainfall and Humidity

The most important feature of the climatic requirement of pistachio is long, hot and dry summer which is essential for the development and maturation of nut. Pistachio trees are known as drought tolerant plants that are able to survive and even produce fairly good yield with very little water (Ferguson, 2003, Goldhamer, 2005). An annual rainfall of at least 300 to 450 mm has been reported as the optimum precipitation for pistachio (Goldhamper, 2005). Inoculation with arbuscular mycorrhiza enhanced the drought tolerance (Abbaspour *et al.*, 2011). The tree flowers and sets fruit freely in England but the nuts fail to mature due to cold and humid weather conditions prevailing in summer. In Iran, where the best pistachio nuts are produced, the mean annual rainfall is 25 to 37 cm and summer temperature is around 31.5°C (Whitehouse, 1957). Being a xerophilous plant, it tolerates the dryness of both soil and air. High humidity is unfavourabale for fruit set and its development, and may also accelerate the incidence of diseases. Sometimes, late spring frosts causes crop losses but usually pistachio escapes frost damage due to comparatively late blooming. Likewise, cool, damp and rainy days during blossoming retard fruit set by interfering in pollen release and its transfer to stigma (Joley, 1969).

## Soil

Although adapted to all kinds of soils it prefers relatively deep, light or dry sandy loam soils with high lime content. Wild trees of pistachio are found growing on the rocky soils of Lebanon, Palestine and the Isle of Cyprus. It prefers 65 to 75 cm of top soil with perfect drainage and is unable to withstand excess water (Popov, 1979). In Greece, pistachio trees raised on *P. terebinthus* rootstocks grow very well in calcareous soils along the seashore of Phaleron and Aigina which are poor and dry. The tree response is favourable to its application of hard or salty water.

Soil having electrical conductivities upto 6ds/m showed no effect of pistachio growth, but increasing it to 8ds/m reduced the yield up to 50 per cent (Sanden *et al.*, 2004). Abbaspour (2012) stated that pistachio plants alleviates the deleterious effect of salt stress, through protection mechanism against oxidative damage by maintaining an inherited and induced activity of antioxidant enzymes and higher proline content that may play a role as an enzyme-stabilizing agent in salt stress. Soil pH of 6.5 to 7.5 is considered suitable for growing pistachio.

Other *Pistacia* species have their own soil preference. In California, *P. terebinthus* grows exceedingly well on dry, gravelly hill sides but poorly on heavy soils underlaid with hardpan and not too well-drained. Thus, the soil preference of the *Pistacia* species rootstock should be taken into account in selecting a site for pistachio orchard (Whitehouse, 1957). On the whole, high yields can be obtained only from the trees grown on deep fertile and well-drained soils. Sandy loam soil is best for high quality nut production (Ferguson, 2003; Goldhamer, 2005).

## Rootstocks

Nine species of *Pistacia* used as rootstock for pistachio are *P. vera, P. atlantica, P. chinensis, P. integerrima, P. lentiscus, P. mutica; P. terebinthus* and *P. palaestina*. The choice of a rootstock is based on disease resistance and its ability to grow under cold (Whitehouse, 1957). *P. mutica* grows on rocky sites of extremely infertile soil in the USSR; *P. khinjuk* is found growing wild on clay soils. *P. terebinthus* is preferred as a stock to *P. vera* in Greece because of its immunity to foot rot and prolonged summer drought. *Pistacia vera* seedlings require a soil of medium texture which is deep, permeable and sufficiently moisture-retentive to prevent critical drying out. *P. atlantica* does well on the sandy North African soils. Pistachio cultivars propagated on *P. chinensis* have been reported to short-lived compared with those on other stocks. Extreme stock-scion incompatibility is observed at the end of the first season's growth in the nursery (Whitehouse, 1957).

In California, the main rootstocks used are *P. atlantica, P. integerrima* and *P. terebitnthus* as they have greater resistance to nematode and *verticillium* wilt than *P. vera* (Joley and Whitehouse, 1953; Crane, 1984). Cultivars budded on these rootstocks initially grow slowly but they pick up growth subsequently and outyield those budded on *P. vera* (Joley, 1969). Crane and Forde (1976) reported significantly greater yield of 'Kerman' when grafted on *P. atlantica* than on three other rootstocks. *viz., P. palaestina; P. terebinthus* and *P. vera*. However, pistachio seedling rootstocks were found better than *P. atlantica* or *P. palaestina* rootstocks (Ibrahim and Nahlawi, 1982).

In New Mexico, *P. terebinthus* showed more cold tolerance than *P. atlantica*. However, *P. atlantica* showed greater per cent take of buds and grow bigger than those of *P. terebinthus* and bears earlier. In California, *P. integerrima* is a marketed as Pioneer Gold which is, however, not winter hardy. The cross between *P. integerrima* and *P. atlantica*, named Pioneer Gold-II is also resistant to *Verticillum* wilt like Pioneer Gold, but more cold tolerant than Pioneer Gold-I. The university of California developed UCB1 (*P. atlantica, P. integerrima*) is a vigorous rootstock and claims to be *Verticillium* wilt resistant as well (Herrera, 1997).

**Table 11.5: Relative Efficiency of different Rootstock***

| *Rootstock* | *Salinity Tolerance* | *Cold Tolerance* | *Disease Tolerance* | | |
|---|---|---|---|---|---|
| | | | *Verticillum* | *Armillaria* | *Phytophthora* |
| *P. terebinthus* | | * | | * | |
| *P. atlantica* | * | ** | *** | *** | * |
| *P. integherrima* | **** | **** | * | *** | *** |
| PG-II | | *** | **** | *** | |
| UGB-I | | *** | * | * | * |

From Most* to Least****.

* Compiled from Ferguson *et al.* (2008).

Atl *et al.* (2011) evaluated five pistachio cultivars on different rootstocks under irrigated conditions in Turkey. They observed best tree vigour, yield and nut quality of cv. Sirit grafted on *P. khinjuk*. The shelling percentages were higher with split nuts than unsplit nuts in all cultivars (Sirit, Kirmizi, Halebi, Uzun and Ohadi) and a positive effect was noted with *P. atlantica* rootstock (Turker and Ak., 2011).

Rootstock selection plays one of the most important roles in orchard management. In Kerman province and in other pistachio growing areas of Iran, pistachio cultivars are grown on pistachio seedling rootstocks of different cultivars such as 'Badami Zarand' and 'Ghazvini'. Sherafati *et al.* (2011) studied different rootstocks and scion combination and reported 'AK bari' cultivar on 'Badami' rootstock caused maximum potassium and zinc acquisition (1.56 per cent and 11.05 ppm respectively). 'Barg-Seyah' scion buded on 'Kalle-ghouchi' resulted in the most amount of iron (241ppm) and copper (12.15ppm). Inter-specific hybrid rootstock of *P. atlantica* x *P. vera* cv. 'Badami-Riz-Zarand' was reported as more tolerant to drought (Kuhbanani and Karimi, 2013). 'Ghazvini' rootstock is tolerant to salinity (Karimi *et al.*, 2009).

Banihashemi (1998) working with 11 cultivars of *P. vera* and different pistachio rootstocks reported that *P. mutica* and *P. khinjuk* were more susceptible to all phytophthora species than *P. vera* cultivars used as rootstock. *P. atlantica* and UCB1 (*P. atlantica* x *P. integerrima*) were found highly resistant to all phytopthora species and did not show any disease symptoms during the experiment and two years thereafter.

## Propagation

### Seed Propagation

The most common method of vegetative propagation is by budding onto seedling rootstock. Seeds of *Pistacia* species are collected in the fall, cleaned and stored dry. Removal of the hull before putting the seeds for germination is necessary since it acts as a germination inhibitor. Seeds of *Pistacia* species vary in their capacity to germinate depending on the hardness of the endocarp. Seed coats can be removed by rubbing over screens. Seeds are usually stratified in a refrigerator (2-4°C) for one to three months before they are induced to germinate. The effect of seed scarification on the germination of *P. terebitnthus* and *P. atlantica* was studied by Crane and Forde (1974). They observed that the seeds of *P. terebinthus* gave 53 per cent germination following scarification with $H_2SO_4$ for 1.5 hours and then by soaking in water for 24 hours. None of the seeds germinated in the control after two weeks from sowing. However, the treated seeds of *P. atlantica* after 2 and 4 weeks gave a germination percentage of 46 and 80, respectively, compared with 26 per cent in the untreated seeds.

Germination of *P. chinensis* seeds ranged from 63 per cent to 92 per cent after 60 days, stratification at 4°C in comparison with 0-24 per cent when sown directly without chilling treatment (Pair and Khatamian, 1982). Ibrahim *et al.* (1984) in Syria studied the germination of seeds of different pistachio cultivars *viz.*, Ascouri, Batouri and Barri and of *P. atlantica*. Germination of seeds was generally high (75 per cent or more) but in the cultivar Barri germination was better when sown in March. *P. atlantica* seeds recorded low germination. Rafieva and Nikolacova (1983) have recommended the use of pistachio seeds without shells for uniform germination and emergence.

Prior to sowing, mechanical removal of the endocarp from the seeds of pistachio cultivar Napoletana advanced germination but eventually lowered the germination percentage (Casini and Conticini, 1979). They also reported that in unshelled seeds of *P. terebinthus*, when immersed in 50 ppm $GA_3$ for seven days, the germination percentage increased from 50 to 79.

Seedling of pistachio species tend to produce taproots in the nursery and suffer considerable transplanting shock. For this reason they are grown in plastic bags or containers and not in nursery rows. Al-Imam and Al-Jubury (2011) studied different growing media on seed germination and seedling growth of cv. Ashoury in Iraq. They reported that media consists of sandy loam soil: peat moss (3:1) significantly increased seed germination, while sandy loam soil: manure: peat moss $(3:\frac{1}{2}:\frac{1}{2})$ significantly increased speed of germination and subsequent seedling growth. Seeds sown in January or February become large enough for budding in September.

## Vegetative Propagation

### Budding and Grafting

T-budding on seedling rootstock is more successful in August than in May (Pair and Khatamian, 1982). Ibrahim and Nahlawi (1982) reported success by T-budding in June. Shield budding in late March and early April resulted in 77 per cent to 95 per cent success. Summer budding proved to be better than spring budding (Olekhnovich and Chernova, 1975). Chip budding, carried out in November gave 90 per cent success (Needs and Alexander, 1982). The method involves the selection of bud-wood from the current season's growth and inserting into the rootstock at 30-35 cm above ground level and removing the top 10 cm of the rootstock immediately. Cleft grafting has been found successful in the month of February, when both the stocks and scions were in dormant condition (El-Sayed *et al.*, 2000). Zaen El-Deen and Abd El-Rhman (2011) from Egypt reported that both cleft and side grafting methods were successful when practiced between January and March. Supplemental irrigation further increased success in January and February grafted plants. Acar *et al.* (2011) suggested chip budding should be carried out in spring at Firal Valley. Unsprout seedlings should be budded in June and the rest in autumn. Thus, budding can be done at 3 different seasons in a year. In California, T-budding is the most commonly used method in commercial pistachio production. T-budding is performed when the rootstock bark is slipping during the early summer and the scion buds buds have matured and hardened.

Ak *et al.* (2013) evaluated different rootstocks (*Pistacia spp.*) on budding success of 'Siirt', 'Kirmizi' and 'Ohadi' pistachio. They reported average bud take rate of 78 per cent in *P. vera*, 68 per cent on *P. atlantica*, 64 per cent on *P. palestina*, 63 per cent on *P. khinjuk* and *P. terebinthus*. The highest T-budding success rate (95 per cent) was obtained from *P. vera* rootstock and 'Siirt' scion and the budding success was much higher in April compared with June and September.

Veneer grafting of the cultivar Kerman, on *P. atlantica* rootstock gave 100 per cent take compared with 83 per cent by bark grafting and 66 per cent by shield budding. Veneer grafting also resulted in the best scion growth. The time of rootstock pruning (at or 25 days after grafting/budding) had no effect on scion growth in veneer-grafted trees but scion development was best in bark-grafted trees with rootstock pruning at grafting and in shield-budded trees with rootstock pruning 25 days after budding (Ruelas, 1986).

## Micrografting

A method of grafting has been developed at Avdat, Israel, for producing light weight, compact rooted and easily transplantable nursery trees for distant locations (Maurice, 1988). The overall propagation system is characterized by establishment of plant material in two successive stages (basket and ridge stages) followed by various grafting methods.

During the basket stage seeds are sown in shallow baskets (60 x 20 cm, 6 cm depth) supported on plastic pipes or stones in order to obtain early lateral rooting. When good lateral roots, have developed, plants are transferred to the ridge stage. The ridges are built on the contour of a 45° hill side underlaid with thick black polyethylene sheeting. This causes the roots to grow down at a slant and on reaching the base of the ridge they burn off by exposure to sun and air accelerating the lateral fibrous root growth. The ridges are covered on the north side by a moss-like ground cover which stabilizes the soil mixture, increases water absorption rates, cools and permits inoculation and development of beneficial soil micro-organisms. Trees are extracted by simply pulling out of the water saturated ridges. This arrangement helps to overcome difficulties related to grafting trees in a restricted volume of soil.

Micrografting in pistachio is best carried out in the autumn of the first season or in spring of the second season but grafting can, in fact, take place at any season when the bark slips. A form of apical wedge grafting on the relatively succulent last growth flush appears to give best results. The wedge of the scion is cut rather long, leaving only a small part near the terminal bud. The wedge is inserted deep into a slit at the apical end of the stock, leaving practically only the terminal bud exposed. The tie should be extended to cover the scion up to the terminal bud. If the apical part of the stock has insufficient diameter, side grafting with terminal bud scions or chip budding are also satisfactory. In budding, care must be taken to choose plump triangular-shaped buds, as found on strong growing flushes.

## Cutting

Pistachio is one of the woody perennial species which shows a tendency to become very difficult to-root when it has passed out of the juvenile phase. Softwood cuttings taken from juvenile shoots of one-year-old seedling and treated with 3000, 10000 and 20000 ppm IBA gave 39 per cent, to 92 per cent rooting but cuttings from older trees were unsuccessful (Pair and Khatamian, 1982). However, Al Barazi and Schwabe (1982) reported that cuttings taken from growing shoots of more than six-year-old pistachio trees when dipped in IBA solution of concentrations ranging from 0-45000 ppm rooted in all the concentrations except control.

## Top Working

Nahlawi *et al.* (1984a) found that eight-year-old pistachio trees when cut back to

20 and 60 cm above ground level and grafted in February gave the most successful result than the trees grafted in March and May. In another trial, old trees were severely pruned in winter and emerged new shoots were successfully T-budded in mid-June.

## Micropropagation

Clonal micropropagation of pistachio using tissue culture technique has been reported in the literature sine 1982 (Barazi and Schwabe, 1982). There are many reports available for successful propagation of *P. vera*; protocols are also available for *P. atlantica, P. terabinthus* and *P. integerrima*. However, micripropagation of rootstock species and hybrids are difficult.

Alderson and Barghchi (1982) and Barghchi and Alderson (1982, 1985) established multiple shoot cultures from apical and lateral buds from aseptically grown seedlings and from greenhouse-grown plants up to two-years-old. An average shoot proliferation rate of 12 shoots per bud was obtained with 4 mg/l IBA on MS medium. Following subculturing, this rate could be increased to as many as 40 shoots per bud. Barghchi and Alderson (1982, 1983a, 1983b, 1985) found that regular subculturing reduced the problem of tip necrosis and apical dominance. It also reduced explant browning and the production of callus at the base of shoots. Cultured shoots rooted readily on MS medium containing 2.5 mg/l IBA. Rooted shoots were successfully established *in vivo* in peat-based compost in pots in the greenhouse. Callus growth was obtained on most tissue explants, on media containing auxin and cytokinin. Sectioning of this callus revealed the presence of embryoids and about l0 per cent of these cultures produced adventitious shoots.

Large-scale propagation of *P. terebinthus* cultivar Tsikoudia, the most popular rootstock for pistachio in Greece was done by using shoot tips (Pontikis, 1984). Shoot proliferation was induced on Anderson medium (Anderson, 1978, 1980). The medium was supplemented (per litre) with 100 mg m-inositol, 0.4 mg thiamine-HCl, 2.5 mg BA, 0.1 mg IBA, 0.1 mg $GA_3$, 30g sucrose and 7 g agar. After a week, explants were transferred to fresh medium to avoid the possible inhibitory effects of polyphenol leached from the explant. Rooting was induced by subculturing plantlets on Murashige and Skoog mineral salts with supplements similar to the shoot proliferation medium plus 89 mg phloroglucinol without BA and IBA at 1 ppm. The rooted propagules were transplanted into pots containing vermiculite and maintained in humid conditions for at least one week before transferring to a glasshouse.

Parfitt and Almedhi (1994) developed a modified medium for superior shoot multiplication of both *P. vera* and *P. integerrima* through manipulation of micronutrient levels and replacement of the sucrose carbon source with $CO_2$ and high light intensity. After multiplication phase, explants were separated and moved to a Gerlite based rooting medium without a sugar carbon source, followed by acclimation in perlite based solid substrate. Tilkat *et al.* (2014) developed a protocol for mass proliferation of *P. khinjuk*, shoot tips and nodal segments of *in vitro* grown axenic regenerates derived from seedlings of *P. khinjuk* stocks were used as material. In both semi-solid and liquid medium, an MS medium supplemented with 1 mg/l

BA and 30g/l sucrose was used. Various immersion frequencies (2,4,8,16 or 20 h) and immersion time intervals (5,10 or 15min) were tested on the temporary immersion bioreactor system (TIS). Repeated trials for immersion frequencies at short intervals (10min/2h, 10 min/4h *etc.*) showed hyper hydrated explants up to 100 per cent, while this problem was not observed in immersions with relatively longer time intervals (10min/20h). The higher hydration rate (8 per cent) were obtained in shoot tips with 10 min immersion frequency every 16h.

An effective *P. vera* micropropogation system was developed by Benmahioul *et al.* (2012) involving rapid axillary bud proliferation and *ex vitro* rooting. The highest shoot proliferation frequency was obtained from nodal explants cultured on MS based salts containing Gamborg (B5) vitamins and supplemented with 4 mg/l BA. The addition of 2 mg/l meta-toplin (mT) generated an optimal number of shoots with suitable morphological features, while kinetin was found unsuitable for shoot proliferation. Microcuttings were rooted *ex-vitro* after being dipped in auxin powder (IBA 2 per cent). Rooted plantlets were transplanted in plastic pots containing a peat : perlite: vermiculite (1:1:1) mixture and then transferred to the greenhouse. After 2 months, 81.5 per cent survival of rooted microshoots was achieved. Tikat *et al.* (2009) developed protocol for direct shoot and plantlet regeneration from *in vitro* regenerated leaf explants of male *P. vera* cv. 'Ar1'. Leaves excised from axenic shoot cultures of pistachio were used to induce organogenesis on a Murashige and Skoog (MS) medium with Gamborg vitamins supplemented with combinations of different concentrations of BAP and IAA. The highest adventitious shoot regenerations in 35 per cent of the explants, with the number of shoots ranging from 2 to 3 per explant, occurred in the explants cultured during the establishment phase in the medium with 1mg/l IAA and 2 mg/l BAP. For shoot multiplication, the highest number of new microshoot/explants (5.76) was obtained in a culture medium supplemented with 1mg/l BAP, but it was not significantly different from the number obtained at 2 mg/l BAP. A high rooting frequency (84 per cent) for microshoots was recorded on a medium supplemented with 2 mg/l IBA. *In vitro* rooted plantlets were transferred to pots filled with a mixture of soil, sand and peat (1:1:1). They were weaned in a growth room and finally moved to a greenhouse.

## Planting

Pistachios should be transplanted to their permanent planting site as early as possible. Transplanting can be done any time during the dormant season; the sooner after leaf fall in December or January it is done, the better. Pistachio roots get injured if exposed to air for more than a few hours. Thorough watering of each plant following transplanting is desirable to reduce losses. Planting is done at a distance of 4 to 5 m in the row and 8 to 10 m between rows adopting the rectangular system of planting. On crowding, alternate plant in the row is removed. In order to ensure adequate pollination, staminate plants are planted throughout the orchard providing one pollenizer for 7 to 8 female plants (Pontikis, 1989).

## Training and Pruning

Apart from the basic cultural managements like fertization; pruning and irrigation in pistachio orchards are not performed regularly which becomes an important factor in low yield, (Okay *et al.*, 2010). It was also indicated that the pruned trees grow more strongly and regularly, the severeness of apical dominance and periodic are decreased, and the yield fluctuation between the years are prevented (Beede and Ferguson, 2005). When the seedling is budded close to the ground and is allowed to grow with little or no pruning or training, it forms a medium-sized, bushy tree with spreading branches. Such trees facilitate harvesting of the clusters of nuts by hand from the ground. Generally, plants are trained to a modified leader system with three or more primary scaffold branches at 30 cm spacing on the trunk, the first scaffold branch on the main trunk being kept about a metre above the ground. In training, new branches more than 75 cm long should be pinched in summer so as to avoid bending and encourage lateral growth (Woodroof, 1979). In order to provide enough light and aeration, centre of the tree should be kept open. A well-trained pistachio tree needs only light pruning subsequently. Since pruning wounds heal very slowly in pistachio, large cuts are to be avoided.

Conventional pruning methods have failed to give desired result in pistachio because of its peculiar bearing behaviour and strong apical dominance. All the lateral buds on a shoot are floral buds except one or two buds near the tip which are vegetative. The latter grow and fruit in subsequent years distancing the bearing area further and further away from the centre and many 8- to 10- year-old branches are observed without any lateral growth. Since there is no lateral vegetative bud in bearing pistachio trees, heading back of such branches becomes unrewarding. Thinning out of 50 per cent branches also fails to initiate new growth due to strong apical dominance of remaining branches (Crane, 1984).

New vegetative growth can be obtained by heading back of all shoots but at the cost of one year's crop. Crane (1984) proposes that researches should be carried out to determine the different degrees of heading back and to test the efficacy of various chemicals in killing the growing points.

Wolpert (1986) observed that severe dormant pruning after an 'on year' generated strong vegetative flush after the 'off year' produced lateral growth with flower buds in the 'on year' and produced a larger crop in the following 'off year' Ferguson *et al.* (1995) demonstrated that two successive seasons of mechanical topping, started before the 'off year', produced changes in shoot growth, trunk growth, and bud retention that mitigated alternate bearing through three biennal cycles, without decreasing yield. They suggested annual hand-pruning to prevent or minimize alternate bearing of pistachio.

Zhang (2014) studied mechanical pruning utilizing a '1-side mechanical slant topping' technique on pistachio trees. This pruning type has been applied in Australia since 2007. The relationship between yield and flower bud numbers show that on-season trees may have more than 3000 flower buds while off-season trees may have less than 200 flower buds. Investigations also show that on- season trees on average have 15.7 nuts per rachis while off-season trees on average have 43.6 nuts per rachis. This style of mechanical pruning prior to an off-season resulted in significantly higher yield in the following off-season (2 years later) compared with trees that were mechanical pruned prior to an on-season.

# Nutrition

## Deficiency and Toxicity Symptoms

Deficiency symptoms of potassium and phosphorus in pistachio have been described by Ashworth *et al.* (1985). The combined effect of P and K deficiencies causes unthriftyness and results in poor yield. Potassium-deficient trees have sparse foliage canopies, small leaves and pronounced shoot die back. The threshold K level of leaves for deficiency appear to be 0.7-0.9 per cent in mid-summer. Phosphorus-deficient trees have normal dense foliar canopies, normal-sized leaves with chlorosis. The chlorosis appears first on leaves terminal to nut clusters, later involving all foliage of the affected trees. Initially, the leaves show interveinal chlorosis but later become bright yellow and desiccate and abscise prematurely. The threshold P level for deficiency in the leaves is 0.09-0.1 per cent. Vegetative growth resumes within 2 months following treatment with P and one year after K.

A leaf scorch symptom is associated with high chlorine levels in leaves (1.9 per cent) while normal leaves without scorch symptoms have 0.4-0.8 per cent Cl during mid-summer. Boron toxicity is characterized by marginal necrosis of mature leaves by the end of growing season. The concentration of boron in the necrosis-free leaf tissues ranges between 220-235 ppm while it is as high as 1000 ppm in necrotic affected tissues. Only trees under severe water stress are free from B toxicity while all trees receiving 4-5 times as much water exhibit symptoms of B toxicity (Ashworth *et al.*, 1985).

## Fertilizer Recommendation

Because of their deep root system pistachio trees do not benefit from the soil-applied fertilizers as much as other tree fruits but responds well to foliar application. Foliar sprays of bayfolar (NPK), heksal (NPK), wuxal-3 (NK), wuxal-5 (NPK) and urea increase leaf N,P,K,Fe and Zn contents (Kaska, 1986). Yields varying from 3.1-9.6 kg/tree were obtained from 25-year-old trees of cultivars Aschouri and Batouri when fertilized with N at 2.5 kg, $P_2O_5$ at 3.0 kg, $K_2O$ at 1.0 kg per tree along with $ZnSO_4$, $MnSO_4$ and Fe-chelate in various combinations (Nahlawi *et al.*, 1984b). According to them, K concentration was higher in bearing branches than in non-bearing branches and yield was generally highest with NK.

Rosecrane *et al.* (1988) noted increase nitrogen, phosphorus, and potassium as well as storage starch pools in pistachio trees due to alternate bearing. Kuru (1993) found that soil in Gaziantep, Turkey, used for pistachio growing had poorer levels of phosphorous than potassium and an application of 60kg P/ha increased the yield at least 3 fold. Weinbaum *et al.* (1995) reported that N fertilization increased yield in the 'on' year, but decreased it in the 'off' year and caused late leaf abscission. Potassium fertilization was reported to increase soil and leaf K-content, nut yield, and quality of pistachio nuts in California (Zheng *et al.*, 2001). Zheng *et al.* (2001) recommended application of 110 to 220 kg/ha K, using potassium sulphate, potassium chlorate or potassium nitrate.

Karimi *et al.* (2006) studied the effects of applied N (ammonium nitrate) and P (triple super phosphate) on 11-13-year-old pistachio trees at Iran. They reported that application of N-fertilizer caused a significant effect on total nut biomass, aborts and blanks and non-split nuts. The highest total nut biomass (3.87kg/tree), split (2.24 kg/tree) and non-split nuts (1.34 kg/tree) were obtained with 900g N/tree. Application of 600g $P_2O_5$ caused the highest nut biomass (3.61 kg/tree), split nuts (2.4 kg/tree) and aborts and blanks (0.351 kg/tree). The interactive effect of N and P showed that the highest total nut biomass (4.74 kg/tree) and split nuts (3.11 kg/tree) were obtained with 900g N and 600g $P_2O_5$/tree/year.

## Soil Salinity and Related Problems

Soil salinity is a serious problem of increasing importance in the principal pistachio growing areas of the world, especially in irrigated fields of arid and semi-arid regions. Continuous cropping together with an excessive use of chemical fertilizers and ill-managed irrigation have turned hundreds of cultivated fertile fields into saline ones. Saline soils contain sufficient soluble salts to suppress plant growth through a series of interacting factors including osmotic potential effect and ion toxicities (Sepaskhah and Maftoun, 1988). The main problem with a high sodium concentration is its effects physical and chemical properties of soils. It contributes directly to the total salinity and toxicity of sensitive crops and has negative effect on the structure of the soil that reduces its capacity to conduct water and air through its profile (Castellanos *et al.*, 2002). To overcome the salinity hazards, Malakouti (2006) suggested application of potassium sulphate and zinc sulphate. Increased concentrations of K and Zn under saline conditions would improve root expansion and elongation, which would increase the surface area contact between tree roots and soil nutrients. Sepaskhah *et al.* (1985) observed that the salinity levels of 3600 and 4800 µg NaCl/g reduced top dry weights of 'Fandoghi' pistachio by 31 per cent and 51 per cent, respectively. The concentration of chloride is well correlated with its concentration in the substrate and is considerably higher than that of sodium. Iron treatments produce a significant increase of this element in the tops.

Sepaskhah and Maftoun (1988) studied the relative salt tolerance of three pistachio cultivars, *viz.*, Badami, Fandoghi and Kaleghoochi using four soil NaCl levels (0, 1600. 3200 and 4800 µg $g^{-1}$). 'Fandoghi' was the least tolerant to NaCl level followed by 'Kaleghoochi' and 'Badami'. A 50 per cent reduction in the top and root dry weights was obtained in all the three cultivars.

## Effect of Micronutrients

Pistachio grown on some salt affected soils faces problems of B excess and toxicity rather than its deficiency (Gupta *et al.*, 1985). While small amount of B is essential for the normal growth of green plants as it is important for cell division and protein synthesis, slightly larger amount may be toxic to many species. Spraying of boron, during flowering (25-50 per cent flower bloomed) alone dissolved in water or in sucrose was reported to increase fruit set and a qualitative improvement of the in shell nut and of the kernel (both fresh and dry weight). Average production increased by 37 per cent in cv. Bianca (Avanzato *et al.*, 2011). To evaluate the response of boron on pistachio, Sepaskhah *et al.* (1988) studied the effects of six soil levels of B (0, 10, 20, 30, 40 and 50µg $g^{-1}$) on the growth and chemical composition of three cultivars, *viz.*, Badami, Fandoghi and Kaleghoochi. The results indicated that B applied up to 20 and 40 µg $g^{-1}$ had no significant effect on shoot dry weight of 'Badami' and 'Fandoghi'. However, higher B levels were associated with a marked suppression of shoot growth of these two cultivars. Addition of B at 10 µg $g^{-1}$ showed a significant increase in the shoot dry weight of 'Kaleghoochi' compared with the control plants. Higher B rates gave no further response. The toxic levels of B in shoots required for growth suppression were 9, 38 and 95 µg $g^{-1}$ dry weight for 'Badami', 'Fandoghi' and 'Kaleghoochi', respectively. Similarly, the B level in the soil saturation extract associated with retarded shoot growth were 0.37, 0.61 and 1.5 µg $ml^{-1}$ for 'Badami', 'Fandoghi' and 'Kaleghoochi', respectively, and those for root growth suppression were 0.24, 0.20 and 0.13 µg $ml^{-1}$, respectively. Therefore, the cultivars 'Badami', 'Fandoghi' and 'Kaleghoochi' are considered as very sensitive, sensitive and semi-tolerant, respectively, to B toxicity. Hossenifard *et al.* (2005) reported a high boron concentration in pistachio orchards (500 mg/l) of Rafsanjan, Iran. The use of fertilizers containing boron is not recommended in this area (Salehi *et al.*, 2010).

In the calcareous soils, foliar application of Zn, Fe and Cu fertilizers is necessary for obtaining better fruit set, yield and quality in pistachio (Soliemanzadeh *et al.*, 2013). A combination of Fe and Zn (each at 1000 mg/l) in the swollen bud stage (middle of March) in Rafsanjan, Iran on 25-year-old 'Owhadi' trees showed increased fruit set. The highest nut splitting rate was reported with Cu (200 mg/l). Spraying of Zn (1000 and 2000 mg/l) showed positive effect on yield (Solemanzadeh *et al.*, 2013). In Zn deficient situation, spraying of Zn three times at 15 days interval (before, during and after blossom) improve tree growth and yield (28 per cent). It also increased (140 per cent compared with control) Zn content of leaves (Kizilgoz *et al.*, 2010).

## Tissue Analysis

Leaf analysis has become a valuable guide for the mineral nutrition of fruits. Leaf sampling technique has been standardized in pistachio by Uriu and Crane (1976). To determine the mineral element status of the plant, the leaflets from the non-bearing shoots should be sampled during one month period prior to harvest. The concentration of mineral elements in the leaves of healthy plants should fall within the following range (Table 11.6).

**Table 11.6: Nutritional Range of Healthy Pistachio Leaves***

| *Element* | *Range (Dry weight basis)* |
|---|---|
| Nitrogen | 2.5-2.9 per cent |
| Potassium | 1.0-2.0 per cent |
| Phosphorus | 0.14-0.17 per cent |
| Calcium | 1.3-4.0 per cent |
| Magnesium | 0.6-1.2 per cent |

| *Element* | *Range (Dry weight basis)* |
|---|---|
| Sodium | 0.002-0.007 per cent |
| Chlorine | 0.1-0.30 per cent |
| Manganese | 30-80 ppm |
| Boron | 55-230 ppm |
| Zinc | 7-14 ppm |

*Uriu and Crane, (1976).

Early in the season the concentration of N, P and Zn in the leaflets is relatively high, thereafter the level drops rapidly during leaf expansion reaching a steady state in early summer. The concentration of manganese is low initially which gradually increases and then remains fairly constant. The trend of K, Mg, Cl and B is similar to Mn but they reach constant levels late in the season. N and P levels are lower in leaves from bearing branches but higher in potassium than the leaves from non-bearing branches.

Elloumi *et al.* (2014) studied the dymanics of macronutrients in leaves and shoots of fruiting and not fruiting branches of 'Mateur' pistachio trees in relation to crop load. They reported that leaf contents of N and P decreased during stages I and II of nut development to reach a constant level during the stage III and finally decline during post-harvest for both 'off' and 'on' trees. Leaf K contents increased during the growing season in fruiting branches. However, K contents were found higher early in the season and then decreased rapidly during stage I and II of nut growth. These dynamics were observed in both fruiting and non-fruiting branches. Zeng *et al.* (2001) stated that pistachio trees exhibited highly fluctuating seasonal leaf K levels. Leaf K concentration was found low (<10g/kg) during spring flush, increased dramatically during fruit development and declined rapidly after harvest. They observed a significant and positive correlation between nut yield and leaf K concentration during nut fill. The critical leaf K value for optimal yield was suggested as 16.9 g/kg. In the month of November, major nutrients (N, P and K) contents in the leaves declined and increased in the buds and stems, indicating a translocation from leaves to perennial organs for winter storage of minerals. Leaf Ca rate decreased gradually from April to November whereas the contents of this element in the buds and stems followed a sinusoidal curve with a minimum in April and maximum in July (Mehdi *et al.*, 2006).

Malakouti (2006) suggested optimum nutrient concentrations in pistachio leaves and nuts as N= 2.00, P= 0.12, K= 1.80, Ca= 2.50, Mg=0.50 and S=0.30 per cent and for Fe, Mn, Zn, Cu and B, the suggested concentrations were 110, 50, 40, 15 and 50 mg/kg. Optimum nuitrient concentrations for pistachio nuts wre recommended to be N=120, P=60, K=150, Ca=5, Mg=5, and Fe, Zn and Cu were 0.30, 0.15, 0.12 and 0.07mg in 100g dry weight, respectively.

## Irrigation

Pistachio trees are capable of withstanding low soil moisture conditions because of their deep -root system. Spiegel *et al.* (1977) found remarkable drought resistance in pistachio trees under run-off farming in the Neger Desert highlands. Male trees are more drought resistant than female trees (Sheremetev, 1984). When there is sufficient rain in winter season, it provides enough soil moisture for growing pistachio during summer months in many regions. Under summer drought conditions, the leaves on unirrigated pistachio trees become yellow and drop down before the maturation of crop whereas on the irrigated trees, the leaves remain green until the normal leaf fall (Whitehouse, 1957). He, therefore, recommended at least two summer irrigations in areas having lower rainfall than 49 cm. In a drip irrigation schedule, irrigation cut-off at 25 per cent and 50 per cent, of cumulative Etc caused significant reduction in the yield and quality of nuts (Phene *et al.*, 1985).

Studies on the effects of water stress and salinity on photosynthesis of the cultivar Kerman revealed that the response of pistachio trees to water stress were typical of those of xerophytic plants (Behboudian *et al.*, 1986). Although photosynthesis declined with decreasing leaf water potential, yet plants continued to photosynthesize until a leaf water potential of -5 MPa was reached. The plants were less efficient in their water use at a lower range of water potential (-6 MPa). Salinity level up to 225 mM Cl had no effect on photosynthetic rate or on stomatal and mesophyll resistance to $CO_2$ diffusion. Irrigation regimes and salinity levels of irrigation water also influence growth and chemical composition of pistachio trees to a great extent. With a 7-day irrigation interval, irrigation water with salinity levels of 4.5 m mhos $cm^{-1}$ caused 40 per cent and 34 per cent reduction in growth of 'Fandoghi' and 'Badami' cultivars after 11 weeks of treatment (Sepaskhah and Maftoun, 1981).

Palma nad Novello (1998) studied the effects of drip irrigation on gas exchange of 'Larnaka' pistachio and found that irrigation with 50 per cent of reference crop evapotransporation (Etc) had the highest water use efficiency and photosynthesis. Mature pistachio trees growing in shallow soil, a reduction in irrigation of 50 per cent of the crop evapotranspiration (Etc) during pit hardening (stage II) had no effect on final yield (Phene *et al.*, 1987). Guerrero *et al.* (2006) recorded significantly higher percentage of split nuts (≈ 65 per cent) by providing irrigation at 100 per cent Etc compared with ~35 per cent under rainfed conditions. Experimental evidence showed that the subsurface irrigation system was superior to flooding system, which save 20-50 per cent water and the higher yield obtained. However, it was found that the application of low water rate (2392 $m^3$/ha) caused decline in growth, yield and increase of early splits (Mohammadi and Panahi, 2006). Different pistachio cultivars showed variations in there physiological behaviour under rainfed conditions. Ghrab *et al.* (2011) working with different pistachio cultivars at southern Tunisia observed best compromise between photosynthesis and transpiration, which maximized water use efficiency in local cultivars 'Meknassy', 'Nouri' and 'Mateur' compared with introduced cultivar 'Libye Rouge'.

Goldhamer and Beede (2004) determined and proposed two stress tolerant period where regulated deficit irrigation (RDI) can be applied: 1) from mid-May through early July, just after full shell development and before the onset of rapid kernel growth; and 2) postharvest. These methods can result in a net reduction of approximately 9 in. (22-23cm) of water per season but have to be carefully managed, timed properly, and monitored with a plant and/or soil based water status techniques to prevent excessive water stress.

## Soil Management

The production of nuts in a pistachio orchard apart from the cultivar and insect damage is dependent largely on soil management which includes the provision of adequate moisture, fertilizer and weed control. Ashworth and Gaona (1982) opined that mulching was effective whether or not the soil was pre-irrigated with sprinkler before mulching in early July. Complete mulching was much more effective than treatment in which polythene sheets were separated by 0.5 or 1.2 m at the rows one year following mulching. Complete mulching of the soil surface of 6-year-old, drip irrigated planting of pistachio trees with clean polythene sheeting for 2 months resulted in elimination of *Verticillium dahliae* to a depth of 120 cm. Zhang (2011) reported increased yield and return by using reflective mulch. Oxyfluorfen plus either glyphosate or paraquat when applied 30 days or more after petal fall to pistachios provided excellent short-term, broad-spectrum control of summer annual weeds (West *et al.*, 1983).

## Flowering, Pollination and Fruit Set

### Flowering

Since pistachio is a dioecious plant, the pistillate and staminate panicles are borne on separate plants. Initially, the buds in the axil of leaves are vegetative for 4-5 years which subsequently develop into floral buds as the plant passes into productive phase. In Northern Hemisphere, depending on position and tree age, axillary buds can differentiate into simple inflorescence buds starting in October and reach their ultimate bud size by late December (Takeda *et al.*, 1979). Terminal buds always remain vegetative. Accurate chilling requirements or desired length of lateral shoots for different pistachio cultivars are not available, except that 1000-1500 hours below 7°C appear to be sufficient in California, USA (Crane and Takeda, 1979). Heavily pruned 'Uzun' pistachio retained significantly more flower buds in their 'on year' which is believed due to increased leaf area (Arpaci *et al.*, 1995). Heavy winter pruning (66 per cent of shoot length headed) also induced a slightly higher kernel/shell ratio and developed new shoots on three-to-four-year old 'Uzun' wood (Arpaci *et al.*, 1995). Rahemi and Asghari (2004) suggested use of volk oil and hydrogen cyanimide as well as their combinations to advance blooming, lateral bud break the percentage flower buds developing into fruit clusters. Parhami and Hokmabadi (2001) while investigating the inflorescence bud dimension on some qualitative and quantitative traits of pistachio cultivars Ouhadi and Ahmadaghaiee at Iran reported that large and medium size buds had more splitting percent when compared with small. As the bud dimensions increases, the fruit set and fruit quality improves.

Exogenous application of polyamine, *i.e.*, Spermine at 0.1 µM significantly increased the flower bud retention, leaf chlorophyll content, leaf and flower bud nitrogen and leaf area. A positive relationship between leaf and flower bud nitrogen content and flower bud retention was noted (Baninasab and Rahemi, 2008). Zahra *et al.* (2009) studied the seasonal changes of peroxidase (POD), polyphenol oxidase enzyme (PPO) activity and phenol content during and after rest in pistachio flower buds of cultivars Kalle-Guchi, Owhadi, Ahmad-Aghaei, and Akbari. They observed increased POD activity in December and February and decreased in January and March in cvs. Kalle-Guchi, Owhadi and Ahmad-Aghaei but increased from November to January and then decreased to March in cv. Akbari. The activity of PPO decreased from November to February and increased in March in cv. Kalle-Guchi. The other cultivars showed irregular upsurge and decrease in different months. However, the activity of both the enzymes from bud swell stage to post-bloom stage, decreased in all cultivars. Total phenol content for all four cultivars decreased from bud swell to post-bloom stage. Askari *et al.* (2011) suggested double spray (February 10 and March 1) with ethephon at 750 mg/l which delay flowering, maturity and increased yield of pistachio cv. Kalleh-Guchi. The treatment is effective to prevent spring frost damage during flowering.

## Pollination

Staminate and pistillate inflorescence may consist of several hundred individual flowers which are apetalous. A pistillate flower comprises a three-carpellate pistil while the staminate flowers have five stamens. Anthesis of both staminate and pistillate flowers should coincide to ensure adequate pollination. Bees visit only the staminate flowers for the collection of pollen. Cross-pollinating agent is wind which necessitates the proper placement of pollenizers in relation to pistillate trees. One pollenizing tree is generally planted for every 8 pistillate trees, the ratio, however, may go up to 1:10 or even 1:12. A solid row of pollenizer trees along the wind direction of the orchard is also planted in addition to those located throughout the orchard. Furthermore, pollen quality of the pollinator is low in some orchards. This situation results in reduction of the production and increase of blank production. The time of pollination of the female flower after first opening has been shown to be a significant determinant of yield (Shuraki and Sedgley, 1994). Crane and Iwakiri (1981) stated that fruit drop can be influenced by the timing of pollination, either in relation to flower age or environmental conditions at anthesis such as temperature. Considering that annual climatic differences affect the phenology of each cultivar also differently, it is suggested that combining different male cultivars with different flowering dates to ensure maximum overlap with the female cultivar (Guerrero *et al.*, 2014).

Until the first year of 21$^{st}$. century, U.S. growers were completely dependent on the 'Kerman' female and 'Peters' male pistachio cultivars. After the release of 'Golden Hills' and 'Lost Hills', the 'Randy' male cultivar was selected for these earlier blooming females. However, 'Randy' flowers about five days earlier than 'Kerman' and is not an effective combination (Kallsen *et al.*, 2014). Atli *et al.* (2006) determine the suitable pollinators for the female pistachio cultivars Uzun, Halebi, Kirmizi, Siirt and Ohadi in Turkey. Fifty selected male types were used to determine the best male. They suggested Type 10 male for 'Uzun' and 'Halebi' Type 39 for 'Siirt' and 'Kirmizi' and Type 79 for 'Ohadi'.

When natural pollination is not sufficient, artificial pollination was reported as much effective as natural pollination (Cagler and Kaska, 1994). However, the effective period for artificial pollination is limited and the success of hand pollination is dependent on the environmental factors such as wind and rainfall. Hand pollination being labour-intensive, spray pollination using aqueous pollen grain solution is preferred. Acar *et al.* (2001) studied the effect of artificial pollination on 'Siirt' cultivar at Turkey. The pollen grains were taken from the early pollen shedding male pistachio trees. The pollen was kept at 4°C for 7-10 days in bottles after drying them using $CaCl_2$ for absorbing humidity. The pollen grains (2 per cent) were mixed with wheat flour (98 per cent) and sprayed at full flowering stage using atomizer connected to a tractor. It caused 11.85 fruits/cluster compared with 9.96 fruits/cluster in natural pollination. Fruit set has increased by artificial pollination (14.13 per cent) compared with natural pollination (9.40 per cent). The blank nut percentage was 47.76 and 55.32 with the artificial and natural pollination, respectively. Artificial pollination using agar and 0.01 per cent boric acid combined with 0.15 per cent (v/v) pollen increased fruit set. The medium containing agar and 100mg/l boric acid combined with 0.05 per cent (v/v) pollen increased fresh and dry weight of Kernal (Zeraatkar *et al.*, 2013).

Spraying pistillate trees with the mineral oil dinitro-*o*-cresol (DNOC) during their rest period advances blooming dates by about three weeks thus coinciding with the flowering date of staminate trees (Procopious, 1973). Unsatisfactory pollination in pistachio has been attributed to varying flowering periods in female trees, inadequate number of staminate trees with different flowering times, particularly late flowering cultivars (Pontikis, 1975). Procopious (1973) recommended the use of DNOC (5-20 per cent) sprays to overcome such problems. Aqueous paclobutrazol at 100-4000 mg/l when sprayed at the end of June (when shoot extension ceases) delays flowering of the staminate cultivars of the following season and coincides with the flowering period of the pistillate cultivars (Porlingis and Voviatzis, 1986). Pontikis (1989) successfully advanced the flowering of female cultivars by about 19 days with the application of hydrogen cyanamide during their rest period which ensured pollination and a good crop.

### Fruit Set

Fruit set in pistachio is determined by a number of factors, such as, winter chilling preceding bloom, weather conditions during anthesis and pollination. On an average, 10 per cent of the flowers generally set fruits, 7.5 per cent containing kernels and 2.5 per cent blanks (Crane, 1986). Unpollinated flowers dehisce within a month after full bloom.

## Xenia, Metaxenia and Blank Production

Pistachio is one of the few plants in which xenia and metaxenia have been observed (Peebles and Hope, 1937; Whitehouse *et al.*, 1964; Pontikis, 1977; Crane and Iwakiri, 1980). Peebles and Hope (1937) comparing the influence of *Pistacia atlantica* and *P. chinensis* pollen on the growth and development of the nuts of Red Aleppo and Trabonella cultivars, observed that *P. chinensis* pollen caused a significant delay in nut maturation and *P. atlantica* increased the splitting of shells of Trabonella nuts.

Falci (1917) observed that the shells of the trees nearest to *P. terebinthus* males had a tendency to split. This is thought to have been due to an increase in size of kernel directly attributed to the influence of *P. terebinthus* pollen. Whitehouse *et al.* (1964) working with 'Sfax', 'Bronte', 'Red Aleppo' and 'Trabonella' as female parents, also found certain kinds of pollen to bring about delayed maturity. Pollens of *P. vera* produce more nuts with split shells and nuts with larger kernels than pollens of four other *Pistacia* species. Pontikis (1977) also noted delayed maturity and increased kernel weight depending upon the type of pollen used.

Crane and Iwakiri (1980) used pollen from five different sources to study the manifestation of xenia and metaxenia in pistachio. None of the five pollens had any effect on fruit size or time of maturity of the cultivar Kerman. An apparent effect of some pollen sources on the shell dehiscence was considered to be due to kernel development and not to the type of pollen. Thus, they, proved that there was no manifestation of metaxenia phenomenon in pistachio. But the reduction in kernel length and dry weight caused by the pollen of *P. atlantica* and hybrid pollen was definitely a xenia effect.

## Blank Production

Blank (seedless) fruit production is widely prevalent in all the species of *Pistacia* and all the regions wherever pistachio is grown. As much as 25 per cent nuts of the cultivar Kerman are blanks but the problem is not so serious in other cultivars. The extent of the problem varies with the cultivar, year and rootstocks. Seed abortion causes 24.5 per cent blanks whereas parthenocarpy is responsible for only 3.2 per cent (Crane, 1986).

## Alternate Bearing

Alternate bearing is a widespread phenomenon occurring both in deciduous and evergreen fruits. Pistachio characteristically is an alternate bearer, but the mechanism differs from that of other alternate bearing tree fruit species (Joley, 1969; Crane and Nelson, 1971; Crane and Al-Shalam, 1977). Biennial bearing in other fruit species is usually the result of greatly reduced flower bud formation during the year of a heavy crop. Whereas in pistachio, abundant floral buds are initiated every year which, however, abscise in large numbers during the summer of a heavy crop. Therefore, bienniality in pistachio is caused by abscission of floral buds during a heavy crop year rather than due to lack of bud formation (Crane, 1984).

Crane and Nelson (1971) reported that abscission generally begins in late June and is most intense during July-August and early September coinciding with the rapid growth and development of the seed. The carbohydrate levels of bearing and non-bearing branches are similar, indicating that carbohydrate depletion is not the cause of bud abscission (Crane and Al-Shalam, 1977). Likewise, the involvement of abscission promoting compound has been ruled out (Takeda and Crane, 1980). Takeda *et al.* (1980) found that most of the $^{14}$C-photosynthate transported from leaves accumulated in developing nuts and poorly in floral buds. They postulated that the carbohydrate deficiency of floral buds themselves may be responsible for the bud-drop phenomenon resulting in alternate bearing.

The nutrient acquisition strategies of alternate bearing pistachio trees were similar to those of biennial plants which store nutrients in the vegetative plant body during the 'off year' and used for reproductive growth during the following year. Richard *et al.* (1998) reported that lightly cropping (off-year) trees stored 7, 14 and 2 times as much N, P and K respectively, as heavily cropping (on-year) trees. Nutrients that accumulated during the vegetative phase in 'off year' trees were used to support reproductive growth during the subsequent 'on year'. Soil nutrient uptake contributed more to storage pools than the leaf nutrient resorption in 'off year' trees, while the reverse was true in 'on year' trees.

Okey *et al.* (2011) observed that IAA, Zeatin and ABA levels in the leaf were high in the 'on year' and was low in the 'off year'. The $GA_3$ levels were however, lower in the 'on year' and higher in the 'of year'. Vemmos (2010) stated that inflorescence buds on defruited branches accumulated twice as much $^{14}C$ photosynthate as those on fruiting ones. A depletion of carbohydrates in flower buds in 'on year' trees was found by some researchers during seed development. Sucrose movement from leaves to flower buds was also reduced, resulting in a decrease of flower bud development on the 'on year' trees which was followed by abscission.

Vermos (2010) listed the following factors that affect bud abscission in pistachio

a) Cultivar: variation between cultivars in percentage of bud abscission has been reported.

b) Rootstock: tree to tree variations in the same orchard have been attributed to genetic differences between the seedlings rootstocks used.

c) Fruit load: the more fruit present on shoots or branches, the more bud drops. Fruit load on a branch can also affect the bud abscission on a neighbouring branch.

d) Soil effect: soil heterogeneity and its capacity for water and nutrients might affect tree growth and yield.

e) Leaf area: Inflorescent bud abscission increased with a decrease in leaf area per branch.

Girdling between inflorescence and shoots before bud abscission showed significant reduction of bud drop (Crane and Nelson, 1972).

Alar at 7500 ppm was found to be effective in reducing shoot growth and increasing flower induction in the following season (Nahlawi *et al.*, 1979). $GA_3$ induces cell division in subapical meristems of quiescent terminal vegetative buds and brings about bud opening (Lin *et al.*, 1984). A single application of PCPA (para-chlorophenoxy acetic acid) maintained inflorescence bud retention for 7 weeks but failed to prevent abscission later. Triple and double applications were more effective and retained the buds for 15 weeks. 2,4-D applied as single, double or triple sprays was as effective as multiple application of PCPA (Gawad and Ferguson, 1987).

## Fruit Growth and Development

Pistachio fruit is a semi-dry drupe consisting of a single seed enclosed in a thin shell. The growth pattern of some of the tissues of the fruit resemble those of drupes while other tissues develop quite differently (Crane *et al.*, 1971 Crane and Al-Shalam, 1974). Growth pattern of fleshy drupaceous fruit, like apricot (Lilleland, 1930) occurs in three distinct periods. The period I has rapid growth separated by a period of decelerated growth which corresponds to period II and the third period again of accelerated growth. Almond, a dry drupe, has only two periods of growth (Brooks, 1939), *i.e.*, a rapid growth phase followed by one of practically no growth. Therefore, almond resembles other fleshy drupes in having similar growth patterns of periods I and II but lacks the period III (phase of rapid growth). Fruit growth pattern in pistachio is similar to that of almond except that second growth cycle (period III) is slightly longer than almond (Crane, 1986). In contrast to almond in which endocarp, nucellus and the integuments of the seed reach their full size in the period I, the ultimate size of only endocarp is attained by the end of period I, and the nucellus and integuments complete their growth in period II when lignification of endocarp also occurs. The endocarp attains its ultimate size by the end of period I (mid-May) while the kernel attains its full size in the last week of July which is signalled by endocarp dehiscence along the ventral suture. Endocarp dehiscence increases further with the physiological maturity which is reached by the middle of September (Crane, 1986).

Significant changes occur in sugar and fat content of the pistachio fruit during growth and development. Initially, sucrose content is very high during the first month of fruit growth, attaining a peak of about 35 per cent in the first week of July. Thereafter it sharply goes down to 14 per cent in the middle of July and finally stabilizes at 6 per cent by the middle of August and remains at that level until harvest. On the contrary, the crude fat is initially only 3.6 per cent in the first week of July and increases rapidly with the rapid decline of sugar reaching a level of about 40 per cent of seed dry weight at maturity. Green pigments disappear from the shells as the nut attains physiological maturity (Crane. 1986). Memi *et al.* (2014) noted continuous shell lignifications throughout the period of fruit growth.

The lack of irrigation supply during stage I and II has shown better results in fruit diameter than a lower input distribution between stage I and III.

## Respiration and Ethylene Evolution

The respiration rate of pistachio fruit is higher during seed growth and development than that during early fruit growth or during maturation (Toumadje *et al.*, 1980). The climacteric peak of respiration observed in many fruits at ripening is indiscernible (Biale, 1960). The blanks have a constantly low respiration rate: about five to six times lower than the fruits with kernels. Ethylene is not involved at any stage of the fruit growth and development from fruit set to maturation as indicated by its steady low evolution rate throughout the period of fruit growth.

## Endocarp Dehiscence

Shell dehiscence is an important character which is taken into account while selecting cultivars for commercial production. *Pistacia vera* is the only species in the genus showing this trait. When the nut matures, the hull covering the endocarp separates easily from the shell which partially exposes the kernel at the apical end.

The bulk of the pistachio crop is marketed in shell, and its shell dehiscence is a desirable trait as it facilitates shell removal by the consumers. Undehisced shells are removed by cracking or by machines before processing and marketing. Starting from the apical end, the dehiscence progresses along the ventral and dorsal sutures and the extent varies due to nut, tree. cultivar and the climatic condition of the year (Crane and Iwakiri, 1981). Shell deshiscence occurs in the month of July when the seed reaches its maximum size, and increases subsequently in August during the period of physiological maturation of the fruit. Nevo *et al.* (1974) did not find any anatomical difference between dehiscent and indehiscent endocarps and they opined that larger volume of seed than that of endocarp and greater shrinkage of the latter upon drying were the factors responsible for the dehiscence. Conversely, Crane (1986) suggests that the dehiscence is not merely a physical phenomenon but is brought about by the anatomical and physiological processes leading to natural abscission. Ethylene is neither involved in hull or shell dehiscence nor in maturation of pistachio fruits but it promotes excessive gum exudation from the trunk and branches, and causes severe flower bud abscission (Crane *et al.*, 1982).

## Maturity

The important indices for determining physiological maturity of pistachio nuts are the easier separation of the hull from the shell and changing of skin (epicarp) from translucent to opaque (Crane, 1978). In addition, kernel dry weight, crude fat percentage, shell dehiscence and change in shell colour should also be taken into consideration in deciding the harvest date. Harvesting of nuts at optimum maturity is necessary because early or late harvests result in inferior quality of the produce (Kader *et al.*, 1982). The nuts do not abscise from trees at maturity, but gradual drying and shrivelling of the hulls for several weeks. The middle to end of September was suggested as the best times of harvest of cvs. Ahmad-Aghaei, Kaleh-Ghouchi, Ohadi and Badami in Rafsanjan, Iran (Mirdamadiha *et al.*, 2005). Zhang and Weir (2006) suggested measurement of a group nut density for prediction of nut maturity of pistachio. The nut density was measured by taking randomly 50 nut samples each week from 13 January until 22 February 2005 (in Australia). Nut density was determined by weighing a container with 1000 ml of water and mesh, used to completely submerge all the nuts (w1) and then adding the 50-nut sample the weight of which had previously been determined (w50). The contair, mesh and nuts along with remaining water was then weighed (w2).

Sample volume = w1-w2+50

Density = w50/sample volume

From mid- January to maturity stage "ripening" both volumes of kernels and densities of nuts increase quickly. They are good indicators to monitor kernel growth and predict maturity at this stage.

The nuts picked a fortnight earlier or later of the optimum date were inferior in quality. The harvesting should be completed within a week. The effect of harvest dates is significant on total and reducing sugars and sensory qualities of nuts, but not on total fat content, fatty acid composition and total protein content (Kader *et al.*, 1982). Total sugars have a pronounced effect on sweetness and overall flavour intensity.

Some nuts rupture their hulls in the orchard prior to harvest and are known as 'early splits'. Even though relatively few nuts become early splits, those nuts are very important factor, because their kernels are frequently decayed by fungi and contaminated with aflatoxin. Unbalanced nutrition was correlated with early splitting which caused *Aspergillus* infection. Deficiencies of K, Fe, P and Zn toxicity are correlated with early split occurrence. A positive and significant correlation between early splitting and Cu content in leaf was also observed (Hosseinifard and Panahi, 2006). Sherafati *et al.* (2012) evaluated 12 pistachio cultivars and recorded the highest (14.8 per cent) early splitting of nuts in cv. Shahpas followed by cvs. Owhadi (2.7 per cent) and Akbari (1.7 per cent). The cv. Pesteh garmeh showed no early splitting while it was only 0.1 per cent in Badami sefid and 0.2 per cent in Abasali.

## Harvesting

In Mediterranean countries, the fruit clusters of small trees are picked by hand, and tall trees are shaken so that the nuts fall on the cloth spread beneath. In California, the early practice of harvesting pistachio nuts by knocking of the trees with long poles or by shaking onto canvass sheets has been replaced by mechanical shaking. The mechanical harvester consists of a tree shaker and a catching frame with a conveyer belt. Mechanical harvesting has the advantage of separating blanks from good nuts, as the force sufficient to remove good nuts from the trees will leave most of the blanks attached to the tree (Crane and Dunning, 1975). Nahlawi *et al.* (1981a) reported that mechanical shakers removed 80-90 per cent of the crop when used at the earliest possible harvest time. The nuts left out after shaking are to be harvested by hand. Ferguson *et al.* (2014) stated that trunk shaking harvester have efficiency of harvesting 93.9 per cent while the commercial harvester have efficiency of 86.2 per cent.

## Yield

Pistachio trees start bearing at the age of seven years and reach the peak production around the 15th year. The yield of nuts in Iran and Turkey vary up to 600 per cent from year to year (Woodroof, 1979). Yields varying from 8-36 kg/tree were reported by Nahlawi *et al.* (1984b) in 10- and 30- year-old trees of 'Batouri' and 'Aschouri', respectively. On an average, a mature tree produces 25 kg marketable nuts per year which amounts to 2.5 to 5 tonnes per hectare depending on the planting density (Woodroof, 1979). 'Ashouri' cultivar of Syria produces up to 200kg of fresh nuts per year (Hadj-Hassan, 2001).

## Dehulling

Hulling should be done as soon as possible after harvest to reduce the chances for fungal growth and to avoid shell staining. Fruits are placed in a large burlap bag on a concrete floor. Roll, beat and press with a piece of wood until most of hulls separated from the nuts.

The dehulled nuts always contain a considerable number of unsplit and blank nuts. Floatation techniques are used to separate these nuts from the split nuts, since immediately after harvesting dehulling the unsplit nuts tend to float. During the floating process, nuts are immersed in a tank of water and agitated for about 10 minutes. Once the agitation is complete, the "sinkers" are primarily the splits and the "floaters" are unsplits and blanks. Since the floatation process is not accurate, some amount of unsplit and low quality nuts always remain. After floatation, the nuts are washed to clean the surface from the residuals left during dehulling.

## Drying

Nuts at harvest contain 40 to 50 per cent water and should be dried to about 5-7 per cent moisture as soon as possible to prevent fungal infection (Kader *et al.*, 1980). The most common methods of drying are sun drying and heated air drying. Sundrying usually takes 3 to 4 days for drying depending on ambient temperature. Heated-air drying is much faster than sun drying. It takes about 7-10 hours to dry the nuts to about 5 per cent moisture by using 70 to 90°C temperature. Different types of dryers such as solar dryer, bin dryer, vertical cylindrical dryer, continuous mobile and steady tray dryer, vertical continuous dryer, funnel cylindrical dryer, drum dryer, continuous belt conveyor dryer, *etc.* are used in different countries. The types of dryer used for drying pistachio nuts are significantly influenced by the drying methods. A combination of bin dryer and sundrying or combination of bin dryer and solar dryer, are recommended as effective dying techniques to prevent adverse quality effect on pistachio nut (Shakerardekani *et al.*, 2011). Kouchakzadeh and Haghighi (2011) developed vacuum-infrared drying technology of pistachio nuts. Using combined vacuum infrared radiation to drying pistachios has a number of advantages. Pistachios can be dried approximately 8 to 10 times faster than dryer that rely on conventional convection and conduction processes only; drying at near room temperature is the most economical way and it seems the quality in the drying process is enhanced.

## Sorting

Consumer demand is for large, in-shell, split pistachio nuts. The United States Department of Agriculture (USDA) standards for pistachio nuts designate size grades (average number of nuts per 28.5g) of 'Extra Large' (20 or less), 'Large' (21-25), 'Medium' (26- 30) and 'Small' (31 or more). The standards also identify as degrading factors the existence of foreign materials, and damaged and unsplited nuts (California Pistachio Commission, 1995). Electro-mechanical devices, optical sorters, and manual sorting are used in processing plants for further grading of the nuts. Shakearardekani *et al.* (2012) suggested segregation or removal of all the yellowish-brown stained nuts from the unstained portion during the sorting process which could contribute to the reduction of aflatoxin content in the final dried pistachio lot.

## Storage

Dried (with 4-6 per cent moisture content) nuts are very stable and can be stored for up to one year at 20°C and 65-70 per cent relative humidity without significant loss in quality attributes (Kader *et al.*, 1982). For longer storage of nuts, low temperatures are recommended. Storage under high concentration of $CO_2$ (98 per cent) and reduced $O_2$ (less than 0.5 per cent) provides good stability in terms of fatty acid loss and formation of peroxide and free fatty acids (Maskan and Karatas, 1998). Using three layer plastic pouches, nuts can be stored for one year in ambient temperature (Raei and Jafari, 2011)

## Pests

### Twig Borer *(Kermania pistaciella)*

The adult moth of twig borers in late March-April, lays its egg in flower and fruit clusters, as well as on new shoots and petioles during high pest populations. The newly larvae directly penetrate fruit cluster. Larvae tunnelled in the pedicel of fruit cluster and branches and began to feed. It causes severe economic damage through fruit dropping, decreasing quality and quantity of the crop yield (Tezerji, 2011). Parasitoids species *Chelonus flavipalpis szepligeti* and *Mirax rufilabris Haliday* were recorded on *K. pistaciella* (Ozgen *et al.*, 2012).

### Pistachio Psylla (*Agonoscena pistaciae*)

Pistachio psylla is an important pest of *P. vera*. In addition to *A. pistaciae*, two other psyllid species, *A. bimaculata* and *Magagonoscena viridis* also reported from Iran (Burckhardt and Lauterer, 1993). The insect hibernates and aestivates in the egg stage, and the psyllid nymph appear in early spring when the new shoots emerge and fresh leaves appear. The infested leaves roll because of the insect feding and the psyllid nymph develop in the shelter. The presence of large populations of psyllid nymph and adults caused severe problems in kernel development, with subsequent bud drop, defoliation and yield loss. (Mehrnejad, 2014). Insecticides Mospilan and Consut were found effective to control the pest (Abbad and Besheli, 2013).

### Gypsy Moth (*Lymantria dispar*)

It attacks pistachio and other nut plantations. Virus preparation virin-ENS at 0.1 l/ha was 93 per cent effective against the moth in late April and early May when the air temperature was higher than 17°C and the larvae were in the 2nd or 3rd instar. When egg masses of the pest were coated with the preparation in early spring or late autumn its effectiveness reached 98.5 per cent over an area of 3000 ha (Tanabaev, 1984).

### Naval Orange Worm (*Amyelois transitella*)

It feeds on the kernel of pistachio nuts and the damage is difficult to detect prior to marketing. The larval infestation can be reduced by spraying carbaryl or azinphos-methyl during the early stage of hull splitting. The entomopathogenic nematode, *Steinernema carpocapsae*, can reduce pesticide reliance by controlling overwintering larvae of the navel orangeworm (Hodson *et al.*, 2012).

### Birds

American crow (*Corvus brachyrhynchos*) has been identified as the most frequent pest, followed by Brewer's blackbird (*Euphagus cyanocephalus*), European starling (*Sturnus vulgaris*) and three other birds. Most of the damage is caused in the 6-8 week period between maturation of first nuts and harvest (Salmon *et al.*, 1986).

### Nematodes

Joley and Whitehouse (1953) reported the infestation of root-knot nematodes (*Meloidogyne* sp.) on pistachio roots. Nematode attack restricts the growth of the tree consequently reducing the nut yield. *P. atlantica* and *P. terebinthus* rootstocks are immune to root-knot nematodes while *P. vera* rootstock is susceptible. Yldz and Elekcioglu (2011) investigate nematodes associated with pistachio orchards at Anatolian region of Turkey. They found nematode genera of Tylenchus, Filenchus, Helicotylenchus, Rotylenchus, Paratylenchus, Pratylenchoides, Trophorus, Geosenamus, Tylenchorhynchus, Criconema, and Xiphinema. Among them, Paratylenchus and Helicotylenchus were the most abundant and frequent groups.

## Diseases

### Verticillium Wilt (*Verticillium dahliae*)

The most serious disease of pistachio tree is the soil-borne verticillium wilt. It has caused considerable mortality in California, especially to the newly planted trees. Ashworth *et al.* (1985) opined that the disease is associated with unthrifty trees having low levels of leaf potassium (0.5-0.8 per cent) during mid-summer. Pre-planting fumigation with a 1:1 mixture of chloropicrin-methyl bromide reduces the inoculum.

Complete mulching of the soil surface of a six-year-old drip-irrigated plantation with polyethylene sheets for two months resulted in elimination of this fungus to a depth of 120 cm (Ashworth and Gaona, 1982). Jamdar *et al.* (2013) reported that *Trichoderma harzianum* 30, 85 and *T. koningi* have inhibitory effect on radial growth of *V. dahliae.*

### Shoot Blight (*Botrytis cinerea*)

Bolkan *et al.* (1984) reported this disease from California. Shoot blighting resulted from initial infection of catkins, racemes and stems. Partial girdling of shoots or twigs led to shrivelled leaves showing necrosis from water stress. Incidence of the disease was high on male trees because of the high susceptibility of catkins. The disease was accentuated by prolonged rainfall and relatively low temperature. Application of fungicide especially benomyl during full bloom of male flowers reduced the number of blighted shoots. In Australia, *Botryosphaeria dothidea* was reported causing shoot blight and cankers. Shoot blight symptoms during late summer prior to harvest included dark lesions on current-season shoots, followed by blackening and dieback. Cankers were observed in older wood, showing wedge-shaped discolouration in cross-sections of these branches.

### Gummosis (*Phytophthora pistaciae*)

Root rot and crown rot or gummosis is the most important disease of pistachio in Iran. The disease occasionally causes yield loss of about 10-12 per cent. Several species of Phytophthora, including *P. citrophthora, P. cryptogea, P. megasperma, P.drechteri, P. nicotianae* and *P. parsiana* have been isolated from diseased trees and their pathogenicity on pistachio trees has been confirmed. The fungus attacks the pistachio trees in Greece but the popular rootstock *P. terebinthus* is resistant to this fungus (Kouyeas, 1952). Poor drainage condition is also favourable to the incidence of crown rot.

In Greece, pistachio is subjected to severe injuries by fungus parasites and the most serious is the fungal parasite of the genus *Phomopsis* which attacks the female flowers in the spring (Anagnostopoulos, 1938). The fungus enters through insect punctures. Control measures consist of burning the infested plant parts. Application of oil sprays in January destroys the overwintering insect.

### Bacterial Dieback (*Xanthomonas translucens*)

Bacterial dieback disease was observed on pistachio trees in Australia in 1992. The disease is characterized by shoot death and dieback, limb and trunk lesions and discolouration of woody tissue in shoots more than one year old (Facelli *et al.*, 2002). Antibiotic streptomycin (1.0mg/l) was found effective in inhibiting the pathogen (Taylor and Edwards, 2000).

### Leaf Brown Spot (*Alternaria alternaria*)

Leaf brown spot is a fungal disease, which mainly infected the leaves and fruits, caused leaf spot and blight. Late spring and early summer, as well as the late summer and early autumn are the two peak period for incidence of disease. Continous rain in summer is the decisive condition for the infestation and outbreak of leaf brown spot (ShuYu *et al.*, 2011). Ash and Lannoselet (2001) however, reported that *Colletotrichum acutatum* causing leaf and fruit excocarp spot in Australia. In New Mexico, French *et al.* (2009) reported Septoria leaf spot of pistachio caused by *Septoria pistaciarum*.

*Alternaria* blight disease caused by *Alternaria* sp. Has become epidemic for some years in southeastern Anatolia region of Turkey (Can *et al.*, 2014). Disease symptoms mainly develop on young fruits and infections continue until harvest. Repeated infestations decrease fruit quality and quantity.

### Viral Disease

In the Badkhyz mountains and Turkmenia regions of the USSR, Kreutzberg (1940) reported the attack of virus on wild pistachio trees resulting in rosette formation of varying sizes with thin branched twigs. Overall growth of the plant is seriously affected. The virus is transmitted by insects and seed. A survey during 1936 and 1937 indicated that the area of infection was confined to the forest tracts of Turkmenia. Uzbekistan and Tadzhikistan as well as to the adjoining regions of northern Afghanistan and northern Iran.

## Breeding

Pistachio breeding programme over the years in California, Turkey and Iran contributed a good number of female and male cultivars in the world. Breeding programme was also initiated in Spain, Australia, Italy and Greece. Rootstock breeding was mainly targeted to develop nematode resistance (*P. atlantica* selections), Verticillium wilt resistance (hybrid of *P. atlantica* and *P. integerrima*-UCB-1), cold resistant (*P. atlantica* and UCB1) *etc.*

The objectives of scion breeding was to develop early bearing nut size, yield, split percentage, nut quality, and early season harvest. Disease resistance, especially resistance to *A. alternate* is also a major objective.

Several quantitatively inherited characters have been described for *P. vera*. Half sib family and parent-offspring regression analysis of flowering and leafing date showed that both are highly heritable (Chao *et al.*, 2003).

Regeneration of plants (2n= 30) from somatic embryos derived from callus of 'Siirt' flowers and from leaf explants of 'Antep' have been reported (Onay *et al.*, 2004, Onay, 2000). Although induction, development, maturation and germination of somatic embryos is being controlled to some extent in pistachio, high frequency regeneration, synchronous development of embryos and transplantation in the field were not very successful (Onay *et al.*, 2011).

## Molecular Markers

Some isoenzymes and genes have been characterized and used for cultivar identification and genetic variability studies in pistachios. Isoenzymic polymorphism in *P. vera* has been reported (Batlle *et al.*, 1998). Fifty cultivars (36 females and 19 males) were analyzed isoenzymatically for seven systems: aconitase (ACO), esterase (EST), glutamate oxalacetate transaminase (GOT), glucose phosphate isomerase (GPI), malate dehydrogenase (MDH), phosphoglucomutase (PGM) and shikimate dehydrogenase (SKD). The result suggested that the GPI reveald the most polymorphic (8 different phenotypes) and the most useful for cultivar characterization (Batlle *et al.*, 1998). Twelve different combinations for female cultivars and 2 for male cultivars were found. Two female cultivars (Bianca and Red Jalap) and 4 male types (A, C-Especial, M-P3 and M-P9) were identified as being of hybrid origin.

Many DNA-based profiling techniques have been developed for the study of genetic variation in pistachios. These technique include: the Random Amplified Polymorphic DNA (RAPD), Amplified Fragment Length Polymorphism (AFLP), Simple Sequence Repeat (SSR) and Inter Simple Sequence Repeat (ISSR). Rovira *et al.* (1995) identified different pistachio spp. by the zymogram phenotypes of ACO, EST, GOT, GPI, MDH, PGM and SKD. This work showed that 2 enzymatic systems, GOT and MDH, were invariable in *P. vera* and that bands on the gel were localized in a different position from other *Pistacia spp.* Hormaza *et al.* (1994) identified a RAPD marker linked to the sex determinant gene. Golan-Goldhirsh *et al.* (2004) assessed polymorphism among the *Pistacia* species of the Mediterranean basin and accessions within the same species using RAFD and AFLP markers.

Esfandiyari *et al.* (2010) used 20 RAPD primers and a pair Sequence Characterized Amplified Regions (SCAR) primer to determine sex in *P. atlantica, P.*

*khinjuk* and *P. vera* var. Sarakhas. Among RAPD primers, only BCI 1200 amplified a specific sex band which was present in female plant. They concluded that SCAR techniques is a reliable technique to identify gender genotypes in seedling stage of *Pistacia* species, that would help to time and expenses in breeding programme. To evaluate the genetic relationship among wild and cultivated *Pistacia* species grown in Iran and the analysis of genetic variation among Iranian pistachio genotypes, two DNA libraries enriched for dinucleotide (AG)n and trinucleotide (ATG)n microsatellite motifs were developed from *Pistacia khinjuk* genome (Arabnezhad *et al.*, 2011). Dendogram constructed on the basis of the Minimum Evolution clustering algorithm revealed that *P. vera* has closer relationship with *P. khinjuk* than with *P. integerrima, P. palestina, P. atlantica and P. mutica.* The dendogram further distinguished the wild Sarakhs pistachio from the rest of the *P. vera* genotypes suggesting that the domesticated *P. vera* was evolved from *P. vera* var. Sarakhs and then this wild genotype likely develops to other local pistachios. The results indicated that microsatellites develop in *P. khinjuk* are distributed in the genome of indigenous pistachio species including *P. vera* genotypes and therefore they will be useful in characterization of Iranian pistachio genotypes.

## REFERENCES

Abbad, M.K.R. and Besheli, B.A. (2013). Int J. Agron. Pl. Prod., 4: 1191-96.

Abbaspour, H. (2012). Jr. Medicinal Pl. Res., 6: 526-29.

Abbaspour, H., Saeidi-sar, S. and Afshari, H. (2011). Jr. Medicinal Pl. Res., 5: 7065-72.

Acar, I., Ak, B.B. and Kuzdere, H. (2001). Xi GREMPA Sem. Pistachio and Almond, Zaragoza, CIHEAM, pp. 145-48.

Acar, I., Arpaci, S., Sarpkaya, K., Karadag, S. and Karayilan, A. (2011). Acta Hort., 912: 239-44.

Ak, B.E., Kandemir, M. and Sakar, E. (2013). Acta Hort., 981: 413-18.

Al-Barazi, Z. and Schwabe, W.W. (1982). Jr. Hort. Sci., 57:247-52.

Alderson, P.G. and Barghchi, M. (1982). In: Plant Tissue Culture (ed. Piyiwara), pp.739-40.

Alekseitsev, G.P. and Tuzov, V.K. (1984). Zashchita Rastenii, 11:34.

Al-Imam, N.M.A. and Al.-Jubury, Y.M.S. (2011). Acta Hort., 912: 245-52.

Al-Saghir, Mohannad, G. and Porter, D.M. (2012). Amer. J. Pl. Sci. 3: 12-32.

Anagnostopoulos, P.T. (1938). Hort. Res., 4: 499-7.

Anderson, W.C. (1978). Comb. Proc. Int. Plant Prop. Soc., 28: 135-39.

Anderson, W.C. (1980). Acta Hort., 112:13-20.

Anonymous (1969). The Wealth of India. Publications and Information Directorate, CSIR, New Delhi. pp. 122-24.

Arabnezhad, H., Bahar, M. and Pour, A.T. (2011). Scientia Hort., 128: 249-54.

Arpaci, S., Tekin, H. and Aksu, O. (1995). Acta Hort., 419: 253-57.

Arreola-Avila, J.G., Nunez-Barrios, A., Vargads-Piedra, G., Valdez-Cepda, R., Borja-de-la-Rosa, A. and Trejo-Calzada, R. (2011). Rivista Chapingo Serie Ciencias Forestales-y-del-Ambienter, 17: 256-65.

Ash, G.J. and Lanoiselet, V.M. (2001). Aust. Pl. Pathol., 30: 365-66.

Ashworth, L.J. Jr. and Gaona, S.A. (1982). Phytopathology, 72: 243-46.

Ashworth, L.J. Jr., Gaona, S.A. and Surber, E. (1985). Phytopathology, 75: 1084-91.

Askari, E., Irani, S. and Razmjoo, K. (2011). Hort. Environ. Biotech., 52: 29-34.

Atli, H.S., Arpaci, S. and Uyger, N. (2006). Acta Hort., 726: 417-22.

Atli, H.S., Aydn, Y., Arpac, S., Acar, I., Karadag, S., Belgel, L., Sarpkaya, K., Kaska, N., Kafkas, S. and Ak, BE. (2011). Acta Hort., 912: 289-94.

Avanzato, D., Vaccaro, A., Meli, M. Delfini, M., Capuani, G., Cocco, M.E., di, Sciubba, F., Tzareva, I. and Teeziev, I. (2011). Acta Hort., 912: 143-49.

Ballistreri, G., Arena, E. and Fallico, B. (2011). Acta Hort., 912: 837-42.

Banihashemi, Z. (1998). Iranian J. Pl. Pathol., 34: (3/4): 63-66.

Baninsab, B. and Rahemi, M. (2008). Hort. Environ. Biotech., 49: 149-54.

Barazi, Z. and Schwabe, W.W. (1982). J. Hort. Sci., 57: 247-52.

Barghchi, M. and Alderson, P.G. (1982). 21st Int. Hort. Congress, Hamburg, Abst. 1244.

Barghchi, M. and Alderson, P.G. (1983a). J. Hort. Sci., 58: 435-45.

Barghchi, M. and Alderson, P.G. (1983b). Acta Hort., 131: 49-60.

Barghchi, M. and Alderson, P.G. (1985). J. Hort. Sci., 60: 423-30.

Batlle, I., Romero, R., Rovira, M. and Vargas, F.J. (1995). Acta Hort., 419: 181-88.

Beede, R. and Ferguson, L. (2005). Pistachio Production Manual. ed. L. Ferguson. University of California, Davis. pp. 97-102.

Beede, R.H., Brown, P.H., Kallsen, C. and Weinbaum, S.A. (2005). Pistachio Production Manual, 4th Ed. Ferguson pp. 76-91.

Behboudian, M.H., Walker, R.R. and Torokjalvy, E. (1986). Scientia Hort., 29: 251-61.

Benmahioul, B., Dorion, N., Kaid-Harche, M. and Daguin, F. (2012). Pl. Cell. Tissue and Organ Cult., 108: 353-58.

Biale, J.B. (1960). In: Handbook der Pflanzenphysioloyie, (ed. W. Ruhland), Vol. 12, Part 2, Springer. Berlin.

Bolkan, H.A., Ogawa, J.M. and Teranishi, H.R. (1984). Plant Dis., 63:163-65.

Brooks, R.M. (1939). Pror. Amer. Soc. Hort. Sci., 37: 193-97.

Brooks, R.M. and Olmo, H.P. (1972). Register of New Fruit and Nut Varieties. University of California Press. Berkeley, pp. 487-88.

Burckhardt, D. and Lauterer, P. (1993). Revue Swisse de Zoologie, 100: 829-98.

Caglar, S. and Kaska, N. (1994). Acta Hort., 419: 55-60.

Can, C. Konukoglu, F., Ozgun, E., Kusek, M., Ozkiline, H., Aktan, A., Citak, M. and Sarpkaya, K. (2014). Acta Hort., 1028: 171-77.

Casini, E. and Conticini, L. (1979). Rivsta di Agricoltura Subtropicale e Tropicale, 73: 233-40.

Castellanos, J.Z., Ortega-Guerrero, A. and Grajeda, O.A. (2002). Terra, 20: 161-70.

Chernova, G.M., Babekova, E. Ya and Olekhnovich, G.S. (1986). Rastitel'nye Resursy, 22:326-33.

Clarke, J.A., Brar, G.S. and Procopious, I. (1976). Quart. Plant Pl. Fds. Hum. Nutr., 25:219-25.

Corazza, L. and Avanzato, D. (1985). Annali dell'Istituto Sperimentale per la Frutticoltura, 16: 49-51.

Crane, J.C. (1978). J. Amer. Soc. Hort. Sci., 103: 332-33.

Crane, J.C. (1984). Fruit Var. J., 38: 74-85.

Crane, J.C. (1986). In: CRC Handbook of Fruit set and Development (ed. S.P. Monselise), CRC Press Inc., Boca Raton Florida, pp. 389-99.

Crane, J.C. and Al-Shalam, I. (1974). J. Amer. Soc. Hort. Sci., 99: 87-89.

Crane, J.C. and Al-Shalam, I. (1977). J, Amer. Soc. Hort, Sci., 102: 396-99.

Crane, J.C. and Dunning, J.J. (1975). Calif. Agric., 29: 6-7.

Crane, J.C. and Forde, H.I. (1974). Calif. Agric., 23: 8-9.

Crane, J.C. and Forde, H.I. (1976). J. Amer. Soc. Hort. Sci., 101: 604-06.

Crane, J.C. and Iwakiri, B.T. (1980). HortSci., l5: 184-85.

Crane, J.C. and Iwakiri, B.T. (1981). Hort. Rev., 3: 376-93.

Crane, J.C. and Nelson, M.M. (1971). HortSci., 6: 489-90.

Crane, J.C. and Takeda, F. (1979). HortSci., 14: 135-37.

Crane, J.C., Bradley, M.V. and Nelson, M.M. (1971). J. Amer. Soc. Hort. Sci., 96: 78-80.

Crane, J.C., Iwakiri, B.T. and Lin, T.S. (1982). HortSci., 17: 383-84.

Elloumi, O., Ghrab, M. and Ben-Mimoun, M. (2014). Acta Hort., 1028: 385-90.

El-Sayed, E.H., El-Sherif, A.H., Said, W.T. and Sari El-Deen, S.A. (2000). Egypt J. Appl. Sci., 15(5): 132-46.

Ermakov, A.I., Popov, R.P. and Yarosh, N.P. (1982). Rastitel'nye Resursy. 18:243-49.

Esfandiyari, B., Nejad, G.H.D., Shahriyari, F.A. and Kiani, M. (2010). Jr. App. Hort., 12(2): 122-24.

Esmaeilizadeh, M., Javanshah, A. and Rahemi, M. (2006). Acta Hort., 726: 279-81.

Facelli, E., Taylor, C., Scott, E., Emmett, B., Fegan, M. and Sedgley, M. (2002). Aust. Pl. Pathol., 31: 95-96.

Falci, R. (1917). II. Pistachio Palermo.

Ferguson, L. (2003). Pistachio production year book. 4th Edn, Ed. L. Ferguson, Univ. California, Davis.

Ferguson, L., Glozer, K., Reyes, H., Rosa, U.A. and Castro-Garcia, S. (2014). Acta Hort., 1028: 377-80.

Ferguson, L., Maranto, J. and Beede, R. (1995). HortSci., 30: 1369-72.

Ferguson, L., Reyes, B., Sanden, B., Grattran, S., Epstein, L. and Krueger, B. (2008). In: Pistachio Production Manual (ed. L. Ferguson), Omnipress, Madison, WI, pp. 71-77.

French, J.M., Heerema, R.J., Gordon, E.A. and Goldbery, N.P. (2009). Pl. Dis., 93:762.

Gawad, H.A. and Ferguson, L. (1987). In: Proc. Plant Growth Regulators Society of America. Lincoln, Nebraska, pp. 63-69.

Ghrab, M., Elloumi, O. and Mimoun, M.B. (2011). Acta Hort., 912: 137-41.

Golan-Goldhirsh, A., Barazani, O., Wang, Z.S., Khadka, D.K., Saunders, J.A., Koatiukovsky, V. and Rowland, L.J. (2004). Pl. Syst.E., 246: 9-18.

Goldhamer, D.A. (2005). Pistachio Production Manual, 4th Edn. Ed. Ferguson L. University of California, Davis. pp. 103-116.

Goldhamer, D.A. and Beede, R.H. (2004). Jr. Hort. Sci. Biotech., 79: 538-45.

Gopalan, C., Shastri Rama, B.V. and Balasubramanian, S.C. (1987). NIN. ICMR, Hyderabad, p' 83'

Guerero, J., Moriana, A. and Couceiro, J.F. (2006). Acta Hort., 726: 513-17.

Guerrero, J., Memi, H., Perez-Lopez, D., Couceiro, J.F., Moriana, A., Martinez, E. and Gijon, M.C. (2014). Acta Hort., 1028: 297-301.

Gupta, U.C., Jame, Y.W., Campbell, C.A., Leyshon, A.J. and Nicholaichuk, W. (1985). Canad. J. Soil Sci., 65:381-409.

Hadj-Hassan, A. (2001). In: Cultivated Syrian Pistachio Varieties. (Eds. S. Padulosi and A. Hadj-Hassan), IPGRI, Rome, Italy.

Hakerardekani, A., Karim, R. and Mirdamadiha, F. (2012). Jr. Fd. Agric. Environ., 10: 459-61.

Halvorsen, B.L., Carlsen, M.H., Phillips, K.M. Bohn, S.K. Holte, K., Jacobs, D.R. Jr. and Blomhoff, R. (2006). The Amer. J. Clinical Nutr., 84: 95-135.

Herrera, E. (1997). Las Cruces, NM. Etrctronic distribution information NMSU and the USDA. pp. 11.

Hodson, A.K., Siegel, J.P. and Lewis, E.E. (2012). Pedobiologia, 55: 51-58.

Hormaza, J.I., Dollo, L. and Polito, V.S. (1994). Econ. Bot., 52: 78-87.

Hosseinifard, J. and Panahi, B. (2006). Acta Hort., 726: 325-28.

Hosseinifard, J., Salehi, M.H., Salehi, F. and Heydarinejad, A. (2005). Acta Hort., 726: 94-96.

Ibrahim, I.H. and Nahlawi, N. (1982). Publication Arab Centre for the Studies of Arid Zones and Dry lands. ACSAD PS/P4. p. 21.

Ibrahim, I.H., Nahtawi, N. and Nahlawi, M. (1984). Publication Arab Centre for the Studies of Arid Zones and Dry Lands. ACSAD/PS/P13, p. 44,

Jacob, M.B. (1951). The Chemistry and Technology of Food and Food Products. Vol. II. Intersci. Publ., New York, p. 970.

Jamdar, Z., Mohammadi, A.H. and Mohammadi, S. (2013). Jr. of Nuts, 4(4): 53-56.

Joley, L.E. (1969). In: Handbook of North Amer. Nut Trees (ed. R.A. Jaynes). The W.F. Humphery Press, Geneva. N.Y., pp. 348-61.

Joley, L.E. and Whitehouse, W.E. (1953). Proc. Amer. Soc. Hort. Sci., 61: 99-102.

Kader, A.A., Heintz, C.M., Labavitch, J.M. and Rae, H.L. (1982). J. Amer. Soc. Hort. Sci., 107:812-16.

Kafkas, S. and Perl-Treves, R. (2001). Theoritical and Applied Genet., 102: 908-15.

Kallsen, C. E. and Parfitt, D.E. (2017). HortSci., 52: 310-12.

Kallsen, C. E., Parfitt, D.E. and Holtz, B.A. (2014). Acta Hort., 1028: 279-83.

Karimi, M., Chaabouni, A.C. and Gargouri, K. (2006). Acta Hort., 726: 367-70.

Karimi, S., Rahemi, M., Maftoun, M., Eshghi, and Tavallali, V. (2009). Australian J. Basic and Applied Sci., 3: 1630-39.

Kaska, N. (1986). In: Foliar Fertilization (ed. A. Alexander). Martinus Nijhoff Publishers. pp.271-91.

Kizilgoz, I., Sakin, E. and Aslan, N. (2010) African J. Agric. Res., 5: 3427-30.

Kouchakzadeh, A. and Haghighi, K. (2011) Agric. Eng. Int., 13(3): 1878.

Kouyeas, V. (1952). Ann. Inst. Phytopath. Benaki. 6: 81-87.

Kreutzberg, V.E. (1940). Acad. Sci., 27: 617-15.

Kris-Etherton, P.M., Hu, F.B., Ros, E. and Sabate, J. (2008). Jr. Nutr., 138: 51746-51.

Kuhbanani, A.M. and Karimi, H.R. (2013). Iranian J. Hort. Sci., 44: 81-83.

Kuru, C. (1993). Dikimden Hasada Antepfisrgr. Ar. Ajans, Gaziantep.

Lilleland, O. (1930). Proc. Amer. Soc. Hort. Sci., 27: 237-15.

Lin, T.S., Crane, J.C. and Ryugo, K. (1984). J. Amer. Soc. Hort. Sci., 109: 39-42.

Malakouti, M.J. (2006). Acta Hort., 726: 293-300.

Maskan, M. and Karatas, S. (1998). Jr. Sci. Food Agric., 77: 334-40.

Mass, J.L., Vanderzwet, T. and Madden, C. (1971). Plant Dis. Rept., 55:72-76.

Maurice, J. (1988). Acta Hort., 227: 93-97.

Mehdi, H., Chaabouni, A.C. and Gargouri, K. (2006). Acta Hort., 726: 367-70.

Mehrnejad, M.R. (2014). Acta Hort., 1028: 191-94.

Memmi, H., Gijon, M.C., Guerrero, J., Couceiro, J.F. and Perez-Lopez, D. (2014). Acta Hort., 1028: 381-84.

Miniati, E. (1981). Fitoterapia, 52: 261-71.

Mirdamadiha, F., Panahi, B. and Talaie, A. (2005). XIII GREMPA Meeting Almond and Pistachio, Zaragoza, pp. 215-19.

Mohammadi, A. and Panahi, B. (2006). Acta Hort., 726: 485-88.

Nahlawi, N., Kotab, M.A. and Meaded, A. (1979). Publ. Arab Centre for the Studies of Arid Zones and Dry Lands. ACSAD/PS/PS. p 32

Nahlawi, N., Kotab, M.A., Nahlawi, M. and Ibrahim, I.H. (1984a). Publ. Arab Centre for the Studies of Arid Zones and Dry Lands, ACSAD/PS/P10, p. 51.

Nahlawi, N., Kotab, M.A., Nseir, P., Hassan, A.H., Kawasma, Y. and Ibrahim. I.H. (1984b). Publ. Arab Centre for the Studies of Arid Zones and Dry Lands. ACSAD/PS/Pll. p.4.-1.

Needs, R.A. and Alexander, D.M. (1982). Australian Hort., 80: 87-89.

Nevo, A., Werker, E. and Sassen, R. Ben (1974). Israel J. Bot., 23: l-13.

Okay, Y., Gunes, N.T. and Koksal, A.I. (2011). African J. Agric. Res., 6: 1161-69.

Okay, Y., Tuna Gunes, N., Ilhami Koksal, A., Koroglu, M. and Alagoz, R. (2010). African J. Agric. Res., J: 3417-26.

Olekhnovich, G.S. and Chernova, G.W. (1975). Subtrop. kultury, 4: 89-91.

Onay, A. (2000). Turkish J. Bot., 24: 91-95.

Onay, A., Piric, V., Tilkat, E., Akturk, Z. and Yildirim, H. (2004). Jr. Hort. Sci. Biotech., 79: 960-64.

Onay, A., Suzerer, V., Yldrm, H., Tilkat, E., Akdemir, H. and Ozden-Tokatl, Y. (2011). Acta Hort., 912: 587-92.

Ozgen, I., Bolu, H. and Beyarslan, A. (2012). Minis Entomol. Zoology, 7: 238-42.

Ozrenk, K., Javidipour, I., Yarilgac, T., Balta, F. and Gundogdu, M. (2012). Fd. Sci. Technol. Int., 18(1): 55-62.

Pair, J.C. and Khatamian, H. (1982). Comb. Proc. Int. Plant Prop. Soc., 32 497-503.

Palma, L.de. and Novello, V. (1998). Acta Hort., 470:317-23.

Parfitt, D.E. and Almedhi, A.A. (1994). Scientia Hort., 56: 321-29.

Parhami, F. and Hokmabadi, H. (2011). Acta Hort., 912: 91-94.

Peebles, R.H. and Hope, C. (1937). Proc. Amer. Soc. Hort. Sci., 34:29-32.

Phene, B.G., Goldhamer, D.A., Menezes, J., Beede, R., Weinberger, G. and Cervantes, Z. (1987). Ann. Rept. Calif. Pistachio Indust. Crop year 1986-1987.

Phene, R.C., Menezes, J.Jr., Goldhamer, D.A., Aitkens, G.J., Beede, R. and Kjelgran, R, (1985). In: Drip/trickle Irrigation in action. Vol. II. St. Joseph. Michigan. USA. ASAE, pp. 805-10.

Pontikis, K.A. (1975). Nea Agrotiki Epitheorisis, 9: 291-98.

Pontikis, K.A. (1977). Anotate Geoponike Shole Athhenon. pp. 1-80.

Pontikis, K.A. (1984). Plant Propagator, 30: 14-15.

Pontikis, K.A. (1989). Fruit Var. J., 43: 125-28.

Popov, K.P. (1979). Rastitel'nye Resursy, l5: 127-31.

Porlingis, I.C. and Voviatzis, D.G. (1986). Acta Hort., 179: 521-27.

Procopious, J. (1973). Jr. Hort. Sci., 48: 393-95.

Raei, M. and Jafari, S.M. (2011). Ann. Food Sci. Technol., 12: 179-85.

Rafieva, M.G. and Nikolacova, M.G. (1983). Byulleten Glavnogo Botanicheskogo Sada, 128: 82-89.

Rahemi, M. and Asghari, H. (2004). Jr. Hort. Sci. Biotech., 79: 823-27.

Richard, C., Rosecrance, S.A., Weinbaum and Brown, P.H. (1998). Ann. Bot., 82: 463-70.

Rosecrance, R.C., Weinbaum, S.A. and Brown, P.H. (1988). Ann. Bot., 82: 463-70.

Rovira, M., Batlle, I., Romero, M. and Vargas, F.J. (1995). Acta Hort., 419: 181-88.

Ruelas, G.S. (1986). Proc. Trop. Region Amer. Soc. Hort. Sci., 23: 50-53.

Salehi, M.H., Hosseinifard, J. and Heydari, M. (2010). Options Mediterraneennes, A. no. 94, XIVGREMPA Meeting on Pistachio and Ahmands. Pp. 79-84.

Salmon, T.P., Crabb, A.C. and Marsh, R.E. (1986). Calif. Agric., 40:5-8.

Sepaskhah, A.R. and Maftoun, M. (1981). Jr. Hort. Sci., 56: 277-84.

Sepaskhah, A.R. and Maftoun, M. (1988). Jr. Hort. Sci., 63:157-62.

Sepaskhah, A.R., Maftoun, M. and Karimian, N. (1985). Jr. Hort. Sci., 60: 115-21.

Sepaskhah, A.R., Maftoun, M. and Yasrebi, J. (1988). Jr. Hort. Sci., 63:743-49.

Shakerardekani, A., Karim, R., Ghazali, H.M. and Nyukling, C. (2011). Jr. Agri. Sci. Toronto, 3(4): 13-21.

Sherafati, A., Hokmabadi, H. and Abbaspour, H. (2012). Acta Hort., 963: 247-49.

Sherafati, A., Hokmabadi, H. and Taheri, M. (2011). Acta Hort., 912: 197-201.

Sheremet'ev, S.N. (1984). Botanicheskii Zhurnal, 69: 149-57.

Shokraii, E.H. and Esen, A. (1988). J. Agric. Food Chem., 36: 425-29.

Shuraki, Y.D. and Sedgley, M. (1994). Jr. Hort. Sci., 69: 1019-27.

ShuYu, Y., Shuchai, Su., GuiMin, Fan., ZhiGang, C. and PingSheng, L. (2011). Pl. Dis. and Pests, 2(4): 13-17.

Smolsky, N.Y. and Smirnov, U.P. (1931). Bull. Appl. Bot. Genet. and Pl. Breeding. XXV, No. 4.

Soliemanzadeh, A., Mozafari, V., Pour, A.T. and Akhagar, A. (2013). South West J. Hortc. Biol. Environ, 4: 19-34.

Spiegel, R.P., Maztgh, D. and Evanri, M. (1977). J. Amer. Soc. Hort. Sci., 102:470-73.

Takeda, F. and Crane, J.C. (1980). J. Amer. Soc. Hort. Sci., 105: 573-76.

Takeda, F., Crane, J.C. and Lin, J. (1979). J. Amer. Soc. Hort. Sci., 104: 229-32.

Takeda, F., Ryugo, K. and Crane, J.C. (1980). J. Amer. Soc. Hort. Sci., 102: 155-58.

Tanabaev, A.T. (1984). Zashchita Rastenii, 11: 23.

Taylor, C. and Edwards, M. (2000). Final Rept. Horticulture Australia Ltd. (Project NT 97002).

Tezerji, Z.S. (2011). Acta Hort., 912: 701-07.

Tilkat, E., Suzerer, V., Ersali, Y., Hoser, A., Kilnic, F.M., Tilkat, E.A., Akdemir, H., Ciftci, Y.O. Onay, A. and Kaplan, A. (2014). Acta Hort., 1028: 145-51.

Tilkat, R., Onay, A., Yadrm, H. and Ayaz, E. (2009). Scientia Hort., 121: 361-65.

Toumadje, A., Crane, J.C. and Kader, A.A. (1980). HortSci., 15:725-27.

Turker, S. and AK, B.Z. (2011). Acta Hort., 912: 253-55.

U.S. Department of Agricciulture, Agricultural Research Service. (2015). National nutrient database for standard reference release 27. Available from http://ndb.nal.usda.gov/ndb/search/list.

Uriu, K. and Crane, J.C. (1976). J. Amer. Soc. Hort. Sci., 102: 155-58.

Vemmos, S.N. (2010). Options Mediterraneans Series-A, Seminaires Mediterraneans, 94: 9-18.

Watt, B.K. and Merrill, A.L. (1963). Agric. Res. Serv., US Dept. Agric., Handbook 8.

Weinbaum, S.A., Brown, P.H. and Rosecrance, R. (1995). Calif. Pistachio Industry Ann. Rept. Crop Year 1995-1995.

West, L.D., Hildreth, R.C. and Schlesselman, J.T. (1983). In: Proc. Western Society of Weed Science, 36: 189-90.

Whitehouse, W.E. (1957). Econ. Bot., 11: 281-321.

Whitehouse, W.E., Koch, E.J., Jones, L.E., Long, J.C. and Stone, C.L. (1964). Proc. Amer. Soc. Hort. Sci., 84: 224-29.

Wolpert, J.A. (1986). Ann. Rept. California Pistachio Industry, Crop Year, 1985-86. pp. 80-85

Woodroof, J.G. (1979). Tree Nuts. AVI Publishing Inc., Westport, Connecticut, pp. 572-603.

Yldz, S. and Elekcioglu, I.H. (2011). Acta Hort., 912: 759-60.

Zaen El-Deen, E.M.A. and Abd El-Rhman, *I.E.* (2011). Res. J. Agric. Biol. Sci., 4: 456-63.

Zahra, P., Majid, R. and Amin, B. (2009). World Applied Sci. J., 6: 1193-99.

Zeraatkar, H., Karimi, H.R., Shamshiri, M.H. and Tajabadipur, A. (2013). Plant Knowledge J., 2(3): 94-98.

Zhang, J. (2011). Acta Hort., 912: 667-74.

Zhang, J. (2014). Acta Hort., 1028: 351-58.

Zhang, J. and Weir, P. (2006). Acta Hort., 726: 337-42.

Zheng, Q., Brown, p.H. and Holtz, B.A. (2001). HortSci., 36: 85-89.

Zohary, M. (1952). Palestine J. Bot., 5(4): 187-228.

# Chapter 12

# Walnuts

The history of walnut cultivation is lost in antiquity. O'Rourke (1969) records its spread after tire last Ice Age from Asia Minor to various parts of the world while Slate (1969) considers it older than man on the basis of fossil testimony. It is said to be cultivated in King Solomon's garden. In ancient Persia, walnuts were even used in barter system (Woodroof, 1979), and there from it dispersed to Egypt, Constantinople, Italy and eventually to the other European countries and the New World.

## Composition and Uses

Kernel of walnut, the edible portion, constitutes about half of the total fruit weight and its major composition as reported by Watt and Merrill (1963) is given in Table 12.1. Walnut has been used globally in human nutrition since ancient times. The high protein and oil contents of the kernel make this nut indispensible for human nutrition. The kernel is consumed fresh, toasted, or mixed with other confectionaries. Walnuts are nutrient-rich nut due to high contents of fats, proteins, vitamins and minerals, they are also good source of flavonoids, sterols, pectic substances, phenolic acids and related polyphenols.

The major components of walnut oil are triacylglycerol (980g/kg oil), in which monounsatured fatty acids (FAs) (mainly oleic acid) and polyunsaturated fatty acids (PUFAs; linoleic and α-linolenic acids) are present in high amounts (Table 12.2). Among vegetable oils, walnut oil has one of the highest amounts of PUFAs (upto 78 per cent of total fatty acid content). Walnuts have high amount of omega-6 and omega-3 PUFA, which are essential dietary fatty acids (Taha and Al-Wadaan, 2011). The protein (14-18 per cent) is mainly composed of glutelins (about 70 per cent of the total seed proteins) together with lesser amount of globulins (18 per cent), albumins (7 per cent) and prolamins (5 per cent) (Martinez *et al.*, 2010).

**Table 12.1: Composition of Walnut Kernel***

| *Content* | *Amount/100 g* |
|---|---|
| Water | 3.5 per cent |
| Protein | 14.8 g |
| Fat | 64.0 g |
| Carbohydrate | 15.8 g |
| Fibre | 2.1 g |
| Ash | 1.9 g |
| Food value | 631 cal. |

* Modified from Watt and Merrill (1963).

**Table 12.2: Nutritional Value of *Juglans regia* L.**

| *Principle* | *Value per 100g* | *Principle* | *Value per 100g* |
|---|---|---|---|
| Vitamins (USDA, 2010) | | Fatty acids (Muradoglu *et al.*, 2010) | |
| Folates | 98 mcg | Unsaturated fatty acids | |
| Niacin | 1.125 mg | Palmitolic acid C 16:1 | 0.77 |
| Pantothenic acid | 0.570 mg | Oleic acid C18:1 | 25.26 |
| Pyridoxine | 0.537 mg | Gadoleic acid C20:1 | 0.05 |
| Riboflavin | 0.150 mg | Total MUFA | 22.37 |
| Thiamin | 0.541 mg | Linoliec acid C18:2 | 57.10 |
| Vitamin A | 20 IU | Lineliec acid C18:3 | 10.34 |
| Vitamin C | 1.3 mg | Total PUFA | 4.29 |
| Vitamin E-y | 20.83 mg | Saturated fatty acids | |
| Vitamin K | 207 mcg | Myristic acid C14:0 | 0.24 |
| Minerals | | Palmitic acid C16:0 | 4.28 |
| Potassium | 441 mg | Stearic acid C18:0 | 1.85 |
| Phosphorus | 346 mg | Archidic acid C20:0 | 0.19 |
| Calcium | 98 mg | Total SFA | 7.21 |
| Magnesium | 158 mg | PUFA/SFA | 9.91 |
| Sodium | 2 mg | | |
| Iron | 2.9 mg | | |
| Copper | 1.5 mg | | |
| Manganese | 3.8 mg | | |
| Zinc | 3.09 mg | | |
| Aluminium | 0.58 mg | | |

Obviously, the walnut is very rich in proteins, fats and minerals and is a concentrated source of energy. These are very high in comparison to other foods as well as fruit nuts it contains a good amount of various vitamins of B group and is the richest in vitamin $B_6$ among all the other nuts such as almond, Brazil nut, chestnut, filbert and pecan (Polansky and Murphy, 1966). The amount of vitamin A and C is, however, very low. Volatile flavour components of walnut constitute about 37 compounds including 6 aldehydes, 6 alcohols and 5 ketones (Clark and Nursten, 1976). Certain carbonyles play a major role in imparting walnut aroma.

Immature fruits can be utilized for preparing various products rich in ascorbic acid such as pickles, chutneys, marmalades, press juice and syrups. The ascorbic acid present in immature fruits, green hulls get readily destroyed at room temperature which can be preserved by pre-treating with sulphur.

Almost all plant parts of walnut are utilized in one way or the other but the fruit and timber have been put to maximum use by man. The fruit has excellent flavour and is mainly consumed as a dry fruit being eaten for table purposes. Its usefulness in cookery is highly acclaimed because it adds desirable aroma, flavour, crispness, tenderness or garnishment to dishes depending upon the preparation. Walnuts have traditionally been regarded as tidbits, a special treat to be added to candy, cakes, cookies, and salted specialities or to be served at special parties or dinners held at auspicious occasions. They are given as gifts alone or along with the other fruits. The nuts are processed by blanching, slicing, chopping, grinding into paste and butter and combining with other foods. They also have unique place in confectioneries and a large number of recipes are available for using nutmeats in candy plants, restaurants and homes. The kernels are also taken as snacks between regular meals.

Walnut shell flour is extensively used as an ingredient in plastic fillers, battery cases, moulding resin forms, industrial tile and as insecticide spreader (Woodroof, 1979). About half of the total production of the flour is consumed by plastic industry. It is also used on large scale as a diluents for agricultural insecticides as lignin and cutin contents of the flour easily absorb and retain active toxic chemicals.

## Area and Production

The walnut is grown extensively in almost all the temperate countries where summers are not too cool. In 2015-16, world production of walnut achieved more than 720,000 metric tons (kernel basis). The major producers are USA (33 per cent), China (33 per cent), Ukraine (6 per cent), Iran (6 per cent), Chile (6 per cent), Turkey (4 per cent) and other countries producing about 12 per cent (Table 12.3). The walnut production increased globally by about 87 per cent from 2005 (384,747 metric tonnes) to 2015-16.

**Table 12.3: Walnut Production in Major Producing Countries (2015)**

| *Country* | *Production (Metric tonnes)* |
|---|---|
| USA | 240,295 |
| China | 240,000 |
| Ukraine | 44,600 |
| Iran | 42,680 |
| Chile | 39,000 |
| Turkey | 30,400 |
| Other Countries | 83,435 |

*Source*: Nuts and Dried Fruits, Global Statistical Review, 2015/16, International Nuts and Dried Fruit Council Foundation, Spain.

USA is the main exporter of walnuts. In 2014, the United states exported 86,000 metric tonnes (shelled) walnut, which is 43 per cent of world exports. China exports only 4.0 per cent (7,991 metric tonnes) of production.

Walnut is one of the important nut crops of India. The area under walnut is about 149,502 hectares.

## Origin and Distribution

The wild seedlings of walnut (*Juglans regia* L.) have been found growing in vast region right from the Carpathian Mountain in eastern Europe, across Turkey, Iraq, Iran, southern USSR and Afghanistan to the north-western Himalayas (Woodroof, 1979; Forde, 1975; Serr, 1969). Researchers (Nekresova, 1936; Berg, 1937; Browicz, 1976; Puri *et al.*, 1983) suggested, *Juglans regia* is native to the mountain range of Central Asia, extending from the Xinjiang province of Western China, part of Kazakhstan, Uzbekistan and Southern Kirghizia in the Central Asian USSR and from the mountains of Nepal, Tibet, Northern India and Pakistan West through Afghanistan, Turkmenia and Iran to portions of Azerbaijan, Armenia, Georgia and Eastern Turkey.

There are about eighteen more species of genus *Juglans* which are native to the North and South America, central China, Manchuria and Japan. Romans liked it so much that they named it in honour of the king of all their Gods, Jupitor's Acorn, or in Latin, *Jovis-Glans* which later on became *Juglans*. The Greeks probably obtained *J. regia* from Persia (Iran). From Greece, seeds were transported to Rome. From Italy the walnut moved to Spain, France, Portugal, and Germany and to the southern USSR and went to China from Tibet. According to Forde (1975), the walnut trees were introduced in England by 1562, where it was not successful because of very cold summer. The American walnut industry was, however, founded on the introduction of French strains to Chile and California but it flourished only when the 'Carpathian' and 'Kaghazi' (thin-shelled) walnuts were also brought into cultivation. It was introduced into Eastern and Southern China from Central Asia about 2000 years ago, where it has since become widely naturalized (Decondolle, 1982, Xi, 1985). In India, walnut remained confined to Kashmir and it appears that it has spread to Himachal Pradesh and Uttar Pradesh hills during the past century only because there has been lack of direct contact among these states. It is still to be tried in the north-eastern Himalayan region and the hills of South India.

The origin of the word 'walnut' has been traced to 'Gaul nut' (O'Rourke, 1969). In the past, France was known as Gaul and the walnut as 'Gaul nut' which later on got converted into 'walnut' in English. Another view is that the name walnut has been derived from 'walnut', an old English word for 'Welsh-nut' (Beede, 1985). The nut is also called as 'Persian walnut', 'English walnut' or 'Carpathian walnut', but the use of term 'Persian walnut' has been considered more appropriate for the nut, because it indicates about the place of its origin (Ramos *et al.*, 1984; Serr, 1969; Woodroof, 1979). 'English walnut' is a misleading term because the nut is in no way connected with England. The transportation of walnuts from Mediterranean countries in English ships is probably the source of the origin of this name (O'Rourke,

**Walnuit Cultivar Hartley**

**Anthracnose Disease**

1969). It was also easier to distinguish it from the American black walnut (Forde, 1975). The walnuts grown in the North America and France were extremely susceptible to severe winters restricting the extension of their cultivation further into the north but the introduction of cold-hardy race of walnut from the Carpathian mountains of Poland greatly solved this problem and the term 'Carpathian walnut', thus, came into existence. According to O'Rourke (1969), 'Carpathian walnut' may be considered as a cold-hardy race of *Juglans regia*, which survives and produces a

**Walnut in Bearing**

**Male Flower Inflorescence**

good crop even when the winter temperature falls to –35°C and it includes all the cold-hardy walnuts from the USSR, Czechoslovakia and Germany.

## Taxonomy

Walnuts belong to the family *Juglandaceae*, and genus *Juglans* L. (Manning, 1978). The genus *Juglans* includes about twenty one species among which *Juglans regia* is the most important being commercially cultivated in many countries (Jaynes, 1969). The other species are grown locally for timber and nuts. A list of the important species along with their common name and native range of availability is given in Table 12.4. The somatic chromosome number (2n) of *Juglans* species is 32.

**Table 12.4: Species of *Juglans****

| *Species* | | *Common Name* | *Range* |
|---|---|---|---|
| Section a. | *Juglans* | | |
| | *J. regia* | English or Persian walnut | Southern Europe, Iran to the Himalayas and China |
| | *subsp. turcomanica* | | |
| | *subsp. Fallax* | | |
| Section b. | *Rhysocaryon* | | Argentina |
| | *J. australis* | | |
| | *J. boliviana* | | Western South America |
| | *J. californica* | Southern California black walnut | California |
| | *J. hindsii* | Northern California black walnut | California |
| | *J. hirsuta* | | Northeastern Mexico |
| | *J. jamaicensis* | West Indies black walnut | West Indies |
| | *J. major* | Arizona black walnut | Southwestern United States |
| | *var. major* | | Northwestern Mexico |
| | *var. glabrata* | | South-central Moxico |
| | *J. microcarpa* | Texas black walnut | Southwestern United States |
| | *var. microcarpa* | | Northwestern Mexico |
| | *(J. rupestris)* | | |
| | *var. stewartii* | | Northern Mexico |
| | *J. mollis* | | Central Mexico |
| | *J. neotropica* | | Northwestern South America |
| | *J. nigra* | Eastern black walnut | Eastern United States |
| | *J. olanchana* | | |
| | *var. olanchana* | | Guatemala |
| | *var.standleyi* | | Southeastern Mexico |
| | *J. pyriformis* | | Southeastern Mexico |
| | *J. soratensis* | | Bolivia |

| Species | | Common Name | Range |
|---|---|---|---|
| | *J. steyermarkii* | | Guatemala |
| | *J. venezuelensis* | | Venezuela |
| Section c. | *Cardiocaryon* | | |
| | *J. ailantifolia* | Japanese walnut | |
| | *(J. sieboldiana)* | | |
| | *var.ailantifolia* | | Japan |
| | *var. cordiformis* | Heartnut | Japan |
| | *J. cathayensis* | Chinese walnut | Eastern China, Taiwan |
| | *J. mandshurica* | Manchurian walnut | Manchuria, Northeastern China, Korea |
| Section d. | *Trachycaryon* | | |
| | *J. cinerea* | Butternut | Eastern United States |

* Arranged according to the classification in Manning (1978).

Xiand Zhang (1994) stated that five species of *Juglans* are native to China. They are- Walnut (*J. regia*), Manchurian walnut (*J. manshurica*), Chinese walnut (*J. cathayensis*), Iron walnut (*J. sigillata*), and Hebei walnut (*J. hopeiensis*). Species such as *J. ailanthifolia, J. cinerea* and *J. mandshurica* are appreciated for their ornamental value and can be found in parks and gardens in some countries. Kutina *et al.* (1991) classified different cultivars of walnut, depending on the shell strength, texture or kernel skin colour in Slovakia. These are:

*Juglans regia* var. *tenera*- a thin smooth shell is called paper.

*J. regia* var. *oblonga,* var. *rotunda,* var. *cylindrica* - have a semi-hard shell called semipaper.

*J. regia* var. *durissima,* var. *connata*- have a hard shell hardly seperable called stone.

*J. regia* var. *maxima,* var. *macrocarpa* - produces very large fruit with a ridged shell called underbrush.

*J. regia* var. *racemosa*- nuts are concentrated in bunches of 8-20.

*J. regia* var. *serotina*- late budding types called St. Johan's nuts.

*J. regia* var. *rubra*- kernel has red and crimson red colour skin.

## Cultivars

For profitable walnut growing, the selection of proper cultivars is required to be done on the basis of microclimate of the site, soil fertility, water supply, bearing habit and nut quality of the cultivars (Hendricks *et al.,* 1985). Chilling, spring frost, rainfall and summer heat are the important climatic factors to be taken into consideration. In most of the European countries, late blooming cultivars are grown to escape spring frost injury. For early cropping, the lateral bearing cultivars are to be chosen. The cultivars should have large nuts and kernels of light colour to meet the market requirement. Pollen shedding and stigma receptivity of the cultivars grown together must coincide to ensure pollination. A vigorous cultivar, Serr, has performed well on the soils of low fertility.

The cultivars differ from region to region. In France, Franquette, Mayette and Parisienne are the cultivars traditionally planted (Charlot, 1987). Sejnovo is the popular cultivar of Yugoslavia which is high in both oil and protein content (Bugarcic *et al.*, 1985). According to McGranahan and Leslie (2006), the ideal walnut cultivars should have a late leaf emergence period, attaining bearing stage in a short period, be of a lateral fruiting type and a vigorous growth to prevent alternate bearing. The cultivar should be resistant or tolerant to walnut blight (*Xanthomonus campestris* pv. *Juglandis*) and anthracnose (*Gnomonia leptostyla*), the shell should be comparatively smooth and the kernel ratio should be at least 50 per cent. Gandev (2017) evaluated two US cvs. Serr and Hartley, two French cvs. Lara and Fernor, Hungarian cv. Tiszacsecsi 83 and two Bulgarian cvs. Sheynovo and Izvor 10 under the climatic conditions of South Bulgaria. The results revealed that cvs. Fernor and Izvor 10 bore fruits on 80 per cent of the lateral buds along the fruiting shoots followed by Lara (45 per cent), Serr (35 per cent), Sheynovo (25 per cent), Tiszacsecsi 83 (15 per cent) and Hartley (10 per cent). The mean fruit weight ranged from 10.63g for Tiszacsecsi 83 to 13.62g for Sheynovo (Table 12.5). The kernel ratio was recorded maximum (55.02 per cent) in Sheynovo (Table 12.6) while the cumulative yield for five years (2009-14) was recorded highest in Serr followed by Fernor and Izvor 10.

All the walnut trees in India are of seedling origin except few seedlings locally named as cultivars or some cultivars introduced from abroad which are grown at the regional research stations. The surveys of the walnut germplasm in all the three states of Jammu and Kashmir, Himachal Pradesh and Uttar Pradesh hills cultivating this fruit have resulted in identifying potential seedlings which can be released as cultivars after further evaluation (Chauhan and Sharma, 1979; Sharma and Chauhan, 1980; Pandey and Sinha, 1984; Qureshi and Dalal, 1985; Woodroof, 1979). Walnut germplasm of Jammu and Kashmir having high bearing potential with better nut and kernel quality needs special attention for exploration, collection and evaluation. Besides the nut yield and quality, other tree characteristics, *i.e.*, plant vigour, ecological tolerance and pest and disease resistance should also be assessed.

Like India, mostly seedlings are preponderant in China. A strain 'grape walnut' is worth mentioning. It derives its name from grape-like cluster bearing habit and each cluster has about 15 fruits (Bien, 1980). The shell is thin and kernel has fragrant flavour and 60 per cent oil content. A large number of cultivars, however, has been evolved in the USA which is the main walnut exporting country of the world. Description of some important cultivars of the world are as follows:

## Hartley

It was selected from seedling in Napa valley by John Hartley in 1909, which is one of the most popular commercial cultivars of California (Brooks and Olmo, 1972). Its desirable traits are light kernel colour, good yield, and tolerance to codling moth and blight. Hartley, however, needs fertile, deep, well-drained soil and irrigation for high productivity. The nuts are large with broad flat base and pointed tip. The shell is light-coloured, thin and seals well. It is primarily a bleaching cultivar sold

**Table 12.5: Yield per Tree, Average Yield per Tree and Total Yield per ha of Walnut Cultivars for the Period 2009-14***

| *Cultivar* | *Yield (kg tree$^{-1}$)* | | | | | *Average Yield (kg tree$^{-1}$)* | *Total Yield kg ha$^{-1}$ (2009-2014)* |
|---|---|---|---|---|---|---|---|
| | *2009* | *2010* | *2011* | *2013* | *2014* | | |
| Izvor 10 | 17.87 b | 23.00 b | 16.77 d | 26.00 a | 12.00 b | 19.13 a | 15,301.33 b |
| Sheynovo | 9.10 d | 15.83 c | 12.83 e | 15.00 c | 4.80 de | 11.51 bc | 9,211.67 d |
| Serr | 20.90 a | 29.73 a | 35.90 a | 8.80 a | 14.70 a | 22.01a | 17,605.33 a |
| Hartley | 17.67 b | 20.67 b | 23.87 c | 13.00 cd | 7.80 c | 16.60 ab | 13,280.00 c |
| Fernor | 12.00 c | 22.00 b | 28.50 b | 22.00 b | 16.00 a | 20.10a | 16,080.00 b |
| Lara | 16.87 b | 20.23 b | 21.67 c | 19.67 b | 7.30 cd | 17.15 ab | 13,717.33 c |
| Tiszacsecsi 83 | 3.93 e | 7.50 d | 5.50 f | 10.00 de | 3.80 e | 6.15 c | 4,917.33 e |

Statistical Significance at *P*= 5 per cent.

* Gondev (2017).

**Table 12.6: Mean Biometric Data* of Walnut Cultivars for the Period 2009-2014****

| *Cultivar* | *Fruit Height (mm)* | *Fruit Width (mm)* | *Fruit Thickness (mm)* | *Shell Thickness (mm)* | *Mean Weight of Fruit (g)* | *Kernel Colour* | *Kernel Ratio (per cent)* |
|---|---|---|---|---|---|---|---|
| Izvor 10 | 41.27 b | 31.29 cd | 31.70 c | 1.1 d | 11.56 c | Pale | 53.75 a |
| Sheynovo | 42.39 ab | 30.97 d | 32.27 bc | 1.2 d | 13.62 a | Dark | 55.02 a |
| Serr | 38.71 c | 33.55 b | 33.04 bc | 1.4 c | 12.53 b | Pale | 53.64 a |
| Hartley | 43.79 a | 33.62 b | 33.58 bc | 1.6 b | 13.36 ab | Pale | 45.50 b |
| Fernor | 41.17 b | 32.80 bc | 33.70 bc | 1.7 ab | 12.55 b | Pale | 46.36 b |
| Lara | 37.14 c | 35.60 a | 35.62 a | 1.4 c | 12.68 ab | Pale | 42.31 bc |
| Tiszacsecsi 83 | 37.52 c | 31.25 cd | 31.35 d | 1.8 a | 10.63 c | Pale | 38.45 c |

*Statistical Significance at *P*= 5 per cent.

** Gondev (2017).

in shell. It is not precocious but very productive when trees are grown up. The tree is moderate to large, and leafs out a fortnight later than Payne.

## Payne

It is the second leading cultivar of California which originated as a seedling in the George Payne orchard in Santa Clara county. The yield potential is very high due to the fruitfulness of lateral buds. Payne can be grown in solid blocks because pollen and stigma mature simultaneously. Nut is medium to small in size with a good shell seal. Payne trees are moderately vigorous, round in shape and require rather heavy pruning to maintain vigour. It can be spaced at 8-10 m. Payne is susceptible to codling moth and blight.

## Franquette

Franquette, an old and leading cultivar of France, is known for its late bud break, thus escaping injury from late spring frost. Lateral buds are not fruitful in Franquette and it reaches full production in 12 years compared with 9 years in other cultivars (Charlot, 1987). It is harvested about a week later than majority of cultivars. Nut is small, shell seal good and kernels light in colour. Tree is large and upright requiring 14-16 m spacing. In California, an open-pollinated seedling of Franquette has been selected and termed as Scharsch-Franquette which is a good pollenizer for medium to late blooming cultivars such as Hartley.

## Serr

It is the result of a cross of Payne x PI 159568. Serr is heavy yielding and well-adapted to warm conditions. The tree is very vigorous and gives poor yield on very fertile soils. It has a tendency for pistillate flower abscission. Its 30 per cent to 50 per cent lateral buds are fruitful. Mid-blooming cultivars Chico, Tehama and others are suitable pollenizers for Serr. Large nuts are of Payne type. Kernel is 59 per cent of nut, light in colour and good in quality. Late spring frost is injurious and codling moth and blight also cause considerable damage.

## Ashley

It resembles to Payne in a large number of characteristics. It is an established high yielding cultivar being fruitful up to 90 per cent on lateral buds and consequently requires heavy pruning to keep the tree vigorous. The large ovoid nut gives about 54 per cent kernel which is high in quality, good in flavour and light tan in colour (Brooks and Olmo, 1972). Early bearing, small trees are planted as fillers. It is, however, unsuitable for high rainfall areas due to problem of blight infection.

## Sunland

This cultivar originated as a cross of Lompoc x PI 159568 in California. It is also a lateral fruiting cultivar having about 90 per cent fruitfulness. The nut is very large, long, oval with a good seal. The shelling percentage is 58 and kernel colour is good. It is highly precocious and tree size is similar to Payne. Sunland is a mid-season, very productive cultivar but susceptible to blight and late spring frost.

## Chico

Selected in 1961 from a cross of Sharkey x Marchetti. Chico is almost cent per cent fruitful on lateral buds. The nut is small but excellent in quality with 47 per cent light coloured kernel. Heavy pruning is required to maintain nut size. It is an excellent pollenizing cultivar and well-suited to hedgerow planting with a spacing range of 4-8 m (Hendricks *et al.*, 1985). It leafs out 2 days after Payne and blooms early but sheds pollen late.

## Vina

It was evolved from a cross of Franquette x Payne in Davis, California. Tree is small with, a lateral bud fruitfulness up to 90 per cent and needs a good pruning programme to maintain its vigour and nut size. Medium size nuts resemble in shape to Hartley but with a less flattened base. The kernels are light in colour, high in quality and shelling percentage is 48. Vina leafs out 7 days after Payne and is a mid-season cultivar both in pollen shedding and fruit harvest.

## Howard

Howard (a sibling of Chandler) is another lateral fruiting, high yielding cultivar adapted to hedgerow planting. In spite of late leafing (15 days after Payne), harvest timing is mid-season. Large nuts are round and smooth-with a good seal. Kernel is light in colour and shelling percentage is 50. Tree is smaller than Vina and Chandler, semi-upright and tolerant to blight (Hendricks *et al.*, 1985).

## Pedro

Pedro has already become a popular cultivar in many walnut growing regions because it attains full bearing age earlier and gives 30 per cent to 50 per cent higher yield than terminal bearing cultivars (Charlot, 1987). The nuts are large, shell heavy, seal fair and kernel 48 per cent of nut. Tree leafs out 18 days after Payne and sheds pollen over a long period, becoming a good pollenizer. High summer temperature adversely affects the quality of shell and kernel.

## Chandler

It has arisen from a cross of Pedro x UC56-224, which is similar in many characters to its parent Pedro in lateral bud bearing, leafing time and harvest period. The nut is large, oval, smooth with a good seal and excellent colour. Kernel percentage is 49. Being an in-shell cultivar, it can supplement Hartley supply in the market. Chandler is moderately vigorous and semi-upright and is considered a very promising cultivar.

## Tehama

Parents of Tehama are Waterloo and Payne. It is a heavy yielding and lateral fruiting (70-80 per cent) cultivar which leafs out 11 days after Payne. Nut size is large but shell seal is not so good because sometimes it cracks at the suture exposing the kernel to damage. Kernel colour is good with a 50 per cent shelling percentage. Tehama is a midseason cultivar mainly serving as a pollenizer. Since the tree is large and vigorous, it requires 13 m spacing.

## New Cultivars

### Cultivars Developed from USA

**Robert Livermore (UC86-11 x Howard)**: A late-bearing walnut with red seedcoat, developed from University of California, Davis. Bears terminally as well as laterally, precocious, protandrous, moderate yielder, well filled nut, average weight 14.1g, uniformly bright red.

**Sexton** (UC85-8) x Chandler): A precocious, high yielder and mid-season walnut developed from University of California, Davis. Nuts are smooth, round solid shells, yield 53 per cent kernel.

**Gillet** (UC76-80 x Chico): Another hybrid developed from University of California, Davis. Large trees; precocious, high yielder and mid season cultivar. Kernel 8.0g, light-coloured, easily removed as halves.

**Forde** (UC61-25 x Chico): Vigorous tree, high yielder and late-season. Nuts are smooth, slightly oval, kernels 8.6g, and plumpy.

**Fredrick** (Payne O.P.): Vigorous tree, harvest one week after payne. Nut 11.4g, smooth and ovate.Kernel 6.6g.

**Ivanhoe** (UC67-13 x Chico): Tree vigour moderate to low, precocious, protogynous, high yielding and matures early. Nut 13.5g and kernel 7.7g.

**Solano** (UC67-13 x Chico): Early-mid season harvesting walnut. Tree moderate vigour, high yielder, nut 14.7g, smooth, oval, kernel 8.0g, early removed in halves.

### Cultivars Developed from China

**Chuanzao 1**(Yunxin 7926 x Xiazao): Early maturity, high yielder, good adaptability to drought, diseases and insects. Nuts is large and round with smooth surface. Kernel percentage is 51.6 per cent and total protein content is 16.80 per cent (GuangLan *et al.*, 2011).

**Xiluo No. 3:** A late fruiting cultivar. High yielder, thin shells, good kernel quality and drought tolerant (Meizhi *et al.*, 1999).

**Liaoning 6** (Selection 10301 x Selection 11001): Trees are hardy, high yielding, resistant to black spot (*Xanthomonas arboricola* pv. *Juglandis*) and anthracnose (*Gnomonia leptostyla*) diseases. Oval shaped nut, round, weighing 12.4g. kernel weight 7.3g with good eating quality (BaoJun *et al.*, 2009).

**Qinquan 1:** High yielder, good resistance to cold and disease. Nut is large and round, smooth surface and thin shell, yellow white kernel of excellent quality (GuoLiang *et al.*, 2009).

**Luzao**: Early ripening (end July -1$^{st}$ week August), precocious and thin shell cultivar. Full kernel, savory and crisp, kernel weight 14.3g (BaoGuo *et al.*, 2008).

### Cultivars Developed from India

Cultivars released from Central Institute of Temperate Horticulture, ICAR, Srinagar, J&K (Ahmed *et al.*, 2013).

**CITH-W-1**: Bold nut (28g), large kernel size (14.21g), good kernel recovery (50.75 per cent) and long trapezoidal in shape.

**CITH-W-2**: Large, ovate nuts and well filled kernel, plumpy, nut weight 16g and average kernel weight 8g.

**CITH-W-3**: Large, attractive nuts, average nut weight 19.3g and kernel weight 9.8g.

**CITH-W-4**: Nuts are large, ovate, average nut weight 19.08g and kernel weight 10.09g.

**CITH-W-5**: High yielder, bigger nuts (19.5g) and kernel (10.16g), good kernel recovery (48.9 per cent) and light shell colour.

**CITH-W-6**: Medium sized nuts, ovate in shape, average nut weight 24g and kernel weight 12.2g.

**CITH-W-7**: Large nuts (24.7g), and kernel (12.4g), nuts are elliptic in shape.

**CITH-W-8**: Nuts having light kernel colour with average weight 20.4g and kernel weight 11.01g.

**CITH-W-9**: Nuts are large (21.2g), kernel weight (10.7g), with kernel recovery 50.5 per cent.

**CITH-W-10**: Its average nut weight 24.5g and kernel weight 12.5g with kernel recovery 51 per cent.

## Cultivars Growing in different Countries

**Argentina:** Criollas, Chandler, Serr, Vina, Howard, Tulare, Trompito, Franquette, Ivarto, Ramillete.

**Australia**: Franquette, Chandler, Lara, Lompoc, Howard, Chico, Ashley, Tulare.

**Azerbaijan**: Sey fi, Sugrai, Araz, Disar, venend, Djar, Qum.

**Bulgaria**:Sheynovo, Silistrenski, Dryanovski, Izvor 10, Lara, Fernor.

**Canada:** Lake, ISU 73-H-24, Combe, North Platte, Coble 2, Broadview.

**Chile:** Paye, Eureka, Serr, Chandler, Sunland, Howard, Tular.

**China:**Bofeng, Bokexiang, Zha343, Luguang, Xiangling, Hanfeng, Wen 185, Liaoning 1, Luguo 1, Liping 2.

**Czech Republic:** Apollo, Bohumil, Jupiter, Lake, Mars, Saturn, Seifersdorfsky, Victoria, Vilem, Sychrov.

**France:** Fernor, Fernette, Franquette, Mayette, Grandjean.

**Germany:** Moselaner, Weinsberger, Weinheimer, Geisenheimer, Kurmarker, Spreewalder.

**Greece:** Hartley, Serr, Franquette, Gustine, Amigo, Pedro, Chandler, Lara, Fernor, Big Top, Ioah, Iliana.

**Hungary:** Milotai 10, Alsoszentivani 117, Tiszacsecsi 83, Milotal kesei.

**India:** Chandler, Howard, Drainovsky, Franquette, Lake English, Opex, Placentia, Wilson, Pratap, Unique, Chakrata Selection, Gobind.

**Iran:** Chandler, Pedro, Hartley, Ronde de Montignac, Jamal (Z63), Damavand (Z30).

**Italy:** Lara, Chandlar, Haward, Tulare.

**Japan:** Oni gurumi, Hime gurumi, Shinano gurumi, Warau, Yourei I gou, Shinrei.

**New Zealand:** Rex, Meyric, Serr, Vina, Tehama, Franquette, NZ purple (Red Danube).

**Pakistan:**Gilgil, Chitral, Leepa, Swat, Serr, Payne.

**Poland:** Albi, Dodo, Leopold, Resovia, Silesia, Targo, Tryumf, K-48, Jacek, Koszycki.

**Romania:** Jupanesti,Geoagiu 65, Mihaela, Roxana, Valcor,Valmit, Franquette, Fernor, Lara.

**Solvenia:** Franquette, Fernor, Fernette, Lara, G-139, Elit.

**South Africa:** Chandler, Vina, Serr, Howard, Sunland, Hartley, Chico, Ashley, Livermore.

**Spain:** Chandler, Howard, Vina, Lara, Fernor, Serr, Hartley.

**Switzerland:** Franquette, Maytte, Parisienne, Rote Donau-Nus, Red Gubler Nut, Clone-II.

**Tunisia:** Franquette, Parisienne, Hartley.

**Turkey:** Maras 18, Yalova 1, Yalova 3, Yalova 4, Sebin, Bilecik, Kaplan 86, Guleteki 1, Sen 1, Kaman 1, Sutyemez 1, Oguzlar 77.

**Ukrain:** Bukovina 1, Bukovina 2, Bukovynsaka bomb, Klishkovasky, Rudkovsky, Chernivsti 1, Chernivsti 2, Yarivsky, Yablunousky.

**United Kingdom:** Broadview, Fernor.

**United States:** Chandler, Howard, Tulare, Robert Livermore, Gillet, Ivanhoe, Solano.

## Climate

### Temperature

Walnuts are highly sensitive to the extremes in winter and summer temperatures as well as to its duration. The plant can tolerate as low as –11°C during deep dormancy without serious damage but as soon as growth commences after dormancy, the temperature even two or three degrees below freezing point (0°C) kills leaves, shoots and flowers and thus resulting into crop failure. Likewise the plants which continue to grow till late fall are subjected to serious foliage damage by frost. There are, however, varietal differences in frost tolerance, the Eureka group being the most susceptible to frost injury (Serr, 1969). Carpathian walnut race, grown in colder parts of the world, is cold tolerant during dormancy but is less productive (Ramos *et al.*, 1984).

High temperature of about 38°C causes sunburning of hulls and shrivelling of kernels resulting sometimes into blank nuts. Severe damages are further aggravated if the humidity is low and temperature exceeds 40°C. Persistently lower summer temperatures are not favourable for proper filling of the nuts. The cool growing season delays harvest causing substantial loss to the grower. Hence cool or short growing season is a limiting factor in walnut production. Like most other temperate nuts, the walnut requires certain amount of chilling in winter to break dormancy in order to make satisfactory vegetative growth and flowering. In the absence of adequate chilling, the bud opening and blossoming are irregular and delayed, resulting into poor crop and die back of shoots. The chilling-hour requirement varies with the cultivars as 700 and 1500 hours are needed for Payne and Franquette cultivars, respectively, at or below 7°C (Woodroof, 1979).

Charrier *et al.* (2011) observed more winter hardiness in timber-oriented genotypes than fruit oriented genotypes. Heat requirement was found significantly different in different genotypes. Frost acclimation changes differed significantly between fruit-and wood-walnuts from January through bud burst. Freeze injury is a leading cause of walnut yield losses in many cold regions of Europe and Asia. Aslamarz *et al.* (2011) studied the cold hardiness and its relationship with proline content in Persian walnut. They stated that cold hardiness of samples increased with accumulated seasonal chilling and decreased as they approached spring bud break. A 1.4 to 2.7 fold increase in proline content was observed in stems from November to midwinter. LT50s (temperature causing 50 per cent lethal injury) ranged from –5.8 to –10°C in November, –12.5 to –19.3°C in midwinter and –5.8 to –9.2°C in early spring. Cultivars and genotypes were classified into three groups based on their cold hardiness: sensitive (Z30 and Serr), semi-hardy (Z53, Z67 and Hartley) and hardy (Lara, Z63 and Pedro).

## Rainfall

Rains and humidity during the late spring and summer increase the incidence of walnut blight but late leafing cultivars generally escape infection (Serr, 1969). An annual rainfall of about 80 cm is considered sufficient for the cultivation of walnut which can be supplemented in drier regions with irrigation, particularly for young plants. In severe drought conditions, the plant may make fair growth but nuts do not fill well. The walnut trees are adaptable to unfavourable conditions and a shrub-like growth habit is observed as an adaptive strategy to unsuitable environment (Iotsovo-Baurenska, 1986).

## Site

In Kashmir, India, the flat land of numerous valleys is used for rice cultivation, around the edges on hilly slopes are apple orchards while higher up the slopes at an altitude of about 2000 m or more are walnut plantations. In such hilly situations, site selection becomes important for successful walnut production. A site, where frequent hailstorms are experienced or cold air settles down in low land making it a frost pocket, is unsuitable for any fruit cultivation. Higher northern sloping land is preferable to southern slopes and low lands as the plants growing there are less subjected to frost injury. On northern slopes growth starts later because these are cooler than southern slopes. Evaporation is considerably less on northern slope and the soil does not lose moisture as quickly as on southern slope.

There is considerable uncertainty regarding the magnitude of potential effects of climate change on walnut. Some studies tend to indicate walnut could be negatively impacted by climate change, while others do not. Walnut may at a disadvantage due to its susceptibility to drought and frost injury in current growing regions given the projected increases in temperature and extreme climatic events. Other regions that are currently considered cold for walnut growth may see increased establishment and growth depending upon the rate of temperature increase and the frequency and severity of extreme climatic events. Gauthier and Jacobs (2011) suggested research investigating a combination of environmental factors, such as temperature, carbon dioxide, ozone, water and nitrogen to understand (a) climate change effects on walnut, and (b) management strategies for walnut acclimation and adaptation to climate change.

## Soil

The suitability of a soil for walnut cultivation should be examined on the basis of top soil, subsoil, aeration, drainage, chemical reaction (pH) and availability of various nutrients. Soil analysis of both top and subsoil will provide useful information. The top soil should be fertile and well drained. Walnut requires soils of neutral range as compared to other temperate fruits which generally do well in a slightly lower pH (Stephens, 1969). Soil pH influences the availability of mineral nutrients, a majority of which are, however, available to the plant in the pH range of 6 to 7. If the soil pH is low which is generally found in cold region, it can easily be raised with the application of lime. Besides the top soil, the condition of subsoil is also important for walnut because its roots are evenly distributed in the soil to a depth of about 3 metres. The subsoil should be free from solid rock, impervious clay or gravel layers which restrict root growth. Tree growing on poor, shallow and dry soils suffers more than from adverse climatic conditions. Soil should also not be infected with root-lesion nematodes and fungal diseases. Walnuts are thought to do best on fertile, deep, well-drained, non-stratified, loamy soil. In deep soil a few roots have been reported to grow to a depth of 3.0m but even in these soils, > 75 per cent of the roots are found in the upper 60-100cm of soil (Probesting, 1943; Catlin and Schreader, 1985).

## Rootstocks

The open-pollinated seed of *J. regia* is used for raising seedling rootstock for walnut. In USA, *J. hindsii*, Paradox (*J. hindsii* x *J. regia*) and Royal (*J. hindsii* x *J. nigra*) seeds are used. In addition, the seedling rootstock of *J. nigra*, *J. mandshurica*, *J. cathayensis* and *Pterocarya stenoptera* are employed to a small extent in other countries (McGranahan and Catlin, 1987: Bien, 1980). Most of the *Juglans* species, *viz.*, *J. cinerea*, *J. major*, *J. hindsii*, *J. microcarpa*, *J. nigra*, *J. ailantifolia*, *J. awtralis* and several hybrids have been reported to be more or less compatible with Persian walnut, exhibiting some signs of incompatibility, such as, overgrowth of scion or stock, swelling at the union, lack of scion vigour, low percentage of union, *etc.* A correct assessment of compatibility is not possible because of the interference of blackline disease.

## Characteristics of Rootstocks

***J. regia*** : Persian walnut is the most common rootstock everywhere because it is readily available besides being fully compatible, resistant to crown rot and tolerant to blackline disease (Hartmann and Kester, 1959; McGranahan and Catlin, 1987). It is, however, unsuitable for waterlogged, saline and oak-root fungus and nematode infected soils. This rootstock is also slow to grow in the nursery.

***J. hindsii*** : It is a common widely used rootstock of Persian walnut in California making a strong graft union. It is resistant to oak-root fungus and root-knot nematode but susceptible to *Phytophthora*, root-lesion nematode and blackline disease. Low availability of phosphorus and zinc in soil or waterlogging hampers the growth of plants on this stock.

***J. nigra*** : It is a satisfactory rootstock for its own cultivars and has also been recommended for *J. regia* in some countries. In comparison to *J. regia* rootstocks, this rootstock is more tolerant to crown gall, *Phytophthora* and oak-root fungus and induces precocity (Delbos *et al.*, 1982). *J. nigra* reduces the tree size both as rootstock and interstock. Lower yield efficiency and susceptibility to blackline disease are the main impediments in its popularity.

***J. microcarpa*** (*J. rupestris*): This rootstock is particularly suitable for high boron, chloride and soil pH conditions where it has given better performance than *J. regia* and *J. nigra* rootstocks (McGranahan and Catlin, 1987).

***J. sieboldiana*** (*J. ailantifolia*): It has been reported as a highly suitable rootstock from the USSR with a wide range of adaptability to varying soil and climatic conditions. It is cold-hardy and resistant to *Phytophthora* (Matheron and Mircetich, 1985, Glonti *et al.*, 1987).

**Paradox** (*J. hindsii* x *J. regia*): Paradox has performed better in California than its parents in respcet of scion growth and tolerance to both nematodes and *Phytophthora* species. It also makes more growth in replanting of old orchards than *J. hindsii* (McGranahan and Catlin, 1987). It is, however, susceptible to blackline disease.

## Other Rootstocks

***J. californica***, It was earlier used in California, has been discarded due to its susceptibility to crown rot, root rot and wet soils. *J. major* may find favour in good and high pH soils but its superiority over *J. hindsii* is still to be investigated. *J. mandshurica* has been used in China in cold region but was unsatisfactory in the USSR (Nedev, 1973).

**Royal** (*J. hindsii* x *J. nigra*) has been tried in California with promising results. The most important rootstock, however, is wingnut (*Pterocarya stenoptera*) which is another member of *Juglandaceae* and, a native of China. It is tolerant to *Phytophthora*, waterlogging and root-lesion nematode but is susceptible to *Armillaria* and blackline disease. Since it is incompatible with many walnut cultivars, it could be used in rootstock breeding programmes (McGranahan and Catlin, 1987).

## New Rootstocks

**RXI** (*Juglans microcarpa* DJUG 29.11 x *J. regia* O.P): A *Phytophthera* resistant hybrid clonal root stock for Persian walnut developed from University of California, Davis. It is less vigorous than typically used *J. hindsii* x *J. regia* seedling hybrids, grows well in waterlogged conditions.

**VX211** (*J. hindsii* PDS96-43 x *J. regia* O.P): A vigorous, nematode tolerant rootstock for Persian walnut developed from University of Californis, Davis and USDA-ARS. Grows well in the presence of nematodes (*Pratylenchus vulnus* and *Meloidogyne* sp.), moderate resistant to *P. citricola*.

**Vlach:** Clonal true Paradox rootstock. Very vigorous rootstock competitive with VX211 for vigour but does not share nematode tolerance.

Among the various walnut rootstocks, *J. nigra* has only reduced tree size by 10-20 per cent but this effect is still to be confirmed (Germain *et al.*, 1975). *J. nigra*, however, definitely dwarfens the tree when used as interstock (Serr, 1968). The dwarf trees are more yields efficient but dwarfing interstock is not recommended for commercial purpose (McGranahan and Catlin, 1987).

## Rootstock Adoption

## Soil

### Aeration

Poor aeration resulting from waterlogging is detrimental to walnut (Catlin *et al.*, 1977), The effect of root-zone flooding varies on damaging young seedlings, the damage being most serious on *J. regia*, medium on *J. hindsii* and the least on Paradox. *P. stenoptera* seedlings are even more adaptive to flooding than *Juglans* seedlings. In another experiment, *J. nigra* was found more tolerant to waterlogging than either of *J. regia* or *J. sieboldiana* (Solignat, 1974). Sometimes, leaves dry and trees wilt suddenly in summer after few days of flood irrigation. This phenomenon is known as 'apoplexy' and no pathogen has been found to be associated with it (Moller and Ramos, 1972). High soil temperature, inhibition of root growth due to inadequate moisture and flooding of root-zone are thought to be the reasons of 'apoplexy'. Paradox roots are supposed to be more tolerant to this malady.

### Salinity

High salt concentration, either in soil or in irrigation water, damages the walnut plants (Catlin and Schreader, 1985). The rootstocks in decreasing order of salt tolerance are *J. hindsii*, Paradox, *J. regia* and *J. nigra* (Serr and Rizzi, 1964; Monk and Wiebe, 1961). Differential capacity of the walnut rootsocks in absorbing and transport of salts has been confirmed in a rootstock trial in California (Browne *et al.*, 1977). An excess amount of chloride was absorbed by Manregian rootstock.

### pH

The chlorosis is usually caused by high soil pH. In such conditions iron and manganese deficiency can occur due to high amount of calcium carbonate present in the soil. Paradox is the least susceptible rootstock to chlorosis (Hendricks *et al.*, 1977; Serr and Rizzi, 1964). Chlorosis also does not occur on *J. microcarpa* or *J. major* rootstocks.

### Cold-hardiness

Adaptability to severe cold in rootstocks is necessary for tree survival in very cold climate. Ozol (1949) graded the *Juglans* rootstock in decreasing order of cold-hardiness as *J. mandshurica, J. cinerea, J. nigra, J. ailantifolia* var. cordiformis and *J. regia*. Among *J. regia*, Carpathian and Manregian types can tolerate more cold than other types (Grimo, 1979; Lagerstedt, 1978).

## Propagation

Walnut can be propagated either by seed or by vegetative methods. The seedlings are highly variable producing the nuts of inferior quality with an average shelling percentage of 30 per cent to 35 per cent as compared to 50 per cent to 60 per cent of good commercial cultivars (Qureshi and Dalal, 1985). The standardization of product is essential for export market which is not possible with the nuts of seedling origin. The need for adoption of vegetative methods for walnut multiplication has been widely realised (Dar, 1987; Pathak and Srivastava, 1975; Thapar, 1963; Woodroof, 1979; Qureshi and Dalal, 1985).

## Seed Propagation

The seedlings are usually raised for use as rootstock. Although walnut germinates without stratification under some conditions, yet proper stratification is desirable for good germination and growth.

The walnut seed fat contains both free and glyceride-bound saturated fatty acids such as palmitic and stearic acids. The non-saturated fatty acids found are palmitoleic, oleic, linoleic, linolenic and myristolenic. During dormancy break a decrease in dry matter and crude fat content and an increase in the number of acids were observed (Kawecki and Kaworski, 1975). The moist nuts are placed in alternate layers with a suitable medium such as vermiculite, peat moss or sand in a container which is then subjected to cold treatment (3-5$^0$C) for a period of two or three months. During this period the nuts should be prevented from drying and should get enough air. For small quantities of nuts, the home refrigerator is adequate for stratification but for large quantities, the nuts can be kept in alternate layers of sand in a protected place in the orchard. The nuts can also be sown directly in the nursery or field in fall if the chances of damage from rodents are not expected, but the soil is to be kept just moist, neither dry nor wet.

A large number of factors such as nut type, size, position of seed, planting depth, soil moisture and temperature and seed treatment influence the germination of seeds. The planting of large nuts at 7 cm depth in soil in vertical position gives better germination and seedling growth (Lal *et al.*, 1984).

Gangwar and Kumar (1975) reported that walnut seeds of medium size (3-4 cm long) gave a germination of 90 per cent compared with only 71 per cent for smaller seeds and 85 per cent for larger ones. Germination was 87.33 per cent in seeds sown in a vertical position and 76.77 per cent in those placed flat in the soil. Chernilevskii (1977) obtained higher percentage of seed germination when seeds were sown in early December and in spring. Low soil moisture and temperature retard the germination and subsequent growth. In a trial of various seed treatments after stratification, such as, 24 hour soaking of seeds in 125 to 250 ppm gibberellic acid, 30 minute soaking in 10 per cent and 20 per cent sulphuric acid and hand cracking, Gaur (1980) found an increase in germination percentage and height of seedlings with gibberellic acid treatment. Hamed and Fathi (1999) evaluated different seed treatment chemicals for raising seedlings of Paradox in Egypt. The seeds were soaked for 24 or 48 hours in gibberellic acid (500 or 1000ppm) or potassium nitrate (0.5 or 1.0 per cent) then stratified at $5^0$C for 3 months before germination. The use of seed treatment followed by stratification increased percentage germination and resulted in earlier and stronger seedling growth. Treatment success followed the order: $GA_3$ at 1000ppm > $GA_3$ at 500ppm > $KNO_3$ at 1.0 per cent > $KNO_3$ at 0.5 per cent > distilled water. The longer soaking period (48h) was found more effective.

Mol'chenko *et al.* (1977) obtained uniform germination and high production of planting materials by immersing the seeds in hot water (80-$90^0$C) for 24 h which were then mixed with 2 parts of peat or sand (1:2) and stratified in wooden boxes. Once a week the mixture was moistened and stirred. The seeds germinated in 35-45 days. This method markedly shortened the period of seed stratification. Treating the seeds with ethrel at 100 to 1000 ppm was found to improve the germination (Casini and Salvadori, 1975, Sinha *et al.*, 1971).

The nursery should be raised at a site having good fertile soil, drainage and irrigation facilities. The nuts are sown 7 cm deep in soil in spring season. The distances between the seeds and rows depend on the period for which the seedlings will remain in the nursery. When the seedlings are to be left in place for about two years or more, a spacing of one metre between rows and thirty centimetres between seeds is a common practice. In areas devoid of rains during summer season, mulches conserve moisture and control the weeds accelerating the germination and growth of the seedlings. Frequent irrigation and removal of weeds are required to obtain maximum growth. In order to ensure the establishment of grafts after transplanting, their taproot is cut off 20-25 cm below the ground level which should be done about a year before actual transplanting.

## Vegetative Propagation

Walnut can be propagated vegetatively by grafting, budding, stooling and micropropagation while rooting of cuttings and layers has not been so successful.

## Grafting

The percentage of success obtained in grafting and budding walnut is definitely lower than in pome and stone fruits thus requiring more precision in grafting operation and handling of stocks and scions (Weber and MacDaniels, 1969). Better success in grafting by cleft, tongue and splice techniques was obtained when grafting was either done in a greenhouse having 28⁰C temperature and 80-100 per cent relative humidity or the grafts after, preparation were placed in the incubator at 25+1⁰C temperature with75±2 per cent humidity (Bein, 1980; Park, 1984; Sharma *et al.*, 1982). Favourable temperature and humidity conditions are essential for callus development in walnut which are usually not available outdoor, resulting in failure of graft unions. Sharma *et al.* (1982) obtained higher survival of grafts when grafting operation was carried out in the middle of January rather than in the last week of December or in the middle of February. They attribute higher success in the middle of January to the favourable temperature and humidity conditions prevailing in the first week of February when the incubated grafts were planted in the field. Avanzato and Tamponi (1988) reported 73 per cent grafting success by heating the unions at 27⁰C as compared with only 6 per cent of the unheated unions. They cleft grafted at the end of January and February and wrapped the graft union area with aluminium foil to maintain uniform temperature and humidity and heated the unions for 24 days.

In a comparison of seven grafting methods during two years, veneer grafting carried out in July gave more than eighty per cent success followed by side tongue grafting done in the late January with fifty per cent (Chauhan and Sharma, 1980).

Chebotar (1974) reported that the optimum period for bench grafting was between December and February. In Yugoslavia, budding on *Juglans regia* rootstocks by the modified Geisenheim method was found most successful between 10 June and 10 July (Simovski and Kuzmanov, 1972). Zachej (1974) opined that the most suitable time for grafting is in March followed by November in Czechoslovakia. Dehghan *et al.* (2009) evaluated three bench grafting methods (side stub, omega and whip tongue) and four walnut cultivars (Z53, Hartley, Pedro and Serr) at Iran under controlled condition in March. The rootstock used was dormant 2-years old Persian walnut seedlings. The plants after grafting were covered with moist sawdust with relative humidity of 85-90 per cent and stored in a humid room at 26-28⁰C for 21 days. Based on the results, the highest grafting success was observed with omega (84.33 per cent) followed by side stub (41.89 per cent) and whip tongue (24.31 per cent) grafting. The cv. Hartley grafted by omega method showed the highest graft take of 88.44 per cent. Achim and Botu (2001) recorded highest grafting by using whip and tongue method followed by callusing at the grafting point with warm water, which provides 26±0⁰C at that point (hot callusing). The grafting success rate was noted > 85 per cent.

In India, generally tongue, cleft and side veneer methods are used. Tongue and cleft methods performed best during February-March. Maximum success was found if the temperature remain around 24-25⁰C during grafting. In Jammu and Kashmir, some refinement has been made with the use of hot callusing cables at graft union by which the temperature is maintained around 25⁰C at graft union (Chandel, 2013).

Ozkan *et al.* (2001) recorded variation among the cultivars in patch budding. Working with four walnut cultivars at Turkey, they noted 90 per cent success with Bilecik, 85 per cent with Yalova 1, 88 per cent with Yalova 3 and 90 per cent with Yalova 4. In India, summer budding (chip method) was standardized. The success was reported between 87 and 92 per cent when performed during the last week of May to first week of June (Chandel, 2013). Pathak and Srivastava (1975) found mid-June is an optimum time of budding. Removal of leaf blades fifteen days before budding operation enhanced bud-take. On the whole, cleft grafting performed in the month of January and thereafter incubating at 26$^0$C and 90 per cent relative humidity for three weeks before planting in the nursery and patch budding in mid-June using double blade knife hold promise as vegetative methods for propagating walnut on commercial scale.

## Stooling

With the use of various plant regulators, such as, IAA, IBA and NAA at high concentrations in combination, Rashid (1978) obtained good rooting and establishment of stool layers. He suggested stooling as a suitable method of walnut propagation but mother stools of the commercial cultivars will have to be raised by cuttings or layers in the first instance. Vahdati and Khalighi (2001) reported higher rooting and longer roots by treating the shoots with equally mixed IBA+IAA+NAA at 5000 and 7500 ppm plus mounding with the mixture of perlite and sand (2:1) and wiring the base of the layers.

## Top-Working

Top-working is usually carried out by modified cleft grafting or bark grafting late in the spring season or when new growth has taken place. The dormant scion wood should be cut from parent tree quite in advance and be stored in refrigerator after proper packing. Bleeding is a problem in walnut top-grafting especially when it is done in early spring which can be avoided by heading back the stock before two weeks of actual operation. In top-working of grown up trees, the frame working or grafting of branches is desirable for quicker healing of wounds and earlier larger yields. After grafting, the open wounds must be covered by grafting wax and, if required, re-waxing may be done. To protect from sun burn, the stem portion needs white washing.

Top-working technique with dehorning has been standardized for higher bud-take success. In this technique, the seedling trees are dehorned by heading back main branches in dormant season (December-January). New branches will emerge from dehorned branches in spring season and these new branches are budded with chip method in second week of May and annular method in first week of July. Chip method showed 85 per cent success while it was more than 85 per cent by annular method (Chandel, 2013).

## Micropropagation

Driver and Kuniyuki (1984) have developed a specific medium, DKW, for the culture of apical and lateral meristems of Paradox walnut. They also reported *in vivo* rooting of tissue culture derived shoots. Heile-sudholt *et al.* (1986) reported quicker growth of axes on solidified woody plant medium than on Driver Kuniyuki's Walnut medium (DKWM), but axillary shoot proliferation and elongation was better on liquid DKWM. They also reported better rooting of axillary shoots in sterile vermiculite following a 15 second dip in 10mM IBA and transferring of micropropagated plants to the greenhouse. Micropropagation technique has been narrated in detail by Kuniyuki and Forde (1985). McGranahan *et al.* (1987) reported successful micropropagation of Paradox, *J. hindsii* and *Pterocarya stenoptera*. Rooting is the main limiting factor in micropropagation of Persian walnut (*J. regia*) due to the poor rooting capacity of this species. Saadat and Hennerty (2001) conducted a number of experiments to identify suitable procedure for rooting of walnut microshoots. Rooting was done in two successive phases: (1) root induction and (2) root development. Root induction was done on DKW medium (1/4-strength micronutrients) containing four different concentrations of IBA or NAA (2, 3, 4 and 5mg/l) in darkness. After one week, for root development, shoots were transferred to a mixture of hormone-free medium and vermiculte (20ml of medium and 25 $M^3$ vermiculite). The maximum rooting (37.5 per cent) was obtained on medium containing 4.0mg IBA/l. In other experiment, the best results were noted by culturing the microshoots on DKW medium (1/2 strength micronutrients) containing 4.0mg IBA/l for 9 days in darkness and transferring to Jiffy-7 pots for root development without auxins. Using this procedure about 83 per cent of surviving shoots in Jiffy-7 pots rooted, which could be used for mass propagation of Persian walnut. On transferring the plantlets to the greenhouse their growth continued and more than 90 per cent of plantlets survived. Jiang *et al.* (2009) studied the *in vitro* culture and plant regeneration of *J. regia* with nodal segments of apical buds excised from the newly-grown shoots of the ten-year old of five elite materials plants at China. Results showed that the DKW supplemented with 6-BA at 1.0 mg/l +IBA at 0.01-0.10 mg/l and polyvinylpyrrolidone at 2 g/l was optimal for segments breaking; while the DKW with 6-BA at 1.0 mg/l +IBA at 0.01 mg/l + 40 g/l sucrose was found optimal for the tube shoot multiplication. The highest rooting frequency was 41.67 per cent by using two steps method. The plantlets with roots were transplanted into the substrate containing peat and sand (1:1), at 20-30$^0$C for 20 days, the survival rate was noted 95 per cent after 40 days of transplanting.

## Planting

To ensure long productive life of walnut trees, planning with regards to selection of site, orchard floor management, irrigation systems, choice of cultivars, planting system, tree density and mechanization should be done before planting because it is very difficult to rectify defects or make adjustments later on the establishment of trees (Wildman, 1985).

### Planting Plan

On flat land, square, quincunx and rectangular system and on sloping land, contour and terrace system of layout are done. In square System, the distance between plants and rows is the same while another tree is planted in the centre of four plants in the quincunx system. More distance is, however, kept between rows than plants in rectangular system. Square system is commonly adopted in walnut because it is easier to layout and convenient for cultural operations and tree thinning. In hill slopes, when the gradient of slope is more than 10 per cent but gradual on a hillock, planting is done in contour design. All the trees in a row are at same elevation but the distance between rows varies, being closer where the slope is steep and further otherwise. The best orientation for the orchard is by following the East-West lines, for better sunlight penetration. A suitable design for a walnut orchard, which includes the selection of the most suitable cultivars based on the agro-climatic situation, can contribute up to 70-80 per cent of the success of the orchard.

## Planting and Care of the Young Trees

If the soil is clayey or not very fertile, it is advisable to dig pits of one square metre dimension and then fill them with the mixture of rotten farmyard manure and soil in 1:2 proportion in the month of November. All the extraneous materials, such as, weeds, stones and plant remnants need to be removed. If there is a hard layer on any side of the pit, it should be broken. The plants should be procured quite in advance of planting from a reliable source and be healed immediately on arrival after laying them in a trench and covering the root portion with moist soil which should never be allowed to go dry in the process.

In late dormant or early spring season, a hole is made just enough to accommodate the entire root system of a plant at each planting site. A plant is then placed in a hole and the roots are spread around in all sides and covered with soil. The soil should be filled in the pit by firming with hands and feet followed immediately by light irrigation. The plant should be topped 40 to 50 cm above the graft union.

The most important operation is irrigation during the first year of the planting in the orchard. Frequent watering is particularly required in the summer season, but water logging is to be avoided in the rainy season. The trunks should be painted white to prevent sunburn. The plant basin should be kept clean by removing weeds as soon as they arise. In windy situations, staking of plants becomes necessary. The young trees need to be protected from the attack of pests and diseases.

## Planting Distance

A number of factors need to be assessed before deciding the tree density. First, a walnut tree takes about 15-20 years to occupy the entire space allotted to it in conventional planting system leaving sufficient space unutilized in early stages. Secondly, in the early years of the orchard, production is directly related to the number of trees planted, *i.e.*, more the tree number higher is the yield. Thirdly, a number of lateral bearing and precocious walnut cultivars are now available for closer planting. Fourthly, the processes of pruning and tree removal have made the high density planting possible. Finally, high density planting requires a higher level of orchard management, *i.e.*, irrigation, fertilization, pruning and control of pests and diseases.

General trend in walnut is towards planting more trees of lateral bearing cultivars per unit area which should be heavily pruned and gradually thinned on overcrowding. But the growth habit and vigour of the cultivars to be grown play the pivotal role in spacing. Sibbet and Osgood (1985) have illustrated this view-point by citing three examples. Chico is a lateral bearing and early fruiting cultivar but is a relatively small and compact tree at maturity. Hence, a spacing of

6.6 x 6.6 m should be kept initially and should be adjusted to 9.3 x 9.3 m finally by pruning and removal of temporary trees. On the contrary, initial spacing of 9.3 x 9.3m to final adjustment of 13.3 x 13.3 m is suggested for Vina, a highly fruitful lateral bearing but vigorous cultivar. Terminal bearing vigorous cultivars like Hartley are, however, to be given 15 m to 20 m spacing, interspace being utilized by Chico fillers in early stages.

Rimos *et al.* (1985) established hedgerows (3.6 x 7.3 m) of Chico cultivar to develop a high density system for walnuts. The trees were mechanically trained into fruiting walls. Hedgerow trees gave as much as double yield in early fruiting years than that of traditionally spaced and pruned trees. They were, however, in favour of further research to determine the best hedging procedures, closest spacing and suitable cultivars. In Iran, the commonly used tree spacing for cv. Chandler is 7m x 5m (285 trees/ha) and 7m x 3.5m (408 trees/ha) for cv. Howard. The objectives are to obtain the earliest possible yield and the mechanization of the maximum number of operations like pruning, harvesting, *etc.* In California, spacing of 5m-9m (within row) and 8m-12m (between rows) are common but growers in the last two decades have been planting in 'hedgerows' at 2.8m-3m (within rows) and 5.6m-6m (between rows). This allows the grower to get a return on investment sooner. Hedgerow orchards are mechanically hedged every 2-4 years.

Pollenizers should be placed in permanent position in order to avoid their elimination during temporary tree removal. Every eighth row of pollenizers be planted perpendicular to the usual direction of the wind.

## Training and Pruning

Pruning of young plants is termed as training which is carried out with two primary objectives, *i.e.*, developing a strong framework capable of supporting heavy crops and to encourage the shoot growth along with the early heavy fruiting. In order to achieve these objectives, it is imperative to know the bearing habit of the cultivars and their nature of growth. Walnuts bear fruits on 1-year-old wood which is produced either terminally or both terminally and laterally. Accordingly, the cultivars are of two types, *viz.*, terminal bearer and lateral bearer. Terminal bearing cultivars (Hartley and Franquette) are thinned but not headed back because vegetative growth remains proliferating in absence of early fruiting. Conversely, lateral bearing cultivars (Vina and Chico) are both thinned and headed back heavily in order to encourage shoot growth which get suppressed due to early fruit production.

Walnut trees are trained according to modified leader system (Serr, 1969). After the first year's growth, the plant is headed back above 2 m of its height. All lateral shoots on the leader are removed leaving one or two shoots at lower level on the trunk. In order to avoid narrow crotches, all necked (primary) buds 1.3 m above the ground should be removed forcing the secondary buds to grow (Freeman and Aldrich, 1985). Both types of cultivars are trained similarly up to two years but a variation of minimum pruning has been brought in modified central leader system for early bearing cultivars (Freeman and Aldrich, 1985). Standard training of young Engligh walnut (*J. regia*) in California consist of removing branches below 1.5-1.8m from the ground, heading the leader and desirable scaffold limbs and leaving some less desirable branches alone for early nut production (Olson *et al.*, 2001).

After selecting the main scaffold limb 1.6 m above the ground, the other primary scaffold limbs are chosen in all directions on the trunk. These limbs should have wider angles and not have arisen just opposite to each other. They should also be more than 30 cm apart vertically on the main trunk. All other vigorous branches should be removed whereas small branches are left undisturbed for fruiting. All the selected limbs should be headed back on early fruiting cultivars but not on late fruiting ones. In later years, secondary scaffold limbs on primary scaffold ones are also allowed to grow by removing the extra vigorous branches. All scaffold limbs are headed back in early fruiting cultivars every year.

Pruning of young bearing trees is again determined by the bearing habit. In the cultivars highly fruitful on lateral buds, the shoot growth is suppressed due to heavy crop load. Therefore, a large number of new shoots should be headed back on the periphery to reduce fruiting and increase vigorous shoot growth throughout the tree's periphery. In contrast, on the terminal bearing cultivars, pruning consists of heading back of selected framework branches and thinning out of competing limbs.

Annual or biennial pruning the young-developing orchard is expensive and time consuming. The use of hedger in place of hand pruning can get pruning done quickly and less expensively than annual hand pruning without sacrificing tree growth or production (Olson *et al.*, 2001).

Hedgerow planting of lateral-bearing cultivars is a method to obtain higher early yield and effectively use mechanized hedging machines for pruning (Ramos *et al.*, 1997). In walnut, hedgerow system, trees are planted with less space in the row than between rows, then trained and mechanically side-hedged on regular intervals to produce a continued wall of shoots and fruiting canopy (Greenfelder *et al.*, 1989). The wall, not the individual tree, becomes the management unit of the orchard. 'Chico', the cultivar originally used for hedgerows, was found productive and has proven to be well adapted to the system. However, nuts were small and kernels were hard to remove from the shell, making it a less desirable cultivar to use in hedgerows system.

Ramos *et al.* (2001) evaluated nine walnut cultivars under a hedgerow management at two locations in California central valley. The cv. Tulare was found well suited to hedgerow planting. The high yield of cv. Tulare was due to its extreme vigour and capacity of rapid development. The cv. Howard exhibited consistently higher yield on Paradox (Table 12.7). Hedging one side every other year has been the standard recommendation for hedgerows and was found effective on precocious laterial bearing cultivars such as Chico, Vina, Tulare and Howard.

## Pruning of Mature Trees

The primary objective of pruning mature walnuts is to obtain regular and high production of quality nuts which is often hampered by overcrowding of branches, insufficient production of renewal wood and inadequate availability of sunlight. Ramos *et al.* (1985) have narrated four reasons of pruning:

**Table 12.7: Influence of Tree Spacing on Yield and Vigor of a Haward Hedgerow Planted on a Silty Clay Loam Soil in Solano County, California***

| *Tree Spacing (m)* | *Density (trees/ha)* | *Rootstock* | *Yield (kg/ha)* | | | | | *Trunk Section ($cm^2$)* |
|---|---|---|---|---|---|---|---|---|
| | | | *1995* | *1996* | *1997* | *1998* | *1999* | *1999* |
| 7 x 3.5 | 407 | *J. hindsii* | 1870 | 4659 | 5739 | 2961 | 6747 | 219.3 |
| 7 x 3.5 | 407 | Paradox | 2265 | 4752 | 6017 | 3375 | 7113 | 190.8 |
| | | | | | | | | |
| 7 x 4.7 | 306 | *J. hindsii* | 1192 | 3312 | 4943 | 2541 | 6646 | 215.4 |
| 7 x 4.7 | 306 | Paradox | 1879 | 3608 | 5021 | 3048 | 6382 | 199.6 |
| Significant: at P 0.05 | | | | | | | | |
| Spacing | | | - | - | - | - | NS | NS |
| Rootstock | | | - | NS | NS | - | NS | NS |

* Ramos *et al.* (2001).

(i) To control tree size.

(ii) To maintain vigour and fruitfulness of spurs.

(iii) To develop renewal wood to replace older, less productive branches.

(iv) To remove weak, dead, or interfering branches and facilitate cultural operations.

Ryugo and Ramos (1979) reported the conversion of vegetative buds into reproductive ones within four weeks of formation in the cultivar Chico on the basis of a trial on heading back of second flush growth to basal buds done at different time of the year. Defoliation or defoliation with etiolation of seven terminal buds in July did not deter the initiation of female flowers for the next season. Thus fruit bud formation is not a problem in walnut.

Shading becomes a problem in mature trees because it reduces photosynthesis thereby weakens spur growth. Hence, selective thinning out of limbs in top and sides of the tree should be initiated well before overcrowding becomes serious. Since the leaves are food manufacturing organs being essential growth and fruiting, not more than 25 per cent of the branch structure be removed at any time (Stoke, 1969). On the basis of experimental evidence, Ramos *et al.* (1985) concluded that benefits of pruning are not immediate, but in time they can result in sustaining yield with increased nut quality and value.

The pruning operations are carried out in the dormant season but early spring is preferable (Maurer, 1976). Further delay causes excess bleeding. All the cut surfaces should be treated with a tree wound dressing.

The best method of pruning young (16 to 17-year-old) and middle aged (34 to 35-year-old) walnut tree is to cut back to 3-year-old shoots; this method also provided 5.5 per cent and 6.5 per cent, more buds for budding, respectively The best method for pruning mature trees (40-year-old) is to cut back 4 to 5 year-old shoots (Novak, 1974).

Foliar spray of dikegulac sodium at 500, 1000 or 2000 ppm in May, June and July, respectively on six cultivars, *viz.*, Franquette, Hartley, Payne, Ashley, Serr and Tehama, resulted in more compact trees and had no influence on nut quality or set (Martin *et al.*, 1980a). They also observed more growth with May and June treatment but not with July treatment (2000 ppm) which caused phytotoxicity.

## Tree Thinning

When closer planting is done, tree thinning is essential to provide adequate light penetration. Tree removal, however, must be undertaken well before the beginning of tree crowding. Two method, 'whiskbrooming' and 'one time' removal are employed. In whiskbrooming, the limbs of temporary trees removed in stages commencing from 10th year onwards, thus minimizing the production losses. But 'one time' tree removal is generally practised because it is more convenient and the production loss is compensated by wood sale. Ramos *et al.* (1985) suggested removal of alternate tree leaving edge trees and alternate rows in place, initially, which can be followed by the removal of additional trees, if required.

## Nutrition

Since walnut trees grow vigorously, they require heavy feeding for proper vegetative and reproductive growth. Leaf and soil analysis are used to determine nutritional demand. The most common nutrients applied are nitrogen, potassium and zinc.

Soil analysis is of limited utility in determining the nutritional requirement of walnuts because the plant deficiencies are not well correlated with soil analysis values (Proebsting and Davy, 1946). Moreover, walnut tree has a very extensive root system trapping nutrients from a large volume of the soil. Soil analysis is more useful in determining the toxicities of various elements, particularly of sodium, chlorine and boron. Information on soil pH is helpful to know about the availability situation of various nutrients. The effects of nutrients on growth and yield are given below.

## Role of Nutrients

### Nitrogen

Nitrogen fertilization is essential for successful walnut production (Drath, 1986). Its application is positively correlated with leaf colour and tree growth but inversely with the severity of disease. Source of N and time of application do not make much difference and the fertilizers can be applied as per convenience (Proebsting and Serr, 1966). Ammonium nitrogen should be applied in soil before the beginning of growth and be supplemented by foliar urea sprays (Proebsting, 1952).

N deficiencies are rarely found and are also difficult to diagnose. Initially, the symptoms are pale colour of foliage, small leaves, reduced shoot growth, low fruit set and yield, and formation of small and poorly filled nuts (Proebsting and Serr, 1966). Simorte *et al.* (2001) working with 3-year-old pot-grown trees of *J. major* MJ209 x *J. regia* and on 10-year-old trees of *J. nigra*, noted reduced leaf and root dry weight in N-deficiencnt plant. Severe deficiency correspond to leaf N content below 17g N/kg DM (mean dry matter), light deficiency between 17 and 24 maximum growth was obtained with 26g N/kg DM. An early application of 175-225 kg N/ha is sufficient to prevent the occurrence of its deficiency (Uriu, 1985).

### Phosphorus

Phosphorus is an energy building nutrient but walnuts do not respond normally to its application. Serr (1961a), however, reported marked improvement in the condition of young trees due to application of superphosphate in winter but the mature trees respond slowly. The improvement in young trees is observed in leaf condition and yielding capacity by the heavy early application of triple superphosphate. Weak growth, small leaves and sparse foliage yellowish rachis are the symptoms of P deficiency. The leaves become yellow and develop necrosis in irregular patches.

### Potassium

Proebsting and Serr (1966) reported the variation in leaf K content from 1.3 per cent to 2.4 per cent but without showing any deficiency symptoms.

K-deficient leaves first become pale and thereafter margins of leaflets fold upward and curl in, and the underside turns greyish (bronzing). In severe cases, leaf size reduces and marginal and tip scorching occurs. A single application of 3.5 tonne potassium sulphate per hectare corrects the deficiency (Proebsting and Serr, 1966).

### Zinc

Zinc is an improtant nutrient controlling enzymatic activities in plants. Deficiency symptoms are recognized early in the season. The bud opening is delayed and leaves are small and chlorotic. The terms 'little leaf' and 'rosette' and 'yellows' are used to describe Zn deficiency symptoms. Terminal die back of shoots is also observed in severe deficiency. Uriu (1985) recommends the foliar spray of zinc sulphate (36 per cent) at 0.5kg/400 l water or Zn EDTA at l kg/400 l water for the correction of zinc deficiency in walnuts. The first spray should be applied just after full bloom and second and third applications after 2-to 3-week interval depending on the severity of deficiency.

### Boron

Deformed leaves, short internodes, chlorosis and die back of twigs and branches are the various symptoms exhibited on the extent of boron deficiency. Soil application of borax (900 -2250 g) per tree every second or third year is the best treatment to prevent B deficiency (Proebsting and Serr, 1966).

Boron toxicity has been reported from many regions. Tip burn and marginal necrosis extending towards the midrib of leaflets appear in mid to late summer. Severely affected trees look as if scorched by fire. The problem can be alleviated by leaching out some B from the soil through heavy irrigation of good quality water.

### Manganese

Manganese plays an essential role in biochemical reactions. In early to mid-summer, interveinal chlorosis develops extending from midrib to margin and subsequently leaves fall early due to Mn deficiency which can be rectified by $MnSO_4$ spray on leaves (Braucher and Southwick, 1941).

### Iron

The entire leaf becomes yellow and in severe cases almost white due to iron chlorosis which begins early in the season, first appearing on terminal leaves. Veins remain green especially adjacent to mid-vein.

### Copper

Copper deficiency appears in the form of necrotic spots, yellowing and premature leaf fall. Small dark brown spots also appear on the shoots around the tip. On the basis of leaf analysis, Brown and Serr (1953) observed that the leaves containing 4 ppm or more Cu do not show deficiency symptoms. They further suggested spray of Bordeaux mixture (10-10-100) in late spring or early summer to cure copper deficiency.

## Magnesium

Magnesium takes part in a number of enzymatic reactions and is usually not available in the soils of low pH. In the situations of Mg deficiency, distinct chlorotic region appears at apical and margins of leaflets in basal leaves (Uriu, 1985). Rachis turned brownish, the symptoms are noticed when the shoots have made ample growth. Mg deficiency can be corrected by spraying magnesium sulphate.

**Table 12.8: Critical Nutrient Levels (Dry-weight basis) in July Walnut Leaf Samples***

| *Element* | *Concentration* |
|---|---|
| Nitrogen (N) | |
| Deficient below | 2.1 per cent |
| Adequate | 2.2 to 3.2 per cent |
| Phosphorus (P) | |
| Adequate over | 0.1 to 0.3 per cent |
| Potassium (K) | |
| Deficient below | 0.9 per cent |
| Adequate over | 1.2 per cent |
| Calcium (Ca) | |
| Adequate over | 1.0 per cent |
| Magnesium (Mg) | |
| Adequate over | 0.3 per cent |
| Sodium (Na) | |
| Excess over | 0.1 per cent |
| Chlorine (Cl) | |
| Excess over | 0.3 per cent |
| Boron (B) | |
| Deficient below | 20 ppm |
| Adequate | 36 to 200 ppm |
| Excess over | 300 ppm |
| Copper (Cu) | |
| Adequate over | 4 ppm |
| Manganese (Mn) | |
| Adequate over | 20 ppm |
| Zinc (Zn) | |
| Deficient below | 18 pm |

*Beutel *et al.* (1983).

## Effects of N, P and K

Abaev *et al.* (1976) obtained good growth and development of walnuts by supplying N, $P_2O_5$ and $K_2O$ at 180:180:60 kg/ha, respectively. But there appears a general consensus on the application of 60 kg/ha each of N, P and K for getting the highest yield and net profit (Kolov, 1977; Thagusev and Caroev, 1971). Kolov (1977) differed only for the dose of K and recommended 30 instead of 60 kg. Ponder and Jones (2001) reported increased nut production and diameter growth by fertilizing the tree with N, P and K. Treatments containing P with N and/or K were found most effective for increasing nut production, and treatments containing N were most effective for increasing diameter growth. Jones *et al.* (1993) reported that late summer fertilization increased nut production by 32 per cent over spring fertilized trees and by 34 per cent over unfertilized trees.

Pramanick *el al.* (2013) from Himachal Pradesh, India stated that a one-year-old seedling requires 500g urea and 750g each of superphosphate and muriate of potash. The fertilizer dose should be increased with the increase in the age of the tree. They also advised to apply full dose of superphosphate and muriate of potash and half dose of urea in October after harvesting the nuts and remaining half dose in the month of February every year. The other recommendations are application of 50kg well decomposed farmyard manure per tree and foliar application of micronutrients like zinc, copper, manganese, boron, iron and magnesium in the month of March.

## Tissue Analysis

Plant analysis or more specifically leaf analysis has been successfully used in diagnosing the nutritional status of the tree. Nutritional standards depicting deficiency, optimum and excess levels of various elements in the leaves have been more or less fixed which serve as reference for comparison with the nutrient position of a given orchard (Table 12.8). The correlation between leaf nutrient level and plant deficiency symptoms has been found excellent for Mg, Mn, P, B. Cl and Na, good for K and N. and fair for Cu, Zn. and Ca (Uriu, 1985).

Leaf sampling period and procedure are very important because considerable variation in leaf nutrient levels occurs due to the sampling season. Abaev and Khvorova (1983) suggested collection of leaves from the middle of the crown during active vegetative growth period in June for the most accurate diagnosis. The terminal leaflets of the compound leaf from spurs or middle of moderately growing shoots should be taken and only 20 to 30 leaflets are sufficient for a sample (Uriu, 1985). However, some authors suggest collecting leaf samples as late as early August (Phares and Finn, 1971). It was suggested to select one or two pair of leaflets from the mid-portion of mature leaves for the sample rather than from the apical or basal end only. The necrotic tissues are, however, analyzed separately to determine the toxicities of elements.

The normal range of N, P, and K has been reported by Serr (1961b) as 2.5-3.25 per cent, 0.12-0.3 per cent and 1.2-3.0 per cent, respectively. However, Abaev and Khvorova (1963) got the highest yield in young plants with 1.54-1.61 per cent N, 0.82-1.02 per cent P, and 2.25-2.56 per cent K. Apparently, the range of nutrient values should be decided locally on the basis of tree performance.

When abnormal symptoms appear on foliage or growth during growing period, these are indicative of some nutritional problems which help considerably in diagnosing the deficiency or excess of nutrient(s). A great deal of experience is, however, required to pinpoint the deficiency or excess toxicity symptoms because the low deficiency level may fail in clear expression of symptoms or the symptoms of more than one element may intermingle creating confusion. Therefore, the symptoms should be observed throughout the growing season and the deficiency or excess should finally be confirmed by leaf and soil analyses.

## Irrigation

Walnut trees are highly sensitive to environmental stresses. Change in water regime, either abundance or shortage, affect the physiological parameters of growth. From a visible morphological point of view, the two stresses seem to cause the same effects-wilting and epinastic leaves appear in trees subjected to both stresses (Belloni and Mapelli, 2001). Ramos *et al.* (1978) compared the unirrigated with irrigated walnut trees of the cultivar Serr, and found increased nut weight from 7.6 to 9.6 g and nut diameter from 40 to 42 mm due to irrigation. Kernel quality as indicated by skin colour and percentage shrivel was also much superior with watering and returns were 4.3 per cent higher. Withholding irrigation for one growing season significantly reduced trunk growth and kernel weight (Martin *et al.*, 1980b). They further obtained heavier kernel weight on the resumption of irrigation in the following year. A water stress in the orchard during the beginning of fruit growing period will result in the production of a large percentage of small nuts. The nut size does not increase by mid-summer or late irrigation after the hardening of nutshell (Aldrich *et al.*, 1976).

Irrigation studies have revealed that walnut roots absorb moisture from the sandy loam soil up to a depth of three metres and 80 per cent of it comes from upper two metres of soil but in heavier soils, the water absorption to a greater proportion is further restricted to the uppermost one metre. Watering is very essential for the establishment of grafts and young plants. The water requirement, however, decreases with the development of roots in adult trees. Shoujia *et al.* (2011) studied the temporal and spatial variations in the water status of walnut trees and the soil in which they were growing were traced by analyzing the differences in hydrogen isotopes during spring and summer in a 7-year-old walnut stand. They observed that the roots were mainly distributed in the upper soil (0-30cm depth) with around 60 per cent of the total root mass in upper soil layer and 40 per cent in deep soil layers (30-80 cm depth). The upper soil layers contributed 68 per cent of the total tree water requirement in the wet season, but only 47 per cent in the dry season. In the wet season, total roots, living roots and new roots were all significantly more abundant than in the dry season. The amount and frequency of irrigation depend on the age of the tree, type of soil, rainfall and its frequency, mulching and soil management system.

The common irrigation systems are flood, furrow, sprinkler and drip irrigation, each having its advantages and disadvantages and none is the best (Prichard, 1985). The choice of an irrigation system is decided on the basis of many factors, such as, water supply, ground level, technological advance and, above all, cost benefit ratio.

Irrigation is generally applied on the soil moisture examination by experience or by using instruments, *viz.*, tensiometer, the gypsum electrical-resistance block or neutron probe. (Schulbach, 1985). The quality of irrigation water should also be assessed because excess salts of boron, chloride, carbonates and bicarbonates make the water unfit for walnut irrigation (Ramos *et al.*, 1984). Regulated deficit irrigation (RDI) has been studied and found economic to many crops. However, experiments with cvs. Chico (Goldhamer *et al.*, 1988) and Chandler (Buchner *et al.*, 2008) showed water stress affected yield and nut quality.

## Weed Control

The herbicides, simazine and diuron, have been successfully used in weed control of young and established walnut groves for four months (Charlot, 1986). Kiryakov and Lukova (1974) found very good weed control in nurseries (second year) by using simazine at 3 kg/ha, which reduced the number of manual weeding from 6 to 3 and was found highly economical. Simazine and diuron at 2-4 lb/acre effectively controlled annual weeds throughout the season. Dichlobenil was found safe for trees on heavier soils but less so on lighter soils, particularly under sprinkler and flood irrigation (Lange *et al.*, 1969). In a 3-year herbicide trial in a young black walnut plantation the best result was obtained with simazine (6 lb/acre) plus pre-planting treatment with paraquat (0.5 pt/acre). This combination showed excellent weed control without causing visible injury to the trees (Roth, 1971).

## Flowering

Walnut seedlings take about eight to ten years to come into bearing, but grafted plants begin fruiting early in about four or five years. Walnut is a monoecious plant, *i.e.*, staminate and pistillate flowers are borne separately on the same tree. Staminate flower are small and densely grouped in catkins, 10-15cm long, borne laterally on 1-year-old wood. Individual flowers lack petal and are characterized by numerous stemens. Each stamen terminates in a pollen bearing anther. Each catkin can release two to four million pollen grains (Impiumi and Ramina, 1967, Westwood, 1993). Pistillate flowers appear in pairs in the form of nutlets at the tip of current season's growth in spikes of typically two to three flowers. However, in some cultivars, female flowers are also produced on the tips of lateral shoots (McGranahan and Leslic, 2009). Fertilization usually occurs five to seven days after pollination. The porogamy and chalazogamy are alternate pathways in walnut flowers (Luzo and Polito, 1991). Though the fruit bud formation of a pistillate flower is initiated during summer, yet its final development occurs shortly before anthesis in the following spring. Peak flowering period of walnut under Shimla, India conditions is mid-April which is prolonged till the first week of May in late blooming cultivars.

Lin *et al.* (1977) studied the initiation of pistillate flowers of 7 walnut cultivars in USA. Flower formation was found during the first 2 weeks of June for cvs. Chico, Vina, Serr, Tehama and Pedro and mid-July for Hartley and Franquette. Initiation of floral parts also occurred at different times, depending on the cultivars. Pistil development began in February for Chico and in April for Franquette. Fertilization occurred throughout April for most of the cultivars but in the middle of May for Franquette. In Bulgaria, the female and male flowers were found to appear at a mean daily temperature of 14.5$^0$C in cvs. Proslavski, Lyubimets and Sheinovo and at 10$^0$C in Dryanovski, Plovdiski and Perushtenski. Mass flowering in all these cultivars was found at a mean daily temperature range of 9$^0$ to 20$^0$C. The phase which lasted for the formation of floral organs until the end of flowering occurred at mean daily temperatures in the range 15.5$^0$ to 19$^0$C and effective temperature sums of 1700$^0$ to 1800$^0$ (Draganov and Nedev, 1973). Draganov and Nedev (1977) further reported that both male and female flowers appeared at a minimum of 50 per cent RH and 4-5 mbar vapour pressure deficit (VPD). Flowering was optimal at 60-70 per cent RH and 6-8 mbar, whereas 75 per cent RH and 10 mbar were detrimental for flowering.

Sladky (1974) stated that the differentiation of staminate catkins in walnut bud primordia occurred in the presence of gibberellin-like substances which controlled the sex differentiation of flowers and stimulated their production. The subsequent development of vegetative buds with leaf primordia was characterized by a reduction in stimulatory substances and by the appearance of inhibitors. The differentiation of pistillate flowers in the terminal buds during the end of growth period was associated with the presence of auxins; these, together with an increased level of inhibitors, regulated the number of pistillate flowers. Using synthetic growth regulators (IAA and IBA) it was possible to change the bud ratio in twigs and regulate the flower numbers in staminate and pistillate inflorescences.

## Pollination

Walnuts are self-fruitful and cross-compatible and in majority of the cultivars, pollen dehiscence and stigma maturity synchronize, hence pollination is not a problem. But the phenomenon of dichogamy occurs in walnut, *i.e.*, in many cultivars pollen shedding takes place before the stigma becomes receptive (protandrous), and in some cultivars stigma maturity precedes the pollen dehiscence (protogyny). Furhermore, the situation is aggravated by short periods of both pollen viability and stigma receptivity (Polito, 1985). Most of the *J. regia* cultivars are protrandrous, only a few cultivars such as Chico, Amigo *etc.* are protogynous. However, high incidence of protogyny is detected in *J. nigra*. Pistillate flower abscission (PFA) is considered as a major non-pathogenic problem of walnut production, reducing final yield. The cvs. Serr, Chandler, Howard and Vina are prone to PFA and the levels of PFA were recorded higher than 90 per cent in Sacramento valley of California (Cartlin *et al.*, 1987). Excessive pollen load has been implicated in cv. Serr in PFA, the loss of female flowers early in the season before fruit drop due to lack of pollination. PFA can be reduced and yield improved in 'Serr' orchards by reducing pollen load. This can be accomplished by pollenizer removal or catkin removal at the beginning of pollen by mechanical shaking. In years of significant bloom overlap between staminate and pistuillate flowers, PFA can be further reduced and yield improved by removing catkins (Krueger, 2000). Rovira *et al.* (2001) studied PFA in 19 cultivars and seletions of *J. regia*. They observed highest average value of PFA (66.1-77.7 per cent) in Spanish selections (MBT-40, MBT-122) and lowest (7.1 per cent) in French cultivars (Franquette, Lara, Mayett, Marbot and Parisienne).Krueger (2000) also opined that previously recommendation of about 10 per cent of the trees in a commercial planting be a cultivar with a pollen shed period overlapping pistillate flower receptivity of the main cultivars appears to be excessive. Two to three per cent is probably adequate to limit losses due to lack of pollination without resulting in excessive PFA. Factors to consider when determining the number of pollinator to plant include: cultivar susceptibility to PFA, walnut pollen load in the area and local pollination and fruit set experience. Walnut flowers are anemophilous representing all the characteristics of wind pollination, *i.e.*, non-sticky, light pollen is produced in abundance, and stigma is protruded, feathery and remains receptive for few days. Effective pollination, however, can be assured by planting the early and late cultivars whose pollen shedding overlaps each other.

In Bulgaria, the cvs. Sheinovo, Dryanovski, Proslavski, Plovdivski, Silistrenski, Slivenski and Zhinovski pollinated each other well and showed a comparatively satisfactory retention of newly set fruits (Nedev, 1975). Generally, cross-pollination with a mixture of pollen from 2-3 cultivars increased fruit set (Tsurkan and Pyntya, 1979).

The crop failures in walnut usually occurs due to adverse weather conditions, such as, snowfall, frost, hail or rain during the flowering period in spring. In case fruit set is not good in spite of good plant health and weather, the inadequate pollination may be a cause of low fruiting. Artificial pollination in walnut is not only easy but is desirable to improve the yield. The catkins can be collected and stored for two or three weeks in a desiccators at 0$^0$C or in a refrigerator provided the humidity is low. This pollen can be used for pollinating the same cultivar or other cultivars at the same location or at a distant location. When stigmas begin to be receptive, about ten catkins, placing each in a small cloth open-mesh bag, are thrown on the top of the trees and stirred two or three times in a day with a pole. A few branches can also be top-worked in the orchard with the pollinating cultivar to improve pollination.

Walnut pollen showed the highest germination capacity when it was first shed. Pollen stored under simulated natural conditions lost its viability in 2-3 days after collection. In some cases freshly collected pollen lost its viability 1-3 hours after exposure to sunlight at 25-30$^0$C. The pollens usually germinate on the stigma within 1-3 days after pollination (Rud and Zhadan, 1972). Hao *et al.* (2011) stated that pollen can be stored for 96 hours under natural conditions; however the viability decreased to 33 per cent after 4 hours. The pollen viability was recorded maximum when walnut anthers shedding, reaching 95 per cent.

## Fruit Growth and Development

The fruit growth begins and continues with the stimulation of pollination and fertilization. The non-pollinated pistillate flowers make some initial growth but ultimately drop down after about three weeks in absence of fertilization.

Nedev and Baev (1973) studied the fruit growth pattern of 5 walnut cultivars in Bulgaria. They found that the mesocarp started to form on 22-25 May, the testa on 3-8 June, the endocarp began to develop on 20-26 June and harden on 5-10 August, and the nuts matured on 15-25 September. The nuts grew most rapidly in length, width and weight from the end of May to the end of June. Labavitch and Polito (1985) have described the development of walnut fruits in about twenty weeks (Table 12.9). In early cultivars and at higher prevailing temperatures, the fruit development does not take a period of twenty weeks to reach harvest stage.

Two rapid periods of walnut fruit growth are interspersed by a period of relatively slow growth. During the first growth period, the fruit gains size and weight due to the formation of new cells and tissues. The fruit attains form and becomes richer in chemical constituents during the second growth period. Predominant sugars present in the developing walnut fruit are glucose and sucrose and fructose in small quantity. Walnut fatty acids are mostly unsaturated which makes them prone to rancidity, Ascorbic, malic and phosphoric acids and various amino acids are other minor soluble constituents of walnut kernel. Akca (2000) studied the fruit growth and development of four Turkish walnut cultivars (Sebin, Bilecik, 32B-18 and 170B-16). He reported that the increase in the kernel weight was slower at the beginning of growth period, and it was rapid between June and August. Rapid nut growth (diameter and length) was observed from May to July. The oil content of the kennel increased rapidly between June and September. Pinney *et al.* (1998) reported that walnuts develop in two distinct stages. First, nuts increase rapidly in both size and weight after pollination. Then, at 8-10 weeks after pollination, the shell hardens and kernel growth becomes the primary process. Oil accumulate as the kernel matures. Maximum total nut weight is attained at approximately 18 weeks after pollination.

**Table 12.9: Approximate Timing of Important Stages in Walnut Fruit Development***

| *Weeks after Bloom* | *Stage of Development* |
|---|---|
| 1 | Fertilization of egg cell |
| 2 | Embryo two to eight cells. Copious endosperm present |
| 3 | Embryo eight to thirty two cells |
| 5 | Globular embryo of up to several hundred cells |
| 7 | Cotyledons begin to grow. Tip hardening begins in shell |
| 8 | Total growth starts to slow, relative to the rapid, early stage |
| 9 | Kernel begins rapid growth as cotyledons start to fill locule |
| 10 | Final nut size attained |
| 12-15 | Kernel growth (weight) resumes at a rapid rate |
| 15 | Maximum total weight |
| 18 | Approximate date of harvest |
| 19-20 | |

* Labavitch and Polito (1985).

YongTao *et al.* (2012) studied the dynamics of fat, protein and sugar metabolism during fruit development of Chinese cv. Xiangling. They divided the development process of fruit in four stages; (a) slow growth [(within 30 days after florescence, (DAF)], (b) fast growth (30 -60 DAF); (c) fat accumulation (60-100 DAF) and (d) fruit maturity (100-140 DAF). Fat content in fruit increased continuously and the maximum increment occurred within 60-90 DAF. It amounted to 50.11 per cent at 100 DAF and maximum (63.13 per cent) accumulation was recorded at 140 DAF. Sucrose and glucose content showed similar trend. They were both low at earlier stage, then gradually increased and reached the peak at 90 DAF, followed by a gradual decrease. Acid invertase (AI, EC 3.2.1.26), neutral invertase (NI, EC 3.2.1.26), sucrose phosphate synthase (SPS, EC 2.4.1.14) and sucrose synthase (SS, EC 2.4.1.13) activity was higher during the early stage and then declined until the minimum point at 80 DAF. There after these enzyme activities turned to ascend the crest value at 90-100 DAF, thereafter kept decreasing until the late maturity stage. Ethylene (ethephon) application hastens fruit abscission but its endogenous production at fruit maturity and drop is still to be ascertained (Labavitbh and Polito, 1985).

### Maturity

For production of quality walnuts, harvesting at proper maturity and subsequent handling of the product are of prime importance. Any delay in picking after maturity of kernels, deteriorates the quality and increases the incidence of mould and pests. Two maturity indices are commonly used for determining the harvesting date of the walnut. A nut is tightly covered with a thick hull which splits on maturity causing the fall of nut on the ground. When about eighty per cent of the hulls have cracked exposing the nuts, it is the time for walnut picking. But the synchrony between hull and kernel maturity is not necessary and kernel often matures earlier than hull splitting. Proper maturity is, therefore, assessed by observing the packing tissue between and around the kernel halves which turns brown on maturity. Depending on the climate, the harvesting is usually done after about two weeks of browning. The kernel becomes darker if the harvesting is further delayed deteriorating the quality.

The use of ethephon (2-chloroethylphosphonic acid) has greatly facilitated the walnut picking (Martin, 197I). The spray of 200-500 ppm ethephon on kernel maturity advances harvesting by about ten days, makes harvest of entire crop possible at a time and promotes hull splitting (Sibbet *et al.*, 1978; Woodroof, 1979). Ethephon treated kernels were of better quality than the control (Martin *et al.*, 1977). The preparation of orchard floor is important for facilitating the harvest. The ground should be levelled and cleared of all external materials, such as, dried leaves, twigs, pebbles, *etc.*

## Harvesting

### Manual Harvesting

Walnut is mostly harvested in the months of September and October, earlier at lower altitudes and later at higher altitudes. Some nuts fall on the ground after splitting of the hull in a natural way while others are forced to drop down by beating with long poles. The total harvest is done in two or three times at few days interval. The fallen nuts are collected in huge heaps in the orchard and covered with leaves and grass for hull fermentation taking about 9-15 days' period. Dehulling, washing and cleaning are carried out manually causing staining of hands, clothes and nuts. Sometimes in order to improve the appearance of nuts, these are bleached with either alkali or acid solution. After cleaning, the nuts are dried under sun for many days by spreading them on a canvass sheet or hard floor. Occasionally the nut drying is not complete due to lesser number of sunny hours and lack of warmth in autumn season. The dried nuts are graded according to the cultivar, size and colour. Ungraded product fetches lower price in the market.

## Machine Harvesting

All operations such as harvesting, dehulling, washing, cleaning, drying, bleaching, grading and packing can also be performed by machines which are desirable for standardizing the product for export. Ethephon is useful for both mechanical harvesting and dehulling. Mechanical harvesting is common in California. The harvesting operation carried out in three steps: (i) shaking of tree to remove nuts, (ii) sweeping of nuts on the ground into windrows, and (iii) picking up of nuts with a harvester. Using modern equipment, one operation can shake three to four trees per minute. The sweeper and harvester can be operated several kilometres per hour.

Processing of nuts involves removing the green hull (hulling) by feeding the nuts through a drum fitted with rotating plates or blades which scape of the hull. Water is sprayed over the nuts. Damaged nuts are manually separated by an air separator. The cleaned nuts are passed through a drying bins using heated air < 43$^0$C and at an airflow of 0.85 $m^3$ $min^{-1}$ $m^3$ to reduce the moisture content of the nut to around 8 per cent. In Iran, a prototype machine was develop to crack walnut. The cracker, which consist of a hopper fitted with a flow rate control device, a cracking unit, a sorter and power system, operates on the principle of attrition using crushing force from a cylinder and helix. The percentage of whole kernels produced was 66.6 per cent and the capacity of the machine was estimated to be about 25.2 kg/hr. (Ghafari *et al.*, 2011). However, all the nuts are not ripen at the same time which makes difficult in mechanical harvesting. Spraying of ethephon at 1000ppm with a turbo pulveriser caused most of the nuts ready to harvest which increased the efficiency of harvester (Yalcin *et al.*, 2014). The hulls are removed by motor driven abrasion hullers; the bleaching is done by dipping the dried nuts in 26 per cent to 28 per cent calcium hypochlorite followed by a thorough rinsing in 1 per cent sulfurous acid. For redrying, warm air at a temperature range of 32-38$^0$C is circulated around the individual nut in a drier till the moisture content is 8 per cent. If the temperature in the drier goes more than 43$^0$C, the kernels will develop rancidity becoming unfit for consumption. After sorting out the small, misshapen, dark coloured, deeply furrowed and off types, the nuts are size-graded. As there is an increasing demand for the kernels, these can be graded for colour and quality with an 'electronic eye' before packing in wooden boxes.

## Yield

Because of differences in tree age, size, soil, fertilization and irrigation practices, a great variation is found in the yield of walnut. A large 40-year-old tree can produce as much as 175 kg nuts while an average yield is only around 40 kg (Metzer, 1977). Full commercial bearing commences after 18-20 years in seedlings and after 8-10 years in grafted trees. The nut yield of about two and half tonnes per hectare is obtained from 8-to 9-year-old trees in the USA under quadruple planting system (Serr, 1969).

# Drying and Storage

## Drying

Delay in drying causes rapid loss in nut quality and makes it susceptible to mould. The nut quality deteriorates even within 9 hours of harvest if walnut remains on orchard floor at a temperature of 32.2$^0$C or more. Walnuts are dried to remove excess moisture from the shell and kernel. Drying stabilizes the product weight, prevents deterioration in kernel quality, makes bleaching efficient and prolongs storage life (Olson and Coates, 1985). The nuts are dried up to 8 per cent moisture level. In the past, sun drying was in practice which was time consuming, labour intensive and often resulted in poor nut quality. Three types of drier, *viz.*, stationery bin, forklift pallet bin and trailer, are now popular for nut drying which is achieved within 24 to 48 hours (Thompson and Rumsey, 1985). Drying air temperature should not exceed 43.3$^0$C because kernels become rancid and inedible at higher temperature.

## Storage

Walnuts are ordinarily stored in gunny bags in small ventilated rooms. Since the nuts are disposed of in the ensuing winter season, the problem of their shelf life is seldom felt. The shelf life of nut meat is examined on the basis of fatty acids, colour and taste. The quality of nut meat, however, deteriorates due to darkening and rancidity which are affected by air, moisture, heat and light. With an antioxidant, kernels at 2.8-4 per cent moisture can be kept for 18 months at 4.4$^0$C with about 3.1 per cent humidity (Woodroof, 1979).

The effect of storage temperature, oxygen permeability of packaging material and variety (Yalova-1 and Yalova-3) on oxidative stability of vacuum-packaged walnut kernels were studied by Bakkalbas *et al.* (2012) at Turkey. They concluded that vacuum-packed walnut kernels in polyamide/polyethylene film pouches [oxygen permeability: 63.4 ± 0.4 ml/m$^2$/24 h (23$^0$C)] with 90 micro m total thickness and storing at 20$^0$C is sufficient to protect the kernel against oxidation for 12 months. Christopoulos and Tsantili (2011) studied the effect cultivars (Chandler, Hartley and Ioli) and storage conditions *e.g.* time (upto 12 months), temperature (1$^0$ and 20$^0$C) and oxygen availability on total antioxidants and colour of kernels during storage. They observed loss of antioxidants with the advance of storage time. After 12 months, maximum loss of total phenolics (TP), FRAP and DPPH were observed in all cultivars stored at 20$^0$C under air. Low temperature (1$^0$C) and packaging in polyethylene-polyamide pouches under $N_2$ or $CO_2$ prevented additively both antioxidant losses and browning of kernel. The decreases in antioxidant of stored walnut kernels are responsible not only for the nutrient loss, but also for the quality deterioration in relation to consumer visual perception.

YanPing *et al.* (2011) reported that treating the fresh walnut (cv. Liohe 4) with $^{60}$CO-gama-ray at 0.5 kGy successfully stored the nuts (in polyethylene bags at 0±1°C) for 90 days. Treatment slowed down the lipid peroxidation process and increased sensory quality.

## Pests

Pest and disease management is a problem in walnut because of the difficulty faced in spraying giant trees, Planting of grafted trees, using dwarfing rootstock, training and pruning and choice of lateral bearing cultivars altogether help in reducing the size of walnut trees considerably making them amenable to spraying and harvesting. In controlling pests and diseases, the sanitary measures can play an effective role but are often overlooked. Sanitary measures involve the planting of healthy trees, precautions against the transfer of diseases and insects through tools used in pruning and cultivation, removal of diseased branches and leaves by pruning, and destroying these by burning (Johnson. 1969). Before acquiring, the plants should be inspected in the nursery for the infection of fungus and incidence of pests. As many as thirty pests have been reported by Butani (1979) in walnut.

### Codling Moth (*Cydia pomonella*)

This insect is a serious pest of apple which also attacks early blooming walnuts but it is confined (Pawar, 1984). Three generations of moth damage the fruits. First generation larvae enter the growing fruit through the blossom end and the damaged immature fruits drop down from the tree becoming breeding sites for naval orangeworm. Second and third generation larvae enter the nut through the side of the husk. The damaged nuts are unmarketable on harvest. Synthetic pheromone traps can be installed in the month of May. Profenophos or Trizophos (2ml/l) should be sprayed to control the pest. Bifenthrin, Cyfluthrin and Spinetoram are also effective.

### Naval Orange Worm (*Amyelois transitella*)

In contrast to the presence of only one larva of codling moth inside an infected nut, several larvae are found in case of naval orange worm infestation. First and second generation larvae attack damaged and mummified nuts during spring and summer but the third generation larvae cause serious damage to nuts on splitting of husks. The control measures include removal of mummy nuts, control of codling moth and walnut blight, and rapid harvest of the nuts after husk split (Van Steenwyk and Barnett, 1985). Two wasps that parasitize navel orangeworm, *Copidosoma (=Pentalitomastix) plethorica* and *Goniozus legneri* showing some larval mortality. It is also suggested to use ethephon to advance husk splitting and harvesting the nuts as early as possible, if worm populations are high.

### Walnut Husk Fly (*Rhagoletis cingulata, Rhagoletis completa* Cresson)

The walnut husk fly causes damage to only mid and late season cultivars feeding on the fleshy tissues of the husk. The nutshell gets stained rendering the nut in shell unfit for sale. The stains cannot be removed through normal bleaching. The infested nuts do not split making harvest a problem. The continuance of fruits on the tree causes the build up and serious attack of naval orange worm subsequently. An early infestation of husk fly damages the kernels which shrivel and turn brown. Late infestations do little damage to the kernels but may stain the shells and make hull removal difficult. The fly can be controlled by timely spray of insecticides based upon catches on green sticky trap spheres. Insecticides mentioned for control of controlling of codling moth are effective to control walnut husk fly. Acetamiprid, Befenthrin, Spinetoram and Phosment are effective to control walnut husk fly.

## Walnut Weevil (*Alcidodes porrectirostris*)

The weevil has been recorded in all the walnut growing regions. Female weevils lay eggs on the fruits which on hatching bore deeper and feed on kernels causing premature fruit drop. After pupation, adult weevils emerge from fruits and feed on petioles and tender shoots. The grub is more destructive because it kills new shoots up to the old wood. Adults are 9 to 12 mm long, black to brown in colour, whereas grubs are 12 to 16 mm long and creamy white in colour (Butani, 1979). The pest can be controlled by spraying carbaryl (0.2 per cent) twice at an interval of 7 days commencing from the time of first attack.

## Indian Gypsy Moth (*Lymantria obfuscata*)

The moth passes through only one generation in a year and overwinters in egg stage. On hatching in the month of March, the caterpillars feed on buds and leaves. Due to nocturnal habit, it hides during day under rocks, stones or clods of soil and attacks the foliage after sunset often resulting in complete defoliation. Masoodi *et al.* (1987) tried a number of control measures and the pest population was most effectively reduced by applying endosulfon (0.05 per cent) and fenvalarate (0.03 per cent).

## Walnut Aphid (*Chromaphis juglandicola*)

A large number of aphids can be seen on the underside of leaves and tender shoots where they suck the sap. The aphids are active from April to September but population is particularly high in late rainy season, when on severe infestation, leaves fall, kernels shrivel and sooty mould develops on honeydew. Aphid has four developmental stages, *viz.*, egg, nymph, wingless adult females, winged adult females and males. The overwintering occurs in egg stage. The granules of disulfotan (5 per cent) should be mixed thoroughly in the soil around the tree trunk to check aphid attack. Spraying of any suitable insecticide such as phosphomidon (0.03 per cent) or dimethoate (0.03 per cent) also controls the pest. Imidacloprid, acetamiprid, and phosmet are also effective to control walnut aphid. The aphid can also be checked biologically by parasitic wasp, *Trioxys pallidus* (Van Steenwyk and Barnett 1985).

## Long Horned Beetle (*Batocera horsfieldi*)

The bark boring beetle is an important pest in India. The grub eats the inner side of the bark making zigzag tunnels and later reaches to the centre of heartwood (Butani, 1979). Infested trees do not die but becomes unfit for timber purpose. The bark is also eaten by adult but it is not of much consequence. The grubs and adults can be hand picked and mechanically destroyed by treating the holes with carbon bisulphide or chloroform-creosote.

### Leopard Moth (*Zeuzera pyrina*)

The young caterpillars feed on inner bark, bore into shoots and branches and pupate inside them. The affected leaves and shoots wilt and die adversely affecting the growth of the tree. The moth completes a single life cycle in about two years. Fully grown caterpillars are 5-7 cm in length and pinkish red in colour whereas moths are white with beautifully designed wings. The pest should be controlled by inserting cotton-wool soaked in carbon bisulphide, para dichlorobenzene or ethyl acetate in the hole followed by sealing with mud (Butani, 1979).

### Storage Pests

Srinath and Gill (1976) found infection of pests in 55 per cent samples of in-shelled walnuts collected from markets. *Tribolium casteneum* and *Oryzaephilus surinamensis* were the commonest pests which could be controlled by fumigation.

## Diseases

### Walnut Blight (*Xanthomonas arboricola* pv. *jugalandis*)

Walnut blight is found throughout the walnut growing regions of the world. In presence of free water, it enters the tree through stomata, wounds, abrasions and insect punctures. The bacteria attack all tissues of leaves, catkins and nuts. Brown and black spots of few millimetre diameters with yellowish green perimeter develop on the leaves. Individual florets die in the distorted infected catkins. Infection of nuts can occur right from early spring to summer. Early infection causes nut drop while late infection results in nut shrivelling and discolouring. Since water is essential for the infection, rainfall, fog, dew and sprinkler irrigation aggravate the blight problem. All the walnut cultivars are susceptible, but early blooming cultivars suffer more damage than late bloomers.

Bacteria overwinter in dormant buds making the control difficult. It is, however, checked by timely sprays of copper containing chemicals and Bordeaux mixture. The first spray is applied on the opening of first female flower which should be followed by subsequent sprays at 7- to 14- day interval according to the weather conditions. The most effective rate of metallic copper is 4.5 kg/ha (Teviotdale *et al.*, 1985).

### Anthracnose (*Gnomonia leptostyla*)

The disease has been reported from Kashmir valley and other walnut cultivating regions in India. It attacks leaves, young shoots, and fruits. Initially small, circular light brown spots develop into large blotches involving a major part of lamina (Pathak, 1980). The fallen leaves should be collected and burnt to prevent the disease spread. The first spray of Zineb or Captan (0.01 per cent) is necessary on leafing to control the fungus which is required to be followed by other two or three sprays at an interval of a fortnight.

### Downy Leafspot (*Microstroma juglandis*)

The white mould appear in geometric patches on the ventral side of leaves which may coalesce turning into large irregular shapes. It is a serious disease in

Kashmir which can be controlled by spraying twice systematic fungicide Bavistin 50WP (0.03 per cent) in June at an interval of 15 days (Puttoo, 1976).

## Walnut Canker or Die Back (*Cytospora juglandis*)

The canker has been recorded in Jammu and Kashmir and Uttar Pradesh hills of India (Bose and Sindhan, 1916; Puttoo and Choudhary, 1984). The swollen or sunken areas of various sizes rupture the bark of branches and trunk. To control this disease, a paste preparation of 50 per cent copper oxychloride (WP) is applied on the diseased portion after scrapping the affected tissues. A foliar spray of copper oxychloride (0.2 per cent) is also effective.

## Deep Bark Canker (*Brennaria* (=*Erwinia*) *rubrifaciens*)

Large longitudinal cracks in the bark of trunk and limbs are the external symptoms of the canker. A mixture of bacterial cells and plant sap oozes from these cracks. This bacterial exudates becomes brown to dark brown on drying giving a bleeding look. The bacteria infect walnut throughout the year except winter. The cv. Hartley is susceptible to canker. The chemical control measure have not proved effective in controlling the problem making the use of preventive measures necessary. Since the canker is associated with the reduced vigour, the maintenance of trees in good condition through cultural practices is essential. Proper water management should be done with the help of tensiometers, (Teviotdale *et al.*, 1985). The injuries from mechanical shakers during harvest need to be avoided.

## *Phytophthora* Root and Crown Rot

A number of *Phytophthora* spp. may cause crown or root rot or both. With the common symptoms of infection, such as, small chlorotic leaves, premature defoliation, decreased yield, die back of terminal shoots and subsequent collapse and death of the tree. The symptoms appear gradually in a number of years depending on *Phytophthora* spp., temperature, soil moisture and relative resistance of rootstock (Mircetich *et al.*, 1985). In the beginning, the crown rot infection occurs at several points which on expansion girdles the tree causing its death. Root rot results in the infection of small and large roots initially which later destroy the entire root system. Even after fumigation, the eradication of *Phytophthora* from infested soil is not possible. Consequently, preventive measures such as better drainage, avoidance of contamination during cultural operations, keeping of graft union sufficiently high while planting and use of Paradox rootstock (more resistant) should be taken to control the rot.

## *Armillaria* Root and Crown Rot

Armillaria root disease affects orchards in all Persian walnut (*J. regia*) growing regions of California. The causal agent is *Armillaria mellea*, which attacks walnut and other woody perennial horticultural crops (*e.g.* Stonefruits) (Baumgartner *et al.*, 2011). The pathogen kills and then decay woody roots, which in turn reduces crop yield and growth, eventually kills infested trees. The primary inoculums consists of mycelium within dead, woody roots left in the soil after infected trees are cleaned.

Viable mycelium has been recovered from 10-year-old or older residual roots (Baumgartner and Rizzo, 2002). Attempts to eradicate inoculums by soil fumigation indicate that penetration by methyl bromide or carbon disulfide is efficacious within the top 0.3 to 0.6 m of soil, but inoculums buried at greater depth is likely to persist (Munnecke *et al.*, 1981). The persistant inoculums and lack of effective preventive or curative methods limit yield throughout the life of an infected walnut orchard (Baumgartner *et al.*, 2013).

### Viral Diseases

All the Persian walnuts on California Black and Paradox rootstocks are susceptible to blackline virus (cherry leafroll virus) which is a widely spread serious disease. Poor terminal growth, die back, small yellow drooping leaves, profuse suckering and small holes and cracks in the bark at graft union are its well-known symptoms. When a small patch of bark is removed at the union, a narrow strip of darkened cambium and phloem tissue at the juncture of rootstock and scion confirms its presence. Blackline virus systematically infects Persian walnut but not California Black and Paradox rootstocks. When the virus moves from the Persian walnut scion to *J. hindsii* or Paradox rootstock, hypersensetive action occurs forming a blackline of the infected cells at the graft union. The virus also dies at the union and stocks remain virus-free (Mirectich *et al.*,1985). They also proved that the pollen was implicated in the natural spread of blackline. It can also be transmitted by seed and scionwood. The following measures are suggested to control this malady:

(i) *J. regia* rootstock should be employed for Persian walnut;

(ii) Only virus-free scionwood should be used for grafting and budding;

(iii) The pollen should also be virus-free;

(iv) *J. hindsii* and Paradox rootstocks can be regrafted with virus-free scion after heading back below the graft union; and

(v) The blackline virus should be detected by using I-ELISA (indirect enzyme-linked immunosorbent assay).

## Breeding and Improvement

Walnut is a heterozygous plant which has been raised from seed in the past resulting into a great amount of genetic diversity. Man has always been utilizing this diversity by selecting the seeds of superior trees for further multiplication and thus has brought enormous improvement in this fruit over a long period. Subsequently, superior seedlings were propagated by grafting and many of them got established as cultivars. This selection process is continuing even now as surveys of thousands of seedling trees are being carried out in various parts of the world (MacDaniels, 1961; Millikan and Cochran, 1966; Valev, 1969).

In its native region, from southestern Europe to the Himalayas, the selection criteria were based on nut yield and kernel quality. The nuts having large size, thin shell, light colour, good flavour were preferred to small size, hard shell, dark colour and poor flavoured nuts. But in other European countries and California, beside the nut traits, the adaptability to local climatic conditions received equal emphasis in selection. In France, late leafing genotypes were identified and multiplied vegetatively and one of such selection is the cultivar 'Franquette'.

Walnut breeding programme at the University of California was initiated in 1948. The programme successfully released 13 cultivars by 1978 of which Vina, Serr, Howard, Chandler and Tulare are popular cultivars in many walnut growing countries. Subsequently, cultivars Robert Livermore (red-skinned), Sexton, Gillet, Forde (all three are lateral bearing) and Ivanhoe (early harvesting), and Solano (big nut size) were developed.

In Iran, the walnut breeding programme started in 1980. Cultivars Jamal (Z63) and Damavand (Z30) were released in 2010 as the first Iranian cultivars (Hassani *et al.*, 2012). Walnut selection breeding in Turkey was initiated in 1971 that released in selection of cultivars Yalova 1, Yalova 3, Yalova 4, Sebin, Bilecik, Kaplan 86, Gultekin 1, Sen 1, Kaman 1, Sutyemez 1, Maras 18 and Oguzlar 77. Sebin and Bilecik cultivars occupy 40 per cent of walnut orchards in Turkey (Akca, 2012).

In India, the Central Institute of Temperate Horticulture (CITH), Srinagar, Jammu and Kashmir released 10 superior selections named after the Institute as CITH 1-10. Some other selections: Govinda, Kotkhai Selection, KW012, Pratap and Unique were made from Himachal Pradesh and Chakrata No. 14 from Uttarakhand. Sharma *et al.* (2014) identified clone GLO109 which has high export quality and recommended for cultivation in different walnut growing areas of Jammu and Kashmir. It produces average nut weight of 20.10g, size 45.45mm x 42.07mm, kernel recovery of 61.40 per cent, oil content of 68.42 per cent and protein content of 15.66 per cent. The clone is also highly resistant to anthracnose disease.

## Objectives for Breeding of Cultivars

### Nut Yield

The yield of nuts is governed by bearing habit, bearing surface, fruit set and nut size. Two types of bearing habits have been observed in walnut cultivars. Firstly, the cultivars like Franquette produce pistillate flowers only on shoots of terminal and subterminal buds which usually do not have more than two flowers. The laterals have negligible pistillate flowers meaning thereby that even a cent per cent set would not give a heavy yield (Forde, I975). Secondly, in 'Payne' type cultivars 80-90 per cent of lateral buds produce pistillate flowers. Further these lateral bearing cultivars bear two to three flowers on most shoots and come to commercial production much earlier than the cultivars fruitful only on terminal or subterminals. Lateral bearing is, however, less important in case of mature trees as they bear on spurs rather than laterally on one-year-old shoots. (Ramos *et al.*, 1984).

Nut size is usually positively related with the kernel yield provided filling is good. It is an important in-shell sales consideration but the increasing trend of sale of shelled produce has put more emphasis on kernel weight which should be about 10g per nut. Consequently, the yield can also be recorded as kilogram of sound kernel per hectare rather than nut yield.

### Nut Quality

Nut quality depends on the colour, shell hardness, smoothness, kernel percentage, kernel quality, shell seal and adherence of meat with shell. The colour (appearance) of shell is still an important character for internal market but not for export where shelled walnut is being sold on large scale. Walnut shell varies in thickness considerably from thick and hard to paper-thin. Thin-shelled walnuts are easily broken and usually have higher kernel percentage varying from cultivar to cultivar and achieving 60 per cent kernel crackout should be an objective of breeding. The kernel percentage is affected by nut size, kernel plumpness, shell fill and shell thickness. Sometimes well-filled shells are too tight to extract the kernel. Hence some space between kernel and shell is required.

Kernel quality has become the most important factor in marketing. Light coloured, sound kernels get traditionally a premium price in the market. The exact kernel colour is a varietal character but the darkening is enhanced by heat damage to nuts directly exposed to sun. Both environmental and genetical factors affect the production of quality kernels.

### Tree Characteristics

The weather prevailing at the time of flowering is a decisive factor for a good or poor crop. A high yielding cultivar will leaf out later in the spring thus escaping from frost injury. Since walnut is a monoecious plant, it is a desirable breeding objective to get a cultivar with a better overlap of pistillate and staminate blooms. As regards the habit and vigour of the tree, Forde (1975) stated that an ideal tree will grow vigorously to moderate size and then slow down and produce nuts in abundance.

## Disease and Pest Resistance

Disease and pest resistant traits in fruit trees are essential to realise the full yield potential augmenting the profit of the grower. Resistance in walnut will, therefore, be desirable against anthracnose, Indian Gypsy moth and blister mite. In the process of incorporating the pest and disease resistant genes, the commercial characteristics should not be overlooked because, howsoever, a resistant cultivar is evolved, it would not be accepted by the growers unless it gives good yield of high quality fruits. PI 159568, an introduction from Afghanistan had never shown symptoms of bacterial blight (*Xanthomonas juglandis*) in California but could not be grown for commercial cultivation due to its shy bearing habit (Forde, 1975).

Rapid genetic improvement in walnut is possible by selecting the parents in hybridization having the traits highly correlated with yield, such as, fruitfulness of laterals, nut size and kernel weight (Forde, 1975). Heritability of various traits was studied by Hansche *et al.* (1972) and they found that leafing date, receptive date of pistillate flowers, first shedding date of pollen and harvest date were highly heritable characters. An easy crossability and variability found among *Juglans* species provide a good scope for walnut breeding (Jaynes, 1969). Hybridization among the various species of the genus and with *Carya* species is being tried to improve the shell thickness, nut size and hardiness in cultivars (Smith, 1976). The interspecific hybrids, however, have been mostly utilized as rootstocks.

## Objectives for Breeding of Rootstocks

A good rootstock should be easily propagable, compatible with most commercial cultivars, capable to produce high yield of good quality fruits, resistant to pest and diseases and adaptable to widely prevalent ecological situations. The need of a drought resistant and dwarfing rootstock for walnut is genuinely felt which might be fulfilled with the use of other *Juglans* species or the interspecific hybrids as rootstocks. The major achievement of rootstock breeding is the development of Paradox (*J. hindsii* x *J. regia*) rootstock from California. However, the offspring from different Paradox mother trees showed variations. Superior clones: VX211 (nematode tolerance) and Vlach (vigorous) of Paradox were selected from California, showing some resistance to *Phytophthora*. Despite the popularity and advantage of Paradox seedling rootstock, the rootstock is susceptible to *Agrobacterium tumefaciens* (crown gall), root-lesion and root-knot nematodes, *Armillaria mellea* (Armillaria root disease) and *Phytophthora* and crown rot. Kluepfel *et al.* (2012) while evaluating > 550 accessions of black walnut, Asian butternut, Persian walnut and wingnut (*Pterocarya*) at National Clonal Germplasm Repository (NCGR) in Davis, CA, USA, observed resistance to *A. tumefaciens* and *Phytophthora* spp. among the *J. hindsii*, *J. major*, *J. microcarpa* and *Pterocarya* accessions and root-lesion nematode resistance in *J. eathayensis*.

## Marker Assisted Breeding

Restriction fragment length polymorphisms (RFLPs) used successfully to investigate the genetic diversity of Persian walnut (Fjellstorm *et al.*, 1994). RFLPs were also used to initiate mapping of walnut genome (Fjellstorm *et al.*, 1994). Randomly amplified polymorphic DNA (RAPD) loci were used to the genetic map (Woeste *et al.*, 1996). LingNa *et al.* (2014) provides a useful protocol for the characterization, identification, and authentication of walnut cultivars. They concluded that SSR markers are useful in cultivar identification and intellectual property protections. Working with 35 walnut cultivars of China, they reported that just four loci (WJR031, WJR069, WJR265 and WJR281) successfully separates 35 cultivars studied, regardless of nut phenotype characters. Dogan *et al.* (2014) studied the genetic relationship of 59 walnut genotypes and cultivars of Turkey by three types of molecular markers. Among the molecular assays tested, SSR provided a very high polymorphism rate of 99.1 per cent while ISSR (inter simple sequence repeat) and RAPD results showed similar to each other but lower polymorphism (71.1 per cent and 69.1 per cent respectively). The dendrogram constructed divided the genotypes into three main groups: Group A included mainly genotypes originated from the Kahramanmaras and Yalova provinces; while Group B contained genotypes originating from the Kreshir, Tokat and Corum provinces of Turkey. Group C contained only cultivars originating from USA and France.

## Genetic Engineering

Gene transfer technology in walnut was first reported in 1988 (McGranaham *et al.*, 1988) and the field trials of mature transgenic trees have been reported (Leslte *et al.*, 2001, Tang *et al.*, 2000, 2001).

The protocol for gene transfer is based on inserting genes of interest into proliferating somatic embrio culture via *Agrobacterium tumefaciens*. Significant progress on the somatic embryogenesis and transformation in walnut has been made in the past 25 years. Somatic embryogenesis has been achieved in five species of *Juglans* (HaoRu, 2000). Cotyledons and zygotic embryos from immature fruits have been routinely used as explants for inducing somatic embryogenosis. Repetitive somatic embryogenesis has been achieved used for the clonal production of *in vitro* plants and for genetic transformation and regeneration of transgenic plants. Using *Agrobacterium* -mediated transformation of somatic embryos, genes of interest have been enigineered into the walnut, and the transgenic plants have been successfully expressed in the transformed genes. Histochemical assessments indicated that more gus-positive somatic embryos, particularly fully gus-positive embryos, regenerated from timentin (500-1000ml/l-containing medium. Transformed embryos have been grown and converted into plants and gus-activity was observed in whole plants (Tang *et al.*, 2000). Polymyerase chain reaction (PCR) assays confirmed the presence of the neomycin phosphotransferase II (nptII) gene in all gus-positive embryos tested (Tang *et al.*, 2011).

Persian walnut somatic embryos were transformed by *Agrobacterium tumefaciens* carrying a gene for synthetic green fluorescent protein (GFP) s65T in order to assess its impact on embryo viability and selection efficiency by Escobar *et al.* (2000). They observed several embryos showing stable fluorescent sectors, allowing efficient and rapid visual selection of primary transgenic embryos. Thirteen stable transgenic embryos and 44 whole plants were regenerated from 2 chimaeric embryos 40 days after transformation. They concluded that the use of GFP as a selective marker can significantly reduce labour, cost and time in walnut somatic embryogenesis-based transformation.

Wu *et al.* (2012) constructed two bacterial artificial chromosome (BAC) libraries, containing a total of 129,024 clones, from *in vitro* grown shoots of 'Chandler' using the HindIII and MboI cloning sites. A total of 48,218 high quality BAC end sequences (BESs) were generated, with an accumulated sequence length of 31.2Mb, representing approximately 5.18 per cent of the walnut genome. Analysis of repeat DNA content in BEEs) revealed that approximately 15.42 per cent of the genome consists of known repetitive DNA, while walnut-unique repetitive DNA identified in this study constitutes of 13.5 per cent of the genome. Among the walnut-unique repetitive DNA, Julia SINE and JrTRIM elements represent the first identified walnut short-interspersed element (SINE) and terminal-repeat retrotransposon in miniature (TRIM) element, respectively; both types of elements are abundant in the genome. They also observed that the density of SSR in the walnut genome analyzed was also slightly higher than that in poplar and papaya. Sequence analysis of BESs indicated that approximately 11.5 per cent of the walnut genome represents a coding sequence.

## REFERENCES

Abaev, M.N. and Khvorova, N.E. (1983). Khimiya V Sel'skom Khozyalstve, 4:31-33.

Abaev, M.N., Lukin, N.F. and Khvorova, N.E. (1976). Agrokhimiya, 12: 113-16.

Achim, G. and Botu, I. (2001). Acta Hort., 544: 503-09.

Ahmed, N., Singh, S.R., Mir, J.I. and Mir, Abid (2013). Proc. Nat. Seminar on Prospects of Walnut in Himalayan States-Impact on Environment and Sustainable Livelihood, 24-25, October, 2013, Uttaraknand, India, pp. 46-52.

Akca, Y. (2000). Turkish J. Agric. Forestry, 24: 349-54.

Akca, Y. (2012). Walnut Growing (In Turkish) 4th Ed. Anit Matbaa. Ankara. p. 328.

Aldrich, T.M. *et al.* (1976). Diamond Walnut News, 58: 33-34

Anonymous (1959). The Wealth of India. Vol. V. CSIR. New Delhi. pp.298-305.

Aslamarz, A.A., Vahdati, K., Hassani, D., Rahemi, M., Mohammadi, N. and Leslie, C. (2011). European J. Hort. Sci., 76(3): 84-90.

Avanzato, D. and Tamponi, G. (1988). Acta Hort., 227: 79-83.

Bakkalbas, E.,Ylmaz, O.M., Javidipour, I. and Artik, N. (2012). LWT-Fd. Sci. Tech., 46: 203-209.

Baojun, Z., GuangPing, L., Yang, S., yongHong, G., ShiHai, W. and DeFu, Y. (2009). China Fruits, 6: 6-8.

BaoJuo, L., GuoHui, Q., SuPing, G. and LiYing, C. (2008). Acta Hort. Sinica, 35: 1088.

Baumgartner, K. and Rizzo, D.M. (2002). Amer. J. Enol. Vitic., 53: 197-203.

Baumgartner, K., Coetzee, M.P.A. and Hoffmeister, D. (2011). Plant Pathol., 12: 515-34.

Baumgartner, K., Fujiyoshi, P., Browne, G.T., Leslie, C. and Kluepfel, D.A. (2013). HortSci., 48: 68-72.

Beede, R.H. (1985). In: Walnut Orchard Management. (ed. D.E. Ramos). Div. Agric. Nat. Res., Univ. of Calif., Publ., 21410: 2-8.

Bein, Ji, (1980). Fruit Var. J., 34:98-99.

Belloni, V. and Mapelli, S. (2001). Acta Hort., 544: 327-33.

Berg, L.S. (1937) Natural regions of the USSR. Translated by O.A. Titelbaum, 1950. NY, MacMillan.

Bose, S.K. and Sindhan, G.S. (1976). AICRP Fruit Res. Workshop. Tech. Doc., 9: 491-99.

Braucher, O.L. and Southwick, R.W. (1941). Proc. Amer. Soc. Hort. Sci., 39: 133-36.

Brooks, R.M. and Olmo, H.P. (1972). Register of New Fruit and Nut Varieties. Univ. of Calif. Press., Barkley and Los Angeles. USA, pp. 613-39.

Browicz, K. (1976). Flora Iranica 121. Akademische Druckund Verlagsanstalt, Graz, Austria.

Brown, J.C. and Serr, E.F. (1953). In: Fruit Nutrition. (ed. N.F, Childers). Rutgers State University. New Brunswick. N. J., p. 268,

Browne, L.T., Brown, L.C. and Ramos, D.E. (1977). Calif. Agric., 3l: 14-15.

Buchner, R.P., Fulton, A.E., Gilles, C.K., Lampinen, B.D., Shackel, K.A., Metcalf, S.G., Little, C.C., Pichard, T.L. and Schwankl, L.J. (2008). Acta Hort., 792: 141-46.

Bugarcic, V., Ogosanovic. D., Korac. M. and Mitrovic, M. (1985). Jugoslovensko Vocarstvo. 19: 281-88.

Butani, D.K. (1979). Insect and Fruits. International Book Distributors. Raipur Road. Dehradun. pp. 325-30.

Casini, E. and Salvadori, S. (1975). Osservazioni e ricerche sull'impiego di alcuni regolatori di crescita nella germinazione der Semi di piante da frutto. La Nuova A.O.P.l. pp.22.

Catlin, P.B. and Schreader, W.R. (1985). In: Walnut Orchard Management (ed. D.E. Ramos). Div. Agric. Nat. Res. Univ., California, pp. 75-81.

Catlin, P.B., Martin, G.C. and Olsson, E.A. (1977). J. Amer. Soc. Hort. Sci., 102: l0l-104.

Catlin, P.B., Ramos, D.E., Sibbett, G.S., Olson, W.H. and Olsson, E.A. (1987). HortSci., 22: 201-205.

Chandel, J.S. (2013). Proc. Nat. Seminar on Prospects of Walnut in Himalayan States-Impact on Environment and Sustainable Livelihood, 24-25, October, 2013, Uttaraknand, India, pp. 16-18.

Charlot, G. (1986). Infor. Centre Technique Interprofessional des Fruits et Legumes, France, No. 86010, p. 50.

Charlot, G. (1987). Infor. Centre Technique Interprolessional des Fruits et Legumes. France. 3-5: 23-24.

Charrier, G., Bonhomme, M., Lacointe, A. and Ameglio, T. (2011). Int. J. Biometrology, 55: 763-74.

Chauhan, J.S. and Sharma, S.D. (1979). Indian J. Agric. Sci., 49:420-22.

Chauhan, J.S. and Sharma, S.D. (1980). Fruit Sci. Rep., l7: 45-48.

Chebotar, E.L. (1974). Sadovodstvo, Vinogradarstvo i Vinodelie Moldavii, 10:32- 34.

Chernilevskii, K.V. (1977). Biol. Ekol. i. Fiziol. Kul'turn. i. Les. Rastenii Kishinev. Moldavian SSR. pp. 58-53.

Christopoulos, M.V. and Tsantili, E. (2011). Scientia Hort., 131: 49-57.

Clark, R.C. and Nursten, H.E. (1976). Jr. Sci. Fd. Agric., 27: 902-908,

Dar, M.A. (l987). Indian Hort., 32: 10-12.

DeCandole, A. (1982). Origin of Cultivated Plants. NY. Noble Printers.

Dehghan, B., Vahdati, K., Rezaee, R. and Hassani, D. (2009). Jr. App. Hort., 11:56-58.

Delbos, R., Kerlan, C., Dunez, J., Lansac, M., Dosba, F. and Germain, E. (1982). Acta Hort., 123: 31-34.

Dogan, Y., Kafkas, S., Sutyemez, M., Akca, Y. and Turemis, N. (2014). Scientia Hort., 168: 81-87.

Draganov, D. and Nedev, N. (1973). Nauchni Trudove, Vissh Selskostopanski Institut Vasil Kolarov, 22: 51-58.

Draganov, D. and Nedev, N. (1977). Gradinarska i Lozarska Nauka, 14:13-18.

Drath, H. (1986). Dtsche Baumsch, 20: 188-90.

Driver, J.A. and Kuniyuki, A.H. (1984). HortSci., 19: 507-09.

Escobar, M.A., Park, Jactn, Polito, V.S., Leslic, C.A., Uratsu, S.L., McGranahan, G.H. and Dabdekar, A.M. (2000). Ann. Bot., 85: 831-35.

Fjellstorm, R.G., Parfitt, D.E. and McGranahan, G.H. (1994). Jr. Amer. Soc. Hort. Sci., 119: 833-39.

Forde, Harold I. (1975). In: Advances in Fruit Breeding, (eds, J. Janick, and J.N. Moore), Purdue Univ. Press, Indiana, Lafayette, pp. 439-55.

Freeman, M.W. and Aldrich, T.M. (1985). In: Walnut Orchard Management, (ed. D.E. Ramos), Div. Agri. Nat. Res., Univ. Calif., Publ., 21410: 57-61.

Gandev, S.L. (2017). Fruits, 72(1): 31-35.

Gangwar, R.P. and Kumar, H. (1975). Haryana J. Hort. Sci., 4: 141-44.

Gaur, N.V.S. (1980). Indian J. Hort., 37: 16-19.

Gauthier, M.M. and Jacobs, D.F. (2011). Ann. Forest Sci., 68: 1277-90.

Germain, E., Jalinat, J. and Marchou, M. (1975). In: Le Noyer (eds. F. Bergougnoux and P. Grospierre) Brochure INVUFLEC. Paris, pp. 53.72.

Ghafari, A., Chegini, G.R., Khazaei, J. and Vahdati, K. (2011). Int. J. Nuts and Related Sci., 2: 11-16.

Glonti, I.Sh., Tutberidze, B.D. and Kalanderishviti, T.L. (1987). Subtropicheskie Kultury, 2: 102-10.

Goldhamer, D.A., Beede, R., Sibbett, S., Dejong, T.M., Ramos, D., Phene, R.C. and Doyle, J. (1988). Walnut Res. Rept., Walnut Marketing Board, pp. 42-52.

Grimo, E. (1979). In: Nut Tree Culture in North America. (ed. R.A. Jaynes), North Nut Growers Assoc. Hamden, CT., pp. 74-83.

Guanglan, Pu, QianWen, X. and LanYing, Z. (2011). Acta Hort. Sinica, 38: 2025-26.

GuoLiang, W., He, L., YanGang, X., Yong, W., PengFei, Z. and XiangYu, S. (2009). Acta Hort. Sinica, 36: 1549-50.

Hamed, M. and Fathi, M.A. (1999). Egyptian J. Agric. Res., 77: 1265-74.

Hansche, P.E., Beres. V. and Nauk, Lozer (1972). J. Amer. Soc. Hort. Sci., 97:279-85.

Hao, L., Ying, G., NingGuang, D., HuZhi, X. and Dong, P. (2011). Jr. Beijing Foresty Univ., 33: 119-23.

HaoRu, Tang, YongQing, Wang and ZhenhLong, Ren (2000). Scientia Silvae Sinicae, 36(3): 102-110.

Hartmann, H.T. and Kester, D.E. (1959). Plant Propagation-Principles and Practices. Prentice-Hall, Inc. Englewood Cliffs, NJ, pp. 493-95.

Hassani, D., Atefi, J., Dastjerdi, R., Keshavarzi, M., Mozaffari, M.R., Soleimani, A., Rahmanian, A.R., Nematzadeh, F. and Malmir, A. (2012). Seed and Plant Improvement J., pp. 525-28 and 529-31.

Heile-Sudholt, C., Huettman, C.A., Preece, J.E., Sambeek, J.W. van and Gaffney, C.R. (1986). Pl. Cell Tissue and Org. Cult., 6: 189-97.

Hendricks, L.C., Griff, R.H. and Ramos, D.E. (1977). Walnut Production in California. Univ. Calif. Div. Agri. Sci. Leaflet 1984.

Hendricks, L.C., McGranahan, G.H., Ramos, D.E., Iwakiri, B. and Forde, H.I. (1985). Rasteniev' dni Nauki, 23: 91-96.

Impiumi, G. and Ramina, A. (1967). Rivista deli'Ortifrutticoltura Italiana, pp. 538-43.

Iotsova-Baurenska, N. (1986). Rasteniev' dni Nauki, 23: 91-96

Jaynes, R.A. (1969). In: Handbook of North American Nut Trees, (ed. R.A. Jaynes), The W.F. Humphery, Press. Inc. Geneva, New York, pp. 376-99.

Jing, F., QingZhong, L., JunLin, Z., Ling, Q., Xin, C. and Jin, W. (2009). Acta Hort. Sinica, 36: 867-72.

Johnson, W.T. (1969.) In: Hanbook of North American Nut Trees. (ed. R.A. Jaynes), The W.F. Humphery Press Inc. Geneva New York, pp. 67-156.

Jones, J.E., Haines, J., Garrett, H.E. and Lowenstein, E.F. (1993). In. Proc. Third North American Agro-forestry Conf. pp. 39-42.

Kawecki, Z. and Kaworski, J. (1975). Fruit Sci. Rep., 2: 17-23.

Kiryakov. D. and Lukova. V. (1974). Rastitelna Zashchita, 22: 6-7.

Kluefel, D.A., Aradhya, M.K., Browne, G.T., McKenry, M.V., Leslic, C.A., McClean, A.E., Moerofelder, J., Velasco, D. and Baumgartner, K. (2012). Acta Hort., 948: 105-11.

Kolov, O.V. (1977). Lensoe Khozyaistro, 1:61-62.

Krueger, W.H. (2000). HortTech., 10: 127-30.

Kuniyuki, A.H. and Forde, H.I. (1985). In: Walnut Orchard Management, (ed. D.E. Ramos), Div. Agri, Nat. Res., Univ. of Calif., Publ. 21410, pp. 38-46.

Kutina, J. and Others (1991). Pomologicaly atlas. Zeemedelske naklandatelstvi Brazda, Praha.

Labavitch, J.M. and Polito, V.S. (1985). In: Walnut Orchard Management. (ed. D.E. Ramos), Div. Agri. Nat. Res., Univ. of Calif., Publ. 21410, pp. 90-94.

Lagerstedt, H.B. (1978). 69th Ann. Rept. North Nut Growers Assoc., pp. 118-21.

Lal, H., Nautiyal, M.C. and Sharma, R.M. (1984). Progressive Hort., 16: 6-8.

Lange, A.H., Elmore, C.L. and Fischer, B.B. (1969). Abst. Meeting of the Weed Sci. Society, America, No. 54.

Leslie, C.A., McGranhan, G.H., Dandekar, A.M., Uratsu, S.L., Vail, P.V. and Tebbets, J.S. (2001). Acta Hort., 544: 195-99.

Lin, J., Shabany, B. and Ramos, D. (1977). J, Amer. Soc. Hort. Sci., 102: 702-05.

LingNa, C., QingGuo, Ma, Yongkun, C., BaoQing, W. and Dong, P. (2014). Scientia Hort., 168: 240-48.

Luzo, J.G. and Polito, V.S. (1988). Bot. Goz., 149: 30-36.

MacDaniels, L.H. (1961). 52nd Ann. Rept. Nut Growers Assoc., pp.26-31.

Manning, W.E. (1978). Ann. Missouri Bot. Garden, 65: 1058-87.

Martin, G.C. (1971). Jr. Amer. Soc. Hort, Sci., 96: 434-36.

Martin, G.C., Irui, K. and Nishijima, C. (1980a). HortSci., 15: 157-58.

Martin, G.C., LaVine, P., Sibbett, G.S. and Nishijima, C. (1980b). Calif. Agric., 34:16-17.

Martin, G.C., Sibbett, G.S. and Drapper, T.M. (1977). Jr. Amer. Soc. Hort. Sci., 102: 805-08.

Martinez, M.L., Labuckas, D.O., Lamarque, A.L. and Maestri, D.M. (2010). Jr. Sci. Fd. Agric., 90: 1959-67.

Masoodi, A.M., Koul, V.K., Bhagat, K.C. and Singh, T. (1987). Indian J. Pl. Protection, 15: 15-17.

Matheron, M.E. and Mircetich, S.M. (1985). Pl. Dis. Reptr., 69: 1039-40.

Maurer, K.J. (1976). Baumschulpraxis, 6: 110-12.

McCranahan, G.H. and Catlin, P.B. (1987). In: Rootstocks for Fruit Crops, (eds. R.C. Rom and R.F. Carlson), John Wiley and Sons. Inc. New York, pp. 411-50.

McGranahan, C.H. Driver, J.A. and Tuleche, W. (1987). In: Cell and Tissue Culture in Forestry, (eds. J.M. Bonga and D.J. Durzan) Martinus Nijhoff Publishers, Boston, pp. 261-71.

McGranahan, G. and Lesile, C. (2009). In: Breeding Plantation Tree Crop: Temperate Species (eds. S.M. Jain and P.M. Pryadarshan), Springer Science, New York., pp. 249-73.

McGranahan, G.H., Leslie, C.A., Uratsu, S.L., Martin, L.A. and Dandekar, A.M. (1988). Bio. Tech., 6: 800-804.

Megranaham, G.H. and Leslie, C.A. (2006). Acta Hort., 705: 117-22.

Meizhi, Z., ShaoTang, G., YuePing, Xu and ChaoBing, Liu (1999). Acta Hort. Sinica, 26: 412.

Metzer, R. (1977). Obstbau, 2: 359-61.

Millikan, D.E. and Cochran, L.H. (1966). 57th Ann. Rept. Nut Growers Assoc., pp.87-82.

Mircetich, S.M.J., Rowhani, A. and Ramos, D.E. (1985). In: Walnut Orchard Management, (ed. D.E. Ramos). Div. Agri. Nat. Res., Univ. of Calif., Publ., 21410:142-52.

Mol'chenko, L.L., Shevchuk, N.S., Yatsyshin, V,M. and Krushel'nitskii, A.F. (1977). Lesnoe Khozyarstvo, 1: 62.

Moller, W.J. and Ramos, D.E. (1972). Diamond Walnut News, 54: 28-29.

Monk, R.W. and Wiebe, H.H. (1961). Plant Physiol., 36: 478-82.

Munnecke, D.E., Kolbezen, M.J., Wilbur, W.D. and Ohr, H.D. (1981). Plant Dis., 65: 384-89.

Muradoglu, F., Oguz, H.,Yeldiz, K. and Yilmaz, H.(2010) Afcrican J. Agric. Sci., 5: 2379–85.

Nedev, N. (1975). Ovoshcharstvo. 54: 25-29.

Nedev, N. and Baev, kh. (1973) Grandinarska i Lozarska Nauka, l0: 29-36.

Nedev, N.V. (1973). Grandinarska i Lozarska Nauka, 10: 7-8.

Nekresova, V.L. (1936). Flora of the USSR. Inst. Of Academy of Sci. USSR.

Novak, Yu. V. (1974). Lesovodstvo i Agrolesomelioratsiya Resp. Mezhved. Temat. Naud. Sbornik.

Olson, W.H. and Coates, W.W. (1985). In: Walnut Orchard Management, (ed. D.E. Ramos), Div. Agri. Nat. Res., Univ. of Calif., Publ., 21410:172-74.

Olson, W.H., Ramos, D.E., Micke, W.C., Yeager, J. and Shawareb, N. (2001). Acta Hort., 544: 437-42.

O'Rourke, F.L.S. (1969). In: Handbook of North American Nut Trees, (ed. R.A. Jaynes), The W.F. Humphery Press Inc. Geneva, New York, pp. 232-39.

Ozkan, Y., Edizer, Y. and Akca, Y. (2001). Acta Hort., 544: 521-25.

Ozol, A.M. (1949.) Dokl. Akad. Nauka. SSR, 66: 725-28.

Pandey, D. and Sinha, M.M (1984). Progressive Hort., 16: 183-87.

Park, K.S. (1984). J. Korean Forest Soc., 66: 95-108.

Pathak, R.K. and Srivastava, R.P. (1975). Progressive Hort., 7: 19-24.

Pathak, V.N. (1980). Diseases of Fruit Crops. Oxford and IBH Publishing Co. New Delhi, pp. 267.

Pawar, A.D. (1984). Pl. Pro. Bull, India, 36: 117-20.

Phares, R.E. and Finn. R.F. (1971). Ann. Rept. North Nut Grow.Assoc., 10: 1513-21.

Pinney, K., Labavich, J. and Polito, V.A. (1998). In: Waalnut Production Manual. Univ. Calf. Pub. No. 3373. pp. 139-43.

Polansky, M. and Murphy, E.W. (1966). J. Amer. Diet. Assoc, 48: 109-11.

Polito, V.S. (1985). In: Walnut Orchard Management, (ed. D.E. Ramos), Div. Agri. Nat. Res., Univ. of Calif., Publ., 21410: 81-86.

Ponder, F. Jr. and Jones, J.E. (2001). Jr. Pl. Nutri., 24: 661-70.

Pramanick, K.K., Kishore, D.K., Sharma, Y.P. and Watpade, S. (2013). Proc. Nat Sem. On Prospects of Walnut in Himalayan States-Impact on Environment and Sustainable Livelihood 24-25, October, 2013, Uttarakhand, India, pp. 37.45.

Prichard, T.L. (1985). In: Walnut Orchard Management, (ed. D.E. Ramos), Div. Agri. Nat. Res., Univ. Calif., Publ., 21410: 35-38.

Probesting, E.L. (1943). Proc. Amer. Soc. Hort. Sci., 43:1-4.

Proebsting, E.L. (1952). In: Fruit Nutrition. (ed. N.F. Childers), Rutgers State Univ. New Brunswick, N.J., p. 262

Proebsting, E.L. and Davey, S. (1946). In: Fruit Nutrition, (ed. N.F. Childers), Rutgers State Univ. New Brunswick, N.J., p. 262.

Proebsting, E.L. and Serr, E.F. (1966). In: Fruit Nutrition, (ed. N.F. Childers), Rutgers State Univ., New Brunswick. N J., pp 262-67.

Puri, G.S., Meher Hormji, V.M., Gupta, R.K. and Puri, S. (1983). Forest Ecology. Vol. 1. Oxford and IBH Publishing, New Delhi.

Puttoo, B.L. (1976). Pest, l0:21-22.

Puttoo, B.L. and Choudhary, K.C.B. (1984). Pest, 18: 38-39.

Qureshi, A.S. and Dalal, M.A. (1985). Progressive Hort., l7: 191-205.

Ram, R.D. and Rathore, D.S. (1987). J. Maharastra Agric. Univ., l2: 113-14.

Ramos, D., Kelley, K., Reil, W., Sibbett, G.S. and Snyder, R. (2001). Acta Hort., 544: 435-43.

Ramos, D.E., Brown, L.C., Uriu, K. and Marangoni, B. (1978). Calif. Agric., 32:5-6.

Ramos, D.E., McGranahan, G. and Hendricks, L. (1984). Fruit Var. J., 38:112-20.

Ramos, D.E., Ryugo, K., Olson, W.H., Osgood, J.W. and Sibbett, G.S. (1985). In: Walnut Orchard Management, (ed. D.E. Ramos), Div. Agri. Nat. Res., Univ. of Calif., Publ., 21410:96-102.

Rashid, A. (1978). Indian J. Hort., 35: 202-206.

Rathore, D.S. (1976). AICP Fruit Res. Workshop Teh. Doc., 9: 1-6,

Rathore, D.S. (1984). Abst. Nat. Symp. Temperate Fruits, Solan, p. 2.

Roth, P.L. (1971). 62nd Ann. Rept. Northern Nut Grower's Assoc., USA. pp. 45-47.

Rovira, M., Ninot, A. and Aleta, N. (2001). Acta Hort., 544: 287-93.

Rud, G. Ya and Zhadan, V.M. (1972). Trudy Kishnjevskogo Sel'skokhozgaistvennogo Instituta, 93: 11 24.

Ryugo, K. and Ramos, D.E. (1979). HortSci., 14: 52-51.

Saadat, Y.A. and Hennerty, M.J. (2001). Acta Hort., 544: 473-80.

Schulbach, H. (1985). In: Walnut Orchard Management, (ed. D.E. Ramos), Div. Agri. Nat. Res., Univ. Calif., Publ. 21410:108-10.

Serr, E.F. (1961a). Calif. Agric. Expt. Sta. Extension Service. Leaflet, 137.

Serr, E.F. (1961b). Proc. Amer. Soc. Hort. Sci., 77: 301-07.

Serr, E.F. (1968). Plant Propagator, 14: 10-13.

Serr, E.F. (1969). In: Handbook of North American Nut Trees. (ed. R.A. Jaynes). The W.F. Humphery Press, Inc. Geneva. New York. pp. 240-63.

Serr, E.F. and Forde, H.T. (1968). Calif. Agric., 22: 8-10.

Serr, E.F. and Rizzi, A.D. (1964). Walnut Rootstocks. Univ. of Calif. Agric. Exr. Serv. Publ. AXT- 120.

Sharma, R.D., Kour, K., Singh, B., Yadav, S., Kotwal, N., Rana, J.C. and Anand, R. (2014). Australian J. Crop Sci., 8: 257-62.

Sharma, S.D. and Chauhan, J.S, (1980). Punjab Hort. J., 20:78-79.

Sharma, S.D. and Chauhan, J.S. (1981). Progressive Hort., 12: 39-42.

Sharma, S.D., Dhuria, H.S. and Chauhan, J.S. (1982). Indian J. Hort., 39:201-204.

ShouJia, S., Ping, M., JinSong, Z. and XianChong, W. (2011). Tree Physiol., 31: 1378-89.

Sibbet, G.S. and Osgood, J.W. (1985). In: Walnut Orchard Management, (ed. D.E. Ramos). Div. Agri. Nat. Res. Univ. of Calif. Publ., 21410: 52-57.

Sibbet, G.S., Martin, G.C. and Drapper, T.M. (1978). Calif. Agric., 32:12-13.

Simorte, V., Bertoni, G., Dupraz, C. and Masson, P. (2001). Jr. Pl. Nutr., 24: 1645-60.

Simovski, K. and Kuzmanov, I. (1972). Jugoslovensko Vocarstvo, 6: 699-704.

Sinha, M.M., Pal. R.S. and Koranga, D.S. (1977). Progressive Hort., 8:69-74.

Sladky, Z. (1974). Preslia, 46: 193-97.

Slate, C.L. (1969). In: Handbook of North American Nut Trees, (ed. R.A. Jaynes), The W.E. Humphery Press, Inc. Geneva, New York, pp. 1-4.

Smith, P.M. (1976). In: Evolution of Crop Plants, (ed. N.W. Simmonds), Longnran, London, p. 310.

Solignat, G. (1974). Pomol. France, 16: 75-77.

Srinath, D. and Gill, J.S. (1976). Sci. and Cult., 42:123-25.

Stephens, G.R. (1969). In: Handbook of North American Nut Trees, (ed. R.A. Jaynes), The W.F. Humphrey, Press Inc. Geneva, New York, pp. 42-54.

Stoke, H.F. (1969) In: Handbook of North American Nut Trees, (ed. R.A. Jaynes), The W.F. Humphrey Press Inc. Geneva, New York. pp. 39-41.

Taha, N.A. and Al-Wadaan, M.A. (2011). African J. Microbiol. Res., 5: 5796-805.

Tang, H., Ren, Z. and KrcZal, G. (2000). Pl. Cell Rept., 19: 881-87.

Tang, H., Ren, Z., Wallbraun, M., Reustle, G. and KrcZal, G. (2001). Acta Hort., 544: 201-205.

Teviotdale, B.L., Schroth, M.N. and Mulrean, E.N. (1985). In: Walnut Orchard Management, (ed. D.E. Ramos), Div. Agri. Nat. Res., Univ. of Calif. Publ., 21410:153-57.

Thagusev, N.A. and Caroev, B.Ju. (1971). Sadovodstvo, 4:30-31.

Thapar, A.R. (1963). In: Fruit Culture in India, (eds. S. Singh, S. Krishnamurti, and S.L. Katyal), Indian Council of Agricultural Research, New Delhi, pp. 342-46.

Thompson, J.F. and Rumsey, T.R. (1985). In: Walnut Orchard Management, (ed. D.E. Ramos), Div. Agri. Nat. Res., Univ. of Calif. Publ. 21410:175-78.

Tsurkan, I.P. and Pyntya, M.A. (1979). Sadovodslvo, Vinogradarstvo i Vinodelie Moldavii, 1:16-19.

Uriu, K. (1985). In: Walnut Orchard Management, (ed. D.E. Ramos), Div. Agri. Nat. Res., Univ. of Calif. Publ. 21410:115-21.

USDA (2010). National Nutrient Database for Standard Reference Rlease 23.

Vahdati, K. and Khalighi, A. (2001). Acta Hort., 544: 531-35.

Valev, K. (1969). Grad. Lozar. Nauka, 6: 3-8.

Van Steenwyk, R.A. and Barnett, W.W. (1985). In: Walnut Orchard Management, (ed. D.E. Ramos), Div. Agri. Nat. Res., Univ. of Calif. Publ. 21410:158-64.

Watt, B.K. and Merrill, A.L. (1963). Agric. Res. Serv., USDA Handbook 8.

Weber, G.I. and Mac-Daniels, L.H. (1969). In: Handbook of North American Nut Trees, (ed. R.A. Jaynes), The W.F. Humphrey Press Inc., Geneva, New York, pp. 18-38.

Westwood, M.N. (1993). Temperate Zone Pomology, Timber Press, Portland, Ore.

Wildman, W.E. (1985). In: Walnut Orchard Management, (ed. D.E. Ramos), Div. Agri. Nat. Res., Univ. of Calif. Publ., 21410: 27-35.

Woeste, K., McGranaham, G.H. and Bernatzky, C. (1996). Mol. Breed., 2: 261-66.

Woodroof, J.G. (1979). Tree Nuts, AVI Publishing Co. Inc. Westport, Connecticut, pp. 616-55.

Wu, J.J., Gu, y.Q., Hu, Y.Q., You, F.M., Dandekar, A.M., Leslic, C.A., Aradhaya, M., Dvorak, J. and Luo, M.C. (2012). Pl. Mol. Biol., 78: 95-107.

Xi, R.T. and Zhang, Y.P. (1994). Chinese Fruit Tree Annals. Chinese Forestry Publishing, Peking.

Xi, S. (1985). Ann. Rept. Northern Nut Growers, 76: 131-34.

Yalcin, M., Acican, T., Alibas, K., Ertura, U., Akca, Y. and Polat, R. (2014). Acta Hort. 1050:323-29.

YanPing, Ma., Huiling, Ma., XingHua, L. and LongBing, Y. (2011). Transactions Chinese Soc. Agric. Machinery, 42(12): 171-76.

YongTao, L., SanMei, M., YongFei, W., XiaLi, X., LiQun, H., QingRong, S. and KeQiang, Y. (2012). African J. Biotech., 11: 1267-76.

Zachej, S. (1974). Ustavu Rastlinnej Vyroby V Piest'arnoch, 12: 247-56.

# Chapter 13

# Strawberries

The cultivated strawberry (*Fragaria* x *ananassa* Duch.) is one of the most important soft fruit of the world. It has adapted well in many different climates-moderate, Mediterranean, subtropical and even at high altitudes of tropical climate. The world production of strawberry has increased considerably over the past years. The production was about 1.1 million tonnes in 1972 which increased to 2.0 million tonnes in 1984 and to 9.2 million tonnes in 2017. With the development of several day-neutral cultivars, the area and production is increasing at a faster rate. A sizeable portion of strawberry is now growing under cover. The development of different mechanical harvesting machinary further reduced the cost of production which is about 35-40 per cent of the total cost of production. Breeders are also able to introduce several cultivars which have large attractive fruits suitable for mechanical harvesting. Some cultivars also have field resistance to some diseases and pests.

## Composition and Uses

The fresh strawberry fruit is a rich source of vitamins and minerals. Strawberry contains pectin (as calcium pectate) about 0.54 per cent. Strawberry fruit is rich source of polyphenols and the concentration varies from 57-133 mg/100g fresh weight (FW) (Aaby *et al.*, 2012). However, cultivars, management practice and harvest time influence phenolic content, ascorbic acid contents and antioxidant capacity (Pincemail *et al.*, 2012). Aaby *et al.* (2012) evaluating 27 cultivars recorded highest variation between cultivars for contents of cinnamoyl glucose (0.6-24.9 mg/100g FW). Concentration of anthocyanins, the most aboundant class of phenolic compounds in the majority of the cultivars, varied from 8.5 to 65.9 mg/100g FW. Flavan-3-ols (11-45 mg/g FW) and ellagitannins (7.7-18.2 mg/100g FW) contributed on average 28 per cent and 14 per cent of total phenolic contents in the strawberry cultivars, respectively. Hernanz *et al.* (2005) stated that anthocyanins (including pelargonidin-3-glucoside, cyanidin-3-glucoside, pelargonidin-3-rutinoside, pelargonidin-3-acetylglucoside) or phenolic acids (including caffeic, ferulic, p-hydroxybenzoic and ellagic acid) were quantitatively more important than those of flavonols (quercetin and kaempferol). Nour *et al.* (2017) evaluated two cvs. Premial and Magic and reported that the cv. Premial (81.62 mg/100g) content twice the amount of ascorbic acid than in cv. Magic (42.15 mg/100g). Total phenols varied from 207.95 (cv. Premial) to 249.98 (cv. Magic) mg GAE/100g while total flavonoid content in fruits of cv. Magic (79.56 mg/100g) was 2-fold higher than the one of the cv. Premial. The quercetin content ranged from 2.72 to 4.66 mg/100g, whereas free ellagic acid ranged from 6.79 to 9.03 mg/100g. The pale shiny fruits of cv. Premial (higher L*, h° and chroma index) had a lower total anthocyanin content (19.07 mg CG/100g) than the bluish fruits of cv. Magic (23.96 mg CG/100g).

Cyanidin-3-glucoside is responsible for the red colour and pelargonidin-3-glucoside is responsible for the orange-brown colour in strawberries (Bakker *et al.*, 1994). The different proportions of these two pigments affect the colour of different strawberry cultivars (Wrolstad *et al.*, 1970).

The flavour of the fruit is attributed to the presence of volatile esters which varies among cultivars. The concentration of the most important aroma compounds, ethyl hexanoate, ethyl butanoate, trans-2-hexenal, 2, 5-dimethyl-4-methoxy-3 (2H) furanone and linalool were found higher in cvs. Senga Sengana, Kristina and the hybrid Lihama-x Senga Sengana (Hirvi, 1983). Dirinck *et al.* (1981) reported methylthiol esters, methylthiol acetate and methylthiol butyrale in strawberries. The mature soft fruit contains about 5.0 per cent total sugar and 0.90 per cent to 1.85 per cent acidity. The major acids being citric and malic with traces of quinic, succinic, glyceric, glycollic and oxaloacetic. Ripe fruit contains slightly more lipid than unripe fruit, and contains more oleic acid and less linoleic acid (Couture *et al.*, 1988). The nutritional composition of strawberry is presented in Table 13.1.

**Table 13.1: Nutritional Composition of Strawberry* (per 100 g of edible portion)**

| | | | |
|---|---|---|---|
| Water (per cent) | 89.9 | Niacin, B-Vit. (mg) | 0.6 |
| Protein (g) | 0.7 | Ascorbic acid, Vit-C (mg) | 59.0 |
| Fat (g) | 0.5 | Calcium (mg) | 21.0 |
| Carbohydrate (g) | 8.4 | Phosphorus (mg) | 21.0 |
| Vitamin A (I.U) | 60 | Iron (mg) | 1.0 |
| Thiamin, B1 (mg) | 0.03 | Sodium (mg) | 1.0 |
| Riboflavin, B2 (mg) | 0.07 | Potassium (mg) | 164.0 |

*Watt and Merrill (1959).

Much of the fruits is eaten fresh and some are processed. The processed products includes canning, jam, conserves, freezing, drinks, *etc.*

## Origin and Distribution

The Romans were believed to cultivate strawberries before the Christian Era. However, the cultivated strawberry (*Fragaria* x *ananassa*), a hybrid of two native

American species; *F. chiloensis* and *F. virginiana*, was first developed in France in the seventeenth century. The wild species *F. vesca* was found throughout the Northern Hemisphere long before the development of cultivated strawberry. Strawberry industry in America was started at the beginning of 18th century and within 25 years several cultivars like Large Scarlet, Hudson Bay, Early Hudson were found growing commercially in New York, Boston, Baltimore and Philadelphia. In China, commercial strawberry plantation began in the early 1950s and from 1980s the acarage and production increased significantly. The strawberry acerage was about 150,000ha in 2015, while the total yield was more than 4000,00 tonnes and the gross output exceeded 40 billion RMB (about 6 billion USD) (Zhang *et al.*, 2017). Most of the countries in Europe had been growing strawberry commercially since the 18th century. In Canada, the main growing regions are Ontario, British Columbia and Quebec. Strawberry is also grown in Israel, Japan, Turkey, Australia and New Zealand. In India, early efforts to popularize its cultivation in Himachal Pradesh and hills of Uttar Pradesh received a set back on account of poor returns per unit area. Later on, several cultivars like Elista, Senga Sengana, Tioga, Torrey, Shasta, Robinson and Fairfax were introduced, and during the last 40 years the area and productivity have increased.

## Production

In 2016, world production of strawberries was 9.2 million tonnes, led by China with 41 per cent of the total and the United States with 15 per cent. Strawberry production in different countries is presented in Table 13.2.

**Table 13.2: Strawberry Production in different Countries**

| *Country* | *Production (millon tonnes)* |
|---|---|
| China | 3.8 |
| United States | 1.4 |
| Mexico | 0.5 |
| Egypt | 0.5 |
| Turkcy | 0.4 |
| Spain | 0.4 |
| World | 9.2 |

*Source*: FAOSTAT (2017).

## Taxonomy

The cultivated strawberry (*Fragaria x ananassa* Duch.), a hybrid between the Scarlet or Virginia strawberry (*F. virginiana* Duch.) and the pistillate South Americam *F. chiloensis* (L.) Duch. is a dicotyledonous, perennial low-growing herb grown in most arable regions of the world. There are about 20 recognized species of strawberries in five chromosome groups (x=7), ten diploids, four tetraploids, one pentaploid, one hexaploid, and four octaploids (Staudt, 1999; Jiajun *et al.*, 2005). The cultivated strawberry is an octoploids (2n= 8x =56). The most common diploid (2n=14), *F. vesca*, is found growing in North America, northern Asia, northern Africa and north temperate Europe. The plants produce bisexual flowers, fruits bright red, long ovate and highly aromatic. Other diploids, *F. daltoniana*, *F. nilgerrensis*, *F. nubicola* have more restricted distribution in Asia and *F. viridis* is found in Europe and central Asia.

Among the tetraploids (2n = 28), *F. moupinensis* is found in eastern Tibet and western China and *F. orientalis* in western Siberia, Mongolia and Korea. Leaves of *F. moupinensis* are trifoliate with small leaflets while the leaves are ovate in *F. orientalis* which also produce slightly aromatic fruit.

The hexaploid (2n=42) species *F. moschata*, is found in northern and central Europe, Russia and Siberia. The plants are vigorous and dioecious; flowers are large and fruits with aromatic flavour. Octaploids (2n=56) including *F. virginiana*, were used by North American Indians to flavour bread and beverages before Europeans settled in America. The plants are tall, produce abundant runners; flowers are imperfect, fruits soft, light to deep red in colour and aromatic. *F. chiloensis* occurs discontinuously along the Pacific coast, commonly on sand dunes, from the Aleutian Islands to California and in Chile; plants are spreading, very vigorous, usually dioecious. Fruit is roundish, dull reddish brown in colour with mild flavour. *F. iturupensis* is another octoploid found in north-east of Japan.

## Cultivars

Strawberries are adapted to different climates: moderate, Mediterranean, subtropical and even at high altitudes under tropical climate. Several cultivars have been bred in different countries and were evaluated under local conditions. Some cultivars, *e.g.*, Senga Sengana, is most suited to conditions of septentrional culture while in meridional region cv. Tioga grows well. Day-neutral strawberries are popular in the northeastern and midwestern United States since the release of Tribute and Tristar cultivars in 1981 (Draper *et al.*, 1981).

Cambridge Favourite, Redgauntlet, Cambridge Vigour, Gorella and Domanil are the important cultivars in UK. The cultivars Yasma, Senga Sengana and Melitopol'Skaya Rannyaya were recommended for cultivation in the south of Ukraine SSR (Rybalov *et al.*, 1985). Chandler, Parker, Santana and Selva were found very promising for protected cultivation in the Metaponto region, Italy (Recupero *et al.*, 1988). Turft, Tioga and Aiko are grown in Kenya (Waithaka and Ngugi, 1985). Smolarz and Zurawicz (1979) in trials with 12 American and 7 European cultivars and 3 local hybrids recorded highest yield from Senga Sengana in central Poland, followed by in Senga Tigajga and Surprise des Halles. Good yields were also obtained from the cvs. Vesper, Pocahontas, Redchief and Midway. Flowers of the American cultivars were generally found more susceptible to frost than the European cultivars or hybrids. Senga Sengana and Senga Gigana showed very good yield in Himachal Pradesh, India (Sharma and Chauhan, 1986). In Czechoslovakia, of the 67 cultivars evaluated, Senga Sengana and Surprise des Halles were found most productive. The cv. Vigerla also cropped well but had rather small fruits with soft consistency (Simanek and Skulcova, 1979).

Senga Sengana, Jonsok, Totem and Bounty are suitable for industrial jam production, and cultivars with small-sized berries are generally more acceptable for freezing than those with large berries (Skrede, 1980). Belrubi, Cambridge Favourite and Seguoia are suitable for freezing (Guegov *et al.*, 1982); the fruits of Senga Sengana are very suitable for freezing and stewing (Lenartowicz and Plocharski, 1986).

Kiprijanovski *et al.* (2012) evaluated newly released Italian cultivars Alba, Maya and Roxana at Republic of Macedonia. The cv. Alba showed earliest flowering and ripening. The cv. Maya producuced higher number of flowers and also produced largest number of marketable fruits (yield 614g/plant). European network for strawberry cultivar evaluation within the COST Action 863 evaluated 23 cultivars (Baruzzi *et al.*, 2012). They reported better performance of cvs. Elegance, Lucy and Zumba in north Europe, cvs. Elegance, Lia, Tela and Zumba in central Europe, and cvs. Aguedillia, Coral® Rociera, Fuentepina, Kilo, Nora and Siba in south Europe.

Descriptions of some strawberry cultivars are given below:

## AC-Yamaska

June-bearing strawberry, developed from Canada (Khanizadeh *et al.*, 1999), vigororous plant, can tolerante temperature below –30°C (with 10 cm straw mulch cover). Fruits are attractive, very large, dark red and good shelf life.

## Amelia

Late season, short-day strawberry developed from East Malling Research Station (EMRS), UK. Fruit medium, red skin, large conical and good flavour. Resistant to powdery mildew and phytophthora crown rot.

## Amiga

Short-day strawberry developed from Spain. Fruit large (26.3 g), long, wedge shaped, aromatic flavour, susceptible to powdery mildew and two-spotted spider mite.

## Arking

It was released from Arkansas Agricultural Experiment Station, USA; originated from a cross of Cardinal x Ark 5431. It produces large firm, medium-red berries late in the season. The plants are vigorous, productive and resistant to strawberry leaf spot, leaf scorch and powdery mildew (Moore, 1982).

## Belrubi

The cultivar was originated in France, a cross between Pocahontas x Redcoat. Fruits are large, conical; flesh red and of good quality. Plants are adaptable to soils with high pH. However, the cultivar is susceptible to mildew and *Botrytis* (Brooks and Olmo, 1978).

## Cabot

It is a late mid-season cultivar with larger berries (17.0 to 25.1g) and typically short conic in shape. Fruit surface is glossy, bright red with lighter interior colour. TSS 5.5 per cent, TA 0.84 per cent and TSS/TA 6.5 (Jamieson *et al.*, 2004).

### Cambridge Favourite

A commercial cultivar of UK raised at Cambridge. Plants are vigorous and spreading and produces runners profusely. Fruits are large, juicy with moderate flavour. The cultivar is not susceptible to mildew and verticillium wilt.

### Cambridge Vigour

This was also raised at Cambridge. Plants are more vigorous than Cambridge Favourite. Fruits are fairly large, orange-red to light red in colour. Flesh is soft, juicy with good flavour. Susceptible to verticillium wilt and mildew.

### Canoga

Canoga was released from the New York State Agricultural Experiment Station in 1979. The plants are vigorous and upright with dark bluish green foliage. The fruits are exceptionally large, oblate in shape, and characterized by an intense, brick-red colour. The flesh is dark red throughout and is very firm. However, Canoga is not resistant to red stele or verticillum wilt (Sanford *et al.*, 1982).

### Charm

It is a June-bearing (short-day) cultivar from USDA-ARS, Corvallis, USA. It is a high yielder, produced medium size fruit (14-18 g), easily capped fruit with excellent fruit quality that are particularly suited to ice cream formulation (Finn *et al.*, 2013).

### Cristal

Short-day strawberry developed from Spain. Fruit large almost cylindrical, glossy firm red flesh, medium sweetness and acidity.

### Delia

Short-day strawberry developed from EMRS, UK. Fruit orange-red skin, large conical, very firm, moderate flavour. Susceptible to powdery mildew and phytophthora crown rot.

### Domanil

The cultivar was originated in Belgium, a very vigorous plant with spreading habit; susceptible to mildew. Fruits are large, round conical. Flesh is firm and of moderate flavour.

### Elegance

Short-day strawberry developed from EMRS, UK. Fruit medium, large, conical, firm, glossy with moderate flavour. Susceptible to powdery mildew.

### Fenella

Short-day strawberry, developed from EMRS, UK. Fruit medium, conical, red skin, conical and good flavour. Moderately susceptible to powdery mildow and good tolerance to rain.

Strawberry Cultivar Cambridge Vigour

Gray Mould

Strawberry Cultivar Camarosa

Strawberry Cultivar Honeoye

Matted Row System

Verticillium Wilt

**Strawberry with Mulching**

## Finesse

Everbearing strawberry, developed from EMRS, UK. Fruit large, conical, uniform, good flavour. Resistant to verrticillium wilt and phytophthora crown rot.

## Florida Elyana

Short-day strawberry for protected cultivation in winter and spring. Developed from the University of Florida. Fruit lagre (24-27 g) conical to wedge-shaped, flesh sweet. Moderately resistant to botrytis rot, anthracnose fruit rot and crown rots.

## Florida Radiance

Short-day strawberry, developed from University of Florida. Fruit bright to dark red skin, medium (21-23 g), orange-red flesh, juicy. Moderately resistant to botrytis fruit rot and anthracnose fruit rot.

## Framura

The plants are very vigorous, erect, dense and produce abundant runners. Fruits are of medium size, conical, slightly angular, medium red, easily detached from calyx (suitable for mechanical harvesting) and suitable for both fresh market and processing. The cultivar is also sufficiently frost resistant and only slightly susceptible to mildew, leaf spot and *Phytophthora* spp. (Ulrich and Sorge, 1981).

## Fuentepine

Short-day cultivar, developed by the Spanish Public Breeding Programme (Soria *et al.*, 2012). Fruit: pleasantly aromatic and tasty, good shelf life and low tendency to form misshapen fruits.

## Galletta

Short-day strawberry, developed from North Carolina State University, USA. Fruit dark purple-red and red blend skin, large (32 g), orange red flesh, firm, early-ripening cultivar. Resistant to crown and fruit to anthracnose.

## Gilbert

Gilbert bears very large, firm, red fruits. The plants are vigorous and produce an abundance of runners which root readily. The cultivar was found winter-hardy and resistant to leaf scorch. It yields well and the fruit is also suitable for freezing (Gilbert, 1982).

## Glory

Fully everbearing strawberry, developed from Italy. Fruit: glossy red, large, flesh firm and sweet. Moderately tolerant to soil born pathogens.

## Gorella

This cultivar was raised in the Netherlands. Plants are moderate in growth with coarse, thick foliage. Fruits are very large, conical and crimson-red in colour. Flesh is juicy with moderate flavour.

## Haruyoi

Haruyoi was produced for protected cultivation in the south of Tokohu, Japan (Honda *et al.,* 1981a). Fruits weigh 11.5-14.4g and are light orange-red. Growth of the plants is vigorous and there is no tendency to dwarfing during forcing.

## Honeoye

Honeoye is a progeny of Vibrant x Holiday. Plants are vigorous and upright. It runners freely and is very winter-hardy. The fruits are large and attractive, uniformly conical in shape and have a bright red glossy appearance. The fruit is moderately firm with relatively tender skin (Sanford *et al.,* 1982).

## Judibell

Very late, short-day strawberry, developed from EMRS, UK. Medium-large conical fruit, medium firm, moderate flavour, resistant to phytophthora crown rot, verticillium wilt and anthracnose fruit rot.

## Lucy

Short-day strawberry, developed from EMRS, UK. Fruit medium, red skin, very large, conical, firm, moderate flavour. Resistant to phytophthora crown rot.

## Mae

Short-day strawberry, developed from EMRS, UK. Fruit large, conical, uniform, firm, moderate flavour and high yielder. Susceptible to powdery mildew and phytophthora crown rot.

## Malling Centerary

It is an early mid-season and short-day type cultivar. It has a moderate yield, very good fruit size (> 70 per cent of 35mm fruit). Berries are firm and attractive with a very regular shape and excellent flavour Developed from East Malling Research Station (EMRS) UK (Whitehouse *et al.,* 2017).

### Malling Opal

Everbearing strawberry, developed from EMRS, UK. Fruit large, blunt conical, uniform, moderately firm flesh and excellent flavour. Moderately resistant to powdery mildew, susceptible to phytophthora crown rot.

### Malling Pearl

Everbearing strawberry, developed from EMRS, UK. Fruit: red skin, large, blunt conical, uniform, moderately firm flesh, excellent flavour. Moderately resistant to powdery mildow.

### Niglu

Very early-ripening and high yielding cultivar developed from China (ZhuangWei *et al.,* 2013). Fruit is conical, red, shiny, uniform with beautiful appearance. Resistance to anthracnose and powdery mildew.

### Palomar

Short-day strawberry, developed from University of California. Fruit: very large (30.2 g); short conic, high yielder. Moderately susceptible to phytophthora crown rot, and common leaf spot.

### Polka

Polka was selected in the Netherlands in 1977 from crosses between Induka (Puget Beauty x Senga Sengana) and Sivetta (Redgauntlet x Gorella).The fruits are fairly large, of good quality and consistent in form and size; they are suitable for fresh market or for canning (Lindeloof and Wassenaar, 1987).

### Prmier

Everbearing strawberry developed by Plant Science, Watsonville, Califormnia, USA. Fruit: medium to large (24.7 g), conic to ovate, good flavour. Moderately susceptible to two-spotted spider mite, Thrips, botrytis fruit rot and angular leaf spot.

### Puget Crimson

Short-day strawberry cultivar developed jointly by Washington State University and Oregon State University, USA (Moore *et al.,* 2011). High yielder, very late-season ripening, large size, and excellent fresh flavour.

### Redgauntlet

This cultivar was raised at the Scottish Horticultural Research Institute and is mainly grown in Kent and Hampshire areas of Engalnd. Plants are tall and vigorous. Fruit large with attractive colour. Flesh firm, juicy with moderate flavour. It has field resistance to some strains of red core and verticillium wilt.

### Saladin

This was also raised at the Scottish Horticultural Research Institute with an aim to produce red core resistant cultivar. Plants are vigorous and flowers about one week later than Cambridge Favourite. It is cultivated mainly for fresh fruit market.

## Sally Bright

Short-day strawberry developed from EMRS, UK. Fruit: deep red skin, conical, good flavour.

## Sasha

Short-day, strawberry, developed from EMRS, UK. Fruit: medium, red skin, large, conical and moderate flavour. Susceptible to powdery mildew.

## Schwartze (Puget Summer®)

It is a short-day strawberry cultivar jointly released by Washington State University and Oregon State University, University of Idaho and USDA-ARS. High yielder, red skin, large, conical and good flavour. Resistant to powdery mildew and phytophthora crown rot.

## Shimei 8

It is a june-bearing cultivar, developed in China. The fruit is conical, bright red and shiny with sepals that are easily separable. First grade fruit weight 42.7 g fruit flesh is orange red, TSS 10.3° Brix, sweet and sour aroma (Yang *et al.*, 2017).

## Senga Sengana

Originated in Hamburg by Prof. Reinhold von Senbusch. Introduced in Europe in 1952. Hybrid between Markee x Sieger. Plant healthy, runners begin to form in late August, skin red, attractive, flesh firmness medium, freezing quality good; calyx comes free easily, frequently remaining on the plant. (Brooks and Olmo; 1972).

## Sonata

Short-day strawberry, developed from The Netherlands. Fruit glossy,light orange-red skin, pleasant aroma and taste.

## Sumas

It was selected from the cross between Cheam x Tioga. The cultivar is winter-hardy and has good tolerance to aphid-borne viruses. Sumas ripens earlier and has good fruit quality, making it potentially suitable both for fresh use and for processing. It is thought to be well adapted to the Pacific northwest regions of the USA and Canada (Daubeny, 1987).

## Summer Evening

Short-day strawberry, developed from Canada, Fruit: very bright, medium-red skin, large, firm flesh, and excellent fresh flavour. Moderately resistant to leaf scorch.

## Sweet Ann

Dayneutral strawberry, developed from California, USA. Fruit: glossy red moderately firm, red flesh and balanced flavour. High yielder.

## Sweet Bliss

June-bearing (short-day) strawberry cultivar developed from USDA-ARS breeding programme, Corvallis, USA (Finn *et al.*, 2011). High-yielding cultivar that produces large-sized fruit with outstanding fresh market and processed quality.

## Teela

Short-day strawberry, developed from Italy. Fruit: bright red skin, firm, high sugar content, good flavour, and long shelf life.

## Terunoka

This cultivar is particularly suitable for forced culture. Fruits are 10-15 g in weight, glossy, scarlet-red and conical. The flesh is firm, making the fruit highly suitable for transporting and handling. Terunoka is highly resistant to *Sphaerotheca humili* (*macularis*) and *Fusarium oxysporum* f. *fragarice* (Honda *et al.*, 1987b).

## Tokun

A aromatic decaploid interspecific hybrid developed from Japan (Noguchi *et al.*, 2011). Vigorous, produces flowers on upright stalks. Pale yellowish-orange fruit and excellent skin gloss. The fruit has aroma components of carmel, coconuts and peach.

## Tyee

This is a Canadian cultivar, resistant to powdery mildew and spotted spider mite. It is also fairly winter-hardy; fruit is smaller and firmer and ripens earlier than cv. Totem. Yields up to 19365 kg/ha were obtained from machine-harvested Tyee in Oregon (Daubeny *et al.*, 1980).

## Vale

An early season, short-day cultivar, developed from Italy. fruit: bright orange-red, firm skin, firm flesh, and faintly aromatic.

## Vantage

Vantage was selected from a cross between Tioga and Veestar. It is productive and suitable for both fresh use and processing. The fruit is bright, firm and has good flavour. The plant is resistant to *Botrytis cinerea* and highly resistant to *Verticillium albo-atrum* (Ricketson *et al.*, 1986a,b).

## Venta

This Lithuanian strawberry cultivar, produced from Senga Sengana x Festival'naya, has moderate vigour and semi-spreading habit. The conical, dark red fruits weigh on an average, 12g. They contain 11.5 per cent dry matter, 4.4 per cent sugar, 1.1 per cent acids and 51.2 mg vitamin C/100 g. The flesh is dark red, juicy and acid-sweet (Anon., 1987).

## Vesper

This is a late ripening American cultivar. Fruits are large, conical and elongated, deep red in colour. The average fruit weight is 9-11 g. The plants are vigorous and are not readily depressed by virus diseases (Popova, 1984).

## Wendy

Short-day strawberry, developed from Canada (Jamieson *et al.*, 2009). Early-season cultivar, firm fruited, attractive fruit, richly flavoured.

## Zijin Jiuhong

Short-day cultivar developed from China (Zhao *et al.*, 2017). High yielder, Fruit is short conical, red shiny, beautiful in appearance and flavour sweet. It has high level of resistance to grey mold.

A list of strawberry cultivars growing in different countries in presented is Table 13.3.

**Table 13.3: Strawberry Cultivars Growing in different Countries**

| | | |
|---|---|---|
| Australia | : | Storey, Naratoga, Redlands Hope, Redlands, DPI Rubygen, Kalinda, Klewa, Lowanna, Millewa, Camarosa, Florida Festival. |
| Brazil | : | Oso Grande, Toyonoka, Campinas, Campi-Dover, Aleluia, Aromas, Albion, Camino Real, Ventana, Diamante. |
| Canada | : | Mira, Jewel, Cabot, Bounty, Sparkle. |
| Chile | : | Camarosa, Albion, Camino Real, Portilas, Sabrosa. |
| China | : | Darselect, Tudal, Cartuno. |
| Egypt | : | Florida Festival, Sweet Charlie, Camarosa, Tamar, Yael, Suzana, |
| France | : | Gariguette, Darselect, Clery, Charlotte, Ciflorette. |
| Greece | : | Camarosa, Florida Fortuna, San Andreas. |
| India | : | Chandlar, Senga Sengana, Redcoat, Tioga, Sweet Charlie, Festival, Camarosa, Belrubi, Winter Dawn, Gorella, Douglas, Ofra, Selva. |
| Italy | : | Sabrosa® Candonga, Sabrina, Florida Fortuna, Eva, Alba, Clery, Roxana, Garda, Nora, Onda, Camarosa, Sabrosa, Milsei. |
| Japan | : | Summer Princess, Summer Engel, 8-9, Flamenco. |
| Latvia | : | Flair, Fleurette, Felicita, Susette, Honeoye, Polka, Elsanta |
| Mexico | : | Florida Festival, Albion, Camino Real, Osceola, Camarillo. |
| Morocco | : | Camarosa, Florida Fortune. |
| The Northernlands | : | Elsana, Sonata, Darselect, Polka, Honeoye, Camarosa, Florence, Flair, Rumba. |
| Norway | : | Frida, Sonata, Florence, Polka, Korona, Iris. |
| Poland | : | Marmolada, Kent, Honeoye, Evita, Elsanta, Selva, Polka, Kastor, Temira, Malling Pandora, Elkat, Vega, Karel. |
| Portugal | : | Sabrosa, Camarosa, Diamante, Albion, Selva, Chandler, Ventana, Chiflon, Irvine, Aromas, Seascape. |
| Russia | : | Stoplight, Belruby, Gera, Vespar, Troubador, Gera. |
| South Africa | : | Camarosa, Aromas, Gaviota, Diamante, Tamar. |

| | | |
|---|---|---|
| Spain | : | Camarosa, Sabrosa, Splendor, Florida Fortuna, Florida Festival. |
| Turkey | : | Sahara, Sabrina, Splendor, Safari. |
| UK | : | Cambridge Favourite, Cambridge Vigour, Gorella, Malling Centenary, Elegence, Milling Opal, Judibell, Lucy, Mae, Finesse, Fenella. |
| Uruguay | : | INIA Guenoa, Earlibrite, Florida Festival, Aromas, Camarosa. |
| USA | : | Totem, Pinnacle, Puget Reliance, Stolo, Sweet Sunrise, Tilamook, Valley Red, Charm, Sweet Sunrise, Hood, Independence, Rainer, Redcrest, Sumas, Whonnock, Florida Radiance, Strawberry Festival. |

## Climate

### Daylength

The cultivated strawberry, *Fragaria* x *ananassa*, is grown extensively in most temperate and in some subtropical countries. Some cultivars are adapted to septentrional culture, *i.e.*, short days in autumn and hard winter, or to meridional conditions, *i.e.*, long days in autumn and moderate winter. The cultivars Gorella, Redgauntlet and Senga Sengana grow under septentrional conditions while Tioga is an important cultivar of meridional regions. However, selection of several day-neutral strawberry cultivars (*e.g.*, Silva, Fern, Hecker, Aptos) have made the cultivation of strawberry possible in different times of the year. Flower buds are not initiated unless the daily dark period is greater than a certain minimum, although this may be less necessary at low temperature (Cooke, 1969). Under normal conditions flowers are, therefore, initiated in autumn, and often the first truss which is initiated under marginally short days is inferior to trusses formed later during subsequent shorter days.

Photosynthesis or $CO_2$ assimilation rates of strawberries are in the range of 15-25 µmol $m^{-1}$ $s^{-2}$ and the light saturation period is between 1000 and 1400 µmole $m^{-1}s^{-2}$ (Carlen *et al.*, 2009; Larson, 1994). The type of strawberry developed from Israel do not required chilling for high yields and initiate flower bud primordial in longer light regimes of 13.5 to 14 hrs than tradiational short day plants (Izhar and Izsak, 1995). Flowering in modern strawberry cultivars is regulated more by sensitivity to high temperature than photoperiod and the short-day cultivars are more sensitive to high temperatures than day-neutrals (Galletta and Bringhurst, 1990). Jonkes (1965) using the cultivars Glasa and Deutsch Evern, found that the extension of natural winter-daylength to 16 hours by means of weak incandescent (tungsten filament) light gave earlier fruiting but reduced the final yield. The cvs. Hokowase and Donner, grown in untreated plastic greenhouse under natural daylength from January to March and after 1300-hour chilling (below 5°C) produced longer petioles, larger leaves, fewer flowers and a lower yield than chilled plants grown under a short photoperiod of 8 hours (Seyama and Takai, 1980).

### Temperature

A day temperature of about 22° to 23°C is considered optimum for maximum growth rate. Growth slows at 30°C with damage occurring above 35°C (Carlen *et al.*, 2009). June-bearing cultivars developed flower buds when the cultivars produce flowers and fruit throughout the year, provided temperatures are between 7 and 15°C. Pyrotis *et al.* (2012) working with polytunnel growing strawberries at two different sites in the Huelva region of Spain, recorded mean temperature of 11-21°C during growth and development, humidity varied between 56-81 per cent and 60-68 per cent, fruit firmness at maturity was 3.2-9.9 N and 5.4-10.4 N at two sites. Increased temperature by 10°C were found to decrease firmness by 0.5 and 0.7 N at eash site. However, humidity when increased more than 75 per cent due to precipitation, reversed the trend. Late-planted strawberries are generally more susceptible to frost than those planted earlier. The cv. Senga Sengana is somewhat resistant to cold than Surprise des Halles. (Zurawicz, 1983). In cold climates, frost damage and winter injury are major constraints for production and maintenance of strawberry planting. Frost damage is of primary concern during the spring as growers take advantage of cultural methods to produce an early crop (Ourecky and Reich, 1976). Blossoms of most June-bearing cultivars are damaged at –2°C (Havis, 1938; Brand and Hawthorne, 1965). Flowers, fruits and flower buds are killed or damaged in several ways; (i) complete killing of the whole primordium, (ii) variable damage within the flower truss, (iii) damage to flower parts, or (iv) injury to part of the vascular tissue of the peduncle (Campbell and Lingle, 1954). The percentage of killed flowers is related to the place of origin of the cultivar, the more northerly cultivars being more tolerant (Bazaichi, 1968). Stancevic (1970) reported significant differences in survival among 12 European cultivars after late spring frost of –3°C.The cvs. Blakemore, Catskill, Fairfax and Dunlap could withstand –1.6°C, but varied in the susceptibility of the flower part damage (Havis, 1938). Ourecky and Reich (1976) classified several strawberry cultivars into hardiness groups based on the temperature to cause a maximum of 10 per cent injury to the flowers (Table 13.4). In comparing the relative hardiness of fruits, flowers and buds, fruits of Sparkle and Guardian showed less tolerance at –4.5°C than the other four cultivars. The large and small buds of Fletcher were extremely frost tolerant. Most cultivars exhibited some tolerance but temperature of –6.5°C or below caused severe damage (Table 13.5). Nikolov (1987) in a study with 2 each of early, mid-early, mid-late and late cultivars in the Plovdiv area of Bulgaria found that Sequoia, the cultivar earliest to flower, required a temperature sum of 148.9-157.9°C to begin flowering, while Shucksun, the last to flower, required 230.8-262.3°C. The optimum temperature for the period between onset of flowering and onset of ripening was 13.8°C for early cultivars, 14.4°C for mid-early, 15.1°C for mid-late and 16.2°C for late; the temperature sum for this period was 268.6-309.4°C for early and 345.5-369.6°C for late cultivars.

Kremenovic (1980) stated that strawberries grown at higher altitude (860 m) had higher ascorbic acid, dry matter and total sugar contents, whereas acid content was higher at the lower altitude (153 m). Plant development was delayed by 8-12 days at the higher altitude compared with lower.

Sulfuric acid rain has been reported to injure strawberry foliage while the yield increased (Cohen *et al.*, 1981). Small, distinct, irregularly shaped, rust-coloured lesions developed on foliage and fruit after a single 1-hr exposure to acid fog at low pH. Fog at pH 11.6 resulted in death of the plants after 8 weeks (Musselman *et al.*, 1988).

**Table 13.4: Classification of Strawberry Cultivars in Relation to Flower Frost Tolerance by Minimum Temperature Resulting in 10 per cent Injury or Less to Open Flowers***

| *Cultivars* | *Per cent injury at 5.5°C* | *Classification (°C)* |
|---|---|---|
| Gala | 100.0 a[z] | |
| Jerseybelle | 100.0 a | 2.5 |
| Robinson | 83.4ab | |
| Fulton | 75.0 b | – 3.0 |
| Sparkle | 74.6 bc | |
| Holiday | 70.4 bc | – 3.5 |
| Redcoat | 67.2 c | |
| Raritan | 61.2 c | – 4.0 |
| Guardian | 60.0 c | |
| Sunrise | 59.3 cd | |
| Midland | 57.8 cd | |
| Surecrop | 55.4 d | |
| Garnet | 44.7 de | – 4.5 |
| Marlate | 44.1 de | |
| Pocahontas | 44.0 e | |
| Redchief | 41.0 e | |
| Catskill | 44.0 e | |
| Darrow | 39.0 e | – 5.0 |
| Fletcher | 39.0 e | |
| Redglow | 38.5 e | |
| Earlidawn | 25.5 f | – 5.5 |

[z]Mean separation by Duncan's multiple range test at 5 per cent level.

*Ourecky and Reich (1976).

## Soil

Strawberry prefers light soil as heavy soil inhibits root development of runners. Most of the roots of strawberry were found in the top 15 cm of soil (Mann and Ball, 1927), while lateral root spread ranged from 30 to 100 cm (Hansen, 1931). Harrold (1955) indicated that the strawberry plants, in the first year, were capable of absorbing water from 45 cm depth, most of the water extracted, however, came from 25 cm surface soil. It is, therefore, essential that the top layer of the soil should be porous and rich in humus. Soil temperature is an important factor influencing root growth. The minimum temperature for active root growth is 5°C and maximum is 31°C (Bowen, 1991) with an optimum temperature ranges between 10 and 23°C (Larson, 1994).

**Table 13.5: Response of Fruits, Flowers and Buds of 6 Strawberry Cultivars to Freezing Temperature***

| *Cultivar* | *Plant Part* | *Per cent Organ Kill* | | |
|---|---|---|---|---|
| | | *– 4.5°C* | *-5.5°C* | *-6.5°C* |
| Sparkle | Fruit | 100.0 | 75.0 | 100.0 |
| | Flower | 12.0 | 85.2 | 85.7 |
| | Buds (> 5 mm) | 0.0 | 38.2 | 36.8 |
| | Buds(<5mm) | 0.0 | 0.0 | 2.5 |
| Redchief | Fruit | 0.0 | 75.0 | 100.0 |
| | Flower | 0.0 | 52.4 | 88.8 |
| | Buds(>5mm) | 0.0 | 0.0 | 66.6 |
| | Buds(<5mm) | 0.0 | 0.0 | 26.4 |
| Fletcher | Fruit | 0.0 | 100.0 | 100.0 |
| | Flower | 0.0 | 40.0 | 60.0 |
| | Buds | 0.0 | 00 | 0.0 |
| | Buds | 0.0 | 0.0 | 0.0 |
| Guardian | Fruit | 100.0 | 75.0 | 100.0 |
| | Flower | 13.8 | 52.2 | 55.0 |
| | Buds(>5mm) | 9.0 | 14.2 | 50.0 |
| | Buds (< 5 mm) | 0.0 | 0.0 | 4.8 |
| Catskill | Fruit | 0.0 | 100.0 | 100.0 |
| | Flower | 0.0 | 54.5 | 100.0 |
| | Buds (> 5 mm) | 0.0 | 14.3 | 71.4 |
| | Buds(<5mm) | 0.0 | 0.0 | 11.5 |
| Earlidawn | Fruit | 0.0 | 28.5 | 100.0 |
| | Flower | 0.0 | 25.5 | 50.0 |
| | Buds(>5mm) | 0.0 | 0.0 | 30.0 |
| | Buds(<5mm) | 0.0 | 0.0 | 7.7 |

*Ourecky and Reich (1976).

Although strawberry is not very sensitive to soil reaction, it, however, prefers a slight acid soil. Clark (1933, 1941) stated that at pH 4.6 and 6.4 (made with calcium nitrate and ammonium sulphate, respectively) the plants made maximum total growth, although the root growth declined in ammonium culture. In humus-deficient loamy sand, raising the pH to 5.4 and 6.3 from 4.0 by liming considerably increased the yield, although the highest yield was recorded with intermediate pH values (Niemann, 1966). At pH 5.3 runner production was found to increase (Baker and Mortensen, 1973). The strawberry does not grow well in poor-drained soil. Waterlogging resulted in death of old roots and new roots failed to develop properly. Planting sites with slopes of about 2 per cent provide drainage without greatly increasing soil erosion. It is essential to improve surface drainage before planting and to clean the ditches and outlets in autumn, if necessary.

Soil heating by placing the cables at a depth of 10 cm to raise soil temperature by 2-4°C was found to advance flowering and harvest by few days in cv. Senga Sengana (Laurinen and Sako, 1980). Lenz (1976) stated that a root temperature of 20° or 28°C was optimum for cv. Senga Sengana grown in pots at an air temperature of 12-36°C. Water utilization was found to increase with root temperature up to 28°C and declined thereafter. However, soil heating to a temperature about 40°C was injurious to strawberries (Dinkel, 1978).

## Propagation

Strawberry is usually propagated by runner. Propagation by seed is not suitable as the seedlings do not come true to type. Nowadays, large-scale commercial propagation by tissue culture has been used widely in the strawberry industry. A million or more plants can be produced in a year from a single meristem-tip (Boxus, 1974).

### Seed Propagation

Strawberry seed if sown immediately after the fruit is picked, germinated with great difficulty. If such seeds are kept for 11 days at 1°C, they germinate quickly (Borgman, 1950). Dimitrov and Topcijski (1957) suggested stratification of seeds at 2-5°C for 60 days. Jonkers (1958) found poor germination after cold treatment at 3-5°C for 16 days, while seed treatment with 96 per cent $H_2SO_4$ resulted in more rapid germination and a higher germination percentage. Treatment with $GA_3$ and thiourea also improve germination (Iyer *et al.*, 1970).

### Vegetative Propagation

#### Runner

Runners produced from stolons of established plants are commonly used for propagation of strawberry. Although, propagation by runners perpetuates all the characters of the parent, viral diseases are quite often transmitted through runners, if the parent plants are already infected. Thus, for collecting runners, a separate runner bed, with virus-tested clonal materials should be maintained. The site and soil where strawberry had not been grown during past few years should be selected for runner beds. Soil samples should be examined to find any contamination of soil-borne virus. In beds, the planting is done at 1.2x1.2 m to 1.8x1.8 m distance.

In everbearing strawberry cultivars, flower bud development continues throughout the greater part of the growing season, and runner production is low. This creates a problem for the nurserymen, not only because of low production of runner plants, but also because inflorescence must be removed by hand throughout the summer in order to prevent fruiting and thereby favour runner initiation (Shoemaker and Marth, 1953). However, suggestion has also been made that flowering and runner production are independent processes (Saha, 1964). Several experiments have reported that $GA_3$ stimulated runner development (Thompson and Guttridge, 1959) and inhibited flowering of strawberry (Denisen, 1959). Dennis and Bennett (1969) reported that spraying of $GA_3$ at 50 to 100 ppm to new and 1-year-old plantings of the everbearing strawberry cv. Geneva in June or double application in June and July inhibited flowering, promoted runner development and increased the yield of marketable runner plants in the following year. Tafazoli and Shaybany (1978), with uniform everbearing cv. Gem, treated N at 60 and 120 ppm, $GA_3$ at 50 and 100 ppm and ethephon at 50 and 100 ppm in the greenhouse. Half of the plants were deblossomed as flowers emerged. The results revealed that both $GA_3$ and N increased runner production and vegetative growth (Table 13.6). Ethephon generally did not affect the growth and runner except that it decreased leaf fresh weight. Rudolph (1986) reported that maximum production of strawberry runners occurred with planting in April, with within-row spacing of 20 cm and between-row spacing of 83.5 cm for cv. Senga Sengana, 125 cm for cvs. Gorella and Fratina and 167 cm for cv. Redgauntlet. Further recommendations include removing the flowers of mother plants, applying 300 $m^3$/ha of organic matter between the rows, intensive irrigation, removal of the runners from between the rows when the mother plants reach 20-25 cm width and using the mother plants for propagation for 2 consecutive years. Taimatsu *et al.* (1985) stated that use of a sawdust bench, slanted at an angle of 2° and with a drainage mat at the lower edge, gave good rooting of strawberry runners which could more easily be lifted and separated from the mother plant than runners rooted in a soil bed. The rate of applied fertilizer had more effect on the production of runners than thickness of the sawdust layer and it is suggested that depth of the layer could be reduced from 7.5-10 cm to 3-5 cm, if additional fertilizer is used. It is also suggested that 60 runners per mother plant (spaced at 40 cm) during early to mid-July would be a satisfactory rate of production.

The runners are mostly lifted during September-October. Sometimes the lifting continues until March to meet the demand for spring planting. However, for spring planting dormant runners are cold stored. For this purpose runners are usually lifted in late November-December and stored until planting, *i.e.*, June-July of the next year.

### Module Plants

Strawberry module plants are fast replacing traditional bare-root transplants as a source of plant as material under protected crop production systems. Module plants offer many advantages over soil grown 'bare root' plants especially the risk of infection by root diseases is reduced, plant nutrition can be totally controlled and higher fruit yield are readily achievable (Beyene *et al.*, 2012).

### Cold Storage of Runners

Cold storage of strawberry plants become a common growing technique for programmed growers of Europe. Plants are usually dug from late November for short storage (1-3 months) to mid December for long storage (3-9 months). Plants have to be exposed to sufficient field chilling and attain a certain level of dormancy before being stored for a long period. Tietz and Gebhart (1979) suggest storing plants for longer period which have accumulated sufficient starch reserve. A prolonged storage considerably decreases the starch content of plant in storage which was correlated with decrease in flower production and yield (Dardi *et al.*, 1996). The late dug (February) plants showed higher survival, vegetative growth as well as yield (Dardi *et al.*, 1998).

**Table 13.6: The Main Effects of Chemical Treatments and Deblossoming on Vegetative Growth, Runner Production and Flowering of 'Gem Strawberries[z*']**

| *Treatments* | *Levels* | *Root Dry Wt. (g)* | *Leaf Dry Wt. (g)* | *Total Fr. Wt. (g)* | *Leaf Fr. Wt (g)* | *No. Leaves* | *No. Runners* | *No. Clusters* | *Avg. Leaf Area ($cm^2$)* | *Leaf N (Per cent)* |
|---|---|---|---|---|---|---|---|---|---|---|
| N | 60ppm | 2.4 | 2.7 | 21.1 | 11.8 | 11.4 | 7.2 | 4.6 | 28.3 | 1.7 |
| | 120ppm | 3.9 | 4.3 | 28.9 | 16.0 | 18.3 | 9.2 | 4.7 | 39.1 | 2.2 |
| $GA_3$ | 50ppm | 2.0 | 2.1 | 16.1 | 8.3 | 9.7 | 6.4 | 4.4 | 23.3 | 1.9 |
| | 100ppm | 4.3 | 5.0 | 34.0 | 19.5 | 20.0 | 10.3 | 4.8 | 44.1 | 2.0 |
| Ethephon | 50ppm | 3.0 | 3.2 | 23.9 | 15.1 | 13.4 | 8.2 | 4.4 | 31.4 | 2.0 |
| | 100ppm | 3.3 | 3.8 | 26.1 | 12.7 | 16.4 | 8.5 | 4.8 | 36.0 | 1.9 |
| Blossom | + | 3.2 | 3.3 | 25.7 | 13.2 | 15.0 | 8.7 | 3.5 | 33.4 | 1.9 |
| | | 3.1 | 3.8 | 24.3 | 14.6 | 14.7 | 8.0 | 5.8 | 34.0 | 2.0 |
| LSD 5 per cent | | 0.7 | 1.2 | 6.4 | 2.0 | 5.2 | 0.9 | 1.4 | 9.4 | 0.4 |

[z]Each figure is the average of 40 observations.

* Tafazoli and Shaybany (1978).

The runners should be fully developed and dormant before storage. The soil should be removed from the roots and packed in bundles of 50 or 100 in polythene bags of a thickness not exceeding 0.07 mm (to permit some diffusion of gases). A temperature of 0°C is suitable for 3-month storage but for longer period the temperature should be –1°C (Linder, 1964). Rosati (1966) advocated stripping of all except 1 or 2 youngest leaves, treating with captan and air drying before storage. As many as 30 000 plants could be stored per $m^3$. Guttridge and Montgomerie (1971) stated that dipping the plants in fungicides before storage, particularly in thiram (0.16 per cent a.i.), improved their survival after storage. Guttridge *et al.* (1965) recorded best growth of plants when runners were cold stored at -1°C. At –2°C young leaves and tissues in the buds were killed, partly or entirely while at 0°C, existing expanded leaves did not survive. However, results from 2 years' trial on the cold storage of Senga Sengana runners showed that a storage temperature of –2°C was more satisfactory than 0°C (Mellenthin, 1963). Control atmosphere storage (2.5-10 per cent $CO_2$ and 2.5-12 per cent $O_2$) showed inhibition of molds infestation during storage which usually is a problem in cold storage runners (Lockhart and Eaves, 1966). However, very low $O_2$ (2 per cent) storage enhance the development of surface molds (Lieten, 1996). After withdrawl from storage the material had to be removed from the bags and planted as soon as possible to avoid injury from waterlogging or over-heating.

## Micropropagation

Experience with strawberry micropropagation indicate that *in-vitro* plants are more uniform, produce higher munber of runners, have better survival in the field and fruit yield is higher than plants propagated by traditional method (Sanchez-Cuevas and Salaverria, 2004). Large-scale commercial propagation by tissue culture has been used widely in the strawberry industry. A million or more plants can be produced in a year from a single meristem tip (Boxus, 1974; Damiano, 1980). Plants have been regenerated from meristematic callus (Jones *et al.*, 1988), anthers (Laneri and Damino, 1980) and immature embryos (Wang *et al.*, 1984). Methods were also developed for efficient shoot regeneration from leaf and runner tissues (Liu and Sanford, 1988). The most widely used media for micropropagation of strawberry are Knop solution, Murashige and Skoog media (MS) and Linsmaier-Skoog (LS) media with agar, sucrose and hormones at desired levels.

It is necessary to avoid the contamination with microorganisms to achieve success in the establishment, incubation and manipulation of the tissue *in-vitro*, since contaminants can destroy the explants, delay their development by competing with them or generate changes in the culture medium that negatively affect their survival and growth. Contamination can come from the explants or can be introduced during the manipulation of plant tissue. Sanchez-cuevas and Salaverria (2004) reported that treatment with 20 per cent chlorine for 20 minutes immersion of shoots, reduced contamination, allowing a greater survival of the explants and higher shoot formation. The treatment that reduced tissue oxidation more efficiently was cystein at 4 g/l, which allowed 100 per cent explants survival.

Abramenko (1983) cultured meristems of strawberry on Knop-Helter (K-H) medium supplemented with BA at 1 mg/l. The meristems regenerated shoots when incubated at 25-27°C and about 3000 lux light intensity in 16-hour photoperiod. Shoots transferred to K-H medium supplemented with IAA at 0.5-1.0 mg/l rooted readily and the plantlets were successfully transplanted. From one meristem 3000 shoot primordia were produced in 6 months and half of these developed into normal plants. These plants were very vigorous and after transplanting in the soil some produced up to 500 new runner plants (10 times that produced by conventional material). The effectiveness of meristem culture in producing virus-free material, however, varied among cultivars (Beloshapkina, 1985). The wilt-resistant cultivars Zenit and Kulon had high regenerative capacity from meristem culture, producing high quality planting materials (Popova *et al.,* 1985). Meristem cultured Senga Sengnna plants was found to produce 2.5 times higher yield than conventionally propagated plants (Dolgikh, 1988). Meristem culture is considered useful in breeding and genetic investigations (Osipova *et al.,* 1986). Gavrilova (1985) successfully regenerated plantlets from anthers of 3 cultivars cultured on MS-agar medium with added benzyladenine and NAA. On an average, Redgauntlet produced 7.2 plants/cultured anther, while Senga Sengana and Festivalnava produced 6.8 and 6.3 respectively. Anthers of cvs. Rabunda and Primella taken from flowers with sepals 5-6 mm long, produced callus in 80-90 per cent cases on media containing 2 mg/l NAA and 1 or 5 mg/l kinetin. Plantlets regenerated in liquid medium with 0.1 mg/l NAA and 2 mg/l kinetin (Laneri and Darnino, 1980). A simple and feasible protocol for micropropagation via somatic embryogenesis from petiole segments has been standardized (Pallavi *et al.,* 2011). The petiole segments of cv. Sweet Charli induced nodulated friable embryonic callus on MS medium containing 1.0 mg/l each of thidiazuron (TDZ) and 2,4-dichlorophenoxy acitic acid (2,4-D). The development and maturation of somatic embryos was achieved on MS medium supplemented with absicic acid (0.5 mg/l). Successful conversion of somatic embryos into plantlets was obtained in hormone free half-strength MS medium. The regenerated plantlets were successfully acclimatized in pre-sterilized cocopeat at primary hardening stage.

A protocol that enables strawberry micropropagation in one step *i.e.,* multiplying shoots and having them rooted in the same culture medium was developed by Debnath (2006). The use of microcuttings, giving both root and shoot growth in a medium containing cytokinin, is emerging as a better choice for micropropagation of strawberries than multiple shoot proliferation, (using a cytokinin supplemented medium) with subsequent rooting of shootlets. *in-vitro* derived strawberry shoots can be proliferated, elongated and rooted on Zeatin-containing medium. Zeatin alone at very low concentration (1-2 μm) produce two to three shoots per explants, averaging 88 per cent rooting incidence in a single medium in 'Bounty' strawberry. Furthermore, the protocol did not use auxin in the culture medium, which lowers the cost and reduces the probability of somaclonal variation among the proliferated plants. The main advantage of this protocol is that the entire shoots tips of the *in-vitro* grown plantlets can be used for shoot proliferation and rooting, whereas basal rooted nodal segments can be transferred to the rooting medium and acclimatized in the greenhouse. The protocol can eliminated stage-II of the micropropagation and can increase both multiplication rate and rooting rate, this translates into a faster micropropagation of strawberry.

# Planting

## Soil Preparation

Thorough preparation of soil is an important task before planting of strawberries. The soil should be prepared by repeated ploughing and removal of weeds, stubbles, *etc.* Since, most of the strawberry roots are confined in the first 40 cm of the soil surface; the soil should be made friable. Deep cultivation increases the crop's absorption of water from the subsoil. Variation in response to depth of cultivation has been recorded in some cultivars. Matted-bed plants of cvs. Cambridge Favourite and Hapil were found to yield more with deeper pre-planting cultivation (25 cm depth) than at 15 cm depth, whereas for Domanil, the reverse was observed (Crisp *et al.*, 1988). Weeds may also be controlled by using herbicides like paraquat, however, certain herbicide require an interval between herbicide application and planting.

## Soil Fumigation

Fumigation of soil is practised where the population of nematodes, soil diseases and pest populations are high. Fumigation should be done in the fall (September or early October) prior to spring planting. Previous crop debris must be cultivated into the soil during the summer so that it is duly decomposed at the time of fumigation. One week prior to fumigation, till the soil to a depth of atleast 25 cm to break clods and loosen the soil. The soil should be moist but not wet. For all fumigation, the soil can be left undisturbed until spring.

Methylbromide has been used for fumigation in the past but has become very expensive and phased out. Medina-Minguez *et al.* (2012) working at two location of Huelva, Spain observed black root rot complex (*Fusarium* spp., *Cylindrocarpon* spp., *Rhizoctonia* spp., and *Pythium* spp.) and nematodes (*Meloidogyne hapla* and *Pratylenchus penetrans*) were the main phytosanitory problems related to the soil, and *Macrophomina phaseolina* was detected occasionally. DMDS (dimethyl disulphide) + chloropicrin (50: 50 w/w), 1, 3 D (dicloropropene) + chloropicrin (65: 35 w/w) and chloropicrin were found effective for soil fumigation. Biosolarization with fresh poultry manure (25,000 kg/ha) increased yield of cv. Camarosa. Biosolarization with fresh poultry manure (7500 kg/ha) and with Brassica pellets (2000 kg.ha) was suggested as a non-chemical soil fumigation for strawberry cultivation (Lopez-Aranda *et al.*, 2012). Plots covered with ethylene-vinylacetate (EVA) film mulch increased soil temperature considerably which was considered lethal for several soil pathogens and could be used as an alternative to 1,3-D+Pic soil per-plnt treatment. The green-EVA film induced both thermal regimes and fruit yield comparable to those obtained under conventional polyethylene (D'Anna *et al.*, 2014).

## Planting

Planting may be done by hand or machine. Care should be taken to prevent damage and drying of roots of the runner. Transplants are usually removed from the soil in the nursery, made into small bundles and often treated until planted in the fruiting field. Some transplants may be planted in the fruiting field immediately after removal from the soil, or placed in a cooler before transplanting. Irrigation should be provided at regular interval to reduce the water stress in the leaf. Defoliation suppresses plant growth, and fruiting may be delayed. Albregts and Howard (1982) stated that subjecting harvested transplants of strawberry cvs. Florida Belle and Dover to excessive wilting increased foliage loss and plant mortality and reduced plant size and yield.

## Planting System

### Matted Rows

In this system of planting, runners are usually planted along the row at about 0.9 m apart and 0.45 m between the plants. After the initial growth of first year, during June-July, the plants are brushed into the row by using short sticks. The width of mats is maintained between 0.40 and 0.45 m. In this system more number of plants can be accommodated per unit area which give a higher yield under suitable condition. However, overcrowding may cause higher percentage of fruit rot and thereby a reduction in yield.

### Spaced Beds

Runners are usually set at 0.3-0.5 m apart in rows with a spacing of 0.9-1 m between the rows. This helps in picking of fruits by moving between the rows. However, an wider spacing between plants reduces yield, particularly in the first year.

### Hill Rows

In this system, the plants are grown either in single or double rows on raised beds. The beds are usually made 15-20 cm high and planting is done 25 to 30 cm apart in a single or double row in each bed. In double hill rows, plants are set alternately. In this system, a large number of plants are set in an unit area and thus makes removing of runners difficult.

Phogat *et al.* (1980), in studies with three different methods of planting of strawberries, recorded highest yield with matted rows compared with ridges or flat beds. Poling and Saflay (1986) stated that annual hill culture is a profitable alternative to the matted row system. The time pattern of the yield response to annual hill culture (annual cropping) *vs* matted row (harvesting limited to years 2 and 3 only) is the major economic advantage of the annual hill system.

## Time of Planting

Although, planting can be done any time from July to following April or May, early planting ensures a good first-year crop. Planting during July-August, however, needs adequate care, particularly from strong sunshine or dry soil conditions. Irrigation is provided to reduce the water stress in the leaf caused by the undeveloped root system of the transplant and the high ambient air temperature. If irrigation is not provided to the transplants during the daytime stress period, the leaves become desiccated and necrotic, and the transplants are essentially defoliated. If defoliation occurs, plant growth, development and fruiting may be delayed (Albregts and Howard, 1972). Planting of runners in October and November, usually, does not require any extra care for their establishment but the yield in the first summer is less compared with early planting.

The results of experiment with cvs. Senga Sengana and Zephyr, showed little difference in yield in the 2nd and the 3rd year of the crop by planting out in previous September, or in the spring at the first cropping year. First-year yield was unprofitable following spring planting and still low when planted in September but Senga Sengana gave an acceptable yield when planted in August. However, August planted crops ripened late in the 2nd year, and on Senga Sengana the incidence of grey mould was greater than on crops planted later (Thuesen, 1983). Spring planting resulted in slightly larger berries in the second harvest than September planting (Thuesen, 1984). Fiedler and Weier (1980a,b) in an experiment in West Germany with 5 strawberry cultivars (Fratina, Fracunda, Senga Sengana, Gorella and Redgauntlet), planted in early April, late May, late July or late October, found that early planting gave significantly higher yield in the first cropping year than late planting. Cultivars differ greatly in response to planting date and yield. The yield effect of planting date declined in the second cropping year. Early planting produced greater mean fruit weight in the first year and fresh planting material produced greater fruit weight than cold stored material. Nestby (1982) also conducted a similar experiment with 4 cultivars in Norway, in which runners were planted on 29 June, 6, 18 or 28 July or 9 August and yields were recorded in the following 3 years. The latest planting date for a satisfactory yield in cv. Glima was 29 June, 6 July for Zephyr and 18 July for Senga Sengana, giving effective temperatures of 1416, 1307 and 1124 day-degree above 6°C, respectively. For cv. Jonsok no clear day-degree requirement was found, but there was a gradual decline in yield for all planting dates after 29 June. For cv. Glima, there was a tendency for berry size to increase as planting was delayed, while for Senga Sengana the largest berries were obtained by planting on 18 July.

Earlier planting dates induced plants to accumulate more biomass. The initial size of planting material affects growth and flowering of cv. Elsanta during autumn season production, with higher drymatter accumulation in tray plants (Palha *et al.*, 2012). Hassell *et al.* (2006) evaluated the effect of different fall planting dates on cvs. Chandler, Bish, Gaviota and Camarosa at South Carolina, USA. Planting was made on 10, 20 and 30 October each year from 1999 to 2003. Early harvest included mature berries harvested from the first harvest through the second full week of April. Main harvest occurred from the third week of April to the first week of June. The most productive, early cultivar was Gaviota yielding the most on all three planting dates. The cv. Camarosa yielded more than Bish and Chandler on the first and second planting dates. All four cultivars planted on 20 October produced the greatest main season yield followed by significantly lower yields from the 10 October planting date, and the lowest yields of all with the 30 October planting date.

## Planting Distance

The planting distance varies in relation to several factors, *e.g.*, cultivar, system of planting, growing conditions, *etc.* In Belgium, France and Italy, the planting densities are usually from 35000 to 45000 plants/ha. Higher plant densities generally increased yield. However, as plant density increased the total amount of foliage and number of crowns were also increased which made harvest more difficult. In addition, as the daily temperature increased in March and April, more fruit-rot was observed with increased plant density (Albregts and Howard, 1978). In Island of Reunion, densities of 40000 to 50000 plants/ha were reported to be the best compromise between cost and productivity in cv. Sequoia (Fournier. 1982). Freeman (1981) in Australia compared the yield and profitability of the strawberry cvs. Torrey and Naratoga for 2 seasons at planting densities ranging from single-row beds at 35000 plants/ha to 3-row beds at 172000 plants/ha. Production per plant at the highest density was found approximately 50 per cent of that at other densities, and the percentage of culled fruit was greater. Production per hectare of marketable fruit at high density was up to 56 per cent greater than with traditional commercial density of 57000 plants/ha. 'Naratoga' out yielded 'Torrey' in most cases and has fewer culled fruits. In both the cultivars leaf area was markedly reduced at high densities but the effect of average berry size was minimal. Profitability was highest at 172000 plants/ha followed by 87000 plants/ha.

Trials conducted at the Purdue University, Indiana with 'Redchief' strawberry revealed that a row width of 45 cm and 1.0 m between row spacing approximately were optimum. Berry size in early season was larger on narrower rows while the size was reduced in wide rows than in narrow rows over the harvest period (Masiunas *et al.*, 1986). In Poland, the wild-type strawberry cv. Baron Solemacher, when palnted on 28 May at 60 x 15 and 60 x 25 cm spacing, produced the highest total in 4-year yield of 24807 and 24719 t/ha, respectively (Malodobry, 1987). In the UK, Wilson and Dixon (1988) planted strawberry cv. Cambridge Favourite on 2 May at 200, 300 and 450 mm intra-row spacing giving 50000, 30000 and 22000 parent plants/ ha, respectively. Once the stolon development ceased, the width of each matted row was reduced mechanically to 245 mm to provide alleys for access. They found highest yield with an intra-row spacing of 200 mm between parent plants and yield was directly related to the number of plants and crowns at fruiting and positively correlated with planting density. Kelin *et al.* (2001) in a plant density trials at China reported negative correlations between plant density and yield per plant while positively correlated with yield per unit area. The maximum yield of 1500-2000 kg/ mu (15mu= 1ha) was recorded with density of 8000 plants/mu in the solar-heated greenhouse, 4400-7600 plants/mu in the arched plastic tunnel and 500-7600 plants/ mu in outdoor. Laugale *et al.* (2012) reported 127 per cent higher yield in the first production year by planting 6.6 plants/m$^2$ compared with 3.3 plnats/m$^2$ which was only 32 per cent higher in the second growing year.

In India, the strawberry cv. Redcoat showed an yield of 33.4 - 40.2 q/ha by planting at 10 x 40 cm spacing (Pandey *et al.*, 1981). Badiyala and Joolka (1983) in a 3-year-trial with cv. Tioga assessed the yield and fruit quality by planting at 5 spacings, ranging from 30x15 to 30x60 cm. In the first year after planting, the highest yield and quality were obtained at the closest spacing and in the following 2 years the yields were found similar under all spacings but the quality was better under wider spacing (30 x 45 and 30 x 60 cm). Ahmad (2009) evaluated the effect of different plant densities (47619, 60606 and 66666 plants/ha on growth and yield of strawberry at Srinagar, Jammu and Kashmir. He recorded highest yield (195.36 g/plant) by planting in double row system (30 x 50 x 60 cm) accommodating 60606 plants/ha and suggested to use double row planting system in this area for strawberry cultivation.

# Nutrition

There is lack of agreement with regard to strawberry fertilization with organic or inorganic manure alone. The plant growth and fruit yield are reported to be influenced favourably by organic manure (Ljones, 1960; Marchal *et al.*, 1982). Annual application of poultry manure at 0, 4, 5, 9, 18 and 36 tonnes/ha showed higher yields each season with incieasing rates of manure up to 18 tonnes/ha (Albregts and Howard, 1981).

## Role of Nutrients

### Nitrogen

Johanson (1963a,b) and Johanson and Walker (1963) reported that the N-deficient plants were stunted and yellowish green in colour. The leaves remained small, while the petioles looked stiff and upright. Margins of older leaves became bright orange-red following which the pigment moved inward. Subsequently, the margins turned brown and dried. Runners were bright red, thick and reduced in number. The yield was greatly affected. The complex relationship between vegetative and subsequent reproductive development are likely to contribute to the variability of response. Sometimes vegetative growth is found antagonistic to reproductive growth (Guttridge, 1960) while an increase in plant growth may or may not be correlated with increased yield (Guttridge and Anderson, 1973). Cultivars also showed variable response with applied N (Voth *et al.*, 1961).

Breen and Martin (1981) in studies with abundant (Hood and Benton) and poor (Olympus) runnering cultivars grown in a milled-bark medium with N from slow-release urea found that growth was poor or normal at low and intermediate rates of N, but high rates were detrimental. Number of leaves, crowns, and runners and subsequent inflorescence and flower production were frequently less than maximum at the highest level of N (4.8 g N/l bark medium). Applied N had a significant effect on the number of crowns/ptant, inflorescence/crown (Table 13.7) and flowers/inflorescence. In Norway, Meland (1986), while studying the effect of N as calcium ammonium nitrate at 50 and 100 kg/daa on 4 strawberry cultivars (Senga Sengana, Zefyr, Jonsok and Glima), found that higher rate of N resulted in fewer inflorescence and blooms per plant and reduced yield. Hynes and Goh (1987) from New Zealand, in trials with increasing rates of N (as urea), reported that in the first and second seasons there were small increases in plant growth and yield with the low rate of N, but higher rates caused marked reductions in both growth and yield. With high N rates, accumulation of high levels of ammonium and soluble salts in the soils during spring of the first season was thought to have initially inhibited plant growth. Following nitrification of the accumulated ammonium, soil acidification occurred and consequently toxicities of Al and Mn limited strawberry growth. Several reports have indicated that N fertilization of strawberries resulted in the development of softer fruit (Darrow, 1931; Greve and Shoemaker, 1930). Fertilization with N has also been reported to result in the production of fruits lower in acidity, total, reducing sugar, (Ricketson and Hill, 1966; Saxena and Locascio, 1968) and increase protein and amino acid contents (YingYan *et al.*, 2003).

**Table 13.7: Effect of Applied N of the Number of Inflorescences in June, 1977 and 1978 in Container Grown 'Hood', 'Olympus' and 'Benton' Strawberries***

| Applied (g/liter of medium) | No. of Inflorescences | | | | |
|---|---|---|---|---|---|
| | June 1977[z] | | June 1978 | | |
| | Per plant | Per crown | Per plant | | |
| | | | Hood | Olympus | Benton |
| 0.3 | 1.9 | 0.74 | – | – | – |
| 0.6 | 5.0 | 1.0l | 6.6 | 7.0 | 4.4 |
| 1.2 | 7.9 | 1.30 | 10.4 | 9.2 | 7.2 |
| 2.4 | 10.7 | 1.30 | 11.6 | 13.6 | 9.4 |
| 3.6 | – | – | 10.4 | 15.8 | 9.4 |
| 4.8 | 12.44 | 1.36 | 8.0 | 11.6 | 7.8 |
| Linear | **** | * | NS | * | NS |
| Quadratic | ** | NS | * | ** | ** |

z: Cultivars (Hood and Olympus) were not significantly different, therefore, pooled values are presented.

y**** = < 0.1 per cent, ** = p < l per cent, * < p < 5 per cent, NS= p > 5 per cent.

*Breen and Martin (1981).

Studies with $N_{15}$-urea, FuTian *et al.* (2006) reported that application of urea at planting and before anthesis showed the maximum N utilization rate (52.5 per cent) while single application at full rate before anthesis had the lowest (34.2 per cent). They also observed that 15 per cent of the N was allotted to the reproductive organs when urea was applied at planting. The percentage rose to 27.8 per cent when urea was applied at planting and before anthesis and increased further to 31.3 per cent when a single application was made before anthesis. The competition of the reproductive organs for $^{15}N$ at the later growth stages of the plant appeared greater than that of the vegetative organs. Split top-dressing of nitrogen at planting and before anthesis with a total rate of 225 kg/ha accelerated blossom bud differentiation and increased early and total yield. Ahmad *et al.* (2011) repored maximum plant height (27.25 cm), number of fruits/plant (14.79), fruit weight (8.11 g) and fruits/plant (120.09 g) by application of N at 150 kg/ha. Andriolo *et al.* (2011) stated that the N concentration used for the strawberry in soilless growing system can be reduced to 8.9 mmol/l without any reduction in fruit yield.

## Phosphorus

In sand culture, P-deficient plants showed reduced growth. The leaves were unusually dark green and curved downwards. Later on, the veins of older leaves turned bluish and the entire leaf looked bluish purple (Johanson, 1963a, b). Application of phosphorus was found effective in increasing the runner production (Taylor *et al.*, 1975). Addition of P in N, K, B and lime mixture increased berry size and weight (Cutcliffe and Blatt, 1984). Odogo *et al.* (2011) suggested application of P at 17 to 34 kg/ha along with farm-yard-manure at 36 to 54 tons/ha for strawberry growing in tropical altitude conditions of Kenya. The treatment caused large-sized and sweet, long-storing berries.

### Potassium

The deficiency of K caused reduction in growth, yield and fruit size and tended to accelerate fruit ripening (Holevas, 1968). Karmer (1964) suggested application of 100-150 kg $K_2O$/ha to soil having a natural $K_2O$ content less than 20 mg/100 g, but for soils with a $K_2O$ content ranging from 20 to 35 mg/100 g of soil, an application of less than 100 kg/$K_2O$ was found adequate. Ahmad *et al.* (2011) reported maximum plant height (23.70 cm), number of fruits/plant (13.93), fruit weight (7.58g) and fruit yield/plant (106.68 g) by application of potassium at 150 kg/ha. Increased supply of K, however, has been found to increase the sugar and acid contents of fruit (Terk-Uile, 1964, Bunemann, 1962). Ricketson and Hiil (1966) indicated that the soluble solids content and total titratable acidity of strawberry fruit increased as the foliar level of K increased from 0.5 to 2.0 per cent. Under saline condition, supplementary K application improved plant growth, yield and fruit quality (Kaya *et al.*, 2002).

### Calcium

Calcium deficiency in strawberry is associated with tipburn of young unopened leaves which subsequently results in deformed and crinkled mature leaves (Johanson and Walker, 1963; Mason and Guttridge, 1974). Examination of tissues affected by calcium stress with a scanning electron micioscope showed cells of irregular structure (Chiu and Bould, 1976). Eavesind Leefe (1962) obtained firmer berries by treating the plants with four sprays of 0.4 per cent calcium chloride. The first spray was given at blossoming time and the subsequent ones at two days' interval. However, the treated fruits developed brown necrotic spots after two days' storage and were less sweet than control fruits. Chiu and Bould (1976) stated that for normal fruit development and for preventing leaf tipbrun, strawberry plants require a small but continuous supply of ionic calcium. Kaya *et al.*(2002) reported that foliar spray of calcium nitrate (9 mmol/l) ameliorated the negative effects of salinity on plant growth under salt-stressed situation. The spraying treatment also increased calcium and nitrogen contents of plants and yield of strawberry cvs. Oso Grande and Camarosa. Bieniasz *et al.* (2012) however, reported that fertilization with calcium preparation did not significantly increase the yield and fruit mass of strawberries. It improved the firmness of fruit. The application of calcium preparations increased the storage time of fruit and increase the possibility of transport especially for cultivars with a very delicate fruit.

### Magnesium

Yellowish green colouration of leaf margin followed by downward and inward curling were recorded due to Mg deficiency. The basal portion of the leaf along the main vein turned yellowish orange while the areas adjacent to mid-rib remained green (Lineberry and Burkhart, 1943). Deficiency of Mg also caused a delay in ripening and colour development of fruit (Bunemann and Gruppe, 1962). Fertilization with Mg increased runner production, yield and improved fruit quality (Niemann, 1966).

### Boron

Deficiency symptoms of B in strawberry are reduced vegetative growth and misshapen or cat-faced fruit, which is due to poor achene development and subsequently reduced tissue enlargment (Johanson, 1980). Excessive borax application to soil can produce toxic effects in strawbberry (Latimer, 1943). Verlinden (1986) reproted that B deficiency in strawberry was associaied with leaf B content of < 25 ppm; excessive B (>200 ppm) induced leaf necrosis resembling K deficiency symptoms. He recommended application of borax at 8 kg/ha in March before planting. High boron rates were found to double the total boron content of plants (Riggs *et al.*, 1987). A combined spray of calcium and boron significantly reduced the incidence of albinism (6.5 per cent) and grey mould (1.2 per cent). Spraying of B alone, however, significantly reduced fruit malformation (Singh *et al.*, 2007).

### Iron

In sand culture, plants grown in Fe-deficient conditions showed dark green venation at 21 days after the trials were started. At 35 days, newly formed leaves showed a slight interveinal chlorosis; at 48 days, the interveinal areas were bright yellow and the veins were dark green. Within another week the oldest leaves showed marginal necrosis (Johanson and Walker, 1963). The Fe chelate EDDHA proved to be the most effective material in supplying Fe to the plants, both in soil and solution culture (Andersen, 1963).

Hosseini and Bahadori (2017) observed that in soil with high pH, the uptake of iron was a problem by strawberry plant. They found that soil application of Fe-EDDHA at 10.5 kg/ha caused the highest fruit yield, average fruit weight and vitamin-C content of fruit. However, spraying of iron sulphate or Fe-HEEDTA was not economical. Soil application of 4.5 kg/ha of sequestrene-138 Fe, was therefore recommended for greenhouse production of strawberry in Jiroft region of Iran.

### Zinc

Deficiency symptoms of Zn appeared as pale green to yellow colour of leaves. The leaf blade tended to be somewhat narrow and concave (Lott, 1946). Foliar spraying of Zn has been reported to increase the yield in strawberry (Krupyshev, 1977).

In order to determine the effect of Zn as foliar spray on strawberry cv. Kordestan under salinity stress conditions, an experiment was conducted by Saadati and Moallemi (2012) in a hydroponic culture in glass house under controlled condition. Three levels of NaCl (0, 15 and 30 mM) and three levels of Zn (0, 1.5 and 3 g/l) were applied. The results showed that vegetative growth, chlorophyll index and leaf relative water content were reduced in the condition of salinity, while foliar application Zn increase them. Electrolyte lekage increased by increased in NaCl concentration and decreased with foliar application of Zn. They concluded that application of Zn could alleviate salinity stress on strawberry.

### Manganese

Mn deficiency resulted in a pale green to yellow colouration in the interveinal areas while the veins remained green (Lott, 1946). Foliar sprays of 0.2 per cent $MnSO_4$ corrected the deficiency (Oydvin, 1968).

### Copper

The younger leaves of Cu-deficient plants turned pale green with red areas between the larger veins. The symptoms became less distinct as the leaves grew older (Lott, 1947). Van der Boon (1961) recorded reduced yield of strawberry when the Cu content of soil was less than 3 mg/kg dry soil.

### Molybdenum

Molybdenum affects fruit quality and nutrient content. Li *et al.* (2017) assessed the effect of different rates of sodium molybdate (0, 67.5, 135, 168.75 and 202.5 g/ha) on strawberry cv. Akihime. Results showed that net photosynthetic rate (Pn) and cholorophill content of the strawberry plants increased with an increase in Mo rate, and the contents of N, Mo, Fe, Cu, total soluble solids, titratable acidity, sweetness, some sugars, organic acids and some volatile compounds in the fruit increased. However, the Mo rates did not significantly affect the concentration of Se, sucrose, lactic acid, acetic acid and some aroma compounds. Fruits sprayed with 135 g/ha Mo had the highest concentrations of six characteristic aroma compounds including methyl butanoate, γ-decalactone, ethyl butanoate, methyl hexanoate, γ-dedecalactone and ethyl caproate.

## Fertilizer Recommendations

Modoran and Gherghi (1958) suggested autumn application of 20 tonnes FYM, 40 kg $P_2O_5$ and 20 kg $K_2O$ per hectare which should be followed by spring application of 20 kg N after the onset of growth but before flowering. After harvest 20 kg N, 40 kg $P_2O_5$ and 20 kg $K_2O$ should be given. On the basis of five-year data, Angelov (1966) suggested pre-planting application of 20 tonnes FYM, 20 kg N, 40 kg $P_2O_5$ and 40 kg $K_2O$ per/hectare, followed by an annual application of 40 kg N, 80 kg $K_2O$ per hectare. Fruit size tended to be larger with pre-planting application of organic and mineral fertilizers than post-planting only. In sandy loam soil, Matvienko (1984) found that 300 tonnes peat and 200 tonnes FYM + 80 kg $P_2O_5$ + 180 kg $K_2O$ per hectare gave yield almost double of the control plants. Ellwanger *et al.* (1984) reported that strawberries treated with a basal fertilizers consisting of 100 kg N, 96.6 kg $P_2O_5$, 66 kg $K_2O$ and 10 tonnes poultry manure markedly increased yield and fruit weight. In trial with cv.Tioga, Joolka *et al.* (1984) found the highest yieid (8.86 kg/3x2 m plot), fruit weight (9.1 g) and TSS (8.13 per cent) and the lowest acidity (0.64 per cent) with N at 150 kg, $P_2O_5$ at 75 kg and $K_2O$ at 50 kg/ha.

Foliar spraying of N (0.5 per cent), $P_2O_5$ (0.2 per cent) and $K_2O$ (0.5 per cent) four times between August and February was suggested by Singh *et al.* (1983). Rosen *et al.* (1988) conducted a field experiment to determine the influence of supplementary

NPK foliar fertilizers (9:8:7 or 16:2:3) applied during flowering, fruit enlargement and/or flower initiation. All plots received soil-applied fertilizers at planting and at the start of the following year according to soil test recommendations. They found that, foliar fertilizers did not significantly increase total yield at any rate or time of application. Jurcev *et al.* (1966) reported that two foliar application of B, Zn. Mn, Cu or B and Zn during flower bud formation increased fruit size, yield and sugar and vitamin C contents of fruits. Cutcliffe and Blatt (1984) found that the first crop yield in cv. Redcoat increased as the rate of applied dolomite increased, where the initial soil pH was 4.9 (Table 13.8). However, leaf tissue K concentration decreased as the rate of applied limestone increased.

**Table 13.8: Effect of Applied Dolomitic Limestone on First Crop Strawberry Yield and Leaf Tissue N, P, K Content at Dromore, Prince Edward Island***

| *Dolomitic Limestone (t/ha)* | *Mean Berry wt. (g)* | *Marketable Yield (t/ha)* | *Leaf Tissue Content (Per cent)* | | |
|---|---|---|---|---|---|
| | | | *Full Bloom* | | |
| | | | *N* | *P* | *K* |
| 0 | 8.9 | 9.6 | 1.6 | 0.20 | 1.0 |
| 2.2 | 9.1 | 10.1 | 1.6 | 0. I8 | 0.9 |
| 4.4 | 9.2 | 10.6 | 1.6 | 0. I8 | 0.8 |
| 8.8 | 8.9 | 10.9 | 1.6 | 0.18 | 0.7 |
| LSD *P=0.05* | NS | 1.3 | NS | NS | 0.2 |

* Cutcliffe and Blatt (1984).

Ganjehi and Golchin (2012) recommended application of N, K and Mg at 110, 120 and 40 mg/l for hyrdroponic culture of strawberry cv. Gavita. Dodman and Amri (2014) in a similar experiment with urea (100 and 200 mg/l/pot), potassium chloride (75 and 150 mg/l/pot) and magnesium sulphate (50 and 100 mg/l/pot) in controlled greenhouse conditions, as hydroponic culture, in cocopeat + perlite (50: 50 v/v) substrate recorded increased number of inflorescence, flowers and fruits with high levels of magnesium sulphate. Low level of urea increased fruit weight and low level of potassium chloride increased soluble solids and fruit firmness.

Strawberry growers at Huaura Valley, Peru applied fertilizer two times and the last dressing usually made 2-3 months before harvest. Casas *et al.* (2017) investigated the effect of a third dressing before 11-12 days of harvest in cv. Aromas. They observed that additional side-dressing increasing yield between 14 and 43 per cent and increased daily harvest for a longer period. The yield increase was from 40.1 tons/ha to 71.6 tons/ha.

## Tissue Analysis

Ballinger and Mason (1960) in their experiment with cv. Albritton opined that leaf blade is the best single plant part for indicating the nutrient status of the plant, though actually the crown was the best indicator for P, the petiole for K and the roots for Mg contents.

John *et al.* (1975) also suggested the use of leaf blade in determining the nutrient status of strawberry. Klossowski and Sobezykiewicz (1966) suggested leaf analysis at the time of fruit ripening while Kalon and Jaskowiec (1982) advocated leaf analysis at peak fruiting period. Bould (1964) concluded that composition at fruiting provided the best indication of the nutrient status. The macroelement composition of strawberry leaves was found to vary greatly from one leaf fraction to another. Kwong (1967) suggested that the youngest mature leaflets appear to be the most desirable leaf fraction for use in making routine foliar analysis for diagnostic work.

Critical levels and deficiency thresholds of different elements as proposed by Bould (1964) at fruiting for N, P, K and Mg; and by Ulrich *et al.* (1980) in recently matured leaves for S, Fe, Mn, Zn, Cu, B and Mo are as follows:

| *Nutrient (Per cent of dry matter)* | *Deficiency Threshold* | *Critical Level* |
|---|---|---|
| N | 2.4 | 2.6-3.0 |
| P | 0.20 | 0.25-0.30 |
| K | 1.0 | 1.5 |
| Mg | 0.10 | 0.15 |
| *Nutrient (ppm of dry matter)* | *Deficiency Threshold* | *Critical Level* |
| S | 300-900 | 1000 |
| F | 5-40 | 50 |
| Mn | 4-25 | 30 |
| Zn | 6-10 | 20 |
| Cu | < 3.0 | 3 |
| B | 18-22 | 25 |
| Mo | 0.12-0.40 | 0.5 |

Papadopoulos (1987) stated that the total N and $NO_3^-N$ in laminae and petioles increased with higher rates of added N. Kaska and Gezerel (1983) conducted an experiment to determine the effect of liquid fertilizers, such as, Bayfolan, Hexal, Wuxal-3, Wuxal-5 and urea on the uptake of nutrients by the leaves of cultivars Tiago, Aliso and Pocahontas in the Cukurova region of Turkey. It was found that N, P, K and Mg were readily absorbed by the leaves, and that the micronutrient absorption and Fe, Zn and Mn contents of the leaves were also increased by these fertilizers. Cutcliffe and Blatt (1984) reported that leaf N concentration in cv. Redcoat at full bloom ranged from 1.4 per cent to 3.2 per cent and application of N as top dressing 6 week after planting had no effect on leaf N of the first crop. However, where N was applied in the planting year and again at renovation following harvest of the first crop, leaf N at full bloom was greater than where no N had been applied. Leaf P concentration were within the range of 0.17 per cent and 0.34 per cent and leaf P at full bloom of the second crop was increased where P was applied in the planting year and again after harvest of the first crop. Leaf K concentrations at full bloom of the first and second crops and after harvest were increased by prior addition of K (Table 13.9).

**Table 13.9: Effects of N, P and K Applications on First and Second Crop Leaf Tissue Nutrient Content***

| *N, P and K Treatments*⁺ | *Leaf Tissue Content (Per cent)* | | | | | | | | |
|---|---|---|---|---|---|---|---|---|---|
| | *Full Bloom* | | | | | | *After Harvest* | | |
| | *First Crop* | | | *Second Crop* | | | *Second Crop* | | |
| | *N* | *P* | *K* | *N* | *P* | *K* | *N* | *P* | *K* |
| kg N/ha | | | | | | | | | |
| 0 | 1.5 | 0.19 | 1.1 | 2.7 | 0.34 | 1.6 | 1.4 | 0.20 | 1.4 |
| 45 | 1.5 | 0.18 | 1.1 | 3.0 | 0.32 | 1.6 | 1.6 | 0.21 | 1.2 |
| 90 | 1.5 | 0.20 | 1.1 | 3.1 | 0.31 | 1.6 | 1.5 | 0.20 | 1.2 |
| i80 | 1.5 | 0.18 | 1.1 | 3.2 | 0.33 | 1.6 | 1.5 | 0.18 | 1.2 |
| LSD *P=0.05* | NS | NS | NS | 0.3 | NS | NS | NS | NS | NS |
| kg P/ha | | | | | | | | | |
| 0 | 1.6 | 0.19 | 1.1 | 3.1 | 0.22 | 1.6 | 1.6 | 0.16 | 1.3 |
| 49 | 1.5 | 0.17 | 1.2 | 3.1 | 0.21 | 1.6 | 1.6 | 0.17 | 1.2 |
| 99 | 1.5 | 0.18 | 1.1 | 3.2 | 0.29 | 1.6 | 1.6 | 0.18 | 1.2 |
| 197 | 1.5 | 0.18 | 1.1 | 3.1 | 0.31 | 1.6 | 1.5 | 0.19 | 1.2 |
| LSD *P=0.15* | NS | NS | NS | NS | NS | NS | NS | NS | NS |
| kg K/ha | | | | | | | | | |
| 0 | 1.4 | 0.18 | 0.9 | 3.1 | 0.31 | 1.2 | 1.5 | 0.20 | 0.7 |
| 28. | 1.5 | 0.19 | 1.0 | 3.2 | 0.32 | 1.3 | 1.5 | 0 19 | 0.8 |
| 56 | 1.4 | 0.18 | 1.1 | 3.1 | 0.32 | 1.4 | 1.5 | 0.20 | 1.1 |
| 112 | 1.5 | 0.18 | 1.1 | 3.2 | 0.32 | 1.6 | 1.6 | 0.20 | 1.3 |
| LSD *P=0.05* | NS | NS | 0.1 | NS | NS | 0.1 | NS | NS | 0.1 |

⁺ N, P and K treatments were applied as top-dress applications 6 wk after planting and immediately after harvest of the first crop

*Cutcliffe and Blatt (1984)

Fattahi and Gholami (2009) studied the changes in nutrient content of strawberry plant parts at different development stages. The plants were grown in a greenhouse in perlite medium and fertigated with Hoagland solution. The result showed that nutrient uptake was variable at different developmental stages. Leaf and petiole were the main sinks for Ca at fruiting stage and also for Mg and K in petioles, Fe in root, Mn in leaf. Nitrogen, calcium and manganese content in leaves have a decisive role during floral transition in strawberry (Eshghi and Tafazoli, 2007).

## Irrigation

Irrigation is essential for strawberries, particutarly in the first year at transplanting and rooting of runner. Irrigation is also useful for utilization of applied nutrients in the soil, preventing frost damage of the blossoms and during the picking period. Strawberry is a shallow-rooted plant and the roots are found within 15-40 cm of the soil surface. The plants thus require more frequent but less amount of water in each irrigation. Increased runner production and early rooting have resulted from irrigation of a newly planted bed (Waldo, 1947). In first year, strawberry plants were found capable of drawing water from 18 inch (45 cm) depth (Harrold, 1955) and formed extensive and deep root systems (Rom and Dana, 1960). Optimum growth of strawberry plants has been found under soil moisture conditions where tension does not exceed 1.0 atmosphere (Rogers, 1935).

Thus, frequent irrigation rather than a few heavy ones were favoured (Mortensen, 1934). Strawberries grown on sandy soil require more frequent irrigation than when grown on heavier soils. Excessive irrigation was, however, detrimental, which encouraged growth of leaves and stolons at the expense of fruit and flower and also increased the incidence of botrytis rot. Results reported from an irrigation trial in Denmark on cvs. Rubina and Dania showed that the ability of berries to swell after rain was affected by growing conditions at earlier stages of development, and that ample watering was necessary during the early development as well as in the final phase of fruit maturation, if berries were to reach a good size (Hansen and Ostermann, 1987). O'Neill (1983) stated that greater susceptibility of older leaves to water stress in the field during summer months appeared to be related to gradients in leaf osmotic potential within the plant and to an age dependency in the ability of leaves to adjust osmotically when challenged by periodic water deficits.

Irrigation is applied in furrows between the rows and alleys are usually cultivated after 2-3 days of each irrigation. Care has to be taken so that water should not wet the leaves as it may aggravate foliage scald. However, use of sprinkler and trickle irrigation are more convenient to regulate water as well as to minimize the amount of water. On an average, the amount of water used for trickle irrigation is 30 per cent less than that required in furrows (Kamber *et al.*, 1986). Tekinel *et al.* (1984), in comparing furrow, sprinkler and drip irrigation methods, found highest yield with drip irrigation and lowest with furrow irrigation. The amount of water used for drip irrigation was 38 per cent less than that used in furrows and 20 per cent less than that used by sprinklers. Application of irrigation significantly increased the yield, fruit size, and quality of strawberry cvs. Redgauntlet, Senga Sengana, Fratina and Gorella. Blasse *et al.* (1984) stated that pre-blossom irrigation in these cultivars was ineffective and the most important irrigation times were during cropping and subsequent flower bud formation in September. The fruits of cvs. Senga Sengana and Fratina, however, showed higher percentage of soluble solids in non-irrigated plots than those from irrigated plots. Albregts and Howard (1985) at Florida evaluated diurnal intermittent sprinkler irrigation intervals (minutes on/minutes off) of 3/17, 5/25, 5/15, 10/20, 15/15 and continuous during the first season and 5/25, 5/15, 5/10 and continuous during the second season for establishing transplants on black polythene-mulched beds. They found that leaf loss during the establishment period generally increased with increased length of the 'off' interval and decreased length of the 'on' interval. Plant mortality was greatest with 3/17 and 5/25 treatments. Seasonal fruit yield was reduced in the first season with the 5/25 treatment.

Strabbioli (1985), in an experiment with virus-free cold-stored 'Aliso' and 'Tuff' strawberries, planted in double rows spaced at 30x35 cm (about 50000 plants/ha) and mulched with black polyethene film, supplied water by sprinkler for 15 days and through labyrinth drippers (4 l/h) under the mulch. Irrigation rates were based on Class A pan evaporation multiplied by four crop coefficients (0.33, 0.66, 1.00 and 1.33). Irrigation was applied every 10 mm of evaporation. Results showed that irrigation increased fruit yield from 345 g/plant (unirrigated control) to 411-554 g/plant, and individual fruit weight from 7.6 to 8.5-10.4 g. The highest values were obtained using crop co-efficient of 1.00. Gehrmann (1985), in trial with cv. Korona grown in sand, induced different levels of water stress by supplying 75, 50 or 25 per cent of the daily water consumption (100 per cent) of the control plants. Plant growth, yield and fruit quality were found to reduce as the water supply was decreased but fruit maturity was accelerated.

Predawn leaf water potential below -0.25 MPa has been suggested as a good criterion for scheduling irrigation of strawberry (Dwyer et.al., 1978). Sruamsiri and Lenz (1986) suggested leaf water potential values below -1.0MPa as the first threshold of drought stress indicater in strawberry. Leaf water use efficiency (WUE) of strawberry depends on the water availability in the growing media and cultivars (KlamkoWski and Treder, 2008). Yuan *et al.* (2004) estimated the WUE of fresh fruit yield is about 1.6g/mm at optimal irrigation. They also reported that in greenhouse produced soil grown strawberry plants, average irrigation water use ranged between 1.3 and 2.1 mm/day at an irrigation of ±25 per cent of pan evaporation.

Deficit irrigation (DI) or non-irrigated strawberry plants showed marked yield reduction compared with irrigated plants (Meland, 1987, Liu *et al.*, 2007). Weber *et al.* (2017) in their experiment with three different irrigation regimes- upper limit of field capacity (UFC), (-12KPa); lower limit of field capacity (LFC), (-33 KPa) and deficit irrigation (DI), (-70 KPA) were established on silty-loam soil and monitored with tensiometers. Deficit irrigation significantly increased the content of sugars, organic acids, their ratio and the content of most identified phenols in cv. Flamenco. Conversly, higher amount of total sugars and organic acids were detected in cv. Eva's Delight at UFC and LFC irrigation. The yield of cv. Eva's Delight decreased under DI. The result suggest superior fruit quality and taste of strawberries grown under minor deficit irrigation for cv. Flamenco.

In sandy soils, irrigation management for strawberry is difficult. GonzalezPerea *et al.* (2017) developed a multiplatform application consisting of a PC and a mobile module for Android devices (APP). It is a useful tool that facilitates the irrigation scheduling. The software takes into account agro-climatic, soil data and information about the required irrigation times. The application has been used in several farms in the Southwestern Spain, achieving water savings which range from 11 per cent to 33 per cent, and highlighting that efficient water use is in fact possible in strawberry production.

## Soil Management

### Mulching

Mulching of strawberries in fall has been recognized beneficial in minimizing winter freezing injury to plants and suppressing early spring growth, thereby reducing fruit bud frost injury (Shoemaker, 1978). Mulching also suppressed weed growth and reduced soil erosion. Mulching for winter crops is done before the plants have been exposed to temprature lower than 6.5-7°C. Clean white straw, marsh hay, pea vines, ground corncobs, sawdust, *etc.*, are used as mulching material. However, use of poor quality straw can introduce weed and/or grain seeds. The most commonly used mulching materials are cereal straws, marsh grasses or evergreen boughs. Tucker (1944) stated that the ideal mulching material must be loose and light; it will not mat down and smother the plants and yet must be heavy enough that it will not be blown away. Wheat straw is the preferred mulch material for strawberries in Kentucky, UK although oat and rye straw also work well. Thickly planted sudangrass also produce a good mulch for strawberries. However, it must be run through a hay conditioner at cutting as it dries very slowly (Masabni *et al.*, 2007). Brierley and Landon (1944) suggested that mulching at the proper time was more important than the ability of the plants to withstand cold and the plants need to be hardened by several light frosts previous to mulching. Time of mulching is critical as early mulching being more deleterious than late application. Brierley and Landon (1941) stated that several light frosts are needed to check growth and respiration and develop cold resistance, and that early mulched plants bleach and fail to harden. At least 10 light frosts previous to mulching are necessary to avoid plant injury and 15 to 20 frosts are optimal (Collins, 1966). Growers using the annual hill system of strawberry culture apply a polyethylene mulch over the beds to maintain a more favourable soil environment (Locascio and Smart, 1968). Mulching with black polythene gave good weed control, reduced labour cost by 25-30 per cent, advanced early cropping of cvs. Festival'naya and Senga Sengana by 4-8 days, increased total yield by 25-34 per cent, improved fruit quality and reduced disease incidence (Nikito-Chkina and Gusev, 1984). Haynes (1987) investigated the effects of white, black and clear polyethylene mulches on temporal fluctuations in protease, sulphatase and phosphatase acitvities and levels of available and biomass N, S and P in raised beds of soil in a strawbberry crop in New Zealand. During spring, summer and early autumn, clear and, to a lesser extent, black polyethylene mulch was found to increase both maximum and minimum mean monthly temperatures over those recorded with white mulch or no mulch. During summer and autumn, soil moisture content was higher under mulched than unmulched conditions. Levels of extractable nitrate and sulphate in soils during autumn followed the order: clear mulch = black mulch > white mulch > no mulch. However, there was no significant difference in the enzyme activity among the treatments. Strawberry fruit yield and vegetative growth followed the order: clear mulch: black mulch > white mulch> no mulch. Laugale *et al.* (2012) noted increased yield and fruit size by using white plastic mulch with black on the lower side. This type of mulch also reduced *Botrytis* incidence compared with unmulched ones.

A disadvantage of using polyethylene mulch is that it must be removed from the field and disposed of at the end of each fruiting season since it does not decompose in the soil. Paper mulches treated with fungicides and with or without a thin polyethylene layer have mulching properties similar to that of black polyethylene but it disintegrates when incorporated into the soil (Everett, 1971). Albregts and Howard (1972) stated that bio-degradeable paper mulch with a polyethylene coating on both sides or with a single polyethylene coating satisfactorily endured 7-month growing season in Florida and gave similar fruit yield and fertilizer leaching result as compared with black polyethylene mulch (Table 13.10). The paper mulches, which are more rigid, were somewhat more difficult to set plants through but it does not have to be removed from the field at season's end. However, escalating costs of labour, machinery and mulch materials have made mulching less economical.

**Table 13.10: Effects of Paper and Polyethylene (PE) Mulches on Marketable Strawberry Yield and Slouble Salts in Solution Extracts****

| *Mulch Types and Placement* | | *Marketable Yield (g/plant)[z]* | | *Soluble Salts (ppm)[y]* |
|---|---|---|---|---|
| | | *1971* | *1972* | *1972* |
| Black paper | (PE down) | 319ab* | 379a | 460a |
| | (Double PE) | 265b | 437a | 374a |
| | (PE down) | | 396a | 605a |
| | (Double PE.) | 302ab | 408a | 535a |
| Gray paper | (PE down) | 330ab | 386a | 530a |
| | (Double PE) | 383b | | |
| Black PE | | 319ab | 413a | 579a |

[z] About 54,000 plants per ha.

[y] Expressed on the basis of soil solution at field capacity.

* Means in any column followed by different letters are significantly different at 5 per cent level.

** Albregts and Howard (1973).

Because of the high financial cost involved for disposal process of polyethelene plastic mulch, biodegradable plastic mulch films, which provide similar agronomic benefits, may be an appealing alternative. Biodegradable plastic mulch can be tilled into the soil or composted at the end of growing season, reducing the labour and environmental costs associated with plastic removal and disposal (Goldberger and Lyons, 2017).

## Weed Control

In strawberry, weed control is considered as one of the most expensive management practices. The weeds pose a great problem and sometimes become unmanageable, specially during wet weather. Different kinds of weeds have been recorded in strawberry fields. In Federal Republic of Germany, the main weed species in strawberry fields were found as *Matricaria chamomilla* [*Chamomilla recutita*] and *Capsella bursa-pastoris* (Seipp, 1987a). The weed species *Ranunculus repens* and *Trifolium repens* were recorded in the Irish Republic (Rath, 1987). *Rorippa silvestris* is considered one of the most serious perennial weeds of strawberry in the Netherlands (Naber, 1987). Prevalent weeds of strawberries in Greece include *Echinochloa crus-galli, Setaria spp., Chenopodium album, Amaranthus* and *Cyperus rotundus* (Giannapolitis, 1987).

Herbicides are used mainly in three different stages for control of weeds in strawberry plantations, namely, (i) perennial weed control before planting, (ii) annual weed control at planting or in young plants, and (iii) annual weed control in established crop.

Cianciara (1979) suggested use of venzar (lenacil) at 1.5-2.0 kg/ha or gramoxone (paraquat) at 3.0 l/ha as pre-planting herbicide. Pre-planting application of treflan (trifluralin) at 1.5 kg/ha was advocated by Marocchi and Grandi (1978). They also found that methyl bromide fumigation before planting gave good control of many weeds. The incorporation of trifluralin at 1.12 kg/ha before planting enhanced the performance of post-planting residual herbicides on spring-germinating weeds and increased persistence (Lawson and Wiseman, 1980). In several soil types, application of aciflurofen-sodium at 0.28-1.12 kg/ha proved very effective when applied before planting strawberries on preformed beds and there were no significant difference in the yield of cv. Chandler, Heidi, Pajaro and Selva. Acifluorofen-sodium remained active against *Medicago polymorpha* and *Erodium moschatum* for 150 days (Agamalian, 1987). Seipp (1987b), however, stated that pre-planting application of stomp (pendimethalin) at 5.0 l/ha, goltix (metamitron) at 5.0 kg/ha, enide (dephenamid) at 14 kg/ha and devrinol (napropamide) at 2.5 kg/ha reduced yields in cvs. Elvira, Induka, Tenira, Elsanta, Tago and Bogota compared with pre-planting application of venzar (lenacil) at 1.5 kg/ha.

Application of napropamide showed good control of broad leaved and annual grass weeds in strawberries, although results varied with season and weed population. A mixture of napropamide and simazine applied in the autumn was found most effective to control broad leaved weeds (Freeman, 1983). The herbicide stomp (pendimethalin) was safe for annual weed control in strawberries. Pendimethalin controls a wide range of weed species but some common weeds, principally groundsel (*Senecio vulgaris*), is resistant (Clay and Davison, 1981). In sandy loam soil, propyzamide at 0.4 kg/ha was found well tolerated by young strawberries but at 0.8 kg/ha some injury was recorded (Clay, 1980). Chloroxuron has long provided effective post-emergence control of small annual broad leaved weeds in strawberries, with minimal crop injury (Schubert, 1969). Smeda and Putnam (1988) found > 90 per cent broad leaf weed control with chloroxuron and > 80 per cent annual grass control with fluazifop-butyl (Table 13.11). Lake and Bennett (1980) reported that a single application of 3, 6-dichloropicolinic acid at 200 g/ha gave good control of creeping thistle (*Cirsium arvense*). Where weed emergence occurred over a longer period, a sequential application of 100 + 200 g/ha was more effective.

Application of nortron (ethofumesate) gave good control of clover (*Trifolium sp.*) in strawberries in various trials in the UK. With established plants of the cv. Cambridge Favourite, 10 l a.i./ha applied between October and December in most conditions were recommended (Clay, 1979).In sandy loam soil, application of isoxaben at 0.8 kg/ha was well tolerated by established strawbberries cv. Cambridge Favourite (Lawson and Wiseman, 1987). Dyachenko (1985) suggested application of 3-4 kg lenacil or 2 kg lenacil + 0.5 kg simazine/ha for control of weeds in established strawberry plantations. A mixture of 1 kg pendimethalin + 2.3 kg napropamide/ha gave excellent control of *Ranunculus repens* and *Trifolium repens*. Metamitron and ethofumesate were well tolerated by cv. Cambridge Favourite but lacked activity against annual weeds (Rath, 1987). Jamwal and Wali (2011) reported good control

**Table 13.11: Mean Visual Injury Rating for Various Broadleaf and Grass Weeds in 1982 and 1983 and to 'Guardian' Strawberries in 1983 from Chloroxuron and Fluazifop-butyl Applications***

| *Treatment[Y] (kg/ha$^{-1}$)* | | *Visual Rating[z]* | | | | | | |
|---|---|---|---|---|---|---|---|---|
| | | *1982* | | | | *1983* | | |
| | | *Application 1* | | *Application 2* | | | | |
| *Chloroxuron* | *Fluazifop-butyl* | *Broadleaf* | *Grasses[d]* | *Broadleaf* | *Grasses* | *Broadleaf* | *Grasses* | *Strawberry* |
| 0 | 0 | 0 c | 0b | 0 c | 0 c | 0 c | 0 b | 0 |
| | 0.07 | | | 0 c | 9.3 b | | 7.4 a | |
| | 0.14 | 0 c | 8.4 a | 0 c | 9.8 ab | 0 b | 8.1 a | 0 |
| | 0.28 | 0.1 c | 8.6 a | | | 0 b | 0 b | 0 |
| 4.48 | 0 | 9.4 a | 0 b | 10 a | 0 c | 9.9 a | | 1.7 |
| | 0.07 | | | 10 a | 9.3 b | | 7.3 a | |
| | 0.11 | 8.5 b | 8.8 a | 9.8 b | 10 a | 10 a | 8.4 a | 2 |
| | 0.28 | 8.7 b | 8.4 a | | | 9.7 a | | 1.7 |

z: A pretransformed rating scale was used with 0 = no injury and 10 = complete control.

y: Means followed by the same letter in the same column are not significantly different at the 0.05 level or probability as measured by Duncan's multiple range test.

d: Broad leaf weeds rated in 1982 included redroot pigweed (*Amaranthus retroflexus* L.) comrmon purslane (*Portulaca oleracea* L.), common lambsquarters (*Chenopodium album* L.), common mallow (*Malvaneglecta* Wallr.) and carpetweed (*Mollugo verticillata* L.). Broadleaf weeds rated in l983 included redroot pigweed, common purslane and carpetweed. Weed height at the time of treatrnent was approximately 7 cm.

* Grass weeds rated in 1982 included large carbgrass [*Digiaria sunguinalis* (L.) Scop], stinkgrass [*Eragrostis cilianensis* (All.) E. Mosher] and foxtails (*Setaria* spp.) Grass weeds rated in 1983 included large carbgrass, stinkgrass and fall panicum (*Panicum dichotomiflorum* Michx.). Weed hieght at the time of treatment varied from 2.5 to 13 cm.

* Smeda and Putnam (1988).

of weeds in strawberry plantations by application of oxyfluorfen + napropamide at 0.60 + 5.0 kg a.i./ha.

## Flowering

The strawberry 'fruit' is not actually a berry fruit instead a receptacle for potential hundreds of tiny fruits, known as achene. The strawberry inflorescence is a modified stem terminated by the primary flower. Arising laterally from this stem are usually two secondary flowers, from which four tertiary flowers may arise, followed by eight quaternary flowers. Individual flowers are typically comprised of ten sepals, five petals, 20 to 30 stamens, and 60 to 600 pistils, depending on the flower order (Hancock, 1999, Darnell *et al.*, 2003). After pollination, the receptacle enlarges and turns red, developing into fruit. The actual fruit is achenes that develop from each carpel, holding one seed each on the receptacle's surface (Himelrick *et al.*, 2002, Martin and Tepe, 2014).

### Effect of Photoperiod

Flowering in strawberry is controlled by photoperiod (Borthwick and Parker, 1952; Pringer and Scott, 1964), with a modifying influence from heat accumulation, particularly on certain cultivars (Smeets. 1979). Strawberries are classified according to their photoperiodic flowering response: short-day (SD) or June-bearing plants initiate flower buds under short days, long-day (LD) or everbearing plants initiate flowers under long-days, and day-neutral (DN) plants are relatively insensitive to daylength with respect to flower induction (Darnell *et al.*, 2003).

Increasing the light intensity for greenhouse-forcing cultivars reduced the time to flower, increased the number of inflorescence, stamen development and fruit set (Smeets, 1979, 1980a,b). The photoperiodic response of strawberry is well recognized in the past, with flower bud initiation under short-days (Arney, 1956a,b; Smeets and Korenberg, 1955), and runner and increased leaf size favoured by long-days (Smeets and Korenberg, 1955, Arney, 1956b). The so-called everbearing strawberry cultivars (*e.g.*, Ozark Beauty, Geneva, Superfection) initiate and produce flowers throughout the growing seasons. Day-neutral strawberries (*e.g.*, Tribute, Brighton, Tristar, Hecker) have gained popularity in different geographical regions. Strawberry fruit production may be limited by low temperatures during autumn, winter or spring in many temperate regions. Low temperatures during bloom and in the fall and winter can destroy flowers and buds (Boyce and Reed, 1983). Flowering in short-day (SD) cultivar may be induced in long-days (LD) so long as the temperature is sufficiently low (Jonkers, 1965, Smeets, 1968); certain cultivars, such as, Deutsch Evern, Talisman and Redgauntlet produce a second crop in the late summer in certain years. Van den Muyzenberg (1942) induced a second flowering in the cultivar Deutsch Evern by SD treatment of 6 to 16 hours. Benott (1977) was able to induce a second flowering in summer in cv. Redgauntlet by SD treatment lasting 20 to 30 days. This SD treatment was most effective when the plants were subjected to it from the beginning of the spring flowering period. Galletta and Bringhurst (1990) have suggested that flowering in modern strawberry cultivars is regulated more by sensitivity to higher temperature than photoperiod, with traditional short-day plants being more sensitive to high temperatures than day-neutrals. Infra short-day types have been developed (Izhar Izsak, 1995) that produce fruits earlier than typical SD types because they lack chilling requirement, and induced by LD (13.5-14.0 hrs) and can tolerate a wider temperature range (10-26°C).

Inflorescence number in DN cultivar increased as light intensity increased from ~200 to ~400 μmol $m^{-2}S^{-1}$ (Dennis *et al.*, 1970). Light saturation curves indicate that strawberry leaf $CO_2$ assimilation increases with increasing photosynthetic photon flux, reaching light saturation at 500 to 700 μmole $m^{-2}S^{-1}$ (Ceulemans *et al.*, 1986). Stamen abortion may occur under very low light intensities (less than from ~10 μmol $m^{-2}S^{-1}$) for several days prior to anthesis (Smeets, 1976).

Control of flowering is important in strawberry production. Lacking an environment condition to off-season fruit production, such as the South Coast region of California, strawberry growers in the mid-Atlantic coast region are using plasticulture, high tunnels to improve spring strawberry yields (Takeda, 2008). Takeda (2012) plugged runner tips of SD cultivar Festival in early July (standard time early August) which flowered in the fall, but less than 30 per cent of the August-plugged transplants flowerd in the fall. Under protected cultivation, July-plugged plants produced fruit in October, November and December. Illumination the crowns of July-plugged transplants with red light in August delayed bloom by several months. In another study, growing transplants under red photoselective nets in August delayed flowering until January. The result suggested that floral bud initiation can be induced even under long photoperiod if the light illuminating the strawberry lacks red and shorter wavelength light whereas excessive red light transmittance either with red LED lights or growing transplants under red photoselective shade net in August delayed flower bud initiation. Changing the time of plugging and altering the light quality for plug plant production can create an opportunity for double-cropping, *e.g.*, fruit production from SD cultivars in fall and early winter and again in the spring.

## Effect of Photoperiod and Temperature

Some cultivars have specific temperature requirements for flower bud initiation (Darrow, 1966). Durner *et al.* (1984) studied the effects of photoperiod and ternperature on day-neutral (Hecker and Tristar), June-bearer (Redchief and Guardian) and everbearer (Ourown and Ozark Beauty) strawberries. They were grown at a constant 21°C under 16 h long-days (LD), 9 h short-days (SD) or 9h with the dark period interrupted (NI) by 3 h of low-level incandescent radiation. Flowering in day-neutrals was unaffected by photoperiod, in June-bearers flowering was inhibited with NI and LD compared with SD, flowering of everbearers was promoted by LD compared with SD. The 3 types were also grown under SD and NI under day/night temperature regimes of 18°/14°, 22°/18°, 26°/22° or 30°/26°C. Significant photoperiod x temperature x type interactions were found on the number of inflorescence and runners and total dry weight/plant.

This same high temperature limitation to flower induction was evident in the LD and DN cultivars under short photoperiods, although the limitation was not as marked as that observed in SD plants. Verheul *et al.* (2007) studied the effects of photoperiod (12, 13, 14, 15 or 16 h), day temperature (12, 15, 18, 24 or 27 °C) and night temperature (6, 9 or 12°C) and their interactions on flower and inflorescence emergence of cvs. Korona and Elsanta. A daily photoperiod of 12 or 13 h resulted in the highest number of plants with emerged flowers. A photoperiod of 14 h or more strongly reduced this number, while no flower emerged at a photoperiod of 16 h. A day temperature of 18°C/or a night temperature of 12°C were found optimal for plants to emerge flowers and resulted in the shortest time to flower. A night temperature of 6°C strongly reduced the number of plants that emerged flowers, especially when the day temperature was also low. Rho-IlRae *et al.* (2007) stated that low temperature and short photoperiod is essential for flower induction but long-day and high temperature is also necessary for flower development. Opstad *et al.* (2011) stated that flowering time of field grown strawberry was mainly determined by thermal relations in the spring and early summer, and accordingly, it was found strongly delayed with increasing latitude and altitude in different geographical locations in Norway. Under high air temperature (D/N 30/27 °C), crown-cooling treatments of 10, 15 and 20°C significantly promoted flower bud differentiation in the first inflorescence in a June-bearing cv. Fukuoka 56. However, continuous cooling treatments of 10 and 15 °C after flower bud differentiation negatively affected anthesis and fruit yield (Hidaka *et al.*, 2017).

## Physiology of Flowering

In one hypothesis proposed by Gooselink (1959), flowering in the strawberry was assumed to be controlled by the relative rates of 2 processes, one promotive to flowering and the other inhibitive. Thompson and Guttridge (1960) reported evidence of the operation of photoperiodic control of flowering in the strawberry through a flower inhibitor system produced in the leaves. A significant reduction in auxin levels was found to have occurred in the strawberry after 21 photoinductive cycles, at which time floral induction had also occurred (Moore, 1961). Moore and Hough (1962) stated that the native auxin level was reduced suddenly and markedly at or near the time of floral initiation. The rapidity of this decrease and the fact that induction in short-day plants is qualitative rather than quantitative was taken as evidence that reduction in auxin level during photoinduction is a consequence of the change from the vegetative to the floral state, and is not in itself a causative factor in floral induction under natural condition. The threshold for photoperiodic induction in 'Sparkle' strawberry was found to lie between 12 and 15 inductive cycles.

Application of GA minimies the effects of LD on vegetative and reproductive growth in SD strawberry. Floral induction under inductive SD conditions is inhibited by exogenous application of gibberellins ($GA_3$ or $GA_{4+7}$). However, application of $GA_3$ applied to field-grown SD and DN strawberry cultivars four weeks after planting in the spring increased yield in the following year (Choma and Himelrick, 1984). Application of $GA_3$ (50 ppm) to latent buds caused 97 per cent bud break of their onset of dormancy (Ahmed and Ragab, 2005).

## Initiation and Development of Flowers

Robertson and Wood (1954) studied the flower-bud initiation and development in early and late formed runners and stated that an young runner plant consists of a short primary axis bearing leaves and leaf initials in spiral succession and terminated by a vegetative growing point. In summer or autumn, the latter develops into a first order, inflorescence, extension growth of the crown being continued by the youngest or uppermost axillary bud. The secondary crown-extension axis so formed may later terminate in a 'second order' inflorescence, and further extension growth again arising from the axillary bud immediately below. This forms a tertiary crown-extension axis, which in turn may end in a third order inflorescence and sometimes further growth may produce a fourth order inflorescence. Thus each inflorescence is terminal and growth habit sympodial. The strawberry inflorescence is of cymose type but shows much diversity of form. Robertson and Wood (1954) distinguished three component units: (i) the terminal flower, (ii) simple branches with from one to three flowers, and (iii) complex branches with more than three but rarely above seven flowers.

Samad *et al.* (2017) reported the mapping of QTLs for flowering time and vegetative traits in a large woodland strawberry (*F. vesca* L.), mapping population that was phenotyped both under field conditions and in a greenhouse after flower induction in the field. The greenhouse experiment revealed additive QTLs in three linkage groups (LG), two on both LG4 and LG7, and one on LG6 that explain about half of the flowering time variance in the population. Three of the QTLs were newly identified and one co-localized with previously characterized *FvTFL1* gene. An additional strong QTL corresponding to previously mapped *PFRlI* was detected in both field and greenhouse experiments indicating that gene(s) in this locus can control the timing of flowering in different environments in addition to the duration of flowering and axillary bud differentiation to runners and branch crowns. Several putative flowering time genes were identified in these QTLs regions. The findings indicate that a few major QTLs may control flowering time and axillary bud differentiation in strawberries. Samad *et al.* (2017) suggested that the identification of causal genes in the diploid strawberry may enable fine tuning of flowering time and vegetative growth in closely related octoploid cultivated strawberry.

## Pollination

Most strawberry cultivars produce hermaphrodite flowers, although few pistillate cultivars (*e.g.* Pegasus, Orleans) were also released. However, imperfect, pistillate or strictly female cultivars require cross-pollination. The anthers open at the sides, sometimes under tension, so that pollen is thrown onto pistils. Pollens are usually at first heavy and sticky but later become dry and are carried by wind (Darrow, 1966). The fertile pollen contents of everbearing cultivars Ostora, Geneva, Mara des Bois and Rapella growing in Poland was reported around 28.5 per cent, while day netural cultivars Selva, Tango, Evita and Irvine was 35.2 per cent (Kaczmarska, 2003). Guitttua *et al.* (2007) stated that the most suitable temperature for pollen germination and tube growth were 25-30°C. They also suggested spraying of boron (0.33 g/l) to increase pollen germination. Pipattanawong *et al.* (2009) observed that sudden rise of temperature (32/27, D/N) from normal temperature (27/22, D/N) after flower bud initiation lowered the ovule fertization rate and increase the unfertilized achene rate, resulting in a high degree of fruit malformation. A long period of high temperature increased achene abortion during early embryogenesis due to low pistil potential, and finally induced fruit malformation. Bees or blowflies are important agents in strawberry pollination. Incomplete pollination may result in developing irregular and poorly shaped berry. Strawberries caged with honey-bees showed increased fruit set, larger berries and less malformed fruit than strawberries caged without bees (Free, 1968). More (1962) concluded that insect pollination is beneficial to strawberry production and that native bee conservation should be practised in the vicinity of strawberry plantation. Sannino and Priore (1979) recorded 34 per cent increased fruit set with bees compared with wind pollination only. Blasse (1981) suggested placement of 2 beehives per hectare for complete pollination. Pion *et al.* (1980) examined the effects of various pollinating agents in muslin cages, which also reduced wind speed by 50 per cent or 20 per cent. The caged plants in which the wind speed was reduced by 50 per cent and 20 per cent resulted in 43-57 per cent and 36-52 per cent malformed fruits, respectively. However, the introduction of honey-bees resulted in almost complete pollination.

Klatt *et al.* (2014) stated that bee-pollinated fruits were heavier had less malformations and reached higher commercial grades. They had increased redness and reduced sugar-acid-ratios and were firmer, thus improving the commercial shelf life. Longer shelf life reduced fruit loss by at least 11 per cent. This was accounting for 0.32 billion USD of the 1.44 billion USD provided by bee pollination to the total value of 2.90 billion USD made with the strawberry selling to the European Union in 2009. A test of strawberry pollination by mason bees (Osmia) in Germay shows variations in the shape of fruits, fruiting and abundance of harvest in different plots depending on the quality of pollination. When pollination was optimal, value added could reach up to 97.5 per cent. In treatment without pollination, the number of deformed fruits was 174 per cent higher than that found in the presence of optimal pollination (Taschner, 2012).

## Fruit Set

Maximum fruits in strawberry develop from tertiary flowers than from primary and secondary flowers. Darnell and Martin (1988), in examining the role of assimilate translocation and carbohydrate accumulation in fruit set of strawberry concluded that fruit set and initial fruit growth are limited by the ability of receptacles to mobilize current source leaf assimilates. The early fruit growth in strawberry is not related to carbohydrate pool sizes. Thus the 'source-sink' hypothesis used to explain flowering and abscission does not appear to explain the process of fruit set in strawberry. However, increased flowering and subsequent fruit set by application of nitrogenous fertilizer have been reported.

Poor pollen performance at high temperature results in low fertilization rates, lead to a reduction in fruit set (Gilbert and Breen, 1986). Anther and pollen degeneration are caused by several factors, including low temperature, low light intensity shortly or before and during flowering and inadequate chilling of plants. Certain strawberry cultivars, cv. Nyoho however, produce heat-tolerant (30/25°C, D/N) pollen, which in turn could result in higher fruit set (Ledesma and Sugiyama, 2005).

Gardner and Marth (1937) reported that an auxin spray could induce fruit set of strawberry flowers. Later, Nitsch (1970) showed that auxins are required for the growth of receptacles from which achenes have been removed and that the achenes, but not the receptacles, are a rich source of IAA.

## Deblossoming

Deblossoming strawberry plant is a beneficial practice for maximizing yield in matted row plantings. Deblossoming June-bearing strawberry plants promoted earlier runner emergence (Jahn and Dana, 1966) and runner number (Barritt, 1974). In everbearing cultivars, deblossoming did not influence runner number but increased production of subsequent inflorescence (Dennis and Benett, 1968). Manual deblossoming is a labour-intensive practice, therefore, different chemicals are used for deblossoming. Choma and Himelrick (1982), in an experiment with greenhouse-grown strawberry cvs. Ozark Beauty (everbearer), Tristar, Hecker and Brighton (day-neutral), applied ethephon [(2-Chloroethyl) phosphonic acid] when the primary flowers opened. Treatment showed an increase in leaf number and decreased petiole length. Both 500 and 1000 ppm ethephon increased the number and percentage of withered inflorescence (≥75 per cent shriveled) and thus eliminated strawberry blossoms.

In a fruit thinning trial with everbearing cv. Rapella, spaced at 100 x 30 cm, Peerbooms (1985) compared the effects of normal flower truss thinning, in which 3 or 4 fruits/stalk were allowed to develop; half thinning, 5 or 6 fruits/stalk and no fruit thinning. Yield of class I fruits amounted to 147, 172 and 157 kg/acre respectively, the average fruit weight was 18.6, 17.2 and 16.3 g, respectively. Lamberts *et al.* (1985), in a similar experiment, thinned the flowers or fruitlets in the autumn so as to bear 1, 3 or 7 fruits/stalk. The treatments were carried out in early November and end of December. Yield and average fruit numbers/plant were found to decline with more severe thinning, but average fruit weight and percentage of class I fruit (weighing more than 15 g) increased. The highest return was obtained by thinning to retain 3 fruits/stalk.

## Parthenocarpy

Most attempts to induce parthenocarpy in strawberries have been cumbersome or only sporadically successful. Tukey (1952) treated 1700 emasculated, unpollinated flowers from 60 strawberry cultivars with various growth regulators but failed to induce parthenocarpic development. However, some investigators were able to induce parthenocarpic strawberry fruit using growth regulators (Nitsch, 1950; Thompson, 1964; Mudge *et al.*, 1981; Beech, 1983). Tafazoli and Vince-Prue (1979) failed to induce fruit set in hermaphrodite flowers of cvs. Redgauntlet and Rabuntla with NAA and concluded that the induction of parthenocarpy was more successrul in male-sterile cultivars than in hermaphrodite ones. Mudge *et al.* (1981) investigated the effect of auxins in a lanolin paste on the flowers of the hermaphrodite cv. Ozark Beauty. They removed the achenes and calyx eight days after hand pollination, and then applied the treatments. Suboptimal NAA concentrations induced parthenocarpic fruit set with limited fruit growth. The induction of parthenocarpy in the hermaphrodite strawberry cv. Redgauntlet was achieved by Beech (1983) by applying auxin and auxin analogues in a lanolin emulsion to emasculated, unpollinated flowers. He found that IAA was the least effective of the five auxins and auxin analogues (lAA, IBA, NAA, NOA and NAAm) used, with no parthenocarpic fruit set at 0.05 and 0.10 mg/l (Table 13.12). At 0.25 and 0.50 mg/l of IAA there were 33 per cent and 35 per cent sets, respectively. Treatment with NAA and NAAm gave 100 per cent parthenocarpic fruit set at 0.05 and 0.10 mg/1. However, abnormalities were observed with NAA and NAAm at 1.0 mg/l included reddening, lengthening and twisting of the pedicels and in some cases splitting of the receptacle tissue around the neck region of the fruit. The results suggest that the optimum concentration for induction of parthenocarpy in strawberry by using NAA or NAAm is 0.05 mg/l.

**Table 13.12: Percentage of Parthenocarpic Fruit Developed after Auxin Treatment***

| *Concentration (mg ml$^{-1}$)* | *IAA* | *IBA* | *NAA* | *NOA* | *NAAm* |
|---|---|---|---|---|---|
| 0.05 | 0 | 22.2 | 100.0 | 55.0 | 100.0 |
| 0.10 | 0 | 88.2 | 100.0 | 39.l | 100.0 |
| 0.25 | 33.3 | 92.9 | 95.8 | 33.3 | 64.0 |
| 0.50 | 35.3 | 93.8 | 85.7 | 83.3 | 81.8 |
| 1.00 | 100.0 | 100.0 | 53.1 | 61.5 | 70.0 |
| SE (Logit) | | | | | |
| Min. | 0.35 | 0.53 | 0.50 | 0.42 | 0.41 |
| Max. | 0.49 | 0.77 | 0.75 | 0.57 | 0.51 |

* Beech (1983).

Mezzetti *et al.* (2004) inserted the *DefHg-iaaM* gene in strawberry (*F. vesca* and *F.* x *ananassa*) to develop parthenocarpic fruit. In emasculated flowers, the presence of the *DefHg-iaaM* gene in the genome of both the species of strawberries plants caused parthenocarpic development of the achenes and some swelling of the receptacle. All emasculated flowers of *DefHg-iaaM F. vesca* plants were able to sustain achene development while 80 per cent (24 out of 30) of emasculated flowers in the control plants did not develop achenes. A similar result was found in *F. x ananassa* where 28 out of 30 *DefHg-iaaM* emasculated flowers developed achenes while only 4 out of 30 emasculated control flowers developed achenes.

## Fruit Growth and Development

The strawberry fruit consisting of the swollen receptacle on the surface of which are embedded the many external achenes or seeds, There is no fleshy mesocarp surrounding the seeds. An aggregate fruit is formed by the simultaneous ripening of a number of separate ovaries, all belonging to a single flower and adhering as unit on a common receptacle (Shoemaker, 1955).

Normal development of the strawberry receptacle is dependent on the release of hormone(s) from the fertilized ovules or achenes (Nitsch, 1950; Mudge *et al.*, 1981).

Thus, the potential weight of berry is a function of the total number of achenes on the surface and the amount of receptacle developed per achene (Strik and Proctor, 1988b). The potential total number of achenes per berry is determined during flower bud initiation and is dependent on berry position on the inflorescence (Janik and Eggert, 1968), and genotype and environment (Webb *et al.*, 1978). Fresh berry weight was found highly correlated to achene spacing and achene number (Abbott and Webb, 1970). Strik and Proctor (1988a,b) studied field-grown genotypes of strawberry in two seasons to determine the relationship of primary berry fresh weight to the total number of achenes per berry and the number of achenes/cm$^2$ of receptacle tissue. Within genotypes, grown in matted rows, berry fresh weight was found correlated with the total number of achenes per berry. In the hill system, berry weight was correlated with the number of achenes and the number of achenes/cm$^2$. The amount of receptacle tissue developed per achene appeared to be related to berry fresh weight among genotypes (Table 13.13). Growth curves and developmental index of fruit length and dry weight were studied for the cultivar Fern (Stutte and Darnell, 1987). There was a strong positive correlation between fruit length and the natural logarithm of fruit dry weight. Secondary fruits showed a 4- to 5-day lag period between pollination and growth, while primary fruits began growing within 24 hours after pollination. Primary and secondary fruits exhibited similar relative growth rates once the development began. Forney and Breen (1985) studied the growth of fruit from greenhouse-grown cv. Brighton. They found that the absolute

**Table 13.13: Mean Values for Yield/Plant, Berry Weight, Number of Achenes/cm$^2$, Total Number of Achenes/Berry, and Weight of Receptacle Tissue Developed per Achene for Strawberry Genotypes Studied in 1985 and 1986***

| *Genotype* | *Yield (g/plant)* | *N*$^z$ | *Primary Berry wt (g)* | *NA/C*$^y$ | *Total No. of Achenes* | *Tissue per Achene (mg)* |
|---|---|---|---|---|---|---|
| Matted row- 1985 | | | | | | |
| 62E55 | 136.74 a$^x$ | 66 | 11.74 a | 20.5 a | 293.5 a | 40 b |
| Redcoat | 109.38 b | 64 | 8.89 b | 20.6 a | 225.1 b | 39 b |
| 71M-59 | 74.67 c | 68 | 10.56 a | 18.8 c | 244.3 b | 43 a |
| Veeglow | 77.50 c | 63 | 8.54 b | 22.3 b | 236.0 b | 36 c |
| Hill system-1986 | | | | | | |
| 62E55 | 737.3 a | 209 | 13.9 a | 14.7 c | 346.7 b | 40 b |
| Redcoat | 662.8 a | 219 | 10.6 c | 13.3 d | 213.8 c | 45 a |
| 83T6 | 651.1 a | 192 | 12.6 b | 17.5 b | 396.1 a | 32 c |
| I 32E57 | 257.0 b | 202 | 8.9 d | 20.6 a | 337.8 b | 26 d |

z: Number of primary berries subsampled.

y: Number of achenes/cm$^2$.

x: Means within columns and years separated by Duncan's multiple range test.

*P*= 5 per cent.

* Strik and Proctor (1988b).

growth rate neared a maximum at the onset of ripening when the rates for primary, secondary, tertiary and quaternary fruits were 106, 53, 33 and 3 mg dry weight per day, respectively. Length, diameter, fresh weight and dry weight were greatest in primary fruit throughout development, and were progressively smaller in fruits at lower rankings. At 15 days after anthesis, fresh weight of secondary, tertiary and quarternary fruits were 55 per cent, 13 per cent and 3 per cent, respectively, of primary fruit weight, The amount and rate of fruit expansion, estimated from achene separation was highest in primary fruit, the maximum rate being over twice that of secondary fruit. Relative growth rate was also highest in primary fruit from anthesis to red-ripe.

The strawberry is generally accepted to be a non-climacteric fruit and, as such, would not be expected to show an increased synthesis of ethylene during ripening. Knee *et al.* (1977), in studies with cv. Prizewinner, stated that cell number per fruit increased three-fold in the first 7 days (Table 13.14), indicating that cell division continued after petal fall (PF). Between 21 and 28 days after PF receptacles became red and softened but there was no change in the rate of ethylene production. Cell expansion continued for at least 28 days. Tubular proliferation of the tonoplast and hydration of middle lamella and wall matrix material had begun 7-14 days after PF but became extreme during ripening. Associated with the hydration of the wall, over 70 per cent of the polyuronide in the wall became freely soluble, and arabinose and galactose residues lost from the wall appeared in soluble fractions. There was no increase in total polysaccharide during ripening and incorporation of $14_C$ glucose into polysaccharides ceased although protein increased and incorporation of $14_C$ proline into wall protein continued. As the berries matured, fruit weight, percentage of soluble solids, ascorbic acid and water-soluble pectins increased whereas total solids, acidity, total phenols, cellulose, protopectin, peroxidase activity and polyphenoloxidase activity decreased (Spayd and Morris, 1981). Total sugar content increased from 2 to 5 mg/g fresh weight between green and red ripeness stages and > 70 per cent of total sugar accumulates in the vacoule (Woodward, 1972, Johm and Yamaki, 1994). Titratable acidity increases between large green and white stages but subsequently decreases from colour turning stage (Woodward, 1972). The fruit contains about 60 mg/100g of ascorbic acid at maturity which accumulates at the later stage of fruit development. The outside of the fruit was reported to content higher amount of ascorbic acid than the inner tissue (Ezell *et al.*, 1947, Spayd and Morris, 1981).Reyes *et al.* (1982), in studies with cv. Benton at three stages (underripe, ripe and overripe) of maturity, found variability in sugar composition due to both degree of ripeness and method of analysis (enzymic, gas liquid chromatography and high performance liquid chromatography) which was found greater for sucrose than for glucose or fructose. Sucrose was almost completely hydrolyzed in the overripe fruit. There was little variation in citric acid levels due to ripeness or method of analysis. Malic acid, however, decreased greatly in overripe fruit. The glucose/fructose ratios for the underripe, ripe and overripe fruits, were 0.86, 0.92 and 0.60, respectively. The ratio of sugars and acids has been related to strawberry flavour (Sweeney *et al.*, 1970).

**Table 13.14: Fresh Weight of Strawberry Fruit Parts and Cell Numbers in the Receptacle during Development***

| *Age of Fruit (Days from PF)* | *Average wt (g)* | | | *Total Cells per Receptacle x $10^{-6}$* |
|---|---|---|---|---|
| | *Calyx* | *Receptacle* | *Achenes* | |
| 0 | 0.084 | 0.035 | 0.032 | 0.99 |
| 3 | 0.120 | 0.096 | 0.049 | I.81 |
| 5 | | 0.178 | 0.065 | 2.17 |
| 7 | 0.120 | 0.35 | 0.107 | 3.15 |
| 14 | 0.140 | 0.72 | 0.090 | 2.95 |
| 21 | 0.154 | 2.41 | 0.130 | 3.42 |
| 28 | 0.180 | 6.15 | 0.140 | 3.37 |
| 35 | 0.154 | 8.68 | 0.120 | 3.32 |

* Knee *et al.* (1977).

The water-soluble pectin levels were very low at 14 days after petal fall and increased to 90 per cent of total pectin levels at red-ripe stage (42 days) (Woodward, 1972). Abeles and Takeda (1990) believed that cellulose (glucanase) (EC 3.2.1.4) is involved in strawberry fruit softening. ChungHyun *et al.* (2011) examined the relationship between fruit firmness and cell wall matrix of three strawberry cultivars. Firmness loss in firm cv. Maehyang was found lower and contained higher amount of ethanol insoluble solids (EIS) at full ripe stage. However, firmness differences were not found at full ripe stage between soft cvs. Akihime and Seolhyang. The water soluble pectin of cv. Maehyang occupied about 54.1 per cent of total pectin whereas it was 63.7 per cent and 62.5 per cent in cvs. Akihime and Seolhyang, respectively. Differences in cellulose contents was not found between cultivars and fruit developmental stages.

Fruit pericarp maturation and ripening were accompained by a decline in hydrogen peroxide ($H_2O_2$) content at the early stage of fruit development, followed by rapid accumulation. An increase in membrane lipid peroxidation (thiobarbituric-acid-reactive substances) coincided with the accumulation of $H_2O_2$, which preceded a rise in ethylene production. In general, the antioxidant activity declined as the fruit matured and ripened. Antioxidant enzyme ascorbate peroxidise (L-ascorbate peroxidise) activity in the fruit pericarp tissue during maturation and ripening was not substantial and thus, may not have a role in alleviating the accumulation of $H_2O_2$ and subsequent events related to oxidative sencence in the fruit pericarp (Woods *et al.*, 2005).

## Harvesting

Strawberry harvesting is one of the most labour-intensive operation. Studies made by University of California in early 2002 show total operating costs around 25,000USD per acre, of which harvesting accourts for about 63 per cent (Rosenberg, 2003). Strawberry fruits are usually harvested when half to three-quarter of the skin develops colour. However, for distant shipments, berries are sometimes harvested while they are green or white and are still hard. A delay in picking usually increases the percentage of overripe and rotted berries. Some cultivar like Cambridge Favourite, however, remains fully coloured for several days in the plant without becoming too soft. Usually the berries should be picked every other day and daily during hot weather and peak seasons. Berries should be picked by nipping off the stalk and not holding the fruit. Aoyagi and Makino (1981) reported that the fruits of cvs. Hokowase and Donner, when harvested at fully-coloured stage and held at 3°C, had higher total sugar and titrable acidity than when fruits were harvested at half-cloured stage (the usual time of picking) and kept at room temperature (15-20°C). Miyazaki *et al.* (1980a) carried out experiments to find the best conditions for hand harvesting calyx-free fruits of cv. America with a single pull. The percentage of calyx-free fruits obtained was generally high early in the season and decreased rapidly as the season progressed. Euparen (dichlofluanid) sprays applied just before harvest combined with $GA_3$ (5 ppm) + 0.5 per cent $KH_2PO_4$ improved the harvesting of calyx-free fruit. Fruits harvested calyx-free showed lower ratio of pectic substances than harvested with calyx-intact (Miyazaki *et al.*, 1980b).

In areas where there is a shortage of pickers, mechanical harvesting is practised. However, the successful operation of a mechanical strawberry harvesting and handling system is dependant upon the optimization of all facets of the production, harvesting, handling and utilization scheme (Morris *et al.*, 1978). Certain plant characteristics are prerequisites for successful mechanical harvesting: a) brittle pedicels, b) easy decapping, c) concentrated ripening, and d) firm fruit (Khanizadeh and Fanous, 1992). Several concepts have been utilized in developing mechanical strawberry harvester (Denisen and Buchele, 1961, Hecht, 1972, Kemp, 1976) with varying degrees of success. Fruit damage by mechanical harvesting, harvester efficiency, and fruit maturity classes of several cultivars after once-over mechanical harvest have been reported (Booster. 1973; Denisen and Buchele, 1967). The cultivars which produce larger number of fruits on fewer long stems are more suitable for mechanical harvesting than those with a larger number of shorter stems carrying less fruit per stem (Fiedler, 1980a). The strawberry cvs. Cardinal and A-5344 are more adapted to machine harvesting, cleaning and sorting without loss of quality (Morries *et al.*, 1974,1978). The cvs. Fratina and Fracunda, which were bred for mechanical harvesting, proved best with 87 per cent and 90 per cent by weight of fruit, respectively (Fiedler, 1980b). The strawberry harvester performs 3 basic functions: fruit is (i) stripped from the plants, (ii) separated from leaves and other foreign materials, and (iii) conveyed to transporting containers. Since most of the mature fruits lie directly on the ground, a unique picking device lifts the fruit from the ground without disturbing the soil surface or damaging the plants or berries (Morries *et al.*, 1978).

Colorio and Glossl (1986) in Italy, using a prototype strawberry harvester, stated that the large-fruited, prolific cv. Hapil proved best suited to mechanical harvesting with 98 per cent of the ripe fruits being lifted (70 per cent of the total crop) on 4 July. The cv. Bogota with large, attractive fruits may also be suitable when harvested at the proper time; 4 July was found too early and only 25 per cent of the crop was harvested. The cultivars Tiago and H.Grande were, however, too soft for mechanical harvesting. Mechanical harvesting of cultivars with concentrated cropping periods, as in cv. Fratina, produced higher yield than hand harvesting (Rudolph *et al.*, 1986).

Kondo and Monta (1999) developed two types of strawberry harvesting robots: one for hydroponic system and other for soil systems. The robot consists of a 3 DOF manipulator, aspirating-type pneumatic end-effector and visual sensors. The pneumatic end-effector grasps the fruit with pressure and then twists the peduncle which results in detachment of fruit. A Cartesian coordinate type 4 DOF manipulator with suction cup and visual sensor to harvest table-top structure was developed by Arima *et al.* (2004). Feng *et al.* (2012) reported a new strawberry harvesting robot for elevated-trough culture using sonar camera sensor and autonomous navigation system. A 6 DOF industrial manipulator was used with a gripping and cutting tool. The successful harvesting rate was recorded 86 per cent.

## Yield

Various components are correlated with yield per plant, including number of crowns (Lacey, 1973; Webb *et al.*, 1974), number of leaves per crown (Nielson and Eaton, 1983), number of leaves per plant (Lacey, 1973), plant size (Olsen *et al.*, 1985), number of inflorescence (Guttridge and Anderson, 1981; Olsen *et al.*, 1985), number of berries per inflorescence (Webb *et al.*, 1974; Guttridge and Anderson, 1973), number of berries per plant (Lacey, 1973), fruit set and total number of achenes per berry (Nielson and Eaton, 1983). Strik and Proctor (1988a) stated that yield per plant within genotypes was mainly dependent on the number of berries per plant, regardless of cultural system. Other variables were correlated with yield, including crown dry weight and leaf area after harvest and number of inflorescence which indirectly affects berry number.

The average yield of strawberry in California is about 30 tonnes/ha, while in Italy it is about 12 tonnes/ha and in Poland which produces mostly processing strawberries, the average yield is only about 6 to 7 tonnes/ha. In India, the cv. Tioga produced an average yield of 5.7 tonnes/ha, Blakemore produced 5.2 tonnes/ha and Shasta showed an average yield of 5.1 tonnes/ha (Sharma and Badiyala, 1980). Skulcova (1980) recorded a yield of 11.82 tonnes/ha in cv. Senga Litessa and 11.77 tonnes/ha in cv. Munchen Kindl. The best cumulative 3-year-yield in New Zealand was found in cv. Brighton (day-neutral) with 79.68 tonnes/ha and Aiko (short-day) with 69.55 tonnes/ha (Stoker and McCarthy, 1986). The average yield of marketable fruit from cultivars Redgauntlet, Korona, Tenira, Senga Sengana and Havelland in German Democratic Republic were 419, 415, 332, 273 and 244 dt/ha, respectively. The proportions of Class A fruit of these cultivars were 73-81, 78-85, 80-84, 54-12 and 65-79 per cent, respectively (Beese, 1986). Ugolik *et al.* (1986) recorded a yield of 30.4 tonnes/ha in Senga Sengana and 25.1 tonnes/ha in Dukat at Poland.

Kargi *et al.* (2012) evaluated five strawberry lines (3,5,6,8 and12), Turkish local cultivars (Osmanli, Ebru, Kaska and Sevgi), American cultivars (Camarosa, Sweet Charlie and Gaviota) and European cultivar (Paros) in terms of yield, fruit weight and fruit quality (Table 13.15). The highest yield was obtained from cv. Paros (365.50 g/plant) followed by Camarosa (299.38 g/plant).

**Table 13.15: Yield, Fruit Weight and Total Soluble Solids Content of different Strawberry Genotypes***

| *Genotypes* | *Average of 4 Years Yield (g/plant)* | *Average Fruit Weight (g)* | *Average TSS (per cent) Contents* |
|---|---|---|---|
| 3 | 213.96 | 8.14 | 7.93 |
| 5 | 215.55 | 8.30 | 8.35 |
| 6 | 281.46 | 9.91 | 7.29 |
| 8 | 281.08 | 11.43 | 7.44 |
| Ebru | 217.17 | 9.26 | 7.91 |
| 12 | 207.21 | 8.24 | 8.68 |
| Kaska | 143.80 | 8.83 | 8.08 |
| Sevgi | 197.99 | 11.18 | 7.21 |
| Osmanli | 83.47 | 3.56 | 9.71 |
| Camarose | 299.38 | 12.57 | 7.52 |
| S. Charlie | 271.33 | 10.64 | 7.62 |
| Gaviota | 177.21 | 10.85 | 7.16 |
| Paros | 365.50 | 11.13 | 7.02 |

*Compiled from Kargi *et al.* (2012).

## Storage

Strawberry are extremely perishable. The short postharvest life is mainly due to susceptibility towards injury during harvesting and postharvest handling, physiological deterioration, water loss and microbial decay. Loss of the quality begins with loss of membrane integrity that generally becomes the leading event in the sensescene (Paliyath and Drollard, 1992).

The successful transport and short-term storage of strawberry fruits is normally dependent on the extent to which decay organisms can be controlled (Woodward and Topping, 1972). Various postharvest treatments based on application of fungicides (Watada, 1971) and heat therapy (Couey and Fallstad, 1966) has been investigated. These methods, however, did not prove effecctive either because of their phytotoxic effects or due to the practical problems involved in their application. Controlled atmosphere storage, on the other hand, has none of these disadvantages, and some control of postharvest decay is possible by raising the $CO_2$ and lowering the $O_2$ concentrations in the storage chamber (Smith, 1957; Hansen, 1967). Woodward and Topping (1972) in their experiment with cv. Cambridge Favourite reported that the output of $CO_2$ from fruits held at 4.5°C in air, or at 1.2 or 5 per cent, $O_2$ fell to minimum after 5 days. Thereafter, the rate increased, more rapidly in air (in which rotting was more prevalent) than in 1 per cent or 2 per cent $O_2$. Fruits stored at 3°C in air or at 5, 10, 15 and 20 per cent $CO_2$ remained in good condition for 10 days, and all concentrations of $CO_2$ reduced rotting due to *Botrytis*. The alcohol content of the fruit increased with the length of storage and with higher concentrations of

$CO_2$. However, 20 per cent, $CO_2$ caused severe injury after 30 days. It was suggested that long-term storage in $O_2$ concentration of 1 per cent or lower may lead to off-flavour, and that the use of higher $CO_2$ concentrations may be restricted to storage for up to 7 days where adequate refrigeration is not available.

Strawberry cv. Oso Grande stored at 10°C under CA with 60 and 90 per cent $O_2$ showed higher appearance rating, preserve fruit quality and reduced decay (Cunha-Jumior *et al.*, 2011). Anthocyanin concentrations increased in both external and internal tissues of 'Selva' strawberry fruits stored for 10 days at 5°C+10 or 20 kPa $CO_2$. The increase was greatest in fruits stored in air at 5°C without $CO_2$ enrichment. Flesh red colour was less intense after $CO_2$ storage than after air storage. (Holcroft and Kader, 1999). Alamar *et al.* (2017) exposed strawberry cv. Sonata to CA of 15 kPa $CO_2$ + 5 kPa $O_2$ at 5°C at early and middle stages during postharvest storage. Cold stored fruit treated with 2.5 d with CA midway through the storage period were firmer and maintained a more vibrant colour despite bursts of increased respiration rate. Furthermore, just 2.5 d of CA was found sufficient to extend the self-life of strawberry (based on disease incidence) by a further 3 d, as compared to control.

Boiecka (1977) stated that generally the fruit of cv. Senga Sengana could not be stored longer than 10 days. The best result with regard to length of storage and fruit quality was obtiined by spraying the plant with bavistin 50 WP at 0.1 per cent during flowering and storage in 20-30 per cent $CO_2$ and 0-3 per cent $O_2$. Lange *et al.* (1978) stored fruits of cv. Senga Sengana up to 12 days at 6°C in normal air at 20 per cent $CO_2$ in air, in 1 per cent $O_2$ + 99 per cent $N_2$ or under low atmospheric pressure (0.1 atm or 0.05 atm). The most efficient treatment for delaying rotting was 20 per cent $CO_2$ in air and it also extended the shelf life of the fruits. Storage under reduced pressure showed no advantage over normal atmosphere cold storage. The time limit for storage of Senga Sengana at 6°C was found to be 8 days with the best treatment (20 per cent $CO_2$). This treatment also had a beneficial effect on fruit texture for up to 8 days. Reduced atmosphere storage, however, was not beneficial to fruit quality. At 6°C, regardless of the treatment, fruit quality could not be preserved for more than 7 days (Plocharski *et al.*, 1978). The fruits could, however, be stored at 6°C with 20 per cent $CO_2$ for 7-8 days without loss of flavour (Plocharski *et al.*, 1979). Hansen and Bohling (1981) reported that 20-30 per cent $CO_2$ inhibited respiration and arrested decay for several days but 50 per cent $CO_2$ resulted in development of off-flavour.

Brackmann *et al.* (2011) stated that duration of low temperature storage of strawberry varies with cultivars and clones. The cv. Guenoa showed low acidity, high TSS/TA ratio and low incidence of decay during storage at –0.5°C for 10 days and 2 days at 20°C. The cv. LBF236.3 γ Vapita and clone LBK 28.1, however showed low TSS/TA, reduction in firmness and increased suspectibility of decay, showing low potential for low temperature storage. Octavo and WeeSim (2017) studied the changes of colour, composition of folate, ascorbic acid and anthocyanin of strawberry cv. Camarosa during refrigerate storage at 40°C for 6 days. They concluded that strawberries should be consumed within a day or two after harvest since the reduction of these three bioactive compounds occurred even after a day of refrigerate storage.

Fruits of Cambridge Favourite, harvested early in the season, could be stored up to 6 days in an ice-bank (about 0.5°C, 97-98 per cent RH) without any reduction in subsequent shelf life at 15°C. Such fruits, when stored for 9 days, spoiled more rapidly on transfer to 15°C than those stored for shorter period (Browne *et al.*, 1982). Browne *et al.* (1984) further reported that fruits picked early in the season were less susceptible to fungal spoilage during storage at 2°C or shelf life at 15°C than fruits harvested late in the season. The commercially treated pallet-packs and 3 per cent, 5 per cent or 7 per cent $CO_2$ (in air) did not consistently retard spoilage of fruit by *Botrytis cinerea* or *Mucor piriformis* during 6 days storage at 2°C or during subsequent shelf life in air at 15°C. Storage at 10 per cent $CO_2$ caused persistent off-flavours. Although $CO_2$-enriched atmosphere did not reduce wastage or extend the postharvest life, the use of polythene covers or pre-cooled strawberries is recommended to keep the fruit cool and reduce moisture loss during distribution. Plocharski (1989) recorded weight loss between 1.6 per cent and 3.6 per cent in fresh fruit of Senga Sengana stored for 6 days and it depended upon the storage atmosphere. When the whole cold storage period was taken into consideration, the losses amounted to about 0.6 per cent per day for the control (normal atmosphere) and no less than 0.26 per cent per day for any of the controlled atmosphere (Table 13.16). The losses due to rotting after 6 days in storage were from 5 per cent in the atmosphere containing 20 per cent $CO_2$, 1 per cent $O_2$ and 79 per cent $N_2$ to about 15 per cent for the normal atmosphere.

**Table 13.16: Weight Losses and Fruit Health of Senga Sengana Strawberries Stored at 6°C in Relation to Controlled Atmosphere and the Time of Fruit Storage in the Cold Room***

| *Atmosphere* | *Weight Losses (per cent)* | | *Per cent of Sound Fruit* | |
|---|---|---|---|---|
| | *6 Days* | *12 Days* | *6 Days* | *12 Days* |
| Normal atmosphere (n) | 3.64 abc | 7.14 e | 85.7 cd | 27.4 a |
| 20 per cent $CO_2$ in the air | 2.30 ab | 4.17 bc | 92.9 de | 81.7 c |
| 20 per cent $CO_2$/24 h->1 per cent $O_2$+99 per cent $N_2$ | 1.64 a | 3.17 abc | 93.2 de | 45.2 b |
| 20 per cent $CO_2$/12 h->1 per cent $O_2$+99 per cent $N_2$ | 2.26 ab | 3.72 abc | 88.9 cde | 43.2 b |
| 20 per cent $CO_2$/24 h-> | 3.49 abc | 7.11 e | 87.8 cd | 31.7 a |
| 20 per cent $CO_2$+1 per cent $O_2$ | 3.47 abc | 6.30 de | 95.2 e | 81.5 c |
| 1 per cent $O_2$+99 per cent $N_2$ | 2.57 ab | 4.98 cd | 89.8 cd | 49.8 b |

* Plocharski (1989).

Wills *et al.* (2000) in their experiment with strawberry cv. Pajaro, fruits were fumigated with nitric oxide immediately after harvest. Fumigation was performed in an anaerobic nitrogen atmosphere from 1.0 to 4000 µl/l and the fruits were then held at 2° or 5°C in air containing 0.1 µl ethylene/l, a concentration prevalent in ambient air at fruit and vegetable markets. Treatment at both temperatures extend the postharvest life of the fruits, the most pronounced effect being obtained with nitric oxide in the concentration range 5-10 µl/l which resulted in >50 per cent extension in self life at 5 or 20°C.

The ability of polysaccharide-based (starch, carrageenan and chitosan) coatings was studied on strawberry (*F. ananassa*) by Ribeiro *et al.* (2007). The effect was studied for 6 days. Loss of fruit firmness was minimum with carrageenan and calcium chloride. The addition of 1 per cent di-hydrated calcium chloride to the coatings reduced microbial growth rate on the fruit. Fruits treated with *Aloevera gel* (1:3 ratio) extend storability up to 16 days when stored at 5°C and 95 per cent relative humidity. The treatment also maintained colour, firmness, and higher fruit quality (TSS, TA, ascorbic acid content) compared to control (Singh *et al.*, 2011). Lutte *et al.* (2013) reported that a combination of 1 per cent $CaCl_2$ + 0.15 per cent phytic acid + 2 per cent citric acid +1 per cent potassium sorbate extend the storage life of strawberry cv. Ail Star at room temperature (20-25°C). The mentioned treatment when combined with chitosan showed restraining the rot of fruits. Garlic oil, cubeba oil, ginger oil composited with carboxymethyl cellulose coating significantly decrease decay index and weight loss, inhibits rate of respiration during the early stage of storage, inhibit reduction of soluble solids and anthocyanins contents and increase of malondialdehyde content in storage (Mingli *et al.*, 2017). Samadi *et al.* (2017) investigated the effect of ascorbic acid, salicylic acid, UV-C radiation and VEG'LYS a natural antimicrobial compound, on strawberry self-life. The results revealed that even 18-days after storage, nearly 80 per cent of fruits treated with VEG'LYS were still not infected by fungi and marketable. Besides showing lower infection levels and better fruit appearance, VEG'LYS treated fruits showed acceptable taste. Combination of candelilla wax edible film and the inoculation with *Bacillus subtilis* HFC103 strain significantly reduced decay percentage (most effective to control *R. stolonifer*) and weight loss (Oregel-Zamudio *et al.*, 2017).

Alginate is a biopolymer with commercial interest for the manufacture of edible coatings due to their colloidal properties. It is a hydrophilic colloidal carbohydrate extracted from several species of brown algae (*Phaeophyceae*). Alginate is useful in forming films due to its gelling property in the presence of polyvalent cations. Postharvest treatment of sodium alginate (1 per cent) in combination with calcium chloride solution (0.4 per cent), glycerol (0.5 per cent), polyethylene glycol 400 (5 per cent), potassium sorbate (0.25 per cent) and chlorine (5 ppm) increased postharvest shelf life of strawberry (Jorge *et al.*, 2013).

Postharvest dipping of fruits in calcium chloride solution (1 per cent) increased the shelf life up to 8 days at 10°C and 80±5 per cent RH (Rahaman *et al.*, 2013). Fruits treated with 1 per cent $CaCl_2$ dip for 1 minutes maintained the fruit firmness with deep red colour on day 8 and 12 at 10°C storage. The fruits also showed no significant fungus infection during 12-days storage (Wan-Mahfuzah *et al.*, 2013).

Gray mold caused by *Botrytis cinerea* led to severe postharvest losses of strawberry industry. In recent years, some studies have shown that postharvest diseases of strawberry can be controlled by using bacterial, fungal and yeast strains. The yeast strain *Hanseniaspora uvarum* significantly controlled postharvest gray mold of strawberry and prolonged the storage time and shelf life (XiaoJie *et al.*, 2017).

## Ripening

Strawberry is considered to be non-climateric. However, it show significant changes in size and quality during ripening (Spayd and Morris, 1981, Abeles and Takeda, 1990). Dennis (1984) classified four fruit ripening stages of strawberry: green, white, pink (or turning) and red. Fruit reach the white stage at about 21 days and are fully red (ripe) with in 30 to 40 days.

Strawberries grown for the fresh market are frequently transported to considerable distances. The time interval involved influences the stage at which this perishable crop is to be harvested, conditions of transport and shelf life in the market. A better understanding of colour development and changes in soluble solids after harvest should lead to improved practices. Smith and Heinze (1958) studied subsequent colour development, texture, flavour and the composition of strawberries picked at various stages of maturity. They reported that at 70°F (10°C) fruit classified as three-quarters-coloured was usually fully coloured in 1or 2 days, half-coloured fruit in 2 or 3 days and quarter-coloured fruit in 3 or 4 days. Austin *et al.* (1960) reported that harvested strawberry fruits will develop full colour at high temperature (75° and 85° F) in complete darkness. When greenish white, up to 10 per cent pink berries were placed at a temperature of 85°F (29.4°C), they developed full colour in 48 hours. At 65°F (18.3°C) about 90 per cent of full colour developed after 96 hours. A low light intensity (25 foot-candles) slightly increased the speed and degree of colouration at high temperature. During ripening, the total cell wall polysaccharides (CWP) of cortical tissues decreased more (1.7 to 1.0 g/100g frozen tissue) than of the pith tissue (1.5 to 1.2 g/100g frozen tissue). Concomitantly, cortical CWP showed greater changes in the extent of solubilisation compared to that of pith. This appeared to be due to high branched pectic substances and the disappearance of highly branched pectic material. For pith, changes in CWP appeared to be associated with a marked decrease in non-esterified pectins. Monosaccharide composition of both tissue types indicatd that losses in highly branched pectic substances were mainly due to a decrease in arabinose residues. The cell walls of both tissue types showed only a small decrease in hemicelluloses during ripening. Cellulose appeared to be minor cell wall component of strawberry fruit and its amount remained relatively constant throughout ripening (TaiHeng and Melton, 2002). In strawberry, quercetin and kaempferol were isolated and identified as the pigment (Williams and Wender, 1952). Given *et al.* (1988) stated that anthocyanin accumulation is one measure of ripening in strawberry. Neither aminoethoxyvinylglycine, an inhibitor of 1-aminocyclopropane carboxylic acid synthase, nor inhibitors of ethylene action (silver, norbornadiene) affected anthocyanin accumulation in ripening fruit. When the achenes were removed from one half of an unripe fruit there was an accelerated accumulation of anthocyanin and induction of phenylalanine ammonia lyase on the de-achened portion of the ripening fruit. The effects of achene removal could be prevented by the application of NAA or 2, 4-D to the de-achened surface. The introduction of NAA into intact unripe strawberry fruits through the peduncles delayed their subsequent ripening, as measured by the accumulation of anthocyanin, loss of chlorophyll and decrease in firmness. These findings suggest that the decline in the concentration of auxin in the achenes as strawberry fruits mature modulates the rate of fruit ripening. Kano and Asahira (1979) studied the effect of endogenous cytokinin in strawberry fruits on maturity. They found that most of the cytokinin activity in fruit was contained in the achenes. The fruits grown at 20°C developed rapidly from 15 days after anthesis and matured in 27 days after anthesis. Cytokinin activity was highest at 10-15 days after anthesis when the achenes were still immature, and decreased gradually as the fruits developed. Ripening of the fruits and achenes grown at 20°C was slower than that of fruits grown at 30°C. At fifteen days after anthesis, cytokinin activity was much higher in fruits grown at 20°C than in those at 30°C. It was concluded that cytokinin is produced in the immature achenes, and that at low temperature ripening was delayed because high cytokinin activity was maintained in the achenes longer owing to their slower development. Inaba and Nakamura (1978) stated that during ripening changes in surface colour, respiration rate, free abscisic acid and ascorbic acid contents were similar in detached fruits and those ripened on plants. Detached fruits lost sugar during ripening and had a higher organic acid content and they looked less acceptable commercially.

Ben-Arie and Faust (1980) stated that endogenous ATP levels declined markedly as the fruit matured from the pink to the dark red stage of ripening. The total phenol content of fruits declined progressively during ripening and could provide a chemical index of ripening (Budini *et al.*, 1980). Sato *et al.* (1986) in Japan studied the varietal difference in fruit quality during ripening of strawberry. Of the 4 cultivars studied, Toyonoka, Harunoka and Reiko required effective cumulative air temperature of 540°C for maturity, Hokowase required 520°C. Fruit firmness of Reiko and Harunoka was high at immature green stage and decreased rapidly thereafter. In the other cultivars, firmness was low at this stage but decreased only slowly. Fruit colour developed more slowly in Toyonoka than in other cultivars. Soluble solids percentage increased rapidly as fruits matured in Reiko, while the percentage was already high at the 420°C stage in Harunoka and Toyonoka, increasing slowly thereafter. Sugar acid ratio was high by the 480°C stage in Toyonoka and Hokowase, and increased rapidly between 420°and 540°C stages in the other cultivars. The decrease in fruit firmness was due to a reduction in HCl soluble pectin in Reiko and to a reduction in hexametaphosphate soluble pectin or crude holocellulose in the other cultivars.

The characteristic strawberry aroma is achieved only when berries are fully ripe. In strawberry cv. Chandlar, Perez *et al.* (1992) found 3-hexenylacetate and hexylacetate comprise 25 per cent of volatiles at green stage which however, decrease at ripe stage. In red berries, ethyl esters are the major volatile components and comprise 60 per cent of the total volatile. Knee *et al.* (1977) stated that under CA storage (high $CO_2$ and low $O_2$), fruit showed a decrease in alcohol acetyl transferase activity and acetate ester production, which may affect fruit flavour.

Strawberry differs significantly from tomato in that the fruit is derived from receptacle rather than ovary tissue and strawberry is non-climacteric. A full length cDNA encoding a ripening-regulated expansion *FaExp2*, was isolated from strawberry fruit. The deduced amino acid sequence of *FaExp2* is most closely related to an expansion expressed in early tomato development and to expansions expressed in apricot fruit rather than the previously identified tomato ripening-regulated expansion, *LeExp1*. Nearly all previously inentified tomato ripening-regulated genes in strawberry were found negatively regulated by auxin. Surprisingly, *FaExp2* expression was largely unaffected by auxin. It was suggested that expansions are a common component of ripening and that non-climacteric signals other than auxin may coordinate the onset of ripening in strawberry (Civello *et al.*, 1999). Using cDNA microarrays, a comprehensive investigation of gene expression was carried out in strawberry fruit to understand the flow of events associated with fruit development and ripening by Aharoni *et al.* (2002). Microarray analysis during and in receptacle and seed (achene) tissues established an interesting parallelism in gene expression between the transdifferentiation of tracheary elements in *Zinnia elegans* and strawberry. One of the genes, CAD, common in both systems and encoding the lignin related protein cinnamyl-alcohol dehydrogenase, was immunolocalized to immature xylem cells of the vascular bundles in the strawberry receptacle. To exanime the importance of oxidative stress in ripening, gene expression was compared between fruit on-vine with free radical generator and untreated fruit. Of 46 genes induced, 20 were also ripening regulated. This suggest that active gene expression is induced to cope with oxidative stress conditions during ripening or that the strawberry ripening transcriptional programmes is an oxidative-stress-induced process. Novel auxin-dependent genes and processes were also identified in addition to transcriptional programmes acting independent of auxin mainly related to cell wall metabolism and stress response.

Nardi *et al.* (2014) suggested that two cDNA (*FaXTH1* and *FaXTH2*) might play different roles on hemicelluloses metabolism and fruit softening in strawberry. *FaXTH1* and *FaXTH2* gene expression was significantly up-regulated when fruit treated with gibberellic acid and ascorbic acid, and when the endogenous source of auxins was removed. On the other hand, the expression of both genes decreased significantly in fruit treated with ethylene while the opposite situation was observed in fruit treated with 1-methylcyclopropene (1-MCP), an inhibitor of ethylene perception.

Medina-Puche *et al.* (2014) characterized the role of *R2R3-MYB10* transcription factor (TF) in strawberry fruit ripening. The expression of this TF takes place mainly in the fruit receptacle and is repressed by auxins and activated by abscisic acid (ABA) in parallel to the ripening process. Anthocyanin was not produced when *FaMYB10* expression was transiently silenced in fruit receptacles. An increase in *FaMYB10* expression was observed in water-stressed fruits, which was accompanied by an increase in both ABA and anthocyanin content. High-throughput transcriptomic analyses performed in fruits with downregulated *FaMYB10* expression indicated that this TF regulates the expression of most of the Early-regulated Biosynthesis Genes (EBGs) and the Late-regulated Biosynthesis Genes (LBGs) involved in anthocyanin production in ripened fruit receptacles. Besides, the expression of *FaMYB10* was not regulated by *FaMYB1* and vice-versa. The observation indicated that strawberry *MYB10TF* plays a general regulatory role in the flavonoid/phenylpropanoid pathway during the ripening of strawberry. Cheng *et al.* (2017) stated that strawberry sucrose synthase gene *FaSS1* plays an important role in the regulation of strawberry fruit ripening, and its expression could be inhibited by abscisic acid and sucrose.

## Pests

### Two-Spotted Spider Mite (*Tetranychus urticae*)

These mites are brick-red in colour, present on underside of the leaves and suck sap from the leaves causing patchy mottled areas. The growth of the plant is reduced and becomes stunted. Organophosphorus pesticides, chlorpyrifos and triazophos were found effective to control red spider mite. The carbamates, aldicarb, thiofanox, formetanate and methiocarb were also effective. The specific acaricides, amitraz, dienochlor, dinobuton, fenbutation oxide, propargite and quinomethionate were usually effective (Goulct and Jessop, 1981). Vacante (1979) stated that fenbutation oxide at 35 g a.i./hl, reduced the percentage of infested leaves from 100 to 12.1 in 10 days. The organophosphate organotin and pyrethroid pesticides were also useful alternative acaricides to the organochlorines, but the organotin cyhexatin was found to increase the yield (McKinlay and Thompson, 1987). Introduction of predator (*Phytoseiulus persimilis*) at the rate of 0.5-2/plant was also found effective to reduce the mite infestation (Scopes and Port, 1981; Port and Scopes, 1981).

Releases of predatory mites can reduce two-spotted mite population especially early in the season when mitecides are not being used. Natural ennimies, *i.e.* the phytoseiid mites *Phytoseiulus persimilis*, *Neoseiulus californicus* and *Amblyseius swiriskii* and the anthocorid *Orius laevigatus* decreased the incidence of the red spider mite, keeping it below the threshold level (Albendin *et.al.*, 2012) However, biological control alone rarely provides complete control.

Hot water treatment (150s at 50°C, 30s at 55°C or 20s at 57°C) can be used to control spider mites on strawberry, without much damage to the plants, but predatory mite, *Neoseiulus californicus* may need to be replaced due to their sensitivity to hot water (Gotoh *et al.*, 2013).

### Blossom Weevil (*Anthonomus rubi*)

The strawberry blossom weevil in Poland and in USSR destroyed 20-50 per cent of strawberry buds (Leska, 1965; Sawzdarg, 1960) and in Finland 50-90 per cent (Vappula, 1965). The yield was found to decrease by 55.6 per cent (Leska, 1965). The weevil lays eggs in the unopened flower buds and then partially severs the individual flower stalks, and the flowers are withered. The metox pl. 30 was recommended as main insecticide to control strawberry blossom weevil in Poland (Labanowska and Rechnio, 1977-78). The synthetic pyrethroids decise 25 EC (deltamethrin) at 0.05 per cent and sumicidin (fenvalerate) at 0.025 per cent gave good control when treated just before flowering and were comparable with the standard insecticide metox at 0.6-1.0 per cent (Labanowska *et al.*, 1981). Ourednickova (2011) obtained good control with insecticide Calyso 480 Sc (thiacloprid) and Spintor (spinosad), where the efficiency was around 90 per cent. The efficiency of NeemAzal-T/S (azadirachtin) was around 50 per cent.

### Vine Weevil (*Otiorhynchus sulcatus*)

The white grubs feed on the root tips in winter and spring, making deep tunnels in the base of the crown and finally the plants completely collapse. Evenhuis and

Alofs (1982) suggested application of carbofuran at 1 kg/acre, which should be worked into the soil to a depth of 2-3 cm. Pommier (1986) advocated application of curater (carbofuran) at 6 to 12 kg/ha around the plants during planting and in case of heavy infestation another application after 60 days. Tamaron (methamidophos) at 0.1 per cent, nexit stark at 0.04 per cent and E605 forte (parathion) at 0.1 per cent applied twice in April or May at 10-day interval at 500 ml/m row, reduced the pest population by 75-77 per cent, 62-65 per cent and 54-60 per cent, respectively, compared with untreated controls (Scherer, 1987b). The nematode, *Heterorhabditis heliothidis* applied to field strawberries at 0.05 or 1.0 million/$m^2$ was found very effective to reduce the pest population. However, a moist soil and temperatures above 12°C were required for good control (Scherer, 1987a).

## Bud Weevils (*Anthonomus signatus*)

Female strawberry bud weevils oviposit in developing flower buds of strawberries. After laying a single egg, weevils girdle or 'clip' the buds at the pedicel, killing the bud and preventing fruit development. This injury is of concern to commercial strawberry growers, who typically assume the loss of one clipped bud is the loss of one average sized fruit (McPhie and Burrack, 2017).

## Leaf Rollers

*Ctenopsenstis obliquana, Epiphyas postvitlana* and *Planotortrix exessana* are the common leaf rollers affecting strawberries. The caterpillars fold the leaves across the midrib and feed within this shelter. One or two early-season (August-September) application of endosulfan at 2.8 l followed by fortnightly application of dichlorvos at 2.4 l and cyhexatin at 400 g/2000 l per hectare from November to February were recommended for control of pests (Baker and Dick, 1981).

## Thrips

Late-season strawberries suffer from fruit distortion leading to downgrading of the fruit and financial loss to growers. Frequent sprays of pyrethroid insecticides (deltamethrin at 2.25 g a.i./ha and cypermethrin at 16.8 g a.i./ha) significantly reduced the percentage of distorted fruits. The distortion was attributed to the feeding of various species of thrips, including *Thrips atratus, T. major* and *Frankliniella intonsa* (Buxton and Easterbrook, 1988). Acetaldehyde (Aa), a volatile compound produced naturally in small quantities by fruits and other plant organs, was found to be toxic to *Frankliniella occidentalis* (western flower thrips) on strawberries (Aharoni *et al.*, 1979). Low levels of $O_2$ (> 1 per cent) or high levels of $CO_2$ (between 50 per cent and 90 per cent) killed over 67 per cent of nymphs and adults of *F. occidentalis* on strawberries held in such atmosphere for 24 to 48 h at 2.5°C (Aharoni *et al.*, 1981). Fumigation with ethyl formate at 0.5 per cent at 30 mm Hg for 1h killed all thrips (Aharoni *et al.*, 1980).

Sampson and Kirk (2013) reported that in semi-protected strawberry crops mass trapping reduced adult thrips number per flower by 61 per cent and fruit bronzing by 55 per cent using blue sticky roller traps. The addition of *F. occidentalis* aggregation pheromone, neryl (s)-2-methylbutanoate, to the traps double the trap catch, reduced adult thrips number by 73 per cent and fruit bronzing by 68 per cent. Rahman *et al.* (2012) suggested application of spinosad followed by release of multi-species predatory mites: *Typhlodromips montdorensis, Neoseiulus cucumeris* and *Hypoaspis miles* which reduced thrips population more than single-species release. Higashida *et al.* (2017) stated that predatory thrips *Haplothrips brevitubus,* is an effective natural enemy for the control of thrips infesting strawberries in the greenhouse. Substantial individual of this predatory thrips on the plants remain even after spraying of spinosad, spinetoram or emamectin benzoate. Vervoort *et al.* (2017) indentified two predatory mites *Amblydromalus limonicus* and *Amblyseius swirskii* which shows ability to combat thrips and multiple cultivation systems. The predatory mite was found scattered over the crop and established itself on the leaves whenever food was available. When thrips levels rise, the predatory mites increased in population and were able to keep thrips at non-damaging levels.

## Nematodes

### Leaf-Nematods (*Aphelenchoides fragariae* and *A. ritzemabosi*)

These nematodes usually feed on the young developing leaves. The leaves become distored, puckered and show greyish or silvery feeding areas. The growth of the plants was severly reduced. Bohmer (1981) reported that the weight of the crown of strawberry cv. Senga Sengana was reduced to 51 per cent by *A. ritzemabosi* and 41 per cent by *A. fragariae*. Fruit yield in the first year was reduced, owing largely to decline in fruit number by 65 per cent and 54 per cent, respectively, by the 2 species. The number of runner production was also reduced considerably. Damage to the plant crowns and reduced yield were related to population density in winter and spring but reduced runner production was due to the summer population density. A five-minute dip at transplanting in 2 per cent aqueous emulsion of thionazin 20E or in isazophos 50 gave 92-99 per cent control of *A. fragariae*. Good results were also obtained when the foliage was sprayed twice in September/October and once or twice in February/March with 5 l/ha of thionazin 20E or 3 l/ha of isazophos 50 (Tacconi, 1980).

### Stem Nematode (*Ditylenchus dipsaci*)

These nematodes usually cause shortening or thickening of the leaf stalks. Use of herbicides simazine and atrazine at 2 or 3 kg/ha, once, lenacil at 3 or 4 kg/ha, once, and simazine at 3 kg/ha, twice (with an interval of 15 days) caused a decrease in the number of *Ditylenchus* on strawberries in field tests in the USSR. Reduction in nematode population was also obtained by using oats, mustard, rye and maize as green manure crops (Andreeva, 1983). Summer treatment of unrooted sucker rosettes with hot water at 48°C for 17.5 minutes followed by mist propagation may be used for clearing the nematodes. At periods of intensive growth and in liberally irrigated plantations, granulated nematicides are preferable to heat treatment (Akinin *et al.*, 1983).

### Root Lesion (*Prathlenchus* sp.) and Root Knot (*Meloidogyne* sp.) Nematodes

Root lesion and root knot nematodes are the most common nematode to parasitize strawberry plants in north-eastern United States (LaMondia, 2002). Eleven different *Pratylenchus* species was recorded in Canada, of these *P. crenatus, P. neglectus, P. penetrans* and *P. pratensis* were reported to parasitize strawberry (Yu, 2008); *Meloidogyne hapla* is the primary root knot nematode species found in strawberry in north-eastern USA and Canda (LaMondia *et al.*, 2005; Trwnshend *et al.*, 1966). Tsolova *et al.* (2017) identified *P. crenatus, P. neglectus, P. thornel* and *Tylenchorluynchus* sp. in the Balkan Mountains in Bulgaria.

The root-knot (*Meloidogyne* sp.), sting (*Beionolaimus longcaudatus*) and stunt (*Tylenchorhynehus* sp.) nematodes were also found in strawberry field in Florida. Khan (2003) conducted a preliminary survey to record the incidence of nematode infestation on strawberry in Himachal Pradesh, India. He recorded *Meloidogyne incognita* was the most predominant species. The other populations recorded were *Tylenchorhynchus mashhoodi, Xiphinema americanum, Pratylenchus pratensis* and *Helicotylenchus dihysteria*. Soil application of nemacur at 6.72 or 13.44 kg/ha was found effective to control all these nematodes (Overman and Howard, 1984). Combined application of neem cake (600g/6m$^2$) and carbofuran (60g/6m$^2$) was suggested in *M. icognita* infected soil (Khan, 2003), Zasada *et al.* (2010) suggested many options to control root lession nematode which include, multiple pre-plant fumigants, soil steaming, soil solarisation, crop rotation and/or treating transplant materials with hot water. The cvs. Glima and Senga Sengana were reported to be resistant to both galling and growth reduction caused by *Meloidogyne hapla* while Talisman. Georg Soltwedel and Koralovaya were susceptilole (Szczygiel, 1981). The cvs. Chandler, Torrey, Ethna and Tiago were indentified as resistant to *M. icognita* (Khan, 2003).

## Diseases

### Gray Mold (*Botrytis cinerea*)

The disease appears as a light brown soft spot on green and ripening fruits, the berries dry out, become tough, and covered by a dusty fungus growth. Infection usually begins on berries touching the soil. However, infection may start in that part of a berry that touches another decayed berry or dead leaf (Swift, 2011). MyeongHyeon *et al.* (2011) working at Korea staed that flowers and red fruits were most susceptible than green fruits. *Botrytis fragariae*, a new species reported from Germany invaded strawberry tissues with an efficiently similar to or lower than *B. cinerea* but showed poor colonization (Rupp *et al.*, 2017). Dowling and Schnabel (2017) first reported that *B. mali* causing gray mold on strawberry in the United States and this species may have fungicide sensitivity spectra different from *B. cinerea*. Fertilizers stimulating excessive foliage development should be avoided to reduce the incidence (Will, 1981). Alofs and Bruyon (1980) obtained good control with euparen M (tolylfluanid), ronilan (vinclozdin) and rovral (iprodione), each applied 4 or 5 times. A combined spraying of ronilan and ridomil 6 or 7 times between 14 May (preharvest) and 26 June (during harvest) was found most effective in Ohio, Nemotodes USA (Tanboon *et al.*, 1983). Freeman (1964) suggested spraying with captan at the appearance of first flower buds and to repeat the treatment every 10 days until the petal fall. Borecka *et al.* (1970) reported that 4 sprays with 0.25 per cent captan, difolatan, euparen and melprex applied at the white bud, beginning of blooming, full bloom and petal fall stages produced satisfactory results. In field trials in Poland under natural infection, Bera (1987) obtained good control with 4 sprays of euparen (dichlofluanid) at 2 or 3 kg/ha between flowering and petal fall.

Incidence of rot was significantly reduced when fludioxonil was applied two times at 1 week intervals from 50 per cent bloom in field trials (MyeongHyeon *et al.*, 2011). After repeated fungicide application, it loose efficiency due to the development of resistant strains of *B. cinerea*. Weiss *et al.* (2014) stated that a simple and effective measure for resistance management is the replacement of chemical fungicides with single side mechanisms by biocontrol agents containing antagonistic microorganisns. *Aureobasidium pullulans* used in Boni Protect forte showed good potential in two different locations in Germany on strawberries and in one year trial in Poland on raspberries. Spray strategies in which Boni Protect forte was used in alternation with chemical fungicides revealed comparable control of *B. cinerea* as the standard treatments with chemical fungicides.

Peng *et al.* (2017) investigated the effect of ultraviolet C (UV-C) treatment on controlling gray mold. They observed significantly reduced lesion diameter and enhanced activities of chitinase (CHI), β-1, 3-glucanase, phenylalanine ammonia-lyase (PAL), peroxidase (POD), and polyphenoloxidase (PPO) in strawberry fruit by UV-C treatment. The activities of antioxidant enzymes including superoxide dismutase (SOD), catalase (CAT), and ascorbate peroxidase (APX) were also higher in UV-C treated fruit. Expression of three defence related genes: *CCR-1allele, CAT, CHI2PPO* and *PLA6* was greatly induced in UV-C treated fruits which suggested that UV-C treatment directly activated disease resistance against gray mold.

## Red Core (*Phytophthora fragariae* var. *fragariae*)

The red core incidence is more prevalent in soils with poor drainage and after a wet autumn and winter. The plants with red core often look slightly wilted. The growth of the plant is checked and the leaves turned dull bluish green in colour. The central tissue of the infected root developed characteristic red discolouration. The fungus lives in the soil and is easily trnasmitted by infected soil implements, infected runners and by irrigation water. The most important aspect to avoid red core is preventing clean land to be contaminated by the fungus. Some cultivars like Tantallon, Troubadour, Saladin showed field resistance. In field trials, aliette [aluminiumtris (ethyl-phosphonate)] proved highly effective in protecting young strawberry plants against *Phytophthora* infection when applied as a pre-planting dip, a soil drench or a foliar spray. Subsequent foliar spray or soil drench programme during the year of planting or the following year ensured good control of *P. fragariae* (Chalandon *et al.*, 1980). Montgomerie and Kennedy (1980) reported that captafol, dichlofluanid or etridiazole (ethazol) applied as soil drench and melataxyl LS 731038 (sodium ethyl phosphite), LS 74783 (aluminium ethyl phosphonate), SN 41703 (prothiocarb) or SN 66752 (propamocarb) as soil drench, root soaks or as foliar sprays decreased the severity of red core. Parikka *et al.* (2017) reported that biofungicide Prestop® (*Gliocladium catenulatum*) was effective and had few red core symptoms during the 10-week trial on *F. vesca*.

The June-bearing cvs. Alistar, Darror, Delite, Earliglow, Guardian, Joliette, Midway, Pathfinder, Redchief, Sparkle, Stelemaster, Sunris and Sure Crop and day-neutral cvs. Tribute and Tristar are resistant to red core (Swift, 2011).

## Verticillium Wilt (*Verticillium albo-atrum* and *V. dahliae*)

This fungus lives in the soil and in dry weather infects the plant. The symptoms of the disease include internal browning of wood at the base of crown, wilting, decrease in runner production, reddening of petioles and stolons late in the season. The older leaves of the infected plants turn brown and yield is severely reduced. Strawberry should not directly follow tomato, potato, pepper, brinjal and raspberry because these plants are also susceptible to *Verticillium* wilt. Njoroge *et al.* (2009) suggested growing broccoli as a successive crop of strawberry compared with postharvest incorporation of broccoli residue as a tool in the management of verticillium wilt in both conventional and organic strawberry production system. Quantitative resistance to *Verticillum* spp. is present in some strawberry cultivars and this resistance originated from wild, non-cultivated species like *F. chiloensis, F. virginiana* and *F. ovalis* (Ebihara *et al.,* 2010). The incidence of *V. dahliae* infections in cvs. Cambridgc Favourite and Cambridge Vigour were higher than in Redgauntlet (McKinley and Talboys, 1979).

## Powdery Mildew (*Sphaerotheca macularis*; *Podosphera aphanis*)

The fungus infects the leaves, specially during dry periods causing upward curling of leaves. The leaves show dark red blotches. The fungus also infects flower stalks, flowers and fruits and covers with white powdery mass. Profic-Ahvasiak *et al.* (1980) obtained best control of mildew with nimrod (bupirimate) at 0.1 per cent, bayleton (triadimefon) at 0.05 per cent, or topsin M (thiophanate-methyl) at 0.1 per cent applied at 10-15 days' interval between May and September. Triadimefon was found very effective for control of mildew. The last treatment, however, should be applied not later than 12 days before commercial harvest (Kepczynska, 1989). Polyversum WP bio-preparation (Oospores of *Pythium oligandrum*) significantly reduced powdery mildew, leaf spot and greymould in strawberries (Meszka and Bielenin, 2010). Myclobutanil, trifloxystrobin and penthiopyrad were registered for control of powdery mildew in Australia. Gomez *et al.* (2017) conducted a series of field trials in Queensland and Victoria to evaluate the efficacy of 19 fungicides and biorational products to manage the risk of fungicide resistance. Treatments were applied weekly starting from December/January for approximately 16 weeks until prior to harvest of runners in March/April. Field trial results identified bupirimate, cyflufenamid, quinoxyfen, proquinazid, azoxystrobin + difenconazole, as the industry standard programme of sulphur, myclobutanil and frifloxystrobin.

Brahm *et al.* (2005) evaluated 10 strwawberry cultivars to powdery mildew under greenhouse at Brazil. Difference in the resistance to powdery mildew among the cultivars was observed. Cultivars Milsei-Tudla showed high resistance. Cultivars Camarosa, Aromas, Oso Grande, Sweet Charlie and Campinas exhibited moderate resistance. The cv.Dover was slightly resistant while Burkley, Vila Nova and Santa Clara were highly susceptible.

### Anthracnose (*Colletotrichum gloesporiodes* and *C. acutatum*)

This disease may appear as a fruit or crown rot, both of which severly reduce yields. Fruit rot appears when fruit begins to ripen in late spring while crown rot can develop in young plants.

Daugovish *et al.* (2009) suggested pre-plant fungicide dips of transplants to reduce/control root rot caused by *C. acutatum.* Plants treated with azoxystrobin, the premixtures of boscalid+pyraclostrobin and cyprodinil+ fludioxonil had 50 to 92 per cent reduction in disease incidence, increased canopy size by more than 100 per cent and produced significantly higher marketable yields. No pretreatments provided protection aganst fruit and foliar infection. Soil drenching with fungicides can manage crown rots; fruit rots can be managed with preventive fungicide applications from bloom through harvest.

### Leaf Spot (*Rumularia tulasnii, Mycosphaerella fragariae*)

Comrnon leaf spot of strawberry is characterized by small white spots with distinct reddish brown border. Severe disease epidemics cause stunting of plants and occasional fruit infection. The cvs. Douglas and Chandler are more susceptible to leaf spot. Development of the disease is enhanced by overhead sprinkler irrigation, rains during late fall, winter and spring and during the period of high humidity. Spraying of anilazine (dyrene) 50 W at 3 lb/acre in 100 gallons of water 5 times between mid-November to mid-January gave good control (Snyder *et al.,* 1987). Mowing the strawberry bed after the plants have finished fruiting helped to reduce infectious materials from field (Swift, 2011).

### Alternaria Spot

*Alternaria sp.* attacks strawberry leaves causing 10 mm diameter purplish brown spots. When midribs, leaf stalks or stolons are infected, extensive necrosis occurs reducing the leaf surface and flower bud formation. The cvs. Senga Sengana, Korona and Induka are resistant while Gorella, Elvira and Redgauntlet are moderately susceptible and Sivetta, Tamella, Tenira, Elista, Bogota and Tiago are very susceptible. Rovral (iprodione) at 0.15 per cent or pomarsol forte (thiram) at 0.5 per cent, which control *Botrytis,* can also be used against *Alternaira* (Soosten, 1981).

### Viral Diseases

More then 30 viruses have been reported to infect strawberries. The main viruses affecting strawberries include strawberry veinbanding virus (SVBV), strawberry crinckle virus (SCV), strawberry mottle virus (SMoV), strawberry mild yellow edge (SMYEV), and strawberry necrotic shock virus (CNSV).

Leaf deformation occur due to strawberry crinkle virus. Chlorosis of leaf is the symptom of strawberry veinbanding virus. Virus-infected plants may be symptomless, or the sympyoms may be confused with caused by bacteria, viroids, insect feeding, herbicide damage, nutritional deficiencies, high temperature, *etc.*

The aphid species found in strawberry crops in Argentina were *Aphis forbesi, Aphis gossypii, Chaetosiphon fragaefolii, C. minor, C. thomasi, Macrosiphum euphorbiae* and *Myzus persicae*. *Chaetosiphon* species have been reported as responsible for most of the strawberry virus trasmissions (Dughetti *et al.*, 2017).

Control of viruses is often difficult because viruses move readily between plants without detection and because there are no 'viricides' or cures for virus-infected plants. Depending on the disease, proper management may include: chemical or biological control of vectors, use of resistant cultivars, use of clean propagation material (*i.e.*, plant material cleaned by meristem culture and heat treatment), and exclusion (plant quarantines) (Moyer *et al.*, 2010).

## Organic Production

Organically grown strawberries command a price premium. Producing strawberries organically entails a system approach to the whole farm. Many of the practices are the same in organic and conventional systems, but the fundamental approach to soil husbandry and pest management may be quite different.

Selection of suitable cultivars is important. Besides determining yields and quality, the cultivar also determines production season and pest control practices. Rhainds *et al.* (2002) suggested growing of 'Honeoye' cultivar which was found very productive and profitable for organic producers. Cultivars Tudia, Tangi and Camarosa showed higher productivity under organic cultivation system (Verona *et al.*, 2005). SanJun *et al.* (2017) suggested growing of cvs. Florida Radiance, Benicia and Camino Real for June-bearing cultivars and Albion and San Andreas for day-neutral cultivars under high-tunnel and organic management in North Carolina.

Soil fertility is to be estimated prior to planting based on soil testing and is primarily maintained with a good rotation of legume and green manures. Different organic sources are used such as guano, feather meal, pelleted chicken manure, green manure crop, compost, vermicompost, *etc.* Use of arbuscular mycrorrhizae fungi (AMF) not only stimulated the plants' growth but also contribute in enhanceing their toleranc to abiotic and biotic stresses such as salinity (Fan *et al.*, 2012). Soil drip-tape application of brown seaweed (*Ascophyllum nodosum*) extract increased yield of strawberry which was due to early growth and crown development (Holden and Ross, 2017). Welke (2004) reported increased yield by foliar application of aerobically-preapared tea compost. It also reduced the incidence of *Botrytis*. Application of vermicompost (10t/ha) significantly increased growth and yield of strawberry (Arancon *et al.*, 2004).

Strawberry (cvs. Chandler and Camarosa) runner can be easily propagated in plugs containing peat and perlite (1:1) under organic conditions and were used successfully for growing strawberries organically (Dolgun, 2007). Organic mulches are used to check weed groth. Straw mulch is used over the winter to minimize cold damage as well. Research in West Virginia indicates that shredded or chopped newsprint makes an excellent and safe mulch (Baniecki *et al.*, 1995). Biodegradable polymers have been shown to be environmentally friendly, thus potentially reducing environmental impacts. A traditional mulching film (non-biodegradable) could be replaced with a biodegradable product. These changes could result in waste production of 0 kg/ha compared to 260 kg/ha of waste for polyethylene (Girgenti *et al.*, 2014).

Organic cultivation like convential production also encounter several insects, mites and diseases problems. They can usually be kept at tolerable levels by selection of resistant cultivars, good site selection, crop rotation and adoption of organic control measures. Biological controls are to be used and they must be deployed before the pests reach critical levels. Release of predatory mites (*Phytoseiulus persimilis, Neoseiulus californicus* and *Amblyseius swirskii*) can reduce two-spotted spider mite population (Albendin *et al.*, 2012). Spraying of NeemAzal-T/S (azadirachtin) reduced the blossom weevil population (Ourednickova, 2011). Predatory thrips *Haplothrips brevitubus* was found very effective for the control of thrips in strawberry (Higashida *et al.*, 2017). Sampson and Kirk (2013) reported that in semi-protected strawberry crops mass trapping reduced adults thrips number per flower by 61 per cent and fruit bronzing by 55 per cent using blue sticky roller traps. Pre-planting soil steaming and soil solarization reduced the nematode population in soil. Biofungicide Prestop® (*Gliocladium catenulatum*) was reported effective against red core diseases of strawberry. Biocontrol agents containing antagonistic microorganisms could be used to minimize gray mold infection (Weiss *et al.*, 2014). MesZka and Bielenin (2010) reported that Polyversun$^{WP}$ bio-preparation (oospores of *Pythium oligandrum*) significantly reduced powdery mildew, leaf spot and gray mould in strawberries.

Coating with 2 per cent cassava starch + 1 per cent chitosan of organically grown harvested strawberry fruit was suggested by Campos *et al.* (2011) to maintain postharvest quality. The treatment showed less than 6 per cent loss in fruit mass. Lower counts of yeast and psychrophilic microorganisms and the best appearance after 9 days storage at room temperature.

There are several reports comparing the convential and organic production systems in strawberry. Roussos *et al.* (2012) reported significantly higher yield under integrated forming system, while both organic and integrated management system resulted in the higher weight of fruits classified in extra category, organically produced fruits showed better colour, higher dry matter, total phenols, crude fibre and carotene contents compared to those produced by the conventional treatments (Abu-Zahra and Tahboub, 2009).

## Protected Cultivation

Many countries are now producing strawberries under protection. China, South Korea and Japan producing strawberries in plastic houses for decades. Most of Spain's strawberry production is in tunnels. More than 90 per cent of strawberry production for supermarkets in Europe is growing under polythene tunnels. The advantages to the grower of using tunnels are (Anon. 2007): (a) extension of fruiting season, (b) improvement in class I grade fruit and reduced wastage. Class I grade (supermarket saleable fruit) for outdoor production is generally 50-70 per cent while under protection it is near to 80-90 per cent, (c) as the fruit is produced in controlled conditions there can be reduced use of pesticides by up to 50 per cent against *Botrytis*, downy mildew and black spot, (d) provision of an improved invironment for the use of biological control, (e) increased opportunities for organic production due to reduced incidence of disease and the enhanced use of predators, and (f) guaranted conditions to carry out production and harvesting operations and no need to stop the pickers in poor conditions or to have long waits for early dew to burn off.

It could greatly benefit growers if berries could be protected from the weather and produce over a long season, into the summer and late fall, as this would extend the season and open up new markets.

Short-day cultivars Carmine, Camarosa, Ventana, Chandler, Darselect, Raritan, Northeaster (Deyton *et al.*, 2009), Florence, Elsanta, Honeoye, Hapil, Symphony (Anon, 2007) and day neutral cultivars Albion, Seascape, Evie-2 (Deyton *et al.*, 2009), Ava, Everest, Elsinor, Jubilee, Camarillo (Lieten, 2005), Alibon (Nicholas *et al.*, 2018) were reported to grow successfully under protected condition.

Strawberries are grown in low tunnels, high tunnels or green houses. Plants are grown in 30-35 cm high-raised beds with a top width of 45 to 55 cm and a bottom width of 55-65 cm, and covered with plastic mulch. Additional specifications include one drip tape per bed, two rows of plants per bed with a distance between rows of 22-28 cm, and within –row plant spacing of 25-35 cm, depending on the vigour of the cultivar. Plant populations range from 50,000 - 70,000 plants/ha. Temperature differences inside and outside high tunnels can vary between 2 and 17°C (Kadir *et al.*, 2006). Santos *et al.* (2008) stated that polyethylene plastic cover of the tunnels reduces the photosynthetic acitive radiation (PAR) by 40 per cent. Yields in high tunnels may be 25 per cent higher than those in the open field (Demchak, 2004) and fruits have better quality (Jett, 2007).

Cropping in substrate culture is becoming increasing popular. Substrate systems include peat bags, container and pots. Raising the plants above the ground in plastic trays or gutter are also in practice. The substrates used are usually peat moss or mix of peat moss and coir. Cropping in soilless systems greatly reduced the use of soil fumigants and fungicides. However, this system requires expertise regarding nutrition and watering. The nutrient solutation should be calculated depending on the soil factors and the needs of the various cultivars (Lieten, 2013). When grown in an inert substrate, plants produced many high order inflorescences, leading to asynchronous production (Savini and Neri, 2004). In substrate containing peat, cvs. Camarosa, Gaviota (short day) and Selva (day neutral) produce the highest number of crowns and leaves compared with sand or perlite without peat (Tehranifar *et al.*, 2007). Temporary suspension of fertigation during flower induction of tray-plants advances the differentiation of terminal flower bud with positive effects on inflorescence development and flower numbers (Savini and Neri, 2004). Higher yield can be obtained by providing a better water supply during flower bud initiation and differentiation, *e.g.*, autumn compared with spring irrigation (Naumann, 1964). Once fruit set has occurred excess use of water may encourage the development of grey mould. Unless plants are established through polythene, crops should be strawed down to ensure the developing fruits are prevented from coming into contact with the soil. It is advisable to apply slug pellets along the rows and then apply straw ensuring it is pushed into the peat plant canopy to provide a barrier between the soil and fruit (Anon., 2007).

Two spotted spider mites can become a problem under low tunnels. These mites prefer hot and dry conditions which the low tunnel environment creates. Exposing plants to cool, damp weather by fully pulling the cover to one side can help to reduce populations, although extended exposure to damp weather can promote gray mould. Predatory mites should work well in low tunnels. In general, gray mold and powdery mildew infection is low under tunnels.

It is a practice among some greenhouse growers of day neutral plant to cut back the plants to their crowns after the first major fruit flush has been harvested and this has resulted in a significant autumn crop of good size (Nicholos *et al.*, 2018).

In New Zealand strawberry yield tend to be about 50 tons/ha. However, in environmentally controlled greenhouse (7m x 7m), over a 6-month harvest period the cultivar Aromas produced about 100-120 tons/ha (Nicholos *et al.*, 2018). Fruits harvested from protected greenhouse showed higher sugar and total phenolic contents (Costa *et al.*, 2012).

## Breeding and Improvement

The objectives of strawberry breeding include high yield, fruit quality, day-nutral cultivar, runner production, extension of fruiting time, adaptability to particular growing environment, resistance to pests and diseases and characters that facilitate mechanical harvest. Modern strawberries, mostly octoploids, have been regrouped as *F. ananassa*. The descendants of the original hybrids are very heterogeneous and produce cultivars with either one or two annual blossoms or perpetual blossoms. Some wild octoploids are used for some special characters in breeding. In USA, *F. ovalis* (*F. virginiana*) has been used in breeding programme because of its resistance to drought and low temperature. Similarly *F. chiloensis* and *F. virginiana* are resistant to red stele, aphids and *Verticillium* wilt. Scott *et al.* (1972) stated that such useful characters can be separated by three or four back crossing. Darrow and Scott (1947) obtained a high percentage of winter-hardy seedlings from crosses of *F. virginiana* x Midland. The fruits of this plant were found to ripen early. However, seedlings from diploid, tetraploid or hexaploid with cultivated octoploid species were mostly sterile or very low in fertility (Bringhurst and Khan, 1963, Jones, 1966). Kataoka and Noguchi (2017) made recurrent selections in sets of one self-pollination and one cross-pollination using 36 cultivars of strawberry. They observed vigorous lines after two or three generations of recurrent selections. The average leaf area of the largest leaf, a measure of plant vigour, was larger than the leaf area of the cultivar 'Kaorino' which is one of the most vigorous strawberry, cultivar in Japan. A line was produced with leaf area twice that of 'Kaorino'. Strawberry is generally predisposed to inbreeding depression, but some self-pollinated lines obtained from recurrent selection had larger leaf area.

Intergeneric hybridization using *F. vesca* x *Potentilla nepalensis; F. vesca* with *P. anserina, P. anglica, P. erecta, P. reptans, P. rupestris; F. moschata* with *P. fruticosa* was also tried (Jones, 1955; Asker, 1970), but the hybrid-seedlings either died or were found sterile. GangJun *et al.* (2017) obtained a synthetic dodecaploid strawberry hybrid γH#15-10 (2n=12x=84) which showed sufficient cold resistance. The cross between *F. x ananassa* cv. Allstar (2n= 8x=56) and YH15-10 produced 13 sudlings, while reciprocal cross failed. Most morphological characters of seedlings were found widely separated among $F_1$ progeny with a high broad sense heritability, which showed that these variations resulted mainly from genetic effect. Some hybrids showed heterosis, especially in growth vigour and runner production. Among the 30 tested hybrids, 28 decaploids (2n=10x=70), one octoploid (2n=8x=56) and one enneaploid (2n=9x=63) were observed. The 63 per cent hybrids showed higher cold resistance than cv. Allstar.

Inheritance of disease resistance had been tried and success was achieved in several occasions. The development of virus-tolerant strawberry cultivar has been an important objective of Pacific Northwest Breeding Programmes for many years (Darrow, 1966, Daubeny *et al.*, 1972). A high degree of tolerance was found in several clones of *F. chiloensis* (Miller and Waldo, 1959). The cultivars Tioga, Totem, Robinson, Tennessee Beauty were reported to be virus-tolerant. Some clones of *F. virginiana* and *F. chiloensis* are highly resistant to *Verticillium* wilt (Varney *et al.*, 1959; Wilhelm, 1955). Cultivars such as Cambridge Favourite, Redgauntlet, Aberdeen, Gala, Gem, *etc.*, showed field resistance to *Verticillium* wilt while Gorella is very susceptible. In Poland, seven strawberry cultivars, which differed in their susceptibility to root-knot nematode (*Meloidogyne hapla*), were used as parents in the hybridization for resistance development. A very high percentage of seedlings (41-78 per cent) in the progeny of cv. Glima were found without any symptom of root galling (Szcygiel and Danek, 1984). The cvs. Gorella, Senga Sengana, Cambridge Favourite, Madame, Moutot, Surprise des Halles and Early Dawn were found resistant to gall nematode (Choleva and Kacharmazov, 1978). Puget Beauty was reported to offer the highest degree of resistance for powdery mildew to its progenies (Daubeny, 1961). In Canada, the June-bearing cullivars Governor Simcoe and Secord of *Fragaria ananassa* (*F. magna*) were derived from crosses between Guardian and Holiday. Holiday being the female parent of Governor Simcoe and the male parent of Secord. Both the cultivars have large, firm fruit with good flavour and are intended for fresh use. Secord is resistant to *Verticillium albo-atrum*, while both cultivars are moderately resistant to *Diplocarpon earliena* (Dale *et al.*, 1986). In the USSR, from 1961, over 12000 forms representing 14 *Fragaria* species were evaluated in the Krasnodar area for resistance to main fungal diseases, namely *Ramularia tulasnei*, *Marssonina potentillae*, *Dendrophoma obscurans*, *Sphaerotheca macttlaris*, *Phytophthora cactorum*, *P. fragariae*, *Verticilliunl dahliae*, *V. albo-atrum*, *Fusarium sporotrichioides* and *Botrytis cinerea*. Those with immunity or resistance to one or more diseases were identified and those suitable for breeding were selected to produce new resistant hybrids and cultivars. These included *F. ananassa* (*F. magna*) cultivars and *F. virginiana* and *F. chiloensis* (Govorova, 1985).

The size of fruit is inherited quantitatively. The cv. Gorella showed high combining ability for fruit size (Oydvin, 1965). The cvs. Francesco (NY 254 x Pocahontas), Savio (Tardiva di Romagna x Md. US 3816) of Italy (Rosati and Faedi, 1980); Veeglow (Redglow x Vibrant) of Canada (Ricketson *et al.*, 1986a); Hinomine (Terunoka x Harunoka) of Japan (Takayanagi and Kawata, 1988), *etc.*, are some of the developed cultivars with large berry size. Studies on commercial cultivars showed variation in soluble solids content and sugars (Wrolstad and Shallenberger,

1981). Some cultivars produce fruits rich in vitamin C. The cvs.Marshall, Clarke, Progressive and *F. chiloensis* were found to transmit relatively high vitamin C content to their seedlings (Hansen and Waldo, 1944). Progenies from strawberry crosses with *F. virginiana glauca* as a common parent showed a significant increase in their fruit quality and nutritional quality, which confirms the interest of the wild genotypes to improve several characters of cultivated strawberry (Diamanti *et al.*, 2012).

## Development of Cultivars in different Countries

### United States of America

First important North American strawberry 'Hovey' (European pin strawberry 'Mulberry' x *F. virginiana*) was developed by Charles Hovey in 1836. Albert Etter developed several cultivars with native *F. chilonsis* of which Ettersburg 80 (1910) was widely grown. Others were Ettersbug 120, Fendalcino and Rose Ettersburg. The everbearer 'Progressive' was introduced by Harlow Rochill in 1912. California breeding programme developed 'Lassen' and 'Shasta' in 1945, 'Tioga' (1964) (was dominant in California in mid 60s to late 70s), 'Pajaro' (1978), Selva (1983), 'Camarosa, (1992), 'Diamante' (1997), 'Albion' (2004), 'San Andreas' (2008), 'Driscoll Ojai' (2008), 'Monterey' (2009), 'Palomar' (2008), 'San Andreas' (2009), 'Portola' (2009), 'Mojave' (2012), 'Merced' (2015). *etc.*

US Depertment of Agriculture, developed 'Blakemore', 'Fairfax', 'Pocahontas', 'Albritton', Surecrop', 'Sunrise', 'Redchief', (1968), 'Earlyglow' (1975), 'Allstar' (1981), 'Tribute' (1981), 'Tristar' (1981), 'Valley Red' (2009), 'Flavorfest' (2012), 'Charm' (2015), 'Sweet Sunrise' (2015), *etc.*

The other important cultivars developed from USA are 'Honeoye' and 'Jewel' from New York, 'Raritan' from New Jersey and 'Sweet Charleie' from Florida. Other cultivars developed from Florida include 'Florida Elyana', 'Florida Radiance' (2009), 'Winter Dawn' (2010), 'Florida 127' (Sweet Sensation®, Sensation™). The North Corolina State University developed 'Galletta' (NCH87-22 x Earliglow).

### Unted Kingdom

In the middle of the 20th century, strawberry breeding programme was intitiated/strengthened in Europe, particulary in Engaland, Scotland, Italy, Germany, Holland and Poland. In UK, the East Malling Strawberry Breeding Programme commenced in 1983 under the leadership of David Simpson, with its first cultivar release Pandora in 1988. Since then the programme has successfully released 42 cultivars primarily for the UK and northern European markets, but also with sales in the USA and South Korea. In 2008 the East Malling strawberry Breeding Club (EMSBC), a public-private partnership, was formed. The EMSBC objectives are to develop a range of high quality cultivars with overlapping seasons, which crop in succession and combine excellent fruit quality high yield, low percentage waste and resistance to fungal diseases. Since 2008, the programme has released five new cultivars (Buddy, Malling™ Centenary, Malling™ Star, Strawbeno, Seremity) with a further two cultivars (Malling™ Glow and Malling™Sunrise) being released by Meiosis Ltd. (Whitehouse, *et al.*, 2017). Three everbearer cultivars [(Sweet Eve (2008), Eves Delight (2008), Velvet (2009)] were released by the Edward Vinson Programme led by Peter Vinson. The Driscoll's Genetics Ltd. Programme, led by Carlos Fear developed 'Jubilee' and 'Camarillo, both suited to protected and open air condition.

**Italy**: Public strawberry breeding programme, CREA-FRF started in Italy at the End of the 1960s and has released more than 30 cultivars until 2016 (Baruzzi *et al.*, 2017). The southern areas of Italy represent 66 per cent of the National strawberry acerage. Winter planting systems with bare root plants and plug plants are adapted. The cvs. Sabrosa® Candonga, Sabrina, Florida Fortuna are the dominant cultivar. In the northern areas, the main cultivars are Eva, Alba, Clery, Roxana and Garda. The main goal of breeding is to produce more and better cultivars for fresh consumption (June-bearing and ever-bearing) well-adapted to the different environmental conditions and to different crop-management practices of the various areas, including fruit quality improvent (taste, aroma, sweetness, flesh firmness, size, regular shape, shelf-life, health promoting properties) and upgrading disease resistance (Baruzzi *et al.*, 2017).

**France:** The *F x ananassa* hybrid 'Pineapple' was growing for many years in France. 'Vicomtesse Hericart de Thury' bred from seed of 'Keens Seedling'by J.L. Jamin was introduced in 1849. The cvs. Due de Malakoff (1854) Docteur Morere (1867) and Mandame Moutot (1906) were the early important cultivars. French-bred cultivars covered approximately 96 per cent planted area in France of which cvs. Gariguette (42 per cent), Derselect (18 per cent), Clery (13 per cent), Charlotte (10 per cent) and Ciflorette (7 per cent) were the most used cultivars in 2010 (Lopez-Aranda *et al.*, 2011). Other cultivars developed from France include 'Donna' (2010), and Amandine (2013).

**The Netherlands:** In The Netherlands, the Fresh Forward breeding programme is focussed improving June-beaing and everbearing cultivars for northern Europe. Genetic research is aimed to develop linkage map for *Fragaria x ananassa*, molecular markers for resistance to *Phytophthora cactorum* and molecular studies on fruit quality, including firmness, flavour volatiles and antioxidant (Simpson, 2014). Breeding programme at The Netherlands produced 'Elsanta' which has been the dominant cultivar of north-west Europe for around 25 years. The other cultivars developed are Sonata, Salsa, Figaro (2007), Rumba (high yielder, good flavour), Vima®Zanta (high yielder, good flavour), Kimberly Elianny (60-day cropping, good yield, flavour and shelf life), Florin (ever-bearing), Florina (high yielder, everbearer, good shelf life) and Flair (June-bearer, excellent flavour, good tolerance to powdery mildew and most root diseases). Other important cultivars developed from The Netherlands include Juspa (1957), Gorella (1960), Korona (1977), Sonata (2007) and 'Flair' (2012).

**Germany:** In early 20$^{th}$ centery Prof Schindler at Pillnitz, Germany bred Oberschlesien (1919) and Frau Mieze Schindler (1933). R. von Sengbusch's produced 'Senga' series of which 'Senga Sengana' (1954) was widely planted for its processing quality. Hansabred GMBH and CO's aim is to develop commercial cultivars for northern Europe with emphasis on improving flavour and disease resistance. Research at the Julius-Kuha Institute includes studies on inheritance of flavour, volatiles and the protential of aroma analysis as a selection tool. Resistance breeding is focussed on *Verticillium dahliae* and *Phytophthora cactorum* (Simpson, 2014). Other cultivars developed from Germany include, 'Malwina' (2012).

**Norway:** The strawbeery breeding programme at Norway has been run by Graminor since 2006. The general objective of their programme is to produce improved cultivars that can withstand extreme, irregular and harsh winter conditions, combined with superior berry qualities and profitable yields. Three June-bearer cultivars 'Gudleif', 'Iris' and 'Blink' were released (Simpson, 2014).

**Turkey:** Atatruk Horticultural Central Research Institute, Turkey has a strawberry breeding programme. Several strawberries have been developed from this programme which includes Yalova-15, Yalova104, Eren-77, Ata-77, Doruk-77 and Bolverim-77. The cultivars Kaska, Sevgi and Ebru were developed from the Cukurova University breeding programme (Serce and Ozgen, 2018).

**Poland:** Strawberry breeding programme in Poland was initiated in 1970s. The objective of breeding was to breed varieties: (a) for growing outdoors, early forcing and evebearing, (b) high yielder, easy to pick, (c) high quality fruit, and (d) good level of disease resistance for *Verticillum* and *Fusarium* wilt, powdery mildew, grey mold. Leaf spot and leaf scorch (Hortynski, 1993). The cv. Grandosa, resistant to powdery mildew and spider mite was selected in 2005, 'Pink Rose'- another cultivar developed which is resistant to common leaf spot, leaf scorch, and powdery mildew.

**USSR:** The strawberry breeding programme in the USSR began at the Moscow fruit growing station in 1924 and some well-known Russian caltivars (Komsomolka, Naryadnaya, Pozdnyaya Zagorya, My-sovka) were bred using the introducing cultivars (Victoria, Luisa, Sharples, Prenner, Deutch Evern *etc.*) as parents (Govorova, 1993).

**Japan:** The early cultivars for stone-wall culture of strawberries in Japan were 'Excelsior' and 'Victoria' (1899) followed by 'Fukuba' and 'Victoria' was the dominant cultivar unil 1960. 'Hottawande' an offspring of 'Fukuba' replace 'Victoria' in 1960 which was furher replaced by 'Kunruwase' in the 1980s. 'Hokowase' an offspring of 'Tahoe' was released in 1960 and was the leading cultivar in southern and western Japan from 1960s to 1988. 'Moriok 16' (Fairfax x Ettersberg) was released in 1967 and 'Kitanokagayaki' (Belle Rouge x Pajaro) in 1996 were dominant cultivars in 1980s and 1990s. After world War II, several American cultivars were introduced and were used for Japanese strawberry improvement. 'Kogyoku' a seedling of 'Fairfax' was released in 1951 and was the leading cultivar of Kanto district from 1960s to 1988. Nagano prefecture developed strawberry cultivars Summer Princes, Summer Engel, 8-9 and Flamenco, which are gaining popularity (Takei, 2010). Recent topics in strawberry breeding include $F_1$ hybrid cultivars grown from seed and interspecific hybrids. The $F_1$ hybrid cultivar 'Chiba $F_1$-gou' was developed from a cross between June- bearing inbred lines (Ishikawa *et al.*, 2007). The decaploid interspecific hybrid 'Tokum', which has a peach-like sweet fragrance was developed from an octoploid cultivar and a diploid wild species (*Fragaria nilgerrensis*) (Noguchi *et al.*, 2011). In addition, DNA markers for genome indetification, disease resistance, everbearing are being developed (Kunisha, 2011) and stuies on cytoplasmic male sterility for $F_1$ hybrid cultivars have been onging since 2010 (Morishita, 2014).

**China:** Although several wild *Fragaria* species were derived from China, culitivated strawberry was introduced in China in the early 20$^{th}$ Century. From the mid-1910s to the late 1940s, strawberry was only planted sporadically in few places. The frist cultivar 'Victoria' was introduced from Russia. In the early 1950s, commercial cultivation was started near big citics and some selections from open-pollinated seedlings were made. In 1980s, the Chinese Ministry of Agriculture established National Strawberry Germplasm Repositories in Beijing and Nanjing. Strawberry breeding started and six cultivars were released in 1980s. In 1990s 15 cultivers, in 2000s 19 cultivars and 18 cultivars between 2000 and 2012 were developed and released (Zhang *et al.*, 2017). 'Benihope', Sweet Charlie', 'Toyonoka' and 'Akihime' are the popular cultivars in China (Zhang *et al.*, 2014). From 2012 to 2016, 30 new strawberry cultivars have been released in China of which 80 per cent bred for forcing/semi-forcing culture. Among the recently released cultivars 'Jingzangxiang', 'Jingtaoxiang', 'Pink Princess', 'Yanli', 'Ningyu', 'Yuexin', Shimei 7' and 'Sigongzhu' are becoming popular (Zhang *et al.*, 2017).

**India:** The major cultivars grown in India are Chandler, Sweet Charlie, Camarosa, Belrubi, Senga Sengana, Selva, Festival, *etc.* Some of the selections and cultivars, developed are Local Jeolikot, Dilpasand, Bangalore Sweet, Pusa Sweet, Pusa Pride, Katrain Sweet, Himalayan Sweet and Shimla Delicious.

**Australia:** Short-day strawberry cultivar 'Rubygem' (2007) suited for mild winter climate was developed from DPI Brisbane and Sydney, Australia and University of Florida, USA. This hybrid (Earlibrite x Carlsbad) is highly resistant to *Fusarium* wilt. 'Sugarbaby' (Coogee x Redlands Joy) a short day partially remontant cultivar was developed from DPI, Queensland, Australia in 2006.

**Israel:** Agricultural Research Organization, Israel developed remontant infra-short day strawberry 'Herut' (Yael x Sweet Charlie) for fall to spring production in Mediterrranean subtropical condition. It has no chilling requirement for flowering. 'Yuval' (Tamar (selfed) x Bella (selfed) is also a remontant infra-short day strawberry, requires no vernalization and produces fruits from November of September planting to the following summer. 'Tamir' (Herut x Hadas) another infra-short day cultivar developed as on alternative to 'Hadas' for Israel, producing around the New year for local and European markets. 'Roteni' was developed in 2010 for low tunnel winter cultivaton and 'Shaked' (2010) for winter cultivation.

**Spain:** Short-day strawberry for late October planting and winter-spring low tunnel production 'Macarena' (2006) was developed from Spain. Other cultivars developed include 'Amiga' (Camarosa x 3-79), 'Cristal' (92-61 x 90-45), 'Sabrina' (2012), 'Deluxe' (2015), 'Dream' (2015) 'Fontanilla, (2015), *etc.*

**Canada:** Agriculature and Agri-Food Canada developed several cultivars such as 'Stolo' (Puget Reliance x Whonnock)- resistant to root weevil, 'Orleans' (L'Acadie x Joliette') for eastern central Canada, 'Nisgaa' (Cavendish x Nanaimo), and 'Laurel' (2012) for eastern Canada.

## Molecular Marker

Molecular techniques have been used in strawberry marker development, mostly based on PCR-driven strategies. A series of molecular marker types have been used by many researchers to fingerprint germplasm and diagram genetic diversity analysis of strawberry. Six cleavage amplified polymorphic sequence (CAPS) markers were developed by Kunisha *et al.* (2013) to identify strawberry cultivars to protect breeders' rights. These wer successfully used to distinguish 14 commercial cultivars. The stability of CAPS analysis makes it a good tool for the identification of strawberries. Martelli *et al.* (1999) investigated the genetic diversity of cultivated strawberry. Nineteen genotypes, including cultivars and breeding partental lines, were analysed with fifteen 10 bp primers. Amplification patterns revealed a high level of polymorphism. The results obtained with one of the primers allowed a characterization of all genotypes. A dendogram representing the genetic retationships among the genotypes was constructed. Tyrka *et.al.*(2002) stated that PstI based amplified fragment length polymorphism (PstIAFLP) method is sufficient for effective identification and useful for assessing the level of genetic diversity in strawberry cultivars and breeding lines.

Hypervariable SSR markers, FxaHGAO2p13 was found capable of distinguishing 72 *Fragaria x ananassa* cultivars or lines. Principal-coordinate analysis detected several groups, which reflect taxon and breeding site (Honjo *et al.*, 2011). The total DNA of 38 strawberry cultivars originating in Europe and America were amplified by 44 pairs of SSR primers (ZhuangWei *et al.*, 2011). Twenty nine cultivars could be distinguished by the fingerprint map constructed by only 4 pairs of SSR primers. Gunaydin and Kafkas (2017) studied the genetic relationship among 34 Turkish and foreign strawberry cultivars and genotypes using CAPS and SSR DNA molecular markers. They concluded that SSR was better than CAPS from the point of view of polymorphism rates and polymorphism information contents, whereas CAPS was better than SSR from the point of view resolving power values, in the characterization of strawberry cultivars and genotypes.

In Japan, a strawberry core collection was established based on SSR and CAPS marker polymorphisms in 119 strawberry cultivars using the 'PowerCore' programme. The core collected consisted of 19 cultivars. The correlation coefficients for the diversity index were found significant between the core collection cultivars and all cultivars. Results demonstrated that the core collection cultivars were suitably selected in terms of reflecting the genetic diversity of all strawberry cultivars (Wada *et al.*, 2017).

## Genetic Transformation

The earlier reports of successful genetic transformation of cultivated strawberry were based on *Agrobacterium*-mediated transformation (Jelenkovic *et al.*, 1986, James *et al.*, 1990, Nehera *et al.*, 1990). The primary focus has been on transformation using genes with potential for pest and herbicide resistance, cold tolerance and ripening. The transformation efficiency varies among cultivar (Debnath and Texixeria da Silva, 2007). James *et al.* (1990) reported transformation frequency of 0.95 in cv. Rapella while Nehera *et al.* (1990) reported 6.5 per cent for cv. Red Coat. Schaart *et al.* (2004) reported an effective method for the production of transgenic plants from which selectable marker genes have been removed.

Khammuang *et al.* (2005) transferred antifreeze protein genes into strawberries. Jimenez-Bermudez *et al.* (2002) developed transgenic strawberry plants incorporating as antisense sequence of a strawberry pectate lyase gene and reported that transgenic lines were significantly firmer than normal fruit.

In addition to *Agrobacterium*, particle bombardment was also used to introduce a gene into strawberry for stable transformation (Wang *et al.*, 2004). Cordero de Mesa *et al.* (2000) stated that coating gold particles with *Agrobacterium* improved the transformation efficiency of leaf explants using the biolistics approach.

The sucrose synthase (Sus) gene is regarded to be involved in strawberry fruit ripening. Hairpin-mediated RNA silencing of *FaSus1* lead to a decrease in sucrose content and inhibit ripening (Hua *et al.*, 2017). Li *et al.* (2017) stated that *miR164* is involved in delaying senescence of strawberry fruit by negatively regulating NAC transcription factor gene under low temperature.

Researchers are working to develop methods to change gene in strawberry by the recent breakthrough technology "gene-editing" (technically known as CRISPR-Cas 9), a process that does not pass on the accessory sequences of genetic engineering that some find objectionable.

## Genome

The cytologically determined C value for *Fragaria vesca* was reported to be 164 Mbp (Akiyama *et al.*, 2001), from which a 206 Mbp C value was extracted (Folta and Davis, 2006). Shulaev *et al.* (2011) reported the *F. visca* genome. The *F. vesca* has a small genome (240Mb) and is the smallest sequenced plant genome other than *Arabidosis* (~147 Mb) and represents a gatway to functional gene studies within Rosaceae. The genome was sequenced to x39 coverage using second-generation technology, assembled *de novo* and then anchored to the genetic linkage map into seven pseudochromosomes. Gene prediction modelling identified 34,809 genes, with most being supported by transcriptome mapping. Genes critical to valuable horticultural traits including flavour, nutritional value and flowering time were identified. Macrosyntenic relationships between *Fragaria* and *Prunus* predict a hypothetical ancestral Rosaceae genome (online version of the paper is available at http://www.nature.com/naturegenetics/.)

Cultivated strawberry (*F x ananassa*) is octoploid and shows allogamous behaviour. Hirakawa *et al.* (2014) studied this octoploid genome through comparison with its wild relatives, *F. iinumae, F. nipponica, F. nubicola,* and *F. orientalis* by *de novo* whole-genome sequencing on an illumina and Roche 454 platforms. The total length of the assembled illumina genome sequences obtained was 689 Mb for *F. Rom x ananassa*, and ~200Mb each for the four wild species. Subsequently a virtual reference genome termed FAN hybrid_r1.2 was constructed by integrating the sequences of the Roche 454 and illumine genome sequence were eliminated. The total length of FANhybrid_r1.2 thus created was 173.2Mb with N50 length of 5137bp. The illumine-assembled genome sequence of *F. x ananassa* and the homoeologous subgenomes of *F. x ananassa* from which heterozygous regions in the four wild species were then mapped onto the reference genome, along with *F. vesca* genome sequence to establish the subgenomic structure of *F. x ananassa*. The genome assembly data, annotation, and gene models of *F. x ananassa* and wild *Fragaria* species are available at Strawberry GARDEN (http://strawberry-garden.kazusa.or.jp).

# REFERENCES

Aaby, K., Mazur, S., Nes, A. and Skrede, G. (2012) Fd. Chem., 132: 86-97.

Abbott, A.J. and Webb, R.A. (1970). Nature, 225, 663.

Abeles, F.B. and Takeda, F. (1990). Scientia Hort., 42: 269-75.

Abramcnko, N.M. (1983). In: Progresivini Smeery U Ovocnarske Vyrobe flolovousy. Czechoslovakia, pp. 209-12.

Abu-Zahra, T.R. and Tahboub, A.A. (2009). Acta Hort., 87: 353-58.

Agamalian, H.S. (1987). Proc. West. Soc. Weed Sci., 40: 150-55.

Aharoni, A., Keizer, L.C.P., Broeck, H.C. vanden, Blanco-Portales, R., Munoz-Blanco, J., Bois, G., Smit, P., Vos, R.C.H.de and Connell, A.P. (2002). Pl. Physiol., 129:1019-31

Aharoni, Y., Hartsell, P.L., Stewart, J.K. and Young, D.K. (1979). Jr. Econ. Entom., 72: 820-22.

Aharoni, Y., Stewart, J.K. and Guadagni, D.G. (1981). Jr. Econ. Entom., 74: 338-40.

Aharoni, Y., Stewart, J.K., Guadagni, D.G. and Mon, T.R. (1980). Jr. Amer. Soc. Hort. Sci., 105:926-29.

Ahmad, M.F. (2009). Indian J. Hort., 66: 132-34.

Ahmad, M.F., Khan, I.A. and Wani, N. (2011). Indian J. Hort., 68: 39-43.

Ahmed, H.F.S. and Ragab, M.I. (2005). Egypt. J. Biol., 5: 35-42.

Akinin, N.I., Gulyaev, A.C. and Metlitskii, O.Z. (1983). Voronezh, USSR; Vserossiiskii N II Zashchity Rastenii. pp. 125-31.

Akiyama, Y., Yamamoto, Y. Ohmido, N., Oshima, M. and Fukui, K. (2001). Cytologia, 66: 431-36.

Alamar, M.C., Collings, E., Cools, K. and Terry, L.A. (2017). Postharvest Biol. Technol., 134: 76-86.

Albendin, G., Garcia, M.C. and Molina, J.M. (2012). Agricola Vergel: Fruiticultura, Horticultura, Floricultura, Citricultura, vid, Arroz, 31(353): 14-22.

Albregts, E.C. and Howard, C.M. (1985). Proc. Soil and Crop Sci. Soc. Florida, 44:197-99.

Albregts, E.E. and Howard, C.M. (1972). HortSci., 7: 569-70.

Albregts, E.E. and Howard, C.M. (1978). Proc. Fla. Sta. Hort. Soc., 91:298-99.

Albregts, E.E. and Howard, C.M. (1981). Jr. Amer. Soc. Hort. Sci., 106: 295-98.

Albregts, E.E. and Howard, C.M. (1982). HortSci., 17: 651-52.

Alofs, W. and Bruyon. J.De. (1980). Fruitteelt., 70 : 340-41.

Andersen, E.T. (1963). Diss. Abstr., 23: 4050.

Andreeva, V.I. (1983). Voronezh, USSR; Vserossiiskii N II Zashchity Rastenii. pp. 111-17.

Andriolo, J.L., Erpen, L., Cardose, F.L., Cocco, C., Casagrande, G.S. and Janisch, D.I. (2011). Hort. Braslleira, 29: 516-19.

Angelov, A. (1966). Gradinarska i Lozarska Nauka. 3: 13-142.

Anonymous (1987). Sadovostvo. Moscow. No.6.

Aoyagi, M. and Makino, H. (1981). Jr. Jap. Soc. Hort. Sci., 49: 583-90.

Arancon, N.Q., Edwards, C.A., Bierman, C., Welch, J. and Metzger, D. (2004). Biresource Tech., 93(2): 145-53.

Arima, S., Kondo, M and Monta, M. (2004). ASAE Ann. Meet. 2004.

Arney, S.E. (1956a). Phyton. 7: 89-102.

Arney, S.E. (1956b). Jr. Exp. Bot., 7: 65-75.

Asker, S. (1970). Hereditas. 64: 135-39.

Austin, M.E., Shutak, V.G. and Christopher, E.P. (1960). Proc. Amer. Soc. Hort. Sci., 75: 382-86.

Badiyala, S.D. and Joolka, N.K. (1983). Haryana J. Hort. Sci., 12: 165-67.

Baker, A.S. and Mortensen, W.P. (1973). Bull. Wash. Agri. Expt. Stn., 776: 5.

Baker, R.T. and Dick, H.B. (1981). NZ. J. Expt. Agric., 9: 377-81.

Bakker, J., Bridle, P. and Bellworthy, S.J. (1994). J. Sci. Fd. Agric., 64: 31-37.

Ballinger, W.E. and Mason, D.D. (1960). Proc. Amer. Soc. Hort. Sci., 76: 359-65.

Barritt, B.H. (1974). HortSci., 9 : 25-27.

Baruzzi, G., Ancay, A. and Others (2012). Acta Hort., 926: 133-38.

Baruzzi, G., Ballini, L., Baroni, G., Birolli, M., Capriplo, G., Carullo, A., D'Anna, F., Funaro, M., Luccchi, P., Magnanni, S., Maltoni, M.L., Sbright, P., Turci, P. and Faedi, W. (2017). Acta Hort., 1156: 179-83.

Bazzocchi, R. (1968). Rivista Ortoflorofruttic., 52 : 394-400.

Beech, M.G. (1983). Jr. Hort. Sci., 58: 541-45.

Beese, M. (1986). Gartenbau, 33: 370-71.

Beloshapkina, O.I. (1985). In. Probt. Vegetat. razmnozheniya V Sadovod, Moscow. pp. 119-22.

Ben-Arie, R. and Faust, M. (1980). Phytochem., 19 : 1631-36.

Benott, F. (1977). Jr. Hort. Sci., 47: 429-39.

Bera, B. (1987). Fruit Sci. Rept., 14: 21-28.

Beyene, G.T., Kehoe, E, E., MacSiurtain, M. and Hunter, A. (2012). Acta Hort., 926: 483-89.

Bieniasz, M., Malodobry, M. and Dziedzic, E. (2012). Acta Hort., 926: 457-61.

Blasse, W. (1981). Gartenbau. 28: 337-38.

Blasse, W., Grittner, I., Bischoff, S. and Eisemann, A. (1984). Gartenbau., 31: 183-85.

Bohmer, B. (1981). Gesunde Pflanzen., 33: 113-17.

Booster, D.E. (1973). Amer. Soc. Agr. Eng., pp. 73-109.

Borecka, H. (1977). Fruit Sci. Rept., 4: 35-41.

Borecka, H., Bielenin, A. and Dzieciol, R. (1970). PR. INST. SAD., 14:273-84.

Borgman, H.H. (1950). Meded. Din. Tuinb., 13: 13-15.

Borthwick, H.A. and Parker, A.W (1952). Proc. 12th Int. Hort. Congr., pp 1-9.

Bould, C. (1964). Jr. Sci. Fd. Agric., 15: 474-87.

Bowen, G.D. (1991). In: Plant roots: the hidden half. (eds. Y. Waisel, A. Eshel and U. Kafkafi). Marcel Dekker, Inc. New York. pp. 309-30.

Boxus, P.H. (1974). Jr. Hort. Sci., 49: 209-10.

Boyce, B.R. and Matlock, D.L. (1966). In. Nut. Fruit Crops. (ed. N.F. Childers) Rutgers Univ., New Jersey, pp. 518-48.

Boyce, B.R. and Reed, R.A. (1983). Adv. Strawberry Prod., 2: 12-14.

Brackmann, A., Pavanello, E.P., Booth, V., Janish, D.T., Schmitt, O.J. and Gimnez, G. (2011). Fd. Sci. Tech., 86: 652-59.

Brahm, R.U., Ueno, B. and Oliveira, R.P. de (2005). Revista Brasileira de Fruticultura, 27: 219-21.

Brand, H.J.Jr. and Hawthorne, P.L. (1965). Louisiana Agr. Expt. Stn. Bull., 591.

Breen, P.J. and Martin, L.W. (1981). Jr. Amer. Soc. Hort. Sci., 106: 266-72.

Brierley, W.G. and Landon, R.H. (1941). Proc. Amer. Soc. Hort. Sci., 38:424-26.

Brierley, W.G. and Landon, R.H. (1944). Minn. Agr. Exp. Stn. Bul., 375: 1-24.

Bringhurst, R.S. and Khan, D.A. (1963). Amer. J. Bot., 50: 658-61.

Brooks, R.M. and Olmo, H.P. (1972). Register of New fruit and Nut varieties. Univ. California Press, Berkley.

Brooks, R.M. and Olmo, H.P. (1978). HortSci., 13: 552-82.

Browne, K.M., Geeson, J. D. and Dennis, C. (1984). Jr. Hort. Sci., 59: 197-204.

Browne, K.M., Geeson, J.D. and Dennis. C. (1982). Grower, 99: 17-19.

Budini, R., Tonelli, D. and Cirotti, S. (1980). Jr. Agric. Fd. Chem., 28: 1236-38.

Bunemann, G. (1962). Proc. 14th Int. Hort. Cong. Brussels, 1 :143-50.

Bunemann, G. and Gruppe, W. (1962). Garlenbauwiss., 27: 166-82.

Buxton, J.H. and Easterbrook, M.A. (1988). Plant Patho., 37:278-80.

Campbell, R.W. and Lingle, J.C. (1954). Proc. Amer. Soc. Hort. Sci., 64: 259-62.

Campos, R.P., Kwiatkowski, A. and Clemente, E. (2011). Revista Ceres, 58: 554-60.

Carlen, C., Potel, A.M. and Ancay, A. (2009). Acta Hort., 838: 73-76.

Casas, A., Calle, R., Castillo, E., Aquino, D., Gamboa, A., Gutierrez, P. and Leon, D. (2017). Acta Hort., 1156: 343-48.

Ceulemans, R., Baets, W., Vanderbruggen, M. and Impens, I. (1986). Scientia Hort., 28: 71-83.

Chalandon, A., Crisinel, P. and Paviot, J. (1980). Mededelingen van de Facultiy Landbouw-Wetenschappen, Rijksuniversiteit Gent, 45 : 207-19.

ChangHyun, L., JeongHo, M., Taeil, K. JinGook, K. and Matsumoto, K. (2011). Hort. Environ. Biotech., 52: 581-89.

Cheng, Z., LiNa, H., XiaoFeng, L., YuZhong, L., YuanYue, S. and JiaXuan, G. (2017). Pl. Growth Regu., 81: 175-81.

Chiu, T.F. and Bould, C. (1976). Jr. Hort. Sci., 51:525-31.

Choleva, B. and Kacharmazov, V. (1978). Gradinarska i Lozarska Nauka, 15: 54-60.

Choma, M.E. and Himelrick, D.G. (1982). HortSci., 17: 773-74.

Cianciara, T. (1979). Prace Instytutu Sadownictwa i Kwiaciarstwa W Skierniewicach. A., 21: 77-82.

Civello, P.M., Powell, A.Z.T., Sabchat, A. and Bennett, A.B. (1999). Pl. Physiol., 121:1273-79.

Clark, J.H. (1933). Proc. Amer. Soc. Hort. Sci., 30: 283-87.

Clark, J.H. (1941). N.J. Agr. Expt. Stn. Bull., p. 691.

Clay, D. (1979). Grower, 92: 30.

Clay, D. (1980). Proc. British Crop Protec. Conf., pp. 345-52.

Clay, D. and Davison. J. (1981). Grower, 95: 20.

Co, H. and Markakis, P. (1968). Jr. Fd. Sci., 33: 281-87.

Cohen, C.J., Grothaus, L.C. and Perrigan, S.C. (1981). Agric. Exp. Stn. Oregon State Univ., Corvallis. Spec. Rpt., 619.

Collins, W.B. (1966). Proc. Amer. Soc. Hort. Sci., 89: 331-35.

Colorio, G. and Glossl, F. (1986). Besseres Obst., 31:230-31.

Cooke, I.J. (1969). Jr. Hort. Sci., 44: 49-55.

Cordero de Mesa, M., Jimnez-Bermudez, S., Pliego-Alfaro, F., Quesada, M.A. and Mercado, J.A. (2000). Funct. Plant. Biol., 27: 1093-1100.

Costa, R.C., Calvete, E.O., Bordignon-Junior, C.L., Francescatto, V. and Reginatto, F.H. (2012). Acta Hort., 926: 509-15.

Couey, H.M. and Fallstad, M.N. (1966). Phytopath., 56: 1345-47.

Couture, R., Willemot, C., Avezard, C., Castaigne, F. and Gossclin, A. (1988). Phytochem., 27:2033-36.

Crisp, C.M., Beech, M.C. and Atkinson, D. (1988). Scientia Hort., 37: 61-70.

Cunha-Junior, L.C., Jacomino, A.P., Trevisan, M.J. and Scarpare-Filho, J.A. (2011). Revisa Brsileira de Fruiticult., 33: 1074-83.

Cutcliffe, J.A. and Blatt, C.R. (1984). Canad. J. Pl. Sci., 64: 945-49.

D'Anna, F., Ipichino, G. and D'Anna, E. (2014). Acta Hort., 10105: 53-58.

Dale, A., Gray, V.P. and Ricketson, C.L. (1986). Canad. J. Pl. Sci., 66: 1031-32.

Damiano, C. (1980). In: Proc. Conf. on Nursery Production of Fruit Plants through Tissue Culture. Applications and Feasibility, USDA-SEA, Agr. Res. Results. ARR-NE-11.

Dardi, R., Faedi, W. and Lavarone, E. (1996). Riv. di. Fruitti. e.di. Ortoflo., 58: 6: 73-76.

Dardi, R., Faedi, W., Bartoletti, L. and Turci, P. (1998). Covenhno, NaZionale La Fragola, Verona, Febbraio, pp. 251-61.

Darnell, R.L. and Martin, G.C. (1988). Jr. Amer. Soc. Hort. Sci., 113: 114-18.

Darnell, R.L., Cantliffe, D.J., Kirschbaum, D.S. and Chandler, C.K. (2003). Hort. Review, 28: 325-49.

Darrow, G.M. (1931). Proc. Amer. Soc. Hort. Sci., 28:231-35.

Darrow, G.M. (1966). The Strawberry. Hoir Winston. New York. pp. 477.

Darrow, G.M. and Scott, D.H. (1947). Proc. Amer. Soc. Hort. Sci., 50:239-42.

Daubeny, H.A. (1961). Canad. J. Pl. Sci., 41 : 239-43.

Daubeny, H.A. (1987). HortSci., 22: 511-13.

Daubeny, H.A., Lawrence, F.J., Martin, L.W. and Barritt, B.H. (1980). Stn. Bull. Agric. Exp. Stn. Oregon State Univ., 645: 40-42.

Daubeny, H.A., Norton, R.A. and Barritt, B.H. (1972). Plant Dis. Reptr., 56 : 792-95.

Daugovish, O., Su, H. and Gubler, W.D. (2009). HortTechnol., 19: 317-23.

Debnath, S.C. (2006). Jr. Hort. Sci. Biotech., 81: 349-54.

Denisen, E.L. (1927). Jr. Pom. Hort. Sci., 6: 104-12.

Denisen, E.L. (1959). Ia. Farm. Sa., 14: 14

Denisen, E.L. and Buchele, W. F. (1967). Proc. Amer. Soc. Hort. Sci., 94 :261-73.

Dennis, F.G. (1984). In. Physiological basis of crop growth and development (ed. M.B. Tesar). Amer. Soc. Agro. and Crop Sci. Soc., Madison, WI. pp. 265-88.

Dennis, F.G. Jr., Lipecki, J. and Kiang, c. (1970). Jr. Amer. Soc. Hort. Sci., 95: 750-54.

Dennis, Jr. F.G. and Bennerr, H.O. (1969). Jr. Amer. Soc. Hort. Sci., 94: 534-37.

Deyton, D., Sams, C., Kopsell, D., Cummins, J., Hayden, T. and Takeda, F. (2009). Presentation at Tenn. Fruit and Veg. Growers Meeting, Jan, 2009.

Diamanti, J., Capocasa, F., Mezzetti, B., Battino, M. and Giampieri, F. (2012). Acta Hort., 926: 101-106.

Dimitrov, P. and Topcijski, S. (1957). Sofia, 4: 103-10.

Dinkel, D.H. (1978). Acta Hort., 16: 91-96.

Dirinck, P.J., Pooter, H.L.De., Willaert, G.A. and Schamp, N.M. (1981). Jr. Agric. Fd. Chem., 29: 316-21.

Dodman, M. and Amiri, M.E. (2014). Jr. Sci. Tech. Greenhouse Cult., 4(16): 111-17.

Dolgikh, V.A. (1988). Sadovodstvo i Vinogradarstvo., 7: 24-25.

Dolgun, O. (2007). Jr. Sci. Food Agric., 87: 1364-67.

Dowling, M.E. and Schnabel, G. (2017). Pl. Dis., 101: 1034-35.

Draper, A.D., Galletta, G.J. and Swartz, H.J. (1981). HortSci., 16: 794-95.

Dughetti, A.C., Kirschbaum, D.S. and Conci, V.C. (2017). Revista de Investigaciones Agropecuarias, 43(1): 36-50.

Dumer, E.F., Barden, J.A., Himelrick, D.G. and Poling, E.B. (1984). Jr. Amer. Soc. Hort. Sci., 109:396-400.

Dwyer, L.M., Stawart, D.W., Houwing, L. and Balchin, D. (1987). HortSci., 22: 42-44.

Dyachonko, L.D. (1985). Visnik Sil's Kogospodars'koi Nauki., 8: 58-60.

Eaves, C.A. and Leefe, J.S. (1962). Canad. J. Pl. Sci., 42: 746-41.

Ebihara, Y., Uematsu, S. and Nomiya, S. (2010). J. Gen. Pl. Pathol., 76: 7-20.

Ellwanger, R.M., Barros, I.B.I. and De, Koller, O.C. (1984). Agronomia Subriogradense, 20: 101-10.

Eshghi, S. and Tafazoli, E. (2007). Int. J. Agric. Res., 2(2): 180-84.

Evenhuis, H.H. and Alofs, W.J. (1982). Fruitteelt, 72: 242-43.

Everett, P.H. (1971). Proc. Fla. Sta. Hort. Soc., 84:124-28.

Ezelle, B.D., Darrow, G.M., Wilcox, M.S. and Scott, H.II (1947). Food Res., 12: 510-26.

Fan, L., Fang, C., Dube, C., Deschenes, M., Dalpe, Y., Tao, S. and Khanizadeh, S. (2012). Acta Hort., 926: 491-96.

Fattahi, M. and Gholami, M. (2009). Jr. Appl. Hort., Lucknow, 11(2): 150-52.

Feng, Q.C., Wang, X., Zeng, W.G., Qui, Q. and Jing, K. (2012). Int. J. Agric. Biol. Eng. 5: 1-8.

Fiedler, W. (1980a). Archiv fur Gartenbau, 28: 279-94.

Fiedler, W. (1980b). Archiv fur Gartenbau, 28: 315-24.

Fiedler, W. and Weier, B. (1983). Archiv fur Gartenbau, 3l: 367-82.

Fiedler, W. and Weier, B. (1984). Institut fur Obstfersch-hung Dresden-Pillnitz. DDr-8057 Dresden.

Finn, C.E., Kempler, C., Moore, P.P., Stirk, B.C., Yorgey, B.M., Martin, P.R. and Galla, G.J. (2011). HortSci., 46: 1701-1705.

Finn, C.E., More, P.P., Yorgey, B.M., Lee, J., Stirk, B.C., Kempler, C. and Martin, R.R. (2013). HortSci., 48: 1184-88.

Folata, K.M. and Davia, T.M. (2006). Critical Reviews in Pl. Sci., 25: 399-415.

Forney, C.F. and Breen. P.J. (1985). Scientia Hort., 27: 55-62.

Fournier, P. (1982). Fruits, 37: 609-15.

Free, J.B. (1968). Jr. Hort. Sci., 43: 107-11.

Freeman, B. (1981). Australian J. Exp. Agric. Anim. Husb., 21:349-53.

Freeman, J.A. (1964). R.A.M. 44. 3: 98.

Freeman, J.A. (1983). Res. Review. Res., Stn., Agassiz. B.C. pp. 4-5.

FuTain, P., Quing, Z., YanMao, J., Yong, P. and Peng, Z. (2006). Pl. Nutr. Fert. Sci., 12: 400-405.

Galletta, G.J. and Brighurst, R.S. (1990). In: Small Fruit Crop Management (eds. G.J. Galletta and D. Himelrick). Prentice Hall, Englewood Cliffs, New Jersey. pp. 83-156.

GangJun, L., li, X., RuiXue, G., WenJi, X, and JiaJun, L. (2017). Euphytica, 213(8): 194.

Ganjehi, B. and Golchin, A. (2012). Jr. Sci. Tech. Greenhouse Cult., 2(8): 71-81.

Gardner, F.E. and Marth, P.C. (1937). Bot. Gaz., 99: 184-85.

Gavrilova, L.I. (1985). In: Probl. Vegetat. razmnozheniya y Sadovod. Moscow. pp. 128,132.

Gehrmann, H. (1985). Acta Hort., 171: 463-69.

Giannapolitis, C.N. (1987). Proc. Meetings E.C. Experts Croup. Dublin. pp. 95-98.

Gilbert, F.A (1982). Fruit Var. J., 36: 114-15.

Girgenti, V., Peano, C., Baudino, C. and Tecco, N. (2014). Science of the Total Environment, 473/474: 48-53.

Given, N.K., Vcnis, M.A. and Grierpon, D. (1988). Planta, 174: 402-06.

Goldberger, J. and Lyons, C. (2017). Performance and Adoptability: Biodegradable Mulch. Biodegradablemulch.org.

Gomez, A.O., Mattner, S.W., Oag, D., Nimmo, P.,Milinkovic, M. and Villalta, O.N. (2017). Acta Hort., 1156: 735-41.

Gonzalez Perea, R., Fernandez Garcia, I., Martin Arroyo, M., Rodriguez Diaz, J.A., Camacho Poyato, E. and Montesinos, P. (2017). Agric. Water Mng., 183: 194-201.

Gooselink, J.G. (1959). Ph.D. thesis. Rutgers Univ, New Jersey.

Gotoh, T., Kitashima, Y. and Sato, T. (2013). Int. J. Acarology, 39: 533-37.

Gould, H.J. and Jessop, N. (1981). Plant Pathol., 30: 171-75.

Govorova, G. (1993). Acta Hort., 348: 45-55.

Govorova, G.F. (1985). Sbornik Nauchnykh Trudov po Prikladnoi Botanike. Genetike r Selektsii. 97 :96-100.

Greve, E.W. and Shoemaker, J.S. (1930). Proc. Amer. Soc. Hort. Sci., 27: 183.

Guegov, Y., Karov, T. and Kalinov, V. (1982). Institut. Internation due Froid, 233-39. Institute of Hort. Plovdiv. Bulgaria.

GuiHua, J., Ming, X., Li, F., HanRong, W., YanJan, W., HuiQin, Z., PuLe, H. and ChongBo, S. (2007). Jr. Fruit. Sci., 24: 234-36.

Gunaydin, S. and kafkas, S. (2017). Acta Hort., 1156: 171-77.

Guttridce, C.G. and Anderson, H.M. (1981). Hort. Res., 21:83-98.

Guttridge, C.C. (1960). Bull. Inst. Agron. Stn. Rech. Cenrbloux. 2: 941-48.

Guttridge, C.C. and Anderson, H. M. (1973). Hort. Res., 13: 125-35.

Guttridge, C.G. and Montgomeric, L.C. (1971). Hort. Res., 11:52-59.

Guttridge, C.G., Mason, D.T. and Ing, E.C. (1965). Exp. Hort., 12: 38-41.

Hancock, F. (1999). Strawberries. CABI publishing, UK.

Hansen, E. and Waldo, C.F. (1944). Food Res., 9:453-61.

Hansen, H. (1967). Erwerbsobstbau., 9: 81-83.

Hansen, H. and Bohling, H. (1981). Besseres Obst., 26: 165-67.

Hansen, H.C. (1931). Amer. J. Bot., 18 : 658-73.

Hansen, P. and Ostcrmann, J. (1987). Frugtavleren, 16: 263-64.

Harrold, L.L. (1955). Jr. Amer. Soc. Agr. Eng., 36:669-72.

Hassell, R.L., Phillips, T.L., Dufault, R.J., Hale, T.A. and Ballington, J.R. (2006). Int. J. Fr. Sci., 6(2): 73-85.

Havis, L. (1938). Bimonthly Bull. Ohio Agr. Expt. Stn., 194: 168-72.

Haynes, R.J. (1987). Biology and Fertility of Soils., 5: 235-40.

Haynes, R.J. and Goh, K.M. (1987). Comm. Soil Sci. and Pl. Anal., 18: 457-71.

Hecht, C. (1972). Mechanical picker for strawberries. U.S. Patent, No 3, 698. 171.

Hernanz, D., Recamales, A.F., Melendez-Martines, A.J., Gonzalez-Miret, M.L. and Heredia. F.J. (2007). Jr. Agric. Fd. Chem., 55: 1846-52.

Hidaka, K., Dan, K., Imamura, H. and Takayama, T. (2017). Environment Control in Biol., 55: 21-27.

Higashida, K. Ohasa, M., Mori, K. and Imai, O. (2017). Ann. Rept. Kansai Pl. Protec. Soc. Japan, 59: 41-45.

Himelrick, D., Powell, A. and Dozier, W.A. Jr. (2002). Alanama Coop. Extn. System, ANR-0633, Auburn, AL.

Hirakawa, H., Shirasawa, K., Kosugi, S. *et al.* (2014). DNA Res. 21: 169-81.

Hirvi, J. (1983). Lobensmittel-Wissenschaft Und-Technologie, 16: 157-61.

Holcroft, D.M. and Kader, A.A. (1999). HortSci., 34: 1244-48.

Holden, D. and Ross, R. (2017). Acta Hort., 1156: 249-54.

Holevas, C.D. (1968). Mem. Gen. II Coloq. Eur. Medit Contr. Fert. Plant Cult. Seville., pp. 701-716.

Honda, F., Iwanaga, Y., Matsuda, T., Morishita, M. and Fushihara, H. (1981a). Bull. the Veg. and Ornamental Crops Res. Stn. C. Kurume, 5: 15-29.

Honda, F., Matsuda, T., Morishita, M., Iwanaga, Y. and Fushihara, H. (1981b). Bull. the Veg. and Ornamental Crops Res. Stn. C. Kurume, 5: 1-13.

Honjo, M., Nunome, T., Kataoka, S., Yano, T., Yamazaki, H., Hamano, M., Yui, S. and Morshita, M. (2011). Breed. Sci., 61: 420-25.

Hortynski, J. A. (1993). Acta Hort., 348: 157.

Hosseini, Y. and Bahadori, M. (2017). Jr. Sci. Tech. Greenhouse Cult., 7: (28) pp. Pel 49- Pel. 160.

Hua, L.N., Zang, M., Wang, S.F., Li, Y.Z., Shen, Y.Y. and Guo, J.X. (2017). Jr. Hort. Sci. Biotech., 92: 107-12.

Inaba, A. and Nakamura, R. (1978). Rept. Faculty Agric. Okayama Univ. Japan, 52: 25-36.

Ishikawa, M., Maeda, F., Itou, Y., Maruo, T. and Koba, T. (2007). Hort. Res. (Japan) 6(Suppl. 2): 216.

Iyer, C.P.A., Chacko, E.K. and Subramanian, M.D. (1970). Curr. Sci. 39:271-72.

Izhar, S. and Izsak, E. (1995). United States Patent # 5,444.179.

Jahn, O.L. and Dana, M.N. (1966). Proc. Amer. Soc. Hort. Sci., 88: 352-59.

James, D.J., Passey, A.J. and Barbara, D.J. (1990). Pl. Sci., 69: 79-94.

Jamieson, A.R., Nickerson, N.L., Forney, C.F., Sanford, K.A., Sanderson, K.R., Prive, J.P., Tremblay, R.J.A. and Hendrickson, P. (2004). HortSci., 39: 1178-80.

Jamieson, A.R., Sanderson, K.R., Prive,J.P. and Tremblay, R.J.A. (2009). HortSci., 44:836-38.

Jamwal, K. and Wali, V.K. (2011). Indian J. Hort., 68: 472-78.

Janik, J. and Eggert, D.A. (1968). Proc. Amer. Soc. Hort. Sci., 93: 311-16.

Jelenkovic, G., Chin, C.K. and Billings, S. (1986). HortSci., 21: 695(abst.).

Jett, L.W. (2007). Growing Strawberries in high tunnels in Missouri, Unversity of Missouri, Columbia, Extn. Leaflet.

Jiajun, L., Yuhu, L., Guodong, D., Hanping, D. and Mingguin, D. (2005). HortSci., 40: 1194-95.

Jimnez-Bermudez, S., Redorndo-Nevado, J., Munoz-Blanco, J., Caballero, J.L., Lopez-Aranda, J.M., Valpuesta, V., Pliego-Alfaro, F., Ouesda, M.A. and Mercado, J.A. (2002). Pl. Physiol., 128: 751-59.

Johansen, F.D. and Walker, R.B. (1963). Proc. Amer. Soc. Hort. Sci., 83:431-39.

Johanson, F. (1980). Hunger in Strawberris. K and H Printers Inc. Everett, Washington. p.36.

Johanson, F.D. (1963a). Wash. Stn. Extn. Bull., p. 561.

Johanson, F.D. (1963b). In: Nut. Def. Symp. Rutgers St. Univ. N.J. pp.75-80.

John, M.K. Daubeny, H.A. and Macelory, F.D. (1975). Jr. Amer. Soc. Hort. Sci., 100: 513-17.

John, O.A. and Yamaki, S. (1994). Jr. Amer. Soc. Hort. Sci., 119: 1024-28.

Jones, J.K. (1955). Ph. D. thesis, University of Reading, U.K.

Jones, J.K. (1966). Sci. Hort., 18: 121-30.

Jones, O.P., Waller, B.J. and Beech, M.C. (1988). Pl. Cell Tissue and Organ Cult., 12:235-41.

Jonkers, H. (1958). Euphytica, l: 41-46.

Jonkers, H. (1965). Meded. Landb Hoogesch., Wageningen, 65-66: 59.

Joolka, N.K., Badiyala, S.D, and Lakhanpal, S.C. (1984). Proc. Nat. Symp. Temperate Fruits, Solan, India, pp. 225-27.

Jorge, T. de S., Soares, A.H., Motta, L.C. do C., Brito, C.F. da C., Silva, M.H.G.E. and Barbosa, W.J. (2013). Acta Hort., 1012: 533-38.

Jurcev, V.K., Velutina, V.A. and Lozuneva, M.F. (1966). Sadovodstvo, 1: 24-25.

Kaczmarska, E. (2003). Acta Scientiarum Polonorum Hortorum Cultus, 2(1): 111-16.

Kadir, S., Carey, E. and Ennahli, S. (2006). HortSci., 41: 329-35.

Kalon, J. and Jaskowiec, J. (1982). Fruit Sci. Rept., 9 : 195-203.

Kamber, R., Eylen, M. and Tok, A. (1986). Res. Rept. Tarsus Res. Inst. Village Affairs, pp. 1-39.

Kano, Y. and Asahira, T. (1979). Jr. Jap. Soc. Sci., 47:463-72.

Kargi, S.P., Kafkas, E. and Baktemur, G. (2012). Acta Hort., 926: 163-67.

Karmer, S. (1964). Archive fur Gartenbau, 12: 255-68.

Kaska, N. and Gezerel, O. (1983). Bache, 12: 29-34.

Kataoka, S. and Noguchi, Y. (2017). Acta Hort., 1156: 197-202.

Kaya, C., AK., B.K., Higgs, D. and Murillo-Amador, B. (2002). Aust. J. Exp. Agric., 42: 631-36.

Kaya, C., Higgs, D., Saltali, K. and Gezeral, O. (2002). Jr. Pl. Nutr., 25: 1415-17.

Kelin, Y., SunJun, L., Chun, Q., Lide, Z. and JingFen, L. (2001). Jr. Southwest Agri. Univ. China, 23: 235-38.

Kemp, I. (1976). New Zealand J. Agr., 132: 57-58.

Kepczynska, E.K. (1989). Fruit Sci. Rept., 16: 77-27.

Khammuang, S., Dheeranupattana, S., Hanmuangjai, P. and Wongroung, S. (2005). Songklanakarin Jr. Sci. Tech., 27: 693-703.

Khan, M.L. (2003). Indian J. Entomol., 33: 83-84.

Khanizadeh, S., Theriault, B., Carisse, O. and Buszard, D. (1999). HortSci., 34: 1286-87.

Khanizadeh, S.D. and Fanous, M. (1992). HortSci., 27: 246-48.

Kiprijanovski, M., Arsov, T. and Gjamovkshi, V. (2012). Acta Hort., 926: 177-84.

Klamkowski, K. and Treder, W. (2008). Jr. Fruit Orna. Res., 16: 179-88.

Klatt, B.K., Holzschuh, A., Westphal, C., Clough, Y., Smit, I., Pawelzik, E. and Tscharntke, T. (2014). Proc. Royal Soc. Londan, Ser. B. Biological Sci., 281(1775): 20132440.

Klossowski, W. and Sobezykiewicz, D. (1966). Inst. Sadown. W. Skier-Niewicach, 10: 213-24.

Knee, M., Sargent, A. and Osborne, D.J. (1977). Jr. Expt. Bot., 8: 377-96.

Knee, M., Sargent, J.A. and Osborne, D.J. (1977). Jr. Expt. Bot., 28:377-96.

Kondo, N. and Monta, M. (1999). ASAE pap No: 99-3071.

Kowng, S.S. (1967). Proc. Amer. Soc. Hort. Sci., 91:257-60.

Kremenovic, G. (1980). Arhiv Za Poljoprivredne Nauka, 41: 509-16.

Krupyshev, P.V. (1977). Referativnyi Zhurnal, 3.55:952.

Kunisha, M. (2011). J. Jap. Soc. Hort. Sci., 80: 231-43.

Kunisha, M., Fukino, N. and Matsumoto, S. (2003). Euphytica, 134: 209-15.

Labanowska, B.H. and Rechnio, H. (1977-78). PR. INST. SAD. 20: 249-54.

Labanowska, B.H., Rechnio, H. and Pala, E. (1981). Fruit Sci. Rept., 8: 91-95.

Lacey, C.N.D. (1973). Euphytica, 22: 546-54.

Lake, C.T. and Bennett, H.A. (1980). Proc. British Crop. Protec. Conf.1980, pp. 315-20.

Lamberts, D., Sohrevens, E. and Lettani, L. (1985). Boer en de Tuinder, 91: 15.

LaMondia, J.A. (2002). Jr. Nematol., 31: 418-23.

LaMondia, J.A., Richar, S.C. and Lorraine, L. (2005). HortSci., 40: 1366-70.

Laneri, U. and Damino, C. (1980). Annali dell Istituto Sperimentale per la Fruiticoltura, 11: 43-48.

Lange, E., Plocharski, W. and Lenartowicz, W. (1978). Fruit Sci. Rept., 5: 39-46.

Larson, K.K. (1994). In: Hanbook of Environmental Physiology of Fruit Crops. Vol. I. Temperate Fruits (eds. B. Schaffer and P.C. Anderson) CRC Press, Boca Raton, FL. pp. 2-17.

Latimer, L.P. (1943). Proc. Amer. Soc. Hort. Sci., 42: 441-43.

Laugale, V., Lepse, L., Strautina, S., Krasnova, I. and Seglina, D. (2012). Acta Hort., 926: 517-23.

Laugale, V., Lepse, L., Strautina, S., Krasnova, I. and Seglina, D. (2012). Acta Hort., 926: 517-23.

Laurinen, E. and Sako, J. (1980). Annales Agric. Fenniae, 19 : 164-72.

Lawson, H.M. and Wiseman, J.S. (1980). Proc. British Crop Protec. Conf. 1980, pp. 353-60.

Lawson, H.M. and Wiseman, J.S. (1987). Proc. British Crop Protec. Conf. 1987. Vol.2, pp.657-63.

Ledesma, N. and Sugiyama, N. (2005). Jr. Amer. Soc. Hort. Sci., 130: 341-47.

Lenartowicz, W. and Plocharski, W. (1986). Prace Instytutu Sadownlictwa i Kwiaciarstwa w Skierniewicach, A. 26: 137-44.

Lenz, F. (1979). Erwerbsobstbau, 2l: 146-48.

Leska, W. (1965). PR. INST. SAD. 9 : 235-44.

Li, J., Lai, T., Song, H. and Xu, X. (2017). Jr. Pl. Physiol., 64: 251-59.

Li, L., Wei, X., Meiling, J., Chao, Y., Ling, L., DongSheng, D. and XiLing, F. (2017). Jr. Integrative Agric., 16: 1502-12.

Lieten, F. (1996). Acta Hort., 439: 623-25.

Lieten, P. (2005). Int. J. Fruit Sci., 5: 89-103.

Lieten, P. (2013). Int. J. Fruit Sci., 13: 84-90.

Lindeloof, C.P.J. Van De, and Wassenaar, L.M. (1987). Fruitteelt, TT: 16-17.

Linder, R. (1964). Fruit Belge, 32: 76-78.

Lineberry, R.A. and Burkhart, L.V. (1943). Plant Physiol., 18: 324-33.

Liu, F., Savic, S., Jensen, C.R., Shahnazari, A., Jacobsen, S.E., Stikic, R. and Anderson, M.N. (2007). Scientia Hort., 109: 113-17.

Liu, Z.R. and Sanford, J.C. (1988). HortSci., 23: 1057-59.

Ljones, B. (1960). Frukt. Og Bacr. 13 : 42-51.

Locascio, S.J. and Smart, G.C.Jr. (1968). Proc. Fla. Sta. Hort. Soc., 81: 147-53.

Lockhart, C.L. and Eaves, C.A. (1996). Can. J. Pl. Sci., 46: 151-54.

Lopez-Aranda, J.M., Miranda, L., Dominguez, P., Soricia, C., Perez-Jimenez, R.M., Zea, T., Talavera, M., Velasco, L., Romero, F., Santos, B de los. and Medina-Minuez, J.J.(2012). Acta Hort., 926: 407-13.

Lott, W. L. (1946). N.C. Agr. Expt. Stn. Ann. Rept., 69: 66.

Lott, W.L. (1947). N.C. Agr. E.xpt. Stn. Ann. Rcpt., 70: 50.

LuHe, Z., BaiHong, C. and Li, A. (2013). Storage and Process, 13: 23-28.

Malodobry, M. (l987). Zeszyty Naukowe Akademii Rolniczci int. Hugona Kollataja W Krakowie. Ogrodnictwo. 220:17-33.

Mann, C.E. (1930). Ann. Bot., 44: 55-86.

Mann, C.E. and Ball, E. (1927). Jr. Pom. Hort. Sci., 6: 104-12.

Marchal, J., Fournier, P. and Munsch, D. (1982). Fruits, 37:773-83.

Marocchi, G. and Grandi, M. (1978). Pudicazione. Instituto di Cloltivazioni Arborec. Italy, 12: 213-20.

Martelli, G., Sunseri, F.,Greco, I., Sabina, M.R., Porreca, P. and Levi, A. (1999). Adv. Hort. Sci., 13(3): 99-104.

Martin, E. and Tepe, E. (2014). Cold climate strawberry farming. Dept. Hort. Sci.e-book, University of Minnesota, St. Paul. MN.

Masabni, J.G., Strang, J.G. and Jones, R.T. (2007). UK. Cooperative Ext. Service, Univ. of Kentucky, UK. HO-16.

Masiunas, J.B., Weller, S.C. and Hayden, R.A. (1986). Abst. 22nd Int. Hort. Congr. California. No.200.

Mason, G.A. and Guttridge. C.G. (1974). Scientia Hort., 2: 299-308.

Matvienko, M.V. (1984). Referativnyi Zhurnal. 55.2 : 763.

McKinlay, R.G. and Thomson, M.E. (1987). Proc. Crop Protec. Northern Britain, Dundee Univ., pp.373-77.

McKinley, R.T. and Talboys, P.W. (1979). Ann. Appl. Biol., 92: 347-57.

McPhie, D. and Burrak, H.J. (2017). Jr. Economic Entomol., 110: 208-12.

Medina-Minuez, J.J., Miranda, L., Dominguez, P., Soricia, C., Perez-Jimenez, R.M., Zea, T., Talavera, M., Velasco, L., Romero, F., Santos, B de los and Lopez-Aranda, J.M. (2012). Acta Hort., 926: 637-44.

Medina-Puche, L., Cumplido-Laso, G., Anil-Ruiz, F., Hoffmann, T., Ring, L., Rodriguez-Franco, A., Caballero, J.L., Schwab, W., Munoz-Blanco, J. and Blanco-Portales, R. (2014). Jr. Exp. Bot., 65: 401-17.

Meland, M. (1986). Giartneryrket. 76 : 67-68.

Meland, M. (1987). Noesk landbruksforkning, 1: 169-73.

Mellenthin, G. (1963). Erwobstd., 5: 67-68.

Meszka, B. and Bielenin, A. (2010). Phytopathologia, 58: 13-19.

Mezzetti, B., Landi, L., Pandolfini, T. and Spena, A. (2004). BMC Biotech., 4(4): 1-7.

Miller, P.W. and Waldo, G.F. (1959). Plant Dis. Reptr., 43: 1120-31.

MingLi, K., JinJun, G., XiaoLei, G. and Min, Q. (2017). Storage and Process, 17: 10-14.

Miyazaki, M., Miva, S. and Yabuuchi, K. (1980a). Jr. Jap. Soc. Fd. Sci. and Tech., 27: 559-63.

Miyazaki, M., Miya, S. and Yabuuchi, K. (1980b). Jr. Jap. Soc. Fd, Sci. and Tech., 27: 564-68.

Modoran, L. and Gherghi, A. (l958). Lucr. Sti. Inst. Cerc. Hortivrtic. pp. 485-89.

Montgomerie, I.G. and Kennedy, D.M. (1980). British Crop Protec. Conf. Pcst and Diseascs. Vo. I. pp. 185-92.

Moore, J.N. (1961). Ph.D. thesis. Rutgers Univ., New Jersey.

Moore, J.N. (1962). Jr. Amer. Soc. Hort. Sci., 94: 362-64.

Moore, J.N. (1982). Arkansas Univ., Fayettevile. AR 727001.

Moore, J.N. and Hough, L.F. (1962). Proc. Amer. Soc. Hort. Sci., 81 :255-64.

Moore, P.P. and Finn, C.E. (2002). HortSci., 37: 230-32.

Moore, P.P., Hoashi-Erhardt, W., Finn, C.E. and Kempler, C. (2011). HortSci., 46: 1698-1700.

Morris, J.R., Kattan, A.A., Nelson, G.S. and Cawthon, D.L. (1978). HortSci., 13: 413-22.

Morris, J.R., Kattan, A.A., Sistrunk, W.A., Cawthon, D.I. and Nelson, G.S. (1974). Ark. Farm Res., 23: 8.

Mortensen, F. (1934). Proc. Amer, Soc. Hort. Sci., 32:424-28.

Moyer, C., Whitaker, V.M. and Peres, N.A. (2010). Florida Coop. Extn. Service, Univ. of Florida, Doc. pp 273. pp. 7.

Mudge, K.W., Narayanan, K.R., Poovaiah, B.W. (1981). Jr. Amer. Soc. Hort. Sci., 106: 80-84.

Musselman, R.C., Sterrett, J.L. and Voth, V. (1988). HortSci., 23: 128-30.

MyeongHyeon, N., HyeonSuk, K., WonKeun, L.Gleason, M.L. and HongGi, K. (2011). Korean J. Hort. Sci. Tech., 29: 151-55.

Naber, H. (1987). Proc. Meeting E.C. Experts Group. Dublin. pp. 71-84.

Nardi, C.F., Villarreal., N.M., Opazo, M.C., Martinez, G.A., Moya-Leon, M.A. and Civello, P.M. (2014). Scinentia Hort., 165:111-122.

Naumann, W.D. (1964). Gartenbau., 29: 21-30.

Nehra, N.S., Chibbar, R.N., Kartha, K.K., Datla, R.S.S., Crosby, W.L. and Stushnoff, C. (1990). Pl. Cell Rept., 9: 10-13.

Neilson, B.V. and Eaton, G.W. (1983). HortSci., 18 : 932-34.

Nestby, R. (1982). Forskning Og Forsok i Landbrusket, 33: 37-42.

Nichols, M.A., Duggan-Jones, D.I. and Christie, C.B. (2018). Acta Hort., 1205: 829-84.

Niemann, J. (1966). Rhe in Mschr. Gemuse, Obstu, Schnittblumen, 54: 391-92.

Nikolov, D. (1987). Rasteniev"dni Nauki. 24: 85-92.

Nikto-Chkina, T.D. and Gusev, A.M. (1984). Izvestiya Timiryazevskoi Sel'skokhozyaistvennoi Akademii, 1: 117-24.

Nitsch, J.P. (1950). Amer. J. Bot., 37: 211-15.

Nitsch, J.P. (1970). In: The Biochem. of Fruits and their Prodts. (ed. A.C. Hulme) Vot. II. Academic 1 Press. NY pp. 427-72.

Njoroge, S.M.C., Kabir, Z., Martin, F.N., Koike, S.T. and Subbarao, K.V. (2009). Pl. Dis., 93: 519-27.

Noguchi, Y., Morishita, M. Muro, T., Kojima, A., Sakata, Y., Yamada, T. and Sugiyama, K. (2011). Bull. Nat. Inst. Veg. and Tea. Sci., 30: 59-67.

Nour, V., Trandafir, I. and Cosmulescu, S. (2017). Erwerbsobstbau, 59: 123-31.

Odongo, T., Isutsa, D.K. and AguYoh, J.N. (2011). Jr. Agric. Sci. Tech., 13: 7-21.

Olsen, J.L., Martin, L.W. and Breen, P.J. (1985). HortSci., 20:74-76.

O'Neill, S.D. (1983). Plant Physiol., 72: 931-37.

Opstad, N., Sonsteby, A., Myrheim, U. and Heide, O.M. (2011). Scientia Hort., 129: 127-34.

Oregel-Zamudio, E., Angoa-Perez, M.V., Oyoque-Salcedo, G., Aguilar-Gonzalez, C.N. and Mena-Violante, H.G. (2017). Scientia Hort., 214: 273-79.

Osipova, L.V., Raslorguev, S.L. and Tyulenev, V.M. (1986). Byulleten Nauchnoi Informatsii Tsentral'noi Geneticheskoi Laboratorii, No. 43, pp. 27-31.

Ourecky, D.K. and Reich, J.E. (1976). HortSci., 11: 413-14.

Ourednickkova, J. (2011). Vedecke Prace Ovocnarske, 22: 213-22.

Overman, A.J. and Howard, C.M. (1984). Proc. Amer. Phytopath. Soc., 39: 98.

Oydvin, J. (1965). Meld. Norg. Landbrtogsh, 4: 1-11.

Oydvin, J. (1968). Gartneryrket, 58: 402-17.

Palha, M.G., Campo, J.L. and Oliveira, P.B. (2012). Acta Hort., 926: 463-69.

Paliyath, G. and Droillard, M.J. (1992). Pl. Physiol. Bichem., 57: 5743-49.

Pallavi, C.M., Rekha, R. and Neelambika, T.M. (2011). Indian J. Hort., 68: 24-27.

Pandey, P., Phoghat, K.P.S. and Shukla, S.N. (1981). Progressive Hort., 12:67-70.

Papadopoulos, I. (1987). Fertilizer Res., 13: 269-76.

Parikka, P.K., Vestberg, M., Karhu, S.T., Haikonen, T. and Hauisalo, J. (2017). Acta Hort., 1156: 751-55.

Peerbooms, H. (1985). Fruitteelt, 75:774.

Peng, J., HuanYu, W., Yu, Z., YuPing, H., Li, W. and YoungHua, Z. (2017). Scientia Hort., 225: 106-11.

Perez, A.G., Rio, J.R., Sanz, C. and Olfas, J.M. (1992). J. Agr. Food. Chem., 40: 2232-35.

Phogat, K.P.S., Pandey, D. and Shukla, S.N. (1980). Progressive Hort., 11:31-33.

Pincemail, J., Kevers, C., Tabart, J., Defraigne, J.O. and Domes, J. (2012). Jr. Fd. Sci., 77(2): C205-C210).

Pion, S., Oliveira, D.De and Paradis, R.O. (1980). Phytoprotection, 61:12-18.

Pipattanawong, R., Yamane, K., Fujishige, N., Bang-SangWoo and Yamaki, Y. (2009). Jr. Jap. Soc. Hort. Sci., 78: 300-306.

Plocharski, W. (1989). Fruit Sci. Rept., 16: 127-40.

Plocharski, W., Lange, E. and Lenartowicz, W. (1978). Fruit Sci. Rept., 5: 47-55.

Plocharski, W., Lenartowicz, W. and Lange, E. (1979). Bull. de I'Institut Int. du Froid, 59: 1152.

Poling, E.B. and Safley, C.E. (1986). HortSci., 21: 1409-11.

Pommier, J.J. (1986). Revue Horticole., 265: 29-33.

Popova, L. (1984). Gradinaarska i Lozarska Nauka, 21 : 31-34

Popova, L.V., Zharkova, I.V. and Zekalashvili, A.U. (198-5). In Probl. vegetal. razrnnozheniya u Sadovod, Moscow. pp. 112-18.

Port, C.M. and Scopes, N.E.A. (1981). Plant Patho., 30: 95-99.

Pringer, A.A. and Scott, D.H. (1964). Proc. Amer. Soc. Hort. Sci., 84: 295-301.

Profic-Alwasiak, H., Szczygiel, A. and Grzymala, L (1980). Prace Instytutu Sadownictwa iKwaiaciarstwa w Skierniewicach, A. 22 : 137-46.

Pyrotis, S., Abayomi, L., Rees, D. and Orchard, J. (2012). Acta Hort., 926: 567-70.

Rahman, M.A., Begum, M.M. and Ahmad, M.R. (2013). Int. J. Postharvest, Tech. and Innovation, 3:230-43.

Rahman, T., Spafford, H. and Broughton, S. (2012). Entomologia Experimentals et Applicata, 142: 258-70.

Rath, N. (1987). Proc. Meeting E.C. Experts Group, Dublin. pp. 107-113.

Recupero, S., Acruti, P. and Magrini, M. (1988). Informatore Agrario., 44: 79-82, 85-88.

Reyes, F.G.R., Wrolstad, R.E. and Cornwell, C.J. (1S82). Jr. Assoc. Official Analytical Chemist, 65 : 126-31.

Rhainds, M., Kovach, J. and English-Loeb, G. (2002). Biol. Agric. and Hort., 19: 333-53.

Rho-IlRae, YongSeop, C., JaeWan, C., HoJeong, J. and HaeBoong, J. (2007). Korean J. Hort. Sci. Tech., 25: 12-16.

Ribeiro, C., Vicente, A.A., Teixeira, J.A. and Miranda, C. (2007). Postharvest Biol. Technol., 44: 63-70.

Ricketson, C.L., Gray, V.P., Cook, F.L, Vandenberg, A.A. and Dale, A. (1986b). Canad. J. Pl. Sci., 66: 1029-30.

Ricketson, C.L. and Hill, R.G. (1966). Proc. 17th Int. Hort. Cong., 1: 418.

Ricketson, C.L., Gray, V.P., Cook, F.L. Vandenberg, A.A. and Dale. A. (1986a). Canad. J. Pl. Sci., 66: 1025-26.

Riggs, D.I.M., Righetti, T.L. and Martin, L.W. (1987). Comm. Soil Scil and Pl. Anal., 18: 1453-67.

Robertson, M. and Wood, C.A. (1954). Jr. Hort. Sci., 29:104-11.

Rogers, W.S. (1935). Rept. East Malling Res. Stn. for 1934, pp. 111-19.

Rom, R.C. and Dana, M.N. (1960). Proc. Amer. Soc. Hort. Sci., 75: 367-72.

Rosati, P. (1966). Atti 11 Congr. Int. Appl. Mat. Plast. Agric. Pisa, pp. 259-61.

Rosati, P. and Faedi, W. (1980). Frutticoltura, 42: 47-49.

Rosen, C.J., Hoover, E.E. and Luby, J.J. (1988). Canad. J. PI. Sci., 68: 277-82.

Rosenberg, H. (2003). Calf. Farmer, August 2003, M1, M2, M9.

Roussos, P.A., Triantafillids, A. and Kepolas, E. (2012). Acta Hort., 926: 541-46.

Rudolph, V. (1986). Gartenbau, 33: 336-37.

Rudolph, V., Makosz, E., Szabo, A., Ivanov, A.J. and Harant, M. (1986). Gartenbau, 33: 340-42.

Rupp, S., Plesken, C., Rumsey, S., Dowling, M. Schnabel, G., Weber, R.W.S. and Hahn, M. (2017). Appl. and Environ. Microbiol., 83(9):, pp.e00269.

Rybalov, L.N., Dymchenko, M.I. and Gnevkovskaya, M.G. (1985). Sadovodstvo, 5: 14-15.

Saadati, S. and Moallemi, N.A. (2012). Iranian J. Hort. Sci., 42: 267-75.

Saha, A.K. (1964). Ph.D. thesis, University of Missouri, Columbia.

Samad, S., Kurokura, T., Koskela, E., Toivainen, T., Patel, V., Mouhu, K. Sergeant, D.J. and Hytonen, T. (2017). Hort. Res., 4: 17020 pp. (http://www.nature.com/articles/hortres201720.)

Samadi, S., Ghasemnezhad, A. and Imani, J. (2017). Acta Hort., 1156: 643-52.

Samchez-Cuevas, M. and Salaverria, J.L. (2004). Revista Cientifica UDO Agricola, 4(1): 21-26.

Sampson, C. and Kirk, W.D.J. (2009). PloS ONE, 8(11): e80787.

Sanford, J.C., Ourecky, D.K., Reich, J.E. and Aldwinckle, H.S. (1982). HortSci., 17:982-84.

SanJun, G., WenJing, G. and Beck, J.E. (2017). HortTech., 27(1): 84-92.

Sannino, G. and Priore, R. (1979). Fruits, 34: 503-06.

Santos, B., Chandler, C.K., Whidden, A.J. and Sanchez, M.C. (2008). Hort. Sci. Dept., UF/IFAS Fla. Coop. Extn. Serv., HS 1134.

Sato, Y., Yamakawa, O. and Honda, F. (1986). Bull. Vege. and Ornamental Crop Res. Stn. Fukuoka, Japan, 9: 23-30.

Savini, G. and Neri, D. (2004). Acta Hort., 649: 169-76.

Saxena, G.K. and Locascio, S.J. (1968). Jr. Amer. Soc. Hort. Sci., 92: 354-62.

Schaart, J.G., Krens, F.A., Pelgrom, K.T.B., Mendes, O. and Rouwendal, G.J.A. (2004). Pl. Biotechnol. Jr., 2: 233-40.

Scherer, W. (1987a). Obstbau, 12: 225-27.

Scherer, W. (1987b). Obstbau, 12: 228-30.

Schubert, O.E. (1969). Proc. Northeastern Weed Sci. Soc., 23: 127-34.

Scopes, N. and Port, C. (1981). Grower, 95: 73.

Scott, D.H., Draper, A.D. and Creeley, L.W. (1972). HortSci., 7: 382-84.

Seipp, D. (1987a). Proc. Meeting E.C. Experts Group, Dublin, pp. 99-106.

Seipp, D. (1987b). Obstbau, 12: 214-17.

Serce, S. and Ozgen, M. (2018). Chronica Hort., 55(3): 21-25.

Seyama, N. and Takai, T. (1980). Bull. Vege. and Ornamental Crops Res. Stn., B. Morioka, 3:39-53.

Sharma, R.L. and Badiyala, S.D. (1980). Progressive Hort., 12: 17-23.

Sharma, S.D. and Chauhan, J.S. (1986). Progressive Hort., 18:126-28,

Shoemaker, J.S. (1978). Strawberries. Small Fruit Culture, AVI. Westport, Conec. pp. 103-187.

Shoemaker, J.S. and Marth, P.C. (1953). Proc. Amer. Soc. Hort. Sci., 62: 255-56.

Shoernaker, J.S. (1955). Small Fruit Culture. McGraw Hill Book Co. Inc., New York. pp. 113-230.

Shulaev, V., Sargent, D.J., Crowhust, R.N. *et al.* (2011). Nature genetics, 43(2): 109-118 doi: 10. 1038/ng.740.

Simanek, J. and Skulcova, E. (1979). Hodnotenic sortimentu jahod Zbornik UVTIZ, Zahradnictvi, 4:43-48.

Simpson, D.W. (2014). Acta Hort., 1049: 107-11.

Singh, D.B., Singh, R., Kingsly, A.R.P. and Sharma, R.R. (2011). India J. Agric. Sci., 81: 407-12.

Singh, R., Sharma, R.R. and Tyagi, S.K. (2007). Scientia Hort., 112: 215-20.

Singh, R.P., Srivastava, R.P. and Phoghat, K.P. (1983). Punjab Hort. J., 23: 64-68.

Skrede, G. (1980). Jr. Sci. Fd. Agric., 31: 670-76.

Skulcova, E. (1980). Vedecke Prace Vyskumneho Ustav Ovocnych a Okrasnych Drevin v Bojniciach, 2:127-34.

Smeda, R. J. and Putnam, A.R. (1988). HortSci., 23: 134-35.

Smeets, L. (1968). Ph.D. thesis. Uitgeverii Ceres-Meppel.

Smeets, L. (1979). Scientia Hort., 12: 19-26.

Smeets, L. (1980a). Scientia Hort., 13: 13-35.

Smeets, L. (1980b). Scientia Hort., 13: 343-46.

Smeets, L. and Korenberg, H.G. (1955). Euphytica, 4: 53-57.

Smith, W.H. (1957). Scientific Hort., 12: 147-53.

Smith, W.L. and Heinze, P.H. (1958). Proc. Amer. Soc. Hort, Sci., 72:207-11.

Smolarz, K. and Zurawicz, E. (1979). Polski. Prace Instytutu Sadownictwa i Kwiaciarstwa W Skierniewicach. A. 21: 83-95.

Snyder, M.I., Paulus, A.O., Voth, V. and Nelson, J. (1987). Calif Agric., 418: 11.

Soosten, R. (1981). Obstbau, 6: 204-06.

Soria, C., Medina-Minguez, J.J., Miranda, l., Dominguez, P., Ariza, M.T., Refoyo, A. and Lopez-Aranda, J.M. (2012). Acta Hort., 926: 139-42.

Spayd, S.E. and Morris, J.R. (1981). Jr. Amer. Soc. Hort. Sci., 106: 101-05.

Sruamsiri, P. and Lenz, F. (1986). Gartenbau., 51: 84-92.

Stancevic, A. (1970). Jugoslovensko Vocarstvo. 4: 111-15.

Standt, G. (1999). Pub. In Bot. Univ. Calf. Pub. Berkley, CA, 81, 122.

Stoker, R. and McCarthy, T.P. (1936). N.Z. Commercial Grower, 41: 36

Strabbioli, G. (1985). Annali dell'Istituto Sperimentale per la Frutticoltura, l6: 75-84.

Strand, L.L. (1994). University of California, Pub. No. 335.

Strik, B.C. and Procror, J.T.A. (1988b). Jr. Amer. Soc. Hort. Sci., 113: 620-23.

Strik, B.C. and Proctor, J.T.A. (1988a). Jr. Amer. Soc. Hort. Sci., 113: 114-29.

Stutte, G.W. and Darnell, R.L. (1987). HortSci., 22: 218-21.

Sweeney, J.P., Chapman, V.J. and Hepner, P.A. (1970). Jr. Amer. Diet. Assn., 57: 432-35.

Swift, C.E. (2011). Colorado State University Fact Sheet No. 2.931.

Szczygiel, A. (1981). Fruit Sci. Rept., 8: 115-19.

Szczygiel, A. and Danek, J.C. (1984). Fruit Sci. Rept., 11:81-85.

Tacconi, R. (1980). Informatore Fitopatologico, 30: 3-8.

Tafazoli, E. and Shaybany, B. (1978). Jr. Amer. Soc. Hort. Sci., 103: 372-74.

Tafazoli, E. and Vince-Prue, D. (1979). Ann. Bot., 43:125-34.

TaiHeng, K. and Melton, L.D. (2002). Postharvest Biol. Technol., 26: 23-33.

Taimatsu, T., Nagamura, S. and Mizula, M. (1985). Bull. Nara Agric. Exp. Stn. Japan, 16:43-50.

Takayanagi. K. and Kawata. J. (l988). Jap. J. Breed., 38: 117-119.

Takeda, F. (2008). Proc. 2007 N. Amer. Strawberry Symp. Ventura, CA. 14-16 Feb. 2007, pp. 32-37.

Takeda, F. (2012). Acta Hort., 926: 237-41.

Takei, M. (2010). The cultivation of strawberry in Japan. Safe vegetable Promotion project in Benguet. Japan. Agric. Exchage. Council. pp. 40.

Tanboon, E.K.P., Grove, G.G. and Ellis, M.A. (1983). Ohio Report, 68:12-13.

Taschner, N. (2012). Obst-und Weinbau, 148: 9-11.

Taylor, B.K., Karoon, K.H. and Pockett, R.J. (1975). Jr. Aust. Inst. Agril. Sci., 41 :207-09.

Tehranifar, A., Postchi, M., Arooei, H. and Nematti, H. (2007). Acta Hort., 761: 485-88.

Tekinel, O., Kaska, N., Dinc, G. and Yurdakul, O. (1984). Doga Bilinr Dergisi., 4: 48-56.

Terk-Uile, J.H. (1964). Fruitteelt. 54: 434-36.

Thompson, P.A. (1964). Jr. Expt. Bot., 15: 347-58.

Thompson, P.A. and Guttridge, C.G. (1959). Nature, 114: 72-73.

Thompson, P.A. and Guttridge, C.G. (1960). Ann. Bot., 24: 482-90.

Thuesen, A. (1983). Statens Planteavl for sog., 85: 3.

Thuesen, A. (1984). Tiddskrift for Planteavl., 58: 69-73.

Tietz, L. and Gebhart, C. (1979). Gartenbau, 26: 365-66.

Townshend, J.L., Olthof, T.H.A. and Staples, J.E. (1966). Pl. Dis. Survey, 46: 144-45.

Tsolva, E., Koleva, L., Kalcheva, S. and Open, De, G. (2017). Proc. Latvian Acad. Sci. 71: 225-88.

Tucker, D.A. (1944). Virginia Fruit, 32: 10,12.

Tuei, R.B. (1952). Ph.D. thesis, Cornell Univ. Ithaca, NY.

Tyrka, M., Dziadczyk, P. and Hortyski, J.A. (2002). Euphytica, 125: 273-80.

Ugolik, M., Kantorowicz-Bak, M. and Orlikowska. J. (1986). Prace Komrsji Nauk Rolniczych iKomisji Nauk Lesnych., 6l: 263-69.

Ulrich, A., Mostafa, M.A. and Allen, W.W. (1980). Strawberry Def. Symp.: a Visiual and Plant Analysis Guide to Fertilization, Div. Sci., Univ. California, Berkeley.

Ulrich, M. and Sorge, P. (1981). Gartenbau, 28: 341-42.

Vacante, V. (1979). Tecnica Agricola, 31: 355-63.

Van den Muyzenberg, E.W.B. (1942). Landb Hogesch. Wageningen, Lab. Tuinpl Teelt, 37: 1-151.

Van der Boon, J. (1961). Fruitteelt, 51: 1129.

Vappula, N.A. (1965). Ann. Agric. Fenn. 1962., Seria Animalia Nogentia. 1/5. 114.

Varney, E.H., Moore, J.N. and Scott, D.H. (1959). Plant Dis. Reptr., 43: 567-69.

Verheul, M.J., Sonsteby, A. and Grimstad, S.O. (2007). Scientia Hort., 112: 200-206.

Verlinden, A. (1986). Boer en de Tuinder, 92: 19.

Verona, L.A.F., Nesi, C.N., Scherer, E.E., Gheller, C. and Grossi, R. (2005). Agropecuaria Catarinense, 18(2): 90-92.

Vervoot, M., Melis, P., Hanssens, J., Craeye, S., Pisman, M., Smagghe, G., Clymans, R. and Belien, T. (2017). Acta Hort., 1156: 833-42.

Voth, V. Proebsting, Sr. E.L. and Bringhurst, R.S. (1961). Proc. Amer. Soc. Hort. Sci., 78: 270-74.

Wada, T., Noguchi, Y., Isobe, S., Kunihisa, M., Sueyoshi, T. and Shimomura, K. (2017). The Hort. J., 86: 365-78.

Waithaka, K., Ngugi, C.W. (1985). Acta Hort., 153: 85-90.

Waldo, G.F. (1947). Ore. Bull., 442.

Wang, D.Y., Wergin, W.P. and Zimmerman, R.H. (1984). HortSci., 19:71-72.

Wang, J. Ge. H., Peng, S., Zhang, H., Chen, P. and Xu, J. (2004). J. Hort. Sci. Biotech., 79: 735-38.

Wan-Mahfuzah, W., Zulkifli, M.S., Latifah, M.N. and Fauziah, O. (2013). Acta Hort., 1012:515-20.

Watada, A.E. (1971). Jr. Amer. Soc. Hort. Sci., 91:177-79.

Watt, B.K. and Merrill, A. (1959). A table of food values. In USDA Yearbook of Agriculture. Washington, Supt. of Documents.

Webb, R.A., Purves, J.V., White, B.A. and Ellis, R. (1974). Scientia Hort., 2: 175-84.

Webb, R.A., Terbalanche, J.H., Purves, J.V. and Beech, M.G. (1978). Scientia Hort., 9:347-56.

Weber, N., Zupanc, V., Jakopic, J. Vebric, R., Mikulic,-Petkovsek, M. and Stamper, F. (2017). Jr. Sci. Fd. Agric., 97: 849-57.

Weiss, A., Weisshaupt, S., Krawiec, P. and Kunz, S. (2014). Acta Hort., 1017: 237-42.

Welke, S.E. (2004). Jr. Sustainable Agric., 25: 57-68.

Whitehouse, A. B., Simpson, D.W., Johnson, A.W., McLeary, K.J., Passey, A.J. and Troop, S.W. (2017). Acta Hort., 1156: 185-88.

Whitehouse, A.B., Simpson, D.W., Johnson, A.W., McLeary, K.J., Passey, A.J., Nellist, C.F., Bates, H. and Harrison, R.J. (2017). Acta Hort., 1156: 69-75.

Wilhelm, S. (1955). Phytopath., 51: 744-48.

Will, H. (1981). Obstbau. 6: 206-07.

Williams, B.L. and Wender, S.H. (1952). Jr. Amer. Chem. Soc., 74:5919-23.

Wills, R.B.H., Ku, V.V.V. and Leshem, Y.Y. (2000). Postharvest Biol. Technol., 18: 75-79.

Wilson, F. and Dixon, G.R. (1988). Jr. Hort. Sci., 63: 221-27.

Wocrdward, J.R. and Topping, A.J. (1972). Jr. Hort. Sci., 47: 547-55.

Woods, F.M., Dozier, W.A. Jr., Ebel, R.C., Himelrick, D.G., Mosjidis, C., Thomas, R.H., Wilkins, B.S. and Pitts, J.A. (2005). Small Fruits Review, 4(3): 85-105.

Woodward, J.R. (1972). J. Sci. Fd. Agr., 23: 465-73.

Wrolstad, R.E. and Shallenbergcr, R.S. (1981). Jr. Assn. Office. Anal. Chem., 64: 91-103.

Wrolstad, R.E., Putnam, T.P. and Varseveld, G.W. (1970). Jr. Fd. Sci., 35: 448-52.

XiaoJie, Q., HongMei, X., Xu, C., HaiLian, Z. and LinYuan, Si. (2017). Fd. Microbiol, 63:205-12.

Yang, L., Li, L., Yang, L., Zhang, J.J. and Dong, H. (2017). Acta Hort., 1156: 189-90.

YingYan, G., YuanMao, J. and FuJian, P. (2003). Jr. Fruit Sci., 20: 475-78.

Yuan, B. Z., Sun, J. and Nishiyama, S. (2004). Biosystem Engin., 87: 237-45.

Zasada, I.A., Halbrendt, J.M., Kokalis-Burelle, N., LaMondia, J., McKenry, M.V. and Noling, J.W. (2010). Ann. Rev. Phytopathol., 48: 311-28.

Zhang, Y., Wang, G., Chang, L., Dong, J., Zhong, C. and Wang, L. (2014). Acta Hort., 1049: 67-71.

Zhang, Y., Wang, G., Dong, J., Zhong, C., Chang, L. and Zhang, H. (2017). Acta Hort., 1156: 7-11.

Zhao, M.Z., Wang, Z.W., Yu, H.M., Wang, Q.L. and Cai, W.J. (2017). Acta Hort., 1156: 131-33.

ZhuangWei, W., Mizhen, Z., Ji, Y., WeiMin, W. and YaMing, Q. (2011). Jr. Fruit Sci., 28: 1032-37.

ZhuangWei, W., MiZhen, Z., YaMing, Q., HongMei, Y., WeiJian, C. and Jin, X. (2013). Jr. Fruit Sci., 30: 1091-93.

Zurawicz, E. (1983). Prace Instytutu Sadownictwa i Kwiaiciarstwa W Skierniewicach. A. 24: 7-17.

## Chapter 14

# Currants

Currants belonging to genus *Ribes* are widely cultivated across temperate Europe, Russia, New Zealand, parts of Asia and to a lesser extent North America. There is an increasing interest in the inclusion of berries, especially currants in human diet mainly health benefits associated with their consumption.

## Composition and Uses

Currents are valuable berry in human diet due to high concentration of bioactive compounds and antioxidant (Hummer and Barney, 2002, Nour *et al.*, 2011, Milosevic *et al.*, 2016). Black Currants posses a high antioxidant activity for which phenolic compounds and ascorbic acid make the major contribution (Lister *et al.*, 2002). Ascorbic acid concentration in black currant surpass many other fruits and berries, and its stability has been described to be more consistent due to the presence of anthocyanins and other flavanoids. Nour *et al.* (2011) evaluated three red and eight black currant cultivars in Romania. The range of total anthocyanins of the tested samples was 12.14-22.06mg/100g (red currants) and 16.17-28.78 mg/100g (black currants). The ascorbic acid content veried between 23.23-44.62mg/100g (red currant) and 161.58-284.46mg/100g (black currants). Citric acid was predominant in all studied black and red currant cultivars, followed by malic acid tartaric acids.

Myricetin was the most abundant flavonol in black currant (8.9 to 24.5mg/100g fresh weight) followed by quercetin (5.2 to 12.2mg/100g) and Kaempferol (0.9 to 2.3 mg/100g). The sum of these major flavonods varied widely among black currant cultivars (Mikkonen *et al.*, 2001).

The black currants are also rich source of vitamins and minerals. The clones of *Ribes aureum* showed on an average 15.74 per cent dry matter, 13.17 per cent sugars, 2.01 per cent acids (Miculka, 1980). The vitamin C content of fruit in black currants varies between 180 and 235 mg per 100 g of fruit while in red currants it varies between 30 and 40 mg. Gubenko *et al.* (1976) found vitamin C content up to 185.6 mg/100 g in black currant cvs. Nadezhdg, Minusinskaya, Biya, Katun, Stakhanovka, Altaya, Golubka and Osennyaya Altaiskaya. The vitamin C content in cv. Ben Nevis was reported to be 151 mg/100 g while it was 153 mg in cv. Silvergieter (Haffner and Heiberg, 1986). Black and red currants have rich mineral composition, especially potassium, calcium and magnesium (Nour *et al.*, 2011). Glutamin and aspartic acids are the principal amino acids of currants. The total amino acids were found higher in white currants while red currants were found to be the richest source of glutamic acids (Dzamic *et al.*, 1975). Shirochenkova *et al.* (1976) found cyanidin-3-rutinoside, cyanidin-3-glucoside, delphinidin -3-rutinoside and delphinidin-3-glucoside in black currant berry. The presence of quercetin, isoquercetin, hyperin and quercitrin were also recorded. *R. nigrum* var. *europeum* was found richer in catechins (Samorodova-Bianki and Babaevskaya, 1981). The composition of black and red currants are presented in Table 14.1.

**Table 14.1: Composition of Currants (per 100 g of edible portion)**

| | *Black Currant** | *Red Currant** | *Dried Black Currants* ** |
|---|---|---|---|
| Water (per cent) | 81.2 | 85.7 | 2.7 |
| Protein (g) | 1.7 | 1.4 | 0.5 |
| Fat (g) | 0.1 | 0.2 | 0.5 |
| Carbohydrate (g) | 13.1 | 12.1 | |
| Vitamin C (mg) | 2.00 | 41 | 1 |
| Riboflavin, B, (mg) | 0.05 | 0.05 | 0.14 |
| Calcium (mg) | 60 | 32 | 130 |
| Phosphorus (mg) | 40 | 23 | |
| Iron (mg) | 1.1 | 1.0 | 8.5 |
| Potassium (mg) | 372 | 251 | |

* Watt and Merrill (1959).

** Mudambi and Rajagopal (1983).

The black currants are not usually used as a dessert fruit because of its strong flavour. However, the cvs. Brodtorp and Goliath are acceptable as dessert fruit due to their mild sweet flavour. Most of the fruits are used for preparation of juice. The other processed product includes jam, jellies, canning yoghourt, liqueur and for pie filling. The black currant cvs. Ben Nevis and Ben Lemond were found suitable for industrial syrup production (Blom and Skrede, 1981).

## Origin and Distribution

Red currants have evolved from several species of *Ribes- R. rubrum*, a native of central and northern Europe, Asia, the Caucasus and Manchuria; *R. vulgare* a native of western Europe; and *R. petraeum*, occurring further south in central and southern Europe and North Africa. The red currant was recognized by the 15th and 16th centuries as of botanical interest and its fruit for use as an appetizer. It seems to have received most attention in Holland and the names of some of the earliest cultivars indicate a Dutch origin. *Ribes petraeum* was found growing wild in woods in 1561 near Berne, Switzerland and red currants increased greatly in popularity in Holland and Britain during the 17th and 18th centuries both as fresh fruit and for making wines and a number of large-fruited cultivars were introduced in Holland.

The black currant was evolved from *Ribes nigrum*, a native of Europe and northern and central Asia, and found growing as far north as Scandinavia and Siberia. Owing to the inherent strong smell of the foliage and the flavour of the fruit, it was less readily accepted as food than red or white currants, although by medieval times black currants were being used for adding colour to wines and gradually gained favour for use medicinally. It was the introduction of a larger berried black currant in Holland in the 16th century that attracted some interest in this fruit, but it was not until the following century that currants were generally accepted in gardens in England (Roach, 1985).

Several species of black currants with sessile, yellow glands originate in South American highlands. Ornamental and golden currents are native to North and Central America (Hummer and Dale, 2010).

## Production

Black curients have been cultivated in northern Europe for more than 400 years. Black currants were introduced in the USA in the seventeenth century. The history of black currant cultivation extends back to 1917 in China, but the industrial production started from 1980s (Qin *et al.*, 2012). The total world production was 655,030 tons in 2016 (FAOSTAT, 2016). More than 50 per cent production was from Russia (395,045 tons). Other major producers are Poland (166, 110 tons), Ukraine (24,500 tones), Germany (13,992 tons), UK (11,353 tons) and New Zealand (10,733 tons).

## Taxonomy and Cytology

The genus *Ribes*, was classified to the *Saxifragacea* family originally, but was later placed within the *Grossulariaceae* on the basis of morphological traits such as the presence of inferior ovaries, syncarpous gynoecium and fleshy fruit (Cronquist, 1981, Sinnot, 1985).

On the basis of morphological characters, the genus has been divided into subgenera, sections or series by many workers.

The European black currant and other black or brown-fruited species with resinous glands have been placed to the sub-genus *Coreosma*. This includes the following species:

*R. nigrum* (European black currant)

*R. americanum* (American black currant)

*R. bracteosum*

*R. petiolare* (the western black currant)

*R. hudsonianum* (the northern black currant)

*R. dikuscha*

*R. ussuriense*

*R. viburnifolium*

The red currants are grouped in subgenus *Ribesia* by Berger. The commercial cultivars are derived from any one or a combination of the following three species:

*R. rubrum* (central and northern Europe and northern Asia), *R. petraeum* (from the mountains of western and central Europe), *R. sativum* (*R. vulgare*) (western Europe). White currants are a colour form of the red currants. *R. multiflorum* from eastern Europe, *R. longeracemosum* from western China and *R. triste* from northern Asia and North America are included in this group. They have crystalline glands and except *R. longeracemosum,* the red fruit.

The other species which are crosses between different species are:

*R. houghtonianum* (*R. rubrum* x *R. sativum* or *R. sativum* x *R. rubrum*) *R. pallidum* (*R. petraeum* x *R. ruburm*)

Black currants are diploid (2x =2n=16) with a choromosome number of 16. The chromosomes are 1.5-2.5mm in diameter, with both mitotic and meiotic processes being highly uniform (Zielinski, 1953). Natural polyploidy are rarely seen (Brennan, 1996).

## Cultivars

There are a large number of black, red and white currant cultivars found in different growing countries of the world. The commercial cultivar of black currants in the UK is Baldwin, though Amos Black, Blacksmith, Boskoop Giant, Cotswold Cross, *etc.*, are also grown in a limited scale. In Poland, the different black cultivars were grouped according to resistance to frost (Holubowicz and Bojar, 1981). The cv. Brodtorp was considered hardy and was found to survive at –35°C while cvs. Roodknop, Baldwin, Blacksmith, Cotswold Cross, Mendip Cross and Silvergieters Black were considered fairly hardy (surviving at –30°C). The cvs. Magnus, Hatton Black. Baldwin, *etc.* are found growing in Australia (Goubran and El-Zeftawi, 1983) and cvs. Seyanets, Golubka, Dochka, Chernoglazaya and Yubileinaya in the USSR (Polikarpova, 1983). In New Zealand, McCarthy (1988) recorded consistently high yields in black currant cvs. Grantly, Ben Nevis and Cotswold Cross. Oosten (1988), in an evaluation trial with 14 black currant cultivars and selections in the Netherlands, found highest yields from Tsema Ben Lomond and M-59-3. The fruits of Boskoop Giant, Laxton's Giant, Blacksmith, Raven and Baldwin are rich in vitamin C than Cotswold Cross, Amos Black, Jet and Ben Nevis.

Among the red currant cultivars, Laxton's No. 1, Red Lake, Jonkheer van Tets, Laxton's Perfection Stanza, *etc.*, are commercially important. In the Netherlands, cvs. Rolan and Rovada showed higher yields. The cultivar Jonkheer van Tets ripens early compared with Rode Rebel and Rovada (Oosten, 1983). The cvs. Red Dutch and Red Castle were considered winter-hardy (Vitkovskij. 1958).

Iftime *et al.* (1989) suggested cultivation of red currant cvs. Rondom, Tydeman's Seedling, Laxton's No. 1 and Rosu Timporium in the north of Moldavian regions in Romania.

## Black Currant Cultivars

### Amos Black

A late ripening cultivar raised from Goliath x Baldwin produces vigorous bush. Fruits are medium-large with good colour and flavour.

### Baldwin

It is a low or medium-vigour cultivar. Fruits are medium in size with good colour, ripens late in the season and is very rich in vitamin C.

### Ben Chaska

Early ripening, cold hardy, large fruited, rust resistant cultivar. Fruit black, round, flavour sweet-acid.

### Ben Como

High yielding, cold hardy, rust resistant black currant for mechanical harvest. Fruit medium-size, round, flavour sweet-tart.

### Blackcomb

Late-flowering, high yielding black currant. Fruit big, flowering late mid-season, better tolerance to late spring frost, high level of field resistance to powdery mildew and rust.

### Ben Nevis

Plants are medium in vigour and a heavy cropping cultivar. Fruits medium-large, taste sweet and have thick and tough skin.

### Blacksmith

Plants are very vigorous and spreading. Fruits medium to large with good flavour and high vitamin C content.

### Boskoop Giant

Produces vigorous bush. Fruits are large, juicy which ripens early and have high vitamin C content. However, the cultivar is susceptible to leaf spot.

### Climax

Plants are vigorous and heavy cropper. Berries are medium to large in size with good flavour.

### Cotswold Cross

Raised at the Long Ashton Research Station, UK, a cross between Baldwin x Victoria. Plants are tall, medium vigorous. Fruits small, uniform with very attractive colour.

**Gofert**

A hybrid of Golubka X Fertodi-1, developed from Poland in 1996 (Pluta and Zurawicz, 2014). Fruit large (≈1.2g) on medium-long strigs, high TSS (18 per cent) and ascorbic acid (208mg/100g) content. Highly resistant to powdery mildew and rust.

**Laxton's Giant**

Plants are very vigorous and spreading, bear large, thick-skinned berries. The fruits ripen early in the season.

**Lunnaya**

This black currant cultivar was bred at the Leningrad Agricultural Institute, USSR from Stakhanovka Altaya x Vystavochnaya Neosypayushchayasya. This mid-season cultivar is highly frost resistant and can stand even up to –45°C, produces large berries with high sugar (8.39-8.86 per cent), acids (3.97 per cent) and ascorbic acid (137.8-165 mg/100 g) contents (Patashova, 1981).

**Mendip Cross**

A combination of Baldwin x Boskoop Giant, flowers in the mid-season. Produces large berries which are slightly acidic.

**Ores**

Midseason black currant, fruit medium, short to medium strigs, firm good for juice, concentrate and freezing, highly resistant to currant gall mite.

**Polares**

It was released from Poland (Pluta and Zurawicz, 2015b), low vigour and moderately productive. Fruit medium, rich in anthocyanin and ascorbic acid. Resistant to powdery mildew.

**Ruben**

Midseason black currant. Developed from Poland. Fruit medium to large, medium strigs, very firm, high content of juice, and resistant to shattering, powdery mildew and blister rust.

**Tahis**

High-yielding, large fruited black current developed from Canada. Very large fruit, tolerate late frost, resistant to powdery mildew.

**Tines**

Early ripening black currant developed from Poland. Fruit large, short to medium strigs, firm and high juice content, winter hardy, resistant to powdery mildew, susceptible to leaf spot, rust and gall mite.

**Tihope**

It is a new disease-resistant black currant cultivar released from Poland (Pluta and Zurawicz, 2015a). Tihope produces large, attractive, and good quality fruit that

Black Currant Bush

Red Currant Bunch

Currant Plantation

can be used for juice, frozen fruit and also for fresh market. Resistant to American gooseberry mildew and rust.

**Titania**

It is a mid-season black currant cultivar resulting from Altajskaja desertnaja x (Consort x Kajaanin Musta) cross developed in Sweden. Fruits are 1.32-2 g in weight with an average yield of 8.4 kg/bush in 4 years. The fruits per raceme vary from 20 to 22 and from 480 to 520 fruits/shoot. The high sugar-acid ratio makes the fruit suitable both for, fresh consumption and freezing. Because of the high pectin content the fruits are suitable for jam, jelly and juice preparation. The flowers can tolerate low temperature. The fruits are firm and well-suited for transport. They

**Black Currant Cultivar Ben Nevis**

**Red Currant Bush**

hang well on the bush and are ideal for "pick-your-own" sales. The plant is very vigorous, upright and suitable for growing by the hedge system (Tamas, 1984).

### Wanfeng (Hanfeng X Heifeng)

It was developed at China in 1990. Currants are black with light green flesh, average cluster weight 7.9g, total soluble solids 14.6 per cent, good eating quality, highly resistant to cold damage, endured –41.5°C in 2000-01 without cold damage (Wenzhi *et al.*, 2004).

### Wellington XXX

It originated as a cross between Boskoop Giant x Baldwin. Plants are vigorous and spreading. The berries are medium in size with tough skin. The cultivar is, however, susceptible to frost.

### Whistler

High-yielding black currant developed from Canada. Small to medium sized fruit, good quality, good tolerance to late spring frost and rust.

Black current cultivars growing in different countries are present in Table 14.2.

**Table 14.2: Black Currant Cultivars Growing in different Countries**

| | | |
|---|---|---|
| China | : | Brodtrop, Heifeng, Hanfeng, Wanfeng. |
| Romania | : | Padina, Geo, Polin, Deea, Abanos, Ronix |
| Poland | : | Bona, Gofert, Tines, Tihope |
| Ukraine | : | Chereshnieva, Yubylejnaia Kopania, Saniuta, Sofievskaia, Vernisazh |
| Siberia | : | Pamiati Potapenko, Shadrikha, Kalinovka, Degtiarevskaya, Podarok Kuminovu, Zonalnaya, Aleandr, Berdchanka, Rakhil, Mariushka, Glarioza |
| Russia | : | Muravushka, Gamma, Gratsiya, Kipiana |
| Lithuania | : | Joninial, Pilenai, Almiai, Laimai, Vyciai. |

## Red Currant Cultivars

### Earliest of Fourlands

It produces vigorous bush, a heavy cropper. Berries are medium which ripens early in the season.

### Fortun

This red currant cultivar was raised in Norway from cross between Raud Flollandsk (Red Dutch) and Jonkheer van Tets. It is an early season dessert cultivar suitable for garden rather than for commercial cultivation (Ovdvin, 1978).

### Jonkheer van Tets

This red currant cultivar was raised in the Netherlands; bushes are vigorous and upright. A heavy cropper, produces large berries with attractive colour.

### Laxton's No. 1

Plants are vigorous and a heavy cropper. Fruits are medium with good quality, ripens early in the season.

**Red Lake**

It makes an upright vigorous bush and is a heavy cropper. Fruits are very large, juicy and bright red in colour.

**Rondom**

It produces vigorous and upright bush, mid-season to late flowering, heavy cropper. Berries are medium in size with thick skin.

**Rosu Timpuriu**

A red currant cultivar, derived from cross between Rouge de Versailles and Heros Cross. It yields almost double that of Heros and the fruits are good in quality and flavour (Botar *et al.,* 1975).

**Viking (*R. petraeum* X *R. rubrum*)**

It is a vigorous, good-yielding, cold hardy cultivar with resistance to North American diseases and pests (Hummer, 2000).

**Margaritar**

It is a white currant cultivar, derived from Rouge de Flollande x Heros Cross; gives very high yield with uniform and compact cluster. It is resistant to leaf diseases (Botar *et al.,* 1975)

**Nortun**

The parentage and origin are the same as that of Fortun. It is somewhat late in ripening and bears larger and more acidic fruits (Oydvin, 1978).

## Climate

The currants can be grown in a wide range of climate but the site should be free from frost. Black currants are particularly sensitive to frosts when flowering. Relatively mild frost during this period can kill a large portion of the flowers and flower buds with consequent effects on fruit yield. One approach to avoid the frost is to select late-flowering cultivar which is likely to escape damaging frosts during the sensitive period. Another possibility is to select relatively tolerant cultivar. Cultivars Zagadka, Tserera, Ben Lomond and Lentyal are the resistant cultivars (Tyurina *et al.,* 2000). Overhead sprinkling or spraying of polyethox polymer (Teric 12A238) during flowering was found effective in increasing frost resistance of flowers (Wilson and Jones, 1983). In the months of March and April, 1977 when repeated night frosts at East Malling caused considerable damage to black currant cvs. Baldwin, Green Black, Tor Cross, Wellington XXX and several other early and mid-season cultivars; the late-flowering cvs. Amos Black, Black Reward, Consort; Malling Jet and Seabrook's Black showed very little damage due to frosts (Keep *et al.,* 1983). Dry cold wind also causes considerable damage. Planting of windbreak to shelter from exposed sites reduces damage from wind and frosts.

Heide and Sonsteby (2011) working with three black currant cultivar: Ojebyn, Ben Torn and Kristein observed growth cessation in short day (SD) conditions, with a critical photoperiod of 16 h. The critical photoperiod for 50 per cent flowering was 16 h in cvs. Ojebyn and Ben Torn, and 17h for Kristein. However, not all plants flowered after exposure to a 10 h photoperiod, and the number of flowers increased by increasing the photoperiod from 10 h to 15 h in all cultivars, while growth cessation was almost immediate in a 10 h photoperiod. They also reported that plants with 15-16 nodes were only marginally responsive to SD induction, buds situated as far down the shoot as the fifth of sixth node were competent to flower. It was suggested that the inability of small black currant plants to flower residues in limitation of leaves to responds to SD and to produce a florigenic signal, while these buds are fully competent to respond to such signal.

To achieve high yield of quality black currants, rich in phenolic compounds, anthocyanins and ascorbic acid, cool summer conditions with ample precipitation are desirable (Woznicki *et al.*, 2015).

Black currants can survive minimum temperatures of –40°C or lower. It has been estimated that black currant cultivars normally require about 800 to 1600h of chilling (0 to 7°C) before buds will break in spring (Barney and Hummer, 2005). Flowering require a base temperature of 5°C and the fruits require a time of 120-140 frost-free days (Harmat *et al.*, 1990).

## Soil

The currants bear fruits on many types of soils but the best production is obtained on deep, loamy soils. Heavy soils, if properly managed, are considered better than sandy soils. Black currants have shown greater tolerance to impeded soil drainage as compared with many other fruits. However, waterlogged conditions for a longer period have proved definitely harmful (Rogers *et al.*, 1936). The currant is heavy feeder, the soils should be supplied with adequate organic matter and be able to retain soil moisture even during drought periods. The management of heavy soil is difficult but sandy soil becomes too dry in summer which restricts the growth and reduces the yield.

The commercial culture of red currant is confined to lighter loam soils, although they can be grown on soils having a wide range of texture. Well-drained soils of good depth which do not dry out and remain cool during summer months are best suited to currant cultivation (Card, 1925). Sako and Laurinen (1979) found black currants to grow more vigorously and yield better on sandy clay soil. An open-textured soil of 0.5 to 0.75 m depth is suitable for black currant cultivation. The root activity of black currant is dependent on soil temperature and moisture content, the root activity in cv. Pamyat Michurina was found to increase at > 3-4°C soil temperature and > 10 per cent moisture. Horizontal root spread was maximum at 18°C and 80 per cent soil moisture (Trunov, 1983). A soil pH of 6.2 was considered optimum for black currant (Jazvicky, 1940).

The optimum soil pH is from about 5.5 to 7.0. On slightly to moderate alkaline soils, application of sulphur reduce the pH to an acceptable level. Total salt and sodium concentration in the soil should not exceed 15,000ppm and 500ppm, respectively (Barney and Fallahi, 2009).

# Propagation

## Cutting

The currants are easy to propagate from hardwood or softwood cuttings. Hardwood cuttings, 20-25 cm in length are usually made from one-year-old shoots. The cuttings are planted during autumn and spring which are rooted by the following winter. It is difficult to detect viruses in one-year-old cuttings, so it is better to use two-year or older cuttings because by that time the infection could be detected and rogued out. McNeill *et al.* (1980) found good rooting by planting cuttings in peat and sand media after storage of cuttings for 33 or 64 days at 0° or 3.3°C. Rooting in single-node dormant cuttings was found very low between October and January and the bud-break was related to rooting in hardwood cuttings. Bud-scale removal is an effective way to force buds to grow and allow cuttings to root though it is time consuming. Soaking the cuttings for 1h in $KNO_3$ solution increased bud-break and rooting (Wainwright *et al.*, 1986). Wainwright and Hawkes (1988) stated that the length of the cuttings significantly influenced the percentage of rooting, amount of new extension growth, mean number of new shoots produced and the dry weight of new growth. They suggested that cutting length may be reduced from the recommended 20 cm to 15 cm without significant loss of productivity. The cultivars Ben Nevis and Ben Lomond showed 89 per cent and 79 per cent rooting, respectively, when 1-year-old dormant shoots were planted direct in the field at a spacing of 15 x 30 cm (Wainwright and Hawkes, 1987).

Barsukov (1982) found that softwood cutting rooted better than hardwood cutting and softwood cuttings of red currants cvs. Houghton Castle and Ruby Castle showed 81.8-89.3 per cent rooting under mist compared with only 3.2-15.3 per cent in hardwood cuttings. Belova (1982) reported that by using softwood cuttings of black currant with a heel of 2-year-old wood, some 200 cuttings/bush were taken instead of the normal 30-40 hardwood cuttings. The softwood cuttings with heel also rooted well and produced up to 86 per cent standard planting material by the end of the year. Salikhov and Oslopov (1985) obtained maximum number of standard 2-year-old transplants by planting cuttings at 20 x 20 cm. Asanica *et al.* (2017) reported average 91.67 per cent rooting of black currant cv. Tinker by using thick cutting which were treated with IBA at 500-1500 ppm IBA and planted in perlite + peat media. The red currant cv. Elite caused an average rooting of 93.33 per cent when thick cuttings were planted in perlite + sand media after IBA treatment at 500 to 1500ppm.

## Micropropagation

Success in micropropagation from shoot meristem on MS medium supplemented with IAA, BA, $GA_3$ and kinetin and/or IBA at different concentrations was reported by Blenda and Kirilenko (1982). The Best results were obtained by callusing on MS medium + 1 mg/l IBA + 0.5 mg/l kinetin followed by transfer for root formation to Gotre medium + 2 per cent sucrose + 1 mg/l IBA and for further growth in Gotre medium without any growth substances. Similarly, Orlikowska (1984) obtained shoot proliferation *in vitro* on MS medium supplemented with BA

(0.25-2.5 mg/l). Removal of apical meristems from proliferating shoots induced considerable elongation of axillary shoots without effecting their number. The addition of ascorbic acid to the medium prevented browning of shoots. Rooting was 100 per cent when base of the shoot was immersed in IBA solution (50mg/l) and then transferred to a liquid medium without growth regulator.

Toshkov and Domozetova (2006) suggested use of 1.0 to 3.0 mm explants from apical and axillary buds for micropropagation. The explants were sterilized with 0.1 per cent $HgCl_2$ for 5 min before placing in MS media supplemented with BAP at 2.0 mg/l. The cultures were cultivated in 22-26°C, 16h photoperiod and 2500-3000 lx light intensity. During the formation of roots, IBA at 2.5 mg/l increased rooting. Sedlak and Paprstein (2012) developed protocol for shoot multiplication of red currant cvs. Detvan, Vitan and Rotte Hollandische in Czech Republic. The explants were actively growing shoots (5-10mm in length) which were surface sterilized with 0.15 per cent $HgCl_2$ followed by serial subculture for three months on MS medium with 1 mg/l BAP.

## Planting

Before planting, the soil should be thoroughly cultivated and all the perennial weeds removed. Perennial weeds may be controlled by using herbicides. Planting is usually done from late September to March when it is dormant. Black currants start growth early in the spring, late planting may reduce growth. Sometimes, *in situ* planting of cuttings is done, where adequate irrigation and other facilities for rooting exist. During planting, care should be taken so that roots of the plants are not damaged or become dry. A continuous row length of 80-90 m is suitable for both hand and machine harvesting. Black currant bushes are usually planted at 0.75 to 1.2 m in rows; 2.7 to 3.3 m between rows; depending on the machineries to be used for spraying, harvesting, *etc.* Trushechiin and Pozadnyakov (1972) suggested planting at 0.6 to 1.0 m in row with 2.5 m between row for machine harvesting. They obtained higher yield by planting 6000 to 9000 bushes/ha. In Poland, Radaczynska *et al.* (1982) recorded highest total 6-year-yield of 40.79 t/ha in cv. Jonkheer van Tets by planting 6700 plants/ha in double row strips and in cv. Erstling aus Vierlanden (46.6 t/ha) from plots with 5000 plants/ha planted in single high density rows. The effects of plant density on growth and yield are variable in different cultivars. Kawecki (1988) in trials with 5 cultivars of black currants spaced at 3 x 1 or 3 x 1.5 m found that Blacksmith bushes were the tallest at both spacing while Brodtorp bushes were the widest. The cvs. Olejbyn and Blacksmith usually produced more shoots, and at wider spacing the number of shoots/bush increased. The cultivar Brodtorp produced higher yield/plant at wide than at close spacing. Neumann (1983) stated that two-row beds at 3.5 + 2.5 m with 0.6 m within the row and a plant width of 1.5 m at right angles to the row had a high yielding capacity/plant and per unit area.

In Kazakhstan, medow-orchards were developed for black currant cv. Katyusha and red currant cvs. Dzhonker, VanTets and Gollandaya Rozovaya. The results showed 2.3-2.7 fold yield increase, while it reduced fruit harvesting time by 1 year (Oleichenko and Titova, 2001).

## Training and Pruning

### Training

The currants are generally trained as bush but there are some other modifications as given below:

#### Single Stem Spindle

Bushes often fruit in the first year on the main stem and in later years on laterals and vertical side shoots. As the branches are well spaced and illumination is good in the bushes trained in this method, fruit quality is better, uniform ripening of berries makes picking easier, and pruning is reduced as compared with bushes grown normally. A double row system of slender spindle bushes produced 20 per cent higher yields (Hillebrand and Stadtmuller, 1983).

#### 4-wire Trellis System

The plants are trained on a 4-wire trellis running at 0.5, 1.0, 1.5 and 2.0 m above the ground. Ten scaffold shoots/plant are retained and the laterals are pruned back to 5-10 cm. The yield rose with planting density. Significant increase in yield was recorded in cvs. Jonkheer van Tets, Hero and Laxton's Perfection by this system and cv. Hero was particularly found suitable for this method. The ease of fruit harvesting has made this system economical despite the high initial cost (Smolarz *et al.*, 1981). Trellis training resulted in higher yields than the conventional bush form of training which was particularly suitable for red currant cv. Erstling aus Vierlanden in Poland (Kawecki *et al.*, 1982).

#### High Stem System

Red currant plants with high stems were produced by wrapping black polythene around the lower 60 cm portion of the shoots, tied so as to form 1 bunch (bouquet shape), 2 bunches (V-shape) or 3 bunches (fan-shaped). A wire support was provided along the rows. The fan-shaped plants had 11-12 bearing shoots and yielded 7941 kg/ha, the V-shaped plants with 9-10 shoots yielding 6416 kg/ha and the bouquet with 6-7 shoots yielding 5033 kg/ha, compared with conventional bushes having 9 bearing shoots and producing 2352 kg/ha (Maric, 1978).

Ancay and Carlen (2016) compare two training system with three red currant cultivars (Red Poll, Rovada and Tatran) in Burson, Switzerland at 1100 m asl. The considered training system for red currant production were: (i) the traditional cordon training system (palmette system) with 3 not removed cordons (main branches) and 15 to 20 medium sized lateral branches per plant, and (ii) the V-training system with constantly renewed cordons, *i.e.*, cordons were removed after 3 years at their base (no permanent wood) and every year, 3 new cordons were established in order to obtain 2-and 3-year-old cordons producing fruits (V-system). The results over 7 years revealed that the V-training system had little influence on yield, cluster attributes and fruit quality, but a very positive incidence on the picking efficiency with 11 per cent more fruits harvested per hour. The economic performance of the new V-system was in average of three cultivars 13 per cent better than the traditional cordon system (palmette). The cvs. Rovada and Tatran were more productive than cv. Red Poll, due to higher number of clusters per plant. The Economic performance was superior for cvs. Rovada and Tatran grown in V-system.

## Pruning

In currants, the fruits are usually borne on l-year-old shoot from previous season's growth. Pruning is usually done to remove older shoots and encourage new shoots. Pruning may be done at any time when the plants are dormant but before the buds start to swell in spring. All the spreading branches close to the ground and weak growths should be removed and heading back is generally done only when it is necessary in high or straggly spreading bushes. After planting, all growth should be cut back at ground level, leaving only one bud per shoot above ground. This will facilitate new growth and prevent fruiting immediately after planting. In the following years, pruning should be done to encourage new shoot growth by removing older branches before they become too thick. However, the intensity of pruning depends on the growth and vigour of the plants. In some black currant cultivars, the yield seemed to be more affected by pruning method than by the amount of pruning. The cultivars Silvergieter and Tinker had the highest yield when all shoots were cut back to 20 cm at planting; weak, low and spreading branches and shoots were removed in the first year, and other branches were removed after 3-4 years. The highest yield was obtained in cv. Ojebyn when the bushes were left untrimmed at planting; 3 new shoots were left each year and old branches were removed after the third cropping (Nes, 1983). Rooijen (1984) in trials with red currants cvs. Jonkheer van Tets and Rotel Red found higher total yields in both the cultivars with light winter pruning; early summer pruning showed no advantage while late summer pruning was found detrimental. Burmistrov *et al.* (1984) reported the effect of rejuvenation pruning of red currants in the USSR in which twelve-year-old bushes of 4 cultivars were cut back to 50 cm above soil level in winter (February), spring (April) or summer (August) and the effects of treatments on growth and fruiting were observed over the subsequent 7 years. They found the largest and smallest number of basal shoot production after summer and winter pruning, respectively. The cultivar Shchedraya produced the longest shoot (mean length 90 cm) followed by Perventz (80 cm). Pruning reduced yields in the first 2 years but increased thereafter, compared with the control, and also increased mean berry weight. Pruning, on any of the dates tested, prolonged the productive life of plantations up to the age of 18 years. In Finland, autumn-pruned black currants gave higher yields (approximately 50 per cent) than spring-pruned bushes. Autumn-pruned plants also had more vertical shoots, making them more suitable for mechanical harvesting.

Zhonghlu *et al.* (2000) suggested pruning techniques for cv. Hanfeng growing in Heilongjiang and Jilin provinces, China. They suggested to prune the shoots at the base leaving 5-7 buds after planting. In the second spring, 4-5 shoots should be headed back, leaving 25 per cent of the length of the shoot, and after the third year, the old fruiting branches should be thinned out.

Modern harvesting equipment may have attachments that allow growers to prune their fields mechanically. This generally entails shaping the bush for harvest efficiency but will also encourage some new growth.

## Nutrition

Currants are gross feeders and require sufficient amount of nutrients to maintain good growth, as flowers and fruits are produced on previous year's shoots. Application of sufficient quantity of organic manure helps in conserving and releasing soil moisture apart from the addition to plant nutrients. Well-rotten bulky organic manure like farmyard manure, sewage sludge, *etc.*, are usually used as pre-planting soil dressing. Among the different nutrient elements, N, P, K, Mg, Ca, Fe and Mn were considered necessary for satisfactory growth and production of currants.

## Role of Nutrients

### Nitrogen

The deficiency of N showed cessation of shoot growth, reduced yield and diminished berry size, in black currant (Wlodek, 1972). The optimum leaf N content for satisfactory growth was estimated as 2.6-3 per cent and the deficiency was developed when leaf N content fall below 2 per cent (Ljones, 1966). The requirement of N for red currant cv. Jonkheer van Tets at 200 q/ha crop level was estimated as 100 kg N/ha while for cv. Red Lake at 150 q/ha level was 50-100 kg N/ha (Papp and Kobzos, 1980). Dalbro and Dullum (1957) recorded yields of 6.5, 10.4, 12.4 and 15.1 kg/sq.m in red currants by application of nitrochalk at 0, 300, 600 and 900 kg/ha, respectively.

Rupp and Trankle (2000) working with red currant cv. Rovada in a loamy soil, recorded highest yield with 150 kg mineralized N/ha. The treatment also increased cluster length, berry weight and colouring of berry. However, increasing the rate to 200 kg/ha decreased the vitamin-C and sugar content of fruits.

### Phosphorus

Kuminov (1961) stated that black currants yielded best when phosphatic fertilizer alone or mixed with compost was applied to planting holes. Kondakov (1973) suggested application of 180 kg P/ha for increased yield and better shoot growth of black currants.

### Potassium

Deficiency of K showed yellowing of leaves and browning of leaf margins. In subsequent stages, the leaves curled downwards resulting in complete withering of the affected bushes (Bosma, 1938). Deficiency of K also decreased the anthocyanin content of berries (Ul'yanova, 1971). Leaf K content between 1.0 per cent and 1.5 per cent was good for growth and yield of black currants (Bould, 1964). Uebel (1982) stated that in soil with adequate K content, additional application of K fertilizer increased yield with greater yield stability and lesser annual yield variation. A fertilizer ratio of 1:1.4 of N:K was suggested for growing currants in Czechoslovakia.

Scibisz *et al.* (1986) found the best fruit set, greater number of fruits per cluster and maximum fruit weight of red currants in Poland with 600 kg $K_2O$/ha. The treatment caused leaf K content of 0.78 to 0.84 per cent.

## Calcium

On highly acidic podzolic soil, liming at 2 g $CaCo_3$/kg soil improved the effectiveness of N, P and K (Popesko, 1968). Ljones (1963) showed that leaf Ca content of black currants varied between 0.99 per cent and 1.61 per cent and different fertilizers like calcium nitrate, potassium sulphate, superphosphate exert secondary influence on leaf Ca content.

## Magnesium

The deficiency of Mg developed reddish purple patches on the leaf surface which at later stages curled downward, dried and abscised (Wallace, 1925). Medel *et al.* (1984) reported that application of N, $P_2O_5$ and $K_2O$ at 60, 30 and 50 kg/ha, respectively to black currants had an adverse effect on leaf Mg content (0.12 per cent). However, in the following year application of MgO at 24kg/ha was found to increase leaf Mg content to 0.30 and 0.32 per cent in red and black currants, respectively.

## Micronutrients

Spraying with B and Zn twice a year was found to increase the yield by 21.4-23.5 per cent and 4.6-9.8 per cent, respectively. Both Zn and B were also effective in increasing the ascorbic acid content of fruit (Krupyshev, 1977). Tumanova (1973) recorded higher total sugar content of berry by spraying Mn at 0.1 to 1.0 per cent. Zn at 0.01 per cent and Co at 0.05 per cent during flower bud development, full bloom, berry formation and at 2 weeks before harvest.

## Fertilizer Recommendations

Kulenova (1977) recommended application of NPK fertilizer in the ratio of 100:50:50 or 50:100:50 kg/ha for black currants. The recommendations of Kawecki *et al.* (1982) were 100-120 kg N, 60 kg $P_2O_5$ and 100-120 kg $K_2O$/ha for red currant cvs. Erstling aus Vierlanden and Red Dutch in Poland. In Norway, Ljones (1984) noted highest yield (9.8 t/ha) in white currant cv. White Dutch by applying 100g $Ca(No_3)_2$, 20 g superphosphate and 40 g $K_2So_4/m^2$. Takac (1988) reported that annual application of 100 kg N, 100 kg $P_2O_5$ and 140 kg $K_2O$/ha or 150 kg N, 100 kg $P_2O_5$ and 200 kg $K_2O$/ha gave average yields of 3.75 and 4.42 t/ha, respectively, compared with 3.63 t/ha in control for black currants and in red and white currants 10.41 and 10.67 t/ha, respectively, compared with 7.31 t/ha in control.

Application of humus and microelements (B, Zn, Cu, Mn, Mo) with NPK increase yield by about 21-54 per cent of black currant cv. Al'go in a frozen meadow chernozem soil in Siberia (Gorokhova *et al.*, 2010). Wojcik and Filipczak (2015) evaluated the effect of mineral fertilizers and swine manure on black currant cv. Tiben in controlled green house conditions. Prior to planting in polythene bags, soil was mixed with 0.7 g P and 2.8g K per plant or with solid swine manure at 300g per plant (=170 kg N/ha). After planting, the plants were fertilized with either ammonium nitrate at 1.3-2.0 g N/plant/year or with swine manure at 300g/plant. The plants showed strongest growth and highest fruit production when P and K fertilizers or swine manure was applied prior to planting, and N fertilizer was applied eash year after planting.

Black currant showed preferentially be fertilized in early autumn, immediately after berry harvest, as this will enhance flower formation and subsequent berry yield without adversely affecting berry concentrations of important fruit quality constituents (Sonsteby *et al.*, 2017).

### Tissue Analysis

Severin and Ermokhin (1979) suggested sampling of leaves from the middle part of 1-year-old shoots at the end of the growing period for tissue analysis. Bould (1964) recorded the optimum value for leaf N in black currants between 2.9 per cent and 3.0 per cent. This value was found in the lamina of the matured leaves taken from the extension shoots before harvesting. The optimum value of leaf K was found between 1.0 per cent and 1.5 per cent and P between 0.25 per cent and 0.30 per cent. Cianciara and Smolarz (1983) suggested that leaf N content of 2.5 per cent in the black currant or 3.0 per cent in the red currants call for immediate N replacement.

## Irrigation

Irrigation is important, particularly for newly planted cuttings or bushes as dry spells may check growth and subsequent cropping. Irrigation during summer months helps promote new shoot growth essential for good cropping. Irrigation may be provided by border strip method; however, sprinkler irrigation system is being followed more conveniently in most commercial currant plantations of Europe. Trials conducted with four levels of irrigation, *i.e.*, i) none, ii) irrigated when soil moisture content of 10 per cent, iii) irrigated at 20 per cent soil moisture content and iv) irrigated at 22.5 per cent soil moisture content on black currant cv. Magnus in New Zealand, using the border strip method of irrigation. The results showed yields of 414, 876, 896 and 988 kg/ha for a total of 0, 2, 12 and 26 irrigations, respectively (Stoker, 1983). Stoker and McCarthy (1986) recorded mean yields of 7.2 and 6.3 tonnes/ha in cv. Magnus by irrigating at 20 and 15 per cent soil moisture content, respectively, compared with 3.0 tonnes/ha for the non-irrigated control. They suggested to apply irrigation at 15 per cent soil moisture content, where water is limited and application cost is high and at 20 per cent soil moisture content where it is plentiful and application cost is low. Treatment at 15 and 20 per cent soil moisture required on an average 5 and 14 irrigations, per season, respectively, to maintain the particular soil moisture level. The average berry weight was found to increase with irrigation frequency but in drier seasons, irrigations tended to reduce total soluble solids content of fruits (McCarthy and Stoker, 1988). Sakshaug and Ingvarsson (1986) in trials with cv. Ojebyn, reported that drip irrigation from the end of flowering to berry colouring enhanced shoot growth and fruit set was increased when irrigated from bud burst until end of flowering. The fruit size and yield increased by applying irrigation from berry colouring until beginning of harvest.

Cerekovic *et al.* (2014) studied the effect of deficit irrigation during flower initiation of two black currant cultivars (Titania and Ben Hope). The greenhouse grown plants were either fully irrigated ($F_1$), with a volume of water replacing the previous day's evapotranspiration, or deficit irrigated (DI) for 3, 4, 5, 7 or 10 days. Deficit-irrigated plants received a volume of nutrient solution corresponding to 50 per cent of the volume of supplied to $F_1$ plants. After full irrigation was continued ($F_1$) or resumed (DI) to allow for 2 and 10 days of recovery for stomatal conductance and leaf area increment, and 7 and 14 days of recovery for root intensity. Full irrigation was continued until the spring. They observed that deficit irrigation reduced the accumulated evapotranspiration (ETc), relative water content (RWC), leaf water potential, osmotic potential and turgor in both the cultivars. In the long term, deficit irrigation reduced the number of strigs and flowers per node in 'Ben Hope' whereas in 'Titania' high temperature during flower initiation seemed more important for the formation of flower initials than water availability.

## Soil Management

### Mulching

Mulching is a common practice in currant orchards. It helps to maintain the soil temperature and moisture and restricts weed growth. The different mulching materials include straw, cutting down of grasses, bitumenous paper, *etc.* Mulching increased yield and improved fruit quality (Kawecki, 1979). Cutting down of grasses to use as mulch sometime create problems by encouraging germination of different weed seeds. Gitinovasov (1984) obtained good results by cutting green manuring crops like peas, oats, winter rye and grass and using as mulch.

During the long dry period, sawdust mulch showed negative influence on growth and yield (Strautina and Kampuss, 2000). Inter-row grassing with perennial herbs and bush row mulching, preventing soil compactness, should be considered during the development of intensive technologies of trellis system in black currant plantations (Pridorogin *et al.*, 2013). Kukhta *et al.* (1999) however, suggested use of black polythene mulch which favourably increase growth and fruit production. This treatment reduced input costs and was the most profitable. Dale (2000) suggested the growers of Ontario, Canada to plant black currants through black plastic and avoid using sod between the rows. Black polythene mulch improved the taste and flavour of berries (Paunovic *et al.*, 2016).

### Weed Control

Traditionally, weed control in black currants were done by cultivating between rows and hoeing between the bushes. Mulching with suitable organic materials along the rows also helps to check the weeds. However, mulching with uncomposted bark (0-80 mm particle size), although showed good control of weeds in red and black currant bushes, it reduced leaf N content and earlier yellowing of leaves in red currant bushes (Grantzau and Scharpf, 1982).

Chemical herbicides are now widely used to eliminate weeds in most commercial plantations. In newly planted black currants terbumeton/terbuthylazine and oxadiazon + simazine gave better control of annual weeds and increased fruit yield in the next season (Rhodes and Tabley, 1983). The rhizomatous and deep-rooted weeds like *Elymus repens* were effectively controlled by 6-8 l fusilade W (fiuazifop)/ha. Lontrel (clopyralid) at 0.5 l/ha gave excellent control of *Cirsium arvense* (Radvany, 1983). Atroshohenko *et al.* (1983) stated that roundup (glyphosate) and goal (oxyfluorfen) decreased population of *Elymus repens* by 92.3 per cent and 84.6 per cent, respectively, and of dicot weeds by 63-85 per cent and 67-75 per cent, respectively. Pendimethalin, napropamide and simazine were found selective for black currants and had no significant effect on photosynthetic pigments, yield or chemical composition of the fruits (Nikolova and Ivanov, 1983). Cianciara (1986) reported good control of weeds in black currant cv. Roodknop and red currant cv. Red Dutch by using simazine, atrazine or gramoxone.

Usually perennial weeds, before planting are controlled by using paraquat (2.0 to 2.2 kg/ha) and simazine (0.8-1.1 kg/ha) after planting while in established plantations weeds are controlled by using paraquat, simazine, glyphosate, oxyflurofen, *etc.*

Napropamide is effective for weed control in newly planted fruit bushes. Clopyralid is a selective herbicide for black currants (Lisek, 2000).

## Flowering

Flower initiation occurs from May to September, depending on cultivar, season and locality (Pisani, 1962; Hardh and Wallden, 1965). Rudloff and Lenz (1961) stated that when flower primordia were developed in June and July, the floral organs differentiated before winter. Initiation of flower starts in the lower middle nodes and then proceeds acropetally. Georgiev and Topchiiski (1972) classified 10 stages of organogenesis. They found flower initiation in late July or early August which continued until late February. The sporogenesis took place in March and both male and female gametophytes developed in April. Flowering, pollination and fruit set started at the end of April and were completed by early May.

In *R. nigrum*, flower primordia are laid down in the axillary buds of the current year's shoot in the year prior to that in which flowering occurs. In *R. rubrum*, flowers are borne mainly on spur shoots which are axillaries of the 2-years or older long shoots. The internal conditions for flower initiation appear to be the release of lateral buds from correlative inhibition and a state of partial dormancy of the buds (Nasr and Wareing, 1961a). Black currant is a short day plant that ceases growing and initiates flowers at photoperiods shorter than a critical length. Surprisingly, however, the flowering response was found to be several-fold stronger in the near-critical photoperiod of 15h than in shorter photoperiods down to 10h (Heide and Sonsteby, 2011). The prompt growth cessation induced by an abrupt change to SD well below the critical length also interferes with flowering by arresting floral initiation at an early stage. Accordingly, the optimum photoperiod for flowering in black currant must be short enough to allow flower initiation and still long enough to prevent premature dormancy (Heide and Sonsteby, 2015). Nasr and Wareing (1961b) stated that when actively growing plants of black currant cv. Baldwin are transferred from long-day to short-day conditions, flower primordia were initiated in the lateral buds, whereas there was no flower initiation in plants maintained continuously under long day. In intact plants, flowers were found to initiate in response to 16 short days and the presence of mature leaves were considered necessary for flower initiation in response: to short days.

The raceme which developed on one-year-old and older wood are pendulous with a primary and one or more secondary racemes (Baldini and Pisani, 1964). Heiberg (1986) in studies with 15 cultivars, reported that the cvs. Jotun, Fortun and Nortun had larger number of flowers per raceme which were recorded as 18.3, 18.2 and 18.0, respectively, the lowest being 12.3 and 12.8 in Sjedraya and Jonkheer van Tets, respectively.

Flowers formed from basal buds are less exposed and form heavier fruits with better nutrition compared with flowers formed at the tip buds. The latter is less insulated and prone to frost damage such that it fails to set fruit.

## Pollination

Most of the cultivars of currants are self-fruitful and thus it is not essential to interplant cultivars for cross-pollination. Some black currant cultivars (*e.g.*, Boskoop Giant), however, have flower structure which prevents self-pollination. It has been noticed that the unfruitfulness in some of the standard cultivars is due to the relative position of the stigma to anthers. If weather during bloom is favourable to insect activity, the pollination is satisfactory and good fruit set is obtained in Boskoop Giant and other similar cultivars. Planting of cross-pollinating cultivars, however, is found to increase the yield. The different cultivars showed differences in self-fertility. Pollens housed in the anther which represent male reproductive part. The stigma is the top most part of the female reproductive part through which pollen grains travel through to fertilization with the ovule in the ovary. Temperature is important in the movement of pollen grain from the stigma to the ovule; during warm weather it may take 48 h to reach the ovule but this may delay to about, 7-9 days. In the USSR, Stepanova (1982), working with 79 cultivars, found low fertility in about 50 per cent cultivars derived from *R. nigrum var. sibiricum,* while three derived from *R. dikwcha* showed good self-fertility. In Czechoslovakia, the white currant cvs. Primus and Blank and hybrid BO 71 showed a high degree of self-fertility as did the black currant cvs. Roodknop and Otelo and hybrid BO 663. The red currant cvs. Titran and Detvan showed a lower degree of self-fertility than cv. Jonkheer van Tets (Mokra, 1988).

Honey-bees are important pollinators for currants. However, low air temperature during flowering sometimes affects movement of bees. Bumble bees and other wild bees are also important in effecting pollination. Experiments have shown that isolation of bushes with gauze caused a reduction in yield compared with bee-pollinated bushes by 47 per cent in red currants and 72 per cent in black currants, mainly as a result of berry drop (Blasse and Hoffman, 1988). Smeekens (1983), working in The Netherlands with black currant cv. Baldwin, obtained 30 per cent less yield by growing in gauze cages than those grown in the open with a beehive. Average bunch weight was also reduced by 28 per cent and berry number/bunch by 19 per cent. A hive density of 4 or 5/ha was recommended for effective pollination.

If the weather is not adequate for insect activity at the time of flowering, supplementing natural populations of pollinating insect may ensure high yields (Fountain *et al.*, 2011). Neira *et al.* (1999) recorded a high degree of correlation between the number of seeds/fruit and the percentage of developed fruits. On average 87 pollen grains of black currant and 1307 pollen grains of other species were noted in the digestive tract of pollinator *Astylus trifasciatus*. Paimetova and Mand (2000) reported the main pollinator of black currant were *Apis mellifera, Bombus lucorum* and *B. lapidarius*. Self-pollination was 3.8 per cent ±0.3 for cv. Bogator, 12.0 per cent ±0.4 for cv. Brodtrop. Bee pollination accounted for 46.7 per cent ±3.3 for cv. Brodtrop and 53.8 per cent ±2.2 for cv. Bogator. The degree of pollination of black currant cvs. Ben Alder, Ben Lomond, Ben Nevis, Ben Tirran, Ceres, Ojebyn, Titania and Triton were estimated in a 4-year study in Poland (Denisow, 2002). The average number of pollen per stigma in the bottom flowers in raceme was 312 grains, 270 in the medium and 133 in the apical flower position. Fewer pollen grains were observed when flowers were subjected to pollination with its own pollen by one bumble bee queen. The smallest number of pollen grains were observed under self-pollination.

In black currants two types of flowers are usually found-one in which the stigma protrudes beyond the anthers, and in the other the anther is just above the stigma (Baldini and Pisani, 1964). In both *R. nigrum* (cvs. Rosenthals Langtraubige and Silvergieters) and *R. rubrum* (cvs. Red Lake and Jonkheer van Tets) abnormal pollen tube growth was observed. In red currants this was manifested as thickening and intensive callose deposition in the tip of the tube whereas in black currants the tubes were twisted. Different forms of ovule degeneration were found at different stages of flower and fruit development, so that even with ample pollination only a small percentage of ovules continued developing (Al-Jaru and Stosser, 1983). Al-Jaru and Stosser (1984a) studied the pollen tube growth and observed that in black currants, the number of pollen tube in the styles of the basal flowers was highest and the rate of growth fastest while in terminal flowers pollen tubes were few and slow growing. In selfed Rosenthals flowers more pollen tubes were found to grow at a faster rate than in flowers cross-pollinated with Goliath. In Silvergieters, however, the reverse was true. In red currants with their shorter styles, no difference was found either in the number of tubes or their rate of growth regardless of flower position and self- or cross-pollination.

In black currants, fruit set was generally higher in the basal and middle parts of cluster than the apical part. A positive correlation was found between berry size or weight and seed number. Berry size and seed number in black currants were generally greater in the basal part than in the middle or apical parts. In red currants, berry size and seed number were mostly independent of position in the cluster. Fruit set did not always depend on seed numbers, but the presence of a minimum number of fertilized ovules was considered necessary for fruit set (Al-Jaru and Stosser, 1984b).

## Fruit Drop and its Control

The fruit drop in different cultivars of black currant ranges between 26 and 60 per cent and the majority of fruitlets drop during the first 3 to 4 weeks after flowering (Verzilov and Plotnikova, 1978). Fruit drop was found lowest near the base of the raceme and increased towards the distal end (Heiberg, 1986). In black currant, Plotnikova and Kirdyashkina (1976) found a direct relationship between berry weight, seed number and position of flowers on the raceme. The terminal berries were smaller had few seeds and were more liable to premature drop than the basal berries. They thought that the number of developing seeds in a fruit, being the auxin source, influence the early drop. Embryo degeneration in the fertilized ovules of the terminal fruit was attributed to hormonal deficiency due to competition from other fruits in the cluster (Plotnikova *et al.*, 1972). Al-Jaru and Stosser (1983) also recorded different forms of ovule degeneration at different stages of flower and fruit development in both *R. nigrum* and *R. rubrum*. Fruit set could be improved by spraying $GA_3$ at 250 or 500 mg/l to flowers before pollination (Cuvasina, 1963). Spraying at full bloom with NAA or $GA_3$ + IAA significantly reduced the abscission of fruitlet. Treatment with 2,4,5-T butyl ester or 2,4,5-TP at 1 to 2 months before harvest reduced fruit drop, increased berry weight, stimulated uniform ripening and increased yield by up to 99 per cent (Verzilov and Plotnikova, 1978).

## Fruit Growth and Development

The seed and fruit development of black currants (cv. Seabrook's Black) was studied by Wright (1956) at the Long Ashton Research Station, UK. He stated that prior to fertilization the ovule grows slowly, mainly by cell division. After fertilization the growth of berry was divided into three main phases :

### First Phase (0-24 days after fertilization)

This is characterized by the rapid swelling of the berry as a result of cell enlargement taking place within the tissues of the pericarp. A cork cambium is found to develop in the apical hypoderm and proceeds to cut off many layers of cells to protect the upper part of the berry as the flower dies.

There is a rapid increase in the size of the seeds. The nucellus cells continue to divide, particularly, at the chalazal end. The triple-fusion nucleus undergoes many divisions, filling the embryo-sac with many free endosperm nuclei before cytokinesis occurs. Many of the cells thus formed divide frequently throughout the remainder of this phase, whilst other cells begin to enlarge. The embryo grows very slowly.

### Second Phase (25-38 days after fertilization)

The volume of the berry increases slowly, but the seeds develop rapidly. The endosperm increases in size, breaking down and absorbing the nucellus; a few of the endosperm cells continue to divide for a while, the rest enlarges and many contain granular inclusions due to the building up of food reserves. The embryo grows slowly.

### Third Phase (39-73 days after fertilization)

The berry volume increases rapidly, largely because of cell enlargement taking place in the placenta. The outer tissues of the pericarp become stretched and pigmented.

The seeds increase in size for a short while. The endosperm cells enlarge and simultaneously accumulate protein reserves which are present as numerous aleurone grains in the mature cell. The embryo undergoes rapid cell division and differentiates into cotyledons, plumule and radicle. When fully grown, the embryo occupies only a small part of the endosperm. At various stages throughout the growth of the fruit one neutral auxin and two acidic auxins were found to occur together with one acidic growth inhibitor. Meristematic activity within the berry and the accumulation of reserve protein in the endosperm of the seed were found to coincide with the auxin peaks. Neumann (1955) studied the relation between hormone content and fruit development in black and red currants.He found that ovule development was dependent on the position of the fertilized fruit in the inflorescence. Inhibition of development and degeneration of the ovules was often noted in the terminal flowers; inhibition began before, as well as after fertilization and resulted in reduced fertility. Presence of 3 hormones in fertilized ovules and fruits were recorded. It was found that terminal fertilized fruits of cv. Rote Hollandische with inhibited ovule development and reduced fertility contained less hormone than did basal fruits. In terminal fruits, IAA was markedly absent. Bouscharain and Bergent (1962) stated that seeds accumulated considerable amount of minerals (Ca, Mg, P, Fe and Cu) during fruit ripening.

Berries from the basal part of the cluster contain higher levels of sugars, organic acids and vitamin C in comparison with berries positioned at apical cluster sections. A similar pattern was observed for phenolic component : berries of the 'Rovada' cultivar contained almost twice the amount of antocyanins in berries from the basal cluster section compared to berries positioned at the apical part of the cluster. The difference were more in black currant cv. Tsema (2.8 fold more anthocyanin in basal cluster section). The basal position of berries in cluster contained from 1.6 to 2.2 fold more hydroxycinnamic acids, from 1.7 to 2.2 fold more flavonol glycosides and from 1.5 to 1.8 fold more flavanols (catechin, epicatechin and procyanidin) (Mikulic-Petkovsek *et al.*, 2016).

During the last 2-4 weeks of fruit ripening significant changes occur in colour parameters, primary and secondary metabolites and radical scavenging activity. The average content of specific phenolic groups and sugar/acid ratio can be used as a simple and reliable maturity index for optimisation of harvest time (Mikulic-Petkovsek *et al.*, 2015).

Li *et al.* (2015) cloned an anthocyanin synthase gene homolog from black currant and its expression at different fruit stages. Anthocyanidin synthase (ANS), a 2-oxoglutarate (2OG) and Fe (II)-dependent oxygenase, catalyzes the penultimate step in anthocyanin biosynthesis, form leucoanthocyanidins to anthocyanidins, the first colored compound in the anthocyanin patway. In this study, a full length, 1427-bp long cDNA named *RnANS1*, which is homologous to the anthocyanidin synthase gene, was cloned from blackcurrant using a homologous cloning strategy. *RnANS1* is highly homologous to other plants *ANS* genes at both the nucleotide and amino acid sequence levels. The deduced protein contains domains conserved in the 2OG and Fe (II) dependent oxygenase, and is phylogenetically closely related to *Paeonia suffruticosa* and *Paeonia lactiflora*. The expression of *RnANS1* was upregulated during fruit maturation, and correlated with the accumulation of anthocyanins and soluble carbohydrates in the fruit.

## Harvesting

The picking period is generally long and may extend from 4 to 6 weeks. In case the fruit is required for jelly making, it is wise to harvest the crop in a few pickings and should be slightly under-ripe when pectin content is high. The harvested fruit is not kept in the sun. The currants should never be picked wet because quality and ability to travel in good condition are adversely affected. The fruit for processing should be just a little more ripen than that for the fresh-fruit market, but it should be still firm. The fruits should be picked with extreme care. Stoker (1985) stated that once the fruits are fully ripe they should be harvested as quickly as possible, preferably within the week, to avoid significant yield loss. The main cluster stem at the base is pinched off, using forefinger and thumb. The hand picking can be carried out by casual piece-work pickers but it is costly as compared to mechanical harvesting. Now many types of machines have been developed for harvesting currants.

The performance of the Finnish advanced type black currant harvester (Junas) and the Russian MPYa-I was recorded. Both machines harvested more than 90 per cent of the crop of cv. Golubka. The machines cleared 0.5 and 0.4 hectare/hour, respectively (Fedotov *et al.*, 1983). Daugaard (1982) selected black currant cultivars on the basis of berry firmness, growth habit, ease of shaking, uniformity of ripening and berry quality and yield. The cultivars most suitable for mechanical harvesting were Topsy, Wallace Seedling, Cotswold Cross, French Janzen, Roodknop and Baldwin. Krotov (1971) studied the suitability of bush for mechanical harvesting and reported that mechanically harvested black currant cv. Ojebyn and red currant cv. Rondom had 0.16 and 0.2 per cent injuries, respectively and did not differ greatly from manually harvested fruits. Dry matter, soluble solids, organic and ascorbic acids and anthocyanins showed the mechanically harvested fruit to be more ripe. This was probably due to machine leaving 5 per cent of mainly unripe fruit to become more ripe on the plants. The quality of frozen fruit from mechanically and manually harvested fruits was similar (Zbroszczyk *et al.*, 1985).

Salamon (2000) studied the mechanical harvesting efficiency of six red currant cultivars and reported that at 0.87/km/h speed, 90 mm amplitude and a 13.7 Hz vibration frequency, 100, 94.2, 90.7, 88.0, 87.4 and 82.7 per cent of cvs, Losan, Rondom, Tatran, Rosetta, Detvan and Maraton fruits were shaken off.

To make the mechanical harvesting efficient, Utkov (2015) suggested the average berry yield should be at least 0.3 t/ha with 90 per cent of simultaneously ripened berries on 3-year-old bushes before harvesting. Bushes should be 1.1-1.8m high with up to 15 perennial branches or up to 40 branches of different age per bush. Branches should not be entangled, reclined more than 50° or lay on the ground. The main load of ripened berries should be positioned more than 0.25m high from the ground. The tearing force of ripe berries should be within 1.2 H and the squashing force should be more than 2 H. The height of the soil heap at the base of bushes should not exceed 0.1m. In these conditions, the harvester KSM-5 would be able to collect 5 tonnes of high quality berries per shift and cover an area of up to 30 ha per harvesting season. Sazonov and Danshina (2016) selected the following black currant cultivars growing in Russia which were considered suitable for mechanical harvesting: Bagira, Kipiana, Partizanka Bryanskaya, Miph, Kudesnik, Selechenskaya 2, Sterlets, Sudarushka, Podarok Veterans, Tamerlan and Debryansk.

Application of ethephon at 480 ppm at 2 to 4 days before harvesting in black currants and at 720 ppm at 4 to 7 days before harvesting in red currants enhanced ripening and abscission. In general, most of the ethephon treated fruits dropped without the string, whereas control fruits fell in bunches (Zatyko, 1975). In similar studies, Zaliwski *et al.* (1974) however, reported the optimum ethephon rate as 500 mg/l and best time for application as 3 to 5 days before harvest for several black and red currant cultivars. Treated fruits could be shaken off by hand separated well from the pedicels and were undamaged and uniformly ripe. The treatment also improved fruit quality.

## Yield

The yield is an important component of production and it greatly depends on the management of the plant. The cultivars which had high yielding capacity showed favourable branch length/plant and bud numbers/m branch, high inflorescence numbers/bud and fruit set/inflorescence (Neumann, 1983). In The Netherlands, the yields recorded in red currant cvs. Rolan and Rovada were 267 and 247 kg/ acre, respectively (Oosten 1983); the yield of white currant cv. Werdavia was 2.58 kg/m$^2$ (Oosten, 1989). Watters (1983) reported that the yields of black currant cvs. Blackdown (9.7 t/ha), Ben Lomond (9.5 t/ha), Jet (8.7 t/ha) and Ben Nevis (8.5 t/ha), in northern Ireland were significantly higher than that of the standard cv. Baldwin (6.5 t/ha). In Poland, the black currant cvs. Ojebyn, Risager, Mendip cross. Roodknop and Blacksmith were considered high yielders (Smolarz, 1982). British cultivars Black Reward and Malling Jet showed a yield of 4211 and 3727 kg/ha, respectively, in Switzerland (Terrettaz and Carron, 1984). In Australia, the cvs. Magnus (4.2 t/ha) and Hatton Black (3.7 t/ha) showed higher annual yield than Baldwin (1.3 t/ha) and White Bud (1.3 t/ha) at a planting density of 4296 bushes/ ha (Goubran and El-Zeftawi, 1983). In Moldavian region of USSR, Polikarpova (1983), after assessing 18 black currant cultivars, recorded higher average yield in cvs. Seyanets, Golubka, Dochka, Chernoglazaya and Yubileinaya.

Ten black currant cultivars were evaluated at the Research Institute of Horticulture, Skierniewice, Poland by Pluta *et al.* (2012). The different cultivars/ clones showed wide variations (Table 14.3). Sasnauskas *et al.* (2009) recorded following yields/bush of different cultivars in Lithunia: (yield :kg/bush, average of four years):- Binar- 1.47, Gagatai- 1.60, Intercontinental- 1.74, Jadrenaja- 1.13, Polar- 1.29, Poezija- 1.69, Storklas- 0.89 and Veloj- 1.83.

**Table 14.3: Yield, Number of Fruits and Soluble Solids Content of Black Currant Cultivars***

| Sl.No. | Cultivar/Clone | Fruit Yield (kg/plot (5 bush) Average of 3 years) | Number of Fruits in Strig | Soluble Solids (° Brix) |
|---|---|---|---|---|
| 1 | Bona | 3.12 | 4.8 | 13.8 |
| 2 | Chereshneva | 3.46 | 13.0 | 14.3 |
| 3 | Yubylejnaiak | 7.19 | 10.5 | 14.8 |
| 4 | Saniuta | 3.53 | 11.6 | 14.0 |
| 5 | Sofievskaia | 2.90 | 9.4 | 14.9 |
| 6 | Vernisazh | 6.65 | 9.7 | 14.2 |
| 7 | Gofert | 5.63 | 10.3 | 18.5 |
| 8 | Tines | 4.96 | 7.8 | 18.0 |
| 9 | Tihope (PC 425) | 9.54 | 9.1 | 16.0 |
| 10 | D13 B/11 | 3.68 | 5.8 | 13.4 |

* Pluta *et al.* (2012).

## Ripening

Anthocyanins are responsible for the final fruit colour of red currants. Both chlorophyll and carotenoids decreased during ripening, chlorophyll disappearing in the ripe fruit, carotenoids were present in low concentration. The carotenoid pattern was that of the chloroplast type, but very complex at the unripe stage, involving a number of minor pigments (Gross, 1982). Skrede (1987) found about 60 per cent anthocyanins in juice and syrups of black currants to be delphinidin derivatives and 40 per cent cyanidin derivatives. Relative amounts of 4 anthocianins namely delphinidin-3-glucoside, delphinidin-3-netinoside, cyanidin-3-glucoside and cyanidin-3-netinoside were detected.

Naichenko and Osokina (1982), in studying the ethylene dynamics during ripening and storage of black currants cvs. Goliath, Junnat and Saunders, could not detect ethylene in green berries some 2 weeks after fruit set but trace amount were found when the berries began to colour and levels increased until harvest. After harvest and during storage, ethylene levels continued to increase until climacteric was recorded and then declined. Sesvecan *et al.* (1987) reported that the sugar content and colour intensity of the juice in red and white currants increased with fruit ripening. They suggested harvesting at full ripeness for optimal juice quality as late harvest resulted in high losses from fruit shrivelling and dropping which were not fully compensated by increase in sugar content and colour intensity. Rubinskine *et al.*(2008) studied the quality changes in black currant cvs. Pilenai, Vyciai, Kriviai and Gagatai berries of different stages of maturity at Lethuania. The highest levels of ascorbic acid were found in berries at the beginning of ripening. Among various cultivars, cvs. Pilenai and Gagatai berries of different maturity had the highest levels of ascorbic acid, 152- 114 mg/100g and 147- 103 mg/100g, respectively. With the exception of cvs. Vyciai and Pilenai larger amounts of pigments accumulated in overripe berries. Mature and overripe berries of cvs. Kriviai and Gagatai had the highest levels of anthocyanin at 387.61- 450.63 mg/100g and 349.79- 393.91 mg/100g, respectively. The highest amount of dry soluble solids were found in overripe berries which was found 14.55-16.95 per cent in overripe berries of cv. Kriviai. The amount of titratable acidity decreased during ripening. Shirochenkova *et al.* (1985), while studying the activity and isoenzyme composition of ascorbic acid oxidase and peroxidase in ripening and storage of black currant cv. Golubka found that in ripening fruits ascorbic acid content decreased and ascorbic acid oxidase increased until maturity and decreased in overripe fruits. In a hot dry year, ascorbic acid content was found more than twice as high at the start of storage (246.7 mg/100g) as in a wet year (109.6 mg/100 g), but changes in both the acid and its enzyme were inconsistent during storage. The isoenzyme spectra of ascorbic acid oxidase and peroxidase varied markedly during ripening.

Zorenc *et al.* (2017) working with three differently coloured cultivars of currants (Jonkheer van Tets, Pink Champagne and Zitavia) observed that the pink fruited cultivar 'Pink Champagne' showed generally lower enzyme activity than the red cultivar 'Jonkheer van Tets'. The white cultivar 'Zitavia' showed very low chalcone synthase/chalcone isomerage activity and *ANS* gene dihydroflavonol 4-reductase activity were detected. The anthocyanin content of the red cultivar increased during the last three weeks of ripening and reached a maximum of 890 mg/kg fresh weight. The pink cultivar show low accumulation of anthocyanins, however, in the last three weeks of ripening their content increased from 14 to 105 mg/kg fresh weight. The white cultivar had no anthocyanins, but the sugar/acid ratios were the highest. In the white and pink cultivars, reduction/lack of anthocyanins was mainly compensated by increased relative concentrations of hydroxycinnamic acids and flavonols.

## Storage

The currants can be stored for a short period in low temperature. Fruits of black currant cvs. Stanza, Rotel, Rosetta and Rondom were successfully stored for 19-21 days at 0° to 1°C (Westerlaken, 1983). The black currant cv. Goliath, stored well for 69, 90-109 and 116-131 days at +2°, 0° and –2°C, respectively (Naichenko and Osokina, 1982). The use of plastic cover (PVC film) reduced weight loss in red currant fruit and improved the overall quality when kept at 1°C and 95 per cent RH. However, a high incidence of decay was observed after 23 days of storage (Hevia *et al.*, 2000). Controlled atmosphere storage is also used in storing currants. Naichenko and Romanshchak (1985) stated that fruits (cv. Goliath) stored at 11 per cent $CO_2$ + 10 per cent $O_2$ + 79 per cent N extended the storage life and quality was improved.

## Pests

### Clearwing Moth (*Synanthedon tipuliformis*)

It is a serious pest of currants. The eggs are laid along the branches by the moth and when hatched the grubs make their way into the stems, suck the sap and cause decay of the branches. Dirimanov and Sengalevich (1972) recorded 3.12 per cent infestation of the branches in a 4-year-old plantation and 41.28 per cent in a 6-year-old plantation. The infestation rate reached 45.09 per cent in Silvergieters, 38.56 per cent in Daniels September and 29.40 per cent in Wellington. Two to three sprays of dimecron 50 or dipterex 80 per cent or heterotex all at 0.2 per cent concentration or a combination of 0.1 per cent dipterex + 0.1 per cent dimecron 50 was suggested to control the moth. The first spray should be applied during 35-40 per cent moth flight and when the eggs start to hatch. Savkovskii and Rybalov (1988a) suggested spraying of 50 per cent or 30 per cent malathion at 1 to 2.6 and 2 to 4.51/ha, respectively, at 20 days after the end of flowering when oviposition usually begins. Frankenhuyzen and Jansen (1983) obtained complete control of clearwing moth by 3 applications of deltamethrin at 0.02 per cent on 15 and 23 May and on 25 June.

Cane borer larvae were not observed in any of the pink or white flowering currants, *R. sanguineum* or in golden currants, *R. aureum* cultivars Crandall, Idaho or Gwens (Hummer and Sabitov, 2004).

## Leaf Midge (*Dasineura tetensi*)

The adult lay eggs in the folds of young leaves and growing points of green shoots. After the eggs begin to hatch, the larvae feeding on the surface of the leaves causing a crinkled and folding like appearance. Young shoots when attacked by midge causing stunting growth and temporary death of shoots (Hellqvist, 2005).

A forcasting model, using air temperature and a logistic function based on developmental rates of a Norwegian population, predicted the time of oviposition with fairly good accuracy (Hellqvist, 2001). In a trial with 6 insecticides, best results were obtained by 0.1 per cent phosdrin, 0.15 per cent nogos 50 or 0.2 per cent folithion which showed over 90 per cent mortality of the pest (Sass, 1972). Fenpropathrin sprays at 50 and 100 g a.i./ha controlled the pest by protecting the plants against oviposition and also by killing the larvae inside small convoluted leaves (Wardlow and Nicholls, 1986).

Harris and Wardlow (1984) tested the effectiveness of sprays of 11 insecticides for the control of larvae of *D. tetensi* on black currant. The combination of azinphos-methyl and demeton-S-methylsulphon at 0.75 per cent and 0.23 g a.i./ ha, respectively; fenitrothion at 0.94 g with or without cypermethrin at 0.04 g and demeton-S-methyl at 0.44 g which gave more than 81 per cent larval mortality in the field. Labanowska *et al.* (1983) conducted several experiments in Poland to control black currant leaf midge. They found satisfactory control with fenitrothion, methidathion and malathion when applied as foliar sprays at standard dosage in high and low-volume sprays but up to 8 applications were found necessary. Soil insecticides chlorfenvinphos at 80 kg/ha of 10 per cent granules, diazinon at 60 and 100 kg/ha of 10 per cent granules or lindane at 80 kg/ha of 2 per cent granules applied around the base of the bushes before blossom time (at the time of emergence of lst generation adults) proved very effective to control the emerging adults or at the cocoon stages, providing protection for 4-6 weeks, which include the blossom period and a part of the preharvest period.

Resistance has been reported from the species *R. americanun* and *R. dikuscha* controlled by a dominant gene *Dt* within the wild species of *Eucoreosma* jointly with some cultivar *e.g.* Ben Connan (Keep, 1985).

## Gall Mite or Big Bud (*Cecidophyopsis ribis*)

This is possibly the most serious pest of black currant. The mite is very small and less than one-hundredth part of an inch in length and can be seen only with a microscope. The mites exist in thousands in every bush when infested. Red and white currants rarely suffer from big bud. Gall mites caused galling of the buds leading to a damaging condition known as 'big bud' (Brennan *et al.*, 2008).

Introduction of predatory mite *Typhlodromus pyri* in the planting of a lower initial infestation level proved to be very effective (Ourednickova *et al.*, 2011).

Thiodan (endosulfan) at 0.2 per cent applied once before and once after flowering showed a reduction in the number of infested buds from 25 per cent to 13 per cent over control (Gorenborg, 1985). Nielsen (1986) suggested the following measures to avoid gall mite infestation and to control the pest; i) use of certified disease and pest-free plants, (ii) not to plant near infested crops, (iii) removal of all infested branches and bushes, (iv) spraying of thiodan emulsion or lannate 10 l (methomyl) at the onset of flowering, at petal fall and again after 14 days, and (v) application of sulphur thiram fungicide sprays during May and June.

Labanowska (2011) suggested use of sulphur once before blooming which reduced the number of affected buds to the level of 40-50 per cent. Ortus 05 SC at 1.5 l/ha applied twice, at the beginning and during the full blossom significantly reduced the number of buds infested by the big bud mite (Labanowska *et al.*, 2015).

Resistance to gall mite is an important genetic trait of *Ribes*. *P*. and *Ce* genes, responsible for gall mite resistance, were established in *Ribes* species and interspecific hybrids using molecular markers. Resistance in *R americanum* is determined by *P* gene and in *R. sanguineum* by *Ce* gene. Both molecular makers were absent in *R. dikuscha* genome. Molecular markers related to *P* and *Ce* genes were identified in the genome of *R. aureum*. Resistance to gall mite in the field conditions in *R. nigrum* x *R. americium, R. nigrum* x *R. aureum* and *R. nigrum* x *R.sanguineum* $F_3$ hybrids fitted and expected Mendelian segregation ratio of 1:1, 3:1 and 1:1, respectively. Seventy five per cent of hybrids with pyramidal resistance to gall mite carrying markers related to *Ce* and *P* genes were obtained in the cross combination *R. nigrum* x *R. aureum* (Mazeikiene *et al.*, 2017). Cultivar Ben Finlay has resistance to gall mite as well as to reversion virus solely due to the *Ce* gene (Mitchell *et al.*, 2011).

## Woolly Currant Scale (*Pulvinaria ribesiae*)

It also causes considerable damage to the currant plantations. This can be effectively controlled by high volume sprays of deltamethrin at 1.7-5 g a.i./100 l water and to a lesser extent by fenitrothion at 47.5 g applied in early June on fruiting black currant bushes of cv. Baldwin (Alford, 1985).

## Weevil

The leaf weevils are black, about 1.25 cm long, having yellowish green or golden scales on their back. They are found feeding on leaves in May and June. But normally it is not a serious pest.

There is another type of weevil known as clay-coloured weevil. This comes out of soil to feed at night. It is brown in colour and speckled with lighter markings and about 0.6 cm long. It starts feeding on leaves in February and lays eggs in summer in the soil and the resulting grubs feed all winter on young roots.

In the UK, spraying of carbofuran at 1.5 kg a.i./ha on 17 August, after harvesting, to an established plantation of cv. Baldwin, was found to control the infestation by *Otiorhynchus sulcatus* (Alford, 1985).

## Red Spider Mite (*Tetranychus urticae*)

This pest in more serious on light soils during dry and hot season. It overwinters as an adult in the rubbish at the base of the bushes and migrates to the leaves and sucks sap as the weather warms up. The migration may take some time and may not be completed until after blossoming. As it feeds on sap, the leaves develop a bronze hue and bush gives a dirty look. The major predators on *T. urticae* are the coccinellid (*Stethorus bifidus*) and the phytoseiid, (*Typhlodromus caudiglans* and *Amblyseius longispinosus*) (Thomas and Burnip, 1984). Phytoseiid mite *Neoseiulus californicus* was shown to be most effective in controlling *T. urticae* in small fruits grown under plastic tunnel (Bylemans *et al.*, 2003). Spraying of chlorpyrifos, demeton-S-methyl, or tetradifon just after flowering was found effective to control the mite.

## Currant Sawfly (*Nematus olfaciens*)

This pest is closely related to the gooseberry sawfly but not the same. The adult sawfly appears in April and May and lays eggs in rows along the main veins on the underside of the leaves. They start feeding in the centre of bush because of their habit of central feeding. There are several broods in a year in May, July and August. If the first brood is controlled immediately, there is less chance of later attacks. Spraying with azinophos-methyl plus demeton-S-methyl sulphone or fenitrothion effectively controlled sawfly.

## Caterpillars

Many types of caterpillars are found during the growing season. These pests range from the common winter moth caterpillars to those of the magpie and tortrix moth. They can be controlled by mixing insecticides in the anti-gall mite spray at grape stage.

## Currant Shoot Borer (*Lampronia capitella*)

The currants shoot borer is one of the major pest in northern and central Europe. The larvae infest the buds in early spring, resulting in the death of developing shoots. Sex pheromones can be used in pest monitoring and in the development of environmentally safe pest control methods (Mottus *et al.*, 2001).

## Aphids

Several aphid species (*Hyperomyzus lactucae, H. pallidus, Cryptomyzus ribis, Nasonovia ribisnigri* and *Aphis spp.*) infest currant (Brennan, 1996). The infested leaves curl downwards. The parts of the upper surface between the veins are usually elevated into large, irregular blisters that are often reddish. The leaves are weakened which ultimately die. Aphids also cause economically significant damage as vector of virus. These aphids can be controlled by winter sprays or by the use of systemic insecticides in the early spring.

A single application of chlormequat chloride at 2500 ppm reduced the density of aphid (*Hyperomyzus lactucae*) on black currants below the level of damage. There was no effect on yield in the year of the experiment, but in the following year bushes which had been treated gave considerably higher yield. This increase was attributed both to the direct action of chlormequat chloride and to the indirect effect through the reduction of aphid population (Singer and Smith, 1976). Hummer and Dale (2010) suggested spraying of pesticide or soap treatment before flowering and after harvest. Calypso 480 SC (thiacloprid) at 0.1 and 0.15 l/ha showed very good control of aphids (*H. lactucae* and *A. schneideri*) (Abanowska, 2004).

## Diseases

### Leaf Spot

The disease is most prevalent in areas of high humidity or rainfall. It causes characteristic irregular, angular and scattered brown spots on the leaves which are first seen in the middle of June. In case of severe attack, the entire leaf shrivels and fruit may dry up. Defoliation in June and July resulted in great loss of crops in that season and also in the following season.

Riordian (1986) reported that due to the wet weather, infection by *Pseudopeziza ribis* was severe and in fungicide trials it reached 82.5 per cent in unsprayed plots in August. In September assessment, the best control of leaf spot was recorded by using fuitriafol. In cultivar assessment trial least infection occurred in Tenah, Tsema and 23832 while Ben Lononel and Black Reward were severely affected compared with the standard cv. Baldwin. Mulholland and Beresford (1989) reported that the leaf spot of black currant caused by *Mycosphaerella ribis* in which maturation of perithecia contained ascospores by early September. Ascospore release always begins in late August or early September, reaches a peak between mid-September and mid-October and decline by mid-November. The onset of ascospore release tended to coincide with bud brust on black currant bushes. In order to prevent primary infection in spring, it is suggested that fungicide should be applied from the time of leaf emergence until the grape stage of flower development. In Poland, spraying with funaben 50 (50 per cent carbendazim), funaben T (20 per cent carbendazim and 45 per cent thiram) and oxafun T (37.5 per cent thiram and 37.5 per cent carboxin) showed good control of mild infection by leaf spot (Rebandel and Mazur, 1981).

Resistance to septoria leaf spot has found in *R. americanum*, *R. dikuscha* and *R. nigrum sibiricum* (Brennn, 2008).

### White Pineblister Rust

The rust caused by *Cronartium ribicola* is seen on the lower surface of leaves as little clusters of orange coloured spots. The leaves become withered and almost black in wet weather and the undersides take on a hairy or woolly appearance. This disease has an alternative host, pine (*Pinus strobus*).

Profic *et al.* (1973) reported good control of rust with benlate (benomyl) at 0.1 per cent applied 8 times between late April and mid-August. The effectiveness of less frequent treatments depended on the disease intensity and the spraying programme. Nielsen and Kirknel (1986a) suggested spraying of triadimefon (bayleton) at 0.05 per cent just after flowering and 10-12 days later. Foliar application of highly refined mineral oils can reduce white pine blister rust in susceptible black and red currants (Picton and Hummer, 2003).

Incorporating of the dominant gene *Cr* from *R. ussuriense*, resistance to *C. ribicola* was achieved. Resistant cultivars Consort, Coronet, Crusader and Titania were developed (Brennan, 2008).

## Mildew

It is not very serious disease but appears when weather is dry. The disease infection causes powdery growth on the leaf surface and young shoots. Scorch like symptoms appear on the leaves that become deformed and dry out on berries, mildew produces a dark brown, felt-like coating that renders the fruit unmarketable (Hummer and Dale, 2010). Nielsen and Kirknel (1986b) suggested spraying of dinocarp (karathane) to control *Sphaerotheca mors-uvae*. Gjerum and Langnes (1984) reported that there was little difference in the effectiveness of triadimefon, bupirimate, chinomethionate, dinocap, dithalimofos, fenarimol and folyfluanid. Three to four sprays were effective than two. An early spraying on a severe attack prevented infection of flowers and subsequent fall of unripe berries. Ionova and Minaaev (1983) reported that spraying of BMK (analogue of carbendazim) reduced mildew incidence on black currants by 81.9 per cent. They suggested spraying at 1.2 kg/ha in a low volume spray of 600 l/ha at fruit set and after harvest. Mineral oil at 8 ml/l (8000ppm) application significantly reduced mildew severity in vegetative growth (Hummer and Picton, 2001). Luna Sensation 500SC containing fluopyram showed high effectiveness in control of black currant fungal diseases (Broniarek-Niemiec, 2015). Potassium bicarbonate treatment significantly reduced the powdery mildew severity in leaves and fruits (Wenneker *et al.*, 2016).

## Anthracnose

It causes small brown necrotic lesions on the leaves, young canes, fruit stems and fruits. The main damage is early defoliation. The disease caused by *Pseudopeziza (Drepanopeziza) ribis* occurs frequently and can be controlled in the summer, by application of Bordeaux mixture at 1 per cent, dithane M-45 at 0.2 per cent or benlate at 0.006 per cent are effective (Zakharieva, 1977). Nowacka *et al.* (1980) obtained good control in black currant cv. Black Smith by applying dithane M-45 at 0.3 per cent during the growing season or 3 times during the first half of season followed by topsin at 0.1 per cent, 3 times during the second half of the season. Spraying with mancozeb 80 at 0.2 per cent and 0.3 per cent cupral 45 proved effective to control *D. ribis* (Burth and Ramson, 1985).

## Die Back

*Alternaria tenuis, Botrytis cinerea, Fusarium avenaceum, Nectria cinnabarina* and *Phoma* spp. were the most common fungi on stems of black and red currant bushes (Tahvonen, 1983). *B. cinerea* infection aggravated die back of branches and inhibited wound healing. In cold autumn, *B. cinerea* infection was increased on damaged branches and subsequent die back was more intensive. This could be counteracted by postharvest spraying with thiophanate-methyl. In a mild autumn maneb and dichlofluanid also reduced die back incidence. Mechanical harvesting resulted more injury and subsequently more infection (Tahvonen, 1979).

## Grey Mould

The grey mould fungus *Botrytis cinera*, causes several problems in association with *Botrysphaeria* on susceptible *Ribes*- dieback of shoots in the spring, premature fruit drop and fruit rot (Hummer and Dale, 2010). Teldor (fenhexamid), Switch (cyprodinil + fludioxonil) and Scala (pyrimethanil) gave good *Botrytis* control. Switch was reported as very effective (Jorg *et al.*, 2003).

## Viral Diseases

### Reversion

It is a viral disease which causes peculiar changes in leaf and shoot until whole bush takes on the appearance of having been reverted to wild state. It is incurable and only remedy is to uproot the bush and burn it. The symptoms first become visible on new leaves formed in June; the end of June is the best time to rogue the bush. It shows reduction in the number of the veins of the leaf. In a reverted leaf, the number of secondary veins between the two points will be reduced to 4, 3, 2 or even 1. In the case of a 1-veined section, the leaf is changed and it resembles an oak leaf. Jones McGarin (2002) characterised the virus as the *black currant reversion virus* (BRV), genus *Nepovirus*, family *Comoviridae*.

Reversion had a significant influence on yield reduction of about 40-50 per cent (Pluta and Zurawicz, 1999). The transmission of the virus is by the eriophyid gall mite. BRV is the agent of black currant reversion disease (BRD), which is economically the most significant viral disease in *Ribes* species. BRV and BRD occur widely in locations where black currant is cultivated commercially (Susi, 2004).

A disease observed on several cultivars designated 'ryabukha' (wild fire) is believed to be of viral origin. It may affect 25-30 per cent of the leaves in June and 80 per cent by August. The most susceptible cultivars are Chernaya Lisavenko, Cheranaya Crozd, Pozdenespelaya and Vystavochnaya (Kalinichenko and Gladkikh, 1973).

## Breeding and Improvement

The breeding work on the improvement of currants is being carried out in many countries. The major objectives in currant breeding are erect growing habit of plant, high productivity, suitability for mechanical harvesting, late flowering with high productivity and shorter maturation period, resistance to late spring frosts and freedom from 'running off', improved fruit qualities acceptable for processing and resistance to insect pests and diseases.

The desirable cultivar should have as many characters mentioned above as possible. Nearly all the characters required occur in the existing cultivars, black currant species or selected hybrids, but it has proved difficult to combine erect habit with strong stout branches and high productivity, late flowering with high productivity, shorter ripening period, resistance to frosts and 'running off'.

The Scandinavian and Russian cultivars are more winter-hardy than West European cultivars. Considerable emphasis has been put on combining intrinsic frost resistance from Scandinavian cultivars and *R. dikuscha* with late flowering for frost escape (Mather *et al.,* 1980). The cultivar Ben Lomond released in 1975 from Scottish Crop Research Institute, has a Swedish, a Finnish and a Canadian parent in its immediate ancestry. They gave it an improved frost tolerance by flowering about 10 days later than Baldwin. Ben More - a hybrid between Goliath and a wild Swedish black currant and Ben Sarek, a sister seedling of Ben More are notably frost tolerant and high yielding (Jennings and Anderson, 1984). The *R. nigrum sibiricum* and *R. dikuscha* were extensively used in breeding for winter hardiness.

Lanham *et al.* (2001) reported the isolation of dehydrin-like genes (dhns) from cold-acclimated (4°C) leaves of black currant cv. Ben Lomond plants using RT-PCR with a guessmer primer based on known dehydrin DNA sequences. Several Dhn like C-DNAS from *R. nigrum* were isolated and one (b8) with an internal lysine-rich coding sequence (K segment) was chosen for further characterization. Northern analysis conclusively demonstrated that b8 was chill-induced in vegetative tissues. The cold induction of dhns was confirmed at the protein level using a polyclonal antibody immunoreactive to the K segments in DHNs.

Late-flowering cultivars usually escape late spring frost damage. *R. petraeum, R. bracteosum, R. divaricatum, R. manschuricum* and *R. multiflorum* are some of the late-flowering species for breeding frost escaping cultivars. Amos Black (Goliath x Baldwin) and Jet (derived from *R. fuscesens*) are very late-flowering cultivars to escape spring-frost damage.

The Scandinavian cultivars, *R. dikuscha* and *R. glutinosum* are being used as donors for leafspot resistance in black currants while cultivars of red currants derived from *R. petraeum* (*e.g.,* Prince Albert, Red Dutch and Viking) showed significant resistance to leaf spot. At East Malling, efforts are being made in combining gall mite resistance (gene Ce from the gooseberry) with other characteristics (Keep, 1984). The gall mite (*Cecidophyopsis ribis*), vector of reversion virus, is widespread on black currant in Britain. The technique of transferring resistance to black currant from initial sterile diploid black currant x gooseberry hybrid was described by Knight *et al.* (1974). In third backcross ($BC_3$) strong resistance controlled by the dominant gene *Ce* was transferred to self-fertile plants. The gene $Sph_3$ from *R. glutinosum* provides strong mildew resistance and from the donors of leafspot resistance, *R. dikuscha* provides the highest level of resistance, conferred by the complementary Senes $Pr_1$ and $Pr_2$ (Anderson, 1972). Some cultivars bred using *R. nigrum* var. *sibiricum* and *R.*

*pauciflorum* showed resistance to *Sphaerotheca mors-uvae*, *Pseudopezizia* (*Drepanopeziza*) *ribis* and *Mycosphaerella ribes* and to reversion virus (Khokhryakova, 1986).

Heritability of resistance to powdery mildew and leaf spot was determined by additive genes. Direct correlation was established between resistance to powdery mildew and resistance to leaf spot. The superior selection D16/1/25 was selected as donor for powdery mildew resistance, cultivar Storklas was selected as donor for leaf spot resistance, and Titania had complex resistance to powdery mildew and leaf spot. D16/1/25 transmitted high productivity and good berry size to their progenies (Sasnauskas *et al.*, 1999). Zhidyokhina and Rodyukova (2000) suggested use of cvs. Tamerlan, Ojebyn and Constance as donors of resistance to powdery mildew. Belorsskaya Sladkaya and Chyorni Zhemehug were considered to be promising parent selections in breeding for resistance to septoriose (*Septoria ribis*). The cv. Titania showed full field resistance to while pine blister rust (Broniarek *et al.*, 2000). Siberian breeding programme initiated in 1930s developed several cultivars of importance. Highly disease resistant and productive cultivars selected were Zonalnaya, Aleandr, Berdskaya, Ranniaya Potapenko, Karachinskaya, Obskaya, Chernaya, Berdsksya Chernaya, Rakhil, Mariushka and Glarioza (Sorokopudov and Sorokopudova, 2011).

Black currant is a non-climactric fruit in its ripening. Qualitative and quantative chages in mRNA populations were found as ripening progressed, leading to the isolation of cDNA clones of five genes showing enhanced steadystate transcription levels in full ripe fruit (Woodhead *et al.*, 1998). Genomic clones of some of these genes have been isolated, and two of the genes (*RIB1* and *RIB7*) showed highly fruit-specific expression. The promoters driving the expression of these two genes may therefore be of considerable value in the future manipulation of ripening processes in transgenic fruit.

The main donor of increased fruit number is the North American wild species *R. biacteosum*. This species have over 60 flowers per raceme (Keep *et al.*, 1982). Madry *et al.* (2000) evaluated 25 $F_1$ hybrids of black currants, originating from diallel cross of five cultivars. A considerable consistence between the values of genetic and phenotypic correlation coefficients was observed. Fruit yield expressed the largest number of significant correlations with other traits. Path coefficient analysis indicated that plant and berry size were the main factors influencing directly fruit yield. It was proved that the number and length of one-year-old shoots via bush size had very large indirect effects on fruit yield per plant. The cvs. Baldwin and Green Black are good donors of high ratio of fruit to vegetative bud (Wilson and Adam, 1967). The juice flavour of *R. nigrum* x *R. petiolare* hybrids was found unacceptable but after one generation of backcrossing it was improved (Anderson, 1977). The cvs. Baldwin, Ben Lomond, Black Reward, Greens Black, Malvern Cross, Roodknop, Seabrooks Black, Wellington XXX, Westwick Choice, *etc.*, are acceptable for juice quality. All these cultivars are pure *R.nigrum* with the exception of Ben Lomond which is a derivative of *R. ussuriense* (Anderson. 1974). The European forms of *R. nigrum* usually showed higher ascorbic acid content in fruit (Knight. 1983). However, Voscilko (1969) noted 200-800 mg ascorbic acid content/100 g in some wild forms of *R. nigrum sibiricum*.

A common breeding programme involving the Swedish University of Agricultural Science, Balsgard, Lithuanian Institute of Horticulture, Latvia State Institute of Fruit Growing and N.I. Vavilov Institute of Plant Industry, Russia were carried out from 1995-2001 (Strautina *et al.*, 2012). Twelve elite hybrids have been selected, which were further evaluated in 2004-2009. Four hybrids were selected which were BRi9502-1A (Lentyai x Intercontinental), BRi9504-5 (65-59-5 x Storkias), BRi9508-3A (Gagatai x Intercontinental) and BRi9508-3B (Gagatai x Intercontinental). In black currant breeding, the main donors used for integrating the different characters are as follows :

| *Characters* | *Main Donors* |
|---|---|
| Regular cropping<br>Late flowering or<br>Frost avoidance | *Ribes bracteosum, R. nigrum* cultivars |
| Frost resistance | Scandinavian cultivars, *R. dikuscha* |
| High yield<br>Fruit size<br>Fruit number | Scandinavian cultivars, *R. dikuscha, R. bracteosum, R. glutinosum, R. nigrun* cultivars. |
| Good setting | Scandinavian cultivars, *R. dikuscha* |
| Fruit quality | *R. nigrum* cultivars |
| Erect habit | *R. bracteosum, R. glutinosum, R. nigrum* cultivars. |
| Resistance<br>*Sphaerotheca mors - uvae* | Scandinavian cultivars, *R. glutinosum, R. dikuscha, R. cereum* |
| *Pseudopeziza ribis* | Scandinavian cultivars, *R. bracteosum, R. dikuscha, R. glutinosum, R. americanum* |
| Reversion virus | *R. dikuscha, R. pauciflorum, R. altreum, R. cereum* |
| *Cecidophyopsis ribis* | *R. grossularia* |

## Genetic Diversity, Molecular Characterization and Chromosomal Mapping

In *Ribes* several molecular markers, such as RAPDs (Korbin *et al.*, 2002, Weijacha *et al.*, 2001) inter-simple sequence repeats (ISSR) (Lanham *et al.*, 2000, Korbin *et al.*, 2002, Yang *et al.*, 2011), anchored microsatellites (Lanham and Brennan, 1998), SSRs (Cavanna *et al.*, 2009, Belaj *et al.*, 2012, Palmieri *et al.*, 2013) expressed sequence tags-SSRs (EST-SSR) and SNPs (Brennan *et al.*, 2008, Russell *et al.*, 2011) were successfully developed to assess the relationship among different species, cultivars and selections and to obtain genetic linkage maps.

Weijacha *et al.* (2001) working with black currant and red currant cultivars and hybrids detected polymorphic bands in these genotypes by the OPQ and OPU primers. They suggested to use these bands to discriminate between members of hybridized pairs. Brennan *et al.* (2002) identified and characterized eleven microsatellite loci in black currant (cv. Ben Alder) and related species. An enriched library was constructed and screened with SSR oligonucleotides. Positive clones were sequenced and primers were designed flanking the repeat motifs. These 11 microsatellites produce amplication products polymorphic from the Eucoreosma section of the genes. The number of alleles varied from 2 to 18 with levels of diversity ranging from (0.18 to 0.91).

Yan *et al.* (2011) analyzed the genetic relationship of 39 black currant cultivars by ISSR. The results showed that the total number of bands (A) amplified by 13 ISSR primers ranged from 3 to 11, the average was 6.77, the mean percentage of polymorphic band (P) was 64.37 per cent. Cluster analysis showed that 39 black currant cultivars were divided in 5 groups at 0.8496 level of similarity coefficient. Cultivars in the group I came from 8 origin countries. The cvs. Ben Lomond and Kantata were farther with other black currant cultivars in genetic distance.

Deep-level second generation sequencing (2GS) technologies are now being applied to non-model species as a viable and favourable alternative to Sanger sequencing. Large-scale SNP discovery was undertaken in black currant using transcriptome based 2GS 454 sequencing on the parental genotypes of a reference maping population, to generate large numbers of novel markers for the construction of a high density linkage map by Russell *et al.* (2011). Over 700,000 reads were produced, from which a total of 7000 SNPs were found. A subset of polymorphic SNPs was selected to develop 384 – SSNPOA assay using the illumina Beadxpress platform. Additionally, the data enabled identification of 3,000 novel EST-SSRs. The selected SNPs and SSRs were validated across diverse *Ribes* germplasm, including mapping populations and other selected *Ribes* species. SNP-based maps were developed from two black currant mapping populations, incorporating 48 per cent and 27 per cent of assayed SNPs respectively. The use of 2GS technology for the development of markers was found superior, in both numbers of markers and biological informativeness of those markers. A proportion of noval genes were discovered across a wide range of putative function and localisation. A small set of SSR markers were found sufficient to effectively describe the *Ribes* germplasm variability and to discriminate between species and accessions, and thus helped increases the genetic understanding of the crop (Palmieri *et al.*, 2013).

## Genetic Map

The first genetic linkage map of black currant was developed by Brennan *et al.* (2008), identified regions where quantitative trait loci (QTLs) for some important traits were located. The analysis was complicated by the fact that the mapping population was found to contain two subgroups, with segregation ratios consistent with these being crossed and selfed offspring. Hacket *et al.* (2010) made QTL analysis based on the trait mean over 3-years and focussed on the crossed offspring. They proposed a mixed model multi-environment approach for this population. The three years were considered as three separate environments, the data from both selfed and crossed offspring were combined and different residual terms were explored to model the correlation between the years. This permits tests for interactions between the QTLs, the year and the type of offspring (selfed or crossed). This was applied to re-analyse two important traits, anthocyanin concentration and bud break. Several additional QTLs were identified, some affecting the trait in both the selfed and crossed offspring, others in just one.

## REFERENCES

Abanowska, B.H. (2004). Bull. OILB/SROP, 27(4): 101-106.

Alford, D.V. (1935). Ann. Appl. Biol., 106: 4-5.

Al-jaru, S. and Stosser, R. (1984a). Erwerbsobstbau, 26: 12-16.

Al-jaru, S. and Srosser, R. (1984b). Obstbau Und Fruchteverwertung, 34: 17-24.

Al-jaru, S. and Stosser, R. (1983). Obstbau Und Fruchteverwetung, 33: 243-52.

Ancay, A. and Carlen, C. (2016). Acta Hort., 1133: 223-26.

Anderson, M.M. (1972). Euphytica, 2l: 510-11.

Anderson, M.M. (1974). Rep. Scot. Hort. Res. Inst. for 1973. pp 39-40.

Anderson, M.M. (1977). Rep. Scot. Hort. Res. Inst. for 1976 pp. 48-49.

Asanica, A., Tudor, V., Sumedra, d., Teodorescu, R.I., Peticila, A. and Iacob, A. (2017). Sci. Paper Sr. B., Horticulture, LXI: 175-81.

Atroschohenko, G.P., Markelov, G.A. and Gitinnovasov, M.M. (1983). In: Intensifikatsiya vozdelpvaniya Produktiynosti Plodovykh i yagodnykh Kul'tur, USSR. pp. 67-69.

Baldini, E. and Pisani, P.L. (1964). Rivesta Ortoflorofruttic. Italiana, 45 : 619-39.

Barney, D.L. and Fallahi, E. (2009). Univ. of Idaho Ext. Bull., 855.

Barney, D.L. and Hummer, K.E. (2005). Currants, gooseberries and Jostaberries-a guide for growers, marketers and researchers in North America. Haworth Press, NY.

Barsukov, N.I. (1982). In Ovoschevodstvo i Sadovodstvo V Zapadnoi Sibri, USSR, pp. 27-30.

Belaj, A., Dominguez, G.M.C., Atienza, S.G., Urdaioz, N.M., De la Rosa, R., Satovic, Z., Martin, A., Kilan, A., Trujillo, I., Valpuesta, V. and DelRio, C. (2012). Tree Genet Genomes, 8: 365-78.

Belova, G.I. (1982). In : Povyshenic Produktivnosti Plodovykh i Ovoshchnykh Kultur USSR, pp.23-25.

Blasse, W. and Hoffmann, S. (1988). Archiv fur Gartenbau, 36: 437-48.

Blenda, V.F. and Kirilenko, F.D. (1982). Fiziologiya i Biokhimiya Kultyurnykh Rastenii, 14: 244-47.

Blom, H. and Skrede, G. (1984). Jr. Sci. Fd. Agric., 35:332-37.

Bosma, B. (1938). Meded. Tuibauw Voorlichtingsdienst, 8: 34.

Botar, A., Palocsay, R. and Dankanits, E. (1975). Lucrarile Stiitifice ale Institutului de Cercetare pentru Pomicultura Pitesti, 2 : 111-25.

Bould, C. (1964). A.R. Long Ashton Res. Stn. for 1963, pp. 83-89.

Bouscharain, H. and Bergent, J. (1962). C.R. Acad. Agric. France, 18: 694-98.

Brennan, R., Jorgenoen, L., Hacker, C., Woodhead, M., Gordon, S. and Russell, J.R. (2008). Euphytica, 161: 19-34.

Brennan, R., Jorgensen, L., Woodhead, M. and Russel, J. (2002). Mol. Ecol. Notes, 2(3): 327-33.

Brennan, R.M. (2006). In: Fruit Breeding. Vol. II (eds. J. Janick and J.N. Mooore) John Wiley and Sons, Inc. New York. pp. 191-95.

Brennan, R.M. (2008). In: Temperate Fruit Crop Breeding: Germplasm to Genomes. (ed. J.F. Hancock) Springer, The Netherlands, pp. 177-96.

Brennan, R.M., Jorgensen, L., Hackett, C., Woodhead, M., Gordon, S.L. and Russell, J. (2008). Euphytica, 161: 19-34.

Broniarek, N.A., Pluta, S. and Bielenin, A. (2000). Bull OILB/SROP, 23(11): 111-16.

Broniarek-Niemiec, A. (2015). Progress in Pl. Protec., 55: 478-82.

Burmistrov, A.D., Razdol'kina, N.A. and Solov'eva, N.A. (1984). Sodovodstvo, 6: 15-16.

Burth, U. and Ramson, A. (1985). Gartenbau, 32: 56-58.

Bytemans, D., Janssen, C., Latet, G., Meesters, P., Peusens, G., Pitoioudis, F. and Wagelmans, G. (2003). Bull. OILB/SROP, 26(2): 37-44.

Card, F.W. (1925). Bush fruits. Macmillaan Co. New York, pp. 265-76.

Cavana, M., TorelloMannoi, D., Becaro, G.L. and Bounous, G. (2009). Genome, 52: 839-48.

Cevekovic, N., Pagter, M., Kristensen, H.L., Brennan, R. and Peterson, K.K. (2014). Scientia Hort., 168: 193-201.

Cianciara, T. (1986). Prace Instytutu Sadownictwa i Kwiaciarstwa Skierniewicach, A., 26:47-58.

Cianciara,T. and Smolarz, K. (1983). Prace Instytutu Sadownictwa i Kwiaciarstwa Skierniewicach A, 24: 49-57.

Cronquist, A. (1981). An integrated system of classification of flowering plants. Colombia Univ. Press, New York.

Cuvasina, N.P. (1963). In : Gibberellins and their effects on plants. USSR Acad. Sci. Moscow. pp. 123-28.

Daltrro, K. and Dullum, N. (1957). Tidsskr. Planteavl, 60: 721-28.

Date, A. (2000). HortTech., 10: 307-308.

Daugaard, H. and Vang-Peterson, O. (1982). Tidsskr. Planteavl, 86: 450.

Denisow, B. (2002). Annales Univ. Mariae Curie Skodowska Sectio EEE Hort., 10: 59-64.

Dirimanov, M. and Sengalevich, G. (1972). Grandinarska i lozarska Nauka, 9: 45-48.

Dzamic, M., Durdevic, B., Velickovic, D., Dzamic, R. and Tesovic, Z. (1975). Archiv Za poljoprivredne Nauke, 28: 37-48.

Fedotov, I.M., Mikheev, N.V. and Zotochkin, N.Y. (1983). Sbornik Nauchnykh Trudov, Vscsoyuznyi Nauchno Issledovatel skii Institut Sadovedstvo imeni D.V. Michurina, 38: 104-107.

Felski, J. and Brzezinska, A. (1988). Acta Hort., 223: 145-48.

Fountain, M., Dean, R. and Cross, J. (2011). 10BC/WPRS-Bull. 70: 151-57.

Frankenhuyzen, A. Van and Jansen, D. (1983). Fruitteelt, 73: 219-21.

Fuller, A.S. (1911). The Small Fruit Culturist. Orange Judd Co. New York, pp. 116, 194.

Georgiev, G.V. and Topchiiski, S. (1972). Gradinarska i Lozarska Nauka, 9: 15-25.

Gitinovasov, M.M. (1984). Sadovodstvo Vinogradarstvo I Vinodelie Moldavii, 7: 35-36.

Gjeium, H.B. and Langnes, R. (1984). Gartneryrket, 74: 376-77.

Gorokhova, O.G., Chevychelov, A.P. and Korobkova, T.S. (2010). Sadovodstvo i Vinogradarstvo, 1: 39-43.

Goubran, F.H. and El-Zeftawi, B.H. (1983). Australian Hort. Res. Newsletter, 35: 97-98.

Granizau, E. arid Scharpf, H.C. (1982). Zeitschrift fur Vegetationstechnik im Landschafts-Und Sportstattenbau, 5 : 64-65.

Gross, J. (1982). Scientia Hort., l8 : 131-36.

Gubenko, A.P., Il'in, V.S. and Chirkova, M.A. (1976). Biol. Aktiv Veshchestva Plov i Yagod. Moscow, pp. 69-71.

Hackett, C.A., Russell, J., Jorgensen, L., Gordon, S.L. and Brennan, R.M. (2010). TAG. Theor. Appl. Genet., 121: 1483-88.

Haffrier, K. and Heiberg, N. (1986). Gartneryrket, 76: 116.

Hardh, J.E. and Wallden, J. (1965). Maatalonst. Aikakausk, 37:67-75.

Harmat, L., Porpaczy, A., Himelrick, D.G. and Galleta, G.J. (1990). In: Small Fruit Crop Management (eds. G.J. Galletta and D.G. Himelrick), Prentice Hall, Englewood Cliffs, NJ. pp. 245-72.

Harris, P.G.W. and Wardlow. L.R. (1984). Ann. Appl. Biol., l04: 8-9.

Hcilubowica,A. and Bojar,F. (1981). Informator o badaniach prowadzonych W. Zakladzic Sadownictwa Akademii Rolniczej W. Poznanich, pp. 171-73.

Heide, O.M. and Sonsteby, A. (2011). Jr. Hort. Sci. Biotechnol., 86: 128-34.

Heide, O.M. and Sonsteby, A. (2015). Scientia Hort., 185: 228-32.

Helberg, N. (1986). Acta Hort., 183: 257-61.

Hellqvist, S. (2001). Acta Agric. Scan. Section-B., Soil and Pl. Sci., 51(2): 86-92.

Hellqvist, S. (2005). Crop Prot., 24: 343-48.

Hevia, F, Lanuza, P., Wilekens, R., Tello, M. and Alvarez, D. (2000). Agro Sur., 28(2): 42-50.

Hillebrand, E. and Stadtmuller, H. (1983). Obstbau, 8 : 270-74.

Hoyle, T. (1955). Rep. 14th Int. Hort. Congr. Netherlands, pp. 342-50.

Hummer, K. and Sabitov, A. (2004). Jr. Amer. Pom. Soc., 58: 215-19.

Hummer, K.E. (2000). Jr. Amer. Pom. Soc., 54: 54-56.

Hummer, K.E. and Dale, A. (2010). Forest Pathol., 40: 251-63.

Hummer, K.E. and Picton, D.D. (2001). HortTech., 11: 445-46.

Iftime, D., Movileanu, A. and Aniculaesci, E. (1989). Cercetari Agronomich in Moldova, 22: 89-95.

Ionova, Z.A. and Minaaev, LA. (1983). Sbornik Nauchnykh Trudov. Vsesoyuznyi Nauchno Issledovatel'skii Institut Sadovodstva imeni L V. lichurina, No. 38, pp. 82-87.

Jamas, P. (1984). Erweebsobstbau, 26: 17-18.

Jazvicky, M.N. (1940). Mod-Jagodn, Kultury, 2: 72-77.

Jerlnings, D. and Anderson, M. (1984). Fruit and Veg. Sci., U.K., February, pp.24-25.

Jones, A.T. and McGavin, W.J. (2002). Pl. Dis., 86: 1333-38.

Jorg, E., Harzer, U. and Ollig, W. (2003). Bull. OILB/SROP, 26(2): 153-60.

KaravoSov, V.T. (1975). Sadovodstvo, 9: 26.

Kawecki, Z. (1979). Txszyty Naukowe Akademii Rolniczo-Tech. W.OIS. Rol., 26:161-73.

Kawecki, Z. (1988). Acta Academia Agriculturae ac Technicae Olstenesis, Agricultura, 46: 135-43.

Kawecki, Z., Rogalski, L. and Ljarbinska, J. (1982). Zeszyry Nankowe Akademii Rolniczo-Technicznej W. Olsztynie Rolnictwo. 35: 59-64.

Keep, E. (1984). Scientific Hort., 35 : 54-71.

Keep, E. (1985). Euphytica, 34: 801-809.

Keep, E., Graflon, W.H., Knight, V.H. and Cumming, L. (1983). Jr. Hort. Sci., 58: 535-40.

Keep, E., Knight, V.H. and Parker, J.H. (1982). Jr. Hort. Sci., 57: 189-96.

Khokhryakova, T.M. (1986). Sbornik Nanchnykh Trudov Prikladnoi Botanike, Genetike i Selektsii, 106: 101-106.

Knight, R.L., Keep, E., Briggs, J.B. and Parker, J.H. (1974). Ann. Appl. Biol., 76:23-30.

Knight, V.H. (1983). Jr. Hort. Sci., 58: 63-71.

Kondakov, A.K. (1973). Referativnyi Zhurnal, 8.55.673.

Korbin, M., Kuras, A. and Zurawicz, E. (2002). Cell. Mol. Biol. Lett., 7(2B): 785-94.

Krotov, A.M. (1971). Sbornik Nau. Rab. Plo. i Yag. Nech. Pol. Vol. III, pp. 180-84.

Krupyshev, P.Y. (1977). Referativnyi Zhurnal, 3.55.951.

Ksliniehenko, A.N. and Gladkikh, V.I. (1973). Zashchita hastenii, 2: 50.

Kukhta, P.N., Kuleshova, I.I. and Adashchik, A.G. (1999). Plodovodstvo, 12: 62-65.

Kulenova, K. (1977). Referativnyi Zhurnal, 3.55.958.

Kuminov, E.P. (1961). Agrobiologija, 4: 500-504.

Labanowska, B.H. (2011). 1OBC/WPRS-Bull., 70: 87-92.

Labanowska, B.H., Bera, B. and Suski, Z.W. (1983). Proc. 10th Int. Congr. Plant Protection, U.K. p. 1013.

Labanowska, B.H., Piotrowski, W. and Gruchala, M. (2015). Progr. In Pl. Protec., 55: 305-11.

Lanham, P.G. and Bresnnan, R. (1998). Theor. Appl. Genet., 96: 917-21.

Lanham, P.G., Kemp, R.J., Jones, H. and Brennan, R.M. (2001). Jr. Hort. Sci. Biotech., 76: 201-207.

Lanham, P.G., Korycinska, A. and Brennan, R. (2000). Jr. Hort. Sci. Biotechnol., 75: 371-75.

Li, X.G., Wang, J. and Yu, Z.Y. (2015). Genetics and Mol. Res., 4: 2726-34.

Lisek, J. (2000). Bull. OILB/SROP., 23(11): 103-108.

Lister, C.E., Wilson, P.E., Sutton, K.H. and Morris, S.C. (2002). Acta Hort., 585: 443-49.

Ljones, B. (1963). Meld. Norg. Landbr. Hogsk., 42: 1-90.

Ljones, B. (1965). Meld Norg. Landbr, Hogsk., 45 : l-44.

Ljones, B. (1984). Meld. Norg. Landbr. Hogsk., 63: 1-7.

Majoros, L. (1983). Kerteszet es Szoleszet, 32: 7.

Maric, Z. (1978). Jugoslvoensko Vocarstvo, 12 : 63-69.

Mather, P.J.C., Modlibowska, I. and Keep, E. (1980). Euphytica, 29: 793-800.

Mazeikiene, I., Bendokas, V., Baniulis, D., Staniene, G., Juskyte, D.A., Sasnauskas, A., Stanys, V. and Siksnianas, T. (2017). Ag. Fd. Sci., 26: 111-17.

McCarthy, T. (1988). New Zealand Comm. Grower., 43: 20-21.

McCarthy, T. and Stoker, R.C. (1988). New Zealand Exp. Agric., 16:73-78.

McNeill, M.M., Smith, R.A. and Cumming, U.G. (1980). Hort. Res., 20:73-78.

Medel, S.F., Torres, M.J.P., Manquian, T.N. and Mansilla, T.R. (1984). Agro. Sur., 12: 100-107.

Miculka, B. (1980). Sbortik UVTIZ Zahradnicvi, 7 : 21-25.

Mikkonen, T.P., Maatta, K.R., Hukkanen, A.T., Kokko, H.T., Torronen, A.R., Karenlampi, S.O. and Karjalainen, R.O. (2001). Jr. Agric. Food Chem., 49: 3274-77.

Mikulic-Petkousek, M., Rescic, J., Schmitzer, V., Stampar, F., Slatnar, A., Koron, D. and Veberic, R. (2015). Fd. Chem., 173: 363-74.

Mikulic-Petkovsek, M., Koron, D. and Vebric, R. (2016). Scientia Hort., 210: 188-96.

Miloseric, T., Milosevic, N. and Mladenovic, J. (2016). Fruits, Paris, 71: 239-48.

Mitchell, C., Brennan, R.M., Cross, J.V. and Johnson, S.N. (2011). Jr. Agri. Forest. Ent., 13: 221-37.

Mokra, A. (1988). Vedecke Prace Vyskumneho Ustavu Ovocnych a Okrasnych Drevin V Bojniciach, 7 : 145-54.

Mottus, E., Kahu, K., Kannaste, A., Liblikas, I., Ojarand, A., Mozuraitis, R., Ovsjannikova, J. and Nikolajeva, Z. (2001). Trans. Estonia Agric. Univ., 2013: 121-25.

Mudambi, S.R. and Rajagopal, M.V. (1983). Food values, Fundamentals of Food and Nutrition. Wiley Easter Ltd., pp. 240.

Mulholland, R.L and Beresford, R.M. (1989). N.Z.J. Crop Hort. Sci., 17: 165-68.

Naichenko, V.M. and Osokina, N.M. (1982). Fiziologiya i Biokhimiy Kul'turnykh Rastenii, 14: 222-24.

Naichenko, V.M. and Romanshchak, S.P. (1985). Fiziotogiya i Biokhimiya Kul'turnykh Rastenii, 17: 81-84.

Nasr, T.A.A. and Wareing, P F. (1961a). Jr. Hort. Sci., 36:1-10.

Nasr, T.A.A. and Wareing, P.F. (1961b). Jr. Hort. Sci., 36: 11-17.

Neira, C.M., Barriga, S.M., Riveros, G.M., Mundaca, B.N., Carriollo, L.R. and Pessot, Z.R. (1999). AgroCiencia, 15: 207-13.

Nes, A. (1983). Forskning Og Forsok i Landbruket, 34: 169-73.

Neumann, U. (1955). Archiv fur Gartenbau, 3 : 274-96.

Neumann, U. (1983). Archiv fur Gartenbau, 3l: 265-76.

Nielsen, S.L. (1986). Frugtavleren, 15: 122-26.

Nielsen, S.L. and Kirknel, E. (1986a). Frugtavleren, 8: 122-26.

Nielsen, S.L. and Kirknel, E. (1986b). Tidsskr. Planteavl, 190 : 377-84.

Nikolova, G. and Ivanov, A. (1983). Gradinarska i Lozarska Nauka, 20: 49-54.

Nour, V., Trandafir, I. and Lonica, M.E. (2011). Fruits, Paris, 66: 353-62.

Nowacka, H., Cimanowski, J., Korolczak, W. and Puchala, Z. (1980). Prace Institutu Sadownictwa i Kwiaciarstwa W. Skin, A. 22, pp. 129-35.

Oleichenko, S.N. and Titiva, I.V. (2001). Sadovodstro i vinogradarstvo, 4: 9-10.

Oosten, A.A. van (1983). Fruitteelt, 73: 200-203.

Oosten, A.A. van (1988). Fruitteelt, 78: 12-14.

Oosten, A.A. van (1989). Fruitteelt, 79: 12-14.

Orlikowska, T. (1984). Fruit Sci. Report., 11: 5-17.

Ourdnickova, J., Knezacek, L. and Silovska, I. (2011). Vedecke Prace Ovocnarske, 22: 173-86.

Oydvin, J. (1978). Gartneyrket, 68: 452-54.

Paimetiva, V. and Mand, M. (2000). Proc. Int. Conf. Fruit. Prod. Fruit Breed., Estonia, pp. 164-67.

Palmieri, L., Grando, M.S., Sordo, M., Grisenti, M., Martens, S. and Giongo, L. (2013). Pl. Omics J., 6(3): 165-74.

Papp, J. and Kobzos, I. (1980). Kerteszeti Egyetem Kozlemenyci, 43:49-57.

Patashova, A.I. (1981). Byulleten'Vsesoyuznogo Ordena Lenina i Ordena Druzhby Narodov Instituta Rastenievodstva Imeni N.I. Vavilova, 112: 79-80.

Paunovic, S.M., Nikolic, M., Miletic, R., Popovic, B., Mitrovic, O. and Kandic, M. (2016). Jr. Mountain Agric. Balkans, 19: 233-46.

Picton, D.D. and Hummer, K.E. (2003). Small Fruits Review, 2: 43-48.

Pisani, P.L. (1962). Rivesta Ortoflorofruttic. Italiana, 46: 3-9.

Plotnikova, I.V. and Kirdyashkina, R.I. (1976). Byrlleten Glavnogo Botanicherkogo Sada, 99: 40-42.

Plotnikova, I.V., Petrovjkaya, B.T. P. and Kirdyashina, R.J. (1972). Fiziologiya. Rastenii, 19:1257-60.

Pluta, S. and Zurawicz, E. (1999). Acta Hort., 505: 167-72.

Pluta, S. and Zurawicz, E. (2014). HortSci., 49: 513-15.

Pluta, S. and Zurawicz, E. (2015a). HortSci., 50: 1096-98.

Pluta, S. and Zurawicz, E. (2015b). HortSci., 50: 1582-84.

Pluta, S., Zurawicz, E. and Pruski, K. (2012). Jr. Berry Res., 2: 23-31.

Polikarpova, L.G. (1983). Sadovidstvo, Vinogradarstvo i Vinodelie Moldavii, 12: 18-20.

Popesko, I.G. (1968). Referativnyi Zhurnal, 5: 55:887.

Pridorogin, M.V., Gldechina, T.V., Gurieva, I.I. and Zudilin, O.E. (2013). Sadovodstvo i Vinogradarstvo, 6: 16-22.

Profic Alwasiak, H., Lenartowicz, T. and Puchala, Z. (1973). Prace. Inst. Sad. W Ski., 17:197-202.

Qin, D., Junwei, H., Sui, W. and ZhuangFei, Z. (2012). Acta Hort., 926: 119-22.

Radaczynska, Z., Tylus, K. and Rejman (1982). Prace Ins. Sad. I Kwia. W. ski., A.23, pp.59-64.

Radvany, B. (1983). Novenyvedelem, 19: 269.

Rebandel, Z. and Maztr, J. (1981). In: Informator O badaniach prowadzonych W Zakiadzie Sadownictwa Akademii Rolniczej W Poznaniu Poznan, Poland, pp.166-71.

Rhodes, P.J. and Tabley, F.J. (1983). Proc. 36th New Zealand Weed and Pest Control Conf., pp. 59-61.

Riordian, F.O. (1986). An. Foras Taluntais. pp. 34-35.

Roach, F.A. (1985). Cultivated fruits of Britain, their Origin and History. Black Well, New York. pp. 295-322.

Rogers, W.S., Kent, W.G. and Pearl, R.T. (1936). Scientia Hort., 4:79-89.

Rooijen, W.J. (1984). Fruitteelt, 74: 1340-41.

Rubinskiene, M., Viskelis, P., Stanys, V., Siksnianas, T. and Sasnauskas, A. (2008). Sci. Works Lithuanian Univ. Agric. Sodininkyste ir darzininkyste, 27(2): 235-43.

Rudloff, C.F. and Lenz, F. (1961). Erwerbsobstbau, 3: 36-38.

Rupp, D. and Trankle, L. (2000). Erwerbsobstbau, 42: 15-20.

Russell, J.R., Bayer, M., Booth, C., Cardle, L., Hackett, C.A., Hedley,P.E., Jorgensen, L., Morris, J.A. and Brennan, R.M. (2011). BMC Pl. Biol., 11(147) (28 October, 2011) 1471-2229/11/147.

Sako, J. and Laurinen, E. (1979). Ann. Agric. Fenniae, 18: 996-1051.

Sakshaug, K. and Ingvarsson, A. (1986). Rep. Inst. Trad. Sve. Lant. 44: 35.

Salamon, Z. (2000). Zeszyty Naukowe Inst. Sadownictwa I Kwiacirstwa w Skierniewicah, 8: 399-406.

Salikhov, M.M. and Oslopov, A.A. (1985). Sadovodstvo, 2: 25-27.

Samorodova-Bianki, G.B. and Babaevskaya, L. (1981). Byulleten Vsesoyuznogo Ordena Lenina i Ordena Druzhby Narodov Inst. Rastenievodstia Imeni N.I. Vavilova, 113: 79-81.

Sasnauskas, A., Stanys, V. and Trajkovski, V. (1999). Sodininkyste ir Darzininkyste, 18(1): 3-16.

Sasnauskas, A., Trajkovski, V., Strautina, S., Tikhonova, O., Siksnianas, T., Rubinskiene, M., Viskelis, P., Lanauskas, J., Valiuskaite, A., Rugienius, R. and Bobinas, C. (2009). Agron. Res., 7(Special issue): 737-43.

Sass, B. (1972). Kert. Kut. Int. Koz., 6: 40-45.

Savkovskii, P.P. and Rybalov, L.N. (1984). Zashchita Rast., 5: 52-53.

Sazonov F.F. and Danshina, O.V. (2016). Sadovodstvo i Vinogradarstvo, 2: 22-27.

Scibisz, K., Niewska, W. and Zajkowska, M. (1986). Abst.22nd Int. Hort. Congress, California, No. 1087.

Sedlak, J. and Paprsteen, F. (2012). Hort Sci. (Prague), 39: 21-25.

Sesvecan, D., Plancher, B. and Gruppe, W. (1981). Erwerbsobstbau, 23: 197-98.

Severin, V.F. and Ermokhin, Yu. I. (1979). Pochvy Zap. Sib. Ikh. Mel.L Ef. Udob. Oms. USSR. pp. 63-69.

Shirochenkova, A.I., Minaeva, V.G. and Khristo, A.A. (1976). Akt. Voor. Botan. Resu. v Sib. Nov. USSR, Nauka, pp. 111-74.

Singer, M.C. and Smith, B.D. (1976). Ann. Appl. Biol., 82: 407-12.

Sinnot, Q.P. (1985). Rhodora, 87: 189-286

Skrede, G. (1987). Norwegian J. Agric. Sci., l: 67-74.

Smeekens, C. (1983). Fruitteelt, 73: 138.

Smolarz, K. (1982). Prace Instytutu Sadownictwa i Kwiaciarstwa w Skierniewicach, 23 : 49-58.

Smolarz, K., Kostusiak. A. and Lachman, J. (1981). Fruit Sci. Rept., 8: 61-68.

Sonsteby, A., Roos, U.M. and Heide, O.M. (2017). Scientia Hort., 225: 638-45.

Sorokopudov, V. and Sorokopudova, O. (2011). Sodininkyste ir Darzininkyste, 30(2): 51-59.

Stepanova, T.I. (1982). Bot. Gen. i Sel., 74: 109-115.

Stoker, R. (1983). New Zealand Comm. Grower, 38 : 29.

Stoker, R. (1985). New Zealand J. Exp. Agric., 13:71-75.

Stoker, R. and McCarthy, T.P. (1986). New Zealand Comm. Grower, 41: 35.

Strautana, S. and Kampus, S.K. (2000). Proc. Int. Conf. Fruit Prod. and Fruit Breed., Tartu, Estonia, pp. 173-76.

Strautina, S., Krasnova, I., Kalnina, I., Sasnauskas, A., Trajkvoski, V. and Tikhonova, O. (2012). Acta Hort., 926: 53-58.

Susi, P. (2004). Mol. Pl. Pathol., 5(3): 167-73.

Tahvonen, R. (1979). Jr. Sci. Agric. Soc. Finland, 5l: 421-31.

Tahvonen, R. (1983). Jr. Sci. Agric. Soc. Finland, 55: 81-90.

Takac, J. (1988). Vedecke Prace Vyskumcho Ustavu Ovocnych a Okrasnych Previn V Bojniciach, 7:183-92.

Tefrettaz, R. and Carron, R. (1984). Suisse dc Viticulture, d'Arboriculture et d'Horticulture, 16: 53-55.

Thomas, W.P. and Burnip, G. (1984). Proc. 37th New Zealand Weed and Pest Control Conf., pp. 224-27.

Toshkov, A. and Domozetova, D. (2006). Agric. Sci., 39(4): 18-25.

Truhsheckin, V.G. and Pozdnyakov, A.D. (1972). Kultura Cher. Smor. SSR Moscow, pp. 132-36.

Trunov, I.A. (1983). Sbornik Nauchnykh Trudov, Vsesoyuznyi Nauchno-Issledovatel Institut Sadovodstva imeni I.V. Michurina No. 38, pp. 66-70.

Tumanova, R.N. (1973). Sadovodstvo, 5: 24-25.

Tyurina, M.M., Volkov, F.A. and Arsent'-ev, A.P. (2000). Sadovodstvo i Vinogradarstvo, 1: 17-18.

Uebel, E. (1982). Archiv fur Gartenbau, 30 : 339-49.

Ul'yanova, D.A. (1971). Sbornik Nauchnykh Rabot, Plodovodstvo i Yagodovodstvo Nechernozemnoi Polosy Vol. 111. Moscow, USSR, pp. 173-79.

Ul'yanova, D.A. (1978). Sadovodstvo, 6: 22.

Utkov, Yu, A. (2015). Sadovodstvo i Vinogradarstvo, 4: 40-44.

Verzilov, V.F. and Plotnikova, LV. (1978). Fitogormony Rosr. Rasr. USSR, pp. 75-82.

Vitkovskij, V.L. (1958). Sad. Ogorod. 6: 54-57.

Voscilko, M.E. (1969). Referativnyi Zhurnal, pp. 219-22.

Wainwright, H. and Hawkes, H. (1987). Grower, 108: 28-29.

Wainwright, H. and Hawkes, H. (1988). Acta Hort., 221: 266-68.

Wainwright, H., Todd, S.F. and Price, D.J. (1986). Acta Hort., 183:331-38.

Wallace, T. (1925). Jr. Pom. Hort. Sci., 5: 1-33.

Wardlow, L.R. and Nicholls, R.F. (1986). British Crop Protection Conference. Pests and diseases Vol. 1 Proceedings of a conference held at Brighton., England, Nov. 17-20, pp. 359-62.

Watt, B.K. and Merrill, A. (1959). A table of food values, In : USDA Yearbook of Agric., Washington, Supt. of Documents.

Watters, B.S. (1983). Agric. in Northern Ireland. 58 : 96-98.

Wenneker, M., Fernandez, G.E. and Hummer, K.E. (2016). Acta Hort., 1133: 515-20.

Wenzhi, Z., Zhonglu, S., Fengzhi, L., Fenglian, Q. and WuJie, Z. (2004). China Fruits, 3: 3-4.

Westerlaken, J. (1983). Fruitteelt, 73: 654-55.

Wiejacha, K., Pluta, S., Golis, A. and Orlikowska, T. (2001). Acta Hort., 560: 185-88.

Wilson, D. and Adam, J. (1967). Rep. Long Ashton Res. Stn. for 1966. pp. 104-11.

Wilson, S.J. and Jones, K.M. (1983). Scientia Hort., 19: 245-50.

Wlodek, L. (1972). Prace. Ins. Sado. i Kwia. Stwa. W. Skier. A., 19: 119-29.

Wojcik, P. and Filipczak, J. (2015). Scientia Hort., 185: 90-97.

Woodhead, M., Taylor, M.A., Brennan, R., McNicol, R.J. and Davies, H.V. (1998). J. Pl. Physiol., 153: 381-93.

Woznieki, T.L., Heide, O.M., Sonsteby, A., Wold, A.B. and Remberg, S. F. (2015). Acta Agric. Scandinavica, Section B-Soil and Pl. Sci., 65: 702-12.

Wright, S.T.C. (1956). Jr. Hort. Sci., 31: 196-211.

Yang, S., Chunyu, Z., ZhiDong, Z., JingHui, W., YaDong, L., Lin, L. and HaiGung, L. (2011). Acta Hort. Sinica, 38: 1747-52.

Zakharieva, T. (1977). Rastitelna Zashchita, 25: 29-30

Zaliwski, S., Lipecki, J., Wicniarska, J. and Selwa, J. (1974). Rock. Nauk. Pol. A. 100: 121-31.

Zatyko, J. (1975). Kert. es Skoleszet, 24: 4.

Zbroszczyk, J., Lenaartowicz, W. and Salamon, Z. (1985). Prace Ins. Sado r Kwia. Stwa. W Skier. A. 25: 207-13.

Zhidyokhina, T.V. and Rodyukova, O.S. (2000). Proc. Int. Conf. Fruit. Prod. and Fruit Breed., Estonia, pp. 162.

Zhongwu, S., Fenglian, Q., Wenzhi, Z. and Fengzhi, L. (2000). China Fruits, 3: 38.

Zorenc, Z., Veberic, R., Koron, D., Miosic, S., Hutabarat, O.S., Halbwirth, H. and Mikulic-Petkovsek, M. (2017). Planta, 246: 217-26.

## Chapter 15

# Gooseberries

Gooseberry belongs to the genus *Ribes*, hence it resembles much with currants in plant characteristics. American cultivars of gooseberry are forms of the native *Ribes hirtellum* Michx. whereas the European forms come under *R. grossularia*, L., Syn. *R. uva-crispa* L. They are grown to a considerable extent in Europe.

Gooseberry is popular in Europe as fresh fruit and as processed product, primarily into pies or preserves. The importance of gooseberry lies partially in its content of phenolic compounds, which are known to have health-promoting effects as antioxidant and anticarcinogens.

## Composition and Uses

Gooseberries are rich in vitamins (A and C) and minerals. The fruit contains about 4.6 to 5.2 per cent total sugar and have about 2.0 per cent acidity. Citric and malic are the major acids found in ripe fruits, however, traces of shikimic and oxalic acids were also recorded. The carotene content is about 0.18-0.20 mg/100 g and about 0.06-0.1 per cent of tanins in juice. The ascorbic acid (vitamin C) content ranged from 10 to 55 mg/100 g (Cvopova, 1984). In general, red-fruited gooseberry cultivars showed higher ascorbic acid content (Patutina *et al.*, 1974). Nikfardjam *et al.* (2013) reported higher ascorbic acid content in cvs. Bekay and Rote Triumph. The main part (> 70 per cent) of the polyphenolic fraction in red gooseberry cultivars is due to anthocyanins consisting of a considerably high amount of acylated compounds (11 to 33 per cent of anthocyanins). Green cultivars show a reduced content of total phenolics as compared to red cultivars. Their phenolic constitution is attributed to relatively high amounts of hydroxycinnamic acid derivatives and flavanols. The phenolic compounds in green gooseberry is quercetin and in red gooseberry it is cholorogenic acid (Pinto *et al.*, 2010). The average content of anthocyanins in gooseberry is 78mg/kg (Mattivi and Vrhovsek, 2010).

The seed oil of gooseberry (*R. grossularia*) contain 8 per cent gamma-linolenic acid which could be used as a source of natural vitamin E. Gamma-linolenic acid is an essential fatty acid for humans with delta-6-desaturase deficiency; it is a precursor of prostaglandins, prostacyclins, and tromboxanes; and has anti-inflammatory and antitumoral effects (Gottman and Galletti, 2001). Composition of gooseberry fruits is given in Table 15.1.

Gooseberries are grown for several purposes such as culinary, canning and jam manufacturing, juice extraction, jelly preparation, wine making and highgrade dessert gooseberries for the fresh-fruit market.

**Table 15.1: Composition of Gooseberry* (per 100 g of edible portion)**

| | | | |
|---|---|---|---|
| Water (per cent) | 88.9 | Calcium (mg) | 18.0 |
| Protein (g) | 0.8 | Phosphorus (mg) | 15.0 |
| Fat (g) | 0.2 | Iron (mg) | 0.5 |
| Carbohydrate (g) | 9.7 | Sodium (mg) | 1.0 |
| Vitamin A (IU) | 290 | Potassium (mg) | 155.0 |
| Vitamin C (mg) | 33 | | |

* Watt and Merrill (1959).

## Origin and Distribution

The gooseberry, as developed in Europe, had its origin from *Ribes grossularis* L. Syn. *R. uva-crispa* L. which occurs wild in many parts of Europe, the Caucasus, North Africa. The gooseberry seems to have been brought into cultivation in Holland and Germany by early Medieval times, and there is a record of its having been imported into England in 1275, although it was not known to the ancient Greeks and Romans. Its cultivation spread by the early 16th century in Britain. Two cultivars, Amber and Hedge-hog, in use in 1665, are still being grown even after more than 300 years. Gooseberries were popular during the 18th century for making wine as well as for cooking. They became quite popular during the later part of that century among the industrial workers of northern England. By the late 1800s, gooseberries were grown commercially in Canada and the United States. In United States, it was grown in Oregeon, Michigan, Washington and Colorado. Commercial production of gooseberries continued and expanded in Europe through the 1900s. Germany, Poland and Russia are the leading gooseberry producer (Barney, 2000). In achieving better fruit size, they bred and selected their own varieties and an enormous number of new named cultivars was introduced. Many of these were only of value for their large berries, but several of the better ones are still being cultivated (Roach, 1985).

## Area and Production

The major portion of gooseberries in Europe is in Germany, Poland, Russia and Britain with lesser amount in Hungary, France, Belgium and Holland. The crop production as reported by Roach (1972) in metric tonns during 1969, is given below:

| Britain | 13 000 (probably not all sold commercially) |
|---|---|
| Poland | 19 000 |
| West Germany | 23 000 |
| Hungary | 4 000 |
| France | 2 000 |
| Holland | 600 |
| Belgium | 1 000 |

In 1999, Europe and Baltic states produced 156,854 tonns from an area of 99,406 ha.

## Taxonomy

*Ribes*, traditionally placed in the polyphyletic family Saxifragaceae in the Englerian classification system (Cronquist, 1981), is now well established into Grossulariaceae based on presence of wholly inferior ovaries, totally syncarpous gynoecium, and fleshy fruits (Sinnott, 1985).

The genus has five subgenera: (1) *Berisia*, European diocious plants; (2) *Grossularia*, gooseberries; (3) *Grossularioides*, thorny currants; (4) *Parilla*, South American Natives and (5) *Ribes*, currants (Kim and Dale, 2010). The genus *Ribes* comprises over 150 species of shrubs of which 10-12 species have been used as the primary gene pool. Gooseberry: *Ribes grossularia* L, syn. *R. uva-crispa* L. is the European gooseberry, and *R. hirtellum* Mich, the American gooseberry (Berger, 1924; Senters and Soltis, 2003). The American gooseberry, *R. hirtellum* are grouped with North American *R. niveum*, *R. oxyacanthoides*, *R. divaricatum*, *R. cynosbati*, *R. leptanthum* and the western Chinese *R. alpestre* and *R. stenocarpum* and the Japanese *R. grossularioides*.

North American gooseberry cultivars are mostly hybrids between *R. hirtellum* and *R. grossularia*.

## Cultivars

There are a large number of gooseberry cultivars grown in various parts of the world.

### White Cultivars

#### Antagonist

An all-round mid-season cultivar produces large, cream-coloured berry of delicate flavour.

#### Careless

A widely grown gooseberry cultivar, heavy cropper of greenish white fruit, mid-season cultivar, used for preparation of jam, canning and for green fruit.

#### Champagne White

Small white berries have excellent flavour.

#### Invicta

It gives high yields. Individual berries about the same size as that of Careless but bushes larger. Invicta is considered a promising cultivar (Williams, 1982).

#### Keepsake (Berry's Early Kent)

Very early and an excellent fresh market cultivar, heavy cropper, berries white, tinged green, plants spreading, however susceptible to frost damage.

#### White Lion

A good cultivar, very late, vigorous and hardy.

#### Whitesmith

An early cultivar, large berries of good flavour and very prolific bearer, fruits moderately large, oval and pale green with yellow tinge.

#### White Sawn

An early cultivar, similar to Whitesmith, fruits slightly sweeter.

### Green Cultivars

#### Angler

The plant makes a neat bush and produces large round berries, having a downy appearance and very early cropper.

#### Bellona

A late-fruiting cultivar, spreading bush, crops heavily, fruits large.

#### Captivator

Medium-sized, smooth, green fruit with good flavour. Canes are moderately vigorous, erect and less spiny.

#### Green Gem

This cultivar was introduced by Laxton Bros of Bedford in 1922, a good, all-round, heavy cropper.

#### Green Ocean

A large and spreading type, bears a large smooth-skinned berry in mid-season.

#### Malakhit

Ripens in late July and produces bright green berries, yields 70.7 centners/ha; frost resistant (Il'in, 1982)

#### Oregon Champion

Small, white to pale green, tart fruit. Canes are large, thornless, vigorous, erect to spreading and quite resistant to powdery mildew.

### Green Walnut

Dark green medium-sized fruits with good flavour. It makes a small neat bush and thus suitable for home gardens.

### Gunner

A valuable mid-season cultivar, fruits large, sweet, olive-green, striking in appearance.

### Heart of Oak

Berry beautiful, pale green in colour, veined deep yellow, oblong, tapering almost to a point.

### Mompakta

Fruits are oval-spherical with a light green, downy skin. Fruit flavour is sweet/ sour and is less prone to cracking during canning.

### Malling Greenfinch

Raised at East Malling Research Station, UK in 1968 from Careless x (Whinham's Industry x Resistenta). A very high yielder, fruits attractive in colour (green berry) with good flavour; bush compact, moderate vigour and fairly spiny (Keep *et al.*, 1985).

### Matchless

A good mid-season cultivar, fruits large, oval in shape, flavour good. Bears a heavy crop and makes a large bush.

### Shiner

A vigorous bush, berry large, almost square, very sweet.

### Surprise

Bears pale green, large and smooth berry with excellent flavour; a heavy cropper.

### Telegraph

A late-season dark green gooseberry cultivar; fruits large with good flavour.

### Thumper

Berry large, sweet, smooth-skinned and appears to be blown up like a tiny ballon.

## Yellow Cultivars

### Bedford Yellow

Berries handsome, golden yellow, streaked red, large, with good flavour, a mid-season cultivar, heavy bearer.

**Broom Girl**

A very early cultivar; fruits medium in size, dark yellow in colour, shaded olive green, attractive appearance.

**Catherina**

A mid-season cultivar; fruits almost orange, oval, large with excellent flavour.

**Champagne Yellow**

A very late cultivar; fruits small with excellent flavour. A neat bush of upright growth.

**Early Sulphur**

A very early cultivar, bears fruit of excellent flavour and attractive primrose-yellow colour.

**Golda**

Raised at Wageningen in 1972; Golda, formerly IVT 72001, derived from Whitesmith x May Duke; a high-yielding cultivar with large, round, yellow fruits ripening a week later than Whinham's Industry and about 10 days earlier than Achilles (Wassenaar *et al.*, 1984).

**Golden Gem**

Berries handsome, deep yellow, having super flavour, mid-season to late, a very heavy cropper.

**Leader**

Bears large smooth delicious fruits of excellent flavour.

**Leveller**

A dessert gooseberry, produces very large, oblong-oval berries with good flavour.

**Pilot**

A less grown cultivar, should be grown where Leveller proves difficult. Not a heavy cropper but similar in other aspects.

**Resika**

Mid-season yellow European gooseberry, developed from Poland. Fruit yellow, medium (3.0g), globose, smooth, good quality for fresh market, processing and freezing. Resistant to powdery mildew.

## Red Cultivars

**Bedford Red**

An early cultivar, a neat upright growing, bearing large fruits, good for small gardens.

**Gooseberry cultivar Invicta**

**Gooseberry cultivar Krusbar**

European gooseberry

Powdery Mildew symptom in gooseberry

### Champagne Red

Small dessert fruit of excellent flavour, bears very late, profusely.

### Conquering Hero

A mid-season cultivar, berries dark coloured, unusually long.

### Guido

Another early cultivar, bears large elongated crimson-red berry; heavy cropper.

### Hinsel

Mid-season red European gooseberry, developed from Poland. Bright red, medium sized fruit (3.0g) for fresh market, processing and freezing. Resistant to powdery mildew.

**Ledenets**

Ripens in mid-July, produces light red berries; yield about 64.2 centners/ha. It is fairly frost and disease resistant (Il'in, 1982).

**London**

The heaviest gooseberry in cultivation, bearing round, smooth fruit in mid-season.

**Lord Derby**

Fruits ripen late, possibly the latest red cultivar; berries large, oval of excellent flavour.

**May Duke**

Earliest to ripe; bush moderately vigorous; fruits medium in size and roundish-oblong in shape.

**Poorman**

Dull-red fruit ripens over a long period. Berries are small to medium, sweet and aromatic, with tough smooth skin. Reasonably free of spines and quite resistant to mildew.

**Pixwell**

One of the most reliable gooseberries in North America. Fruit small to medium in size, green and hangs in cluster below the canes. Canes are vigorous, erect to spreading, and resistant to mildew.

**Molodezhnyi**

Ripens in July and produces small light red berries; yields, on an average, 68.6 centners/ha. Fairly frost and disease resistant (Il'in, 1982).

**Rosko**

Derived from Gladde Gele x Whinham's Indirstry. A high yielder, produces round, purplish red fruit of good flavour, ripening a few days earlier than Golda (Wassenaar *et al.*, 1984).

**Samaryanin**

Ripens in late July and produces light red berries. Yields, on an average, 99 centners/ha and fairly frost and disease resistant (Il'in, 1982).

**Senena**

Berries small to medium, light red, yielding 61.7 centners/ha. Moderately frost resistant but highly resistant to *Sphaerotheca mors-uvae* (Il'in, 1982).

**Warrington**

Good for preparation of jam, berry vivid crimson, small, firm and very late.

A list of gooseberry cultivars growing in different countries is presented in Table 15.2.

**Table 15.2: Gooseberry Cultivars Growing in different Countries**

| | | |
|---|---|---|
| Canada | : | Captivator, Pixwell, Welcome, Pembina pride, Invicta. |
| Czech Republic | : | Astrid, Dekor, Mistral, Produkta, Roman, Skvost, Viking, Rolonda, Rokula, Invicta. |
| Finland | : | Hinnonmaki Gelb, Hinnonmaki Rot. |
| Germany | : | Rochus, Macurines, Rolonda, Tixia |
| Italy | : | Careless, Early Sulphur, Green Finch, White Smith |
| Lativia | : | Koknese, Kursu D Zintars, Anna Viksne, Citronu, N-17-15, 25-F2-60 |
| Lithuania | : | Nesluchovskyj, Kirdeikiai, Kursu, Dzintars |
| Netherlands | : | Achilles |
| Poland | : | White Smith, Captivator, Invicta, Pixwell, Pax, Laskovij, Rochus, Rolonda, Ruskos, Spine Free |
| Russia | : | Krasnoslavianski, Misorskij, Puszkinskij, Laskovij, Ruskos. |
| Ukraine | : | Niesluchovski, Kamieniar. |
| United Kingdom | : | Invicta, Pax, White Smith, Pixwell, Malling Greenfinch,Spine Free. |
| United States | : | Captivator, Oregon, Champion, Poorman, Pixwell, Shefford, Speedwell. |

## Climate

### Temperature

Gooseberries are hardy fruits, grow best and produce fruits in a cool, somewhat humid climate and thus do not perform well in dry and warm climate. The most suitable climate for gooseberry cultivation is in northern Europe, northeastern United States and Canada where there is no danger of spring frost. High temperature during summer and inadequate supply of moisture at fruit swelling limit the production. The gooseberries are highly susceptible to frost and not only the flowers but also the fruits at their initial stage of development are severely damaged. Intense sunlight combined with temperature can cause berries to sunscald. Production requires 160 to 200 growing days (5°C base temperature) and a 120 to 140 days frost free growing season (Harmat *et al.*, 1990). Late frosts cause a typical damage to gooseberries. This consists of brown mark as the berries are scorched and distorted berries are developed. Light frosts also cause considerable damage to berries. Therefore, gooseberries must be planted in frost-free sites.

### Wind

Wind is also an important factor to be kept in mind while selecting a site for successful cultivation of gooseberries. The reasons, why gooseberries appear to like being grown as an undercrop below orchard trees, are their disliking for intense sunlight and need for protection from cold.

The plants should be protected by planting rows of windbreak trees or choosing a site alongside other vigorous soft fruits or orchard trees.

## Soils

A wide range of soils are used for gooseberry cultivation, varying from heavy clay to sand but the soil must be deep, fertile and well drained. Poor drainage is extremely harmful and bushes are killed readily by water logging. The shallow soils dry out quickly with a consequent reduction in plant growth and fruit size. In warm regions, bushes perform better in heavier soils that retain moisture and keep cooler. An organic mulch helps protect the shallow roots. Gooseberries are tolerant to a fair degree of soil acidity and are relatively resistant to chlorosis on calcareous soils. Liming has been reported to increase the yield of gooseberry only if added simultaneously with mineral fertilizers in amount equal to half of the hydrolytic acidity. Lime alone decreased the yield to some extent if used in large quantity (Gomolyako, 1940).

The optimum soil pH for gooseberries has been suggested in the range of 4.6 to 4.8 (Jazvicky, 1940). In New Zealand, Watt (1950) reported that gooseberries perform best on a soil of somewhat higher pH. He recommended that the pH be maintained in the range of 5.5 to 7.0 by liming.

The root systems of gooseberry is relatively close to the surface and extremely sensitive to changing soil environment.

## Propagation

### Cutting

Gooseberry can be best propagated by hardwood cuttings. Twenty centimetre cuttings from one-year wood are made any time during the dormant season. They are tied in bundles and placed in a box of moist sand or peat in a cool basement. In early spring, cuttings are set to the nursery at a depth of 15 cm with a spacing of 15 cm in rows at 90 cm apart. Cuttings should not be taken from bushes or stools infected with American gooseberry mildew and preventive spraying for this disease should be given to the nursery rows.

Raspopova (1986) produced cuttings by removing branches from the base of the stock plant and cutting them into 2-internode sections with a green lateral shoot when most of the laterals on the branches attained growth between 12-15 cm. Rooting up to 94.7 per cent was obtained with this method. Siksnianas and Sasnauskas (2006) recorded 72.9 per cent rooting of soft-wood cuttings when treated with IAA at 50 mg/l water solution for 20 hours before planting for rooting. Hybrid of blackcurrant and gooseberry cv. Josta showed better rooting of hard wood cutting in autumn than in spring (Cejka *et al.*, 2013). Aleshkevich *et al.* (1982) studied the differences in tissue structure of goosberry leaves in relation to rooting capacity with easy-to-root cvs. Rozovyi, Russkii and Kolobok and with difficult-to-root cv. Finik. They correlated the size and number of chloroplasts with rooting capacity. In the leaves of Finik chloroplasts were large but few, while in easy-to-root Kolobok they were smaller but numerous.

### Layering

Gooseberries can be propagated by bending branches downward in the fall and partly covering them with soil. It can also be propagated by mound layering. All the branches are cut back to about 8-10 cm above ground in early spring. In July, soil is mounded around the vigorous shoots, covering half their length. The rooted shoots are removed in fall or spring and set to the nursery for one year before planting them in the field.

## Grafting

Increase in yield was found by grafting on some rootstocks. Harmat (1975) reported that the gooseberry cv. Zold Orias when grafted on 6 selected clones of *R. aureum,* produced on an average, 20 per cent higher yield than on a general population of *R. aureum*. Pallagi 2, selected from Brecht was the best rootstock producing about 75 per cent higher yield.

Zika and Kricnarova (1987) observed that NAA at l-10 g/l when applied to the stems of *Ribes aureum* rootstock immediately after grafting in January-February or the application made in the following spring when grafted in August, prevented lateral budbreak from the stems. Root sucker development was also reduced. Treatment with NAA at 1 and 2 g/l, reduced the number of laterals by 50 per cent and 75 per cent respectively.

## Micropropagation

Tissue culture is a new method of multiplication of gooseberries. Welander (1985) developed a rapid method of micropropagation of gooseberry cv. Hinnonmaki Yellow. The optimum concentration for shoot production was 2.2 μM BA and 0.05 μM IBA in MS media. The Lepoivre salt medium with no hormone was used for elongation of shoot and in this medium about 50 per cent of the shoots formed roots which had 100 per cent survival in the field. Apical meristems of cv. Russkii cultured on MS media produced shoots which were rooted in a perlite-sand mixture treated with half-strength MS media (Mironova, 1985).

Wainwright and Flegmann (1985), in studies with shoot cultures of cvs. Invicta and Careless, found that addition of 0.1 μM NAA to a MS initiation media containing 1 μM BA reduced the degree of basal shoot browning, increased the percentage of explants survival and shortened the time of shoot initiation.

Micropropagation protocol for five gooseberry cultivars was developed by Skovorodnikov (2012). Experiments were conducted with various growth regulators in MS media. Explants comprised stem segments 2-4 cm long with 1 bud taken from cuttings. Response of cv. Severnyi Kapitan explants to 5 growth regulators showed that medium with CPPU (forchlorfenuron) added at 0.2 mg/l gave the best survival and plantlet regeneration. This medium was subsequently used for the culture of explants of cvs. Snezhana, Asiroma, Nesluhknyanskii and Karpaty. At the proliferation stage, addition of BA at a concentration of 1 mg/l increased the average number of shoots production per explants by 3.4 times. Increases in plantlets height were achieved by adding $GA_3$ at 0.5-1.0 mg/l. Plantlets were rooted successfully in a non-sterile peaty soil by treatment with Kornevin, an auxin-based product.

## Planting

Gooseberry is a long-lived plant and careful preparation of land will be well repaid. Good drainage is essential. Gooseberry is normally grown on a leg and should not be planted too deeply. All suckers and buds on the roots and lower stem must be pulled off and it may be necessary to remove a few of the upper roots from the main stem before planting. There is a need for generous soil preparation and initial manuring before planting of gooseberries. Soil test before planting is important because gooseberries have a comparatively high potash requirement. A heavy base dressing, ploughed in before planting, followed by annual dressings in the form of a mulch round the bushes, are ideal for this crop.

Gooseberries may be planted at any time after the leaves have fallen, provided the soil is in reasonable condition for working. The planting should be firm and it is always wise to check the newly planted bushes after spells of severe winter frost and replant them properly, if they have been lifted out of the ground. Most growers plant two-year-old bushes, though the planting of three-year-old bushes is also practised. The planting is usually done in square system, however, rectangular system is also followed.

### Square Planting

The square planting is reasonably good because two-way cultivation can be practised. The distance varies with the cultivar. Most growers allow 2 m between bushes for strong growing and 1.8 m for the weaker cultivars.

### Leveller Planting

The specialized Leveller gooseberry production has shown that a much closer planting distance is suitable to this cultivar. Although of somewhat spreading habit, Leveller makes a small bush and it is planted at 4 ft (1.3 m) between bushes and 5 ft (1.6 m) to $5^1/_2$ ft (1.8 m) between rows. In the UK, gooseberries cvs. Careless, May Duke, *etc.*, are usually planted in rows at 2.4-2.7 m apart. Rybalov (1985) reported that cv. Finik, when planted at 2.5 x 1.2.5 x 0.75 or 2.5 x 0.5 m, produced an average annual yield (6 cropping years) of 1 13.2, 132.8 and 139.7 centners/ha, respectively; the berry weight over 6 years average was 4.1 g. Average shoot extension length for the 3 planting densities were 21.4, 22.4 and 21.8 cm, respectively. Davlebaeva (1985) obtained the highest average yield with a 3x0.6 m planting.

## Training and Pruning

Gooseberries can be very difficult to harvest if they are of a thorny variety. Cordon training offers the advantage of opening up the plant and leaving the fruit accessible. A single branch is selected and trained up a stake to a height of 1.5-2.0 m. Only new, well spaced, medium-sized branches are left at the end of the growing season. Poorly spaced, small branches, and branches that bore fruit, are removed (McKay, 2005). In Hungary, small grafted trees are attached to wire-trellis, which helps in harvesting and spraying for plant protection (Soltesz *et al.*, 2007).

Gooseberries are often pruned similarly as currants, although they tend to bear heavier on one-year canes than do currants and from spurs on older wood. However, bearing wood older than three years is seldom profitable. Since gooseberry is susceptible to mildew, it is important to keep the bush open for good air circulation. Also, the gooseberry has very sharp spines and an open bush greatly facilitates harvest. The fruiting wood should be developed horizontally. The young plant is cut back to several strong shoots at planting time. Usually 3-6 shoots are selected and trained to form an open bush. Shoots developing near the ground or tending to crowd the centre should be removed. Pruning becomes a thinning out process, the severity being determined by the density of growth. Heavy pruning is suggested for weak growth; light pruning for strong growth. Renewal of main branches from sucker-like basal shoots is necessary from time to time. In order to produce large, high-quality dessert berries, a system of hard-spur pruning should be adopted.

The pruning for quality and weight consists of normal leader tipping and branch selection. Strong growing leaders are only reduced by a third, weak ones by two-thirds, medium ones by half. All the weaker side shoots which are not more than 15 cm in length are left untouched unless they grow inwards and crowd the bush. All the remaining side shoots are either removed if over-vigorous or crowded or spurred back to leave them about 8-10 cm long. Light pruning to trim the bush is advocated for an early crop. Old branches that have stopped growing and are tending to give poor crops of small berries can be replaced by training in new vigorous shoots from time to time, provided the bush itself is still growing vigorously.

Porter (1979) in trial with cv. Yorkshire Champion grown from cuttings, 20 or 30 cm long, spaced 10, 20 or 30 cm in the row, cut out hedge centres to various degrees to facilitate mechanical harvesting. The highest yield (18 t/ha) of machine-collected fruit was from bushes with no centre cutting out.

The pruning is mainly carried out in dormant season. It is also believed that in some cases where there is a danger of bud-pecking by small birds in the winter, the pruning should be delayed until after the hard frosts are over. If pruning is carried out early in winter there is a risk of leaving the bushes wide open to bird attack. A limited form of summer pruning is practised by some specialist Leveller growers. This consists of shortening back the longest of the side shoots to 4 or 5 leaves where they are causing congestion or unduly shading the fruit. The leaders are, of course, never touched.

## Nutrition

Gooseberries respond to the application of farmyard manure (FYM) and use of adequate manure before planting followed by mulch round the bushes give beneficial effect. Ljones (1966) obtained about 53 per cent increase in gooseberry yield by the application of FYM.

## Role of Nutrients

### Nitrogen

Deficiency of N restricted growth, blossom development and reduced fruit size (Wallace, 1925). Ljones (1966) suggested application of calcium ammonium nitrate at 25 g/m$^2$. Wlodek *et al.* (1978) reported that plants of cvs. White Triumph and Resistenta when received ammonium nitrate or urea each at 80, 110, 200 or 260 kg/ha or calcium ammonium nitrate or ammonium sulphate at 200 kg N/ha, did not improve yield or fruit weight in either cultivar with increasing N rates, but a slight positive effect on plant growth was recorded in cv. Resistenta. High yield of good quality fruits was obtained with the lowest N rate and ammonium nitrate gave slightly better result than other sources.

### Phosphorus

Phosphorus application showed no significant effect on the yield of gooseberry (Grunnet and Vendelboe, 1952; Johansson, 1957). However, a basal application improved root growth.

### Potassium

Gooseberries have a high requirement for potassium. The deficiency of K resulted in brittle shoots and marginal leaf necrosis (Hadj-Hassan and Bunemann, 1974; Ljones (1966) observed that yield of gooseberry increased by the application of 25 g $K_2SO_4$/m$^2$. In sand culture, higher K level increased the total sugar content and titratable acid in fruits. In ripened fruits, malic and citric acids as well as glucose and fructose were increased by increasing K rates (Hadj-Hassan and Choureitah, 1973).

Gooseberry is very susceptible to chloride damage. The toxicity symptoms appear as necrosis of leaf margins and subsequent leaf abscission. Quast (1982) suggested use of $K_2SO_4$ in place of KCl, as the plants are more sensitive to the latter.

### Magnesium

The deficiency of Mg showed leaf blotch, similar to leaf scorch symptoms under K deficiency. The need for a balance between potassium and magnesium in the soil and also in fertilizer practices to be considered for supply of both the nutrients. A favourable response to the application of $MgSO_4$ at 30 lb per acre was recorded (Power, 1945).

### Boron

On course-textured soils with low B availability, annual fertigation (over 12 weeks at 7 day intervals, beginning at the stage of bud break at 0.25 kg/ha) or soil application (at the bud burst stage at 1 kg/ha) effectively improve growth, fruit yield, and the processing quality of the fruit (Wojcik and Filipczak, 2015).

## Fertilizer Recommendations

Johansson (1957) made the following recommendations for fertilizing gooseberries on soil not too poor in organic matter: 4 kg sulphate of ammonia or similar other N source, 1-2 kg superphosphate and 3 kg of 50 per cent potassium sulphate per 100 m$^2$ per year. Cuhareva (1967) suggested application of a mixture of 20 tonnes/ha of organic manure along with $P_2O_5$ and $K_2O$ each at 240 kg/ha to the bushes at the time of planting.

The fertilizers are applied during dormant season but nitrogen can be applied in split doses; one in dormant and another before fruit-setting stage.

## Irrigation

The gooseberries like other bush fruits respond well to drip irrigation. The water requirements are high during summer when there is no rainfall or drought occurs. During that time water application gives beneficial effect and enhances the yield to a considerable extent.

Water is critical during berry development through harvest, and during bud formation. The amount of water depends upon the type of soil, drainage and weather, so frequent testing of soil moisture is recommended.

Sava (2012) working with gooseberry cultivars Kolobok, Captivator, Grushenka, Sadko, Sema and Severni Captan, reported increased fruit weight form 2.0 to 3.2 g from non-irrigated field to 2.1 to 3.9 g on irrigated field. The average yield of studied gooseberry cultivars were also increased from 1.2 to 14.6 t/ha in non-irrigated field to 7.4 to 19.5 t/ha in irrigated field.

## Soil Management

Gooseberries, like strawberries, enjoy mulch, and peat or straw mulch was found excellent. The mulching restricts the growth of all annual weeds, helps to retain summer moisture and also makes picking during wet weather much easier. As with all soft fruits, shallow cultivation also suppresses the growth of weeds.

In an experiment on soil management variants, in 7-year-old plantation of gooseberry cultivars Resistenta and Robustenta, the treatments included bare soil maintained with gesatop 50 (simazine) at 3 kg/ha or gesatop 50 at 6 kg/ha, mulching with bituminous paper, clean cultivation or permanent sward. In the cv. Resistenta the best result with regard to plant growth and cropping were obtained with gesatop 50 at 6 kg/ha. Robustenta responded best to paper mulching and clean cultivation. (Kawecki, 1978). A double application of nitrogen fertilizer is needed with mulch to avoid nitrogen deficiency.

## Weed Control

Applying mulch around plant basin reduce weed emergence. Gooseberries have shallow and fibrous roots, hence deep cultivation near the plants should be avoided. A spring application of pre-emergent herbicide, such as dichlobenil or oryzalin will minimize annual weeds (Maughan and Black, 2014).

Herbicides have been found very effective to control the weeds in gooseberry plantation. Cianciara and Smolarz (1973) obtained good control of weeds by applying gesatop 50 (simazine) at the rate of 3 or 5 kg/ha, treatment with 3 kg/ha gave better results. Kohary and Harmat (1976), in trials with 8 herbicides, found that sinbar (terbacil) at 1 or 1.5 kg/ha gave the best general weed control. In the third year, afalon (linuron) + ramord (propachlor) at 5+5 kg/ha; afalon at 5 kg/ha or maloran (chlorbromuron) at 6 kg/ha were also effective in controlling the weeds.

# Flowering, Pollination and Fruit Thinning

## Flowering

The gooseberry takes about 2-3 years to come into bearing depending on the management of the bush. In the fourth year a crop of around 25 tonnes/ha can be expected.

The fruiting habits of gooseberry resemble those of the apricot more closely than those of any other tree fruit. Fruit buds are borne laterally, contain flower parts only and give rise to inflorescence without leaves or if the leaves are present they are reduced in size. Flower buds are formed laterally on long shoots or on spurs. The flower differentiation usually occurs in the first week of August.

## Pollination

Most of the cultivars of gooseberries are self-fruitful; hence there is no need of interplanting cultivars for cross-pollination. In a study conducted by Jefferies *et al.* (1982) on some commercial plantations of gooseberry cv. Careless at Stafford, poor crops of small fruits with few seeds were recorded, although the flowers appeared normal and there was an effective pollination period of at least 11 days. Hand pollination resulted in a good set of normal fruits. Few pollinating insects were recorded on a typical affected site, but on trouble-free sites in Hereford and Worcester, insects, particularly honey-bees (*Apis mellifera*) were plentiful and effective in transferring pollen. On trouble-free sites, excluding pollinating insects by caging bushes resulted in a typical crop failure, but caging had little effect at the problem site. Introducing honey-bee colonies to the affected site increased fruit set and seed number. Il'in (1986) reported that fruit set after selfing was 1.8-60.3 per cent, being highest in Chelyabinskii Zelenyi, Samarodok, Afrikanets and Avenarius Chelyabinskii (35.3-60.3 per cent). Low set under isolates was obtained from Krasnyi Krupnyi and Kaskad 21-24 (1.7-4.6 per cent). Fruit set was higher in all cultivars under open-pollination than with selfing.

## Fruit Thinning

The thinning of berries at early stages is rarely practised but some form of thinning of crop is essential if large berries are to be marketed, and this is usually done by marketing a portion of crop for early sale, when a size of 3/5 inch (1.5 cm) has been attained generally in mid-May. If the early prices are not considered good enough, thinning is sometimes delayed until a better size and weight have been achieved, and the crop is then thinned and the thinning are sold as green, semi-mature fruits.

# Maturity and Ripening

The berry weight increased until full maturity. Gooseberry harvested at full maturity weighed 30-40 per cent more than those harvested at physiological maturity and had a better quality, due to continued soluble solids and dry matter accumulation. The crop is considered matured when it has attained proper size, colour and quality. The pectin content in gooseberry fruit increases until technical maturity and declined thereafter (Strelnikova, 1971).

Kalousis *et al.* (2005) studied the physio-chemical characteristics of five cultivars. Fruit was harvested when 20 per cent of the fruit on a bush were ripe. Harvest dates varied from 15 June (Rokula) to 8 July (Greenfinch). The cv. Invicta gave the largest fruit but was the least sweet, cv. Leveller had the smallest fruit and cv. Greenfinch the sweetest. Fruit colour change with fruit size, especially for red cultivars Rokula and Pax. However, fruit size was not a good indicator of fruit sweetness. Selecting fruit on the basis of firmness and colour rather than fruit size was found as a more accurate indicator of fruit sweetness and therefore consumer acceptability. Famiani *et al.* (2009) working with gooseberry cv. White Smith reported that the concentrations of glucose, fructose and sucrose were relatively low up to 60 days after full bloom (DAFB), with glucose and fructose levels being higher than that of sucrose. All the sugars then increased up to harvest (approx: 110 DAFB), and sucrose become the most abundant soluble sugar. Malate and citrate concentrations increased up to 70-80 DAFB, then decreased. Phosphoenolipyruvate carboxykinase (PEPCK) increased during fruit ripening and was most abundant when the malate and citrate contents decreased. The results suggested that PEPCK plays a role in the catabolism of malate and citrate in gooseberry flesh during ripening.

Deposition of the cuticular membrane (CM) per fruit increased rapidly up to 41 DAFB, thereafter CM mass per fruit remained constant. Elastic strain of CM at maturity averaged 23.8 per cent. Micro cracks in the CM were recorded at 64 DAFB (Khanal *et al.*, 2011).

## Harvesting

A judgement is required to decide when and how to pick gooseberries. A lot will depend upon the type of market for which the grower caters and price at that time.

### Picking for the Early Market

The bushes are picked over taking the largest berries as they reach marketable size. The picking continues until the market price drops. After that the crop is left to mature. For green gooseberry, reasonable size and quality are easily obtained as a result of thinning.

### Picking for Green Gooseberry

Green gooseberries for the fresh fruit market are picked at the same stage of maturity as for processing. They are left on the bushes till they have attained full size. But once the ripening has commenced, the fruits are unsuitable for processing. It is better to pick early than too late. The fruit for processing is generally picked in one picking or at the most two, but it is uneconomical to harvest fruit in more than two pickings.

### Picking for Dessert Fruit

The ripe dessert fruit should not be so soft which squashes in picking or during transport to market. For this some judgement is required so that the berries reach the consumers at exactly the right stage for eating.

## Manual Harvesting

The gooseberries are usually picked from the plants by hand. The thorny nature of the bushes makes picking difficult. In larger plantations special scoop is used. The berries for general market should be picked more carefully. The grain fanning mill is helpful to clean leaves in case of large crops. The green-coloured gooseberries are most popular, though some markets prefer the pink or red types. Gooseberries are kept in shade to prevent sun scald.

## Mechanical Harvesting

It is desirable to introduce mechanical harvesting for a viable commercial gooseberry crop. The important requirements for mechanical harvesting are that the bushes must be robust and relatively upright in habit with firm berries that can withstand rough handling. Additional requirements are high yield, suitable dry matter and acid contents and acceptable flavour, together with resistance to gooseberry mildew. Out of 6 green-fruited cultivars examined, Ivicta, Riesen Von Coethen and Whitesmith were recommended for mechanical harvesting. Among 5 red-fruited cultivars, the most suitable appeared to be Rote Orleans and Whinham's Industry (Daugaard, 1983).

Shaking of gooseberries from bushes show that the working speed, amplitude and vibration frequency of the shaking fingers have a significant influence on the shaking efficiency and damage to shoots and fruits. Tests were carried out using a combine gathering fruits from the whole row of bushes at 3 forward speeds from 0.5 to 1.5 km/h, 5 amplitudes of vibration from 40 to 100 mm, and 7 vibration frequencies from 11.7 to 21.7 Hz (Salamon, 1999). As working speed increases, the accuracy of shaking decreased and damage to the fruits and shoots increased. Increasing the amplitude and vibration frequency resulted in increased accuracy of shaking but also increased damage. Lower amplitudes were more beneficial for harvesting gooseberry fruits. At a speed of 0.5 km/h, amplitude of 40 mm, and frequency of 20 Hz, the harvesting efficiency was 91.5 per cent, the number of shoots knocked off was 2.3 per cent, broken shoots 0.6 per cent, and damaged fruits 2.5 per cent.

# Yield

The yield of a particular cultivar greatly depends on the training system, planting distance and management conditions. The cv. Malinovyi yielded 26.2 centners/ha (Zaletilo, 1979). The cv. Yorkshire Champion, when spaced at 10, 20 or 30 cm apart, and pruned to hedge for mechanical harvesting, produced a yield of 4.4 to 6.9 tonnes/ha (Anon., 1978). At the National Fruit Trials, Brogdale, Kent, the cumulative yield from 1977 to 1983 for cv. Malling Greenfinch was 129 tonnes/ha compared with 54 tonnes/ha for cv. Careless (Keep *et al.*, 1985). At Moldova, average yield of more than 6.0 t/ha was recorded in cvs. Severnii Captan, Donetski krupnoplodnai, Donetski pervent, Puskinskii and Colobok (Sava, 2010).

# Storage

Gooseberries should be refrigerated as quickly as possible after harvest. Harvesting should be made in small, shallow containers to improve air circulation around the fruit and reduce bruishing. With proper cooling and storage, fresh gooseberries can be stored for up to three weeks. Harb and Streif (2004) in an experiment with gooseberry cv. Achilles, fruits were stored at 1°C in regular air and under six various controlled atmosphere (CA) conditions. Air storage was found not appropriate to handle gooseberries due to a drastic reduction in fruit firmness (lower by 25 per cent than CA storage), darkening of fruit colour, mainly texture, and lower acidity level. Low oxygen storage in combination with high $CO_2$ concentrations (18 per cent or more) also adversely affected fruit quality, mainly due to the development of off-flavours in fruits. Harb and Streif (2004) recommended storage at 12-15 per cent $CO_2$ combined with 18 per cent $O_2$ which can store the fruits up to 7 weeks. Pre-treatment of harvested fruit with ozonated vapour maintained higher ascorbic acid and anthocyanin contents in storage (Kampuse *et al.*, 2015).

## Pests

### Gooseberry Sawfly (*Nematus ribesii*)

It is a serious pest of gooseberry, and the caterpillars cause the damage. The adult sawfly appears in April and May and lays eggs under leaves. The caterpillars are fully grown in three weeks and start feeding on the leaves in the centre and base of the bush. Later, they spread on the entire bush. The caterpillars are green in colour with black dots on each segment and a black head. They can be controlled by spraying fenitrothion, malathion, derris or chlorpyrifos in mid-May.

Ionova (1971) obtained effective control of gooseberry-sawfly with DDVE (dichlovos), nogos, bromophos, gardona and C-9491 (iodofenphos) of all 3 generations and DDVF at 0.2-0.5 per cent was safely used 10 days before harvest.

### Aphids

The different species of aphid found to damage gooseberry include *Aphis grossulariae*, *Nasonovia ribisnigri* which suck the sap until the shoots become stunted. The trouble really begins in late autumn when the eggs are laid and bushes should, therefore, be washed during mid-winter with tar-oil preparation. Spraying of dimethoate, formothion or endosulfan was found effective to control aphid.

### Red Spider (*Bryobia ribis*)

A severe attack of red spider can do enormous damage to the bushes within a week. The adult mite is not always red in colour but it varies from whitish green to grey and rusty red and is visible to naked eye. In April the adults start to feed by sucking the sap of the leaves. Spraying of demeton-S-methyl, dimethoate and formothion should be done as soon as the mites are observed on the opening buds.

### Gooseberry Fruit Worm

Gooseberry fruit worm (*Zophodia convolutella*) causes serious damage to the gooseberry plantation. Larvae feed on multiple fruit and spin silken webbing, joining fruit and leaves together. Lokhman (1974) recommended the following cultural and biological methods to control this pest: (i) tilling of soil around the bushes and mulching with a 6-8 cm thick peat layer before flowering to hinder the flight of the moths and to encourage eventual pupation of the caterpillars in that layer (ii) spraying with a 1x $10^6$ spore suspension of *Paecilomyces fumosoroseus* per ml water + 0.02-0.03 per cent chlorofos in the first half of June. If fruit damage after these treatments exceeds 5-10 per cent, additional spraying of the soil with spore suspension + chlorofos in late June or early July is suggested. Borg (1981) obtained good control by spraying acephate at 0.1 per cent applied just after flowering.

## Diseases

### Die Back

Die back is caused by the ever-present grey mould fungus (Botrytis). In this case both shoots and fruits are attacked. The first sign of attack is yellowing of the foliage. Infected leaves turn almost white and drop off, leaving the twigs bare. The infected branches ultimately die and other branches are infected by the powdery grey spores. Excessive nitrogen application causes succulent growth of the plants which become more susceptible to the disease infestation. The trouble can be controlled by removing all decayed wood whenever it is noticed. Shelter from strong wind and cultivation help to prevent the disease appearance. Dieback of gooseberry is an increasing problem in commercial filed in the Netherlands. The casual agent was diagnosed as *Eutypa lata,* based on morphological characteristic and r-DNA-ITS data (Wenneker *et al.,* 2011).

### Anthracnose (*Colletotrichum acutatum*)

The symptoms on fruit surface were round, brown, shrivelled, sunken spots (Vichova *et al.,* 2013). Spraying of copper hydroxide, propiconazole, and basic copper sulphate gave good control.

### Collar Rot (*Phytophthora cactorum*)

The affected plants showed brown, water-soaked lesions appear mostly at the base of the stem, typical for collar rot. Resistant *Ribes aureum* rootstock may be a good solution because it keeps susceptible tissue away from infected soil. Good results in disease control were obtained after use of metalaxyl and fosetyl-Al twice during the season (Meszka and Bielenin, 2011).

### Leaf Spot (*Pseudopeziza ribis*)

It appears after a wet, cold summer and especially where the soil is not well drained. The leaves are covered with brown spots which cause them to fall prematurely. Control is obtained by spraying with Bordeaux mixture; spraying of benomyl, carbendazim, zineb, *etc.,* are also effective to control the leaf spot.

Profic-Alwasiak (1980) tested some fungicides for control of *Alternaria alternata* on the gooseberry cv. Whitesmith. Captafol, moncozeb or metiram each at a kg/ha, applied shortly after bloom and 3 times more at 8-10 days' interval gave good control.

### Cluster Cup Rust (*Puccinia pringsheimiana*)

Red or orange blotches are first seen on the fruits and leaves. The berries may be entirely covered in extreme cases. These blotches turn into small wart-like growths which burst to reveal clusters of cupped growths. Spores released from these cluster cups infect the gooseberry bush. The disease can be prevented by correcting drainage, scrupulous weed control and avoiding overcrowding or any other factor which promotes a wet or stagnant atmosphere.

### American Gooseberry Mildew (*Sphaerotheca mors-uvae*)

American gooseberry mildew attacks gooseberry causing serious infections. The fungus appears as a white powdery growth on the young leaves, stem surfaces and berries. Both side of the leaf can be covered with white mycelium and spores which turn rusty brown with age. Powdery mildew infestions may result in a significant reduction in marketable fruits. Severe infection reduces shoot growth and adversely impacts the following year's crop (Wenneker, 2016).

Brockamp (2010) suggested one or two sprout applications with sulphur and if required weekly, preventive spraying with a combination of fennel oil and sodium bicarbonate. Luna Sensation 500Sc containing fluopyram and trifloxys trobin at 0.6 and 0.8 l/ha effectively control mildew (over 96 per cent) (Broniarek-Niemiec, 2015). Applications of potassium bicarbonate (formulated as Armicarb/Karma) were found effective in reducing the incidence and severity of American powdery mildew in gooseberry. However, potassium bicarbonate might cause serious phytotoxicity (Wenneker, 2016).

### Vein-Banding Virus

It causes translucent, pale yellow banding along the main veins. The first leaves that are coming in spring showed banding in all of the veins. Leaves with vein-banding are often yellowish and asymmetrical. Vein banding associated virus is a distinct species within the genus Badnavirus. The sequence of a 1.5 kb DNA fragment amplified from viral genomic DNA was similar to those of a wide range of badnaviruses and contained motifs characteristic of the ribonucleases (RNase) H domain of the badnavirus open reading frame (ORF) III polyprotein (Jones *et al.*, 2001).

Aphids which spread vein-banding include *Aphis grossulariae, A. schneideri, Hyperomyzus lactucae* and *Nasonovia ribisnigri*. Control of aphids is worthwhile because of the direct damage they cause.

## Breeding and Improvement

### Genetic Resources

In order to improve the characterization and conservation of North Europian *Ribes* germplasm by means of coordinated trans-national achievements, a genetic resource project entitled "RIBESCO-Core Collection of Northern European Gene Pool of *Ribes*" was launched in 2007. The four-year project was initiated in 2007 by nine partners in eight EU countries within the Baltic Sea Region. In the project, a decentralised core collection of cultivated *Ribes* species was established, including blackcurrant, red and white currant and gooseberry. The germplasm accessions maintained in the phenotypic characterization using trans-nationally defined descriptors, thereby bringing together the phenotypic information of approximately 1400 *Ribes* accessions. Genotypes were selected into the core collection partly on the basis of agronomic, historical or other cultural values. The quality of berries have one selection criteria. Germplasm representing a high level of genetic variation was selected. Conservation of the selected core collection, *ex situ* field collection was established and cryopreservation for long-term storage was developed. The project promoted both the preservation and use of genetic material (Karhu *et al.*, 2012).

In Latvia, more than 70 genotypes of gooseberry originated in Latvia, have been collected and described at the Latvia Institute of fruit growing. It included local cultivars (Koknese, Kursu Dzintars and Anna Viksne) and breeding clones. Genotypes potentially useful for both commercial growing and breeding were recognized and described, as well as characterized by their spinelessness, winter hardiness, spring frost tolerance, resistance to pest and disease and high berry quality. Differences between ripening times of early ripening accessions (5-14, 6-4) and the late-ripening ones (Koknese 55, 448-10, 536-9, 433-3, 323, 250-F2-60) were upto a month, depending on season. Accessions with erect plant habit (5.2) and no spines (250-F2-60, 250-F2-71 and 448-10) were recorded. However, very few genotypes had a fruit weight over 4g (484-18,39). The most productive accessions were 4-14, 3-10, 410-15, 435-1, 5-2, Anna Viksne and Koknese 55. The highest soluble solid contents of fruit of 16.1 and 15.8° Brix were found in breeding clones 424 and 323-1, respectively. Accessions with uniform fruit ripening time (Kursu Dzintars, N-17-15) and fruit size (410-15, 433-26, 6-4. Citronu) were found (Kampuss *et al.*, 2012).

At Michurin, All-Russian Research Institute of Horticulture, between 2011and 2015, 22 new gooseberry germplasm was added to the collection. The cvs. Compromiss, Orphey and Uralskiy Rozoviy were recommended for cultivation (Zhidyokhina *et al.*, 2016).

The *Ribes* collection is maintained at the National Clonal Germplasm Repository (NCGR) in Corvallis, Oregon and its worksite at the Arctic and Subarctic Plant Gene Bank (ASPGB) in Palmer, Alaska. *Ribes* collection and public access for requesting accessions is available online through the Germplasm Resources Information Network (GRIN). GRIN is available at http://www.ars-grin.gov/npgs.

## Breeding

The genus *Ribes* (Grossulariaceae) is a diverse group of woody flowering plants in which hybridization is known to occur in both natural and control conditions. Agricultural uses of gooseberries and currants have led to breeding experiments that have produced several hybrids. Numerous fertile hybrids from control crosses developed by Keep (1962) show correlation between morphological similarity and the absence of strong reproductive barriers, suggesting that these barriers are weaker between closely related species.

There are about 15 active *Ribes* breeding programmes carried out worldwide with most of them focused on releasing black currant cultivars. The main breeding centres are located in Scotland, Poland, New Zealand, Sweden, Estonia, Latvia, Lithuania, Romania, Russia, Serbia and US. The breeding effect on gooseberries are small and mostly located in eastern Europe and US. The fruit production of gooseberry is mainly destined for processing and freezing industries with less fruit going to the fresh market. Interspecific hybridization within the *Ribes* species used in some countries particularly in Russia, Poland and Lithuania (Pluta, 2012).

The main goals of *Ribes* breeding programmes are common and similar in different breeding centres. These are (1) breeding for resistance to main pests and diseases, (2) breeding for fruit quality and sustainability for processing and fresh market, and (3) good adaptability to local environmental conditions (winter hardiness, spring frost tolerance, chilling requirements, and machine fruit harvest). Besides agronomic traits such as yield, resistance to pests and diseases are also the focused objective of breeding.

Zazuliana (1999) evaluated gooseberry cultivars, hybrids and selections of diverse origin. Crosses were also made between certain genotypes, and their progenies were evaluated. Two new cultivars, Belovezhskii and Korall, which combined filed resistance to American powdery mildew, with high yield, attractive appearance and excellent in flavour were developed.

In Latvia, gooseberry breeding continued from the 1930s to 1970s. the varieties selected by local breeders (Koknese, Veldze, Perse, Kursu Dzintars, Parsla) still form the basis of the gooseberry assortment, having good resistance to mildew, as well as good productivity and winter hardiness (Strautina and Lacis, 2000).

The genus *Ribes* has many sources of resistance to mildew. *Ribes hirtellum, R. divaricatum, R. cynosbati, R. irriguum, R. niveum, R. oxyacanthoides* show high resistance to mildew (Keep, 1970) and *R. niveum, R. gracile, R. rotundifolium, R. irriguum* have high resistance to leaf spot (Keep, 1974). Two high-yielding mildew-resistant cultivars, Malling Invicta and Malling Greenfinch, which derive mildew resistance from North American *R. divaricatum* via the resistant German cultivar Resistenta (Bauer, 1955), were released from East Malling Research Station, UK in 1979 and 1984 respectively (Keep *et al.*, 1980, 1985).

Breeding for spinelessness using the Canadian *R. oxyacanthoides* derivative, Captivator as donor, has involved repeated backcrossing to European *R. grosularia* cultivars, in an attempt to improve yield and fruit size at East Malling. However, even in late backcrosses, spineless or near spineless segregants yielded less than their spiny sibs (Keep, 1976). An alternative breeding line was also started at East Malling in 1968 when a totally spineless hybrid of gooseberry x *R. sanguineum* was backcrossed to gooseberry. A low proportion of plants in this and subsequent backcrosses were almost or completely spineless and some had high yield potential (Keep, 1984).

## Molecular Markers

Molecular markers have been developed for use in genetic diversity studies within the genus *Ribes*. Genetic variation in the gooseberry germplasm was examined using RAPD, ISSR and AFLPs. AFLP markers separated closely related genotypes which were indistinguishable with RAPDs or ISSRs. RAPD markers were, however, useful to identify mislabelled plants in field trials (Lanham and Brennan, 2001).

Tanovic *et al.* (2012) developed protocols for genetic fingerprints of standard cultivars through molecular marker analysis using SSRs.

## REFERENCES


Aleshkevich, L.V., Kochetova, N.I. and Kochetov, Yu, V. (1982). Vestinik Sel'skokhozyarstvennoi Nauki, 10: 50-53.

Anonymous,(1978). New Zealand Commercial Grower, 33: 29.

Bauer, R. (1955). Rpt. 14th Int. Hort. Congr., Scheveningen. pp. 685-96.

Berger, A. (1924). Bull. NY. Sta. Agric. Exp. Sta., 109.

Borg, A. (1981). Kemisk bekampning i suarta vinbar Vaxtokyddnotiser. 45: 186-89.

Brockamp, L. (2010). Proc. 14th Int. Conf. Organic. Fruit Growing, Germany pp. 376-79.

Broniarek-Niemiec, A. (2015). Progr. In Pl. Protec., 55: 478-82.

Cejka, B., Matejickova, J., Matejicek, A. and Kaplan, J. (2013). Vedecke Prace Ovocnarske, 23: 139-43.

Cianciara, T. and Smolarz, K. (1973). Prace. Ins. Sed. w. Skier., 17: 181-87.

Cronquist, A. (1981). An Integrated System for Classification of Flowering Plant. New York. Columbia Univ. Press. pp. 555.

Cuhareva, N.V. (1967). Sadovodstvo, 1: 6.

Cvopova, E. (1984). Vedecke Pra. Vys. Ust. Ovo. Okras. Drev. Bojni., 5: 153-62.

Daugaard, H. (1983). Frugtavleren, 12: 430-35.

Davlebaeva, *E.G.* (1985). Int. Voz. plod. i. yogod. Kul, USSR. pp.45-51

Famiani, F., Baldicchi, A., Battistelli, A., Moscatello, S. and Walker, R.P. (2009). Jr. Hort. Sci. Biotech., 84: 249-54.

Gomolyako, L.G. (1940). Biokhim Kultur Bastenu, 79: 372-74.

Gottman, F.D. and Galletti, S. (2001). Jr. Agric. Fd. Chem., 49: 349-54.

Grunnet, H.O, and Vendelboe, B. (1952). Tidsskr. Planterrvl, 55: 591-620.

Hadj-Hassan, A. and Bunemann, G. (1974). Gartenbauwiss., 39 : 521-52.

Hadj-Hassan, A. and Choureitah, A. (1973). Gartenbauwiss., 38 : 315-25.

Harb, J. and Streif, J. (2004). Jr. Hort. Sci. Biotech., 79: 329-34.

Harmat, L. (1975). Cyumolastermesztes., 2 : 69-75.

Harmat, L.A., Porpaczy, D.H. and Galletta, G. (1990). In: Small Fruit Crop Management (eds. G. Galleta and D. Himlerick) pp. 245-272, Prentice Hall, Englewood, Cliffs, NJ.

Il'in, V.S. (1982). Sadovodstvo, No. 11, pp. 26-27.

Il'in, V.S. (1986). Sadovodstvo, No. 1, pp. 27-28.

Ionova, Z.A. (1971). Sbor. Nau. Rahot Vses. Nau. Issled. Inst. Sadov. Imeni, I.V. Nich. No. 16, pp.326-28.

Jazvicky, M.N. (1940). Plod. Jagodn. Kultury, 2 : 72-84.

Jefferies, C.J., Atwood, J.G. and Williams, R. R. (1982). Scientia Hort., 16: 147-53.

Johansson, E. (1957). Medd. Tradgardsfors, Alanarp., 104: 14.

Jones, A.T., McGavin, W.J., Geering, A.D.W. and Lockhart, B.E.L. (2001). Pl. Dis., 85: 417-22.

Kalousis, N.K., Cullum, J. and Wainwright, H. (2005). Small Fruits Rev., 4(1): 33-41.

Kampuse, S., Muizniece-Brasava, S., Kurma, Z., Sabovics, M., Savri, S. and Ruse, K. (2015). Acta Hort., 1071: 405-12.

Kampuss, K., Strautina, S., Kampuse, S. and Krasnova, I. (2012). Acta Hort., 926: 39-45.

Karhu, S., Antonius, K., Rantala, S., Kaldmae, H., Pluta, S., Rumpunen, K., Ryliskis, D., Sasnauskas, A., Schulte, E., Strautina, S. and Toldam, A.T.B. (2012). Acta Hort., 926: 27-32.

Kawecki, Z. (1978). Rocz. Nauk. Rolniczych. A. 103 : 39-48.

Keep, E. (1962). Genetica, 33: 1-23.

Keep, E. (1970). Rep. East Malling Res. Stn. for 1969, pp. 133-37.

Keep, E. (1974). Ann. Appl. Biol., 7: 131-35.

Keep, E. (1976). Acta Hort., 60: 123-28.

Keep, E. (1984). Scientific Hort., 35: 54-71.

Keep, E., Parker, J.H. and Knight, V.H. (1980). Rep. East Malling Res. Stn. for 1979. pp. 215-16.

Keep, E., Parker, J.H. and Knight, V.H. (1985). Rep. East Malling Res. Stn. for 1984, p 221.

Kenny, A. (1971). Fm. Fd. Res., 2: 8-9.

Khanal, B.P., Grimm, E. and Knoche, M. (2011). Scientia Hort., 128: 289-96.

Kohary, E. and Harmat, L. (1976). Kertgazdasag, 8 : 29-35.

Lenham, P.G. and Brennan, R.M. (2001). Acta Hort., 546: 135-37.

Ljones, B. (1966). Forskn. Fors Landbr., 17 : 227-40.

Lokhman, S.A. (1974). Izves. Timi Sel. Aka., 1: 134-43.

Mattivi, F. and Vrhovsek, U. (2010). Obstbau und Fruchteverwertung, 60: 442-48.

Maughan, T. and Black, B. (2014). Utah State Univ. USDA-RMA Publication. Horticulture/Fruit/2014-6.

McKay, S.A. (2005). NY Fruit Quart., 13(2): 29-32.

Meszka, B. and Bielenin, A. (2011). Progs. Pl. Protec., 51: 1184-87.

Mironova, O.A. (1985). Probl. Vege. Yaz. Y. Sad. Moscow, USSR, pp. 102-106.

Nikfardjam, M.P., Kopp, M., Hempfling, K. and Engel, K.H. (2013). Obstbau und Fruchteverwertung, 63: 213-22.

Patutina, I.G., Saumjan, K.V. and Sljapnikova, A.S. (1974). Land. aftfi Zen. 19: 2857-58.

Pinto, M. da. S., Kwon, Y.I., Apostolidis, E., Lajolo, F.M., Genovese, M. and Shetty, K. (2010). Jr. Fd. Biochem., 34: 639-60.

Piofic-Alwasiak, H. and Szczygiel, A. (1980). Fruit Sci. Rept., 7: 35-43.

Pluta, S. (2012). Acta Hort., 946: 27-36.

Porpoczy, A. (1988). Acta Hort., 224: 361-67.

Porter, L.A. (1979). New Zealand Commercial Grower, U: 23.

Power, W.L. (1945). Science, 101 : 301.

Quast, P. (1982). Mitteilungern des Obstbauversuchsringes des Alten Landes, 37:123-26.

Raspopova, G.I. (1986). Sadovodstvo, 3: 23-24.

Roach F.A, (1985). Cultivated fruits of Britain, their Origin and History. Blackwell, New York, pp. 295-322.

Rybalov, L.M. (1985). Sadov. Vino. i Vin. Mol., 9: 28-29.

Salamon, Z. (1999). Inzynieria Rolnicza, 5: 249-56.

Sava, P. (2010). Seria Horticultura, 53: 235-40.

Senter, A.E. and Soltis, D.E. (2003). Taxon., 52: 51-66.

Siksnianas, T. and Sasnauskas, A. (2006). Agronomijas Vestics, 9: 135-39.

Sinnott, Q.P. (1985). Rhodora, 87: 187-286.

Skovorodnikov, D.N. (2012). Vestnik Orelgau, 6: 24-27.

Soltesz, M., Szabo, Z. and Nyeki, J. (2007). Acta Hort., 732: 397-403.

Strautina, S. and Lacis, G. (2000). Acta Hort., 538: 469-72.

Strelnikova, V.N. (1971). Sbo. Nau. Rab. Vses. Nau. Issl. Inst. Sad. I.V. Michurina, 15: 181-86.

Tanovic, B., Hummer, K.E., Robertson, N.L. and Gilmore, B.S. (2012). Acta Hort., 946: 73-76.

Vichova, J., Jilkova, B. and Pokorny, R. (2013). Pl. Dis., 97: 1249.

Wainwright, H. and Flegmann, A.W. (1985). Jr. Hort. Sci., 60: 215-21.

Wallace, T. (1925). Jr. Pom. Hort. Sci., 5: 1-33.

Wassenaar, L.M., Lindeleof, O.P.J. and Van De. (1984). Fruitteelt, 74: 1102.

Watt, B.K. and Merrill, A. (1959). A table of food values. In : USDA Yearbook of Agriculture, Washington, Supt. of Documents.

Watt, J.H. (1950). Bull. New Zealand Dept. Agr., No. 282.

Welander, M. (1985). Scientia Hort., 26 : 267-72.

Wenneker, M. (2016). Acta Hort., 1133: 515-20.

Wenneker, M., Raak, M.M.J.P. van., Brouwershaven, I.R. van, Martin, W. and Kox, L.F.F. (2011). European J. Pl. Physiol., 131: 441-49.

Williams, R.F.V. (1982). Hort., 32 : 30-32.

Wlodek, L., Potocka, W., Jackiewicz, A., Piatkowski, M. and Kemula. F. (1978). Prace. Inst. Sad. W. Ski. A., 20: 101-109.

Wojcik, P. and Filipczak, J. (2015). Scientia Hort., 56: 171-74.

Zaletilo, A.P. (1979). Trudy ser Kaz., Ges. Skh. Kho. St., 8: 34-44

Zazulina, N.A. (1999). Plodovodstvo, 12: 100-103.

Zhidyokhina, T.V., Koveshnikova, Ye. Yu., Bryksim, D.M., Rodyukova, O.S., Khromov, N.V. and Guryeva, I.V. (2016). Sadovodstvo I Vinogradarstvo, 1: 12-19.

Zika, J. and Kricnarova, L. (1987). Vedecke Prace Ovoc., 11: 117-25.

# Chapter 16

# Raspberries

Raspberries are grown as a high input/high value fresh products in some regions while in other regions raspberries are grown as a processing crop. In recent years the demand for fresh raspberries all the year round in Western Europe and North America has resulted in the production season being extended considerably by using a range of cultivars and production systems (Knight, 2004).

## Composition and Uses

Raspberry contains almost all important nutrients without being very rich in any particular ingredient of nutrients (Table 16.1), The total sugar content is 8-9 per cent, mostly consisting of glucose and fructose with little or no sucrose (Anon., 1972; Kichina and Ivanov, 1977). The acidity is due mainly to citric acid (97 per cent) rest being malic acid (3 per cent). The good jellying quality of raspberry is ascribed to its calcium pectate (0.4 per cent). The raspberry flavour components are 35 alcohols, 23 acids, 34 carbonyls and 19 esters, the major components being geraniol, linalool, α-terpineol and α-and β-ionoes (Broderick, 1976; Latrasse, 1982). Kichina and Ivanov (1977) reported 164-177 mg and 109-126 mg tannin content per 100 g in dark-coloured and light-coloured raspberry fruits, respectively.

Raspberry fruit contents abundant dietary phtochemicals such as flavonols, phenolic acids, ellagitannins, vitamin C and E, folic acid and β-sitosterol (Bowen-Forbes *et al.*, 2010). Anthocyanins and other phenolic compounds such as ellagitannins and ellagic acid, which distinguish raspberry from other berries, occur in a high content, and are mainly responsible for their broad health beneficial properties (Rao and Snyder, 2010). Ellagitannins possess several biological activities, including anticancer properties (Nowak *et al.*, 2017).

**Table 16.1: Nutritional Composition of Raspberries (per 100 g edible portion)***

| *Constituents* | *Red Raspberry* | *Black Raspberry* |
|---|---|---|
| Water (per cent) | 84.2 | 80.8 |
| Calories | 57 | 73 |
| Protein (g) | 1.2 | 1.5 |
| Fat (g) | 0.5 | 1.4 |
| Carbohydrates (g) | 13.6 | 15.7 |
| Vitamin A (1.U.) | 130 | t |
| Thiamine (mg) | 0.03 | 0.03 |
| Riboflavin (mg) | 0.09 | 0.09 |
| Niacin (mg) | 0.9 | 0.9 |
| Ascorbic acid (mg) | 25 | 18 |
| Calcium (mg) | 22 | 30 |
| Phosphorus (mg) | 22 | 22 |
| Iron (mg) | 0.9 | 0.9 |
| Sodium (mg) | t | t |
| Potassium (mg) | 168 | 199 |

t = traces.

* Westwood (1978).

Kula *et al.* (2016) working with 11 red and black raspberries recorded significant differences in the levels of major anthocyanins in most red raspberry cultivars, of which cyanidin 3-*O*-sophoroside was the most prevalent one. Black raspberries showed 4-11 times more anthocyanin content compared to red raspberries, with the dominating anthocyanins being cyanidin 3-*O*-rutinoside and cyanidin 3-*O*-xylosyl-rutinoside. Sanguiin H-6 was the main ellagitannin identified in all varieties. A number of other compounds belonging to flavonols, flavan-3-ols and phenolic acids, were found at much lower levels than anthocyanins and ellagitannis, with the presence of procynidins $B_1$ and $B_2$ being reported for the first time in the fruits of black raspberries. Andrianjaka-Camps *et al.* (2016) reported 98 to 225 toatl phenolic content (mg gallic acid equivalent per 100g fresh weight) among 16 red raspberries cultivars (Table 16.2). Miletic *et al.* (2015) studied the chemical and antioxidant properties of matured raspberry fruits of cvs. Willamette and Meeker. They reported very high content of ellagic acid (from 27.5 to 29.2 mg/100g fw for 'Willamette', from 21.0 to 25.1 mg/100g fw for 'Meeker') and cyanidin (from 38.7 to 43.2 mg/100g fw for 'Willamette', from 21.3 to 30.4 mg/100g fw for 'Meeker'). Significant amounts of quercetin (up to 1.1 mg 100g fw), apigenin (up to 0.4 mg/100g fw), gallic acid (up to 6.1 mg/100g fw), and pelargonidin (up to 5.1 mg/100g fw) were also detected.

Raspberries are consumed fresh, used in making processed products, such as, jam, jelly, syrup, yoghurt, flavourings, and are also canned and frozen (Jennings,

1976). Wine can also be prepared from the berries. Valset (1976) evaluated the raspberry cultivars for jam quality and reported that Camenzind and Veten cultivars imparted the best flavour, Veten and Malling Jewel, the best colour and Zeva, the best consistency to the jam. In a study on the suitability of raspberry cultivars for freezing, Lenartowicz *et al.* (1980) gave the highest ratings to Szopska, Allen, Canby, Glen Clova and the hybrid NY 588. Roots, young shoots, leaves (fresh and dried) are used for medicinal purposes (Anon., 1972).

**Table 16.2: Total Phenolic Contents (mg gallic acid equivalent (GAE) per 100g Fresh Weight) of different Red Raspberry Cultivars**

| *Cultivar* | *Phenolic Contents* | *Cultivar* | *Phenolic Content* |
|---|---|---|---|
| Amira | 154 | Kweli | 200 |
| Autumn Bliss | 188 | Polka | 139 |
| Brillance | 150 | Sugana | 144 |
| Jan J | 225 | Regina | 98 |
| Erika | 141 | Rubifall | 106 |
| Heritage | 167 | Versalle | 86 |
| Himbo Top | 162 | Chambord | 113 |
| Imara | 180 | Deauville | 102 |

Compiled from Andrinjaka-Camps *et al.* (2016).

## Origin and Distribution

Raspberries are natives of the temperate regions of northern Hemisphere, the genetic diversity of various *Rubus* species being located in the different centres of diversity, *i.e.*, Chinese, Japanese Centre, Indochinese, Indonesian Centre, Hindustan Centre, European-Siberian Centre, South American Centre and North American Centre (Zeven and Zhukovsky, 1975). The subgenus *Idaeobatus* has originated in eastern Asia whereas *R. occidentalis* is native to North America (Ourecky, 1975). Raspberry species from the *Rubus* subgenera Cylactis (Arctic raspberry), *Idaeobatus* (raspberries), *Anaplobatus* (flowering raspberries) and *Orobatus* are native of northern hemisphere (Funt and Hall, 2013).

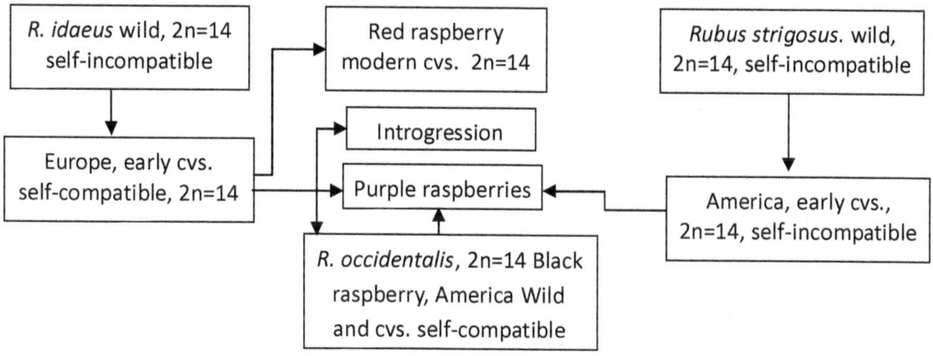

Jennings (1976) has traced the origin of cultivated raspberries to three diploid species or subspecies, *i.e.*, *R. idaeus* var. *vulgatus*, *R. idaeus* var. *strigosus* and *R. occidentalis* as under.

The raspberry is grown in many countries of the world, *viz.*, USSR, USA, Canada, Belgium, Bulgaria, Czechoslovakia, Denmark, Finland, France, Germany, Hungary, Ireland, Italy, Netherlands, Norway, Poland, UK, Yugoslavia, Australia and New Zealand. *Rubus* subgenus *Idaeobatus* is distributed principally in Asia but also to East and South Africa, Europe and North America. In contrast, subgenus *Eubatus* is mainly distributed in South America, Europe and North America (Jennings, 1988).

Black raspberry (*R. occidentalis*) is mainly grown in Oregon, Ohio, Pennsylvania and New York in the United States and in Korea and Poland.

## Area and Production

The world raspberry production was 795 thousands tonnes in 2016 of which nearly 25 per cent production was from countries of European Union followed by Russia and United States (Table 16.3).

**Table 16.3: Raspberries Production in 2016**

| *Country* | *Production (000 tonnes)* |
|---|---|
| European Union | 210 |
| Russia | 165 |
| United States | 136 |
| Poland | 129 |
| Mexico | 113 |
| Serbia | 62 |
| World | 795 |

Source: FAOSTAT (2016).

## Taxonomy

*Rubus* is a variable genus contains 14 subgenera and over 400 species out of which 57 are found in India (Anon., 1972). Four sections of the genus are horticulturally important. *Idaeobatus* section includes the raspberries in which ripe fruits separate from the receptacle causing it to be hollow on the underside. *Eubatus* section consists of blackberries and is characterized by the adherance of druplets of the ripe fruits to receptacle both detaching from the calyx. Herbaceous, north-circumpolar or alpine species are included in *Cylactis* section. The flowering raspberries have been placed in *Anoplobatus* section.

The main species bearing edible fruits have been listed by Ourecky (1975) are: *R. idaeus* var. *vulgatus* (European red raspberry), *R. occidentalis* and *R. leucodermis* (black raspberries), *R. odoratus* (thimbleberry), *R. spectabilis* (salmonberry), *R. idaeus* var. *strigosus* (red raspberry), *R. phoenicolasius* (wineberry), *R. ellipticus* (golden evergreen raspberry), *R. nivens*, *R. coreanus*, *R. parvifolius*, *R. macraei*, *R. hawsiiensis* and *R. glaucus*.

All the present raspberry cultivars have been developed form the hybridization of *R. idaeus* var. *strigosus* and *R. occidentalis*. The botanical variety *vulgatus* produces dark and long conical fruits with few or no glandular hair whereas the fruits of the variety *strigosus* are round and light red with numerous hair. The purple raspberry (*R. neglectus*) is the hybrid derivatives of black raspberry (*R. occidentalls*) with *R. idaeus* var. *strigosus* (Ourecky, 1975).

## Cultivars

A large number of raspberry cultivars are grown which can be divided into many groups, *viz.*, early, medium and late depending on the harvest period; summer and autumn (Primocane) fruiting; red, yellow, black and purple on the basis of fruit colour.

The cultivars are also available for specific purposes, such as. winter-hardy, insect-and disease-resistant and suitable for protected cultivation, mechanical harvesting, processing, canning, juice making and freezing.

### Autumn Bliss

This cultivar was raised at East Malling Research Station, UK (Keep *et al.*, 1984). Autumn Bliss is of complex parentage including in its ancestry an early autumn-fruiting *Rubus idaeus* var. *strigosus* selection from New York State Agricultural Experiment Station, the herbaceous *R. articus*, the American black raspberry *R. occidentalls*, and the red raspberry cultivars Malling Landmark, Promise, Lloyd George, Pyne's Royal, Burnetholm and Norfolk Giant. It ripens earlier than the standard Heritage and Zeva Herbsternte, overlapping in mid-August with summer-cropping cultivars Malling Leo and Malling Joy and continuing up to October. The fruits are large, oval-conical, fairly easy to pick and slightly dark red, with a pleasant rather mild flavour.

### Autumn Treasure

A mid-season, primocane-fruting red raspberry for fresh production, suitable for low input systems and protected cultivation. Fruit: medium to large, firm, pleasant flavour, bright red and good shelf life. Highly resistant to *Phytophthora* root rot.

### Cascade Bounty

It is highly productive cultivar producing large quantities of medium-size, bright coloured and tart fruit. It is winter hardy and extremely tolerant to *Phytophthora* root rot.

### Chilcotin

This cultivar has been released to replace the famous cultivar Willamette (Daubeny, 1978). The fruit is glossy red and can be transported well for fresh market. It is good for pick-your-own system and freezing. Its yield is higher than Willamette.

## Chinook

A primocane fruit raspberry released from USA as an alternative to Autumn Bliss. Fruits round, firm. It has bright red colour with good flavour, field tolerance to root rot.

## Chilliwack

Chilliwack has been evolved from Skeena x BC64 -10 -198 (Summer x Carnival). The high quality fruits are suited for both fresh market and processing, and are resistant to *Botrytis cinerea* and *Rhizopus stolonifer* (Daubeny, 1987). The plant is resistant to *Amphorophora agathonica* and to some extent to *Phytophthora erythroseptica*.

## Comox

Comox has been derived from Skeena x BC 64-9-81 (Creston x Willamette) producing large firm fruits of high quality (Daubeny, 1987). The fruits are resistant to *B. cinerea* and *R. stolonifer* and to *Amphorophora agathonica*. The cultivar is self-fertile. It has given significantly more yield than Meeker, Willamette and Skeena.

## Fengmanhong

A selection of local Chinese cultivar Hongshumei. It can withstand winter temperature of -40°C. Fruits are large, glossy, bright red colour, average fruit weight 6.4 g, with soluble solids content of 12.8 per cent.

## Glen Clova

It has been introduced by D.L. Jennings in Scotland which is a seedling of complex origin (Brooks and Olmo, 1972). Glen Clova is the best early cultivar with bright red fruits ripening over a long period (Oosten and Van, 1987a). Fruit is of medium size, firm flesh and good flavour. The bush is erect, vigorous and heavy cropper. It is also winter-hardy (Gwozdecki *et al.*, 1984/85).

## Heritage

A primocane fruiting red raspberry, selected in 1960 from a cross between New York 463 and Durham and released in 1969. It is a high-yielding, reliably productive, has good fruit quality, and is firm enough to ship. Heritage has been good parent with Caroline, Polana, Redwing and Ruby as direct offspring and has produced a number of yellow-fruited sports such as Kiwigold and Graton Gold (Volk *et al.*, 2013).

## Korere

An early, floricane-fruiting red raspberry producing good yields of high quality fruit suited to fresh and processing markets. Resistant to raspberry bushy dwarf virus (RBDV).

## Lewis

A floricane fruiting red raspberry. It is an outstanding fresh-market cultivar with large, glossy, attractive and firm fruit. In the pacific North West, yields of Lewis were similar to cv. Meeker, whereas the fruits were consistently large and firmer than those of Meeker (Finn *et al.*, 2001).

## Lyulin

The cultivar was derived from Elite No. 11824 (Newburg x Bulgarski Rubin) and released in 1962 (Khristov, 1983). The bush has upright short canes (1.25-1.30 m), requiring no support. Hence, the cultivar is suitable for mechanization. It is comparatively resistant to *Didymella applanta*. Three-year-old plants yield 9829 kg fruits/ha.

## Malling Admiral

The parentage of Malling Admiral includes Burnetholm, Preussen, Norfolk Giant and Malling Promise. A late-ripening cultivar with good quality. Fruits are large, conical, dark red in colour with good flavour. It is moderately resistant to mildew, spur blight and cane *Botrytis*.

## Malling Bella

A mid-season primocane that is well adopted to production in the UK and southern Europe. It is suitable for double cropping and long cane production and has large mid-red berries. It has a traditional raspberry flavour.

## Malling Charm

It is an early primocane cultivar with large berries that are sweet and juicy and a long shelf life. It comes into production in mid-August, covering the gap in the market until mid-to late September.

## Malling Delight

The Malling Delight is of complex parentage involving Burnetholm, Preussen, Malling Promise, Baumforth A, Lloyd George and Pyne's Royal. Fruits are very big and a heavy cropper; however, it needs picking at slightly unripe stage as berries soften quickly and bleed. It has excellent dessert quality but not suitable for freezing or canning (Keep, 1975). It is susceptible to *Botrytis* and spur blight, but resistant to the aphid *Amphorophora idaei*, strains 1 and 3.

## Malling Joy

A very late-ripening cultivar derived from the American black raspberry Cumberland and number of old and well-known cultivars with the addition of Malling Promise. Fruit is large, conical, dark and very firm. It is susceptible to spur blight and cane *Botrytis* but is resistant to all four strains of virus-carrying raspberry aphids.

## Malling Leo

It is derived from the American black raspberry Cumberland and includes the red raspberries Malling Landmark, Lloyd George, Pyne's Royal, Norfolk Giant and Burnetholm in its ancestry. A late-season cultivar; fruits are round, firm, bright orange-red in colour. Leo is susceptible to cane spot but resistant to *Botrytis* and spur blight.

### Malling Promise

The first raspberry to ripen in June. Fruit is of high quality being suitable for good jam but not for bottling. The vigorous plant is resistant to frost but takes few years to attain its full potential for heavy yield. The cultivar is also virus tolerant (Brooks and Olmo, 1972).

### Malling Minerva

An early ripening floricane fruiting raspberry. Fruit: attractive shape, good colour, good flavour and shelf life.

### Meeker

A hybrid of Willamette and Cuthbert. It has long cluster of fruit, tall canes, with long, strong, well-connected and well-spaced laterals. The fruit is late-ripening, large, conic, smooth and regular. It is presently used almost exclusively for processing (Volk *et al.*, 2013).

### Moutere

An early ripening floricane-fruiting raspberry, developed from New Zealand, producing high yields of attractive fruits. Fruit very large, glossy dark red, round, firm, pleasant aroma and adopted to machine harvest.

### Norfolk Giant

It is a late cultivar suitable for biennial cropping (Waister and Cormack, 1975). It is a hardy, and vigorously growing cultivar, blossoming late, thus avoiding frost even in cold regions. The fruit has strong but fair flavour. It is suitable for mechanical harvesting.

### Perron's Red

Perron's Red was selected in 1977 from a cross between Zeva Herbsternte and Pathfinder, having inherited the best qualities of both the parents (Huber, 1987). Fruits are conical and excellent for fresh market, freezing, jam, jelly and syrup. It is a vigorous primocane-fruiting raspberry for roadside market, pick-your-own and home gardens.

### Royalty

Royalty, tested as NY 632, is a cross between a purple and red raspberries, and ripens from red to purple (Sanford and Ourecky, 1982). The fruit is large (about 6 g), flavour and appearance good and has long shelf-life. The plant is hardy, being immune to *Amphorophora agathonica*, *Aphis rubicola* and *Byturus rubi*.

### Titan

It was selected from Hilton x NY 589 cross and tested as NY 883 (Sanford *et al.*, 1985). Bright red fruits are large (4-6g), mild in flavour and are suitable for jam making. The canes are glabrous with sparse basal spines. Titan is susceptible to most raspberry diseases and is unsuitable for mechanical harvesting.

**Raspberry Plantations**

**Raspberry cultivar Malling Admiral**

**Yellow Rust disease of raspberry**

**Raspberry cultivar Meeker**

### Tulameen

Developed from Canada, a cross of Nootka x Glen Prosen. It is recognized for its bright glossy medium red-coloured fruit with a quintessential raspberry flavour. The firm fruit texture results in good shelf life and suitability for long distance transport. It was widely planted in Europe, the pacific Northwest of North America, in South America, particularly Chile, and in Australia and New Zealand (Volk *et al.*, 2013).

### Veten

This Norwegian cultivar has been selected from a cross between Asker x Lloyd George (Boicheva, 1985). The fruits are moderately large (about 3 g), dark red, elongated and have good flavour. It is stout and high yielder. Veten is an early cultivar with fairly good winter-hardiness.

### Vintage

A high-yielding, glossy, bright-coloured, primocane-fruiting raspberry for fresh market with excellent flavour. Yield similar or higher than Heritage.

## Willamette

Willamette is one of the most important red raspberry cultivar being ideally suited for processing and juice making (Daubeny *et al.*, 1989; Bergamini and Ramina, 1980). The dark purple-red fruits are easily removed on ripening and thus can be harvested mechanically. The plant is moderately vigorous and canes have the least amount of branching. The cultivar is precocious, heavy yielder and heat tolerant. Willamette is resistant to cane botrytis, cane spot, powdery mildew and to crown gall, but susceptible to spur blight and bacterial blight.

## Yellow Antwerp

It is a mid-season cultivar having the finest and sweetest yellow fruits. The fruits are large, round, with good keeping quality. The vigorous plants are highly productive and fairly tolerant to pests and diseases. The cultivar is, however, not very popular.

The commercial evaluation of the raspberry cultivar depends not only on the high yield but also on size, colour, firmness and chemical composition of the berries (Wieniarska *et al.*, 1982). The fruit traits are related more to the genotype of the cultivar than to the environment (Potter, 1972). The cvs. Newburgh, Phoenix, Rubin. Cayuga, Malling Exploit, Malling Promise, Taylor and Viking showed good yields of 8.1-9.8 t/ha in Romania (Tudor, 1979). These cultivars are good in frost resistance, berry size and sucker formation. In a trial at northern Ireland, to compare the performance of Glen Clova, Malling Delight, Glen Isla and Malling Admiral; Malling Delight significantly out-yielded the other cultivars over 3 cropping seasons with an average annual yield equivalent to 18.5 t/ha (Watters, 1983). In New Zealand, the cultivars Sentry, Citadel, Avon and Southland showed good yield in the third harvest season. However, Sentry had unacceptable fruit characters. 'Citadel' showed promise, 'Avon' had a very concentrated harvest and 'Southland' produced a crop in autumn as well as in summer (Smale and Brewer, 1983). Janda and Gavrilovic (1983), in evaluation trial with 8 Yugoslovian and 32 introduced cultivars of red and black raspberries, recommended cultivation of cvs.Malling Exploit, Willamette and Gradina in the Zapadna Morava valley of Yugoslavia. However, the cvs. Podgorina, Gradina, Willamette and Malling Exploit were recommended for general commercial cultivation in most parts of Yugoslavia (Bugarcic *et al.*, 1984). In Poland, Wieniarska *et al.* (1982) found high yields in cvs. Glen Clova and Canby. Fruits were found larger in cvs. Veten, Malling Seedling Z and Glen Clova while the fruits of Glen Clova, Malling Admiral, Malling Orion and Newburgh showed higher vitamin C content (Table 16.4).

# Climate

Raspberry is essentially a crop of cool climate but some cultivars can be grown satisfactorily under subtropical conditions. In comparison with other temperate fruits, it requires mild winter and summer seasons. Williams (1960) studied the dormancy and flower initiation under controlled environment and found that the cultivars Malling Promise and Lloyd George needed at least a 3-week low-temperature (3.3°C) period of dormancy while high summer temperatures prolonged it (Mage, 1975). Duration of dormancy depends on the cultivar, geographical source and position of bud on the cane.

**Table 16.4: Yield, Fruit Weight and Vitamin C Content of Fruit of 10 Raspberry Cultivars at Felin Field Station of Poland***

| *Cultivar* | *Total Yield (kg/plot) for 3 Years (1978-80)* | *Fruit Weight (g)* | *Vitamin C Content (mg per cent)* |
|---|---|---|---|
| Canby | 110.9 | 2.9 | 23.0 |
| Glen Clova | 116.3 | 3.7 | 31.2 |
| Malling Admiral | 92.6 | 3.3 | 26.3 |
| Malling Jewel | 73.2 | 3.1 | 17.3 |
| Malling Orion | 64.1 | 3.4 | 26.3 |
| Malling Seedling | 96.3 | 3.8 | 18.5 |
| Newburgh | 68.4 | 3.0 | 25.9 |
| Norna | 101.2 | 3.2 | 17.5 |
| September | 65.1 | 2.5 | 18.6 |
| Veten | 85.3 | 3.8 | 26.1 |

* Wieniarska *et al.* (1982)

In regions where sufficient chilling units are lacking, the budbreak can be induced by several chemicals, *viz.*, DNOC (5 per cent), $KNO_3$ (5 per cent), thiourea (1 per cent) or cyanamide (4 per cent). Snir (1983) obtained the best effect with cyanamide which also increased fruit size, earliness and yield.

Raspberries frequently suffer from winter injury and the extent of damage is determined by the cultivar, poor shoot maturation, low autumn and winter temperatures, insufficient or excessive snow cover, premature thaws and changes in shoot water content (Elsakova and Yaroslavets, 1978). Redalen (1986) found the least winter injury in cvs.Asher, Malling Leo, Lloyd George and Nootka. To protect from winter injury the plants are covered with soil or bent to ground securing in position with wire wickets. Soil is also placed on the tip.

Protected cultivation of raspberry in soilless system is gaining popularity. The root systems of plants in pots are exposed to lower winter temperatures than the ones grown in soil. As the roots are usually more susceptible to frost than the canes, they may suffer winter injury in areas with low temperatures. Palonen *et al.* (2016) reported that the roots of the Canadian cultivars Ottawa and Muskoka and the Finish cultivar Maurin Makea were most frost hardy and could tolerate almost 10°C lower temperatures than the roots of the Scottish cultivar 'Glen Ample'.

Since the dormant period is short in raspberry, the yield is often affected by frost. The cultivars Malling Exploit, Malling Promise and Zeva have been found frost-hardy, whereas cvs.Tragilo, Glen Clova and Malling Jewel were found frost-susceptible (Terrettaz, 1978). Frost hardiness was improved with the application of ABA (abscisic acid), and BA (benzyladenine) (Zarly, 1978).

According to Arnaudova and Angelov (1974), the raspberry cultivars begin growth as the mean daily temperature rises above 5°C and effective temperature sums (from the start of growth to start of ripening) are: Malling Enterprise 694.1°; Malling Promise 707.9°; Malling Exploit 723.1° and Newburgh 743.9°. The combination of high temperature and high nitrogen enhances both vegetative growth and flowering (Locksin and Elfving, 1981). Hot and dry summer and severe winter are not conducive to good performance of raspberry.

Pre-planting primocane fruiting root treatment by *Bacillus subtilis* showed better adaptations to winter frost in western Siberia (Belyaev *et al.*, 2017).

## Soil

Raspberries grow successfully in a wide range of soil types from sandy loam to clay when rich in organic matter. The character of subsoil is more important than the type of surface soil. The subsoil should be deep and well-drained. If the root growth is restricted due to impervious subsoil or high water table, the plants may be damaged during a drought. Moderately acidic soils (pH 5.8-6.7) are desirable for majority of cultivars. Best growth occurs on moderately well-drained soil with good water holding capacity and high organic matter content that is greater than 3 per cent (Watson *et al.*, 1980). In Troyan area of Bulgaria, raspberries showed good growth on deluvial soils but poor growth on pseudopodzolic soils (Kirov, 1983). The unfavourable profile structure of pseudopodzolic soil with temporary surface water-logging make them unsuitable for growing red raspberries (Pironkova and Hristov, 1998).

## Propagation

Raspberries are commercially propagated vegetatively, and seed propagation is adopted only in breeding.

### Seed Propagation

Newmex (1985) reported that seeds did not have embryonic dormancy but the germination was physically prevented by the testa/or endosperm. When the seed-coat was removed, injured or scarified, 100 per cent germination was obtained within 15 days at 20°C. The cultivars differ in their temperature requirement for seed germination. Scarification is done by using concentrated sulphuric acid (Jennings and Tulloch, 1965). Pre-germination treatment of seeds with $GA_3$ (500 ppm) also increased germination (Jennings and Tulloch, 1965). Seeds from fully mature fruits germinate as readily as from ripe, fruits (Ourecky, 1975).

### Vegetative Propagation

Vegetative propagation is done by root suckers, root cuttings and tip layers.

#### Suckers

Red raspberries produce suckers in abundance during growing season which are used for planting. The suckers are separated along with a piece of old wood and planted in spring. Sucker production can be encouraged by inserting a spade deeply at interval injuring the roots of old plants or by mulching (Hartmann and

Kester, 1959). Actively growing young suckers are used for planting in moderately rainy weather.

## Cutting

Raspberry can be propagated from detached segments of roots used as root cuttings (Hudson, 1954; Heydecker and Marston, 1968; Torre and Barritt, 1979).

Red raspberry cuttings have long been found difficult or impossible to propagate from above ground shoots (Stoutemyer *et al.*, 1933; Hartmann and Kester, 1983). Red raspberry cuttings with etiolated bases arising from root cuttings or obtained by etiolating above-ground shoots of both English and American cultivars are capable of rooting (Williams and Hudson, 1956; Williams and Norton, 1959). Howard *et al.* (1987), in their experiment with leafy cuttings of the red raspberry cvs. Malling Admiral, Malling Delight, Malling Leo and Malling Orion, taken from axillary shoots of the current year's new cane, reported that the relatively stress-free environment of a ventilated high humidity 'fog' system enabled etiolated, blanched and normal (control) cuttings of the relatively free-rooting Malling Leo to root within the range of 62 to 77 per cent. In a potentially stressful 'mist' environment, needing heavy shade, cuttings from etiolated shoots gave 73 per cent rooting but normal cuttings failed to root. The difficult-to-root Malling Admiral required the combination of pre-etiolation and fog to root at 50 per cent.

Black and purple raspberries are propagated as a rule by tip-layering. These berries have arched canes that root at the tips, the black rooting more easily than the purple (Sudds, 1935).

## Micropropagation

Planting of healthy stock is essential for the successful growing of raspberries. Sufficient information is available on their micropropagation elucidating the medium, quantity of nutrients, shoot portion to be used, light and temperature requirements and heat treatment to get rid of viruses (Anderson, 1980; Baumann, 1981; Hedtrich *et al.*, 1980; Snir, 1981; Welander, 1985). Snir (1981) reported the *in vitro* method for propagating 50000 plantlets in a year from one bud tip. The superiority of Anderson's medium over MS medium was reported by Anderson (1980) and Avita and Barrientos (1985) but contradicted by Desjardins and Gosselin (1987). Desjardins and Gosselin (1987) obtained higher proliferation rates of the red raspberry cv. Madawaska on MS medium with 1mg/l BAP and 50 mg/l ascorbic acid. The *in vitro* rooting of red raspberry was reported as being genotype depended (James *et al.*, 1980). Rooting can be achieved spontaneously in cultures or on rooting medium with reduced to half MS medium macronutrients and the addition of 0.1-1.6 mg/l IBA (Welander, 1985). Addition of 0.1 μm phloroglucinol improve root formation (James *et al.*, 1980). Kiss *et al.* (1984) found a modified nutrient medium containing Morel macronutrients, MS nutrients and BA suitable for shoot proliferation. Nitsch and Nitsch medium with 0.1-1.0 mg/litre IBA was good for rooting. Better proliferation is obtained with nodal segments than shoot tips (Welainder, 1985). Sakin *et al.* (1990) stated that the modified MS medium can be successfully used in *in vitro* propagation of raspberry. They suggested using 0.5 + 0.5 mg/l of IBA + BA during the initial growth stage and 0.5 + 1.0 mg/l of IBA +

BA during the proliferation stage. Hedtrich *et al.* (1980) suggested long days of 16 hour at 24°C day and 20°C night temperatures for *in vitro* propagation. Plantlets grown under fluorescent lamps showed better responses concerning the number of shoots and leaves and shoot length, whereas those grown under LED light showed higher dry weight and longer root length (Poncetta *et al.*, 2017).

For production of 100 per cent virus-free plants, a combination of meristem culture and weekly heat treatment is required (Baumann, 1981). Pyott (1981) gave 8 weeks' heat treatment at 37°C to shoot explants for eliminating the virus. He got more than 75 per cent rooting in plantlets when dusted with rooting powder (0.1 per cent IBA) and placed under intermittent mist.

Ruzic and Lazic (2004) multiplied raspberry cv. Willamette on MS media containing 2 x $FeSO_4$ x $7H_2O$. In the rooting phase, half-strength of MS media with 1 and 2 mg IBA, 0.1 mg $GA_3$ and 1.0g activated charcoal/l were used. Rooted plantlets were successfully (96 per cent) aeclimatized under a mist system. Popescu and Isac (2000) reported high frequency shoot regeneration from leaf-derived callus in raspberry. They induced with high efficiency by indirect and direct organogenesis, when cultured on Lee and Fossard medium. The percentage of leaf cytokinin/auxin ratio and a significantly higher frequency of shoot regeneration was obtained from leaf derived cali, reaching 100 per cent when 2,4-D at 2.2 μM and BA at 13.3 μM were present in the culture medium, as compared with 46.7-73.3 per cent for direct organogenesis. Not only the frequency of regenerating explants, but also the number of shoots formed per calls (with an average of 3.8) was significantly higher in the case of indirect organogenesis as compared with direct organogenesis. Dziadczyk *et al.* (2013) stated that *in vitro* culture conditions influence the initiation of raspberry cell suspension culture. They used established callus culture obtained from leaf explants as inoculums for cell culture initiation. Different combinations of IAA, NAA, BAP and 2, 4-D were added to MS medium in order to get the callus culture suitable for initiation of a cell suspension. The best callus (vigorously growing, healthy and friable) was obtained on the medium supplemented with 4.0 mg/l IAA and 1.0 mg/l BAP. The best culture medium for induction of raspberry cv. Nawojka cell suspension was observed when supplemented with 1.0 mg/l 2,4-D. Also the medium with 8.0 mg/l IAA and 1.0 mg BAP was similarly efficient.

## Planting

A thorough preparation of the soil is essential for raspberry planting which is achieved by ploughing to a depth of 15 cm in early spring followed by disking and harrowing immediately before setting the plants. An insecticide can be mixed in the soil in case of a danger of grub's damage. Pinkerton *et al.* (2002) suggested soil solarisation between planting for the management of soil born plant pathogens. In Pacific Northwest, soil solarisation from mid-July to mid-September produced more numbers of healthy primocanes in the second year of 'Qualium' red raspberry compared to non-solarized plots. Red raspberry plants growing in solarised soil yielded more fruit than plants growing in non-solarized soil in the third year after solarisation. Planting distance in raspberry depends on the training system and the type of cultivating equipments. Two training systems, *viz.*, hill (stool) and hedgerow (solid row) are used for raspberry.

Summer-bearing red raspberries can be grown in hill systems on raised beds or flat ground. The canes are trained by tying them together in bundles, which then require some form of support. Hills usually are established by setting traditionally propagated or nursery-matured planting stock 75 cm apart in rows and 2.0 to 2.5 m between rows. Confine hills to a diameter of about 30 to 45cm by removing all suckers that develop between hills or in aisles. Young plantations of cv. Tulameen when grown on raised beds, the use of black plastic films helped to reduce soil water content in the rooting zone. Fruit yield per cane was 110 per cent higher on raised beds covered by black plastic film, compared to uncovered ones (Neuweiler and Heller, 2000). In hill system, the plants are spaced sufficiently apart in rows so that cultivation can be done on all four sides. Each hill consists of 5 or 6 vigorous canes that are tied firmly to a central stake. All suckers arising between the hills are removed. This system is recommended for small planting of black and purple raspberries.

In the hedgerow system, the plants are grown in a continuous row which can be cultivated on two sides. In this system, the competition between plants is greater than in the hill system. Hedgerows are best established by planting rooted plants or root cuttings about 60 cm apart in rows. Primocane fruiting raspberries typically are grown in 30 to 45 cm wide hedgerows, whereas summer-bearing hedgerows are kept about 30 cm wide. The hedgerows width is maintained through hand pruning, mowing, chemical cane burning, or rototilling along the edges of the rows.

In the valley of Mexico, a density of 40 canes/$m^2$ was suggested for multi-annual hedge plantations as they ensure the highest yields (199 kg/$m^2$), without affecting fruit quality or the vegetative growth of the cane (Alvarado-Raya *et al.*, 2016).

The suckers arising in the rows are not removed. The distance between rows is decided by the equipment used for cultivation. In a comparison of the hedgerow and stool systems, Mason (1981a) could not find significant yield differences.

Terrettaz and Carron (1987) have recommended spacing of 2.40 m between rows and 60 cm within the row. Kingston and Epenhuijsen (1987) concluded that an interrow spacing of about 1.60 m gave the highest yield but that a minimum spacing of 2 m was required for mechanical harvesting. Initially, closer spacing gives higher yield which subsequently levels out (Jankovic *et al.*, 1981).

Jordan and Ince (1985) got 7.4, 4.2 and 3.6 tonnes fruit yield per hectare from planting of the mother plants with multicane bases and large crowns, immature suckers and mature suckers, respectively. These yield variations were caused by the differences in the production of fruiting canes.

Raspberries are usually planted in spring but can also be set during rainy season. The planting stock needs to be heeled to protect the roots from drying. The plants are headed back to about 15 cm before setting. For bare-root plants, set the plant so that the highest point of attachment of roots to cane is 2.5 to 5 cm below the ground level.

## Training and Pruning

Hill planted raspberries are staked in the second year after setting the plants.

The canes are tied to the stakes at a point halfway between the ground and tips of the canes. However, hedgerow is more common and canes are trained to wire trellises which are strung between posts, set 5 to 10 m apart in the rows. In a 4-year pruning trial, the highest yield was produced by American 'V' at 45° system and the lowest by the conventional hedgerow system (Watters, 1981). Keepel (1984) has described four systems of cane support. The vertical single wire frames are used with cane growing between them, but new shoots make harvesting difficult. Wire trellis with movable wire girds are used in biennial cultures doing away with the tying in of canes. V-cultures using angled posts with the canes tied to wires singly or in bundles of four, provide protection for the- new shoot growing in the middle.

Raspberries are pruned (topped) to keep them from growing too tall by cutting back the ends of the canes as they reach proper height. Kollanyi (1975) did not find any significant yield difference by reducing plant height from 3 per cent to 39 per cent, but berry weight increased in proportion to pruning severity. Terrettaz and Carron (1980) got 47 per cent more yield by increasing plant height from 120 to 180 cm and a decrease of 60 per cent in individual fruit weight. When cane density was increased from 6 to 10 canes per metre, the yield also increased by 28 per cent, with a simultaneous 10 per cent reduction in fruit weight. The cane height of 180 cm adversely affected the ease and speed of picking. Rebandel and Przysiecka (1981) studied the effect of pruning and $GA_3$ on the growth and cropping of raspberry. The number of fruits per cane was highest on pruning the cultivar Malling Promise at 90 cm and treating with $GA_3$ (200-250 ppm in early June), in Latham at 100 cm and in cultivar Malling Exploit at 90 cm without $GA_3$ treatment.

Raspberry canes are biennial as they grow in the first year, fruit in the second and then die. Fruiting canes are cut down after harvest. New canes grow from buds on the base of old canes and suckers arise directly from the roots. By removing weak new canes and suckers, and keeping about 7 strong canes per hill, the yield increases from 1.5 to 2.2 times (Sidorovich, 1985). Sidorovich (1985) compared mannual, chemical and mechanical methods of removing young shoots and found that the most effective mechanical method was the high pressure, (100-200 atmosphere) water jet when the shoots were 15-20 cm high or by repeated cuttings (3 or 4 times). Rebandel and Przysiecka (1981) reported the highest yield on the removal of young shoots on 22 May than either on 10 May or 12 June. The cutting down of all young canes in the first year than leaving 8 canes per plant resulted in better yield in the following year (Rebandel *et al.*, 1985).

Nehrbas and Pritts (1986) examined the relationship between light distribution primocane growth and floricane yield components in Royalty purple raspberry and Titan red raspberry under 3 pruning regimes. They found that canopy light penetration was positively associated with flower number per lateral and yield in 'Royalty', and with flower number per lateral, berry weight and earliness of ripening in 'Titan'. Early primocane suppression resulted in an increase in berry size and yield in both the cultivars relative to controls, although 'Titan' was found more responsive. Fruiting plants in the alternate row sowing system had more floricanes and produced greater yield than controls, but berry size was reduced. This reflects the positive direct effect of cane number on yield but negative effects on all other yield components. Jordan and Ince (1986) tipped (cane apex removed), the primocane of field-grown raspberry cv. Heritage during early growth in the first and second season after planting in annual production system. Tipping times, at 3- to 4-week interval, were compared with untipped plants. Early tipping (late November) in the first season was found to increase the total yield from 5.44 to 7.51 t/ha. Later tipping did not affect total yield, but delayed the production peak and improved late season's yields. Second season's total yield was reduced from 16.5 to 9.5 t/ha by the latest tipping (mid-November and early December).

Yield per cane increased with planting distance, but yield per hectare was reduced if either the minimum cane density was not reached or the increased yield per cane was not sufficient to compensate for the wider planting distance (Stirk and Cahn, 1999). In North-America, Buszard (1986) studies cane densities as high as 10 canes/m$^2$ for three cultivars grown in a hedgerow, and obtained maximum yield with $\approx$ 3.3 canes/m$^2$. Lopping produced higher yields than topping, but slightly reduced fruit size and cold hardiness. Hedgerows are more difficult to train. Biennial bearing may occurred in the hedgerow systems, due to greater competition between primocanes and floricanes for light (Stirk and Cahn, 1999).

In dense raspberry canopies, the simultaneous existence of both primocanes and floricanes results in strong competition for light between the canes. This may cause increased disease incidence because of reduced air circulation, slower drying, and limited penetration of pesticides, as well as lower picking efficiency. Vanden Heuvel *et al.* (2000) applied trellising and cane density treatment to vigorous raspberry cv. Titan plants in Southern Ontario. The trellising treatments were compared with hedgerow (control), V-trellis, and single-sided shift trellis. Cane densities (9,16, 23 and 30 canes/m$^2$) were established before bud break. In the second season, plants trained on the V-trellis yielded more than did those in a hedgerow system, while those trained on a shift-trellis had the lowest yield because of a small fruiting framework. Fruit quality was lower on the shift-trellis than in other two systems. These differences were related to light penetration into the three canopies. The maximum cane density (30 canes/m$^2$) produced the highest yield, through an optimum cane density was not found for any system. Yield potential per cane decreased as cane density increased, however, the increase in cane numbers compensated for this decrease, resulting in an increase in total yield per unit area. Fruit quality also decreased as cane density increased, but cane density did not affect canopy microclimate.

Oliveira *et al.* (2004) working with raspberry cv. Autumn Bliss recorded higher yield with cane densities of 16 and 24/m row. They stated that light condition is the most important environmental conditions affecting productivity in greenhouse grown raspberry. Summer pruning of primocane-fruiting red raspberry is an effective technique to induce late season production under non-heated greenhouse in mild-winter climates. In Southern Portugal, 'Autumn Bliss' yielded most when pruned at 15 nodes. Late pruning (15 August) in the second year reduced fruit yield without affecting cane growth. Although longer pruning shortened the cropping cycle, pruning early (16 July) is crucial to avoid yield loss due to the onset of low-light autumn conditions. Longer pruning induced a higher number of shorter laterals.

This study made by Oliveira *et al.* (1999) confirms the possibility of achieving a second crop in November-December when primocanes are cut immediately after the harvest of the normal summer crop.

## Biennial Cropping

In view of the difficulties encountered during mechanical harvesting due to the presence of both vegetative and fruiting canes, the biennial cropping has been tried successfully as an alternate raspberry production system (Clark, 1984). The advantages of biennial system are the elimination of selective hand pruning and decreased harvest cost while higher establishment cost is the main disadvantage of the system. In the cultivar Glen Clova, Wright and Waister (1982) compared the growth and yield of raspberry fruiting canes in annual, partly biennial and biennial systems. They reported the highest yield in biennial system and associated it with the increased number of fruits per cane. The fruit size remained unaffected. Increased leaf area and better light condition in biennial system also accounted for the higher yield. Clark (1984) has described the establishment of full biennial system by using dinoseb-in-oil [12 l DNBP: (20gl$^{-1}$) + 6 l summer spraying oil 100 l$^{-1}$ ha$^{-1}$] on four occasions during the growing season. He demonstrated 50 per cent increase in yield over the annual system (Table 16.5).

**Table 16.5: Fresh Weight of Raspberries Harvested from 1981/82 Trial***

| *Production System and Cane Density* | *Mean Fresh Weight of Fruit (kg m$^{-1}$ of row)* | *l.s.d (0.05)* |
|---|---|---|
| Trellised annual | 1.53 | 0.45 |
| Trellised fully biennial | 2.28 | |
| 5 canes m$^{-1}$ | 1.21 | 0.35 |
| 10 canes m$^{-1}$ | 1.84 | |
| 15 canes m$^{-1}$ | 2.38 | |
| 20 canes m$^{-1}$ | 2.12 | |

| *Production System* | *Fresh Weight of Fruit (kg m$^{-1}$ of row) Cane density* | | | |
|---|---|---|---|---|
| | *5* | *10* | *15* | *20* |
| Trellised annual | | | | |
| Block I | 0.77 | 1.17 | 1.72 | 2.00 |
| Block 2 | 1.00 | 1.36 | 1.92 | 2.14 |
| Block 3 | 0.95 | 1.66 | 1.87 | 1.73 |
| Sub-plot means | 0.91 | 1.40 | 1.84 | 1.96 |
| Trellised fully biennial | | | | |
| Block 1 | 1.66 | 2.40 | 3.06 | 1.86 |
| Block 2 | 1.44 | 2.49 | 3.08 | 2.95 |
| Block 3 | 1.81 | 1.93 | 2.61 | 2.06 |
| Sub-plot means | 1.64 | 2.27 | 2.92 | 2.29 |

* Clark (1984)

## Nutrition

### Role of Nutrients

Raspberry requires annual fertilizer application for good vegetative growth and fruiting.

#### Nitrogen

For good vegetative growth and fruiting, raspberry requires nitrogen application every year. Its deficiency causes uniform yellowing of older leaves which later turn bright yellow with necrotic tips and margins before falling. The cane growth declines and red colouration of canes and leaves may occur (Davis, 1942). The growth of raspberry plants reduces considerably when leaf N content falls to 2.0 per cent (Ljones, 1966, Naumann, 1961). Barring few reports of low yield, N application has generally increased the yield and total growth of canes (De Gomez *et al.*, 1986; Papp *et al.*, 1982). The yields were lower both in absence and excess application of N (Quast, 1982). Higher N application delays fruit ripening, reduces sugar and ascorbic acid content of berries, decreases frost resistance and increases the incidence of spur blight (Goode, 1970; Papp, 1973).

Raspberries require a majority of their N early in the growing season. Application of N in early spring as soil surface bands of granular fertilizer (ammonium nitrate) enhances cane growth and N concentrations in fruiting clusters, laterals and leaves of fruiting canes (Kowalenko *et al.*, 2000). Laszlovszky and Smolarz (2000) reported higher mean fruit weight and yield of cv. Polana by application of 100 and 150 kg of N/ha. Primocane-fruiting (fall-bearing) raspberries positively respond to higher nitrogen doses in comparison to summer-bearing raspberries (Strik, 2008).

#### Phosphorus

As a result of P deficiency, unusually dark green leaves appear initially which subsequently turn to coppery purple. The leaves become small, stiff, curl and, in severe cases, die. The roots darken and the plant growth decreases (Naumann, 1961; Ramig and Vandeceveye, 1950). Raspberries do not respond to P application normally (Ljones, 1965).

#### Potassium

Reddish brown spots in the mesophyll region of the leaves, marginal leaf scorching and reddening of the canes are the usual symptoms of K deficiency in raspberry (Wallace, 1938; Harris, 1940). The reduction in plant growth is, however, not pronounced (Naumann, 1961). Acidic and dry soils may cause K deficiency even on N, P and K application (Medel and Vargas, 1983).

#### Calcium

The Ca deficiency symptoms include reduction in leaf size, hardening, curling and creamy white bands on either side of midrib (Davis, 1942: Ramig and Vandeceveye, 1950). Ca deficiency also causes dying back of the root tips and formation of swollen areas behind the dead portion forming a compact mass of short roots. Ramig and Vandeceveye (1950) reported 0.2 per cent Ca content in petiole as optimum for normal growth of the plants.

Calcium application along with NPK results in good plant growth and high yield (Smolarz and Mercik, 1983a,b). Ca sprays increase berry size and prolong their shelf life (Eaves *et al.*, 1972). It also reduces fungus rot during storage. However, Vance *et al.* (2017) working with raspberry cvs. Tulameen and Vintage and with different Ca formulations (Ca chloride ($CaCl_2$), $CaCl_2$ + boron, Ca silicate, Ca chelate and Ca, acetate) concluded that foliar application of calcium do not improve quality or shelf life of raspberry.

**Magnesium**

In sand culture experiments, Mg deficiency showed reduction of leaf pigments and interveinal chlorosis with puckering and rolling of leaf margins (Davis, 1942; Hinkle and Eisenegen,1950). Mg application increased Mg content in tissues but decreased tissue Ca and K content and sometimes P and N contents also (Borgs *et al.*, 1987). Besides increasing the yield, Mg also accentuates the ascorbic acid content of the berries (Powers, 1945).

**Boron**

The characteristic symptoms of B deficiency are dying back, delayed or complete failure of bud development on fruiting canes, distorted leaves with unusually long petioles, or leaves becoming small, indented and feathery (Askew *et al.*,1951a; Mont, 1955). Kowalenko (1981b) obtained 23 per cent higher yield with the application of 1 kg boron per hectare. In association with N, it also increased berry size and number. Askew *et et al.* (1951a,b) reported that B treatment improved the sugar content of the berries. B content of raspberry leaves is adversely affected by calcium nitrate application (Ljones, 1963). Inconsistent results of B application on yield have been reported by Chaplin and Martin (1980).

**Zinc**

Very little information is available on zinc requirement of raspberries except that fruit dry matter and carbohydrate of red raspberries increase with the foliar application of zinc sulphate (Askew *et al.*, 1951b). Leont'eva *et al.* (2015) reported maximum Zn and Cu accumulation in "Sputnitsa' raspberry roots (524±0.59 and 405±0.44 ppm dry weight, respectively). Under N90, P90 and K90 application Zn and Cu contents in the roots were found positively higher than without fertilization.

**Iron**

Marginal chlorosis of raspberry leaves has been reported in sand culture due to iron deficiency which can be corrected by Fe injections (Davis, 1942; Yan Haarlem, 1948). On an average, dry fruits contain about 62 ppm iron (Beeson, 1941).

**Manganese**

In case of Mn deficiency, interveinal chlorosis occurs. Beginning near the margins, it extends towards the midrib on lower leaves and later upper leaves are also affected besides die back of canes in winter (Wallace, 1951; Lamb, 1956). The deficiency can be corrected by spraying 2 per cent manganese sulphate twice, first in April and again in July. Beeson (1941) reported 46 ppm average Mn content in raspberry fruits on dry weight basis. Among the four microelements used, Mn had most pronounced effect on yield, carbohydrate, vitamin C and dry weight of berries (Harris, 1944) (Table 16.6).

**Table 16.6: Effect of Boron, Copper, Manganese and Zinc on Raspberries****

| *Treatments* | *Yield* | *Total Carbohydrate* | *Vitamin C* | *Dry Weight* |
|---|---|---|---|---|
| Boron | $0^+$ | $0^+$ | + | + |
| Copper | 0 | 0 | 0 | 0 |
| Manganese | + | + | + | + |
| Zinc | 0 | + | 0 | + |

+ Significant increase, 0 No significant increase.

$0^+$ Increase but not quite significant at 5 per cent level.

** Harris (1944).

## Molybdenum

In combination with chlormequat, Mo reduced leaf chlorosis, increased yield and improved fruit quality (Simonov and Kulikova, 1984).

## Chloride Toxicity

Similar to K deficiency, chloricle ions cause leaf scorch in raspberries which is enhanced by continuous use of KCl (Rogers *et al.*, 1936). Quast (1982) suggested application of $K_2SO_4$ in place of KCl to avoid chloride toxicity in raspberries.

## Fertilizer Recommendations

A number of fertilizer trials have been carried out in raspberries but without consistent results. Generally, N, P and K are applied together and sometimes Ca, Mg and B are also added to fertilizer mixtures. Raspberry responds to nitrogen and potassium fertilization (Stanton, 2013). Buskiene and Uselis (2008) obtained higher yield from cv. Polana by application of 120 kg N and 180 kg K/ha. Krawiec (2016) reported more vigorous plant growth, higher yield and better fruit quality with fertigation programme of 17-18 weekly applications starting from spring and ending at harvest. The rates used were 76-80 kg N, 19-21 kg P and 81-87 kg k/ha. Holubowicz and Rebandel (1975) obtained highest yield, from the cultivars Latham and Malling Promise by applying 40 t/ha organic mulch + 50 kg N, 25-50 kg $P_2O_5$, 75-100 kg $K_2O$ and 100 kg Mg/ha in autumn. 'Malling Promise' produced high yield (10.35 t/ha) when received N, $P_2O_5$, and $K_2O$ at 37.5:37.5:37.5 kg/ha + wuxal foliar spray (Subotic and Boskovic, 1979). In a long term fertilizer trial of three cultivars of raspberry, application of N, $P_2O_5$, $K_2O$ and Ca at 60:48:72 kg and 18 q/ha, respectively, resulted in good plant growth and high yield (Smolarz and Mercik, 1983b). The fertilizers should be applied in early spring (Pahoto, 1956).

## Tissue Analysis and Nutritional Status

Raspberries have a relatively low nutrient requirement. Research on the fate of applied $^{15}N$ (a naturally occurring isotope of N) has shown that primocane rely primarily on fertilizer N for growth, whereas floricane growth is highly dependent on stored N in the over-wintering primocanes, crown, and roots from 30 per cent to 40 per cent of stored N was allocated to new growth. Plants receiving higher rates of N fertilizer took up more N, often leading to higher N concentration in the tissues, including the fruit. Reallocation of N from senescing floricanes and primocane leaves to canes, crown and roots has been documented. Accumulation of other macro-and-micronutrients in plant parts usually preceded growth. Primocanes generally contained the highest concentration of most nutrients during the growing seasons, except Ca, Cu and Zn, which often were more concentrated in roots. Roots typically contained the highest concentration of all nutrients during winter dormancy (Stirk and Bryla, 2015).

Dierend and Stocker (2001a) studied the uptake of N, P, K, Mg and Ca of 4 raspberry cultivars in a field experiment at Germany. They reported average uptake (per hectare) of N, P, K, Mg and Ca of 82, 14, 79, 13 and 46 kg, respectively. Nutrient uptake was much higher for the vigorous growing cultivars Schoneman and Resa than for the lesser growing Himboqueen and Meeker. One-year-old shoots showed higher nutrient uptakes than 2-year-old shoots and fruits. Nutrient uptake was found higher from spring until the end of harvest. If the growers do not remove the bearing shoots and the surplus one-year-old ones, the nutrient removal from the field amounted to 36 kg N, 6 kg P, 17 kg K, 3 kg Mg and 10 kg Ca/ha/year. They (2001b) suggested that if the N min-reserve is more than 40 kg N/ha, then no N fertilizer is needed. If pruned shoots remain in the planting, only 10-15 kg $P_2O_5$, 20-30 kg $K_2O$ and 40 kg MgO/ha, need to be applied to maintain adequate supply of mineral nutrients. John *et al.* (1976) found that the nutrient status of the leaves was greatly affected by genotype, cane age and sampling time, and the maximum change in the leaf composition occurred during the periods of high metabolic activity in spring, late summer and autumn. They, however, recommended the sampling during the early-harvest when leaf composition was the most stable. Hughes *et al.* (1979) reported the least flux of nutrient concentrations in the last two weeks of August. According to them, the best time to sample leaves should coincide with the least internal nutritional status. The stability also varies with the element. On the basis of biweekly analysis, Kowatenko (1981a) demonstrated that leaf K on new canes was fairly stable between 10 and 15 August, P between 1 and 20 September and-Ca between 10 July and 15 August. He (1981c) further stated that relatively constant leaf tissue Mg concentration occurred between 10 July and 15 August and K concentrations between 15 August and 25 September which were promising dates for leaf sampling. The leaf-lamina samples taken from non-fruiting canes from late June to early August were better indicators of the nutritional status of the crop (Bould, 1968). Ramig and Vandeceveye (1950) found that the analysis of leaf blades was good for assessing N and Mg levels whereas petiole analysis was a better indicator of P, K and Ca status.

Galindo-Reyes *et al.* (2011) studied the tissue concentration and accumulation dynamics of N, P and K in the primocane pruning of raspberry cv. Malling Autumn Bliss grown under polyhouse and managed to produce two harvests per year. The leaf was the organ that showed the highest concentrations of N, P and K, with 52, 53 and 37 per cent of the total amounts extracted by the plant, and also showed the highest seasonal variation in N concentration across the year, thus suggesting the root as the main storage organ for N in raspberry. The vegetative organs of the plant (leaves, stems, crown and root) extracted 127, 30 and 50 kg/ha of N, P and K, respectively.

On the basis of nutritional survey, Medel and Vargus (1983) reported that raspberry leaves in Chile contained average levels of N (3.08 per cent), Fe (174.23 ppm), Cu (8.88 ppm), Zn (22.88 ppm), Mn (109.72 ppm), high levels of P (0.59 per cent) and low levels in K (0.5 per cent), Ca (034 per cent) and Mg (0.18 per cent). Optimum levels of various elements in leaves as reported by various investigators are N, 2.7 to 3.3 per cent; K, 1.2 to 1.6 per cent; Ca, 0.48 to 1.11 per cent; Mg, 0.23 to 0.47 per cent and B, 25 to 37 ppm (Papp *et al.*, 1982; Ljones, 1963, 1965, Wright, 1942).

## Irrigation

Irrigation of raspberry plantations is essential in dry regions and is often profitable even in humid regions. Irrigated plants make more vegetative growth and yield fruit over a longer period than unirrigated plants. Kongsrud (1976) created drought conditions by withholding water supply at different periods, and found that drought during ripening and picking reduced berry size and thereby yield, and also restricted cane height. From the results of the experiments on irrigation carried out for seven years, Bjurman and Ingvarsson (1985) suggested the following:

i) Irrigation during spring and summer is necessary only on light soils or during extremely dry conditions.

ii) Irrigation 3-4 weeks before harvest can be beneficial to berry development.

iii) Irrigation after harvest counteract dormancy and can result in winter damage.

iv) The weak growing cultivars respond better to irrigation than vigorous cultivars.

Terrettaz and Carron (1985) reported higher yield from sprinkler and drip irrigation than with no irrigation. Irrigation can be applied once a week during the fruiting season and once every 2 or 3 weeks during the rest of the dry season. Light sandy soil needs more frequent irrigation than heavier soils. DeBoer *et al.* (1983) in comparing the trickle and sprinkler irrigation systems over 3 years, observed that both the systems gave an average fruit yield of 1660 g/m row of the cv. Boyne. Total water requirement between May and September averaged 68 cm with an average irrigation requirement of 34 cm. Ivanov (1984) reported that drip irrigation increased yield by 2.4-19.2 per cent in comparison with sprinkler irrigations but fruit quality (mean fruit weight, solids and sugar contents) improved only slightly. Drip irrigations in raspberries, as compared with sprinkler irrigation, resulted in 16.0-20.8 per cent reduction in water requirement and 7.3-14.5 per cent less production costs per 100 kg fruit. Drip irrigation at 3-day interval and a 90 per cent rate of ETM (maximal evapotranspiration) gave the best result.

Buskiene and Petronis (2000) investigated drip irrigation efficiency during 1998-99 for autumn raspberry cultivar Polana, allowing raspberry to crop only on annual stalks (in autumn). The year 1998 was moderately wet and the crop was irrigated 3 times, while 1999 was dry and the crop was irrigated 7 times. In wet years, irrigation was found inefficient, while in dry years, the yield increased 3-fold, average berry mass 47.1 per cent, stem height 18.3 per cent and diameter 24.09 per cent. Cropping stem length was 3.2 times bigger than without irrigation. Kireva and Petkov (2003) also recorded 3-fold yield increase of cv. Lyulin under drip irrigation than without irrigation at Sofia, Bulgaria. Reduction of irrigation by 20 and 40 per cent decreased the yield by 11 and 22 per cent, respectively. Irrigation wings from pipelines and holes spaced at 3.0 cm providing water flow of 2.7 l/ha, ensured sufficient water quantities, uniform surface and soil distribution. Costs of establishing the plantation under microsprinkler irrigation or drip irrigation could be recovered in three years (Kireva and Petkov, 2003, Petrova-Branicheva, 2016).

## Soil Management

Lawson and Wiseman (1976) reported that dense weed cover reduced cane growth in spring-planted raspberries and caused considerable mortality of planting material if weeds remained beyond late June. In the second year, weeds similarly decreased the fruit yield. Since regular cultivation for controlling weeds is not possible due to shallow roots of the plants and cost factor, proper soil management becomes essential. Studies have demonstrated the benefits of monocot ground covers for weed control and erosion control (Freyman, 1989). Chamberlian and Crandall (1979) observed that a cover of red fescue or ryegrass between raspberry rows caused a decrease in cane growth and fruit yield one year but not the next, after irrigation and fertilization regimes were adjusted to compensate for competition from the cover crop. Clay and Davison (1976a) compared the effect of different soil management systems, *viz.*, cultivation, non-cultivation, straw mulching and the use of residual and a contact herbicide on the growth and cropping of raspberries. Straw mulch resulted in considerably higher yield and less cane production. They suggested that a non-cultivation system combined with careful regulation of sucker growth should lead to improved yield. Organic matter and P contents were highest at the surface of non-cultivated soils (Clay and Davison, 1976b). Pine bark mulching provides total control of weeds and raspberry suckers (Romero-Franco *et al.*, 2001). Use of pine bark mulching in acid soils could have positive effects on reduction of aluminium toxicity (Romero *et al.*, 2001). Biodegradable and compatible materials available in the trade name Mater-$B_1$® (produced using vegetable starch) that show mechanical properties similar to those of traditional plastic could be used as an alternative of plastic films (Girgenti *et al.*, 2012). Permanent sward, non-tillage, straw mulch and regular cultivation systems were also compared by Terrettaz and Catzeflis (1978). Mulching increased the fruit weight by 12 per cent but led to increased frost damage and fungal attack. Permanent sward reduced growth, improved cane ripening and decreased the incidence of frost damage and diseases. The other systems gave intermediate results.

Weedicides are very useful in controlling weeds in raspberry plantation. The yield, number and length of cane were more improved by chemical weed control than by hand weeding (Reithmeier *et al.*, 1978). They reported better weed control with spring application of domatol (simazine + aminotriazole) at 5 kg/ha in 10001 water than 2 or 3 applications per year of gramoxone (paraquat) at 3 l/ha in 6001 water. Terbacil and diuron were found to be promising herbicides for use in raspberry plantation in Hungary. Most weeds, except *Convolvulus* spp. and *Cirsium arvense* were killed by this herbicides, while sufficient, healthy raspberry shoots were formed. In old, weed-infested plantations terbacil at 2-3 kg/ha + diuron at 2 kg/ha + paraquat at 0.4-0.5 per cent were effective in weed control (Sirkany, 1981).

Nikolova and Petrov (1985) reported the best growth and production of new canes by season-long weed control with 1.32 kg pendimethalin + 1.5 kg simazine/ha. The fruit quality as measured by total sugars, titratable acidity and ascorbic acid contents was not impaired by the use of weedicides. In newly planted raspberries on clay soil, Hartley and Loader (1985) obtained good weed control by 2 kg terbacil, 3 kg simazine or 3 kg terbuthylazine/ha, but terbumenton + terbuthylazine was injurious to the plants. The rate of weedicides increased slightly on heavy textured soils. The weedicides should be sprayed before the emergence of weeds and new canes in early spring. The use of weedicides can be supplemented with mechanical cultivation, if required.

A large percentage of North American processing red raspberry is produced in maritime areas of western regions in Washington, British Columbia and Oregon. Most of these raspberry plantings are caneburned, they are treated with herbicides in early spring to remove the first-emerging primocanes in an attempt to reduce their competition with floricanes during flowering and fruiting and thereby increase fruit yield. Yushan *et al.* (2017) studied the effect of carfentrazone and oxyfluorfen applied for primocane management. They observed that caneburning provide early season weed control, however, the effect fell below 70 per cent in commercial field of cv. Meeker. Oxyfluorfen provided longer suppression of primocane growth than did carfentrazone, particularly in the plots of cv. Cascade Bounty. Oxyfluorfen, improved 3-year 'Meeker' average yield by 43 per cent from non-treated field, compared with a 29 per cent increase with carfentrazone.

## Flowering and Pollination

Two groups of cultivars with different growth-cycles are commonly recognised in red raspberry flowering: (a) 'annual-fruiting'-the entire cycle of vegetative growth, flowering and fruiting is normally completed in a single growing season (Carew *et al.*, 2000), and (b) 'biennial-fruiting'-the shoots (canes) have a 2-year life-cycle during which they pass through a sequences of seasonal phases involving vegetative growth, flower formation and fruiting (Sonsteby and Heide, 2008). However, a third group, 'tip flowering' type, is also sometimes considered (Dale, 2008).

The annual-fruiting cultivars are commonly referred as primocane-fruiting cultivars, and terms as autumn-fruiting, autumn-cropping, everbearing, and tip-fruiting are also variously used (Keep, 1988), whereas biennial-fruiting cultivars, sometimes referred as summer-cropping cultivars, are now often referred to as floricane-fruiting cultivars (Dale *et al.*, 2005). Sonsteby and Heide (2009), however, suggested to use the terns annual-and biennial-fruiting which adequately explain the essential difference, namely a 1-year or 2-year life-cycle.

Low temperature and/or short days are required for flowering in biennial-fruiting cultivars, the annual-fruiting cultivars flower freely at temperature as high as 30°C (with some exception *e.g.* in cv. Autumn Treasure in which flowering was delayed and reduced by temperature above 20°C). In some annual-fruiting cultivars also, flowering is advanced by long-day conditions. In biennial-fruiting cultivars, floral initiation is accompanied by growth cessation and bud dormancy, and since dormancy released requires exposure to low temperature or winter conditions, this confers a 2-year life cycle upon the shoots. Shoots originating from root adventitious buds in biennial cultivars are juvenile and must develop 15-20 leaves before they can be induced to flower. In contrast, the juvenile phase is absent in annual-fruiting cultivars which respond to a vernalization treatment already at the 5-leaf stage. In these cultivars, floral initiation starts at the tip and is then progressing basipetally, but only some buds on the cane will sprout and flower in the current season, while the remaining buds become dormant. In most cultivars, the number and proportion of annual-flowering buds increase with increasing temperature, being much larger at temperatures above, than below 20°C. The distinctive physiological characteristics of annual-fruiting raspberries are the ability to initiate flower buds under the high temperature and long-day conditions of summer and the absence of dormancy in a certain proportion of the initiated buds, leading to flowering in the first year (Sonsteby and Heide, 2012).

Neri *et al.* (2012) studied the flower differentiation of both primocane-fruitings (cvs. Erika, Dolomia, Lagorai and LO3) and floricane-fruiting (cvs. Tulameen and Violet) in Northern Italy. They observed that in primocane fruiting raspberry cultivars, the flowers differentiation quickly in July while in the floricane-fruiting cultivars the buds became reproductive in October. Notably, the cv. Lagorari showed an intermediated behaviour with a few buds differentiating in July and a high number in October.

Dale *et al.* (2012) stated that flower bud initiation in red raspberry is controlled by three systems: (1) temperature and day length; (2) a period of chilling; and (3) juvenility. They postulated that in the first system, flower bud initiation is primarily controlled by the maximum temperature become higher in longer day length, Within the species there appears to be continuous variation for the maximum temperature-being highest in primocane-fruiting genotypes, progressively lower in tip-fruiting, and early floricane-fruiting genotypes, and lowest in late floricane-fruiting genotypes. A period of chilling appears to irreversibly condition the primocane to initiate flower buds when grown below the maximum temperature at which initiation will occur, once the juvenile period is over. Hence, chilling is epistatic over juvenility, and juvenility is epistatic over temperature and daylength.

The inflorescence of red raspberry is a cyme, in which terminal flowers develop first, followed by the sequential development of flowers further down the inflorescence axial. The uppermost lateral buds produce only one-to-three inflorescence, the complexity of the flowering laterals increases gradually in buds at lower positions (Sonsteby and Heide, 2008, 2009; Heide and Sonsteby, 2011).

Red raspberries are self-fertile and do not require any cross-pollinizing cultivars for fertilization. However, it require insect transfer of pollens. The most active pollinators were found as honebees although bumblebees and hoverflies (Sprhidae) visited as well. The open shape of flowers also enables short-tongued pollinators to reach the nectar. Although flowers are considered self-fertile, the lack of entomophilous pollination resulted in a lower fruit set, higher fruit drop and lower fruit size (Prodorutti *et al.*, 2003). Robbing of flower buds by a highly invasive bumble bee (*Bombus terrestris*) reduces nectar availability for managed honey bees (*Apis mellifera*) and that could potentially affect honey production (Saez *et al.*, 2017).

Zurawicz (2016) recently reported that cross-pollination increased the number of drupelets in the fruits of red raspberry. Working with ten red raspberry cultivars (Canby, Glen, Ample, Laszka, Polana, Polka, Radziejowa, Schoneman, Vetan and Willamette) at Poland, the cultivars were self-and control cross-pollinated in a green house. The study revealed overwhelming advantages of cross-pollination over self-pollination. The average number of druplets in the fruits from cross-pollinated flowers ranged from 48.3 in 'Glen Ample' (76.9 per cent more druplets than in self-pollinated 'Glen Ample') to 92.9 in Radiziejowa (96.8 per cent more druplets than in self-pollinated 'Redziejowa'). Bees are more attracted to flowers of cultivars which simultaneously offered ample nectar and some pollen than to flowers with a rapid rate of pollen shedding, but low nectar yield (Schmidt *et al.*, 2015).

Each raspberry flowers contains 100 to 125 pistils, with each pistil representing one potential druplet. About 75 to 85 druplets must from in order to produce a normal-looking fruit. Insufficient pollination or fruit set within a flower leads to the development of small, misshapen, and crumbly fruit (Welch, 2007).

Pollen germination studies on 18 cultivars showed that the average germination was 75 per cent, the highest being 89 per cent and 88 per cent in cvs. Glen Clova and Schoneman 2-4, respectively, and the lowest of 58 per cent in cv. Norma. (Redalen, 1976). He further reported that a solution containing 15-20 per cent sucrose and 50-100 ppm boric acid in distilled water was the most satisfactory medium for raspberry pollen germination. In pollination experiments, the cultivar Willamette has been found highly fertile with an average fruit set of 84 per cent and showed no difference between open-self- and cross-pollination (Kuhan, 1987).

## Effect of Growth Regulators

Application of $GA_3$ and gibberesol ($GA_3$+ $GA_2$) at 50 ppm resulted in good development of fruiting laterals and increased the number of berries per inflorescence, fruit set and yield (Rebandel, 1981; Redalen, 1981). Chlormequat (CCC) at 1-1.2 per cent was sprayed to increase the strength of canes for growing without any support which reduced the cane height and increased the number of leaves and productivity (Koziovana *et al.*, 1986; Kuryata *et al.*, 1985). The ascorbic acid content was slightly reduced but fruit size was not affected, and no adverse effect was observed in the year following the application.

Treatment with ethephon (500, 1000 or 2000 ppm) at early stage of cane growth reduced cane diameter and height without affecting the number of buds per cane and lower concentrations also increased the number of flowers and buds (Redalen, 1986). However, ethephon in raspberry is mainly used to regulate the timing and intensity of fruit ripening, thereby decreasing the number of total pickings (Joliffe, 1975). After a week of ethephon (150 to 600 ppm) treatment, the amount of ripe fruits increased at harvest in cultivars Malling Jewel and Glen Clova (Knight, 1976). When ethephon (0.1 per cent) was applied four days before starting of harvest, it accelerated ripening and increased the yield without reducing the following year's crop (Kuryata *et al.*, 1988). Knight (1976) also reported uniform ripening in the cultivar Malling Jewel after 7 and 11 days of ascorbic acid (0.5 per cent) spray. Treatment with methyl jasmonate (MJ) developed an intense red colour in red raspberry cv. Autumn Bliss after 2 days of harvest prior to anthocyanin accumulation (Moro *et al.*, 2017). Preharvest MJ treatment increase flavanoids, particularly ellagic acid and quercetin. It can be useful to food industry to obtain berries with enhanced health promoting properties (Flores and Ruiz Castillo, 2015).

## Harvesting

Sjulin and Robbins (1987) stated that the strength of fruit adhesion to receptacle tissue decreases as fruits change colour and ripen. Visual classification of fruit colour according to the Natural Colour System (NCS) chart and by physical measurement of fruit adherence to the receptacle or fruit compression resistance may be used to identify the optimal harvest time. The light red colour stage corresponding to NCS code 3060-Y90R was identified as the optimal harvest stage for commercial fresh marketing of the 'Glen Ample' cultivar. Fruit harvest at this stage developed the same chemical and sensory quality as *in situ* matured fruits (Stavang *et al.*, 2015). Mason (1980) determined the efficiency of a mechanical harvester on the basis of berry ripeness which was estimated by berry's titratable acidity, weight and age. As the acidity decreased with age, the percentage of berry removal by harvester increased. The percentage of berries removed rose from 5 per cent for 10-day-old berries to 77.97 per cent for 36- to 46-day-old-berries and declined to 20-33 per cent for older ones. The berries should be picked when they are fully coloured and readily separate from the stem. Harvesting of firm, ripe and sound berries is important for a better return from the raspberry plantation. Both manual and mechanical harvesting are in practice but the latter is gradually replacing the former. Kingston and Pallesen (1986) reported on the basis of yield that Willamette cultivar was most suitable for mechanical harvest and Nootka, Glen Clova and Skeena for manual harvest.

The berries are handpicked by using the thumb, index finger and middle finger and are generally placed in the basket without holding in the hand. All the overripe, injured and decaying berries are discarded. The firm ripe fruits should be separated from the very ripe fruits at the time of picking. The firm fruits are suitable for marketing and fully ripened fruits for canning and freezing. Picking should be carried out twice per week but during hot and wet weather, it should be done on alternate days.

## Machine Harvesting

Matrtin, (1985) has described the mechanical harvesting and advocated cultural modifications for its improvement. Minimum field slope, uniform plant growth and long row are required for efficient mechanical harvesting. The trellis should be higher for mechanical picking than for hand picking which can be up to 2 m. Topham and Mason, (1991) demonstrated maximum amount of harvest when the first picking was delayed until 33 per cent of the potential crop had ripened and picking was done at an interval of 2 days. The harvesting machines have the container storage, pneumatic cleaning and inspection belt. The berries are detached by vibrating tines attached to the outer surface of rotating drums on each side of the row. Mason (1985) suggested the use of relatively low vibration frequency and short harvesting interval for improving the efficiency of the harvester. A prototype harvester for use in 6 ft spaced rows was developed at the Scottish Institute of Agriculture Engineering (Ramsay, 1984). Field trials with correctly adjusted shaking drum diameter, stroke and frequency, and shaking finger diameter, length and distribution pattern, achieved machine-harvested yields of 3.58, 3.50 and 2.44 tons/acre from 4-year-old Glen Clova, Glen Prosen and Malling Jewel, respectively. Working at 2 miles/h, the prototype was capable of harvesting 1 acre/h.

Canumir *et al.* (1999) reported that mechanical harvest outperforms manual harvest on areas bigger than 7 ha with yields over 10 t/ha per season. The machine harvester (Korvan 930 Koevette, used in Chile) showed 70 per cent efficiency and a fruit loss of 10 per cent. The investment towards machine harvester could be recovered in 3 to 4 years. Summer-bearing red raspberries are machine harvested every 2 or 3 days to minimize loss of ripe fruit. Thus, there may 10-15 harvest per season depending on climate and cultivar. In general, machine harvest caused 15 per cent loss of total yield compared with 5-7 per cent when the berries were handpicked. Rabcewicz *et al.* (2017) carried out a study to determine the possibility of harvesting raspberries with a pulsating air steam that causes the shoots of plants to vibrate. Two tractor-driven test devices were developed. The first one was a high capacity compressor, and the stream of air produced by it was directed onto one side of a row plants. The second one was a device producing two counter flowing air streams colliding in the middle of a row of plants. The frequency of the air pulses was adjusted steplessly by varying the rotational velocity of the shutters closing and opening the outlets of the fans. In the field trials with the device based on a compressor-generated air stream, two air pulse frequencies were used: 500 and 540 pluses per minute (8.3 and 9 Hz). The mean detachment force was 0.727 N. About 50 per cent of ripe raspberries were removed from the bushes. The results achieved during harvesting with the device operating on the principle of colliding two pulsating air streams were markedly better and put the harvesting effectiveness within the range achieved in the trials with combine harvester with mechanical finger shakers. For two pulse frequencies (8 and 9 Hz), almost 62 per cent of fruits were harvested.

## Yield

The yield in raspberry is governed by the cultivar, planting density, site, fruiting year, cropping season, cropping pattern and method of harvesting. The yield potential of each cultivar is different from the other. 'Malling Delight' is one of the high yielding cultivar (Keep, 1975; Dijkstra and Oosten, 1980). Higher yields are expected from high planting density, summer fruiting, annual cropping pattern and hand harvesting than low planting density, autumn (primocane) fruiting, biennial cropping pattern and mechanical harvesting. The yield in the second cropping year will be more than first cropping year. The cultivar Lyulin produced 6232 and 9829 kg/ha fruit in the first and second cropping year, respectively (Khristov, 1983). Winiarska *et al.* (1986) reported the yields of cvs.Canby, Malling Admiral, Norna, and Malling Seedling cultivars at full bearing as 13.8, 13.7, 13.0 and 9.20 tons/ha. Yield of 11 primocane fruiting raspberry cultivars were reported by Andrianjaka-Camps *et al.* (2016) is presented in Table 16.7.

**Table 16.7: Marketable Fruit Yield and Fruit Weight of 11 Primocane Fruiting Raspberry Cultivars Tested in a Production System on Substrate (2 pots m$^{-2}$) in 2012 in Conthey, Switzerland. All the cultivars produced fruits before end of August.**

| *Cultivar* | *Fruit Yield Pot$^{-1}$* | *Fruit Yield m$^{-2}$* | *Fruit Weight (g)* |
|---|---|---|---|
| Amira | 990$^{ab}$ | 1980$^{ab}$ | 5.24$^{a}$ |
| Autumn Bliss | 1018$^{ab}$ | 2036$^{ab}$ | 3.13$^{c}$ |
| Brillance | 449$^{b}$ | 898$^{b}$ | 4.85$^{abc}$ |
| Joan J | 899$^{ab}$ | 1799$^{ab}$ | 4.41$^{abc}$ |
| Erika | 1063$^{ab}$ | 2126$^{ab}$ | 5.21$^{a}$ |
| Heritage | 1118$^{ab}$ | 2237$^{ab}$ | 3.16$^{c}$ |
| Himbo Top | 1082$^{ab}$ | 2164$^{ab}$ | 5.16$^{a}$ |
| Imara | 1179$^{ab}$ | 2358$^{ab}$ | 5.44$^{a}$ |
| Kweli | 1504$^{a}$ | 3009$^{a}$ | 5.05$^{ab}$ |
| Polka | 743$^{b}$ | 1487$^{ab}$ | 3.68$^{bc}$ |
| Sugana | 1261$^{ab}$ | 2523$^{ab}$ | 5.00$^{ab}$ |

Means (n=3) followed by the different letters indicate a significant difference between the cultivars ($p<0.05$, Tukey test).

Adopted from Andrianjaka-Capms *et al.* (2016).

## Ripening

Red raspberry fruit have been classified as climacteric (Blanpied, 1972; Jennings, 1988) and non-climacteric (Kader, 1985; Veazie and Nonnecke, 1992) relative to colour development. Ethylene emanation and internal ethylene concentration in raspberry fruit from floricanes increased as fruit colour developed (Burdon and Sexton, 1990). Jennings (1988) reported that respiration of floricane produced raspberries decreased during ripening. Veazie and Nonnecke (1992) detected ethylene only after red pigment developed in cv. Heritage. Ethylene-forming enzyme activity commenced in drupes and receptacle tissue from fruit at the yellow and mottled stages, respectively. Ethylene also was involved in fruit softening (Robbins *et al.*, 1989). Ethylene production increased after soluble solids content increased and titratable acidity decreased substantially, indicating a lack of involvement of ethylene in these ripening changes (Veazie and Nonnecke, 1992).

Enhance fruit ripening was reflected by decreased values for titratable acidity, total phenolics and increased concentration of total anthocyanins and soluble solids (Kivi *et al.*, 2014). Progression to the red ripe stage (cvs. Glen Clova and Glen Prosen) was found accompanied by a reduction in the ordered cellulose and a dramatic reduction in pectin content and the degree of methyl-esterification. The softer fruit (cv. Glen Clova) exhibited greater reductions in both parameters, implicating increased pectin hydrolysis, as one of the main factors contributing to the differences in firmness between the cultivars (Stewart *et al.*, 2001).

During ripening (cv. Glen Clova) endo-β-1, 4-glucanase specific activity (per g fresh weight) increases approximately 15-fold. Highest activity was found associated with the surfaces of the receptacles where the weakening abscission zones are located (Iannetta *et al.*, 2000).

Amrita-Kumar and Ellis (2001) stated that in raspberry development of fruit colour and flavour are critically dependent on products of phenylpropanoid pathway. They reported that L-phenylalanine ammonia-lyase (PAL) is encoded in raspberry cv. Meeker by a family of two genes (*RiPAL1* and *RiPAL2*). RiPAL1 shares 80 per cent amino acid sequence similarity to *RiPAL2*, but phylogenetic analysis places *RiPAL1* and *RiPAL2* in different clusters within the plant PAL gene family. *RiPAL1* was associated with early fruit ripening events, whereas expression of *RiPAL2* correlated more with later stages of flower and fruit development.

Yellow raspberry fruits have reduced anthocyanin contents. Anthocyanidin synthase (*Ans*) catalyzes the conversion of leucoanthocyanidin to anthocyanidin, a key committed step in biosynthesis of anthocyanins. Molecular analysis of the *Ans* gene enabled to identify an inactive *ans* allele in a yellow fruit raspberry ("Anne"). A 5 bp insertion in the coding region was identified and designated as $ans^{+5}$. The insertion creates a premature stop codon resulting in a truncated protein of 264 amino acids, compared to 414 amino acid wild-type ANS protein. This mutation leads to loss of function of the encoded protein that might also result in transcriptional downregulation of *Ans* gene as a secondary effect, *i.e.*, nonsense-mediated mRNA decay. Further, this mutation results in loss of visible and detectable anthocyanin pigments. Functional characterization of raspberry *Ans/ans* alleles via complementation experiments in the *Arabidopsis thaliana idox* mutant supports the inactivity of encoded protein through $ans^{+5}$ and explains the proposed block in the anthocyanin biosynthetic pathway in raspberry (Rafique *et al.*, 2016).

## Storage

Raspberries are very perishable and have a very short life after harvest (2 to 5 days) due to their natural soft texture and sensitivity to mould and other pathogens. Rapid cooling (5°C) within an hour of harvest followed by refrigerated storage (-0.5 to 0°C and 90-95 per cent relative humidity) can keep the fruits 10-12 days.

Fresh fruit of raspberry changes colour during storage as anthocyanins develop and pH increases (Robbins and Sjulin, 1989). This colour change is a characteristics that detracts from good postharvest appearance. Robbins and More (1990) observed that during storage for 16 days at 0 or 4.5°C or storage for 8 days at 20°C, fresh raspberry (cvs. Chilcoin, Meeker, Nootka and Willamette) fruit become darker, less red, and more blue as recorded in L*a*b* CIE coordinates. Rates of change for cultivars during storage did not differ. Colour changes depended on temperature, with rates of change fastest at 20°C, especially during the first 4 days. Fruits stored at 0°C for 16 days were redder and less blue than stored at the 4.5°C. Nunes *et al.* (2003) stated that development of off-flavour is the primary limiting factor for raspberry stored at 10 or 15°C, and objectionable aroma is the limiting factor when stored at 20°C.

The response of three raspberry cultivars (Meeker, Chilliwack and Qualicum) to modified atmosphere (MA) conditions was studied by Toivonen *et al.* (1999). Three levels of atmospheric modifications were attained in polyethylene packages; (i) high 10 per cent $CO_2$, 5 per cent $O_2$, (ii) moderate 6 per cent $CO_2$, 10 per cent $O_2$, and (iii) slight 3 per cent $CO_2$, 15 per cent $O_2$. The three cultivars all showed a similar response for soluble solids and titratable acidity retention, which was not affected by MA over 7 days of storage at 1°C. Ripening, as measured by fruit colour, was delayed with increasing level of atmospheric modification. There was no difference between cultivars for delay of ripening with MA. High MA levels enhanced loss of firmness in 'Qualicum' and 'Chilliwack'. Volatile analysis of the juice from 'Qualicum' and 'Meeker' after 7 days storage showed that 'Qualicum' was prone to greater acetaldehyde and ethyleacetate accumulation than 'Meeker'. Ethyl acetate accumulation has been linked to off-flavours in MA-packaged raspberries. Moor *et al.* (2014) evaluated the different passive modified atmosphere package (MAP) (30μm LDPE bag) (Estiko, Estonia). The Xtend® (Stepac, Israel) and an active MAP (30μm LDPE bag, flushed with a gas mixture containing 10 per cent $O_2$ and 15 per cent $CO_2$) on the postharvest quality of raspberry cv. Polka. Fruits were initially stored for 3 days at 1.6°C, then a half of the bags were removed into simulated retail conditions (6°C) and a half remained at 1.6°C for another 24 hours. The raspberries stored at 1.6°C, showed the best quality in passively modified LDPE bags, since the fruit had higher soluble solids content and titratable acidity. The Xtend® bags was found most suitable for maintaining fruit quality.

Modified atmosphere packaging of fruits with the biodegradable non-commercial film showed lesser levels of acetaldehyde, ethyl acetate and ethanol evolution and maintain fresh weight in storage at 1±1°C for 5 days followed by 20±1°C for 3 days (Briano *et al.*, 2016).

Storage in control atmospheres (CA) (10 per cent $O_2$ + 15 per cent $CO_2$, 10 per cent $O_2$ + 31 per cent $CO_2$) suppressed rotting significantly. Raspberry cultivars with a lighter red colour with less blue like Malling Admiral and Glen Lyon kept a better colour under CA compared with a dark and more bluish cultivar like Veten. Total soluble solids were unchanged after one-week under CA storage while titratable acids decreased (Haffner *et al.*, 2002). Forney *et al.* (2016) assessed nine red raspberry cultivars and advanced selections for their potential market-life when held at 1°C in air or controlled atmosphere (CA, 12.5 kPa $CO_2$, 7.5 kPa $O_2$) with 95 per cent RH. Fruits decay of all genotypes was found suppressed by CA storage. The log of physiological breakdown (PB), which was expressed as juice leakage, ranged from 6 to 28 days in air-stored fruit and was increased by -1.0 to 9.7 days in CA-stored fruit depenting on genotype. Fruits of genotypes with the longest market-life when held in air or CA included 'AAC Eden', 'KO3-9', 'Glen Ample' and 'KO2-15' and those with the shortest market-life were 'Encore', 'Nova', 'KO2-14' and 'KO6-1'.

Postharvest fruit coating with chitosan (1 or 2 per cent) in solution for 5 minutes was found effective in maintaining titratable acidity and retarding respiration, ethylene production, weight loss and decay incidence of fruits at storage (0°C, 90 per cent RH) for 12 days. (Tezotto-Uliana *et al.*,2014). Fruits coated with *Aloe vera* gel and stored at 4°C maintained higher levels of antioxidant capacity, total anthocyanin and antioxidant enzymes during storage for 8 days (Hassanpour, 2015).

Pre-harvest spraey of *Bacillus amyloliquefaciens, Curtobacterium pusillum* or *Saccharomyces cerevisiae* suspensions were reported as potential alternative to control post-harvest decay caused by *Botrytis* and *Rhizopus* of strawberry (Antoniolli *et al.*, 2011). Pre-harvest spray of Shemer® (commercial formulation of the antagonistic yeast *Metschnikowia fruiticola*) allowed a good control of grey mould (Prodorutti *et al.*, 2008).

Brohier *et al.* (1983) stated that spraying of calcium chloride at 0.25 per cent and 0.50 per cent increased berry firmness and reduced ethylene levels at harvest. Sprayed fruit also remained firm when stored for 13 days at 3.5°C. Spayed *et al.* (1984) stored hand-harvested red raspberry cv. Mecker, at 4°, 14° or 25°C up to 12 days using fast release $SO_2$ generating pads (paper impregnated with sodium bisulphite which reacts with moisture from the fruit to form $SO_2$) or dual release pads (fast release pads and slow release pads with sodium bisulphite in plastic-lined paper pouches). Visual mould ratings were found lower for both $SO_2$ generators than control and fruits stored at 4°C with the generators were brighter with better red colour than controls but irregular bleaching of anthocyanins occurred at 14° or 25°C.

## Pests

### Raspberry Cane Midge (*Resseliella theobaldi*)

The eggs are laid and larvae feed under the epidermis of new canes with natural splits in the bark or where the bark has been damaged (Anon., 1978). The midge damage is usually associated with spur blight, cane blight and *Fusarium* spp. which results in wilting and yield reduction (Dalman and Malkki, 1986). Monitoring of pest with pheromone traps is possible as the sex pheromone has been identified and synthesize, though the sex pheromones is only attractive for males (Cross and Hall, 2006). The males are active in the morning (from 8 to 12 am) and after 5pm. Adult males normally do not fly into the traps either at night or on wet days (Sipos and Penzes, 2011). Both chemical and cultural control measures are undertaken. The insecticides, parathion methyl (0.1 per cent), phosalone (0.2 per cent) and fenitrothion (0.1 per cent) give good control. Neonicotinoid thiacloprid (Calypso) is effective against eggs and small larvae of the raspberry cane midge (Jaastad, 2007). Dalman and Malkki (1986) suggested mechanical removal of young canes at a height of 10-20 cm. The cultivar Glen Clova has been recommended for cultivation because it produces non-splitting canes in the second flush. A chalcidoid species of parasitoid, *Aprostechus epicharmus* was reported as the principal parasitiod of *R theobaldi* (Vetek *et al.*, 2006).

## Raspberry Beetle (*Byturus tomentosus, B. unicolor*)

The beetle causes considerable damage in United States, Canada, and eastern Scotland by larval feeding of fruits. They occasionally feed on developing druplets, causing deformed and discoloured fruit. A semiochemical trap (floral attachment) developed by Scottish Crop Research Institute, Scotland were found very attractive and efficient could be a useful tool in integrated pest management (Baroffio, 2012). The adult beetles and larvae can be destroyed by spraying zolone (0.2 per cent) or parathion methyl (0.15 per cent) between petal fall and first pink fruit stages (Taylor and Gordon, 1975; V'ltov, 1984). In Italy, spraying with deltamethrin (1.2 g/hl) was found to reduce the percentage of infested fruits from 1.4 per cent in untreated control to 0.2 per cent (Suss and Fava, 1981). Mospilan 20 SP showed good control (Abanreska *et al.*, 2000).

The potential use of volatile-enhanced while traps to "lure and kill" adult raspberry bettles is proposed as an alternative to insecticidal control (Woodford *et al.*, 2003)

## Raspberry Sawfly (*Monophadnoides geniculatus*)

The flies lay eggs on leaf tissues during flowering and larvae feed on underside of the leaves making elongated holes in between the veins. There are usually two generations in a year. The larvae should be destroyed by spraying 0.1 per cent parathion methyl or 0.2 per cent phosalone.

## Wingless Weevil (*Otiorhynchus* spp.)

Wingless weevils are widespread in red raspberry in several regions of Europe. In Scotland, the main species associated with raspberry is the clay-coloured weevil (*O. singularis*), but vine weevils (*O. sulcatus*) were also detected. In Italy, seven *Otiorhynchus spp.* were identified. The most abundant were *O. apenninus, O. sulcatus, O. armadillo, O. ovatus* and *O. globus*. Crop scouting and nocturnal beating were considered as the best methods to detect adult weevils (Gordon *et al.*, 2003).

The adult weevil damages raspberry plants by feeding during night. Gordon and Woodford (1986) tested the efficacy of various insecticides in controlling the weevil and recommended two night sprays of azinophos-methyl plus demeton-S-methylsulphon at 280 g a.i. or of fenitrothion at 550 ml a.i./ha.

The strawberry blossom weevil (*Anthonomus rubi*) is an important in open field raspberry crops of Switzerland. In certain fields more than 30 per cent of flowers are attacked by weevil and damaged flower buds which subsequently do not bear fruits (Linder *et al.*, 2011). Systematic insecticides are used to control the pest.

## Aphids (*Amphorophora rubi, A. sensorita, A. agathonica*)

Aphids are vectors for viruses which are serious problems in raspberry culture. To combat aphid attack, breeding of resistant cultivars has been taken up after identifying numerous dominant genes. A series of allels, $A_1$, $A_2$, $A_3$, $A_4$, *etc.*, have been found to confer resistance against various strains of *A. rubi* (Keep *et al.*, 1970). Kennedy *et al.* (1973) did not find immunity to *A. rubicola* in the cultivars tested by him but cvs.Canby, Lloyd George and New York 632 were immune to *A. agathonica*.

## Mites

Different species of spider mites: two spotted spider mite (*Tetranychus urticae*), McDaniel spider mite (*T. mcdanieli*), yellow spider mite (*Eotetranychus carpini*) and European red mite (*Panonychus ulmi*) were reported on red raspberry from United States. Leaf and bud mite (*Phyllocoptes gracilis*) causing serious damage on both leaves and fruits (Trandem *et al.*, 2011). Two spotted spider mite (TSSM) is a major pest on soil less floricane-fruiting raspberries grown in plastic tunnel. The very favourable climatic conditions in tunnels lead to big TSSM outbreaks. TSSM significantly reduced yield and fruit quality (Linder *et al.*, 2011). Population development of TSSM and predatory mite species was studied in Finland, Italy and Switzerland (Tuovinen *et al.*, 2000). The most important phytoseiid species reported were *Phytoseius macropilis* in Finland, *Amblyseius andersoni* in Italy and *A. andersoni*, *Typholodromus pyri* and *Euseius finlandicus* in Switzerland. Most of the mentioned predatory mites provide effective control. Tuovinen and Lindqvist (2015) reported that the releases of predatory mites *Neoseiulus cucumeris* reduced the numbers of *P. gracilis* on cv. Glen Ample.

The spider mite (*Neotetranychus rubi*) is another phytophagous mite common on raspberries in North-eastern Europe. It was also reported from north-eastern Italy on commercial raspberry. The predatory mite *Amblyseius andersoni* can keep *N. rubi* populations at low levels during spring but not during summer when it prefers TSSM (Malagnini *et al.*, 2009).

Single postharvest application of wettable sulphur and of the miticide spirodiclofen was both very effective and provides excellent control of leaf and bud mite. The timing of application does not pose a risk of pesticide residues (Linder *et al.*, 2008). Postharvest spraying with vegetable oils (rapseed oil emulsified with soft soap) or sulphur targeting overwintering females in the buds was reported as a very valuable component of a management strategy for the leaf and bud mite (Trandem *et al.*, 2011).

## Nematodes

Nematodes are harmful both directly and indirectly. Directly, they reduce plant growth and yield, and cause mortality by root rot. Indirectly, they transmit virus diseases. *Pratylenchus penetrans* caused 24 per cent plant mortality (McElory, 1977), a reduction of 66 per cent and 43 per cent in plant height and 47 per cent and 26 per cent in yield by *P. penetrans* and *Xiphenema bakeri* infection, respectively. McElory (1977) obtained 30 per cent increase in growth and yield with either soil injection of dibromochloropropane or foliar sprays of oxamyl applied 4 times at 2-week interval starting at full-leaf stage. Trudgill and Alphey (1976) reported good control of *Longidoris elongatus* and transmission of raspberry ring spot virus by applying dazomet and dichloropropane nematicide at 336 kg a.i./ha.

Root zone treatment of the cultivar Willamette by fenamiphos (8 and 16 kg a.i./ ha) through trickle irrigation system reduced the nematode population significantly and increased plant vigour and fruit yield. Application of organic mulches increase soil food web structure and the abundance of nematode antagonists (*e.g.*, predacious nematodes), resulting in reduced crop damage by *P. penetrans* (Forge and Kempler, 2009). Application of oxamyl reduced root *P. penetrans* population densities and was considered an important addition to current control methods used to manage *P. penetrans* (Zasada and Walters, 2016).

## Diseases

Diseases are more harmful to raspberry than insects and the diseases attacking the canes through mechanical wounds are more destructive. The incidence of diseases can be reduced by planting healthy stocks, choosing disease-resistant cultivars, selecting frost-free sites, protection from wind, removing old canes after harvest, destroying seriously diseased plants and using fungicides when needed. The cultivars Puyallup, Chilcotin, Royalty, and Brandywine are highly resistant to cane diseases (spur blight and grey mould) whereas Malling Admiral and Glen Moy are fairly resistant (Seemuller, 1987). The cultivars Glen Clova, Malling Delight, Malling Orion, Zeva 2 and Sirus are highly susceptible to these diseases.

### Spur Blight (*Didymella applanata*)

Spur blight causes brown-violet discolouration around the bud and lesions at the base of canes (Kuhne, 1980). Leaf shows brown lesions with yellow margins which may shrivel and drop. Infested buds fail to survive in winter. Swait (1980) reported its incidence from nil in the first year to more than 50 per cent in the 6th year. With a programme of 4 or 5 sprays of dichlofluanid (0.1 per cent) commencing from the emergence of new canes, Burchil and Swait (1975) were able to control spur blight in a plantation of Malling Jewel. Lime sulphur application before new growth resumes in spring is suggested to decrease incidence. Broad-spectrum multi-site fungicides mancozeb, chlorothalonil, copper hydroxide and dithianon were found effective against spur blight. The efficacy of these compounds was between 64.4 and 81.7 per cent. All fungicides were applied prevently, four times until the beginning of harvest and once after harvest (Stevic *et al.*, 2017). Biocontrol agents *Bacillus subtilis* and *Pseudomonus fluorescens* showed antagonistic activity towards *D. applanata* (Shternshis *et al.*, 2016).

### Grey Mould (*Botrytis cinerea*)

Bud failure, inhibitation of lateral shoot emergence and yield reductions are caused by grey mould infection (Rebandel *et al.*, 1985). The infected buds die even due to mild frost. The increase in incidence of grey mould is, however, slower than spur blight during 6 years of observations (Swait, 1980). Kolbe (1981) reported that spraying every 5-6 days with 0.2 per cent dichlofluanid during bloom increased fruit set by 6 per cent, yield by 39 per cent and individual fruit weight by 4 per cent. Malling Promise and Seiwka Gromiszowej are the most susceptible cultivars and Malling Jewel the most resistant to this disease (Borecka, 1973). Eight active substance: boscalid, pyraclostrobin, cyprodinil, fludioxonil, fenhexamid, iprodion, pyrimethanile and thiram are in use to protect raspberries against grey mould in Poland. At least 3-4 treatments are made from flowering, every 5-7 days. Frequent repeating treatments with the same substances lead to resistance of grey mould. Substance with different mechanism 'laminarin' (β 1-3 polysaccharide), obtained from brown seaweed (*Laminaria digitata*) is recommended as preventive treatment, used every 10-14 days, to protect strawberries against grey mould, powdery mildew and leaf spot (Pluciennik *et al.*, 2014). Laminarin can be successfully included into commercial protection of the raspberry plants against grey mould in the programmes mixed with the chemical fungicides. With the use of such a strategy, there are opportunities to reduce the risk of *B. cinerea* resistance to commonly used fungicides and to decrease pesticide residues in fruits (Krawiec *et al.*, 2016).

### Cane Spot (*Elsinoe veneta*)

Occurrence of purple spots and lesions on internodes from mid-April to late July are the typical symptoms found on young canes. The lesions are predominant in the fruit bearing region of cane 46-120 cm above soil level. The disease on Malling Jewel raspberries was effectively controlled by spraying dichlofluanid (0.1 per cent) before harvest in a 4-spray programme (Munro *et al.*, 1988).

### Anthracnose (*Leptosphaeria coniothyrium*)

The symptoms on canes include stunting and loss of vigour. Affected bark becomes desiccated and often cracks. It may penetrate deeper into the canes and is also associated with raspberry midge (Kuhne, 1980). It can be controlled by spraying fungicides at bud-break before flowering and during full bloom (Tsbnkovski and Paneva, 1978). Cuprocin (0.4 per cent), benomyl (0.1 per cent), chlorothalonll (0.2 per cent) and dithane M-45 (0.2 per cent) have been found effective in reducing the disease incidence.

### Yellow Rust (*Phragmidium rubi-idaei*)

Yellow rust is a serious foliage disease. In early spring, greenish-orange, glossy, pimple like structure (about 1 mm in diameter) appear on the upper surface of young leaves. Later on rust pustules appears on the lower surface of the leaves from late spring through fall. Infected leaves on primocanes die prematurely and the developing berries fail to ripen. Anthony *et al.* (1987) reduced the disease incidence on young canes by removing all fruiting canes and old cane stubs but removal of successive flushes of replacement canes did not reduce the infection of lateral shoots on fruiting canes. The production of basidiospores can be inhibited *in vitro* by dinoseb-in-oil at 1 mg a.i./ml. The cultivars Latham, Malling Delight, Malling Jewel, Mecker and Glen Clova are susceptible in varying degrees (Anthony *et al.*, 1985).

### Root Rot (*Phytophthora fragariae* var. *rubi*)

The symptoms usually appear in the mid-summer to early fall. Plants showing reduced vigour and leaves turned yellow, red or orange and being scorching on the edges. The plant wilts and dies. The diseases is frequently observed in areas with high rainfall and poor drainage.

It is suggested to grow raspberries on raised beds in high rainfall areas. Heiberg (1999) reported that plastic mulch and raised beds (30 cm) both individually and when combined reduced the severity of rot. Treatment with oxadixyl (125 a.i./kg) also reduced soil moisture and incidence of rot. Wilcox *et al.* (1999) suggested planting in raised beds and application of metaloxyl. Application of compost covering the ridges reduced the spread of the pathogens (Stirnimann, 2000)

In a field evaluation of 99 raspberry clones for susceptibility to root rot, most of the clones were found highly susceptible but no infection was observed in WSU 585 and Cherokee and 10 per cent or less in Sunrise, MN and O, Durham, Pathfinder and Lathan (Barritt *et al.*, 1981).

Cultivars Cascade, Bounty, Jaclyn, Newburgh, Cascade Delight, Prelude, Summer, Moutere, Ukee, and Josephine were identified as resistance to the disease (Moore and Hoashi-Erhardt, 2012). Clones that are derivatives of *Rubus strigosus, R. niveus, R. lasiostylus* and *R. innominatus* showed resistance to *Phytopthora rot* (Kempler *et al.*, 2012).

## Viral Diseases

Viruses are serious problem in raspberry culture. After reviewing the literature on raspberry viruses, Converse (1977) has grouped the viruses into 4 categories according to the mode of their transmission:

i) Insect transmitted viruses and mycoplasma, ii) Nematode transmitted viruses, iii) Pollen transmitted viruses, and iv) Viruses transmitted by unknown methods.

Jones (1981) has listed 4 aphid-borne viruses, *viz.*, BRNV (black raspberry necrosis virus), RLMV (raspberry leaf mottle virus), RLSV (raspberry leaf spot virus), and RVCV (raspberry vein chlorosis virus), and 4 nematode-borne viruses, *viz.*, RRV (raspberry ring spot virus), TBRV (tomato black ring virus). AMV (arabis mosaic virus) and SLRV (strawberry latent ring spot virus). Pollen transmits RBDV (raspberry bushy dwarf virus). Transmission of viruses also occurs through cultural practices, such as, pruning, irrigation, *etc.*

The symptoms of infection vary with the virus involved and viruses latent. Jones (1979) studied the effect of RRNV and RBDV on growth and yield of the cultivar Lloyd George in pots. The plants infected with both viruses were stunted and produced fewer and shorter canes than virus-free controls. The virus infection delayed ripening, decreased the berry weight and deformed the fruits but there was no significant reduction in yield. However, Murant (1987) stated that when infested RBDV alone, plants of many cultivars exhibit no foliar symptoms. Several cultivars develop a veinal or general chlorosis on the leaves in addition to crumbly fruit. Daubeny *et al.* (1978) described the symptoms of RBDV as reduction in yield, fruit size and enhancement in drupelet abortion resulting in crumbly fruit. Leaf abnormalities were also observed.

Since virus infected plants cannot be cured by chemical and cultural practices, planting of virus-free plants are the only remedy which is generally followed in all the raspberry growing regions. The tip cuttings are made virus free by heat treatment technique before raising plants from them. Aphid-borne viruses can be controlled by breeding aphid-resistant cultivars and significant progress has been made in this regard (Keep *et al.*, 1970; Kennedy *et al.*, 1973). The nematode-borne viruses need to be eliminated by controlling the nematodes.

Quito and Martin (2012) reported a novel Reovirus, Raspberry latent virus (RpLV) in addition to RBDV and RLMV causing severe crumbly fruit in red raspberries. In Europe, raspberry plants are commonly infested with a complex of aphid-transmitted viruses that together caused raspberry mosaic disease (RMD). By grafting and vector transmission to a range of red and black raspberry cultivars, these viruses have been loosely characterized and identified as RLSV, RLMV, BRNV, and Rubus yellow net virus. An additional, very widespread virus RVCV, is spread by a different aphid vector (MacFarlane and McGavin, 2010). Lightle *et al.* (2014) reported that *Amphorophora agathonica* is the primary vector of aphid-transmitted viruses in red raspberry in the Pacific Northwest region of US. The lower developmental threshold of *A. agathonica* was estimated to be 2.7°C. In the field, apterous and alata aphid populations began rapidly increasing at ~800 growing degree-days and peaked at 1050 growing degree-days.

Spak and Kubelkova (2000) stated that vegetatively propagated plants are the main source of RBDV virus spread in cultivated raspberry, than naturally infested wild Rubus populations. Surveys in two important raspberry growing regions (Kent and the West Midlands, UK) in 1995 and 1996 identified RBDV infected plots of Autumn Bliss, Dinkum, Leo, Glen Moy, Malling Admiral and Malling Delight (Knight and Barbara, 1999). Mavric *et al.* (2004) detected RBDV in Autumn Bliss, Fall Gold, Golden Bliss and Meeker in Solvenia.

McGavin *et al.* (2010) obtained the complete nucleotide sequence of a UK strain of the sadwavirus Black raspberry necrosis virus (BRNV) by amplification and sequencing of virus RNA from infected plants. The RNA1 was 7,572 nucleotide (nt) in size and RNA2 was 6,350 nt in size, each excluding the 3′ poly-A tail. The RAN1- and RNA2-encoded polyproteins are predicted to be processed into (RNA1) a protease cofactor, an RNA polymerase and an A1kB protein, and (RNA2) a movement protein and to capsid proteins.

Raspberry leaf blotch virus (RLBV) is a recently described (-)ssRNA virus (genus *Emaravirus*) associated with a long-known, severe raspberry leaf blotch disorder (RLBD). The virus is presumably transmitted by the raspberry leaf and bud mite (*Phyllocoptes grandis*). RLBV is widely distributed in the cultivated and wild raspberries in Finland and shows considerable genetic variability (Dong *et al.*, 2016). RLVB has also recently been reported to be associated with red raspberry cv. Glen Ample showing severe symptoms of raspberry leaf blotch disorders in Scotland, England, Finland, Bulgaria and Italy (Zindovic *et al.*, 2015).

## Physiological Disorder

### Sunscald or White Drupelet Disorder

The exposure of red raspberry fruits at green to white to pink maturity stages to high temperature and solar radiation causes white drupelet disorder (Renquist *et al.*, 1987). This solar injury occurs at 42°C and higher temperature at 4 or more hours of ultraviolet radiation in laboratory. Renquist *et al.* (1987) also reported that 40 per cent fruits of the cultivar Heritage were affected. The amount of injured fruits was reduced by 8 per cent to 16 per cent by covering the plants with three types of covers (30 per cent and 60 per cent with black fabric and 20 per cent with white polyester) one to three weeks before harvesting. Watering the raspberry planting on a weekly basis (in hot, dry weather) may reduce the incidence of sunscald and/ or white drupelet dirorder.

## Breeding and Improvement

Different aspects of raspberry breeding, *i.e.*, objectives, techniques, systems and inheritance of characters have been reviewed by various workers (Darrow, 1937, 1967; Jennings, 1976; Ourecky, 1975). Early improvement began in Europe in the 16th century with the selection of superior chance seedlings but planned breeding commenced only during late 19th or early 20th century in Canada, UK, USA and USSR (Darrow, 1937; Hedrick, 1925). Many superior cultivars of red, black and purple raspberries were released and some of them are under cultivation even now (Brooks and Olmo, 1972).

## Breeding Objectives

Emphasis is laid on the following objectives in the breeding of raspberry cultivars:

### High Productivity

In all breeding programmes, the productivity always remains the main criterion in selecting a breeding material for releasing as a cultivar. The number of fruits per cluster, number of cluster per plant and plant hardiness play important role in the productivity (Ourecky, 1975). Among 28 cultivars, Malling Seedling gave the highest yield at all sites in Poland (Gwozdecki *et al.*, 1983).

### High Fruit Quality

The cultivation of a high yielding cultivar will be of no use if its fruit quality is poor. Raspberry fruits must be large in size, firm in texture, bright in colour, rich in soluble solids, and very good in flavour. The drupelets should be large and compact.

### Improved Shelf Life

In view of the perishable nature of the fruit, even small improvement in its keeping quality will be highly desirable.

### Suitability for Processing

Jennings (1976) advocated breeding of raspberries which meet the specialized requirement of processing. Small bright coloured seeds are required for attractive jams.

### Prolonged Ripening Period

A long ripening season is needed for 'pick-your-own' operations as well as for full utilization of harvester. According to Keep (1976), the harvesting season can be extended by breeding extra early and extra late summer fruiting types and early autumn fruiting selections.

### Suitability for Mechanical Harvesting

The development of mechanical harvester has given a boost to raspberry culture. Hence breeding of the cultivars suitable for mechanical harvesting has been suggested by the breeders (Jennings, 1974; Keep, 1975; Ramsay *et al.*, 1985). The characteristics suggested by them for this purpose are: firm fruit, superior fruit quality, ease of fruit abscission; flexible pedicels and fruiting laterals, peduncles of unripe fruit that do not break young cane that does not damage easily and self-supporting shorter plants.

In raspberry breeding programme worldwide it has been a challenge to select for fruit firmness, resistance to mechanical damage and extended shelf-life. Components of fruit firmness which need to be considered are skin strength, tissue strength or integrity and the attachment of druplets to each other (O'Donoghue and Martin, 1988). Fruit firmness has been selected in seedling populations from the raspberry gene pool which has been derived from *R. idaeus* and *R. strigosus* (Hall and Stephens, 1999).Improvement in fruit firmness has been achieved through the use of black raspberry (*R. occidentails*) using the firm fruited cultivar Cumberland as a source of parent for fruit firmness (Keep and Knight, 1986). Breeding evaluation of remontant forms of raspberry for berry firmness was carried out from 2005-09 in Moscow, Russia. The firmest berries were produced by cv. Atlant and the interspecific elite from 15-146-11, indicating the suitability of these genotypes for mechanical harvesting. Cultivars Gerakl, Atlant, Bryanskoe, Divo, Evraziya and Progenies 47-18-4 and 13-39-11 were identified as useful donors for breeding for high resistant to crushing (Evdokimenko, 2010).

### Dwarf Compact Plant

The maintenance cost can be reduced by restructuring the plant which has erect sturdy canes requiring no trellising and negligible topping (Ourecky, 1975).

### Wider Adaptability

The adaptation to both climatic and edaphic factors is required. The selection should be able to tolerate low temperature and frost in colder regions and resistance to leaf diseases and fluctuating winter temperatures in warmer regions. It should withstand wet and dry soil condition as well.

Winter hardiness is a major objective of raspberry breeding of Bulgaria, Latvia, Poland, Rumania, Russia, Norway, Sweden and Eastern Canada (Knight, 2004). Wild *Rubus* species *R. idaceus vulgates, R. arcticus, R. crataegifolius, R. pungens* are used as a source of hardiness. Released winter hardy cultivars include: 'Boyne', 'Nova', 'Killarney', 'Ottawa', 'Muskoka', and 'Jenkka'.

Raspberry breeding in the southern USA was initiated in the late 1920s and early 1930s and involved state agricultural experiment stations in Mississippi, North Carolina, Tennesse and Virginia. The USDA was also involved through collaboration with the North Carolina programme from the 1930s through the early 1960s. These five raspberry breeding programme released a total of 20 cultivars including 12 involving interspecific hybridization with heat tolerant Asian raspberry species. The most significant of the heat tolerant raspberry cultivars involving Asian germplasm are 'Southland' from the USDA/North Carolina programme and 'Dormanred' from the Mississippi programme (Ballington, 2016).

## Disease and Insect Resistance

Breeding pest and disease and pest resistance into crops offers major hope for reduction in high pesticide use and increasing organic production (Stephens *et al.*, 2002). Resistance to *Bortrytis cinerea* is desired to increase the self-life of fresh fruit and reduce grey mould in the riper fruit harvested by machine. Several sources of resistance to *B. cinerea* in raspberry have been found including *R. pileatus* (Jennings and Williamson, 1982, Jennings and Brydon, 1989), *R. occidentalis* (Jennings and Williamson, 1982, Knight, 1980), *R. crataegifolius* (Jennings and Williamson, 1982) and *R. coreanus* (Keep *et al.*, 1977). Knight (1980) found that 2182/5, a third generation back-cross from *R. occidentalis*, had strong resistance to grey mould. Stephens *et al.* (2002) found best shelf life in clone that had both *R. pileatus* and *R. occidentalis* in its ancestry.

Root rot (*Phytophthora rubi*) disease is a problem of raspberry cultivation in North America, Europe, Australia and many other countries. Wild rubus species (*R. strigosus*, *R. ursinus*) have been identified potential source of resistance. The cvs. Latham and Winkler's Samting were identified as having significant resistance. Resistance or tolerance to root rot has been a major objective in the breeding programme of James Hutton Limited (JHL), Scotland since 2002 (Jennings *et al.*, 2016). Highly tolerant 'Glen Erich' was identified at JHL. A mapping population segregating for root rot resistance (Glen Moy x Latham) was screened under both field and glasshouse conditions over a number of seasons. Two QTL associated with resistance were identified (Jennings *et al.*, 2016). For marker assisted breeding the SSR marker Rub118b, a 110-bp allele from 'Latham' was found in resistant germplasm but was not found in any of the susceptible germplasm (Graham *et al.*, 2011). Using the resistance source from 'Cherokee', 'Latham' and accessions of *R. occidentalis* and *R. strigosus*, several *P. rubi* resistance cultivars: Cascade Delight, Cascade Dawn, Cascade Bounty, Prelude, Heritage and Taylor were developed.

Fire blight (*Erwinia amylovora*) outbreak are less common in *Rubus* species but can reach economically damaging levels under certain conditions (Braun *et al.*, 1999). The raspberry cv. Willamette appeared to be highly susceptible with average 50 per cent disease severity ratings while cv. Royalty, was reported as a moderately resistant with an average 25 per cent disease severity rating (Bastas and Sahin, 2014). No cultivars of *Rubus* are truly resistance to spur blight, however, some cultivars like Brandywine, Festival, Hilton, Killarney, Latham, Madawaska and Newburgh are less prone to spur blight.

Raspberry bushy dwarf virus (RBDV), a pollen-borne virus is the most common virus of raspberry worldwide. A number of RBDV resistance cultivars have been developed from England, Scotland, USA, Finland, New Zealand, Canada and Romania. Some of the resistant cultivars are Malling Promise, Malling Exploit,

Malling Admiral, Malling Jewel, Malling Joy, Malling Orion, Glen Clova, Glen Magna, Bristol, Octavia, Marcy, Sunrise, Latham, Lewish, Newburgh, Phoenix, *etc.*

Resistance to the large raspberry aphid (*Amphorophora idaei*) in red raspberry, using single major genes or polygenic major genes, has proved successful in controlling this virus vector aphid in Europe (Birch *et al.*, 2003). Surveys conducted in the early 1990s found that >75 per cent of the UK *A. idaei* population consisted of biotypes with the ability to break the most used resistance gene $A_1$. Since then growers in England have reported intermittent breakdown of the formerly strongest resistance gene, $A_{10}$. Genetic analysis of raspberry aphid populations, based on rDNA 1GS DNA patterns, has shown that *A. idaei* populations in UK are genetically very variable within and between the 5 known *A. idaei* biotypes. This is attributed to alate migrations of parthogenetic asexul females in summer and sexual males in autumn. This means that resistance-breaking genes are readily exchanged between raspberry aphid populations (Birch *et al.*, 2003). Several aphid-immune cultivars are available, including: Haida, Nootka, Skeena, Qulicum, Malahat, Cowichan, Canby, Esquimalt, Saanich *etc.* East Malling, UK released 'Autumn Bliss' in 1984 which has several genes, including the aphid resistance from *R. occidentalis*. Similarly, 'Autumn Byrd', released in N. America has aphid resistance gene from *R. occidentalis*. $A_{10}$ gene from *R. occidentalis* was incorporated to develop Malling Joy, Malling Leo, Gaia, Glen Rosa, Glen Doll and Glan Fyne (Birch *et al.*, 1994, Jennings *et al.*, 2008).

Wild *Rubus* species, *R. coreanus*, *R. craetigifolius*, *R. occidentalis*, and *R. phoenicolaesius* has been used as source of resistance to raspberry beetle (*Byturus tomentosus*) in raspberry breeding programme (Briggs *et al.*, 1982). Their derivatives and the purple raspberry cultivar Glen Coe (Jennings *et al.*, 2003) showed some level of resistance to larvae-induced damage. In Estonia, late-flowering cultivars 'Glen Ample' and 'Glen Magna' showed lowest damage compared with 'Ivars' and 'Novokitaivska' (Arus *et al.*, 2013).

## Development of High Yielding Primocane Fruiting Cultivars

The cultivation of primocane fruiting cultivars is becoming more and more popular in the world because of high returns, low costs, avoidance of winter killing and suitability for mechanical harvesting (Keep, 1988; Lawrence, 1980; Ourecky, 1978).

Several clones of *R. strigosus* were used in New York State breeding programme to develop primocane fruiting cultivars. 'Ranere' was used to develop 'Cherokee' at Virginia Polytechnic, USA which has resistance to root rot and primocane fruiting. Using raspberry related species as donors of useful characters for over 50 years. East Malling, UK has released 'Malling Leo' the first red raspberry cultivars with *R. occidentalis* in its ancestry, in 1976, 'Autumn Bliss' which has genes from primocane fruiting. Lloyd George' *R. strigosus* and *R. arcticus* and for aphid resistance *R. occidentalis* was released in 1984. 'Autumn Bliss' has been used widely to develop early-ripening parent in primocane breeding programme worldwide (Knight, 2004). USDA-ARS, Oregon breeding programme releases primocane-fruiting 'Summit' and 'Amity' and floricane-fruiting 'Willamette' and 'Canaby' which are popular in many countries.

## Breeding System

The basic chromosome number of raspberry is seven (x=7), as has been determined by a number of workers (Kumarov and Juzepczuk, 1939; Einset, 1947; Darlington and Wylie, 1955). Almost all the cultivars are self-fertile, but self-incompatibility has been observed in 11 of 23 *Rubus* species studied. The growth of self-pollen tubes was restricted in the style indicating the reaction due to 57 and 52 incompatibility alleles (Keep and Knight 1986). Lloyd George and Preussen cultivars have been used most frequently as donors in raspberry breeding.

## Inbreeding, Outbreeding and Backcrossing

All the three systems have been used in breeding of raspberry cultivars. Inbreeding helps in the expression of many desirable recessive characters and in developing homozygosity. For the incorporation of desirable genes into breeding line, outbreeding is generally resorted to producing $F_1$ hybrids. General and specific combining ability for plant vigour and winter-hardiness have been studied by Ure (1963) and he found low vigour in progenies due to common parentage. Keep and Knight (1968) successfully introduced many characters like resistance to virus vectors and cane spot, primocane fruiting, dwarfing habit, good flavour and fruit firmness into desirable genotypes.

## Interspecific Hybridization

Hybridization between *R. idaeus* var. *vulgatus* and *R. idaeus* var. *strigosus* is considered as subspecific, almost fully fertile $F_1$ hybrids. A wide range of sterility in other specific crosses and their use in breeding have been reported by Ourecky (1975). Many purple cultivars have been evolved as a result of crosses between red and black raspberries. Rietsema (1955) reported successful crosses between *R. idaeus* and *R. illecebrosus, R. lasiostylus,* and *R. phoenicolasius, R. lasiostylus* and *R. phoenicolasius*.

## Polyploidy and Mutation

All commercial raspberry cultivars are diploids and autopolyploidy decreases pollen and ovule fertility (Topham, 1967). Spontaneous mutants are very rare and little work has been done on induced mutations.

## Inheritage of Specific Characters

Ourecky (1975) has given the detailed account of inheritance of characters and listed the genes identified for them and donor cultivars in *Rubus*, subgenus *Idaeobatus* in tabular form on the basis of researches conducted in the UK.

Adaptation to both severe winters and hot summers is required in raspberries. The Asiatic species, *R. parvifolius* is used as a parent in breeding cultivars for tropical and subtropical regions. Winter-hardiness has been reported as a dominant character (Jennings *et al.*, 1961). The inheritance of yield is complex, and its contributory factors are large fruit size, number of fruits per lateral and high frequency of nodes bearing more than one fruiting lateral and sturdy dwarfness is recessively controlled by gene $d_1d_2$ (Keep, 1969).

Lewis (1941) reported the inheritance of primocane fruiting as governed by a single recessive gene. But on the basis of work done at East Malling, Keep (1961) postulated that it was largely additive and associated with parental seasons of primocane fruiting. Later, the predominantly additive inheritance of primocane fruiting was also suggested by Ourecky (1976) and Fejer (1977).

Large fruit size (a desirable character for 'pick-your-own' cultivation), drupelet number, drupelet weight, seed weight, percentage of light seed and percentage of seediness are quantitatively inherited (Toyoma, 1961). 'Newburgh' and 'Heritage' cultivars have been good sources of fruit firmness character in breeding. Firmness of Heritage fruit is associated with its small, compact and coherent drupelets (Ourecky, 1969). The inheritance of fruit colour is qualitative and under polygenic control (Lawrence, 1966).

Since aphids are vectors for many viral diseases, the breeding of aphid-resistant cultivars is of paramount importance. Keep *et al.* (1970) have reported in detail the relationship between ten major genes ($A_1$, $A_2$, $A_3$..... $A_{10}$) for resistance to four strains of *Amphorophora rubi*. $A_2$ is a dominant gene giving full resistance by its presence. The genetics of resistance against American strain of raspberry aphid (*A. agathonica*) has been described by Daubeny (1966). Kennedy *et al.* (1973) reported cvs.Canby, Lloyd George and New York 632 as immune and cvs.Washington and Malling Exploit as resistant to *A. agathonica*. Three genes $Sp_1$, $Sp_2$ and $Sp_3$ govern the resistance to powdery mildew which is of epistatic nature (Keep, 1968a,b).

The raspberry breeding programme has been quite successful as many good cultivars have been evolved over the years in the UK and the USA, possessing good plant and fruit attributes and adaptability to ecological factors. Red raspberry cultivars developed from different countries are listed in Table 16.8.

## Molecular Markers

Molecular markers are used to distinguish between cultivars and to group cultivars of similar origin. The development of similar sequence repeat (SSR) markers (Grahm *et al.*, 2002, Stafne *et al.*, 2005) has allowed the development of a raspberry genetic linkage map. This facilitates the development of diagonostic markers for polygenic traits and the identification of genes controlling complex phenotypes. A number of marker-based technologies have been set up to fingerprint each genotype at any stage of development and independly from environmental factors that may influence the phenotype. An applicative scheme for raspberry fingerprinting has been developed in order to verify the putative pedigree of several accessions derived from a breeding programme in which some genotypes showed the same allelic pattern: Tulameen and Tulameen Gialla Sugana and Sugana Gialla (Venturi *et al.*, 2012).

Expressed sequence tags (ESTs) are a source of SSRs that can be used to develop markers to facilitate more basic research across genera. To accelerate development of EST-SSRs in the genus *Rubus*, Bhushakra *et al.* (2015) generated 1149 and 2358 cDNA sequences from red raspberry and black raspberry, respectively. The cDNA sequences were screened using rigorous fittering criteria which resulted in the identification of 121 and 257 SSR loci for red and black raspberry, respectively. Primers were designed from the surrounding sequences resulting in 131 and 288 primer pairs, respectively, as some sequence contained more than one SSR locus. Sequence analysis revealed that the SSR-containing genes span a diversity of functions and share more sequence identity with strawberry genes than with other Rosaceae species. Development of SSR markers that are transferable between red and black raspberry for marker-assisted selection, evaluation of genome collinearity and to facilitate comparative studies in *Rubus* will be more effective using SSR markers developed from black raspberry sequences by Dossett *et al.* (2015).

**Table 16.8: Red Raspberry Cultivars Developed from different Countries**

| | | |
|---|---|---|
| Australia | : | Dinkum, Alkoopina, Bogoing, Glen Yarra. |
| British Columbia | : | Chilcotin, Chilliwack, Qualicum, Malahat, Skeena, Comox, Haida, Algonquin, Chemainus, Cowichan, Esquilnalt, Nanoose, Rudi, Sannich, Tulameen, Ukee, Algonquin. |
| Bulgaria | : | Essenna Poslata, Ljlin, Samodiva, Raliza. |
| Canada | : | Saanich, Regency, Jeanne d'Orleans, Rudyberry, Ukee, Red River, Cowichan, Boyne, K-81-6, Killamey, Nova, Regal, |
| China | : | Fengmanhong, Boyne, Zaohong. |
| Czech Republic | : | Granat |
| Denmark | : | Zenith |
| England | : | Joan Irene, Malling Minerva, Valentina, Malling Hestia, Marcela, Autumn Treasure, Bethan, Octavia, Autumn Bliss, Autumn Britten, Brice, Octavia, Joan Squire, Julia, Malling Delight, Malling Charm, Malling Bella, Aurtumn Byrd, Maiiling Jewel, Malling Juno, Autumn Cascade, Autumn Cygnet, Gaia, Julia, Malling Augusta, Malling Joy, Valentina. |
| Estonia | : | Aita, Alvi, Helkel, Tomo |
| Finland | : | Jatsi, Jenkka, Ville. |
| France | : | Comtessse, Favorite, Galante, Meco, Princess, Wawi |
| Germany | : | Resa, Rubaca, Rumiloba, Rusliva, Weirula. |
| Hungary | : | Julesi, Fertodi Zamatos, Fertodi 103, Fertodi Aranyfurt, Fertodi Carmen. |
| Italy | : | Dolomia, Lagorai, Vajolet, Amira, Erika. |
| Latvia | : | Arta, Dita. |
| Mexico | : | Anita, Gina. |
| Netherlands | : | Marwe, Holland, Dwarf, Marve. |
| New Zealand | : | Moutere, NRY, Raspberry Shortcake(TM), Adele, Awaroa, Korere, Rakaida, Selwyn, Waiau, Waimea, Korpiko, Wakefield, Ruby Beauty®, Motueka, Tadmor, Clutha. |
| Norway | : | Baldar, Borgund, Frosta, Hitra, Stiora, Tamber, Varnes, Vene. |
| Poland | : | Benefis, Laska, Polesie, Beskid, Nawojka, Polana, Pokusa, Poranno Rosa. |
| Romania | : | Citria, Ruvi, Star, Opal, Gustar, Elite 89-15-3. |
| Russia | : | Sentyabr'sKaya, Latam, Solnyshko, Stolichnaya, Novokitaevskaya, Orbita, Nezhnost, Ozarenie, Babje Leto, Brygantyna, Gusar, Kredo. |
| Scotland | : | Glen Doll, Glen Fyne, Glen Cally, Glen Ericht, Glan Ample, Glen Lyon, Glen Prosen, Glen Clova, Glen Magna, Glen Moy, Glen Rosa. |

| | | |
|---|---|---|
| Sweden | : | Ariadne, Boheme, Carmen, Elektra. |
| Switzerland | : | Frutafri, (Tula Magic®), Sugana, Elida, Framita, Himbo-Star, Himbo-Top. |
| USA | : | Chinook, Driscoll Cardinal, Driscoll Carmellina, Driscoll Dulcita, Driscoll Francesca, Driscoll Madona, Driscoll Maravilla, Georgia, Jaclyn, Cascade Bounty, Cascade Dawn, Explorer, Nantahala, Pequot, Cascade Gold, Crimson Giant, Pacific Deluxe, Vintage, Cascade Harvest, Crimson Night, Double Gold, Niwot, Grandeur, Radiance, Heritage, Meeker, Cascade Delight, Coho, Cardine, Prelude, Encore, Lewis, Canby, Cascade Nectar, Latham, Lauren, Mandarin, Taylor, Titan, Willamette, Anna Maria, Gloria, Godiva, Hollins, Holyke, Isabel, Joe Mello, Lawrence, Sweetbriar, Tola, Wilhelm, Nordic, Redwing, Lauren, Titan, Amity, Summit, Esta, Emily, |

Miyashita *et al.* (2015) stated that amplified fragment length polymorphism (AFLP) markers are a reliable technique for assessing genetic diversity and studying phylogenetic relationship in *Rubus*. In their studies with 81 accessions, valuable information on the availability of unique fragments in different accessions were recorded which may be useful for the development of improved genotypes through conventional breeding and marker-assisted selection.

## Linkage Map

Genetic linkage map was constructed utilizing a cross between phenotypically diverse European red raspberry' Glen Moy' and the North American 'Latham' which is 789cM (Graham *et al.*, 2004). SSR markers were developed from both genomic and cDNA libraries from 'Glen Moy'. These SSRs, together with AFLP markers were utilised to create a linkage map. In order to test the utility of the genetic map for QTL analysis, morphological data based on easily scoreable phenotypic traits were collected. The segregation of cane spininess, and the root sucker traits of density and spread from the mother plant, was quantified in two different environments. These traits were analysed for significant linkages to mapped markers using map QTL and were found to be located on linkage group 2 for spines and group 8 for density and diameter.

Sources of aphid resistance have been identified in wild germplasm and are used to develop mapping populations to study the inheritance of these valuable traits. Bushakra *et al.* (2017) have constructed a genetic linkage map using single-nucleotide polymorphism and transferable (primarily simple sequence repeat) markers for F1 population ORUS 4305 consisting of 115 progeny that segregate for aphid resistance. The linkage map of seven linkage groups representing the seven haploid chromosome of black raspberry consists of 274 markers on the maternal map and 292 markers on the parental map including a morphological locus for aphid resistance.

## Genetic Transformation

The *Agrobacterium tumefaciens* non-oncogenic strain GV2260 carrying the pBin 19 binary plasmid containing the *nptII* gene was used for the transformation of raspberry cv. Ruby. The *DefH9-laam* gene caused perthenocarpic fruit development. Under standard cultivation conditions, the *DefH9-laaM* gene increased fruit yield which was attributed to the increase in fruit weight, fruit number per inflorescence and number of inflorescences per plant (Mezzetti *et al.*, 2004a). The total IAA (auxin) content of young flower buds expressing the *DefH9-laam* gene indicates that auxin plays a role in plant fecundity in raspberry (Mezzetti *et al.*, 2004b). Georgieva *et al.* (2008) achieved transformation using *A. tumefaciens*. Leaf petioles and leaf segments of red raspberry cv. Elit-1 were infected with *Agrobacterium* strains EHA101, LBA4404 and LBA4404 carrying 35s and 2x35s promoters. The regenerants were obtained on MS nutrient medium enriched with 0.3 mg/l IBA, 0.01 mg/l, 2,4-D and 2 mg/l TDZ. The *Agrobactenium* growth was inhibited by cefotaxime. Kanamycin was used as a selective agent for the transformants. The transformation efficiency was within the range of 0.54-2.08 per cent for Elit-1 transformed with *Sac B* and *Cod A* gene. The integration of the marker genes *npt* and Hygromycin in the putative transgenic plant was confirmed by PCR analysis using primers, the nucleotide sequence of which was found complementary to these genes.

## Genome

Raspberry chromosomes are small, varying from 1.8-3.1 μm in length (Jinno, 1957). The genome compliment of raspberries is only 275 Mbp. This has made chromosome studies in raspberries difficult in terms of getting good view of chromosome structure and morphology (Hall and Kempler, 2011). VanBuren *et al.* (2016) reported a 243 Mb draft genome of black raspberry selection ORUS4115-3 using an illumina-only approach with genome anchoring from a high-density genetic linkage map. The black raspberry genome is largely collinear to the diploid woodland strawberry (*Fragaria vesca*) with a conserved karyotype and few notable structural rearrangements. Centromeric satellite repeats are widely dispersed across the black raspberry genome, in contrast to the tight association with the centromere observed in most plants. Among the 28005 predicted protein coding genes, Vanburen *et al.* (2016) identified 290 small-scale gene duplicates enriched for sugar metabolism, fruit development, and anthocynin related genes which may be related to key agronomic traits during black raspberry domestication. The genome assembly and annotation are available on the Genome Database for Rosaceae (https://www.rosaceae.org). Jibran *et al.* (2018) stated that draft genome for black raspberry (ORUS 4115-3) lacks assembly of scaffolds to chromatin scale. Jibran *et al.* (2018) used high-throughput chromatin conformation capture (Hi-C) and Proximity Guided Assembly (PGA) to cluster and order 9650 out of 11,936 contigs of this draft genome assembly into seven pseudo-chromosomes. The seven pseudo-chromosomes over ~97.2 per cent of the total contig length (~ 223.8Mb). Locating existing genetic markers on the physical map resolved multiple discrepancies in marker order on the genetic map. Centromeric regions were inferred from recombination frequencies of genetic markers, allingment of 303 bp centromeric sequence with the PGA, and heat map showing the physical contact matrix over the entire genome.

## REFERENCES

Abanowska, B.H., Gajek, D. and Niemczyk, E. (2000). Bull. OILB/SROP, 23(11): 137-40.

Alvarado-Raya, H.E., Avitia-Garcia, E. and Castillo-Gonzalez, A.M. (2016). Revista Mexicana de Ciencians Agricolas, 7(1): 17-29.

Amrita-Kumar and Ellis, B.E. (2001). Pl. Physiol., 127: 230-39.

Anderson, W.C. (1980). Acta Hort., 112: 13-20.

Andrianjaka-Camps, Z.N., Wittemann, M.S., Ancay, A. and Carlen, C. (2016). Acta Hort., 1133: 345-51.

Anonymous (1972). The Wealth of India, Pub. and Inf. Direc., CSIR, New Delhi, pp. 85-90.

Anonymous (1978). Tech. Note, Scottish Agric. College, No. 20, p. 4.

Anthony, V.M., Shattock, R.C. and Williomson, B. (1985). Pl. Path., 34: 521-27.

Anthony, V.M., Williamson, B. and Shattock, R.C. (1987). Ann. Appl. Biol., 110: 262-73.

Antoniolli, L.R., Silva, G.A. da, Alves, S.A.M. and Moro, L. (2011). Pesquisa Agropecuaria Brasileira, 46: 979-84.

Arnaudova, T. and Angelov, A.S. (1974). Gradinarska i Lozarska Nauka, 11: 32-39.

Arus, L., Kikas, A., Kaldmae, H., Kahu, K. and Luik, A. (2013). Biological Agric. Hortic., 295(4): 227-35.

Askew, H.O., Chittenden, E.T. and Monk, R.J, (1951a). Jr. Hort. Sci., 26:268-84.

Askew, H.O., Chittenden, E.t. and Watson, J. (1951b). N.Z.J. Sci., Sec. A., 32:13-26.

Avita, G.E. and Barrientos, F.P. (1985). Hort. Mexicana, 1: 41-49.

Ballington, J.R. (2016). Acta Hort., 1133: 13-21.

Baroffio, C.A. (2012). Acta Hort., 926: 615-17.

Barritt, B.H., Crandell, P.C. and Bristow, P.R. (1981). Fruit Var. J., 35: 60-62.

Bastas, K.K. and Sahin, F. (2014). HortSci., 49: 1492-97.

Baumann, G. (1981). Obstbau, 5 : 88-89.

Beeson, K.C. (1941). USDA Misc. Pub., 369: 131-51.

Belayaev, A.A., Shternshis, M.V., Chechenina, N.S., Shpativa, T.V. and Lelyak, A.A. (2017). Environ. Sci. and Pollution Res., 24: 7016-22.

Bergamini, A. and Ramina, A. (1980). Informatore Agrario, 36: 12031-37.

Birch, A.N.E., Fenton, B., Malloch, G., Jones, A.T., Phillips, M.S., Harrower, B.E., Woodford, J.A.T. and Cattey, M.A. (1994). Insect Mol. Biol., 113: 567-78.

Birch, A.N.E., Jones, A.T., Fenton, B., Malloch, G., Geoghegan, I., Gordon, S.C., Hiller, G. and Begg, G. (2003). Bull. OILB/SROP. 26(2): 51-54.

Blanpied, G.D. (1972). Pl. Physiol., 49: 627-30.

Blurman, B. and Ingvarsson, A. (1985). Rappot Ind. Trad., Sveriges Lantbruk Suniverulet, 31: 30.

Boicheva, R. (1985). Rasteniev'dni Nauki, 22 : 80-83.

Borgs, M., Krzpvanski, Z. and Czech, M. (1987). Revista di Agri. Subtropic Tropicale, 81: 165-71.

Boricka, H. (1973). Prace Inst. Sada. W. Skierniewicach, 17: 209-16.

Bould, C. (1968). Jr. Sci. Fd. Agric., 19: 457-64.

Bowen-Forbes, C.S., Zhang, Y. and Nair, M.G. (2010). J. Food Comp. Anal., 23: 554-60.

Braun, P.G., Hildebrand, P.D. and Jamieson, A.R. (1999). Acta Hort., 505: 369-72.

Briano, R., Giuggioli, N.R., Baudino, C. and Peano, C. (2016). Acta Hort., 1133: 397-404.

Briggs, J.B., Danek, J., Lyth, M. and Keep, E. (1982). J. Hort. Sci., 72: 831-62.

Broderick, J.J. (1976). Intern. Flavours and Fd. Additives, 7: 27-30.

Brohier, R.L., McFarlane, F. and Dooley, L.B. (1983). Aust. Hort. Res. Newslerter, 55: 100.

Brooks, R, M. and Olmo, H.P. (1972). Register of New Fruit and Nut Var., Univ. of California Press, Berkeley, pp. 528-47.

Bugarcic, V., Misic, P., Milutinovic, M., Subotic, R. and Gacesa, B. (1984). Jugoslovensko Vocarstvo, 18: 23-32.

Burchill, R.T. and Swait, A.A.J. (1975). Ann. Appl. Biol., 81 :341-47.

Burdon, J.N. and Sexton, R. (1990). Scientia Hort., 43: 95-102.

Bushakra, J.M., Bryant, D.W., Dossett, M., Vining, K.J., VanBuren, R., Gilmore, B.S., Lee, J.M., Mockler, T.C., Finn, C.E. and Bassil, N.V. (2017). Theor. Appl. Biol., 128: 1631-46.

Bushakra, J.M., Lewers, K.S., Staton, M.E., Zhebentyayeva, T. and Saski, C.A. (2015). BMC Plant Biology, 15: 258, pp. http:/www. Biomedcentral.com/1471-2229/15/258.

Buskiene, L. and Petronis, P. (2011). Proc. Int. Conf. Fruit Prod. and Fruit Breed., Estonia, 12-13 September, 2000. pp. 177-80.

Buszard, D.J.I. (1986). Acta Hort., 183: 175-82.

Canumir, V.J.A., Moreno, M.C. and Hetz, H.E. (1999). Agro Ciencia, 15: 249-56.

Carew, J.G., Gillespie, T., White, T., Wanwright, H., Brennan, R. and Battey, N.H. (2000). J. Hort. Sci. Biotech., 75: 495-503.

Chamberlian, J.D. and Crandall, P.C. (1997). Proc. Western Washington Hort. Assoc., 69: 142-44.

Chaplin, M.H. and Martin, L.W. (1980). Comm. Soil Sci. Pl. Anal., 11: 547-56.

Cheng, Y., Geng, W., Yang, B. and Liu, J. (2017). Russian J. Pl. Physiol., 64: 566-75.

Clark, R.J. (1984). Scientia Hort., 24: 315-21.

Clay, D.V. and Davison, J.G. (1976a). In : Proc. 1976 British Crop Prot. Conf. Weeds, London, UK, 1: 241-48.

Clay, D.V. and Davison, J.G. (1976b). In: Proc. 1976 British Crop Prot. Conf. Weeds, London, UK, 1: 249-52.

Converse, R.H. (1977). HortSci., 12: 471-76.

Cross, J. and Hall. D. (2006). IOBC/WPRS Bull., 29(9): 105-09.

Dale, A. (2008). Acta Hort., 77: 219-23.

Dale, A. Oliveira, P.B. and Valdivieso, T. (2012). Acta Hort., 926: 215-19.

Dale, A., Pirgosliev, S., King, E.M. and Sample, A. (2005). J. Hort. Sci. Biotech., 80: 346-50.

Dalman, P. and Malkki, S. (1986). Ann. Agric. Fenniae, 25:232-41.

Darlington, C.D., and Wytie, A.P. (1955). Chromosome Atlas of Fl. Pls., Allen and Unwin, London.

Darrow, G.M. (1937). In: Better Plants and Animals 2, USDA Yearbook of Agri., pp. 496-533.

Darrow, G.M. (1967). Amer. Hort. Mag., 46 :202-318.

Daubeny, H.A. (1966). Proc. Amer. Soc. Hort. Sci., 88: 346-51.

Daubeny, H.A. (1978). Canad. J. Pl. Sci., 58: 219-82.

Daubeny, H.A. (1987). HortSci., 22: 1343-45.

Daubeny, H.A., Lawrence, F.J. McGregor, G.R. (1989). Fruit Var. J., 43: 46-48.

Daubeny, H.A., Stace Smith, R. and Freeman, J.A. (1978). Jr. Amer. Soc. Hort. Sci., 103: 519-22.

Davis, V.E. (1942). MS thesis, Maryland University.

DeBoer, D.W., Peterson, R.M. and Evers, N.P. (1983). HortSci., 18: 930-31.

DeGpmez, T.E., Martin, L.W. and Breen, P.J. (1986). HortSci., 21: 441-42.

Desjardins, Y. and Gosselin. A. (1987). Canad. J. Pl. Sci., 67: 863-69.

Dierend, W. and Stocker, I. (2001a). Erwerbsobstbau, 43: 43-49.

Dierend, W. and Stocker, I. (2001b). Erwerbsobstbau, 43: 114-19.

Dijkstra, J. and Oosten, A.A. Van (1980). Erwerbsobstbau, 22: 164-67.

Dong, L., Lemmetty, A., Latvala, S., Samuilova, O. and Valkonen, J.P.T. (2016). Ann. Appl. Biol., 168: 122-32.

Dossett, M., Bushakra, J.M., Gilmore, B., Koch, C.A., Kempler, C., Finn, C.E. and Bassil, N.V. (2015). Jr. Amer. Soc. Hort. Sci. 140: 243-52.

Dziadczyk, E., Domaciuk, M., Dziadczyk, P., Pawelec, I., Szczuka, E. and Bednara, J. (2013). Annals Univ. Mariae Curie Sklodowska Sectio-C., Biologia, 68(2): 15-24.

Eaves, C.A., Lockhart, C.L., Stark, R. and Craig, D.L. (1972). Jr. Amer. Soc. Hort. Sci., 97: 706-07.

Einset, J. (1947). Gentes Herb., 7: 181-92.

Elsakova, S.D. and Yaroslavets, E.T. (1978). Byulleten Vsesoyuznogo Ordena Lenina Inst. Rast. N.I. Vavilova, 82: 75-79.

Evdokimenko, S.N. (2010). Sadovodstvo-i-Vinogradarstvo, 1: 30-34.

Fejer, S.D. (1977). Canad. J. Gen. Cytol., 19: 1-13.

Finn, C.E., Lawrenc, F.J., Langford, G., More, P.P., Yorgey, B. and Stirk. B.C. (2001). HortSci., 36: 1155-58.

Flores, C. and Ruiz Castillo, M.L. del (2015). J. Food Com. and Analysis, 39: 55-61.

Forge, T.A. and Kempler, C. (2009). Canad. J. Pl. Pathol., 31: 241-49.

Forney, C.F., Jamieson, A.R., Pennell, K.D.M., Jordan, M.A. and Fillmore, S.A.E. (2016). Acta Hort., 1120: 57-64.

Freyman, S. (1998). Acta Hort., 262: 349-56.

Funt, R.C. and hall, H.K. (2013). Raspberries. CAB international, UK. pp. 282.

Georgieva, M., Kondakova, V., Djilyanov, D., Badjakov, I. and Yancheva, S. (2008). Universitatea Din Carova Univ. of Cariova, 13: 5-14.

Girgenti, V., Peano, C., Giuggioli, N.R., Giraudo, E. and Guerrini, S. (2012). Acta Hort., 926: 571-76.

Goode, J.E. (1970). Pl. Path., 19: 103-110.

Gordon, S.C. and Woodford, J.A T. (1986). Crop Res., 26: 111-19.

Gordon, S.C., Woodford, J.A.T., Grassi, A., Zini, M., Tuovinen, T., Lindquvist, I. and McNicol, J.W. (2003). Bull. OILB/SROP, 26(2): 55-60.

Graham, J., Hackett, C.A., Smith, K., Woodhead, M., Mackkenzie, K., Tierney, I., Cook, D., Bayer, M. and Jennings, N. (2011). Theor. Appl. Biol., 165: 318-28.

Graham, J., Smith, K., Mackenzie, K., Jorgenson, L., Hackett, C. and Powell, D. (2004). Theo. Appl. Genet., 109: 740-49.

Graham, J., Smith, K., Woodhead, M. and Russell, J. (2002). Molecular Ecology Notes, 2: 250-52.

Gwozdecki, J.A., Korsak, A. and Smolaroz, K. (1984/85). Prace lnst. Sad. i Kwiaciarstwa w Skierniewicach, 23 : 67-72.

Gwozdecki, J.A., Radacyznska, Z., Rechnio, N., Sikora, J. and Smolarz, K. (1983). Prace Doswiadezalne Zakresu Sadownictwa, 24: 77-89.

Haffner, K., Rosenfeld, H.J., Skrede, G. and Wang-Laixin (2002). Postharvest Biol. Technol., 24: 279-89.

Hall, H.K. and JoStephens, M. (1999). Acta Hort., 505: 93-99.

Hall, H.K. and Kempler, C. (2011). Fruit Veg. and Cereal Sci., Biotechnol., 5: 44-62.

Haltvick, E.T. and Struckmeyer, B.F. (1965). Proc. Amer. Soc. Hort. Sci., 87: 234-37.

Harris, C.H. (1944). Proc. Amer. Soc. Hort. Sci., 48: 300-302.

Harris, G.H. (1940). Sci. Agric., 20: 344-54.

Hartley, M. J. and Loader, D.L (1985). Proc. N.Z. Weed and Pest Control Conf., Harting, pp.244-47.

Hartmann, H. and Kester, D.E. (1983). Plant Propagation, 4th Ed. Prentice-Hall Inc., p. 631.

Hartmann, H.T. and Kester, D.E. (1959). Plant Propagation, Prentice Hall, Inc., Englewood, Cliffs. NJ, P. 491.

Hassanpour, H. (2015). Fol. Sci. Technol., 60: 495-501.

Hedrick, U.P. (1925). The Small Fts. of New York, J.B. Lyon, Albany.

Hedtrich, C.M., Feucht, W. and Schimmelppeng, H. (1980). Erwerbsobstbau, 22: 159-60.

Heiberg, N. (1999). Acta Hort., 505: 249-55.

Heide, O.M. and Sonsteby, A. (2011). J. Hort. Sci. Biotech., 86: 433-42.

Heydecker, W. and Marston, M.E. (1968). Hort. Res., 8: 142-46.

Hinkle, D.A. and Eisenengen, W.S. (1950). Soil Sci., 70: 23.

Holubowicz, T. and Rebandel, Z. (1975). Prace Komisii Nauk Polni. Komis, Nauk Las, PTPN, 41 : 107-16.

Howard, B.H., Tal, E. and Mitra, S.K. (1987). Jr. Hort, Sci., 62: 485-92.

Huber, T. (1982). Ft. Var. J., 41: 104.

Hudson, J.P. (1954). Jr. Hort. Sci., 29: 27-43.

Hughes, M., Chaplin, M.H. and Dixon, A.P. (1979). HortSci., 14:46-47.

Iannetta, P.P.M., Wyman, M., Neelam, A., Jones, C., Taylor, M.A., Davies, H.V. and Sexton, R. (2000). Physiologia Plantarum, 110: 535-43.

Ivanov, A. (1984). Gradinarska i Lozarska Nauka, 21: 35-37.

Jaastad, G. (2007). Norsk frukt og boer, 1: 15

James, D.J., Knight, V.H. and Thurbon, J.J. (1980). Scientia Hort., 12: 313-19.

Janda, L. and Gavrilovic, J. (1983). Jugoslovensko Vocarstvo, 17: 3-11.

Jankovic, R., Bugarcic, V. and Subotic, R. (1981). Jugoslovensko Vocarstvo, 15: 415-21.

Jenninggs, D.L. and Williamson, B. (1982). Ann. Appl. Biol., 100: 375-81.

Jennings, D.L. (1974). Bull. Scottish Hort. Res. Inst. Assoc., 8: 34-37.

Jennings, D.L. (1976). In : Evoluation of Crop Plants. ed. NW. Simmonds, Longman, London, pp.251-54.

Jennings, D.L. (1988). Raspberries and blackberries: their breeding, diseases and growth. Academic, New York.

Jennings, D.L. and Brydon, E. (1989). Ann. Appl. Biol., 155: 507-13.

Jennings, D.L. and Tulloch, B.M.M. (1965). Jr. Exp. Bot., 16: 229-40.

Jennings, D.L., Tulloch, B.M.M. and Topham, R.B. (1967). 13th Ann. Rep. Scot. Hort. Res. Inst., p. 45.

Jennings, S.N., Brennan, R. and Gordon, S.C. (2003). 10BC/WPRS Bull., 26: 67-72.

Jennings, S.N., Ferguson, L. and Brennan, R.M. (2008). Acta Hort., 77: 203-206.

Jennings, S.N., Graham, J., Ferguson, L. and Young, V. (2016). Acta Hort., 1133: 23-28.

Jibran, R., Dzierzon, H., Bassil, N. *et al.* (2018). Hort. Res., 5(8): pp. 11.

John, M.K., Daubeny, H.A. and Chucah, A.N. (1976). Jr. Sci. Fd. Agric., 27: 877-82.

Joliffe, P.A. (1975). Canad. J. Pl. Sci., 55: 429-37.

Jones, A.T. (1979). Jr. Hort. Sci., 54: 267-72.

Jones, A.T. (1981). Bull. Scot. Hort. Res. Inst. Assoc., 19: 15-25.

Jordan, D. and Ince, J. M. (1986). Abst.22nd Int. Hort. Congress, California, No.850.

Jordan, D.T. and Ince. J.M. (1985). N.Z. Comm. Grower, 40: 31.

Kader, A.A. (1985). Coop. Exten. Univ. of Calif. Div. Agri. and Natural Researches, Spec. Pub. 3311. pp. 3-11.

Keep, E. (1961). Jr. Hort. Sci., 36: 174-85.

Keep, E. (1968a). Euphytica, 17: 417-38.

Keep, E. (1968b). Canad. J. Gen. Cytol., 10: 253-62.

Keep, E. (1969). Euphytica, 18: 256-76.

Keep, E. (1975). Garden, 100: 419-21.

Keep, E. (1976). Acta Hort., 60: 123-28.

Keep, E. (1988). Jr. Hort. Sci., 63: 1-18.

Keep, E. (1988). Jr. Hort. Sci., 63: 1-18.

Keep, E. and Knight, R.L. (1968). Rep. E. Malling Res. St. for 1967, pp. 105-107.

Keep, E. and Knight, R.L. (1986). Grower, 106: 27-30.

Keep, E. and Knight, V.H. (1986). Rept. East. Malling Res. Stn. For 1986. pp. 105-107.

Keep, E. Knight, R.L. and Parker, J.H. (1970). Rep. East Malling Res. St. for 1969, pp. 129-31.

Keep, E., Knight, V.H. and Parker, J.H. (1977). Euphytica, 26: 505-10.

Keep, E., Parker, J.H. and Knight, V.H. (1984). Rep. East Malling Res. Stn. for 1983, p. 191.

Keepel, H. (1984). Bessers Obst., 29: 113-15.

Kempler, C., Muehlichem, A.M. and Forge, T.A. (2012). Acta Hort., 926: 59-64.

Kennedy, G,G., Schaefers, G.A. and Ourecky, D.K. (1973). HortSci., 8: 311-13.

Khlistov, L. (1983). Gradinarska i Lozarska Nauka, 20 : 32-38.

Kichina. V.V. and Ivanov., S.V. (1977). Polosy, 10: 175-80.

Kingston, C. and Epenhuijsen, K. Van (1987). N.Z. Comm. Grower, 42: 18-19.

Kingston, C. and Pallcsen, J. (1986). N.Z. Comm. Grower, 41: 26-27.

Kireva, R. and Petkov, P. (2003). Selskostopanska Tekhnika, 40(2): 47-51.

Kirov, S. (1983). Pochvoznanie i Agrokhimiya, 18 : 17-20.

Kiss, F.J., Zatyko, J. and Radics, I. (1984). Kertgazdasag, 16: 63-68.

Kivi, A.R., Sartipnia, N. and Nikmanesh, L. (2014). Jr. Biodiversity and Environ. Sci., 4: 421-25.

Knight, J.N. (1976). Acta Hort., 60: 99-112.

Knight, V.H. (1980). Acta Hort., 112: 127-34.

Knight, V.H. (2004). Jugosl. Vocar., 38: 23-38.

Knight, V.H. and Barbara, D.J. (1999). Acta Hort., 505: 263-71.

Kolbe, W. (1981). Erwerbsobstbau, 23: 108-110.

Kollanyi, L. (1975). Kerteszeti Kutao Intezet Kozlemenyei, 5: 23-29.

Kongsrud, K.L. (1976). Forskning og Forsok i Landbruket, 27: 73-83.

Kowalenko, C.G. (1981a). Comm. Soil Sci. Pl. Anal., 12: 795-809.

Kowalenko, C.G. (1981b). Comm. Soil Sci. Pl. Anal., 12: 1151-62.

Kowalenko, C.G. (1981c). Comm. Soil Sci. Pl. Anal., 12:1163-79.

Kowalenko, C.G., Keng, J.C.W. and Freeman, J.A. (2000). Canadian J. Pl. Sci., 80: 363-71.

Koziovana, N.A., Solovei, E.P., Yaroslavets, E.I. and Kondrashov, G.I. (1986). Referativnyi Zhurnal 5.55.860.

Krawiec, P. (2016). Acta Hort., 1133: 305-309.

Krawiec, P., Krawiec, M. and Szot, I. (2016). Acta Hort., 1133: 469-72.

Kuhan, B.F. (1987). Tidsskr Planteavl. 91: 85-88.

Kuhne, H. (1980). Erwerbsobstbau, 22: 161-63.

Kula, M., Majdan, M., Gold, D. and Krauze-Baranowska, M. (2016). J. Food Comp. Anal., 52: 74-82.

Kumarov, V.L. and Juzepczuck, S.V. (1939). Flora of USSR Vol.9, Acad. Scientiarum, Moscow.

Kuryata, V.G., Rmenyuk, G.L. and Prokopeuko. L.S. (1985). Fiziologia itliokh. Kul'turunykh Rastennii, 17: 366-72.

Kuryata, V.G., Sogur, L.N. and Dabizhuk, T.M. (1988). Fiziologia i Biokh. Kul'turnykh Rastennii, 20: 296-302.

Lamb, J.G.D. (1956). Jr. Dept. Agric. Dublin, 52: 117-130.

Laszlovszky, Z.A. and Smolarz, K. (2000). Zeszyty Naukowe Instytute Sadownictwa I Kwiaciarstwa w Skierniewicach, 8: 239-42.

Latrasse, A. (1982). Lebensmittel Wissenschaft Technologie, 15: 45-51.

Lawrence, F.J. (1966). Diss. Abstr., 26: 6272.

Lawrence, F.J. (1980). A.R. Oreg. Hort. Soc. for 1980, 71: 102-04.

Lawson, H.M. and Wiseman, J.S. (1976). Weed Res., 16: 55-62.

Lenartowicz, W., Plocharski, W. and Zbroszczyk, J. (1930). Fr. Sci. Rep., 7: 125-34.

Lender, C., Baroffio, C.A. and Mittaz, C. (2011). 10BC/WPRS Bull, 70: 107-11.

Leonf'eva, L.I., Kornilov, B.B., Prudnikov, P.S. and Leonicheva, E.V. (2015). Vestnik Oreal GAU, 6: 65-70.

Lewis, D. (1941). Proc. 7th Intern. Genet. Congr., Edinburg, 190.

Lightle, D.M., Quito-Avila, D., Martin, R.R. and Lee, J.C. (2014). Environmental Entomol., 43: 467-73.

Linder, C., Baroffio, C.A. and Mittaz, C. (2008). 10BC/WPRS Bull., 39: 85-87.

Linder, C., Baroffio, C.A. and Mittaz, C. (2011). IOBC/WPRS-Bull., 70: 165-69.

Ljones, B. (1963). Meld. Norg. Landr. Hogsk., 42: 190.

Ljones, B. (1965). Meld. Norg. Landr. Hogsk., 44: 1-13.

Ljones, B. (1966). Meld. Norg. Landr. Hogsk., 45: 1-44.

Ljones, B. and Skshang, K. (1967). Meld. Norg. Landr. Hogsk., 46: 1-19.

Locksin, L.S. and Elfving, D.C. (1981). HortSci., 16: 521-28.

MacFarlane, S.A. and McGavin, W.J. (2010). Julius Kuhn Archiv., 427: 39-40.

Mage, F. (1975). Meld. Norg. Landr. Hogsk., 54 : 2-24.

Malagnini, V., Grassi, A., Maines, R., Angeli, G., Ioriatti, C. and Duso, C. (2009). 10BC/WPRS Bull., 50: 53-55.

Martin, L.W. (1985). HortSci., 20: 1014-15.

Mason, D.T. (1977). Canad. Agric., 22: 20-22.

Mason, D.T. (1980). Jr. Hort. Sci., 55: 165-70.

Mason, D.T. (1981a). Hort. Res., 21: 149-58.

Mason, D.T. (1981b). Jr. Hort. Sci., 56: 193-98.

Mason, D.T. (1985). Jr. Agric. Eng. Res., 31: 223-30.

Mavric, I., Virscck-Marn, M. and Koron, D. (2004). Proc. Zbornik-Reteratov, 24-26 March 2004. pp. 437-41.

McGavin, W.J., McMenemy, L.S. and MacFarlane, S.A. (2010). Archives of Virology, 155: 1897-99.

Medel, S.F. and Vargas, V.H. (1983). Agr. Sur., 11: 82-92.

Mezzetti, B., Costantini, E., Chinochetti, F., Landi, L., Pandolfini, T. and Spena, A. (2004a). Acta Hort., 649: 107-10.

Mezzetti, B., Landi, L., Pandolfini, T. and Spean, A. (2004b). BMC Biotech., 4(4): (15 March 2004).

Miletic, N., Leposavic, A., Popovic, B., Mitrovic, C. and Kandic, M. (2015). Acta Hort., 1099: 211-18.

Miyashita, T., Kunitake, H., Yotsukura, N. and Hoshino, Y. (2015). Scientia Hort., 193: 165-73.

Mont., R.J. (1955). N.Z.J. Sci., Rech. Sec. A., 36: 610-13.

Moor, U., Poldma, P., Tonutare, T., Moor, A. and Starast, M. (2014).Agron. Res., 12: 745-52.

Moore, P.P. and Hoashi-Erhardt, W. (2012). Jr. Amer. Pomological Soc., 66: 28-33.

Moro, L., Hassimotto, N.M.A. and Purgatto, E. (2017). J. Plant Growth Regu., 36: 773-82.

Munro, J.M., Dolan, A. and Williamson, B. (1988). Pl. Path., 37: 390-96.

Murant, A.F. (1987). In: Virus Diseases of Small Fruits (ed. R.H. Converse) USDA, Washington, D.C.

Naumann, W.D. (1961). Gartenbauwiss, 26: 32-56.

Nehrbas, S.R. and Pritts, M.P. (1986). Abst. 22nd Int. Hort. Congress, California, No. 849.

Neri, D., Massetani, F., Zucchi, P., Giacomelli, M. and Savini, G. (2012). Acta Hort., 926: 243-50.

Neuweiler, R. and Heller, W. (2000). Prol. 13th Int. IFOAM Scientific Conf. Basel, Switzerland. 28-31 August, 2000, 136.

Newmex, J. (1985). Canad. J. Pl. Sci., 65: 135-38.

Nikolbva, G. and Petrov, P. (1985). Proc. 1985 British Crop Prot. Conf. Weeds, 3: 1053-57.

Nowak, A., Sojka, M., Klewicka, E., Lipinska, L., Klewicki, R. and Kolodiziejczyk, K. (2017). Jr. Agric. Food Chem., 65: 2947-55.

Nunes, M.C.N., Emond, J.P. and Brecht, J.K. (2003). Acta Hort., 628: 599-606.

O'Donoghu, E.M. and Martin, W. (1988). Jr. Hort. Sci., 63: 217-20.

Oliveira, P.B., Oliveira, C.M. and Monteiro, A.A. (2004). HortSci., 39: 520-24.

Oliveira, P.B., Oliveira, C.M., Monteiro, A.A. and Lopes-de-Fonseca, L. (1999). Acta Hort., 505: 101-105.

Oosten, A. and Van, A. (1987). Fruitteelt, 77: 12-13.

O'Rourke, A.D. and Brose, M.D. (1989). Est. Bull. Coop. Ext. College of Agric., Washington St. Univ., 1521: 17.

Ourebky. D.K. (1969). Fruit Var. Hort. Dig., 23: 78.

Ourecky, D.K. (1975). In : Adv. in Fr. Breeding, (eds. J. Janick, anal.I.N. Moore), uit Purdue Univ. Press, West Lafayette, Indiana, pp. 98-129.

Ourecky, D.K. (1976). Acta Hort., 60: 135-44.

Ourecky, D.K. (1978). Fruit Var. J., 32: 50-57.

Paholo, L.T. (1956). Sad-i-Ogorod, 9: 56-58.

Palonen, P., Tommila, T. and Rantanen, M. (2016). Acta Hort., 1133: 239-45.

Papp, J. (1973). Kertezzeti Egyerem Koz., 37: 151-58.

Papp, J., Papaine, I. and Nagy, J. (1982). 21st. Int. Hort. Cong., Hamburg, Absr. No. 1059.

Petrova-Branicheva, V. (2016). Jr. Mountain Agric. on the Balkans, 19: 124-43.

Pinkerton, J.N., Ivors, K.L., Reeser, P.W., Bristow, P.R. and Windom, G.E. (2002). Pl. Dis., 86: 645-51.

PironKova, D. and Hristov, S. (1998). Pochvoznanie Agrokhimiya Ekologiya, 33(6): 86-88.

Pluciennik, Z., Bronlarek-Niemiec, A. and Masny, S. (2014). Paper presented at the 57th Fruit Plant Protection Conf., Ossa, Poland.

Poncetta, P., Iroatti, D. Mignani, I. and Gionogo, L. (2017). Acta Hort., 1155: 369-74.

Popescu, A.N. and Isac, V. (2000). Acta Hort., 538: 667-70

Potter, J.M.S. (1972). Rep. on Curivar Trial, 71: 1-16.

Powers, W.L. (1945). Science, 101: 301.

Prodorutti, D., Belletti, P.A. and Frilli, F. (2003). Notiziario-ERSA, 16(4): 34-37.

Prodorutti, D., Ferrari, A., Pellegrini, A. and Pertot, I. (2008). 10BC/WPRS-Bull., 39: 189-92.

Pyott, J.L. (1981). HortSci., 16: 308-09.

Quastm, P. (1982). Mitt. des. Obst. des Altan Landes, 37: 126-29.

Quito, D. and Martin, R.R. (2012). Acta Hort., 926: 593-99.

Rabcewicz, J., Bialkowski, P. and Konopacki, P. (2017). Jr. Hort. Res., 25: 61-66.

Rafique, M.Z., Carvalho, E., Stracke, R., Palmieri, L., Herrera, L., Feller, A., Malnoy, M., Martens, S. and Frontiers Media, S.A. (2016). Frontiers in Pl. Sci., 7. pp.1892.

Ramig, R.E. and Vandeceveye, S.C. (1950). Plant Physiol., 25: 617-29.

Ramsay, A. (1984). Fr. and Veg. Sci., February, pp. 10-11.

Ramsay, A.M., Cormack, M.R., Mason, D.T. and Witliamson, B. (1985). Agric. Eng., 40: 2-9.

Rao, A.V. and Snyder, D.M. (2010). J. Agric. Food Chem., 58: 3871-83.

Rebandel, Z. and Przysiecka, M. (1981). Akademia Rolnicza (poland). pp. 159-62.

Rebandel, Z., Przysiecka, M. and Cofta, H. (1985). Roc. Akad. Ratni W. Peznonii, ogrodnictvr, 156: 177 -90.

Rebrindel, Z. (1981). Akademia Rolnicza (potand)., pp. 163-65.

Redalen, G, (1976). Metd Norg. Landr., Hogsk., 55: 1-11.

Redalen, G. (1981). Meld. Norg. Landr., Hogsk., 60: 8.

Redalen, G. (1986). Acta Hort., 183:199-206.

Reithineier. K., Kess, H. and Junker, H. (1978). Erwerbsobstbau, 18: 65-67.

Renquist, A.R., Hughes, H.G. and Regoyski, M.K. (1987). HortSci., 22: 396-97.

Rietsema, I. (1955). Rep. Hort. Cong., 14: 712-15.

Robbins, J., Moore, P.P. and Patterson, M. (1989). Acta Hort., 262: 311-17.

Robbins, J.A. and Moore, P.P. (1990). HortSci., 25: 1623-24.

Robbins, J.A. and Sjulin, T.M. (1989). HortSci., 24: 980-82.

Rogers, W.S., Kent, W.G. and Pearl, R.G. (1936). Sci. Hort., 4: 79-89.

Romero, R., Rigueiro, A. and Fernandez, J. L. (2001). World Congr. on Conservation Agric. Madrid, Spain. 1-5 October, 2001. Vol. 2. pp. 457-62.

Romero-Franco, R., Fernandez-Lorenzo, J.L. and Riguiro-Rodriguez, A. (2001). Actas Congresso 2001 Sociedal Espanola de Malherbologia, Spain. pp. 221-24.

Ruzic, S. and Lazic, T. (2004). Jegoslovensko Vocarstvo, 38: 109-17.

Saez, A., Morales, C.L., Garibaldi, L.A. and Aizen, M.A. (2017). Basic and Applied Ecology, 19: 26-35.

Sakin; S., Agaogl, Y.S., Kabak, K. and Sakin, M. (1990). Abst. 23rd Int. Hort. Con. Firenze, Italy. No. 1272.

Sanford, J.C. and Ourecky, D.K. (1982). N.Y. Fd: and Life Sci. Bull., 97: 200.

Sanford, J.C., Ourecky, D.K. and Reich, J.E. (1985). HortSci., 20: 1133-34.

Sarkany, L. (1981). Kerteszet es Szoleszet, 30: 1-5.

Schmiolt, K., Filep, R., Orasz-Kovacs, Z. and Farks, A. (2005). Jr. Hort. Sci. Biotech., 90: 47-56.

Seemuller, E. (1987). Obstbau, 12: 356-59.

Shternshis, M., Shpatova, T. and Belyaev, A. (2016). Int. J. Agron., Vol. 2016, pp. Artixle ID 2797125.

Sidorovich, A.S. (1985). Referativnyi Zhurnal, 955.743.

Simonov, I.N. and Kulikova, S.N. (1984). Referativnyi Zhurnal, 10. 55.699.

Sipos, K. and Penzes, B. (2011). IOBC/WPRS Bull., 72: 3-5.

Sjulin, T.M. and Robbins, J. (1987). Jr. Amer. Soc. Hort. Sci., 112: 481-87.

Smale, P. and Brewer, L. (1983). N.Z. Comm. Grower, 38: 26.

Smolarz, K. and Mercik, T. (1983a). Prace. Das. I. Zark. Sand, 24: 19-100.

Smolarz, K. and Mercik, T. (1983b). Prace. Das. l. Zark. Sand, 24: 101-108.

Snir, I. (1981). Scientia Hort., 14: 139-43.

Snir, I. (1983). HortSci., 18: 710-13.

Sonsteby, A. and Heide, O.M. (2008). Scientia Hort., 117: 249-56.

Sonsteby, A. and Heide, O.M. (2009).Jr. Hort. Sci. Biotech., 84: 439-46.

Sonsteby, A. and Heide, O.M. (2012). Acta Hort., 926: 221-28.

Spak, J. and Kubelkova, D. (2000). Jr. Phytopathol., 148: 371-77.

Spayd, S.E., Norton, R.A. and Harynen, L.D. (1984). Jr. Fd. Sci., 49: 1067-69.

Stafne, E.T., Clark, J.R., Webber, C.A., Graham, J. and Lewers, K. (2005). Jr. Amer. Soc. Hort. Sci., 130: 722-28.

Stavang, J.A., Freitag, S., Foito, A., Verrall, S., Heide, O.M., Stewart, D. and Sonsteby, A. (2015). Scientia Hort., 195: 216-25.

Stephens, M.J., Alspach, P., Hall, H.K. and Kempler, C. (2002). Acta Hort., 585: 349-53.

Stevic, M., Pavlovic, B. and Tanovic, B. (2017). Pesticidi I Fitomedicina, 32(1): 25-32.

Stewart, D., Iannetta, P.P.M. and Davies, H.V. (2001). Phytochem., 56: 423-28.

Stirk, B.C. (2008). Acta Hort., 777: 403-10.

Stirk, B.C. and Bryla, D.R. (2015). HortTech., 25: 452-59.

Stirnimann, W. (2000). Revue Horticole Suisse, 73: 121-25.

Stoutemyer, V.T., Maney, T.J. and Picket, B.S. (1933). Proc. Amer. Soc. Hort. Sci., 30: 278-82.

Strik, B.C. and Cahn, H.K. (1999). HortSci., 34: 611-14.

Subotic, R. and Boskovic, I. (1979). Savremena Potjoprivreda, 27: 351-56.

Sudds, R.H. (1935). Proc. Amer. Soc. Hort. Sci., 32: 401-06.

Suss, L. and Fava, C. (1981). Informatore Fitopatologica, 31: 35-40.

Swait, A.A.J. (1980). Jr. Hort. Sci., 55: 133-37.

Taylor, C.E. and Cordon, S.C. (1975). Jr. Hort. Sci., 50: 105-12.

Terretraz, R. and Carron, R. (1987). Revue Swisse Vit. Arbo. etd' Hort., 19: 151-53.

Terrettaz, R, and Catzeflis, J. (1978). Revue Swisse Vit. Arbo, etd' Hort., 10: 189-92.

Terrettaz, R. (1978). Revue Swisse Vit. Arbo. etd' Hort., 10: 233-35.

Terrettaz, R. and Carron, R. (1980). Revue Swisse Vit. Arbo etd' Hort., 12: 269-72.

Terrettaz, R. and Carron, R. (1985). Revue Swisse Vit. Arbo. etd' Hort., l7: 153-55.

Tezotto-Uliana, J.V., Fargoni, G.P., Geerdink, G.M. and Kluge, R.A. (2014). Postharvest Biol. Technol., 91: 72-77.

Toivonen, P.M.A., Kempler, C., Escobar, S. and Emond, J.P. (1999). Acta Hort., 505: 33-38.

Topham, P.B. (1967). Ann. Bot., 31: 687-97.

Topham, P.B. and Mason, D.T. (1981). Hort. Res., 21: 29-39.

Torre, L.C. and Barritt, B.H. (1979). Jr. Amer. Soc. Hort., Sci., 104: 28-31.

Toyoma, T.K. (1961). Diss. Abstr., 22: 384.

Trandem, N., Vereide, R. and Bothun, M. (2011). 10BC/WPRS Bull., 70: 113-19.

Trudgill, D.L. and Alphey, T.J.W. (1976). Pl. Path., 25: 15-20.

Tsonkovski, K. and Paneva, V. (1978). Rastitelna Zashchita, 26: 15-18.

Tudor, T.A. (1979). Lucrari Stiintifice, Institutal Agronomic "Nicolae Balcescu" Horticultura, 22: 95-100.

Tuovinen, T. and Lindqvist, I. (2015). 10BC/WPRS Bull., 109: 123-28.

Tuovinen, T., Lindqvist, I., Grassi, A., Zini, M., Hohn, H., Schmid, K., Gordon, S.C. and Woodford, J.A.T. (2000). Proc. Int. Conf. Pest and Diseases, Brighton, UK, 13-16 Nov. 2000, pp. 333-38.

Ure, C.R. (1963). Diss. Abst., 20: 2671.

Valset, K. (1976). Landbruksforlaget, pp. 13-17.

Van Haarlem, J. R. (1948). In : Temperate to Tropical Fruit Nutrition (ed. N.F Childers)., Sommerset Press Inc., Somrnewilla, New Jersey.

VanBuren, R., Bryant, D., Bushakra, J.M. *et al.* (2016). The Plant J., 87: 535-47.

Vance, A.J., Jones, P. and Stirk, B.C. (2017). HortSci., 52: 382-87.

Vanden Heuvel, J.E., Allan Sullivan, J. and Proctor, J.T.A. (2000). HortSci., 35: 1215-19.

Vasilaka-Kis, M.D., Struckmeyer, B.E. and Dena, M.N. (1979). Jr. Amer. Soc. Hort. Sci., 104: 61-62.

Veazi, P.P. and Nonnecke, G. (1992). HortSci., 27: 331-33.

Venturi, S., Galassi, E., Dondidni, L., Tartarini, S. and Costa, G. (2012). Acta Hort., 926: 97-100.

Vetek, G., Thuroczy, C. and Penzes, B. (2006).Bull Entomol. Res., 96: 367-72.

V'ltov, B. (1984). Rastitelna Zeshchita, 32: 25-26.

Volk, G.M., Olmstead, J.W., Finn, C.E. and Janick, J. (2013). HortSci., 48: 4-12.

Waister, P.D. and Cormack, M.R. (1975). Grower, E4 : 505-06.

Wallace, T. (1938). Jr. Pom. Hort. Sci., 15: 3-13.

Wallace, T. (1951). The Diagnosis of Mineral Deficiencies in Plants, His Majesty's Stationery Office, London.

Wastson, L.E., Parker, R.W. and Polster, D.F. (1980). Manual of plant species suitability for reclamation in Alberta. Vol. 2. Edmonton, AB: Land Conservation and Reclamation Council, p. 537.

Watters, B.S. (1981). Agric. Northern Ireland, 56: 127-30.

Watters, B.S. (1983). Agric. Northern Ireland, 58 : 61-63.

Welander, M. (1985). Jr. Hort. Sci., 60: 493-99.

Welch, T. (2007). Commercial Red Raspberry production in the Pacific Northwest. A pacific Northwest Extension Pub., PNW 598 pp. 10.

Westwood, M.N. (1978). Temperate Zone Pomology. W.H. Freeman and Co. San Francisco, pp. 280-81.

Wieniarska, J., Smolarz, K. and Lipecki, J. (1982). Fruit Sci. Rep., 9:73-82.

Wilcox, W.F., Pritto, M.P. and Kelly, M.J. (1999). Pl. Disease, 83: 1149=54.

Williams, I.H. (1960). Jr. Hort. Sci., 35: 214-20.

Williams, I.H. and Hudson, J.P. (1956). Gardeners' Chronicle, I39: 34-35.

Williams, M.W. and Norton, R.A. (1959). Proc. Soc. Hort. Sci., 74: 401-06.

Winiarska. J., Lipecki, J., Smolarz, K. and Gwozdecki, J. (1986). Fruit Sci. Rep., 13: 167-72.

Woodford, J.A.T., Birch, A.N.E., Gordon, S.C., Griffith, D.W., McNicol, J.W. and Robertson, G.W. (2003). Bull. OILB/SROP, 26(2): 87-92.

Wright, C.J. and Waister, P.D. (1982). Jr. Hort. Sci., 57: 442-48.

YuShan, D., Walters, T.W. and Miller, T.W. (2017). HortSci., 52: 692-99.

Zarly, B. (1978). Acta Hort., 81: 128-35.

Zasada, I.A. and Walters, T.W. (2016). Jr. Nematology, 48: 177-82.

Zevgn, A.C. and Zhukovsky, P.M. (1975). Dictionary of Cultivated Plants and their Centres or Diversity Centre for Agri. Publ. and Doc., Wageningen.

Zindovic, J. Marn, M.V. and Pelsko, I.M. (2015). Jr. Pl. Pathol., 97(2) pp. 398.

Zurawicz, E. (2016). Acta Hort., 1133: 145-51.

# Chapter 17

# Blueberries

Blueberry is a relatively new fruit crop steadily gaining popularity in many parts of the world. With the current trend of growing interest in its commercial exploitation, the crop has become an economically important segment of the fruit industry. Blueberries obtained either from the wild or its plantations which have become one of the most popular small fruits in the United States, Canada, Chile and some of the European countries.

## Composition and Uses

Blueberries have a high nutritive value being rich, particularly, in vitamins and minerals (Table 17.1). Stanisavljevic and Jokovic (1986) studied 8 high bush blueberry cultivars (Berkeley, Bluecrop, Earliblue, Concord, Grower, Gold Traube, Herbert and Stanley) and recorded the following constituents (average of eight cultivars):

| Dry matter | 14.28 per cent | Iron | 0.300 mg |
|---|---|---|---|
| Soluble solids | 10.91 per cent | Copper | 0.0406 mg |
| Total nutrients (100 g fresh weight) | 174.85 mg | Zinc | 0.0938 per cent |
| | | Boron | 0.344 per cent |
| Nitrogen | 75.739 mg | Manganese | 0.752 per cent |
| Phosphorus | 10.991 mg | Total sugar | 9.85 per cent |
| Potassium | 77.264 mg | Reducing sugar | 8.22 per cent |
| Calcium | 12.834 mg | Total acids | 1.5 per cent |
| Magnesium | 6.172 mg | pH | 3.00 |

**Table 17.1: Composition of Blueberries (per 100 g edible portion)***

| *Constituent* | *Amount* | *Constituent* | *Amount* |
|---|---|---|---|
| Water (per cent) | 83.2 | Niacin (mg) | 0.5 |
| Calories | 62.0 | Ascorbic acid (mg) | 14.0 |
| Protein (g) | 0.7 | Calcium (mg) | l5.0 |
| Fat (g) | 0.5 | Phosphorus (mg) | 13.0 |
| Carbohydrates (g) | 15.3 | Iron (mg) | 1.0 |
| Vitamin A (I.U.) | 100 | Sodium (mg) | 1.0 |
| Thiamine (mg) | 0.03 | Potassium (mg) | 8l.0 |
| Riboflavin (mg) | 0.06 | | |

*Watt and Merrill (1963).

The major sugars present in the fruits are fructose and glucose but there is divergence of opinion about the predominant acid which is citric acid according to Kushman and Ballinger (1967) and quinic acid according to Patten *et al.* (1989b). The hybrids of *Vaccinium corymbosum* and *V. australe* have high sugar content (up to 10.9 per cent), carotenoids (up to 1.36 mg/100 g fresh fruit) anthocyanins (1992 mg/100 g), catechins (258 mg/100 g) and 340 mg/100 g of triterpenic acid (Shapiro *et a1.*, 1984). The ripe fruits of *V. myrtillus* were found to have 6.4-9.1 per cent sugars, 0.8-1.2 per cent organic acids and 96-250 ppm vitamin C (Stoyanov, 1983). The main acids were reported as citric acid, malic acid, quinic acid and tartaric acid, and different cultivars showed remarkable difference in organic acid composition (Hui *et al.*, 2011). The astringent substances in leaves of blueberry were found as tannins belonging to catechin groups. In highbush blueberry it ranged from 1.2 per cent to 6.76 per cent. (Krstic-Pavlovic and Milutinovic,1982). Blueberry is a nutrient-rich fruit due to its high levels of the secondary metabolites such as anthocyanins, flavonoids, alkaloids and terpenes (Chen *et al.*, 2017). Triterpenoids are important due to their high medicinal value (Liu, 2005); Bunea *et al.* (2011) working with cultivated and wild blueberries in Romania reported total polyphenol contents of 424.84 to 819.12 mg GAE/100g fresh weight, total flavonoids from 84.33 – 112.5 mg QE/100g and total anthocyanins from 100.58 -300.02 C3GE/100g. In *V. myrtillus*, anthocyanins petunidin-3-glucoside and delphinidin-3-glucoside have the highest contribution while in *V. corymbosum*, peonidin-3-glactoside represent the major anthocyanin. The total phenolies were higher in *V. myrtillus* than in the cultivated blueberries (Milivojevic *et al.*, 2012). The major volatiles emitted by ripening blueberries are ethylene, acetaldehyde, methyl acetate, methyl alcohol and ethyl alcohol (Kushman and Ballinger, 1967). Chun Yu *et al.* (2009) identified 67 kinds of volatile compounds in different cultivars of highbush blueberry. The character impact components (CICs) of cv. Duke were ethyl butanoate, damascenone, ethyl 2–methylbutyrate, ethyl butanoate, D-limonene, and 2- butanone. The unique flavour of highbush blueberry cultivars was formed because of interaction between these CICs.

Blueberries are consumed both fresh and processed. Besides a table fruit, it is a good additive to meat or cereal garnish. Its red-purple juice is very tasty which, on fermentation, becomes a good table wine or, on fortification, excellent cordial.

Because of good cooking and baking properties, the fruits are processed into a large number of products such as jams, jellies, sauces, pies, cakes, toppings and pastries (Galletta, 1975). Canning, drying, freezing and dehydrofreezing methods are used for preserving the blueberries. Besides giving fruits, this versatile plant meets the requirements of ornamental plantings, reforestation, wildlife and soil management programmes (Galletta, 1975). Almost all plant parts, leaves, branches and fruits are of high ornamental value because their showy colours vary with the stages of plant growth and development.

## Origin and Distribution

North America is undoubtedly the home of cultivated blueberries but its wild relatives are widely distributed in the North temperate zone and the mountains of the tropics, chiefly in North America, Europe and eastern Asia (Camp, 1942). Interestingly, majority of *Vaccinium* species have tropical origin. The wild germplasm of current cultivars is diverse but only 3 selections, *viz.*, Brooks, Sooy and Rubel have contributed more of the genes in present-day cultivars (Hancock and Siefker, 1982). Systematic collections of blueberry species have been made to expand the ecological range of commercial cultivation.

Blueberries have been domesticated in the twentieth century and are one of the most recent fruit crops to be brought under cultivation. Their domestication was probably delayed because of the availability of wild berries in abundance. Commercially grown blueberries have been categorized into three types, *viz.*, lowbush blueberries, highbush blueberries and rabbiteye blueberries. The lowbush blueberries were probably the first to come under cultivation (Eck and Childers, 1966). The credit for domesticating highbush blueberries goes to Dr. F.V. Coville of the U.S. Department of Agriculture who began his domestication work at Whitesbog, New Jersey, in 1906 (Coville, 1910). M.A. Sapp of Crestview, Florida, apparently started rabbiteye blueberry cultivation in around 1893 after he transplanted selected wild seedlings into his field (Moore, 1966).

Since the beginning of blueberry cultivation in the United States, its plantation has simultaneously spread to some other countries as well. Apart from the United States, New Zealand and Japan, blueberry has been introduced in Chile, Finland, Netherlands, Poland, USSR, Austria, Denmark, Bulgaria, Poland and some other countries where preliminary experiments on screening of cultivars are in progress. With a steady expansion in its cultivation, blueberry promises to become a fruit of worldwide interest.

## Production

The vast majority of commercial blueberry crops can be divided into two categories: highbush and lowbush. More than three quarters of the world's blueberry production is of the highbush blueberries, with the United States, Canada and Chile accounting for 76 per cent of this production. Over 90 per cent of worldwide lowbush blueberry production is in the United States and Canada, and 97 per cent of this production is in processed wild blueberries.

In 2016, world highbush blueberry production grows by 92,000MT to 655,000MT. Processed highbush production remains primarily of North American origin, as 76 per cent of processed highbush blueberry production was located in North America in 2016. Lowbush production grow by an even larger percentage than that of highbush, increasing by 24.5 per cent over 2014 and adding 41,000 MT to total 206,500 MT in 2016. Europe and China are the only other sources of these wild blueberries outside United States and Canada at 6 per cent and 3.6 per cent of global production, respectively.

Apart from USA, Canada, Chile and China, the other important blueberry producing countries are Poland, Germany, Mexico, France, Netherland, Spain, Sweden and New Zealand.

## Taxonomy

Blueberries have been placed in the family Ericaceae, genus *Vaccinium* and subgenus *Cyanococcus* (Camp, 1945). The predominant types of cultivated blueberries are hihgbush (*Vaccinium corymbosum* L. and complex interspecific hybrids with wild plants such as *V. darrowii* Camp.), rabbiteye (*V. virgatum* Ait.) and lowbush blueberries (*V. angustifolium* Ait.). Highbush blueberry is further divided into northern highbush (predominantly *V. corymbosum* L. and *V. darrowii* Camp) (Retamales and Hancock, 2012). The three types of blueberries mainly differ in plant stature though other differences have also been observed and are given in Table 17.2.

**Table 17.2: Some Characteristics of the Three Types of Blueberries***

| *Character* | *Lowbush* | *Highbush* | *Rabbiteye* |
|---|---|---|---|
| Plant height | 0.30-0.60 m | 1.8-4.0 m | Up to 6.0 m |
| Fruit weight | < 1g | 3-4 g | l0-20 per cent less than highbush |
| Fruit yield (high) | 5.0-7.2 t ha$^{-1}$ | 2-5-30t ha$^{-1}$ | 15-17 t ha$^{-1}$ |
| Ploidy level | diploid (2n=24) | tetraploid (2n=48) | hexaploid (2n=72) |
| Soil requirement | poor rocky | sandy loam to clayey | sandy loam |
| Chilling hours | not determined | 800-1000 for northern highbush (< 300 for southern highbush) | 250-600 |
| Winter-hardiness | most hardy | - | least hardy |
| Drought- tolerance | High | - | High |

* Based on the information of Hancock and Draper (1989) and Retamales and Hancock (2012).

## Cultivars

The cultivars should be selected on the basis of their adaptability to the soil and climatic conditions under which they are to be grown. As a result of selection and hybridization, a large number of cultivars have been released for cultivation. Modern trend is towards planting early cultivars. In Yugoslavia, the large-fruited cvs. Earliblue, Berkeley, Bluecrop, Blueray, Coville, Darrow and Herbert, and the small-fruited Burlington, Goldtraube, Rancocas and Rubel were recommended for production (Oblak *et al.*,1984). The cvs. Bluecrop, Herbert and Jersey were recommended for planting in Kanagawa prefecture of Japan (Katano and Shigeta, 1984). The cvs. Bluecrop, Jersey and Goldtraube were found to grow well in Poland, even on soil unsuitable for blueberries (Gruca, 1981). Cultivar Sharpblue, Misty, Diana and Baldwin perform well in Zhejiang, China (HuaJiang *et al.*, 2013). In the Netherlands, for fresh market, Bluetta (very early-ripening); Patriot, Bluejay and Spartan (early-ripening); Bluecrop, Berkeley and Coville (mid-season-ripening); Dixi (late-ripening) and Elliot (very late-ripening) were recommended (Wijsmuller, 1988a,b). The low-chill southern highbush blueberries grown in Florida were developed by crossing northern highbush cultivars, which had large fruit, high fruit quality, and early ripening but were poorly adapted in Florida (Lyrene, 2002).

**Aron**

This highbush blueberry, a Finnish cultivar, was selected from the backcross of *V. cormybosum* cv. Rancocas x (*V. uliginosum* x Rancocas), resembles *V. corymbosum* vegetatively and reaches a height of 1m (Hiirsalmi and Lehmushovi, 1982). It is more resistant to cold. It yields about 30 kg of medium-sized, good quality berries per 100 m$^2$ a year; 100 berries weigh 100-130 g. The berries may be used fresh or frozen, retaining their shape on thawing.

**Berkeley**

In a varietal trial in northern Italy, Berkeley gave the highest yield among 14 cultivars and produced fruits of good quality (Eynard *et al.*, 1985). It has been evolved from Stanley x (Jersey x Pioneer) cross. Light blue, large, firm-flesh berries of medium dessert quality are very attractive. This late-season bush is easy to propagate.

**Bluecrop**

This northern highbush cultivar was evolved from (Jersey x Pioneer) x (Stantey x June) and released in 1952. It is an early-mid-season cultivar producing large, light blue, firm and subacidic berries. The fruits have good flavour and are resistant to cracking (Brooks and Olmo, 1972). The bush is vigorous, consistent producer and drought resistant. Bluecrop has given highest yields in Italy and USSR as well (Bounous *et al.*, 1985; Butkui and Butekenc, 1987).

**Blue Ribbon (Toro x G-344)**

High-yielding, northern highbush blueberry, self-fertile, suitable for fresh market. Fruits large to medium, light blue, typically sweet with juicy and crispy texture and good post-harvest storage life.

**Blue Silk (Elliott x Puru)**

Highbush blueberry. Fruit medium to large, medium blue, firmness medium and medium sized plant.

**Bobolink™ (FL03-291) (FL00-28 x FL98-363)**

Southern highbush cultivar introduce in 2009 by university of Florida. Large fruit, light blue colour, and sweet flavour. Vigorous plant and requires cross-pollination.

**Bounty**

This cultivar is a result of Murphy x G 125 cross (Ballington *et al.*, 1989). The plants are moderately vigorous with a spreading growth habit and require extensive training during the initial years. It is a consistent producer of large fruit with above-average flavour and excellent colour, stem-scar and firmness. The fruits have long lasting flavour and, therefore, the cultivar is suitable for fresh market. It is a stem blight-tolerant and cane canker-resistant, early mid-season to mid-season cultivar.

**Brightwell**

This rabbiteye blueberry was selected in 1963 from a cross between Tifblue x Menditoo (Austin and Draper, 1983). It is productive, early-ripening, has medium-sized fruit of excellent colour, stem-scar, firmness and pleasant flavour. Due to upright growth habit and ease of berry separation, it is suitable for mechanical harvesting. However, it requires cross-pollination for fruit setting.

**Choice**

Lyrene *et al.* (1985) evolved this late-ripening rabbiteye blueberry for North Florida, from the seeds of Tifton 31. The berries are small, firm, light blue, having good flavour and small dry picking scar. The bush is vigorous, high yielder and resistant to leaf diseases. For cross-pollination, Powderblue and Bluebelle cultivars are suitable. This late-ripening cultivar has been recommended for the fresh market and pick-your-own operation.

**Cosmopolitan (Nui x Darrow)**

Highbush blueberry, suitable for home garden. Fruits large, medium to dark colour and medium firmness. Highly productive.

**Coville**

Coville is a descendent of (Jersey x Pioneer) x Stanley which was introduced by F.V. Coville in 1949 (Brooks and Olmo, 1972). The berries of this late-ripening cultivar are extra large, firm, aromatic, good in dessert quality and resistant to cracking. This vigorous and productive bush can easily be propagated.

**Croatan**

A popular northern highbush blueberry cultivar of North Carolina. Highly productive and modest winter hardy. Fruits are soft with mild flavour. Resistant to stem canker

**Duke**

It is a standard early-season cultivar developed from Stanley x Weymouth in 1986. 192-8 (E 30 x E 11). Plants of Duke are vigorous, upright and consistently high yielding. Canes are numerous, stocky and moderately-branched, exposing the fruit in all part of the plant. The berries are medium in size, have small dry picking scars, good colour and firmness. Fruit ripens early, and the flavour is mild but becomes more aromatic after several hours in cold storage (Draper *et al.*, 1987). It is recommended for commercial cultivation in areas where high-chilling highbush blueberries are grown successfully. Resistant to mummy berry.

**Earliblue**

It is a standard early season cultivar developed from Stanley x Weymouth in 1952. The berries are large, oblate, light blue, very firm and good in flavour. The plant is upright, hardy and productive (Brooks and Olmo, 1972).

**Emerald**

Highbush blueberry cultivar. Large berries, firm, medium blue with excellent flavour. Resistant to cane canker and stem blight.

**Fundy**

This lowbush blueberry cultivar is a selection from the open-pollinated seedlings of Augusta (Hall *et al.*, 1988), The bush is upright growing and gives consistently high yield. The attractive berries make the cultivar suitable for pick-your-own culture.

**Fuku-berry (Fuku= happiness) (Woodard x Tifblue)**

Rabbiteye blueberry developed from Japan in 2008. Large fruit, round-oblate, very light blue colour, good for fresh market, require cross-pollination and summer pruning.

**Georgiagem**

Georgiagem is a new low-chilling cultivar having the cultivars Earliblue and Bluecrop and *V. darrowii* in its ancestry. It is an early-ripening, moderately productive cultivar producing fruits of medium size, good colour, small stem scar, good firmness and pleasant flavour. Plants are moderately vigorous and semi-upright. Outstanding characteristics of Georgiagem are easy-rooting, early-ripening and very good fruit quality. It is recommended for areas where low-chilling blueberries are grown successfully as it requires chilling of at least 350 hours below 7.2°C (Austin and Draper, 1987).

**Jersey**

This is an important cultivar of USA, which is a cross between Rubel x Grower, having been released in 1928. Medium-sized berries are round-oblate, blue, firm, but lack aroma (Brooks and Olmo, 1972). The bush is erect, vigorous, hardy and productive. It is a late-ripening cultivar.

**Lielogu (Delite x Woodard)**

Cold resistant, rabbiteye blueberry, developed from Latvia and introduced in 2010. Large, light blue fruit, medium firmness, high-yielding, resistant to anthracnose and fire blight

**Northblue**

Northblue has originated from a cross of B 10 x US 3, made in 1968 (Luby *et al.*, 1986). It is a high-yielding cultivar with large fruit size. The plants are moderately vigorous, usually 1.0 m tall and more heavily branched than most other highbush cultivars. The fruits are more acidic, a characteristic that has been partially responsible for longer storage life. Northblue is comparable in firmness with many commercial highbush cultivars. The cultivar is particularly suitable for colder regions.

**'O'Neal**

Developed from North Carolina. Large sized berries, firm sweet and medium blue colour. Resistant to stem canker.

**Pearl**

A southern highbush blueberry. It is highly productive with good fruit quality. It ripens earlier then rabiteye cultivars in southeastern US and enable producers to participate in the lucrative early US markets (Stringer *et al.*, 2012).

**Premier**

A vigorous and productive rabbiteye cultivar. Fruits are large with good flavour, dark blue colour. Susceptible to gall midge.

**Powderblue**

Another vigorous and productive rabbiteye cultivar. Fruits are medium to large, good firmness and flavour. Fruits are resistant to cracking.

**Romero (FL95-3 x Star)**

Early ripening cultivar developed by the University of Florida and introduced in 2009. Fruit dark blue and pleasant acid blueberry flavour. It is a self-fertile and low chilling cultivar.

**Ruilan**

Developed at China from Northblue. The fruit is oblate with good flavour and blue colour. It shows high bearing and resistant to cold and multiple diseases.

**Star**

A popular cultivar of highbush blueberry, developed by the University of Florida. Fruits are large, medium blue, good firmness and flavour. It has excellent postharvest fruit quality.

**Tifblue**

Tifblue, a rabbiteye blueberry cultivar, was derived from Ethel x Clara cross at Beltsville (Brooks and Olmo, 1972). The fruit is light blue, firm, good in flavour and has very small dry scar. Dozier *et al.* (1989) reported the highest average gross income from Tifblue over a 6-year period. Susceptible to rain cracking.

Anthracnose disease symptom

Blueberry plantation

Blueberry cultivar Tifblue (Rabbiteye)

**Blueberry cultivar Jersey**

**Blueberry plant showing symtoms of Shock Virus**

**Mummy Berry Symptoms**

### Titan™ (T-959) (T-460 X FL80-11)

High yielding rabbiteye blueberry, developed from University of Georgia and introduced in 2011. Fruit very large (3.0 to 3.5g), medium-to-light blue, excellent firmness, good flavour, suitable for fresh market and machine harvest.

## Climate

Different types of blueberries differ considerably in the climatic requirements, particularly with regard to winter temperature and chilling. Cold injury to plants can occur by early fall freezes before cold acclimation, by severe midwinter freezes that exceeds the limits of the plants tolerance, or complete declamation. Ideally, blueberry cultivars for temperate regions should acclimate to cold quickly in the fall, have a high midwinter-hardiness, and deacclimate late and or slowly during spring or during unseasonably warm spells in winter, and do all of this without adversely delayed time of fruiting (Ehlenfeldt *et al.*, 2012). The cultivars Chippewa, Polaris, Northblue, Blueray and Patriot showed highest winter hardiness. The medium winter hardy cultivars were reported as Northland, Spartan, Jersey and Bluejay, and the least hardy cultivar was Duke under Latvian condition (Ehlenfeldt *et al.*, 2012). Lowbush can survive at a winter temperature lower than –40°C because of snow cover protection. Highbush types are more winter-hardy than rabbiteye as their flower buds can tolerate temperatures of –30° to -35°C while rabbiteye may be injured by temperature lower than –25°C (Hancock and Draper, 1989). Cold-hardiness is one of the most important factors affecting blueberry industry in the north of eastern Poland. Winter temperature in 1978-79 and 1984-85 were –33.4° to –34°C which killed most of the cultivars, *viz.*, Herbert, Darrow, Blueray, 13-14 A, Collins, Weymouth, Heerma, Earliblue and Lateblue. However, the cvs. Bluecrop and Berkeley were found relatively hardier (Pliszka, 1986). Hancock *et al.* (1987), in studies with 18 cultivars of highbush blueberry at East Lansing, USA, found significant correlation between bud position on the shoot and percentage of brown ovaries due to frost. Significant differences in the percentage of brown ovarics were also noted among cultivars, the values ranging from 25 per cent in Eliot to 94 per cent in Bluehaven. However, most of the variation was found associated with differences in the, stage of bud development, the least developed buds being the most hardy. Fear and Lawson (1987) observed that floret survival after unreasonably low temperature in winter was higher in Northblue, Jersey and Northsky than in Collins. The varietal differences to cold damage of flower buds in rabbiteye cultivars has also been reported as it was the least in Tifblue and maximum in Bluegem (Clark *et al.*, 1986). Fluctuating late, winter temperatures and spring frosts are detrimental to blueberry production. The regions affected with these situations should grow the late-blooming cultivars having high bud number per shoot.

Blueberries require chilling temperature for regular leaf and bloom development which varies with the type and its cultivar. Most of the highbush cultivars require 650 to 850 chilling hours below 7.2°C. The low-chilling highbush cultivars need only about 300 chilling hours. Austin and Bondari (1987a) reported a range of 250-650 chilling hours for eight rabbiteye cultivars. The rest requirement of the rabbiteye cultivar Briteblue has been determined as 600 h (Austin and Millinix, 1984). Flower bud expansion of all types is significantly influenced by the levels of accumulated chilling (Austin and Bondari, 1987b).

Blueberries needs high or low chilling hours depending on the type and cultivar. Northern highbush blueberry (NHB) cultivars are well adapted to low temperature over winter (below–2°C), developing easily with 800-1000 chill hours. Southern highbush blueberry (SHB) cultivars do not grow well at very low temperature during winter and prefers less than 550 chilling hours. Compared with NHB cultivars, SHB tend to grow longer in autumn, to have less cold resistance and to start flowering earlier in spring (Lyrene and Ballington, 2006). SHB cultivars Emerald, Jewel, Misty, Sebring, Springhigh, Sweetcrisp requires less than 300 hours chilling while cvs.

Arlen, Bladen, Reville needs 800-900 hours of chilling. Oliveira *et al.* (2017) stated that the effect of chill on precocity and yield is cultivars-dependent. Working with three highbush blueberry cultivars in Portugal, they suggested to grow cultivar Legacy in south of Portugal, with a warm climate, the cv. Duke is best for places where plants can achieve 1200 chilling hours and cv. Elizabeth in place where at least 1000 hours chilling hours is available.

Very high summer temperature impairs the fruit quality whereas severe drought can restrict the plant growth, killing them eventually. Berry-splitting and harvest-delay may result from heavy and continuous rains during harvest season.

A control environment study conducted by Banados and Strik (2006) using container grown blueberry cvs. Duke, Bluecrop and Elliott in Oregon, USA reported that plants under long-day (LD, 16h of light) had three to six flushes of growth and did not initiate flower buds or enter endo-dormancy. Bud break occurs after 5 to 15 days in green house (16h light, 22°C). Plants under short-day (SD, 8h of light) had two flushes of growth and developed flower buds after two weeks of SD and dormancy after four weeks of SD. The number of flower buds and the degree of dormancy increased with time of exposure to SD. After cold storage, bud break and bloom occurred after 6 and 26 days, respectively. Fruit harvest started 65, 95 or 130 days after bud break in cvs. Duke, Bluecrop and Elliott, respectively.

## Soil

The basic soil requirements of blueberries for healthy plant growth are acidic in nature (pH 3.5-5.5), acid peat or sandy loam with high organic matter and capacity to supply moisture constantly throughout the growing period (Brown and Draper, 1980; Spiers, 1978). Most species of blueberry thrive naturally only on acidic soils with an abundant supply of moisture during the growing season (Cain and Eck, 1954). However, Cain (1952) stated that blueberries can be grown successfully on soils above pH 6 with $(NH_4)_2SO_4$ fertilization and sawdust mulch. Spiers (1984a) stated that rabbiteye blueberries could be successfully grown over a soil pH range of 3.9 to 6.1 but leaf N content decreased above soil pH of 5.0. Rabbiteye blueberry plantings are nevertheless established on upland soil characterized by good drainage, low nutrient and water- holding capacity and low organic matter content (Cain and Eck, 1954). Haynes and Swift (1985), in a glasshouse experiment with blueberry grown between pH range of 4.0-7.0 recorded poor growth with higher pH. The plant growth was best at pH 4.3-4.6.

Unlike many fruit crops, blueberry lacks root hairs but the entire root system is very fine, fibrous and hair-like. The fine roots, however, cannot push their way into a heavy compact soil. For this reason blueberry requires open and porous soil for easy penetration of roots. The fine and fibrous nature of the roots make it necessary to have a constant supply of moisture throughout the growing period. The failure of blueberries in many orchards has been attributed to the fact that blueberries do not tolerate heavy calcareous and dry soil conditions (Cain and Eck, 1954). Soil organic matter is of special importance to blueberry soils for improving soil structure and the retention of moisture and nutrients. Johnson (1948) stressed the value of organic matter in commercial planting and recommended the addition of peat to sandy soils. Mycorrhizal infection, nutrient uptake and root activity in blueberry were positively influenced by the addition of both organic materials and sulphur, although sulphur alone was detrimental to growth (Li *et al.*, 2006). Under good nutrient supply, roots of vigorous plants have higher percentage of cortical cells with fungal hyphae than plants in poor sandy soils low in organic matter and subject to dryness (Eynard and Czesnik, 1989). Spiers (1986) stated that in a sandy, well-aerated soil, the major factors influencing root distribution appears to be the, soil moisture.

In New Jersey, USA, the major soil series Sassafras, Pocomoke, Berryland and Downer used for blueberry production often have soil pH levels much lower than the soil pH range of 4.0 to 5.2 that is considered satisfactory for blueberry. Many conventional blueberry farms in New Jersey have been evaluated with very low pH in the 3.3 to 3.9 range. The recommended target is 4.8 so as not to pose a serious risk of over-liming. The lime requirement for blueberry producing soils at 70 days of incubation with calcium carbonate show a general lime requirement of 112 kg/ha for each 1/10 soil pH unit increase desired would elevate pH of each of the soils to within a range pH 4.3 to 5.0 (Heckman *et al.*, 2002). Mulching resulted in uniform root distribution near the crown area. Incorporating peat moss concentrated roots at 30 to 45 cm depth, while mulching tended to concentrate roots at the upper 15 cm of the soil. Blasing *et al.* (1984) found maximum yield of cv. Berkeley in soil hiving a 20 cm moderately humid layer overlying a 5 to 10 cm much less porous layer which prevented rapid drying out. They opined that good soil aeration in the upper layer was important for a good crop.

## Propagation

Commercial propagation of highbush and rabbiteye blueberries is done mostly by hardwood cuttings (Childers, 1983; Hancock and Draper, 1989). Lowbush types generally consist of wild plants and thus are rarely propagated. Blueberries can also be multiplied through seed, grafting and tissue culture.

## Seed Propagation

Stratification of seeds at 3-4°C for a period of one month is essential for good germination (Butkene and Butkus, 1980; Sherestenikina and Zaranchuk, 1984). The germination is further increased by the provision of alternate temperature and light (Gulletta, 1975, Butkene and Butkus, 1980). The seeds should be sown in acidic and organic medium. The germination percentage of mature, large, filled seeds is higher than immature, small seeds which can be further stimulated by treatment with $GA_{4+7}$ at 100-500 ppm (Ballington *et al.*, 1976). Stratified seeds of *V. vitis-idaea* germinated well in light at 4-5°C but the survival of planted-out seedlings fluctuated between 10 per cent and 67 per cent of the seeds sown (Targonskii *et al.*, 1984).

## Vegetative Propagation

### Cutting

Both hardwood and softwood cuttings can be used but hardwood cuttings are preferred for propagation. In a comparison of thin, medium and thick cuttings,

Bates (1979) obtained the highest rooting percentage (70 per cent) with thick, long cuttings. In early spring, 10 to 15 cm long cuttings are prepared from the previous season's growth and inserted in a medium of peat and sand (1:1). Peat and soil mixture medium has been recommended to reduce root breakage in subsequent transplanting (Shelton and Moore, 1981a). They (1981b) further compared 14 rooting media for highbush blueberry cuttings and found vermiculite to be the best. The cuttings did not benefit from the treatment of root promoting substances. Mycorrhizal inoculation, however, promotes root formation (Powell and Bagyaraj, 1984). In trials with blueberry cultivars Blueray and Coville, Badescu *et al.* (1985) reported that hardwood cuttings taken in early December and stored at 2° to 5°C in wet sand rooted better than cuttings taken in mid-January or March. Blueray showed better rooting than Coville and best rooting was obtained in perforated polythene bags.

Zee *et al.* (2001) working with highbush blueberry cultivars Berkley, Bluecrop, Blueray, Coville, Dixi, Early Blue and Herbert obtained best rooting by using one-year-old undamaged and healthy branches (6-12 mm thick and 120-150mm long) detached from the mother bush during January and treated with 2g IBA/1 for 20 seconds. A mixture of acid peat and perlite (2:1) substrate, 75-90 per cent relative humidity and 18-20°C were reported as the optimum rooting conditions. The rooted cuttings ranged from 67.1 per cent (Coville) to 86.7 per cent (Blueray). Fisher *et al.* (2013) suggested use of mini-cuttings (four-node segments with 3.2cm length and 2.17mm diameter) which allowed higher use of propagative material. Mini-cuttings of cv. Powderblue with two cut leaves caused 98 per cent rooting in 90 days when treated with IBA at 2000mg/l for 15 seconds and planted in January in a commercial substrate (Plantmax$^{Reg.}$) and washed sand at 3:1 ratio. Bottom heat increase the percentage of rooting (Marangon and Biasi, 2013).

The cuttings strike roots by June, but are left as such till next spring to make further growth. The cuttings are then transplanted in the nursery, 15-25 cm apart and 50 cm between rows, where they are grown for another one year. Fertilizers can be applied to the plants and they should be protected from pests and diseases.

Softwood cuttings are made from the currently growing shoots in summer season by retaining two leaves per cutting. These are planted in perlite under mist. Almost 100 per cent, rooting occurs and application of growth substance promotes root formation (Childers, 1983). Softwood terminal cuttings of Woodward and Tifblue blueberry (*V. ashei*) rooted, better in media containing milled pine bark alone or mixed with equal part of perlite than in sphagnum peat moss or mixed with perlite (Pokorny and Austin, 1982). Badescu *et al.* (1985) suggested use of peat + sand or sawdust (50 per cent +50 per cent) or peat +sand +soil (50 per cent +25 per cent +25 per cent) as rooting media for softwood cuttings of blueberry cvs. Dixi, Bluecrop, Heerma II and Heerma I. The recommendations for a propagation medium of rabbiteye blueberry in Japan are the mixture of peat moss and Kanumatsuti (a volcanic ash deposit). However, addition of coal ash (clinker ash) improved root dry weight (Ban *et al.*, 2009).

Litwinczuk *et al.* (2005) compared highbush blueberry (cv. Herbert) plants propagated by softwood cuttings (HT) and obtained by micropropagation (TC) of axillary (AX) and adventitious (AD) shoots of 1-year-old *in-vitro* cultures or 11-year-old cultures (SH) in Poland. They observed cutting-derived HT plants grow more slowly produced significantly less and shorter shoots and were more variable than micropropagated plants. However, majority of the HT plants developed flowers 1-year earlier, flowered more abundantly, bore significantly larger berries than TC plants. There was no clear difference between AX and AD plants. SH-derived plants had smaller berries with the fewest seeds compared to AX and AD obtained plants. This reveals that culture age had more significant influence than shoot source for the variation observed among micropropagation systems.

## Micropropagation

Significant advances in blueberry micropropagation have been achieved along with tremendous progress in plant tissue culture techniques and their acceptance globally by commercial growers. *In-vitro* culture, using semisolid gelled media for lowbush (Debnath, 2009), highbush (Cao and Hammerschlag, 2000) and half-high blueberries (Graham *et al.*, 1996) have been developed. Although *in-vitro* propagation can be via shoot proliferation from pre-existing buds, adventitious shoot regeneration through organogenesis, somatic embryogenesis, or axillary bud development is most commonly used as it is the most reliable method to get true-to-type micropropagules (Debnath, 2017). Woody plant medium (WPM) or modified WPM is mostly used as basal medium with the addition of cytokinin and zeatin for shoot proliferation. Direct *ex-vitro* acclimatization and rooting in hydroculture showed good results (Vescan *et al.*, 2012).

Automation in large-scale liquid culture can get rid of most manual handling of using semisolid gell media in micropropagation and can reduce the propagation costs significantly. Although liquid media in a bioreactor systems enables control over microenvironmental conditions (aeration, agitation and dissolved oxygen) of *in-vitro* culture (Paek *et al.*, 2005) and can be used an alternative to gelled media to speed up and reduce costs of micropropagation (Levin and Vasil, 1989). The newly developed platform bioreactor (Welander *et al.*, 2014) functions well for both large-scale micropropagation and basic research (Welander *et al.*, 2017).

## Planting

The field is prepared for planting by ploughing to eliminate or incorporate the existing vegetation and to improve the internal drainage. Cover crops can be grown to improve the soil structure and organic matter content. Before planting, the rows are stocked as per layout plan.

Planting is done in early spring, early fall and late winter but early spring is the best planting period. The planting distance depends upon the blueberry type to be grown. Lowbush blueberries are raised in solid plantings as they are naturally seeded and spread by rhizomes. The small plants do not need more space. Highbush plants are given a distance of 1.2 to 1.5 m between plants and 2.4 to 3.0 m between rows. Rabbiteye blueberries are spaced 1.8 m apart in rows and 3.6 m between rows (Hancock and Draper, 1989). In high density planting, spacings of 0.6 m and 0.7 m are provided for highbush and rabbiteye types, respectively (Shelton and Freeman,

1989). High density, planting results in higher production, more profit and quicker return of investment and thus compensates for extra initial cost of additional plants.

Traditionally highbush blueberry plants were planted at 1.35m between plants in the row and 2.70m between the rows (Eck, 1988) which was subsequently revised to 1.20 X 3.0m (Eck *et al.*, 1990). Rabbiteye blueberry usually planted at 1.5 to 2.4m between plants in row and from 3.7 to 4.3m between rows (Eck *et al.*, 1990). Decreasing the plant spacing from 1.35m to 0.91m in a row increased 14 per cent yield of highbush blueberry (Eck, 1988). Austin and Mullinix (1980) working with rabbiteye blueberries recorded 25 per cent and 40 per cent yield increase from plants spaced within rows at 1.2m when compared with 1.8 and 2.4m, respectively. Moore *et al.* (1993) studied the influence of in-row plant spacing on the yield and fruit size of highbush blueberries cvs. Blueray (erect growing) and Bluecrop (spreading). Over the 5-year harvest period, plots with 0.61m plant spacing within a row produced a cumulative total yield of 17.24 t/ha more than plots with conventional 1.2 to 2m spacing. Plant spacing did not affect fruit size. Stirk and Buller (2005) reported total dry weight per hectare of wood, crown, and roots (but no leaves) was 218 per cent to 562 per cent higher when plants were at 0.45m than at 1.2m within rows

## Pruning

At planting, bare-root plants generally have one-third to two-thirds of the branches removed to balance top growth with the roots. The remaining one to three strong canes per plant are headed back 50 per cent to a strong bud. Any flower buds (round/plump) should be rubbed or clipped off. Early flower buds should be removed during the first two years after planting.

During the second year of establishment, pruning should consist of removing weak, damaged, or diseased wood and flower buds. In the third year, flower buds may be left on the most vigorous shoots. By the fourth season a light crop may be harvested, but flower buds should be thinned to prevent over-fruiting and bending of the shoots (Bratsch, 2009).

Blueberry bears fruit on the terminal 2-3 inches of previous season's growth. Most cultivars tend to overbear making it necessary to thin out the bushes and prune off the excessive buds to encourage new growth for the next year's crop. Pruning depends on the growth habit of the cultivar. Cabot and Pioneer cultivars have spreading growth habit requiring pruning of the lower drooping branches, whereas cultivars like June, Rubel and Concord need more thinning at the centre. Efforts are made to develop an open upright plant. The shoots are thinned, leaving only five or six branches at the centre. The criss-crossing shoots are also removed. Pruning is generally carried during dormant period but sometimes old or misplaced wood is removed immediately following harvest.

Pruning comprises the removal of weak and broken branches and the wood older than 4 years in order to increase the berry size, fruiting shoots of the cultivars Cabot. Pioneer and Sam should be cut back to three to five fruit buds per shoot (Eck, 1986). Amount of cutting back varies with the number of fruit buds per shoot as determined by growing conditions and the cultivar. Siefker and Hancock (1987) found a significant correlation between the percentage of basal area removed in pruning and berry weight. The basal area of bush can be removed up to 40 per cent without affecting the yield. In trials with 3-year-old mother plants of cv. Bluecrop, Smolarz (1985) recommended hard pruning to 5 cm above the ground level which stimulates the production of many good quality cuttings.

Stirk *et al.* (2003) compared the effect of pruning intensity on berry weight, yield and harvest efficiency in mature 'Bluecrop' and 'Barkley' for five years. They compared (a) conventional pruning with removal of the most unproductive canes, thinning of 1-year-old shoots from the top of bushes, (b) speed pruning- where only one or two of the most unproductive canes were removed at their base, and (c) unpruned control. The unpruned trees produced highest yield while the conventionally pruned plants produced fruits that were 27 per cent larger and could be harvested in about half the time. Fruit began to ripen on the conventionally pruned bushes about 5 days earlier than on the unpruned ones. The speedily pruned bushes were intermediate for all these characters.

Lowbush cultivars are generally pruned by mowing or burning on biennial basis. Hall *et al.* (1986) reported increased branching of shoots following pruning with a modified burner. In lowbush blueberry plants, growth response to mechanical pruning at various heights was compared with that achieved by the conventional burning practice (McLeod and Misener, 1988). The plant growth response indicated that elevated mechanical pruning was not feasible. However, the crop can be mechanically pruned at ground level if the remaining stem sections are burned later. Lyrene (1984), working with rabbiteye blueberry at Florida found that removing the terminal 10 cm from shoots on 27 October induced flower buds below the cut, even though no new vegetative growth occurred. The average number of new flower buds/twig ranged from 1.3 for Delite to 6.1 for Climax, with an average of 3.0 for all the clones tested. Unpruned twigs of the same clones averaged only 0.5 new flower buds on comparable twig sections. Mainland (1986) found significant increase in yield of rabbiteye blueberry by combined summer and winter pruning. In addition to winter pruning, southern highbush berries are also commonly hedged at 100-120cm after the fruiting season.

## Nutrition

The nutrient requirement of blueberries is comparatively low than many fruit trees. However, several studies proved that regular fertilization is required for a sustainable yield and good fruit quality. Traditional granular fertilizer is being replaced by liquid and controlled release fertilizer to minimize nutrient leaching (Vargas and Bryla, 2015; Smith and Harris, 2017).

### Role of Nutrients

Tamada (1989) studied the nutrient deficiencies of highbush and rabbiteye blueberries in water and soil culture experiments. The plants, receiving all the essential mineral nutrients were completely healthy with large leaves of normal green colour but exhibited deficiency symptoms in the absence of nutrients.

### Nitrogen

Deficiency of nitrogen in blueberry caused stunted growth, low vigour, chlorotic leaves and decreased yield (Hanson and Hancock, 1996). Nitrogen application increases yield, fruit size, branching, percentage of long shoots (> 20 cm) and leaf N, P and K contents but decreases leaf Ca and Mg contents (Austin and Bondari, 1986; Eck and Stretch, 1986; Blasing, 1989). High rates of N fertilizers, however, reduce the plant growth, yield and frost resistance (Patten *et al.*, 1988a; Smolarz and Mercik, 1989). Martin and Pelofske (1983) obtained the highest average yield (47.6 kg ripe fruits per eight plants) of cv. Bluecrop from plots receiving 140 kg N/ha, with a leaf N level of 1.96-2.06 per cent and from soil of pH 4.6, At soil pH of 3.0-3.5, Kruger and Naumann (1984) recorded the maximum yield in cv. Berkeley with 30 kg N/ha, higher doses (60, 90 or 120 kg N/ha) were found to reduce the yield. Nitrogen source for fertilization of blueberry has been extensively researched. Use of ammonical source of N was suggested by Poonnavhit and Darnell (2004). If the soil pH is below 5.0, urea is suggested and if above 5.0 pH, ammonium sulphate is the preferred source of N (Krewer and NeSmith, 2006). Peterson *et al.* (1988) found $NH_4$ to be a better nitrogen source for highbush blueberry than $NO_3$ because the absorption of $NH_4$-N is more rapid. The plants receiving $NH_4$-N were twice in size and dry matter yield than those treated with $NO_3$-N. Brown *et al.* (1988) stated that sulphur-coated urea is an acceptable N source for blueberries compared with ammonium sulphate, but sulphur-coated urea is less effective in maintaining low soil pH.

The use of liquid fertilizer through drip irrigations system is now in practice in blueberry production. Soluble ammonical compounds provides soil acidification, slow movement through the soil profile and are the preferred N source for blueberry production. Smith and Harris (2017) working with southern highbush blueberries cv. Star reported adequate plant growth and superior fruit quality by using liquid fertilizer at 63.5 per cent $NH_4$-N or controlled release fertilizer at 66 per cent $NH_4$-N.

Mycorrhizal infection of blueberries plays a key role in N nutrition as it increases the yield by enhancing N uptake (Bajwa and Read, 1986; Read, 1987). A high level of mycorrhizal colonization is found in acid soil with good nutrient supply than in poor sandy soil (Eynard and Czesnik, 1989).

### Phosphorus

Phosphorus is essential for plant growth. P deficiency symptoms are slight purpling on the leaves, dull leaf colour in rabbiteye blueberries and yellowish green colour in highbush blueberries (Tamada, 1989). Spiers (1984b), in a study with rabbiteye blueberries grown in sand culture with different levels of P (10-27 mg/l), found that phosphorus concentration in the leaves increased linearly but that of Cu and Mn decreased in response to increasing level of P fertilization. As leaf P decreased below 0.09 per cent, P deficiency symptoms were found.

Hart *et al.* (2006) advised application of P if soil P level is less than 50ppm and leaf P is below 10 per cent. At 26-50ppm soil P and 0.08-0.01 per cent leaf P, up to 45kg $P_2O_5$/ha is recommended which should be increase to 45- 67kg $P_2O_5$/ha when soil and leaf P are below 25ppm and below 0.07 per cent, respectively.

### Potassium

In absence of K, rabbiteye and highbush blueberry plants made the least growth proving its essentiality (Tamada, 1989). K deficiency causes interveinal chlorosis on the young leaves, while at later stages scorching and defoliation occur. In rabbiteye blueberries visual K deficiency symptoms became increasingly evident as leaf K content decreased below 0.25 per cent (Spiers, 1984a). Hancock and Nelson (1988) did not find any correlation between soil K and yield, and in only one year leaf K was significantly associated with yield. The optimum leaf K level is about 0.43 per cent (Eck, 1977). Eck (1983) obtained the highest yield (up to 7.95 kg/plant) of blueberry cv. Bluecrop with 40 kg K/ha, which resulted in leaf K content of 0.33 per cent to 0.59 per cent.

Krewer and NeSmith (1999) suggested one or two application of K in light textured soil where leaching of nutrient is high. Hart *et al.* (2006) suggested application of 84 to 112kg/ha $K_2O$ when soil levels of K is between 0-100ppm and leaf K is less than 2 per cent.

### Calcium

The role of calcium is similar to K on growth and defficiency symptoms of blueberry plants (Tamada, 1989). The green region of leaf is narrower than in K deficiency and the leaf surface uneven. Deficiency symptoms of Ca in rabbiteye blueberries were more prevalent when leaf Ca decreased below 0.20 per cent (Spiers, 1983). They are highly efficient in Ca uptake as they grow in low Ca environment (Twigg and Link, 1951). The defficiency, however, adversely affects the plant growth in high pH condition due to excessive Ca absorption.

Stuckrath *et al.* (2008) studied the effect of foliar application of Ca (0.78g, 4.68g and 7.8g $Ca_2^+$) in pre-harvest, at three different conditions (tunnel, mesh, and ambient) on texture and pectin cotent. Differences ($P <=0.05$) in Ca content were found in tunnel cultivar fruit, during the period of cell expansion toward harvest, at Ca foliar level of 5ml/l (7.8g). A linear correlation between Ca concentration and texture was established and a good correlation between low methoxyl pectin and Ca content was obtained with the high dose of Ca (5ml/l).

### Magnesium

Mg deficiency results in distinct reddening between the veins followed by necrosis (Tamada. 1989). The leaf colour is more dull in rabbiteye blueberry than in highbush blueberry. Magnesium deficiency can be corrected by applying dolomite at 1 ton/ha (when the soil pH is bellow 4.0 or 4.5) and the application should be made in autumn (Hart *et al.*, 2006).

### Iron

Interveinal chlorosis appearing first on younger leaves is a typical symptom of iron deficiency (Tamada, 1980). Severe chlorosis was cured by applying 4 oz FeEDTA (10 per cent Fe) to 3-year-old plants which increased Fe content and reduced Ca content (Clark *et al.*, 1987a). Hanson and Hancock (1996) stated that iron deficiency symptom is the indicator of high soil pH (>5.5). At high pH, Fe becomes unavailable (Hart *et al.*, 2006). Foliar application of iron chelate (10 per cent Fe) at

1kg/400 l water per hectare was recommended to overcome the deficiency (Krewer and NeSmith, 1999).

### Manganese

Because of lower soil pH in which blueberries grow, Mn absorption is higher but the plants do not exhibit the toxicity symptoms at foliar level up to 15000 µg/g (Korcak, 1988, Miller, 1987). The critical deficiency levels of Mn in leaves are 20 to 30 µg/g dry weight. The concentration of Mn in roots increases in the presence of Al at higher levels in the soil solution (Korcak, 1988).

Peterson *et al.* (1987) reported poor plant growth and higher levels of extractable aluminium when the plants grew in soils treated with aluminium sulphate than in soils treated with sulphur alone. But the addition of sawdust was essential for good plant growth regardless of either S or Al application.

### Boron

Deficiency of B results tip dieback. Leaves were mottled and cupped shaped. Shorter internode and leaf chlorosis also appears (Hanson and Hancock, 1996). Hart *et al.* (2006) suggested application of borax (11 per cent B) at 11-22kg/ha in B-deficient plants. In acid podzol soil where low availability of B tends to keep foliar B concentration of less than 24ppm standard in lowbush blueberry, treatment with DAP (89kg P/ha), $ZnSO_4$ (3.3kg Zn/ha) and either borate 2.2 kg B/ha or solubor at 0.74kg/ha, increase B content of leaf and yield (Smagula and Litten, 2002).

## Tissue Analysis

Hanson (1987) found weak correlations between soil and leaf nutrient levels for P, K, Ca and Mg in fully grown plants of highbush blueberries. He further stated that soil analysis could be helpful in monitoring the nutrient status of young plants. On the basis of a study of seasonal variation in nutrient content in the leaves of highbush blueberry cultivar Bluecrop, Clark *et al.* (1959) demonstrated that least variation in leaf values for most elements occurred in the sampling period following harvest, *i.e.*, between mid-July and mid-August. Ca, Mg and Zn levels in the leaves did not vary much with the sampling dates, year, mulch or location. The matured mid-shoot leaves on fruiting canes were sampled for analysis. They further opined that the optimum ranges of specific nutrient elements in blueberry leaves were still to be determined.

After consolidating the leaf analysis data from several sources, Cain and Eck (1954) suggested the following deficiency levels for blueberry below which the plants might respond to nutrient application (Table 17.3).

**Table 17.3: Estimated Deficiency of Leaf Nutrient Levels in Blueberries on Dry Matter Basis***

| N | P | K | Ca | Mg | Fe | Mn | Zn | Cu | B |
|---|---|---|---|---|---|---|---|---|---|
| (per cent) | | | | | (ppm) | | | | |
| 1.80 | 0.07 | 0.40 | 0.30 | 0.08 | 60 | 20 | 10 | 10 | 10 |

* Cain and Eck (1954).

In a complete nutrient treatment in water and sand culture experiments on rabbiteye and highbush blueberries, the leaf values recorded for various nutrients are given in Table 17.4 (Tamada, 1989).

**Table 17.4: Nutrient Levels in the Leaves of Tifblue rabbiteye Blueberry on Dry Matter Basis***

| Experiment | N | P | K | Ca | Mg | Fe | Mn |
|---|---|---|---|---|---|---|---|
| | (per cent) | | | | | (ppm) | |
| Water culture | 1.67 | 0.24 | 0.78 | 0.34 | 0.290 | 70 | 74 |
| Sand culture | 1.14 | 0.113 | 0.55 | 0.55 | 0.306 | 105 | 75 |

* Tamada (1989).

Studies with 6-year-old blueberries cv. Tiflbue and Premier showed that N,P,K, Mn and Zn contents tended to decrease before harvest, stabilize at harvest and increase slightly after harvest. Mg, Ca and B increased as the growing season progressed, while Cu decreased. Fe showed high variability without a clear tendency during the season. The period of greatest stability of macro- and micronutrient demand of both cultivars per ton of fruit production was 4.7, 0.35, 5.2, 2.8 and 0.74 kg for N,P,K,Ca and Mg, respectively. The demand for micronutrients fluctuated between 6 to 60g per ton (Vidal *et al.*, 1999). Lafond (2013) determined the minimum and maximum leaf micronutrient concentration in wild lowbush blueberry (*V. angustifolium*) under the climatic and edaphic conditions of the Saguenay Lac Saint Jean region, Quebec, Canada. Minimum and maximum leaf concentrations were (mg/kg) for Al (26.2-73.5), B (32.2-52.9), Cu (3.2-6.5), Fe (27.8-61.4), Mn (873-1394) and Zn (11.0-17.3). The determination of these new minimum and maximum leaf micronutrient concentrations established sufficiency ranges for the growing conditions in the region. Leaf analysis need to be interpreted with knowledge of genotype, growing environment, berry size and yield (Stirk and Vance, 2015). Separation of species for leaf sampling and determination of nutrient sufficiency range seems warranted (Barlow *et al.*, 2017).

## Fertilizer Recommendations

Usually, blueberry plants are fertilized with ammonium sulphate or urea under field conditions and are fertigated with other nutrients as needed, based on leaf tissue analysis (Bryla and Stirk, 2015).

Warman (1985,1987,1988) could not get enhanced yield of lowbush blueberry with any farmyard manure or fertilizer treatment but green manuring increased Mn level with a concomitant decrease in Mg level of leaves probably because of pH reduction. Childers (1983) recommended 68 kg N, 705 kg $P_2O_5$ or 578 kg of complete fertilizer (8-8-8) per hectare for highbush blueberry to give the best result. K should be derived from potassium sulphate instead of potassium chloride as the latter is injurious to young plants.

The growth response, yield and fruit size of rabbiteye blueberry remained unaffected with the rate and source of fertilizers applied (Austin and Bondari,

1986, Spiers, 1987a, b). Peterson *et al.* (1988), however, got increased plant growth with 112 and 224 kg N/ha over control but 336 kg N/ha decreased plant growth. On comparing various N sources, they obtained the best plant growth by sulphur-coated urea followed by ammonium sulphate.

In highbush blueberry, fertigation with ammonium sulphate resulted more growth and fruit production than fertigation with urea (Vargas and Bryla, 2015b).

## Irrigation

Adequate water availability is essential for good growth and fruiting of blueberries as both drought and water logging conditions are harmful. The absence of root hair makes the establishment of blueberry planting unsuccessful under dry soil conditions. Typically, 25 to 50mm of water per week is recommended for blueberry although actual water requirements will likely vary not only with environmental conditions, but also with the cultural practices under which the crop is grown (Bryla and Strik, 2007). It is recommended to use four drip laterals per row for commercial production of highbush blueberry in sandy soil (Holzapfel *et al.*, 2015). Trickle irrigation has been found very useful in such a situation. Since blueberries have minimal lateral transport of water and nutrients, even distribution of water throughout the root zone is essential (Abbott and Gough, 1987). In a split root water application experiment, the highbush plants on unirrigated site had poor vegetative growth and fruiting (Abbott and Gough, 1987). Byers and Moore (1987) have shown that 57 per cent to 75.5 per cent irrigation water could be saved during the growing season by a moisture method without affecting the highbush plants, yield, berry size, berry quality and vegetative growth. Seymour *et al.* (2004) working with lowbush blueberry recorded 43 per cent higher yield with irrigation than rainfed plots. However, the rainfed berries had greater firmness, higher Brix and lower moisture content than irrigated berries.

The highbush bluebeery plants are injured under conditions of high transpirational demand and low soil moisture. Cameron *et al.* (1989) studied the influence of soil moisture stress on the growth and gas exchange characteristics of highbush plants and reported significant reduction in gas exchange rates, leaf number and whole plant leaf areas due to stress. Benoit *et al.* (1984) reported that greenhouse lowbush blueberry plants grown under water treatments of 0.2 (about field capacity), 0.6, 1.0 and 5.0 bars did not differ significantly in number of stems per plant or flower buds per stem. A significant increase in the number of flowers per stem was found for the 0.2 and 0.6 bar treatments over 1.0 and 5.0 bar treatments. Established field plants showed no significant difference in the number of flowers between the unirrigated control and the 0.6 or 0.2 bar treatments. The number and total weight of berries in the crop-year increased significantly with increased water availability during the vegetative growth-year but individual berry size remained the same. SuJin *et al.* (2004) reported soil water potentials in well-water (daily irrigation in 4 l pots) and water stressed (irrigation at 7-day interval) were about -0.22 and -0.99Mpa, respectively, where as leaf water potential in well-watered and water-stressed highbush blueberries (cv. Rancocas) were -1.07 and -1.79 Mpa, respectively. The relationship between net $CO_2$ assimilation rate and stomatal conductance was linear ($r^2$ =0.96**) in well-watered blueberry, but curvilinear ($r^2$ =0.90**) in water stressed blueberry, demonstrating the involvement of non-stomatal factors in photosynthesis in water stress condition. Erb *et al.* (1988) advocated growing of drought resistant plants in dry areas to combat moisture stress conditions.

Holzafel *et al.* (2004) evaluated the response of highbush blueberry (Bluetta) to different levels of water application, from 20 to 133 per cent of reference evapotranspiration, under microjet and drip irrigation over a period of 7 years at Chile. They stated that water requirement of blueberry is a functions of canopy cover and type of irrigation method used. With drip and microjet irrigation, fruit yield increased monotonically with higher levels of water supply. During the first two years of harvest, plants under drip irrigation produced higher yields than microjet. However, in the fourth and subsequent season microjet surpassed drip irrigation. For the last season, a 7-year-old blueberry had the highest yield of 10300kg/ha for microjet irrigation, with a level of water replacement of 6200 $m^3$/ha, compared with 6800kg/ha for drip irrigation with the same amount of water application.

Bryla and Strik (2007) investigated plant water requirements in three northern highbush blueberry cultivars, Duke, Bluecrop, and Elliott, grown either at a high-density spacing of 0.45m apart within rows or a more traditional spacing of 1.2m, spacing between rows was 3.0m, high-density planting significantly increased total canopy cover and water use per hectare. However, although canopy cover (often considered a factor in water use) increased up to 246 per cent, water use never increased more than 10 per cent. Because of more canopy cover at high-density, less water penetrated the canopy during irrigation (overhead sprinkler) or rain, reducing both soil water availability and plant water potential in each cultivar and potentially reducing water use. Among tested cultivars, water use was highest in 'Duke' which used 5-10mm/day and lowest in 'Elliott' which used 3-5mm/day. Peak water use in cultivar was during fruit development, but water use after harvest declined sharply.

A large number of studies indicate that blueberries are sensitive to flooding. Abbott and Gough (1987) reported delayed flowering and fewer flowers in blueberry cultivar Bluecrop following flooding. Davies and Flore (1986, 1987) noted reduced carbon assimilation in rabbiteye blueberry after only 1-2 days of flooding. Root conductivity was also reduced in flooded plants (Davies and Flore, 1986; Crane and Davies, 1987). The response of 3-year-old cultivar Woodard (*V. ashei*) plants to 0, 5, 15, 25 and 35 days of flooding during spring and summer was studied by Crane and Davies (1988). Almost all plants survived initial flooding, but plant survival decreased to 83 per cent after 25 days of spring and 67 per cent after 5 days of summer flooding.

## Soil Management

Soil management and irrigation are important for the establishment of blueberry plants. Patten *et al.* (1989a) studied the effect of orchard management practices and irrigation on the growth and yield of rabbiteye blueberry. The plants were grown for 4 years with or without mulch, with cultivated or sodded row middles, and under various irrigation treatments. The results are given in Table 17.5. Plant establishment and growth were optimum with 360° low volume spray emitters.

**Table 17.5: Influence of Orchard Management Practices and Irrigation Treatments on the Growth and Yield of Rabbiteye Blueberry Plants***

| *Treatment* | *Canopy Volume ($m^2$)* | | | *Yield (kg/plant)* |
|---|---|---|---|---|
| | *1984* | *1985* | *1986* | *1985 and 1986* |
| Sod | 0.10 | 0.27 | 0.69 | 2.54 |
| No sod | 0.09 | 0.26 | 0.61 | 2.68 |
| Mulch | 0.10 | 0.28 | 0.67 | 3.02 |
| No mulch | 0.90 | 0.24 | 0.62 | 2.39 |
| One drip emitter | 0.08 | 0.23 | 0.59 | 2.51 |
| Two drip emitters | 0.07 | 0.21 | 0.59 | 1.90 |
| 360° low volume spray emitters (LVSE) | 0.15 | 0.37 | 0.87 | 3.29 |
| 40° LVSE | 0.10 | 0.26 | 0.53 | 2.76 |

* Patten *et al.* (1989a).

Dale *et al.* (1989) obtained the highest yield and vigour of highbush blueberry plants with the- incorporation of peat before planting, use of sawdust mulch and application of high level of irrigation.

## Mulching

Since, blueberries do not survive in prolonged drought conditions and in view of the specific characteristics of the roots, mulching is necessary for better growth (Eck and Childers, 1966; Cain and Eck, 1954). It aids in controlling weeds and maintaining uniform soil moisture. Straw chaff, peanut husk or sawdust may be used for mulching. Legume hay mulch has been reported to be injurious and should be avoided (Shoemaker, 1955). Patten *et al.* (1988b) reported increased growth of drip irrigated plants mulched with sawdust. Salt induced leaf chlorosis and necrosis were evident on plants with no mulch whereas mulched plants showed healthy growth. In mulched soil K, Na, Mg and Cl levels were several times lower and uniform throughout the soil profile compared with the non-mulched treatments. In comparison with control, mulch increased the plant volume and yield of the cultivar Tifblue (Patten *et al.*, 19989a). Lareau (1989) concluded that sawdust mulch is superior to any other factors which contribute to the growth and productivity of highbush blueberries grown on mineral soils. If optimum levels of nutrients are available, the sawdust mulch improves soil moisture content and aeration.

In coastal Columbia and adjacent area of the Pacific Northwest, beds of blueberry are usually mulched with conifer sawdust (SD), which is very effective at deterring weed growth, moderating soil temperature and moisture, and maintaining pH in the range optimal for blueberry. However, the high C/N ratio of SD mulch can stimulate microbial immobilization of nitrogen (N), necessitating the application of fertilizers in excess of actual crop requirements in order to avoid crop N deficiency (Forge *et al.*, 2013).

The increasing use of mulching in modern agriculture has occurred with the advent of plastic materials. However, there are several major concerns about using mulch films which linked mainly to the end of life of the plastic mulch films: removal and disposal of the used plastics at the end of crop cycle and environmental issues, when the plastic materials are not collected or properly disposed of. The availability of new materials such as the biodegradable and compostable materials such as the trade mane Mater-Bi® (produced using vegetable components such as starches, vegetable oils) that show mechanical properties similar to those of traditional plastics are an interesting solution. Mater-Bi® mulch film shows positive effects on the moisture content in the soil during the cultivation period (Girgenti *et al.*, 2012). Woven black weed mats are successful in weed control, protection of soil from erosion and hand picking disturbance. It also allow rainfall and irrigation water to permeate once the material has aged (Ireland and Wilk, 2006).

### Weed Control

The lowbush blueberry fields pose a serious weed problem. Although some weeds are controlled by cutting and burning, many are controlled by chemical treatments. Hexazinone, a broad spectrum herbicide, provides control against many herbaceous and woody species (Jensen *et al.*, 1981; Yarborough and Ismail, 1985). Jensen (1986) reported increased yield of blueberries by 85 per cent and 75 per cent following application of atrazine and hexazinone, respectively. Increased yield due to greatly improved weed control were related to increase in the number of flower buds per shoot and in the density of fruiting shoots. Yarborough (1985) compared the efficiency of hexazinone, 2.4-D, dicamba and glyphosate for controlling *Betula* sp., *Acer rubrum* and *Salix* sp. Dicamba and glyphosate were more effective than hexazinone and 2,4-D. Glyphosate controlled aspen (*Populus trernuloides*) most effectively but caused some injury to blueberry plants (Yarborough and Ismail, 1985). Post-emergence application of 1.1 kg sethoxydim + 1 per cent oil and 1.1 kg fluazifop-butyl + 0.5 per cent wetter showed 88 per cent and 75 per cent control, respectively of *Digitaria sanguinalis* (Wallinder *et al.*, 1982). Hanchar *et al.* (1985) reported that 1.1-4-5 kg hexazinone/ha would be economically feasible for weed control in lowbush plants, whereas hexazinone at the rate of 3.3 and 2.3 kg/ha maximized yield and profit, respectively. Weed management is difficult in berry production, especially in organic system. Organic foliar herbicides are often expensive and only partially effective, although the successful use of various mulches for weed management has been reported (Forge *et al.*, 2013).

## Flowering, Pollination and Fruit Set

### Flowering

The shoot growth pattern of blueberry is episodic and sympodial. In the first growth flush, most shoots come from axillary buds on the previous year's growth (Gough *et al.*, 1978). Flower buds are produced on primary, secondary and tertiary shoots that are produced in the first, second and third growth flushes (Lindberg, 2013). The reproductive buds of blueberries consist of terminal and later flower buds. The induction of flower buds from vegetative buds begins at the shoot apex and moves basically in response to a short day length with appropriate temperature (Darnell, 1991). Fruits are produced on one-year-old woods with reproductive buds in the apical area.

Floral bud initiation occurs under short day length in late summer and early fall (northern hemisphere). Six weeks of 8-h photoperiod in the fall promoted floral bud initiation in rabbiteye blueberry (Darnell, 1991); southern highbush blueberry initiate floral buds at 8-h photoperiod but not in 16-h photoperiods (Spann *et al.*, 2003). In early-mid-season, and late-season northern highbush blueberry cultivars, floral bud initiation only occurred under short day (at constant temperature 22°C) with number of flower buds correlated to the length of exposure (Stirk, 2012). Plants grown under short day for four or more weeks ceased growth and entered endo-dormancy, whereas those grown under long day make continuous growth and did not initiate flower buds or go dormant (Banados and Strik, 2006).

Summer pruning of vigorous growth in highbush blueberry leads to the production of lateral shoots, depending on fruiting date. Longer lateral length was positively correlated with number of flower buds (Bandos *et al.*, 2009). Rabbiteye blueberry (Tifblue) when pruned in early summer (June, in Japan) stimulated secondary shoots and later in autumn, terminal flower buds of this stimulated secondary shoots produced fruits in the following year (Karimi *et al.*, 2017).

Blueberry flowers are perfect and are generally borne with the corolla opening inverted. Individual flowers consist of a five-lobed white of whitish pin urn-shaped corolla that are 0.6-1.2cm long and upto 8mm wide (Eck and Mainland, 1971). Rabbiteye flowers generally have narrower corollas than highbush flowers. The style is multi-lobed and hollow, with at least five stylar canals (Gough and Shutak, 1977). During stigmatic receptivity nectar is produced at the base of the corolla and pollen is shed through pores at the end of each another.

## Pollination and Fruit Set

Among the three important cultivated species of *Vaccinium*, only *V. corymbosum* (highbush) is self-compatible. *Vaccinium angustifolium* (lowbush) and *V. ashei* (rabbiteye) are self-incompatible, and hence require cross-pollination for a good fruit set (Stushnoff, 1972; Stushnoff and Palser, 1969). A good commercial crop is ensured only when about 80 per cent of the flowers set and fruits mature (Eck, 1986). High fruit set can be achieved only through cross-pollination by insects as the pollen grains of *Vaccinium* sp. are relatively heavy, thereby making wind pollination difficult. Also, the structure and position of the flower do not permit ready self-pollination. Flowers hang in pendant position with the pistil extended beyond the stigma. The sides of the stigma are angled and act like an inverted funnel, thereby preventing the pollen from falling on to the stigma. The flowers are well adapted for insect-pollination because of well-developed and fragrant corolla and location of nectaries at their bases.

For commercial blueberry, production, autogamy and self-compatibility may act as an insurance policy against total crop failure during acute pollinater shortage. Selfing and cross-pollination in the same blueberry cultivar can produce extraordinary fruit sets of between 70 per cent and 80 per cent (Sampson and Spiers,

2002). A longer pistil extending beyond the mouth of a flower increases opportunities for pollinator contact and stigmatic loading (Kearns and Inouye, 1993). Therefore, selecting cultivars with squatter blooms and longer styles could profoundly increase blueberry yield (Tamada *et al.*, 1977).

Adequate fruit set in rabbiteye blueberry (cv. Bluebell) is expected to occur after pollination of flowers up to at least 5 day after anthesis (DAA) (Young and Sherman, 1978). Moore (1964) assessed the length of flower receptivity in highbush blueberry by recording fruit set after hand pollination at time intervals ranging from 0 to 5 DAA. Fruit set of 'Bluecrop' declined sharply at 5 DAA.

Pollen quantity and viability of five rabbiteye blueberry cultivars were determined by Qin *et al.* (2013) to provide the scientific basis for selecting appropriate pollinizers. The pollen quantity and pollen viability of 'Premier' was 600.34 and 79.84 per cent, significantly higher than other four cultivars. They suggested use of cvs. Premier, Tifblue and Baldwin to use as pollinator for rabbiteye blueberry.

Among the insects, honey-bees are the most important pollinators and their use is regarded as a necessary practice in maximizing blueberry yield (McGregor, 1976; Eck, 1988). Honey-bees comprise about 95 per cent of the insect visitors to this crop and their foraging activity makes them efficient pollinators. Hensels (1983), in an experiment, grew blueberry bushes for 2 years in gauze cages, in some of which a small population of bees was placed just before flowering. In the first year average yield amounted to 5600 g/bush for plots with bees compared with 375 g/bush for plots from which bees and other pollinating insects were excluded. In the second year, the corresponding average yields were 4350 and 1050 g/bush, respectively. Goodman and Clayton (1988) reported that plants visited by bees gave a higher fruit set (61 per cent vs 18 per cent), increased fruit numbers (1156 vs 144) and more berry weight (1.6 kg vs 0.2 kg) than did plants caged to prevent access to bees and large insects. In self- and cross-pollination studies with blueberries, Lyrene (1989) reported only 37 per cent fruit set under self-pollination compared with 74 per cent to 91 per cent fruit set following cross-pollination.

Blueberry are native to North America, where there are no native honeybees and they are pollinated in the wild mostly by bumblebees (*Bombus* sp.). Zajmi *et al.* (2012) studied honeybee vs. bumblebee pollination in highbush planting in Kosovo. They observed that honeybees placed in the experimental plot visited the blueberry flower for a short period of time and did not enter the flower. In contrast bumblebees from commercial hives placed in the plots visited flower for longer periods of time. Hicks (2011) studied the activity of native and introduced bees in pollinating the lowbush blueberry in Newfoundland, Canada. Native bees have been shown to be important to blueberry pollination in this province. The diversity and abundance of native bees was greatest in managed blueberry plots compared to unmanaged plots but this did not manifest into greater fruit set. The supplementation of blueberry plots with imported bees *Bombus impatiens* and *Apis mellifera* did not increase fruit set and the diversity of native bee species was decreased in the supplemented plots.

The efficacy of Japanese honeybee [*Apis cerana japonica* (ACJ)] and European honeybee [*Apis mellifera* (AM)] on rabbiteye blueberry was investigated in Japan

(Maki *et al.*, 2000). ACJ was able to work under cooler condition than AM, even in the dusk after sunset. The fruit set ratio of flowers which ACJ pollinated was high enough and reached at the similar level of those by artificial pollination.

Gibberellic acid ($GA_3$) treatment increase fruit set and yield of rabbiteye blueberry (NeSmith and Krewer, 1992). $GA_3$ can be used to salvage the crop following a moderate freeze during bloom (NeSmith *et al.*, 1995).

## Fruit Growth and Development

Blueberry fruit is a true berry, consisting of a mullti-carpellate ovary with 100 or more ovules. The endocrap consists of a single layer of stone cells which constitute the lining of the five compound locules (Yarbrough and Morrow, 1947).

Growth of blueberry in terms of fruit diameter measurement exhibits a cyclic growth pattern like that of stone fruits (Young, 1952; Edwards *et al.*, 1970; Spiers, 1981). Stage I is a period of rapid pericarp development characterized by cell division. The duration of stage I of first and second harvest fruits lasts to about 30 days in the lowbush and highbush Species whereas in later harvests the stage I lasts up to 60 days (Young. 1952). Edwards *et al.* (1970) found that the duration of stage I can be increased by low temperatures and decreased when a late bloom is followed by rise in temperature.

Stage II is characterized by rapid endosperm and embryo development. This stage can vary (5-56 days) considerably between blueberry species and between cultivars of a species as well. Young (1952) reported stage II to be shorter in the lowbush species than in the highbush cultivars.

Stage III is the period of final rapid mesocarp growth due principally to cell einlargement. Apart from attaining maximum size, it builds up sugar content and develops colour from light purple through dark purple and finally to blue. After attaining dark blue colour the berry does not increase much in size except for slight increase in sugar content and flavour (Eck, 1986).

Shimura *et al.* (1986) studied the seasonal fruit growth curve of the cultivars Weymouth and Jersey (*V. corymbosum*) and Woodard and Tifblue (*V. ashei*) and observed a double sigmoidal pattern. They divided the fruit growth into three stages: (I) a period of rapid growth, (II) a period of slow growth, and (III) a second period of rapid growth which lasts until the fruit ripens. The stage II period of the early cv. Weymouth was found to be the shortest and that of Tifblue, a late cultivar, the longest. Pericarp colouration in Weymouth occurred between stages II and III. In Tifblue, it occurred during the later part of stage III. Fruit respiration increased between I and III or during the early part of III. Ethylene evolution peaked at the same time or a little earlier than respiration depending on the cultivar.

NeSmith (2006) studied the fruit development period of several rabbiteye blueberry cultivars in Georgia. He calculated the fruit development period (FDP) for the different cultivars as a function of real time (days) and thermal time (HU). Depending on cultivar and chill hours at a location, flowering time varied as much as 24 days from year to year. Average FDP for the cultivars recorded were: Alpha 75.3 days; Austin 81.7 days; Premier 82.3 days; Climax 84.4 days; Brightwell 87.4

days; Tifblue 92.2 days and Ochlockonee 93.9 days. The shortened fruit development period of cv. Alpha resulted in a similar ripening time with cvs. Climax and Premier, even through flowering time of cv. Alpha was 7 to 10 days later. Using HU instead of days generally reduced the coefficient of variation associated with FDP.

## Maturity

Maturation of the fruit begins with the onset of stage III of the growth curve. During this period the fruit undergoes biochemical and physiological changes and attains its full size and maximum edible quality. The soluble solids increase rapidly. The final part of maturation is marked by the onset of ripening, characterized by a rapid increase in the rate of respiration. Fruit acidity decreases sharply but soluble solids continue to increase. There is a rapid fall in the chlorophyll content while anthocyanin starts accumulating.

Taking into consideration the days from full bloom to harvest, the fruit growth period of blueberries ranges from 50 to 90 days in highbush and lowbush and from 60 to 130 days in rabbiteye (Eck, 1986). The length of period is greatly influenced by environmental factors, temperature being the major factor influencing berry growth. The degree of pollination is the other factor influencing the length of growth period, since higher seed counts are associated with larger fruit and shorter maturity cycle (Tamada *et al.*, 1977). Within the flower clusters, individual flowers will open over a period of 2 weeks contributing towards a protracted harvest period that may require as many as three to four pickings to remove the entire crop.

JiaBo *et al.* (2011) reported increased total sugar content during fruit development, which was Duke > Bluecrop > Elliot; while the contents of total acid decreased as the fruits reached full-maturing was Duke < Bluecrop < Elliot. The contents of vitamin C were higher at young fruit stage while the contents of pigment reached to the highest at fruit ripening stage, which was Elliot > Duke > Bluecrop.

The intensity of colour was found to increase in the first six days (in cv. Jersey) and then stabilized. The percentage of total sugar increased for 9 days after colour and then stabilized (Woodruff *et al.*, 1960).

## Ripening

Currently, there is no consensus on whether blueberry is a climacteric fruit or not. Frankel (1972) could not find any surge in respiration or ethylene production during ripening of highbush blueberry, or treatment with accteldehyde or ethylene (Janes *et al.*, 1978). However, Lipe (1978) noted increase ethylene production at the redberry stage in rabbiteye blueberry.

Ripeness of blueberries is determined by their blue colour which reaches up to 100 per cent when the berry is fully ripe. Ripening date may be affected by several component stages of development between flower bud break and ripe fruit. Lyrene (1983) found that ripening date was better correlated with flowering-to-ripening interval than with flowering date in rabbiteye blueberries. Gupton (1984) in estimating fruit ripening date of fourteen rabbiteye blueberry cultivars, stated that the time required to complete a specific stage of fruit development differs among cultivars. Date of bud break was found to have little effect on ripening date (Table 17.6). Interval between bud break and anthesis was not a good predictor of ripening date. The cv. Menditoo completed this stage as rapidly as Climax and Premier. Though its buds began to swell earliest, Powderblue fruit progressed most rapidly from anthesis to ripe fruit on early shoots, followed by Climax, Centurion and Beckyblue.

In cvs. Bkomidon, Northcountry and Northland, berry development from blue-pink to fully mature required about 3-6 days (Lin *et al.*, 2012). Forney *et al.* (2012) studied the compositional changes in blueberry fruit during ripening. The primary sugars were glucose and fructose, which increased from about 2 to 6 per cent as fruit ripened from white to blue. Citric acid comprised 60 to 80 per cent of the organic acids in blueberries and declined by 60 per cent as fruit ripened from white to blue. Additional acids present consisted of quinic, malic, and small amount of succinic and tartaric acids. Total blueberry phenolics in turning fruit were 60 per cent less than in white fruit and 12 per cent less than in blue fruit. Anthocyanins increased as colour developed. Antioxidant capacity decreased as fruit ripened from white to turning. Fruit firmness decreased about 80 per cent as fruit ripened from white to blue.

Abundance of phosphoenolpyruvate carboxykinase (PCPCK) was noted in the flesh at different stage of development. In addition, changes in the amounts of citrate, malate, soluble sugars, isocitrate lyase, NADP-malic enzyme, phosphoenolpyruvate carboxylace, pyruvate, and orthophosphate dikinase in the flesh of ripening fruit were detected. Dissection of blueberries provided evidence that both PEPCK and phosphoenolpyruvate carboxylase were present in the same cells (Famiani *et al.*, 2005). Xylose was found to be the most aboundant noncellulosic neutral sugar associated with fruit walls, and the observed high Xyl/Glc ratio suggested that xylans, which are usually a minor hemicellulosic fruit wall component are abundant in blueberry. The pectic matrix showed increased solubilisation at early and intermediate stage of ripening, but no changes were detected at late ripening. Little reduction in pectin polymer size occurred during blueberry ripening. In contrast, hemicellulose levels decreased as ripening progressed, and a clear depolymerisation of these components was observed (Vicente *et al.*, 2007).

Goto *et al.* (2013) working with highbush blueberry cvs. Earliblue, Berkeley and Jersery noted increased IAA concentration during seed enlargement. During the matururation stage, the change in the ABA concentration resemble that of ethylene evaluation. Ethylene evaluation peaked at the same or on colour stage earlier than ABA concentration. The ABA concentration increased simultaneously with ethylene evaluation in 'Earliblue' and 'Barkeley' and the ABA concentration increased later than ethylene evolution in 'Jersey'.

Temperature seems to be a major factor influencing the ripening of blueberries (Shutak *et al.*, 1956). At a suitable temperature, berries develop blue colour, either light or dark. Berries ripen faster at 27°C and slower at 16° to 21°C (Shutak *et al.*, 1956). Gorchov (1984) stated that summer-ripening plants have more asynchronous ripening than autumn-ripeners. Ripening asynchrony was mostly due to variability in the time taken for fruit development. Fruits with large seeds developed faster.

**Table 17.6: Classification of Cultivars by Estimated Time Required to Complete Specific Stages of Bud and Fruit Developed on Early and Late Developed Shoots in 1982***

| *Cultivar* | *Julian Date of Bud Break* | | *Days from Bud Break to Anthesis* | | *Julian Date of Anthesis* | | *Days from Anthesis to Ripe Fruit* | | *Days from Ripe Fruit to 50 per cent Ripe* | | *Days from Anthesis to 50 per cent Ripe* | | *Julian Date 50 per cent Ripe (Sample Shoots)* | | *Julian Date 50 per cent Ripe (Whole Plant)* |
|---|---|---|---|---|---|---|---|---|---|---|---|---|---|---|---|
| | *Early* | *Late* | *Early* | *Late* | *Early* | *Late* | *Early* | *Late* | *Early* | *Late* | *Early* | *Late* | *Early* | *Late* | |
| Aliceblue | 29 | 52 | 47 | 37 | 76 | 89 | 74 | 64 | 19 | 32 | 93 | 96 | 169 | 185 | 173 |
| Beckyblue | 36 | 54 | 38 | 31 | 74 | 85 | 72 | 68 | 13 | 12 | 85 | 80 | 159 | 165 | 162 |
| Bluebelle | 35 | 38 | 37 | 42 | 72 | 80 | 94 | 76 | 20 | 33 | 114 | 109 | 186 | 189 | 196 |
| Briteblue | 20 | 35 | 55 | 47 | 75 | 82 | 94 | 76 | 17 | 25 | 95 | 96 | 170 | 178 | 175 |
| Centurion | 46 | 58 | 43 | - | 89 | - | 70 | - | 29 | 45 | 99 | - | 188 | 191 | 195 |
| Climax | 36 | 50 | 40 | 33 | 76 | 83 | 70 | 63 | 9 | 13 | 79 | 76 | 155 | 159 | 160 |
| Delite | 31 | 49 | 44 | 36 | 75 | 85 | 84 | 69 | 8 | 34 | 92 | 103 | 167 | 188 | 176 |
| Homebell | 32 | 52 | 41 | 31 | 73 | 83 | 80 | 76 | 17 | 15 | 97 | 91 | 170 | 174 | 170 |
| Menditoo | 52 | 57 | 33 | 33 | 85 | 90 | 81 | 68 | 20 | 33 | 101 | 101 | 186 | 191 | 195 |
| Powderblue | 45 | 53 | 37 | 37 | 82 | 90 | 64 | 56 | 18 | 26 | 82 | 82 | 164 | 172 | 170 |
| Premier | 40 | 58 | 37 | 34 | 77 | 92 | 76 | 61 | 8 | 7 | 84 | 68 | 161 | 160 | 160 |
| Southland | 32 | 48 | 41 | 36 | 73 | 84 | 73 | 67 | 18 | 35 | 91 | 102 | 164 | 186 | 171 |
| Tifblue | 41 | 55 | 38 | 41 | 79 | 96 | 80 | 70 | 6 | 17 | 86 | 87 | 165 | 183 | 169 |
| Woodard | 32 | 44 | 41 | 37 | 73 | 81 | 73 | 72 | 18 | 20 | 91 | 92 | 164 | 173 | 170 |

* Gupton (1984).

Blueberries do not ripen uniformly thus extending the harvesting time by 3-7 weeks. Uniform ripening can be achieved by the application of ethephon (240 to 1920 ppm). A single foliar spray applied two weeks before the first harvest increases the percentage of the total yield of ripe fruits during first picking (Eck, 1970; Warren *et al.*, 1973; Dekazos, 1978). In order to minimize the loss of fruit firmness, multiple application of low ethephon concentration (500 ppm) is more desirable (Eck, 1986). Poll and Wood (1983) reported that spraying with ethephon at 500 or 1000 ppm advances ripening. With the higher rate, only 2 harvests were necessary to remove most of the fruits.

## Harvesting

As agricultural labour has become scarce and expensive, the use of machines in blueberry production has gained importance. Blueberry picking starts around 10 July in Michigan; 20 May in North Carolina and 20 June in Jersey (Shoemaker, 1955). Gernerally, 3-7 pickings are made at weekly interval to remove the entire crop. Only fully ripe berries are picked. Mechanical harvesting of ripe blueberries results in a substantial loss of berries falling to the ground. Losses accounting up to 16 per cent have been reported in Michigan (Cargill and Ledebuhr, 1983) and more than 20 per cent in North Carolina (Mainland *et al.*, 1975). It has been reported that machine-harvested blueberries have poorer quality than hand-harvested ones (Miller and Smittle, 1987). Decay of machine-harvested blueberries during postharvest handling is a major problem of the industry (Ballinger *et al.*, 1978; Cappellini *et al.*, 1982). Capellini *et al.* (1982) reported that about 15.2 per cent of the blueberries in New York markets were defective and about two-third of this was due to decay. In spite of this drawback, machines continue to be in common use for blueberry harvesting. In West Germany, Heermann (1984), in comparing the hand and mechanical harvesting of highbush blueberries, stated that the first, harvesting may be by hand followed by machines having drums with fingers, or adjustable frequency beaters (4 pairs). At a speed of 2-4 km/h, 3 workers can harvest 1 tonne/h, compared with hand picking at 30-50 kg/day for one worker. Patten *et al.* (1988c), in studying the efficiency of machine harvesting at different times of day with rabbiteye blueberries, reported that the number of mature fruits remaining on the plants after harvesting decreased with later harvest times during the day. Thus, an increase in harvest efficiency corresponded to decreased leaf-water potential. The cultivars recommended for mechanical harvesting are Ama, 13/16A and Goldtraube 71 (Wijsmuller, 1988b).

Over-the-raw (OTR) has been used to harvest blueberry destined for processing and freezer since early 1960s (Dale *et al.*, 2010), however research has shown that blueberries harvested by OTR machines have significantly reduced shelf life (Takeda *et al.*, 2013), and as much as 25 per cent of fruit is lost to the ground, thus reducing the packable yield (van Dalfsen and Gaye, 1999). Hu *et al.* (2017) proposed a platform-based harvest-aid system which offers a promising solution. The platform-based, harvesting-assist system will use hand-held, pneumatic shakers such as the "Golia' model. The platform will be larger enough to accommodate two workers on each who will operate a hand-held shaking devise. Compared with existing harvesting methods (OTR harvester, hand-harvest, and the handle shaker with portable fruit catch frame), the semi-mechanical harvesting systems will have the benefits (a) it can significantly improve harvest efficiency compared with hand-harvest by about 10-20 times, (b) harvest aid system will significantly reduce fatigue because the workers do not need to hold the shaker all time, (c) ground losses are expected to be reduced greater due to a new design of the conveyance system, and (d) the system will be scale neutral: its performance will be appealing to large and small growers alike, whereas its cost will be substantially lower than a regular OTR harvester and be affordable to small-and-medium-sized growers.

Takeda *et al.* (2017) demonstrated that pneumatic shakers could remove 3.5-15 times more blueberries (g/minute) compared with hand harvest. Differences in picking rates were observed across cultivars. Soft fruit catching surfaces reduced the impact forces and bruise damage to blueberries. The blueberry producers have great concerns with labour costs and availability and are open to exploring mechanization as a way to mitigate the need for hand-harvest labour (Gallardo *et al.*, 2018).

Several attempts have been made to develop a rapid, quality-based method for sorting mechanically harvested blueberries (Ballinger and Kushman, 1970; Hamann *et al.*, 1973; Wolfe *et al.*, 1974). Another method based on bulk density has been developed for sorting out mechanically harvested blueberries (Patten *et al.*, 1989b). Freshly harvested blueberries are separated into four bulk density classes based on sink/float determination in 2, 4 and 6 per cent NaCl brines. A 4 per cent brine removes all green, 90 per cent of red fruit and 10-20 per cent of blue fruit. A solution of 6 per cent brine floats the remaining red berries and 50-60 per cent of the blue fruits. The berries that sink in 6 per cent brine are over mature and mechanically damaged. This method provides a basis for the development of separation techniques for mechanically harvested blueberries.

## Yield

Hancock and Draper (1989) reported marked variation in the yield of three types of blueberries. The yield is furthermore determined by the climatic and cultural factors than with the genetic potential of a cultivar. According to them, as high as 7.2 t/ha yield has been recorded in lowbush plants but the commercial yield is only 1.8 t/ha. The highest yield of highbush plants is 30 t/ha, the commercial average being 5.5 t/ha. Likewise, rabbiteye blueberry gives a maximum yield of 17 t/ha with a commercial average of 3.0 t/ha.

Maodobry *et al.* (2001) recorded total yield in 11 years at southern region of Poland for four highbush blueberry as Bluecrop -40.6kg, Heerma -30.2kg, Crotan -24.7kg and Darrow -21.7kg per bush. In northern Spain, the highbush blueberry cv. Ivanhoe produced 6.44kg and Berkeley 6.36kg/tree (Ciordia *et al.*, 2004). The rabbiteye blueberry cultivars yield about 6.0kg/plant. The average yield of cvs. Austin and Brightwell were 6.0 kg/plant. Late season cv. Ochlokonee produced 8.1 kg/plant compared with 4.3 kg/plant in cv. Tifblue (NeSmith, 2006).

## Storage

Blueberries were less firm and exhibited more postharvest decay when harvested by machine than harvested by hand (Ballinger *et al.*, 1973; Austin and

Williamson, 1977). Fruit decay could be reduced by rapid cooling of highbush blueberries, especially when combined with fungicides and controlled atmosphere (Ceponis and Cappellini, 1983a, 1985). Blueberries that are destined for shipping should be cooled to 1°C within 4h of harvest (Sargent *et al.*, 2006). Fruits cooled at 0°C to removed field heat increase eight-to-tenfold shelf life than non-cooled fruits (Beaudry, 1992). The recommended cold storage requirement for blueberries is to store at 0-1°C and 85-95 per cent relative humidity (Viente *et al.*, 2005), which can keep the fruit in acceptable condition for 2-3 weeks (Viente *et al.*, 2005; Schotsmans *et al.*, 2007). Temperature during storage is a critical factor in the development of decay in blueberries. Ballinger *et al.* (1978) recorded that berries held at 22.2°C developed an unacceptable 20 per cent decay within 1 to 5 days. Plump, firm and uniformly coloured blueberries can be held in acceptable and marketable condition for 2 weeks at 0°C if stored in film-capped baskets (Hruschka and Kushman, 1963). Bunemann *et al.* (1957) also reported that blueberries could be held in acceptable eating quality for 4 to 6 weeks at 0°C. Pre-cooling of blueberries to 2°C reduced the decay when they were held for 24 hours at 21°C after a three-day simulated transit period at 10°C (Hudson and Tietjen, 1981). Miller (1986) stated that fruits of rabbiteye blueberries (cvs. Climax and Woodard) that were harvested by machine were softer and had high moisture loss and decay after 1, 2 or 3 weeks' storage at 3°C than hand-picked berries.

Bunemann *et al.* (1957) tried several modified atmospheres ranging from 5 per cent to 25 per cent $CO_2$ and 2 per cent to 15 per cent $O_2$. Off flavours developed in all the modified atmospheres. Smittle and Miller (1988) compared the storage life of rabbiteye blueberry cvs. Climax and Woodard harvested by hand and by machine and stored in air, 100 per cent $N_2$ or in 10, 15 or 20 per cent $CO_2$ with 5 per cent $O_2$. Storage of fruits in high $CO_2$ atmospheres resulted in greater percentages of marketable and firm fruits and better sensory ratings than did storage in air. Fruit quality characteristics of hand-harvested fruits stored for 42 days and machine-harvested stored for 21 days in 20 per cent $CO_2$ with 5 per cent $O_2$, remained like those of fresh fruits. The $CO_2$ atmosphere was found more effective in retarding decay in the rapidly cooled berries than in those cooled to 2°C for 48 or 72 h (Ceponis and Cappellini, 1983b).

Modified atmosphere packaging (MAP) was reported effective in extending the storage life of blueberries. Highbush blueberry (cv. Lateblue) packaged in micro-perforated (1mm diameter) and non-perforated film maintained better quality attributes throughout the cold storage (4°C for 15 days). Non-perforated film packaging reduced pathogen-induced decay during cold storage (Chiabrando and Giacalone, 2008). Moggia *et al.* (2014) reported that MAP increased the proportion of sound fruits (80-90 per cent) compared to control (<60 per cent). MAP also retained fruit firmness and lower the proportion of dehydrated fruits, which amounted to 4-10 per cent in MAP vs. 20-30 per cent for control.

Harb and Streif (2004) recommended storage of highbush blueberry cv. Duke at 12 KPa$CO_2$ + 18 KPa$O_2$ at 0-1°C storage, which maintained the fruit firmness on acceptable values. Incidence of *Botrytis* was also efficiently controlled at this atmosphere storage for six weeks. Rabbiteye blueberries (cvs. Centurion and Maru) stored at 2.5 $KPaO_2$ + 15 $KPaCO_2$ at 1.5°C, showed only half as much blemished fruit compared with storage in regular air. Additionally, fungal development in 'Maru' fruit was minimised by CA storage (Schotsmans *et al.*, 2007).

Fungicide sprays and dips, hot fungicide dips and hot water dips reduce postharvest decay and maintain quality during storage (Burton *et al.*, 1979). Dekazos (1980) reported that blueberries, sprayed before harvest with daminozide and/or ethephon at specific stage of berry development and stored at 30°C and 95 per cent to 100 per cent RH, effectively maintained storage quality for more than two months. Rabbiteye blueberry fruits are more resistant to fungal decays than highbush fruits. Postharvest decay of Jersey bluberries (*V. corymbosum*) could be controlled by dipping in 10 g captafol/l water for 30 min or 100 g captafol/l for 3 min (Ballinger, 1983).

Gray mold is a major disease during storage of blueberry. A large number of studies focusing on postharvest technologies in blueberry have been reported with attempts to extend storage and shelf life and preserve quality during storage and shipment (Saito and Xia, 2017a). Controlled atmosphere prevented mold infection during storage and had little effect on quality (Hancock *et al.*, 2008). Dual release $SO_2$ generating pads or slow release $SO_2$ generating pads in combination with MAP effectively controlled fruit rot in blueberries and reduced fruit-to-fruit spread of gray mold during cold storage (1°C for 5 wks) (Saito and Xia, 2017a). Chitosan coatings containing essentials oils (carvacrol, cinnamaldehyde and trans-cinnamaldehyde each at 0.5 per cent) are effective in extending the shelf life of fresh blueberries. Adding essential oils with chitosan substantially decreased bacterial and yeasts/molds on the fruit (Sun *et al.*, 2014). Blueberry leaf extracts (BLE) had greater variety of phenolic acids. Chitosan coating (2 per cent W/V) and 8 per cent or 12 per cent BLE blueberry fruits with MAP (3$KPaO_2$+12$KPaCO_2$) and stored at 2±°1C and 95±2 per cent RH for 35 days storage and 3 days at room temperature showed effective control of fruit decay, extend shelf life and maintain high nutritional value of fresh blueberries during postharvest storage (GuiYun *et al.*, 2014).

## Pests

### Gall Midge (*Dasineura oxycoccana*)

Blueberry gall midges over-winter as larvae or pupa in the soil below the blueberry bushes, with adult flies generally emerging in late January to early February in Florida. The female gall midge lay eggs into expanding floral buds as well as tender new vegetative tissues, at a time when the floral buds are beginning to separate (Lyrene and Payne, 1995). Egg hatch occurs within a few days, giving rise to larvae. These larvae develop in the buds by feeding on the nutritive plants juices using enzymes that destroy tissue within the proximity of the oviposition site (Gagne, 1989). Developing floral buds or shoots can die within a week of egg deposition. High infestation in rabbiteye blueberries cause up to 80 per cent yield loss (Lyrene and Payne, 1995).

Native natural blueberry gall midge include larval predators like *Toxomerus marginatus* and parasitiod include *Aprostocetus* sp., *Aphanogmus* sp., *Caraphron pallidiventris*, *Quadrastichus* sp. and *Platygaster* sp. (Sampson *et al.*, 2006). Clear sticky sheets are an effective and practical tool for monitoring blueberry gall midge and should hung with their bottom edge about 5cm from the ground. If bucket traps are used, they should be placed as close to blueberry bushes as possible to ensure accurate emergence monitoring (Rhodes *et al.*, 2017). Diazinon, malathion, and spinosad spray are effective to control gall midge in blueberries (Liburd, 2004).

## Flower Thrips (*Frankliniella* spp.)

In southeastern united States, three species of thrips-*Frankliniella bispinosa*, *F. tritici* and *F. occidentalis* were reported (Turner and Liburd, 2007). First two larval instars and the adult fed on pollen, styles, ovaries petals and fruit which resulted reduced fruit set and deformed or softened berries. Arevalo (2006) suggested use of acetamiprid and spinosad to control flower thrips.

## Blueberry Aphid (*Illinoia pepperi*)

The aphids suck sap on new succulent leaves and stem tips. Several species of parasitic wasps (Pnion and Aphidius sp.) and predatory insect attack aphids and their eggs (Isaacs *et al.*, 2008).

This aphid is a vector of blueberry shoestring virus and is found in most commercial plantings. Five morphological forms have been recognized during its seasonal phenology (Elsner and Kriegel, 1989). Vigorous new shoots are excellent habitat for aphid populations. The aphid has been controlled successfully with pirimicarb at 828 g a.i./ha and acephate at 1106 g a.i./ha without affecting its predators' population (Whalon and Elsner, 1982).

## White Grub (*Cyclocephala longula*)

During the past few years, white grubs become recognized as a pest of southern highbush blueberries in California. White grubs feed on plant roots causing the plants to be stunted. In some cases, plant death has occurred when large grub attack newly planted fields. Entomopathogenic nematode *Heteterorhabditis bacteriophora* and insecticide imidacloprid provide good control (Haviland and Hernandez, 2012).

## Blueberry Maggot (*Rhagoletis mendax*)

The blueberry maggot is a significant pests in North America and if left unchecked, infestations can reach 100 per cent. Larvae feed internally, and infested fruit can be difficult to identify without destructive sampling, making it possible for infested fruit to be missed in the sorting process (Burrack and Littlejohn, 2012). Placement of baited traps using the mixture of ammonium acetate and protein hydrolysate effectively captures flies when placed on the top of the canopy (Teixeira and Polavarapu, 2001). Insecticides imidacloprid, organophosphates and thiocloprid provides good control.

## Spotted-wing Drosophila (SWD: *Drosophila suzukii*)

It is an invasive fly pest of thin-skinned fruits from Asia that has become established throughout the United States, Canada and Europe. The host range of

SWD includes grape, strawbwrry, caneberries and cherries. The female SWD has evolved a modified ovipositor that is selerotized and serrated, which unlike more other drosophilid flies, allows her to attack fruits as they are ripening (Atallah *et al.*, 2014). The larvae develop inside the fruit, causing rapid decay and increased risk of secondary infections by fungal and bacterial pathogens. (Iglesias and Liburd, 2017).

Biorational insecticides, *Chromobacterium subtsugae*, sabadilla and pyrethrins + azadirachtin can be used to control adult SWD in organic blueberry production. This biorational insecticide has minimal effects on parasitoids *Aphelinidae* and *Pteromalidae* (Iglesias and Liburd, 2017). Lee *et al.* (2017) suggested use of bait trap consisting of new blueberry wine which was most effective, specific, inexpensive and most cost-effective bait for trapping SWD.

## Lepidoptera Larvae (*Operoptera bruccata, Cheimophila salicella, Spilonota ocellana*)

These larvae damage flower and fruit clusters more frequently than leafrolls during April, May and early June causing production losses up to 7 per cent (Gillespie, 1985). Malathion treatment reduces larvae populations and increases the yield.

## Scarabacid Larvae (*Scitala sericans, Aphodius tesmaniae*)

These two species are more common among the eight scarabacid species found in blueberry plantations (Osmelak and Clayton-Greene, 1988). The reduction in scarabacid population has been obtained by applying chlorpyrifos (500 g a.i./l) in February.

## Sharp-Nosed Leafhopper (*Scaphytopius magdalensis*)

This is a stunt virus vector which preferably attacks highbush blueberry. Reduced oviposition, growth rate and high mortality of nymphs have been found on rabbiteye foliage. (Etzel and Meyer, 1986). The hopper can be controlled by malathion sprays.

## Birds

Conover (1985) and Tobin *et al.* (1988) reported the following birds causing damage to blueberries:

1. European starlings (*Sternus vulgaris*)
2. American robins (*Tordus migratorium*)
3. House finches (*Carpodacus mexicanus*)
4. Northern mockingbirds (*Mimus polyglottos*)
5. Grey catbirds (*Dumetella carolinensis*)
6. Northern orioles (*Icterus glabula*)

Fire crackers, covering with screen or netting and use of mesurol as deterrents have been suggested to check the bird damage (Childers, 1983).

## Nematodes

Five nematodes, *viz.*, *Xiphinema americanum*, *Meloidogyne incognita*, *Tetylenchus christiei*, *Hernicycliophora similis* and *Trichodorum* sp., have been reported to infect blueberries out of which *X. americanum* is the most harmful (Clark *et al.*, 1987b).

## Diseases

### Mummy Berry (*Monilinia vaccinii-corymbosi*)

The fungus is very common in all blueberry growing regions. It attacks shoot tips and blossoms in humid weather and later the spores infect the young fruits. The cream-coloured mummified berries generally remain attached to the bush. Thorough raking of the soil beginning from early spring until after bloom is recommended at weekly interval. The mummified berries from the plant crowns should be removed (Childers, 1983). In Canada, spraying of funginex 1.9 E (triforine) at 1.7 l/ha on 4 and again on 14 May showed very good control of twig and blossom blight of blueberry (*V. corymbosum*) and increased the yield from l.2 t/ha in the control to 2.9 t/ha (Lockhart, 1982).

The use of hive-dispersed biocontrol product such as Serenade (a commercial formulation of the bacterium *Bacillus subtilis* which has shown activity against flower infection by *M. vaccinii*- corymbosi) as a supplement during pollination can reduce the risk of mummy berry disease. This may be a prudent practice that optimizes the benefits to pollination of high bee densities while reducing the associated disease-vectoring risk (Dedej *et al.*, 2004).

### Botryosphaeria Stem Blight (BSB) (*Botryosphaeriacea* family)

Botryosphaeria stem blight is caused by fungi in the *Botryosphaeriacea* family including *B. dothidea* (Witchera and Clayton, 1962), *Neofusicoccum ribis*, and *Lasiodiplodia theobromae* (Wright and Harmon, 2010).

BSB is a destructive disease of blueberries, losses can be severe in young fields where plants often become infected and die within two years. In older fields, stem blight is most commonly seen as a dieback of one or more canes; however mature plants of susceptible cultivars may die when the stem blight pathogen infects the crown (Smith, 1997).

Biofungicides Fracture® (a.i., Banda de Lupinus albus doce, BLAD; Serenade MAX® (a.i., *Bacillus subtilis* strain ST 713) in rotation with conventional fungicides Fracture®, Switch® (a.i., cyprodinil and fludioxonil) were found to adequately control the Botrytis blossom blight (Percival *et al.*, 2017).

### Canker and Twig Dieback (*Pestalotiopsis calvispora*, *P. neglecta* and *Truncatella* (=*Pestalotia*) *angustata*)

The symptoms appear as light-brown canker lesions, surrounded by a reddish halo, and shoot dieback. Espinoza *at al.* (2008) stated that cvs. O'Neal and Brightwell were the most susceptible and Bluecrop and Misty the least susceptible. *P. clavispora* was reported as highly sensitive to fludioxonil and pyraclostrobin. Elfar *et al.* (2013)

observed that at least four species of Diaporthe were primary pathogens, capable of causing stem canker symptoms on blueberry. They were identified as *D. ambigua, D. neotheicola, D. passiflorae* and *D. australafricana*. The symptom developed due to Diaportha includes necrotic reddish-brown cankers on blueberry shoots and stems. In Spain, *Neofusicoccum parvum* and *N. australe* were reported to cause stem canker and diback in blueberry (Castillo *et al.*, 2013).

## Anthracnose (*Colletotrichum acutatum, C. gleosporiodes*)

Disease symptoms occur on all above ground plant parts, however fruit infection is the most serious. Incidence of disease is more under hot and humid condition during flowering, just before or during harvest promoting fruit rotting. Meszka and Bielenin (2012) observed cultivar Bluetta was most susceptible to anthracnose, showing over 50 per cent of infected fruits. The cvs. Nelson, Herbert, Toro, Bluerey, Bluecrop, Patriot and Bluegold showed medium susceptibility. Cultivars Chandler, Darrow and Ellitt showed a low level of susceptibility while cvs. Bridget, Sierra, Berkley and Boniface were not infected.

The control of anthracnose requires several fungicide treatment between before flowering to fruit harvest. Good efficacy in the control was reported by spraying cyprodinil + fludioxonil and piraclostrobin + boscalid and biological compound Polyversum, which reduced disease severity by 70-80 per cent (Meszka and Bielenin, 2012).

## Phytophthora Root Rot (*Phytophthora cinnamomi*)

The root rot is the most serious among the diseases affecting the plantations of *V. corymbosum* (Cheah and Hunt, 1988). The underground runners lack feeder roots, show discrete lesions and a distinctive olive-brown discolouration. Above ground symptoms include discoloration of leaves (yellowing or redding) defoliation, and stunted growth (Milholland, 1995). Sterne (1982) stated that planting on peat and on poorly drained soil may be exacerbating the disease. The control measures suggested by Caruso (1989) are improvement of drainage in affected spots, sanding, application of ridomil and fertilization of stressed plants. Rabbiteye blueberry cultivars Tifblue, Callaway, Southland, Menditoo and Delite are completely resistant (Cheah and Hunt, 1988). Smith and Miller-Butler (2017) suggested planting on raised bed and treated with phosphorus acid fungicides four times a year. Highbush blueberry cvs. Emerald, SantaFe, Primadonna and Springhigh have resistant to *Phytophthora* root rot.

## Blueberry Rust (*Thekopsora minima*)

The disease initially appears as small lesions on the lower leaf surface but often goes unnoticed until yellow coloured pustules form on the underside of the leaf. Corresponding red-brown lesions develop on the abaxial surface. Under high disease pressure, infected leaves drop and severe defoliation can lead to reduced production in the following season (Simpson *et al.*, 2017). Fungicides that effectively control blueberry rust including propiconazole, mancozeb and copper (Lazar *et al.*, 2006). Simpson *et al.* (2017) demonstrated that fungicides fenbuconazole, tebuconazaole, azoxystrobin and dithianon are effective in controlling blueberry rust.

## Fruit Rot

Blueberry (*V. corymbosum* and *V. ashei*) fruits are susceptible to infection by *Colletotrichum gloeosporioides* (*Glomerella cingulata*) at all stages of development but did not show symptoms until ripening. The cvs. Powderblue and Morrow were found most resistant and Jersey, Harrison and Blueray the most susceptible to anthracnose fruit rot (Daykin and Milholland, 1984). The blueberry fruit rot is also caused by *Phomopsis vaccinii*. The fungus apparently penetrates the fruit at all stages of development, remaining latent until maturation. Infected fruits are typically soft and often split, causing juicy leakage; infected tissue is reddish brown and mushy (Milholland and Daykin, 1983).

## Viral Diseases

### Blueberry Shoestring Virus (BBSSV)

It is the most serious disease which moves both in the phloem and xylem (Urban *et al.*, 1988). The virus is difficult to eradicate because it is widely prevalent in native stands of blueberries and there from transmitted by the aphid vector to the commercial plantations (Hancock *et al.*, 1986). Typical symptoms of BBSSV are strap-like leaves and red-stem-streaks, remain latent. The removal of all blueberry plants in the vicinity of commercial fields has been suggested to control this virus.

### Blueberry Stunt Virus (BBSV)

BBSV is common to all regions and appears mainly on susceptible cultivars Cabot, Pioneer, Concord and Scammell. Rancocas is a highly resistant cultivar (Childers, 1983).

### Blueberry Shock Virus (BlShV)

The virus was first observed in the pacific North West in 1987. With BIShV infection, the newly emerging flowers and new growth become necrotic but the plant will subsequently put on new growth and the symptoms are not known to reoccur on the same part of the plant (Martin *et al.*, 2012). The virus is pollen-borne so control of the vector is impractical (Finn *et al.*, 2017). While transmission and movement are not well understood, the virus movement, and subsequent expression of symptoms varies by genotype (Bristow and Martin, 1999). At Oregon State University, North Willamette Research and Extension Center (NWREC), cultivars Legacy, Toro, Bluecrop, Baby Blues, Darrow and all rabbiteye cultivars tested negative for BIShV for over 10 years, while cvs. Berkeley, Bluegold, Brigitta Blue, Nui and Spartan tested positive in the year following their 1$^{st}$ bloom (Finn *et al.*, 2017).

### Blueberry Red Ringspot Virus (BRRSV)

The virus caused red ringspots or blotching on one year old stems or older. In mid-to late summer reddish brown spots develop on older leaves. In some blueberry cultivars, circular areas of light colour appear on fruit surface or fruit deformations may occur (Plesko *et al.*, 2010).

### Blueberry Scorch Virus (BlScV)

In 2004, it has been found in open fields in Europe (in Piedmont, Nort-West of Italy) though it was already present in North America. Symptom expression varies according to the cultivar and the virus strain. Susceptible varieties usually show a rapid flower and twig blight (scorch). Some highbush showed symptoms associated BlScV and a number of samples tested positive in specific DAS-ELISA for BlScV. The virus was reported to cause damage on *V. corymbosum* and *V. ashei* (Prodorutti *et al.*, 2011). In 2001, BlScV was first reported from Connecticut and Massachustts. The symptoms including leaf chlorosis and necrosis, blossom blight, tip die back, or a general decline in vigour was recorded (DeMarsay *et al.*, 2004).

### Blueberry Necrotic Ring Blotch Virus (BNRBV)

The virus was reported on southern highbush blueberries in several south-eastern states in the USA (Quito-Avila *et al.*, 2013). Symptoms included irregularly shaped circular spots or blotches with green centres on the upper and lower surface of leaves. Sequence analysis showed that the virus possesses protein domains conserved across RNA viruses in the alpha-virus-like supergroup.

## Organic Production

Worldwide highbush blueberry area increased from 42,000 ha in 2005 to 77,290 ha in 2010. In 2010, the USA had the largest planted area with 46 per cent of the world, followed by Chile (17 per cent) Canada (12 per cent), and Argentina, Poland and China (with 4-5 per cent each). Certified organic blueberry area in the USA increased from 194 ha in 2003 to 1,665 ha in 2011 (Stirk, 2014).

The main challenges for expansion of area under organic production include high input cost, limited options for disease or pest control, and reduced yield in organic system (Julian *et al.*, 2011).

Application of organic sources of nutrients affected macro-and micronutrients in soil and the leaves. Application of feather meal increase soil Ca and B significantly than fish emulsion. However, fish emulsion increased soil and leaf K and P. Application of fish emulsion at high rates reduced soil pH. Fish emulsion appeared to be an acceptable fertilizer source for blueberry during establishment (Larco *et al.*, 2013). Root colonization by ericoid mycorrhizae and dark sepate endophytes were generally higher in organic production system (Sadowsky *et al.*, 2012). Speciality fertilizers (fish meal, feather meal, or seed meals) decompose rapidly (>50 per cent decomposition in 1 month) and release plant N rapidly (Gale *et al.*, 2006).

Weed management was identified as one of the most challenging aspects of organic blueberry production. Application of surface mulches such as saw dust have been shown to improve blueberry production by reducing weed pressure, maintaining soil moisture, and increasing plant growth (White, 2006). Effective weed control was reported by using weed mat plus a mulch of saw dust (Larco *et al.*, 2013). The use of woven polyethylene ground covers, or weed mat has been approved for organic production in USA. Julian *et al.* (2012) reported that weed mat was the most economical method of weed control during establishment of northern highbush blueberry, which increase net income of 13 and 22 per cent in cvs. Liberty and Duke, respectively (Stirk and Vance, 2017).

Organic fungicides for suppression of mummyberry include Actinovate (*Streptomyces lydicus*), Serende Max (*Bacillus subtilis*), Serenade Aso (*Bacillus subtilis*) and Sonata (*Bacillus pumilis*). Actinovate, Serenado Aso, Sonata and Pre-Stop (*Glicladium catenulatum*) may be used to minimize Botrytis fruit rot. Adult of Spotted wing Drosphila can be control by Spinosed (*Saccharopolyspora spinosa*). Soft bodied insects such as aphids, whiteflies, and mealy bug are more susceptible to potassium laureate, commonly referred to Neem pesticides work best against immature insects, and need to be applied when temperatures are worm. A list of insecticides approved for use in organic cultivation is presented in Table 17.7. Sulphate is frequently used to prevent foliar diseases. Copper sulphate (often referred to as bluestone, vitriol, Bordeaux mixture, or basic copper sulphate) was one of the first organic fungicides ever used. It should be applied before the disease attacks, as they only work on germinating fungal spores. Extracts of citrus, giant knot weed and *Bacillus subtilis* are effective to control mummy berry disease in lowbush blueberry (McGovern *et al.*, 2012).

**Table 17.7: List of Insecticides Used in Organic Blueberry Cultivation with Trade Name, Active Ingredient, and Application Rate**

| *Insecticide* | *Active Ingredient* | *Rate* |
|---|---|---|
| Entrust SC | Spinosad | 6 fl oz acare$^{-1}$ |
| Grandevo | *Chromobacterium subtsugae* | 3 lb acre$^{-1}$ |
| Venerate XC | *Bukholderia* spp. | 2 qt acre$^{-1}$ |
| Venerate XC | *Bukholderia* spp. | 2 qt acre$^{-1}$ |
| Veratran D Lo | Sabadila alkaloids | 8lb acre$^{-1}$ |
| Veratran D Hi | Sabadila alkaloids | 15 lb acre$^{-1}$ |
| OxiDate 2.0 | Hydrogen dioxide, Peroxyacetic acid | 64 fl oz 100 gal $H_2O^{-1}$ |
| Azera Entrust SC | Pyrethrins + azadirachtin<br>Spinosad | 2.5 pt acre$^{-1}$<br>6 fl oz acre$^{-1}$ |
| Azera Grandevo | Pyrethrins + azadirachtin<br>*C. subtsugae* | 2 lb acre$^{-1}$ |
| Azera Venerate XC | Pyrethrins + azadirachtin<br>*Burkholderia* spp. | 2 qt acre |

Wang *et al.* (2008) reported that highbush blueberry cv. Bluecrop grown from organic culture yielded significantly higher sugars (fructose and glucose), malic acid, total phnolics, total anthocyanins, and antioxidant activity (ORAC) than fruit from the conventional culture. In organically cultured fruit, the average values for the ORAC, total anthocyanins, and total phenolic content were 46.14 micromol of Trolox (TE/g of fresh weight (fwt), 131.2 mg of fwt and 319.3mg/100g fwt, respectively compared with 30.8 micromol of TE/g of fwt, 82.4 mg/100g fwt and 190.3mg/100g fwt, respectively.

## Breeding and Improvement

Blueberries were earlier harvested from the wild plants but their improvement through selections from wild types and hybridization paved the way for commercial cultivation (Galletta, 1975). Blueberry germplasm is being continuously assembled and evaluated for various horticultural traits, such as, plant height, flower bud number, soil pH tolerance, drought tolerance, upland soil adaptations and stem blight resistance (Erb, 1987; Erb *et al.*, 1988; Garvey and Lyrene, 1987; Moon *et al.*, 1987). The genetic variability among wild species of blueberry and their crossability have made it possible to evolve superior cultivars.

## Objectives of Breeding

Breeding objectives of blueberries have changed over the years. In the past, the emphasis was on large fruit size, good blue colour, good flavour, high yield and plant vigour but now ecological tolerance and disease resistance have become more important objectives.

Blueberries have been bred for more than a century with multiple introgressions between *V. corymbosum* (VC), *V. angustifolium* and *V. virgatum* sym. *V. ashei*, as well as other *Vaccinium* species. This introgression from the secondary gene pool, included species from sections *Cyanococcus, Polycodium, Pyxothamnus, Vaccinium* and *Batodendron* contributing to blueberry improvement (Ballington, 2009). In the University of Florida, blueberry breeding programme, VC has been adopted to lower latitudes by the incorporation of genes from wild *Vaccinium* species in section *Cyanococcus* native to Florida and the south US. These crosses ultimately yielded, what is now referred as southern highbush blueberries (Chavez and Lyrene, 2010). In the last two decades, one of the major focus are for these wild introgressions has been the use of sparkleberry (VA; *V. arboreum*) (Lyrene, 2013). VA has several adaptive characteristics, such as single trunk architecture, open racemes, deeper root system, and higher fruit soluble solids (Rodriguez-Armenta *et al.*, 2017). Chavez and Lyrene (2010) studied to understand whether crosses of diploid *V. darrowii* and diploid *V. fuscatum* (both section *Cyanococcus*) with diploid *V. arboreum* (section *Batodendron*) could be made and to assess the possibility of transferring traits from *V. arboretum* into cultivated blueberries of section *Cyanococcus*. They observed many successful crosses. Crosses were more productive when *V. arboreum* was the pollen parent. Pollen stainability of F1 (*V. darrowii* x *V. arboreum*) hybrids were low, ranging from 0.9 o 23.7 per cent. Numerous meiotic abnormalities were found in the F1 hybrids. This led to a high percentage of incomplete and unstained tetrads and probably to the production of unreduced gamets.

### Edaphic Adaptation

Modern blueberry, cultivars should adapt to the less acidic soils poor in organic matter and low in moisture content (Galletta, 1975). Hybrid plants with *V. angustifolium* in their ancestry have been shown to be capable of absorbing and tolerating relatively high levels of magnesium. However, the hybrids with the least *V. corymbosum* gemplasm produced the most vigorous growth (Korcak, 1986) on upland soil conditions, while *V. ashei*, *V. atrococcum* and *V. darrowii* have been identified as potential germplasm sources for adaptability to upland soils (Galletta, 1975).

### Climatic Adaptation

In order to attain a wider climatic adaptability, a blueberry cultivar is required to be cold tolerant in winter, frost tolerant in spring (late-bloomer) and heat tolerant in summer seasons. Hancock *et al.* (1987) reported that late-blooming cultivars normally escape the frost damage but they are also late maturing. The need is to combine the late-blooming with early maturity. The low-chilling cultivars should be bred for growing in the warmer regions. The USDA-ARS blueberry breeding programme has pursued the goal of incorporating the cold-hardiness and late bloom *V. constablaei* into a form suitable for northern growers. Combining *V. constablaei* with *V. virgatum* (rabbiteye) allowed many of the best aspects of both germplasm to be used into a usuable form. Such hybrids derive late bloom and cold-hardiness from *V. constablaei*, and many aspects of vigour from *V. virgatum*. The first commercial product in the USDA programme of this introgression is the cultivar "Nocturne' a hexaploid mixed-species hybrid which achieves the majority of the goals desired in combining these germplasm (Ehlenfeldt, 2017).

The hybrid *V. darrowii* x *V. corymbosum* called 'US75' played a critical role in the development of southern highbush cultivars with low chilling requirements like 'Cooper', 'Georgiagem', 'Gulfcoast', 'O'Neal' and 'Sierra' which includes gene from five species (Hancock *et al.*, 1995).

### Precocity

Early commercial harvests have become a necessity of modern fruit cultivation. Efforts should be made to evolve precocious cultivars giving commercial crops in only two to three years.

### Fruit Quality

A good fruit quality will always remain the primary objective of any fruit breeding programme depending on the fruit use either for fresh market or for processing. In general, improvement in fruit size, colour, texture, flavour, nutritive value and processing quality is desired. Ballington *et al.* (1984) working with 11 species reported *V. stamineum* was the best in terms of soluble solids, large fruit size and firmness; *V. elliottii* was easily harvested and *V. corymbosum* had a favourable balance between soluble solids and acids.

### Extending Harvest Season

In order to make blueberries available for longer duration, both to fresh market and processing industry, early as well as late-ripening cultivars should be developed. Galletta (1975) stated that fruits ripening very early or late are often poor in quality but fruit maturity period can be extended from 2 to 4 months in any locality with the inclusion of available species germplasm.

## Suitability for Mechanical Harvesting

The compact plant habit and uniform fruit ripening are the two main characters which facilitate mechanical harvesting. Uniform ripening reduces the yield and exposes the fruits to weather and pest hazards, such as, high temperature, intense rains, hails and severe attack of any pest (Galletta, 1975). But the mechanical picking of blueberries has become a common practice, requiring due attention in breeding. Concentrated ripening involves three parameters, *i.e.*, short flowering period, uniform ripening and the capability of the ripe berries to remain on the plant in good condition for a longer period.

## Resistance to Pests and Diseases

With the expansion in blueberry culture and problems in chemical control, the breeding of cultivars resistant to serious pests and diseases has become very important. Although, it is not possible to incorporate resistance in a cultivar against many pests and diseases; yet resistance or tolerance against major pests and diseases has to be secured. Insect problem is serious in some blueberry growing regions; blueberry maggot particularly causes great damage to the crop. Botrytis stem and blossom blights, *Phytophthora* root rot, mummy berry and blueberry shoestring virus are serious diseases of the crop. Hancock (1989) suggests the use of biotechnological approaches to develop stronger resistance.

Wild *V. corymbosum* showed resistance to blueberry stem canker. 'Echota' was the first highbush blueberry cultivar with genes for stem canker resistance. The selection 'US41' a colchiploid *V. corymbosum* has been used as a source of resistance to *Phytophthora cinnamomi*. *V. angustifolium* has proven to be a valuable source of genes for resistance to *Botryosphaeria* sp.

## Systems of Breeding

Both intraspecific and interspecific hybridizations have been successfully used in blueberry breeding programmes. The progenies have been carried over 4 to 5 generations based on phenotypic selections achieving remarkable success (Galletta, 1975). A potential parent must exhibit phenotypic superiority for the traits to be incorporated in the progeny but should not transmit undesirable traits at the same time. It should be tested by selfing and crossing to tester clones.

Interspecific hybridization has made significant contribution in blueberry breeding progress involving crosses from *V. corymbosum* x *V. angustifolium* and *V. darrowii* x *V. corymbosum* (Hancock, 1989). The winter-hardy cultivars Northland, Northblue, Northsky and North Country have been released for cultivation. *Vaccinium* species with similar ploidy are completely interfertile but different chromosome levels reduce their interfertility. To determine the crossability, the mixed pollen should be used on several female plants (Galletta, 1975). Crosses between *V. corymbosum* and *V. ashei* are more successful than reciprocal crosses.

*V. corymbodendron* (section *Pyxothamnus*) is a tetraploid species native to high altitude locations in Peru, Colombia, and Venezuela, and is of considerable interest because it flowers at times when night time temperatures drop below freezing (Luby *et al.*, 1991). It also is notable for its profuse, concentrated flowering, and monopodial plants structure both of which may be useful in breeding for mechanical harvest. Initial hybridization experiments of 4x *V. corymbodendron* with a range of 2x species from section *Cyanococcus* found virtually all hybrids were tetraploids. 4x xX 2x cross in conventional *Vaccinium* germplasm typically result in an almost absolute block of triploid offspring. The triploids expressed very low fertility as both males and females, none the less, several were used successfully in crosses. A single hybrid from 4x *V. corymodendron* x 2x *V. vitis-idaea* (lingonberry, selection *Vitis-idaea*) was unexpectedly found to be tetraploid and was highly fertile. These crossing results suggest *V. corymbodeandron* either possesses no ploidy barriers to hybridization or possesses genomic dosage factors that differ from section *Cyanococcus* species (Ehlenfeldt and Ballington, 2017).

Inbreeding has a limited scope in blueberry breeding because of the problems of self-incompatibility, seed germination and inbreeding depression. It is only useful in developing homozygous parent stocks. Lowbush and rabbiteye blueberries are generally self-incompatible whereas highbush blueberry cultivars vary in self-fruitfulness (Czesnik *et al.*, 1989; Galletta, 1975).

Lowbush blueberries have been hybridized with *V. corymbosum* to produce half-high blueberry cultivars. The half-high produce higher yields and larger fruit than lowbush blueberry. Some of the cultivars of half-high blueberry are Northblue, Northsky, Northcountry, Northland, Polaris and Chippewa.

Polyploidy in blueberries has been induced by seed and shoot treatments with colchicine (Aalders and Hall, 1963; Rousi, 1966). It is useful in increasing the fertility of the sterile hybrids and transferring gene from diploid to tetraploid species.

## Inheritance of Characters

Enhanced knowledge about blueberry germplasm diversity and genetic relationship among plant breeding materials could be an important support to crop improvement strategies (Giongo *et al.*, 2017). Heterozygosity of *Vaccinium* species precludes the possibility of simple inheritance but simultaneously provides enormous opportunities for large scale selection and improvement. Most of the characters are quantitatively inherited requiring partitioning into their component parts and assigning due weightage to each part. For instance, yield is determined by plant size, fruiting shoots per cane, self-fruitfulness, pollination, fruit size and incidence of pests and diseases. Interestingly, more than one aspect of a trait is desirable. Large fruits are good for fresh market and hand picking but small fruits are considered better for flavour and mechanical harvesting.

Inheritance studies in blueberries have been rather limited (Czesnik and Madry, 1989). They reported the inheritance of some characters in $F_1$ hybrids of *V. corymbosum* x *V. uliginosum* on the basis of general combining ability and specific combining ability effects.

An exhaustive list of prepotent and potential donors for many blueberry characters with probable mode of inheritance has been prepared by Galletta (1975).

A number of screening methods for physiological traits and disease resistance have been developed.

Parental choice is a key process in breeding, with particular regard to traits that are still poorly defined in blueberries like flavour and aroma. The capacity to choose the most efficient parents to be able to develop outstanding progenies depends on fruit texture and aroma properties (Giongo *et al.*, 2017).

In blueberry, three major dehydrins of 65, 60 and 14 Kda have been found to increase with cold acclimatization and decreases with deacclimatization. A comparison of these dehydrin levels among various blueberry cultivar selections has revealed their level of accumulation to be closely associated with cold hardiness level. The sequence of a 2.0 kb CDNA clone, which encodes the 60KDa blueberry dehydrin, was used to map a dehydrin-related gene to linkage group 12 of the *V. caesariense* testcross population. A preliminary comparison of the segregation pattern of the dehydrin-related gene to that of the cold hardiness trait suggests that the marker does not segregate with cold hardiness (Panta *et al.*, 2004).

## Biotechnology

### Indexing Blueberry Cultivars by Molecular Marker

The development of DNA-based markers has facilitated species distinction and cultivar identification. Sequencing of the chloroplast *mat* K gene and the nrITS (Kron *et al.*, 2002) has allowed researchers to study phylogenetic relationship between closely related species and clarify the understanding of the taxonomy of the *Vaccinieae.* Random amplified polymorphic DNA sequence (RAPD) markers have been used to distinguish cultivars and wild accessions (Burgher *et al.*, 2002).

The RAPD analysis was able to identify distinct field clones in lowbush blueberry. Burgher and Mackenzie (2004) collected field clones (from Nova Scotia, Canada) with varying phenotypes to estimates genetic relationship in lowbush blueberry. The RAPD analysis showed 68 per cent genetic similarity among field clones reflecting expected genetic variation. Approximately 15 per cent of the field clones were not related. JeeHwa *et al.* (2014) worked to construct a DNA marker database for blueberry cultivars growing in Korea using simple-sequence report (SSR) markers. A set of 49 SSR primer pairs was tested to select polymorphic SSR markers in 6 cultivars. The genetic relationship among 34 cultivars by using 17 SSR marker. Out of 17 SSR markers, a set of 3 minimum SSR markers enabled that these SSR markers may be used for identifying blueberry cultivars and alternative choice to distinctness, uniformity, stability (DUS) examination in blueberry.

Garriga *et al.* (2013) developed an inter-simple sequence repeat (IISR) marker based system to identify blueberry cultivars. Ten cultivars of *V. corymbosum*: Bluecrop, Bluegold, Duke, Elliott, Legacy, Misty, Nelson, O'Neal, Sierra and Toro and three *V. ashei*: Climax, Permier and Tifblue were analyzed. The sensitivity and reliability of this molecular marker system was compared with identification by SSR. Six IISR primers were used and high levels of polymorphism among the cultivars (80 per cent of polymorphic loci) were detected with high repeatability. Using pairs of ISSR primers provided greater reliability in cultivar identification. They concluded that IISR-based technology is a simpler, faster and has less expensive alternative to SSRs for genotyping blueberry cultivars and can be used in genetic traceability studies as well as genetic improvement programme.

## Genetic Transformation

Although Vaccinum cultivars have generated exclusively through the traditional methods of controlled hybridization and deliberate selection, genetic transformation could provide a powerful approach to supplement conventional breeding methods, for Vaccinum by introducing genes of interest (Song and Hancock, 2012). Graham *et al.* (1996) reported transformation of half-highbush blueberry cv. Northcountry using disarmed *Agrobacterium tumefaciens* strain LBA4404. *Agrobactirium* –mediated transformation is the major gene delivery method; and effective selection has been reported using either the *npt II* or the bar gene as the selective marker (Song and Hancock, 2012). In 2004 stable transformation was reported for four highbush blueberry cultivars (Aurora, Bluecrop, Brigitta Blue and Legacy) using the *Agrobacterium* strain EHA105, with a transformation frequency ranging from 5.0-15.3 per cent (Song and Sink, 2004). In 2006, the first field trial of transgenic blueberry with herbicide resistance was performed (Song *et al.*, 2008). In 2009, a blueberry C-repeat binding factor (CBF) gene (GenBank AF234316) was transformed into southern highbush blueberry cultivar Legacy in order to improve cold tolerance by elucidating the CBF-regulated network in blueberry. This progress has demonstrated that blueberries can be improved through genetic engineering (Song and Hancock, 2012).

## Transcriptomics

Application of transcriptomics approaches can greatly enhance understanding of bluberry physiology. The success of transcriptomics approaches is dependent on the extraction of high-quality RNA which is complicated by the abundance of polyphenolics and polysaccharides in blueberry (Vashisth *et al.*, 2011).

The *bZIP* (basic leucine zipper) genes are an important family of transcription factors in plants encoding transcriptional and physiological process. The conserved domain of the *bZIP* gene in *V. corymbosum* was cloned (Liu *et al.*, 2017) using two degenerate primers and the 5′ terminal and 3′ terminal of the gene were amplified by using rapid amplification of cDNA ends (RACE) medhods. The complete cDNA sequence termed as *SabZIP* was obtained. Sequence analysis showed that the nucleotide sequence of *SabZIP* was 1451 bp containing an open reading frame (ORF) of 1077 bp and encoding a protein with 358 amino acids. The alignment analysis indicated that the predicted protein sequence contained two typical Leucine Zipper structure domain and basic domain DNA-binding domain. Real-Time quantitative polymerase chain reaction (Real-Time PCR) analysis showed that *SabZIP* gene expression was increased 2 h after being cold induced (4°C) and reached a peak after 8h, and then decreased. These results provided more information about *bZIP* type transcription factors and had an important insight into the molecular adaptation mechanism of cold stresses in blueberry (Liu *et al.*, 2017).

Vashisth *et al.* (2011) extracted RNA of high quality from eight and six organs of rabbiteye and southern highbush blueberry, respectively, using a modified cetyltrimethl ammonium bromide based method. The expression stability of 12 reference genes were evaluated. Ubiquitin-congugating Enzyme (UBC28), RNA Helicaselike (RH8), Clathrin Adapter Complexes Medium Subunit Family Protein (CACSa), and Polyubiquitin (UBQ36) were the mostly stably expressed genes across multiple organs in both blueberry species. Further, the expression stability of the reference genes in the branch abscission zone following treatment with fruit abscission-inducing compounds were analyzed, CACSa, RH8 and UBc28 were most stable expressed genes in the abscission zone under abscission-inducing conditions.

The blueberry genomic database (BBGD454) was developed to provide the research community with valueable resources to identify genes that play an important role in flower bud and fruit development, cold acclimation and chilling accumulation in blueberry. The database was developed using SQL Server 2008 to house 454 transcript sequences, annotations and gene expression profiles of blueberry genes. BBGD454 can be accessed publically from a web-based functionalities to allow scientists to access and search the data in order to correlate gene expression with gene function in different stage of blueberry fruit ripening, at different stages of cold acclimation of flower buds, and in leaves (Darwish *et al.*, 2013). It can be accessed from http://bioinformatics.towson.edu/BBGD454/

## Genetic Map

The costruction of the first genetic map in autoteraploid blueberry has been made possible by the development of a new single nucleotide polymorphic (SNP) markers developed using genotyping by sequencing in mapping population created from a cross between two key highbush blueberry cultivars, Draper x Jewel (McCallum *et al.*, 2016). In total, 1794 SNP markers and 233 simple sequence repeat (SSR) markers exhibited segregation model for meiosis in an auto tetraploid. Of these, 700 SNPs and 85 SSRs were utilized for construction of the 'Draper' genetic map, and 450 SNPs and 86 SSRs for the 'Jewel' map. The 'Draper' map comprises 12 linkage groups (LG), associated with the haploid chromosome number for blueberry, and totals 1621 cM while the 'Jewel' map comprises 20 LG totalling 1610 cM. Tentative alignments of the two parental maps have been made on the basis of shared SSR alleles and linkages to double-simplex markers segregating in both parents.

## REFERENCES

Aatders, L.F. and Hall, I.V. (1963). Canad. J. Gen. Cytol., 5: 380-83.

Abbott, J.D. and Gough, R.E. (1987). HortSci., 22: 40-42.

Arevalo, R.H.A. (2006). Ph. D. thesis, University of Florida, Gainesville, Fl. USA.

Attallah, J., Teixeira, L., Saalazar, R., Zaragoza, G. and Kopp, A. (2014). Proc. Biol. Sci., 281(1781), 20132840. Pubmed http://dx.doi.org/10.1098/rspp.2013.2840.

Austin, M.E. and Bondari, K. (1986). Jr. Amer. Soc. Hort. Sci., 111:840-44.

Austin, M.E. and Bondari, K. (1987a). Scientia Hort., 31:71-79.

Austin, M.E. and Bondari, K. (1987b). HortSci., 22:1247-48.

Austin, M.E. and Draper, A.D. (1983). HortSci., 18: 252.

Austin, M.E. and Draper, A.D. (1987). HortSci., 22: 682-83.

Austin, M.E. and Miltinix, B.G. (1984). Plant Gr. Regu. Bull., l2: 8-10.

Austin, M.E. and Mullinix, B.G. (1980). Jr. Amer. Soc. Hort. Sci., 105: 111-14.

Austin, M.E. and Williamson, R.E. (1977). Jr. Amer. Soc. Hort. Sci., 102: 454-56.

Badescu, G., Bortez, M., Badescu, L. and Enache, E. (1985). Acta Hort., 165: 189-95.

Bajwa, R. and Read, D.J. (1986). Trans British Mycol. Soc., 87:269-77.

Balling, J.R., Ballinger, W.E., Swallow, W.H., Galletta, G.H. and Kushman, I.J. (1984). Jr. Amer. Soc. Hort. Sci., 109: 684-89.

Ballinger, W.E. (1983). Jr. Amer. Soc. Hort. Sci., 108:422-24,

Ballinger, W.E. and Kushaman, L.J. (1970). Proc. Highbush blueberry Mech. Symp., N.C. State Univ., Raleigh.

Ballinger, W.E., Kushaman, L.J. and Hamann, D.D. (1973). Jr. Amer. Soc. Hort. Sci., 98: 583-87.

Ballinger, W.E., Maness, E.P. and McClure, W.F. (1978). Jr. Amer. Soc. Hort. Sci., 103: 130-34.

Ballington, J.R. (2009). Acta Hort., 810: 49-60.

Ballington, J.R., Galleta, G.J. and Pharr, D.M. (1976). HortSci., 11:410-11.

Ballington, J.R., Mainland, C.M., Duke, S.D., Draper, A.D. and Galletta, G J. (1989). HortSci., 24: 161-62.

Ban, T., Kitazawa, H., Matsumoto, S., Kobayashi, N., Tokumasa, K., Kobatake, M. and Asao, T. (2009). Jr. Appl. Hort., 11: 54-55.

Banados, M.P. and Strik, B. (2006). Acta Hort., 715: 65-71.

Banados, M.P., Uribe, P. and Donnay, D. (2009). Acta Hort., 810: 501-07.

Bandos, M.P. and Stirk, B.C. (2006). Acta Hort., 715: 65-71.

Barlow, P., McCurran, A. and Miller, S. (2017). Acta Hort., 1180: 161-68.

Bates, P.M. (1979). New Zealand Comm. Grow., 31: 23.

Beaudry, R.M. (1992). Ann. Rept. Michigan Sta. Hort. Soc., 122: 140-45.

Benoit, G.R., Grant, W.J., Ismail, A.A. and Yarborough, D.E. (1984). Canad. J. Pl. Sci., 64: 683-89.

Blasing, D. (1989). Acta Hort., 241: 213-20.

Blasing, D., Hartge, K.H. and Naumann, W.D. (1984). Erwerbsobstbau, 26: 266-68.

Bounous, G., Songia, E., Mazzitelli, M. and Vai, M. (1985). Acta Hort., 165: 61-65.

Bratsch, A. (2009). Specialty Crop Profile: Blueberries. Virginia Cooperative Extension Bull. Pub. 438-503.

Bristow, P.R. and Martin, R.R. (1999). Phytopathol., 89: 124-30.

Brooks, R. M. and Olmo, H.P. (1972). In: Reg. New fruit and Nut var., Univ. Calif. Press, Berkeley, pp.168-79.

Brown, G.R., Wolfe, D. and Rasnake, M. (1988). HortSci., 23: 314-15.

Brown, J.C. and Draper, A.D. (1980). Jr. Amer. Soc. Hort. Sci., 105:20-24.

Bryla, D.R. and Stirk, B.C. (2007). HortSci., 132: 270-77.

Bryla, D.R. and Stirk, B.C. (2015). HortTechnology, 25: 464-70.

Bunea, A., Rugina, D.O., Pintea, A.M., Sconta, Z., Bunea, C.I. and Socaciu, C. (2011). Notulae-Botanicae, Horti-Agrobotanici, Cluj-Napoca, 39(2): 70-76.

Bunemann, G., Dewey, D.H. and Watson, D.P. (1957). Proc. Amer. Soc. Hort. Sci., 70: 156.

Burgher, K.L., Jamieson, A.R. and Lu, X.W. (2002). Jr. Amer. Soc. Hort. Sci., 127: 98-103.

Burgher, M.K. and MacKenzie, K. (2004). Small Fruit Review, 3: 295-305

Burrack, H.J. and Littlejohn, K. (2012). Int. J. Fr. Sci., 12: 106-13.

Burton, C.L., Tennes, B.R. and Brown, G.K. (1979). Proc. 4th North Amer. Blueberry Res. Worker Conf., Univ. of Arkansas, Fayetteville.

Butkene, Z.P. and Butkus, V.F. (1980). Trudy Akad, Nauk Lit SSR. C No. 3/91: 45-55.

Butkus, V.F. and Butkene. Z.P. (1987). Lietuvos TSR Mokslu Akademijos Darbai, C (Biologijos Moksiai). 2: 28-36.

Byers, P.L. and Moore, J.N. (1987). HortSci., 22: 52-54.

Cain, J.C. (1952). Proc. Amer. Soc. Hort. Sci., 59: 161-66.

Cain, J.C. and Eck, P. (1954). In: Fruit Nutr. (ed. N.F. Childers), Somerset Press, Somerville, NJ.

Cameron, J.S., Brun, C.A. and Hartley, C.A. (1989). Acta Hort., 241:254-59.

Camp, W.H. (1942). Brittonia, 4:205-47.

Camp, W.H. (1945). Brittonia, 5: 203-75.

Cao, X. and Hammerschlag, F.A. (2000). HortSci., 35: 945-47.

Cappellini, R.A., Ceponis, M.J. and Koslow, G. (1982). HortSci., 17: 55-56.

Cargill, B.F. and Ledebuhr, R.L. (1983). Proc. Int. Symp. Fruit, Nut and Veg. Harv., Tei Aviv, Israel,

Caruso (1989). Acta Hort., 241: 307-11.

Castillo, S., Borrero, C., Castano, R., Rodriguez, A. and Aviles, M. (2013). Pl. Dis., 97: 1112.

Ceponis, M.J. and Cappellini, R.A. (1983). Plant Dis. Rptr., 67: 169-71.

Ceponis, M.J. and Cappellini, R.A. (1985). HortSci., 20: 228-29.

Chavez, D.J. and Lyrene, P.M. (2010). Euphytica, 171: 262-72.

Cheah, L.H. and Hunt, A.W. (1988). Orchardist New Zealand, 61:60-61.

Chen, L.S., Liang, G.Z, Liu, H.R., Tan, Y. and Zhu, D.Z. (2017). Acta Hort., 1180: 71-76.

Chiabrando, V. and Giacalone, G. (2008). Fd. Sci. and Tech. Int., 14(3): 199-205.

Childers, N.F. (1983). Mod. Fruit Sci. Hort. Publ. Rutgers Univ., New Brunswick. NJ

ChunYu, Z., YaDong, Li., Xuesen, C., ZhiDong, Z., HaiGuang, Liu and Lin, Wu. (2009). Acta Hort. Sinica, 36(2): 187-94.

Ciordia, M., Garcia, J.C. and Diaz, M.B. (2004). Acta Hort., 649: 301-304.

Clark, J.R., Dombek, D.G. and Buckley, J.B. (1987a). Arkansas Farm Res., 36: 4.

Clark, J.R., Marx, D.B. and Dombek, D.G. (1989). Bull. Arkansas Agril. Exp. Sta., No. 920.

Clark, J.R., Moore, J.N. and Baker, E.C. (1986). Arkansas Farm Res., 35: 3.

Clark, J.R., Riggs, R.D. and Moore, J.N. (1987b). HortSci., 22:408-09.

Conover, M.R. (1985). Appd. Animal Behaviour Sci., 13 :383-86.

Coville, F.V. (1910). USDA Bureau of Pl. Ind. Bull. No. 193.

Crane, J. H. and Davies, F.S. (1987). HortSci., 22:1249-52.

Crane, J.H. and Davies, F.S. (1988). Jr. Amer. Soc. Hort. Sci., 113: 180-84.

Czesnik, E. and Madry, W. (1989). Acta Hort., 241:50-55.

Czesnik, E., Bounous, G. and Giffre, D. (1989). Acta Hort., 241: 56-63.

Dale, A., Cline, R.A. and Ricketson, C.L. (1989). Acta Hort., 241:120-25.

Dale, A., Hanson, E.J., Yardborough, D.E., McNicol, D.E., Stang, E.J., Brennan, R., Morris, J.R. and Hergert, G.B. (2010). Hortic, Rev., 16: 255-382.

Darnell, R.L. (1991). Jr. Amer. Sc. Hort. Sci., 116: 856-60.

Darwish, O., Rowland, L.J. and Alkharouf, N.W. (2013). Bioinformation, 9(17): 883-86.

Davies, F.S. and Flore, J.A. (1986). Physiol. Plantarum, 67: 545-51.

Davies, F.S. and Flore, J.A. (1987). Jr. Amer. Soc. Hort. Sci., 111:565-71.

Daykin, M.E. and Milholland, R.D. (1984). Pl. Dis., 68 : 948-50.

Debnath, S.C. (2009). *In-Vitro* Cell. Dev. Biol. Plant, 45(2): 122-28.

Debnath, S.C. (2017). Acta Hort., 1180: 111-15.

Dedej, S., Delaplane, K.S. and Scherm, H. (2004). Biological Control, 31: 422-27.

Dekazos, E.D. (1978). Proc. Fla. Sta. Hort. Soc., 91: 168-70.

Dekazos, E.D. (1980). Proc. Fla. Sta. Hort. Soc., 93: 145.

DeMarsay, A., Hillman, B.I., Peterson, F.P., Oudemans, P.V. and Schloeman, S. (2004). Pl. Disease, 88(5): 572.

Dozier, W.A. Jr., Knowles, J.W., Caylor, A.W., Powell, A.A. and Akridge, R. (1989). Bull. Alabama Agr. Exp, Sta., Auburn- Univ., No. 598, p. 19.

Draper, A., Galletta, G.J., Jelenkovic, G. and Vorsa, N. (1987). HortSci., 22: 320.

Eck, P. (1970). HortSci., 5: 23-25.

Eck, P. (1977). Jr. Amer. Soc. Hort. Sci., 102: 816-18.

Eck, P. (1983). Jr. Amer. Soc. Hort. Sci., 108: 520-22.

Eck, P. (1986). In: Handbook of Fruit Set and Dev. (ed. S.P. Monselise), CRC Press, Inc. Boca Raton, Florida, 75-85.

Eck, P. (1988). Blueberry Sci. Rutgers Univ. Press, New Brunswick, NJ.

Eck, P. and Childers, N.F. (1966). Blueberry Culture, Rutgers Univ. Press, New Brunswick, N.J.

Eck, P. and Stretch, A.W. (1986). HortSci., 21 : 249-50.

Eck, P., Gough, R.E., Hall, I.V. and Spiers, J.M. (1990). Blueberry management In: Small Fruit Crop Management (eds. G.J. Gallettea and D.G. Himelrick) Prentice Hall, Englewood, Cliffs, NJ. pp. 273-333.

Eck. P. and Mainland, C.M. (1971). HortSci., 6: 494-95.

Edwards, T.W., Sherman, W.B. and Sharpe, R.H. (1970). HortSci., 5:274-75.

Ehlefeldt, M.K., Rowland, L.J., Ogden, E.L. and Vinyard, B.T. (2012). Jr. Amer, Soc. Hort. Sci., 137: 31-37.

Ehlenfeldt, M.K. (2017). Acta Hort., 1180: 263-67.

Ehlenfeldt, M.K. and Ballington, J.R. (2017). Acta Hert., 1180: 257-61.

Elfar, K., Torres, R., Diaz, G.A. and Latorre, B. A. (2013). Pl. Dis., 97: 1042-50.

Elsner, E.A. and Kriegel, R.D. (1989). Acta Hort., 241:330-32.

Erb, W.A. (1987). Ph.D. Diss. Univ. of Maryland, College Park. USA.

Erb, W.A., Draper, A.D. and Swartz, H.J. (1988). HortSci., 23: 312-14.

Espinoza, J.G., Briceno, E.X., Keith, L.M. and Latorre, B.A. (2008). Pl. Dis., 92: 1407-14.

Etzel, R.W. and Meyer, J.R. (1986). Jr. Econ. Entom., 79: 1513-15.

Eynard, I., Gay, G. and Savino, P.G. (1985). Acta Hort., 165:53-60.

Eynard, L. and Czesnik, E. (1989). Acta Hort., 245: 115-19.

Famiani, F., Culterera, N.G.M., Battistelli, A., Casulli, V., Proietti, P., Standardi, A., ZhiHui, C., Leegood, R.C. and Walker, R.P. (2005). Jr. Exp. Bot., 56: 2959-69.

Fear, C.D. and Lawson, V.F. (1987). Fruit Var. J., 41 :148-49.

Finn, C.E., Mackey, T.A., Postman, J.D. and Martin, R.R. (2017). Acta Hort., 1180: 423-29.

Fischer, D.L. de, O., Fachinello, J.C., Antunes, L.E.C., Fischer, C. and Giacobbo, C.L. (2013). Revista de la Faculted de Agronomia La Plata, 112: 1-5.

Forge, T., Temple, W. and Bomke, A. (2013). Acta Hort., 1001: 369-74.

Forney, C.F., Kalt, W., Jordon, J.A., Vanqvist-Tymchuk, M.R. and Failmore, S.A.E. (2012). Acta Hort., 926: 331-37.

Frankel, C. (1972). Pl. physiol., 49: 757-63.

Gagne, R.J. (1989). The plant-feeding gall midges of North America. Cornell Univ. Press, Ithaca, NY.

Gale, E.S., Sullivan, D.M., Hemphill, D., Cogger, C.G., Bary, A.I. and Myhre, E.A. (2006). J. Environ. Qual., 35: 2321-32.

Gallardo, R.K., Stafne, E.T., DeVetter, L.W., Zhang, Q., Li, C., Tadeda, F., Williamson, J., Yang, W.Q., Cline, W.O., Beaudry, R. and Allen, R. (2018). HortTechnology, 28:10-16.

Garriga, M., Paarra, P.A., Caligari, P.D.S., Retamales, J.B., Carrasco, B.A., Lobos, G.A. and Garcia GonZales, R. (2013). Canadian J. Pl. Sci., 93: 913-21.

Garvey, E.J. and Lyrene, P.M. (1987). Jr. Amer. Soc. Hort. Sci., 112: 856-58.

Gatleta, G.J. (1975). In: Advances in Fruit Breeding, (eds. J. Janick and J.N. Moore). Purdue Univ. Press. West Lafayette, Indiana, pp. 154-96.

Gillespie, D.R. (1985). Canad. Entom., 117: 949-53.

Giongo, L., Grisenti, M., Khomenko, I., Poncetta, P., Loretti, P., Ajelli, M., Biasioli, F. and Farneti, B. (2017). Acta Hort., 1180: 197-204.

Girgenti, V., Peano, C., Giuggioli, N.R., Giraudo, E. and Guerrini, S. (2012). Acta Hort., 926: 571-76.

Goodman, R.D. and Clayton-Green, K.A. (1988). Aust. J. Exp. Agric., 28:287-90.

Gorchov, D.L. (1984). Bull. Eco. Soc. Amer., 65: 103.

Goto, R., Watanabe, M., Murkami, M., Sagawa, S., Komri, S. and Suzuki, A. (2013). Hort. Res. Japan, 12(2): 165-71.

Gough, R.E. and Shutak, V.G. (1977). Rhode Island Agr. Exp. Bull., 423.

Gough, R.E., Shutak, V.G. and Hauke, R.L. (1978). Jr. Amer. Soc. Hort. Sci., 103: 94-97.

Graham, J., Greig, K. and McNicol, R.J. (1996). Ann. Appl. Biol., 128: 557-64.

Gruca, Z. (1981). In: Informator O badaniach prowadzonych w Zakladzi Sadownictwa Akademii Rolnicsej w Poznaniu Poznan, Poland, pp. 107-09.

GuiYan, Y., Jin, Y., XinCheng, G., BingJun, Q., HuaJun, W., Yun, D. and YanYun, Z. (2014). Postharvest Biol. Technol., 92: 46-53.

Gupton, C.L. (1984). Jr. Amer. Soc. Hort. Sci., 109: 803-06.

Hall, I.V., Brrows, J.M. and Hildebrand, P.D. (1986). Canad. J. Pl. Sci., 66: 1033-35.

Hall, I.V., Jamieson, A.R. and Brydon. A.D. (1988). Canad, J. Pl. Sci., 68: 553-55.

Hamann, D.D., Kushman, L.J. and Ballinger, W.E. (1973). Jr. Amer. Soc. Hort. Sci., 98:572-76.

Hanchar, J.J., Skinner, S.P., Yarborough, D.E. and Ismail, A.A. (1985). HortSci., 20: 404-05.

Hancock, J., Callow, P., Serce, S., Hanson, E and Beaudry, R. 92008). HortTechnology, 18: 199-205.

Hancock, J.F. and Draper, A.D. (1989). HortSci., 24: 551-56.

Hancock, J.F. and Nelson, J. (1988). HortSci., 23: 857-58.

Hancock, J.F. and Siefker, J.H. (1982). HortSci., 17: 363-66.

Hancock, J.F., Morimoto, K M., Pritts, M.P. and Ramsdell. D.C. (1986). HortSci., 21: 1059-60.

Hancock, J.F., Nelson, J.W., Bittenbender, H.C., Callow, P.W., Cameron, J.S., Krebs, S.L., Pritts, M.P. and Schumann, C.M. (1987). Jr. Amer. Soc. Hort. Sci., 112:702-06.

Hanson, E.J. (1987). Jr. Pl. Nutri., l0 : 1419-27.

Hanson, E.J. and Hancock, J.E. (1996). Managing the Nutrition of Highbush blueberries. Bulletin E-2011, Michigan State University Extension, East Lansing, Michigan.

Harb, J.Y. and Streif, J. (2004). European J. Hort. Sci., 69(2): 66-72.

Hart, J., Stirk, B. White, I. and Yang, W. (2006). Nutrient Management for Blueberries in Oregon. Publication No. EM8918. Oregon University Extension Service. Corvallis, Oregon.

Haviland, D.R. and Hernandez, N.M. (2012). Int. J. Fr. Sci., 12: 114-23.

Haynes, R.J. and Swift, R.S. (1985). New Zealand Com. Grow., 40: 21.

Heekman, J.R., Pavlis, G.C. and Anastasia, W.L. (2002). Hort Technology, 12:220-22.

Heermann, D. (1984). Erwerbsobstbau, 26: 274-275.

Hensels, L.G.M. (1983). Fruitteelt, 73 : 535.

Hicks, B.J. (2011). Jr. Cadian Entomol. Soc., 7(15): 108-18.

Hiirsalmi, H. and Lehmushovi, A. (1982). Ann. Agricuturae Fenniae, 21: 151-54.

Holzapfel, E., Jara, J. and Coronata, A.M. (2015). Agric. Water Management, 148: 207-12.

Holzapfel, E.A., Hepp, R.F. and Marino, M.A. (2004). Agricultural Water Management, 67(3): 173-84.

Hruschka, H.W. and Kushman, L.J. (1963). U.S. Dep. Agric. Mark. Res. Rep., No. 612.

Hu, B., Yang, W.Q., Andrews, H., Li, C. and Takeda, F. (2017). Acta Hort., 1180: 335-39.

HuaJiang, C., WeiPing, C., JianHua, Li, WeiDong, Z. and XueDan, C. (2013). Acta Agriculturae Zheiangensis, 25: 777-81.

Hudson, D.E. and Tietjen W.H. (1981). Proc. Amer. Soc. Hort. Sci., 38 : 239.

Hui, W., HuiQin, Z., ChangNian, Ma and Ming, X. 2011). Jr. Fruit Sci., 28: 1045-49.

Huncock, J.F., Erb, W.A., Goulart, B.L., Scheerens, J.C. and Gough, R.E. (1995). Jr. Small Fruit and Vitic., 3: 1-16.

Iglesias, L. and Liburd, O.E. (2017). Acta Hort., 1180: 283-91.

Ireland, G. and Wilk, P. (2006). Blueberry Production in the northern NSW, Australia. Primefact, 195.

Janes, H.W., Chin, C.K. and Frenkel, C. (1978). Bot. Gaz., 139: 50-52.

JeeHwa, H., YongSham, K., Young, K., EunJu, K., EunHee, S. and KeunJin, C. (2014). Korean J. Breed. Sci., 46: 58-65.

Jensen, K., Doohan, D. and Thomson, J. (1981). Proc. Northeast Weed Sci. Soc., 35: 147.

Jensen, K.J.N. (1986). HortSci., 21 : 1143-44.

JiaBo, P., XiaoYan, L., JiaHui, W., YaDong, L., Jun, K., XiangYi, W. and ChengWu, M. (2011). Dongbei Nongye Daxue Xuebao, 42(10): 76-79.

Johnson, S. (1948). Mich. Agr. Exp. Sta. Tech. Bull. No. 205.

Julian, J.B., Stirk, B.C. and Yang, W. (2011). Ore. St. Univ. Pub., AEB0022.

Julian, J.B., Stirk, B.C., Laro, H.O., Bryla, D.R. and Sullivan, D.M. (2012). HortSci., 47: 866-73.

Karimi, F., Igata, M., Baba, T., Noma, S., Mizuta, D., Kim, J.G. and Ban, T. (2017). The Hort. J., 86: 300-304.

Katano, Y. and Shigeta, T. (1984). Bull. Kanagawa Hort. Exp Stn., 31:16-23.

Kearns, C.A. and Inouye (1993). Techniques for Pollination Biologists. Univ. Press of Colorado, Niwot, Co, USA.

Korcak, R.F. (1986). Jr. Amer. Soc. Hort. Sci., 111: 816-28.

Korcak, R.F. (1988). Jr. Amer. Soc. Hort. Sci., 113: 189-93.

Korcak, R.F. (1989). HortSci., 24 : 573-78.

Korn, K.A., Powell, E.A. and Luteyn, J.L. (2002). Amer. J. Bot., 89: 327-36.

Krewer, G. and NeSmith, D.S. (1999). Blueberry Fertilization in Soil. Frt. Publication No. O1-1, University of Georgia Extension, Athens, Georgia, USA.

Krewerk, G. and NeSmith, D.S. (2006). Blueberry cultivars for Georgia. Southern Region Small Fruit Consortium.

Krstic-Pavlovic, N. and Milutinovic, M. (1982). Jugoslovensko Vocarstvo. 16: 27-33.

Kruger, E. and Naumann, W.D. (1984). Erwerbsobstbau, 26: 133-47.

Kushman, L.J. and Ballinger, W.F. (1967). Proc. Amer. Soc. Hort. Sci., 92:290-95.

Lafond, J. (2013). Int. J. Fruit. Sci., 13: 345-55.

Larco, H., Stirk, B.C., Bryla, D.R. and Sullivan, D.M. (2013a). HortSci., 48: 1484-95.

Larco, H., Stirk, B.C., Bryla, D.R. and Sullivan, D.M. (2013b). HortSci., 48: 1250-61.

Lareau, M.J. (1989). Acta Hort., 241:126-31.

Lazar, E.E., Morris, S., Treverrow, N. and Hayward, S.E. (2006). Final Rept. New Dept. Primary Indust. Narara, Australia.

Lee, H., Krewer, G. and Weibelzah, E. (2017). Acta Hort., 1180: 523-29.

Levin, R. and Vasil, I.K. (1989). *In-Vitro* Cell. Dev. Biol. Plant, 25: 21-27.

Li, Y., Xuedong, T., Lin, W. and Zhang, Z. (2006) Acta Hort., 715: 375-80.

Liburd, O.E. (2004). Berry/Vegetable Times, 4: 2-4.

Lin, Wu, LiLi, S., Xiangxin, L., YaDong, L. and ZhiDong, Z. (2012). Acta Hort., 926: 357-61.

Lindberg, W. (2013). M. Sc. Thesis. Mich. State Univ., East Lansing.

Lipe, J.A. (1978). Jr. Amer. Soc. Hort. Sci., 103: 76-77.

Litwinccizk, W., Szcxzerba, G. and Worna, D. (2005). Scientia Hort., 106: 162-69.

Liu, J. (2005). Ethonopharmacol, 100: 92-94.

Liu, Q.Z., Chen, X., Xu, L., Zhong, X. and Wang. J. (2017). Acta Hort., 1180: 17-22.

Lockhart, C.L. (1982). Canad. Agric., 27 : 39-41.

Luby, J.J., Ballington, J.R., Drapar, A.D., Pizza, K. and Austin, M.E. (1991). Acta Hort., 290: 391-456.

Luby, J.J., Wildung, D.K., Stushnoff, C., Munson, S.T., Read, P.E. and Hoover, E.E. (1986). HortSci., 21:1240-42.

Lyrene, P. (2002). Jr. Amer. Pomol. Soc., 56(2): 79-85.

Lyrene, P. and Ballington, J. (2006). In Blueberries for Growers, Gardeners, Promoters (eds. N. Childers and P. Lyrene). Gainsville, USA: AgScience, pp. 26-37.

Lyrene, P.M. (1983). HortSci., 18 : 221-23.

Lyrene, P.M. (1984). HortSci., 19 : 98-99.

Lyrene, P.M. (1989). Jr. Amer. Soc. Hort. Sci., 114: 995-99.

Lyrene, P.M. (2013). HortSci., 48: 146-49.

Lyrene, P.M. and Payne, J.A. (1992). Proc. Fla. Sta. Hort. Soc., 105: 297-300.

Lyrene, P.M., Sherman, W.B and Crocker, T.E. (1985). Circ. Agric. Expt. Sta., Univ. of Florida, S-320: 3.

Mainland, C.M. (1986). Abst. 22nd. Int. Hort. Cong. California, No. 852.

Mainland, C.M., Kushman, L.J. and Ballinger, W. (1975). Jr. Amer. Soc. Hort. Sci., 100: 129-34.

Maki, K., Ishikawa, S. and Yoshida, T. (2000). Honeybee Sci., 21(4): 159-63.

Maodobry, M., Lech, W., Dziedzic, E. and Kowalski, J. (2001). Folia Hort., 13(1): 73-82.

Marangon, M.A. and Biasi, L.A. (2013). Pesquisa Agropecuaria Brasilera, 48: 25-32.

Martin, L.W. and Pelofske, P.J. (1983). Comm. Soil. Sci. Pl. Anal., 14: 131-42.

Martin, R.R., Polashock, J.J. and Tzanetakis, *I.E.* (2012). Viruses, 4: 2831-52.

McCallum, S., Graham, J., Jorgensen, L. and Others (2016). Molecular Breed., 36(41):1-24.

McGovern, K.B., Annis, S.L. and Yarborough, D.E. (2012). Int. J. Fruit. Sci., 12: 188-204.

McGregor, S.E. (1976). Agri. Handbook, 496, USDA, Washington DC.

McLeod, C.D. and Misener, G.C. (1988). Applied Eng. Agri., 4: 118-21.

Meszka, B. and Bielenin, A. (2012). Progress in Pl. Protec., 52: 88-91.

Milholland, R.D. (1995). In Compendium of Blueberry and Cranberry Diseases (eds. F.L. Crauso and D.C. Ramsdell) APS Press, USA. pp. 7-8.

Milholland, R.D. and Daykin, M.E. (1983). Pl. Dis., 67: 325-26.

Milivojecic, J., Maksimovic, V., Maksimovic, J.D., Radivojevic, D., Poledca, M. and Ercisli, S. (2012). Turkis J. Biol., 36: 738-45.

Miller, P.M. (1987). Canad. J. Bot., 65: 1643-46.

Miller, W.R. (1986). Abst. 22nd. Int. Hort. Cong., California, No. 123.

Miller, W.R. and Smittie, D.A. (1987). Jr. Amer. Soc. Hort. Sci., 112: 487-90.

Moon, J.W., Jr. Hancock, J.F.Jr., Draper, A.D. and Flore. J.A. (1987). Jr. Amer. Soc. Hort. Sci., 112: 170-73.

Moore, J.N. (1964). Proc. Amer. Soc. Hort. Sci., 85: 295-301.

Moore, J.N. (1966). In: Blueberry Culture. (eds. P. Eck, and N. F. Childers). New Brunswick: Rutgers Univ. Press. pp. 45-74.

Moore, J.N., Brown, M.V. and Bordelon, B.P. (1993). HortSci., 28: 1162-63.

Morggia, C., Lobos, A. and Retamales, J.B. (2014). Acta Hort., 1017: 153-58.

NeSmith, D.S. (2006). Acta Hort., 715: 133-36.

NeSmith, D.S. (2006). Acta Hort., 715: 137-42.

NeSmith, D.S. and Krewer, G. (1992). HortSci., 27: 316-18.

NeSmith, D.S., Krewer, G., Rieger, M. and Mullinix, B. (1995). HortSci., 30: 1241-43.

Oblak, M. Petronijevic, M., Krgovic, L. and Milutinovic, M. (1984). Jugoslovensko Vocarstvo, 18:51-58.

Olivera, P.B., Pinto, R.M., Mota, M. and Oliveria, C.M. (2017). Acta Hort., 1180: 511-16.

Osmelak, J.A. and Clayton, G.K. (1988). Pl. Protc. Quarterly, 3: 55-57.

Pack, K.Y., Chakroborty, D. and Hahn, E.J. (2005). Pl. Cell. Tissue Org. Cult., 81: 287-300.

Panta, G.R., Rowland, L.J., Arora, R., Ogden, E.L. and ChonChong, Lim. (2004). Jr. Crop. Improvement, 10: 37-52.

Patten, K.D., Neuendorff, E.W., Nimh, G.H., Peters, S.C. and Cawthon, D.L. (1989a). Jr. Amer. Soc. Hort. Sci., 114:728-32.

Patten, K. D., Scheuerman, M.R. and Burns E.E. (1989b). Acta Hort., 241:354.

Patten, K.D., Neuendroff, E.W. and Nimr, G. (1988c). Jr. Amer. Soc. Hort. Sci., 113: 953-56.

Patten, K.D., Haby, V.A., Leonard, A.T., Neuendorff, E.W. and Davis, J.V. (1988a). Comm. Soil Sci. Pl. Anal., 19: 1065-74.

Patten, K.D., Neuendoiff, E.W., Leonard, A.T. and Haby, V.A. (1988b). Jr. Amer. Soc Hort. Sci., 113:4-8.

Percival, D.C., Abbey, J., Lu, H. and Harris, L. (2017). Acta Hort., 1180: 241-47.

Peterson, D.V., Mullins, C.A., Lietzke, D.A. and Deyton, D.E. (1987). Jr. Amer. Soc. Hort. Sci., 112: 612-16.

Peterson, L.A., Stang, E.J. and Dana, M.N. (1988). Jr. Amer. Soc. Hort. Sci., 113: 9-12.

Plesko, I.M., Marn, M.V. and Koron, D. (2010). Julius Kutin Archiv, 427: 204-205.

Pliszka, K. (1986). Abst. 22nd. Int. Hort, Cong. Califomia, No. 1259.

Pokorny, F.A. and Austin, M.E. (1982). HortSci., l7 : 640-42.

Poll, J.T.K. and Wood, F.H. (1983). New Zealand Comm. Grow., 38:29.

Poonnachit, U. and Darnell, R. (2004). Ann. Bot., 93: 339-405.

Powell, C.L. and Bagyaraj, D.J. (1984). NZ J. Agri. Res., 27:467-71.

Prodorutti, D., Turina, M., Bragagna, P., Profaizer, D. and Angeli, G. (2011). 10BC/ WPRS Bull., 70: 65-67.

Quin, Y., youJiang, L. YongQuan, R., Dorgwei, T., Chen, C., ZhorgChao, L., XingYuan, L. and Wenttua, Z. (2013). Guizhou Ag. Sci., 3: 4-6.

Quito-Avila, D.F., Brannen, P.M., Cline, W.O., Harmon, P.F. and Martin, R.R. (2013). Jr. General Virol., 94: 1426-34.

Read, D.J. (1987). Angewandte Botanik, 61: 25-37.

Retamales, J., Matus, F. and Gonzalez, P. (1990). Abst. 23rd. Int. Hort. Cong., Firenze, Italy, No.4327.

Retamales, J.F. and Hancock, J.F. (2012). Blueberries. Wallingford, CABI, UK.

Robbins, J.A.A. and Doughty, C.C. (1984). HortSci., 19: 100-02.

Rodes, Z.M., Benda, N. and Liburd, O.E. (2017). Acta Hort., 305-12.

Rodriguez-Armenta, H.P., Olmstead, J.W. and Lyrene, P.M. (2017). Acta Hort., 1180: 435-43.

Rousi, A. (1966). Zuchter Gen. Brdg. Res., 36: 352-59.

Saaito, S. and Xiao, C.L. (2017). Acta Hort., 1080: 123-28.

Sadowsky, J.J., Hanson, E.J. and Schilder, A.M.C. (2012). Int. J. Fruit Sci., 12: 169-87.

Sampson, B.J. and Spiers, J.M. (2002). Acta Hort., 524: 53-61.

Sampson, B.J., Rinehart, T.A., Liburd, O.E. Stringer, S.J. and Spiers, J.M. (2006). Ann. Entomol. Soc. Am., 99: 113-20.

Sargent, S.A., Brecht, J.K. and Forney, C.F. (2006). In: Childrs, N.E. and Lyrene, P.M. (eds). Blueberries for Growers, Gardeners and Promoters. Childers Pub. Florida, USA. pp. 139-51.

Schotsmans, W., Molan, A. and Mackay, B. (2007). Postharvest Biol. Technol., 44: 277-85.

Seymour, R.M., Starr, G. and Yarborough, D.E. (2004). Small Fruits Review, 3: 45-56.

Shapiro, D.K., Kudinov, M.A., Narizhnaya, T.I., Kononovich, T.V. and Starkova, N. Yu. (1984). Rastitel'nye Resursy, 20: 396-400.

Shelton, L.L. and Freeman, B. (1989). Acta Hort., 241: 250-53.

Shelton, L.L. and Moore, J.N. (1981a). HortSci., 16: 320-21.

Shelton, L.L. and Moore, J.N. (1981b). Arkansas Farm Res., 30: 14.

Sherstenikina, A.V., Zaranchuk, L.G. (1984). Botanika: Iledovanie, 26: 166-67.

Shimura, I., Kobayashi, M. and Ishikawa, S. (1986). Jr. Jap. Soc. Hort. Sci., 55: 46-50.

Shoemaker, J.S. (1955). Small Fruit Culture. McGraw Hill Book Comp., Inc., pp. 341-82.

Shutak, V.C., Hindle, R. and Christopher, E.P. (1956). Proc. Amer. Soc. Hort. Sci., 68: 178-83.

Siefker, J.A. and Hancock, J.F. (1987). HortSci., 22: 210-11.

Simpson, M., Wilk, P., Colcins, D., Robertson, D. and Daniel, R. (2017). Acta Hort., 1180: 105-10.

Smith, B.J. (1997). Acta Hort., 446: 457-64.

Smith, B.J. and Miller-Butler, M.A. (2017). Acta Hort., 1180: 53-60.

Smith, E.D. and Harris Jr. G. (2017). Acta Hort., 1180: 341-46.

Smittle, D.A. and Miller, W.R. (1988). Jr. Amer. Soc. Hort. Sci., 113: 723-28.

Smolarz, K. (1985). Acta Hort., 165: 185-88.

Smolarz, K. and Mercik, S. (1989). Acta Hort., 241: 171-74.

Song, G.Q. and Hancock, J.F. (2012). Int. J. Fruit. Sci., 12: 316-32.

Song, G.Q. and Sink, K.C. (2004). Pl. cell. Rept., 23: 475-48.

Song, G.Q., Sink, K.C., Callow, P.W., Baughan, R. and Hancock, J.F. (2008). Jr. Amer. Soc. Hort. Sci., 133: 605-11.

Spann, T.M., Willianmson, J.G. and Darnell, R.L. (2003). HortSci., 38: 192-95.

Spiers, J.M. (1978). Jr. Amer. Soc. Hort. Sci., 103: 705-08.

Spiers, J.M. (1981). HortSci., 16 : 175-76.

Spiers, J.M. (1983). Jr. Pl. Nutr., 6 : 1073-86.

Spiers, J.M. (1984a). Jr. Amer. Soc. Hort. Sci., 109: 559-62.

Spiers, J.M. (1984b). Jr. Pl. Nutr., 7: 1567-81.

Spiers, J.M. (1986). Jr. Amer. Soc. Hort. Sci., 111:877-80.

Spiers, J.M. (1987a). Jr. Amer. Soc. Hort. Sci., 112: 600-03.

Spiers, J.M. (1987b). Jr. Pl. Nutr., 10: 2247-61.

Stanisavljevic, M. and Jokovic, P. (1986). Abst. 22nd Int. Hort. Cong. California, No. 155.

Sterne, R.E. (1982). Pl. Dis., 66: 604-05.

Stirk, B. and Buller, G. (2005). HortSci., 40: 1998-2001.

Stirk, B., Buller, G. and Hellman, E. (2003). HortSci., 38: 196-99.

Stirk, B.C. (2012). Acta Hort., 926: 197-214.

Stirk, B.C. (2014). Acta Hort., 1017: 257-68.

Stirk, B.C. and Vance, A.J. (2017). Acta Hort., 1180: 347-53.

Stirk, D.R. and Vance, A.J. (2015). HortSci., 50: 1453-66.

Stoyanov, D. (1983). Nauchni Trudove, Vissih Lesotekhnicheski Institut, Sofiya, Gorsko Stopanstvo, 27/28: 127-32.

Stringer, S.J., Draper, A.D., Spiers, J.M. and Marshall, D.A. (2012). Int. J. Fruit Sci., 12: 246-48.

Stuckrath, R., Quevedo, R., Fuente, L.de.la, Hernandez, A. and Sepulveda, V. (2008). Jr. Pl. Nutr., 31: 1299-1312.

Stushnoff, C. (1972). HortSci., 7: 10-13.

Stushnoff, C.L. and Palser, B.F. (1969). Phytomorphology, 19: 312-21.

SuJin, K., DukJun, Y., JiHyo, K., Taechoon, K., Byoungyil, L. and HeeJae, L. (2004). Jr. Korean, Soc. Hort. Sci., 45: 143-48.

Sun,X.X., Narciso, J., Wang, Z., Ference, C., Bai, J.H. and Zhou, K.Q. (2014). Jr. Fd. Sci., 79(5): M955-M960.

Takeda, F., Krewer, G., Li, C., MacLean, D. and Olmstead, J.W. (2013). Hort Technology, 23: 430-36.

Takeda, F., Yang, W.Q., Ferivalds, Li, A., Sung, K., Xu, R., Hu, B., Williamson, J. and Sargent, S. (2017). Agronomy (Basel), 7:1-18.

Tamada, T. (1989). Acta Hort., 241: 132-38.

Tamada, T., Iwagaki, H. and Ishikawa, S. (1977). Acta Hort., 61: 335-41.

Tamada, T., Iwagaki, H. and Ishikawa, S. (1977). Acta Hort., 61: 335-41.

Targonskii, P.N., Bogdanova, G.A. and Sakova, V.G. (1984). Rastrtel'nye Resursy, 20: 29-35.

Teixeira, L.A.F. and Polavarapu, S. (2001). Florida Entomol., 84: 363-69.

Tobin, M.E., Woronecki, P.P., Dolbeer, R.A. and Bruggers R.L. (1988). Wildlife Soc. Bull., 16:300-03.

Turner, J.C. and Liburd, O.E. (2007) Insect management in blueberries in the eastern United States. University of Florida, I Go 70.

Twigg, M.C. and Link, C.B. (1951). Proc. Amer. Soc. Hort. Sci., 57:369-75.

Van Dalfsen, K.B. and Gaye, M.M. (1999). Appl. Eng. Agric., 15: 393-98.

Vargas, O.L. and Bryla, D.R. (2015). Acta Hort., 50: 479-85.

Vashisth, T., Johnson, L.K. and Malladi, A. (2011). Pl. Cell Rept., 30: 2167-76.

Vescan, L.A., Pamtil, D., Clapa, D., Fira, A., Sisca, C.R., Pop, I.F., Petricele, I.V., Ciuzan, O. and Pop, R. (2012). Romanian Biotech. Letters, 17: 6893-6902.

Vicente, A.R., Ortugno, C., Rosli, H., Powell, A.L.T., Greve, L.C. and Labavitch, J.M. (2007). Jr. Agric. Fd. Chem., 55: 4125-30.

Vidal, I., Amaro, J. and Venegas, A. (1999). Agricultura Tecnica Santiago, 59: 309-18.

Vuente, A.R., Cirello, A.M., Martinez, G.A., Powell, A.L.T., Labavitch, J.M. and Chaves, A.R. (2005). Stewart Postharvest Review, 4: 1-9.

Wallinder, C.J., Talbert, R.E. and Caviness, D.M. (1982). Proc. 15th Ann. Meet. Southern Weed Sci. Soc., p. 139.

Wang, S.Y., Chen, C.T., Sciarappa, W., Wang, C.Y. and Camp, M.J. (2008). Jr. Agric. Food Chem., 56: 5788-94.

Warman, P.R. (1985). Proc. of the 5th Int. Symp. Agri. Wastes, 16-17 Dec., 1985, Chicago, Illinois, USA.

Warman, P.R. (1987). Pl. Soil, 101: 67-72.

Warman, P.R. (1988). Comm. Soil Sci. Pl. Anal., 19: 1031-40.

Warren, J.M., Ballinger, W.E. and Mainland, C.M. (1973). HortSci., 8: 504-07.

Watt, B.K. and Merrill, A.L. (1963). Comp. Foods, USDA Agr. Handbook, 8.

Welander, M., Sayegh, A., Hagwall, F., Kuznetsova, T. and Holefors, A. (2017). Acta Hort., 1180: 387-91.

Whalon, M.E. and Elsner, E.A. (1982). Jr. Econ. Entom., 75:356-58.

White, L.D. (2006). M. S. thesis, Corvallis, USA.

Wijsmuller, J. (1988a). Groenten en Fruit, 43: 70-71.

Wijsmuller, J. (1988b). Fruitteelt, 78: 23-25.

Witcher, W. and Clayton, C.N. (1962). Phytopath., 53: 705-12.

Wolfe, R.R., Chan, W.Y. and Cobianchi, A. (1974). Trans. Amer. Soc. Agri. Engineers, 17:1117-20.

Woodruff, R.E., Dewey, D.H. and Sell, H.M. (1960). Proc. Amer. Soc. Hort. Sci., 75: 387-401.

Wright, A.F. and Harmon, P.F. (2010). Pl. Dis., 94: 966-71.

Yarborough, D.E. (1985). In: Proc. 39th Ann. meeting of the Northeastern Weed Sci. Soc. pp. 204-206.

Yarborough, D.E. and Ismail, A.A. (1982). Proc. Northeasrern Weed Sci. Soc., 36: 193-98.

Yarborough, D.E. and Ismail, A.A. (1985). HortSci., 20 : 406-07.

Yarbrough, J.A. and Morrow, E.B. (1947). Proc. Amer. Soc. Hort. Sci., 50: 224-28.

Young, M.J. and Sherman, W.B. (1978). HortSci., 13: 278-79.

Young, R.S. (1952). Proc. Amer. Soc. Hort. Sci., 59: 167-72.

Zajmi, R.A., Braha, S., Shehaj, M. and Rexhepi, A. (2012). Acta Hort., 926: 559-62.

Zec, G., Colic, S., Todorovic, R. and Marinkovic, D. (2001). Jugoslovensko Vocarstvo, 35(3/4): 117-23.

## Chapter 18

# Cranberries

Cranberry is widespread throughout the cool temperate northern hemisphere, including northern Europe, northern Asia and northern North America. The cranberry fruit is used as fresh or processed as juice, jam, sauce, and also as medicine. Because of its rich contents of polyphenols and other phytochemicals, the demand of cranberries has increased many fold in the last 20 years.

## Composition and Uses

Chemical composition of cranberries showed that they contain: water- 86-87 per cent, dry matter- 13.2 per cent, sugar -3.65 per cent, pectin - 0.77 per cent, protopectin -0.35 per cent, fiber -1.80 per cent and titratable acid - 2.5 per cent. The main part of soluble dry matter in carnberries is sugar, which is represented exclusively by hexoses – glucose and fructose. Fructose predominates in cranberries (1819 mg/100g) and a good amount of sorbitol (382 mg/100g). Citric acid is the main organic acids, and there are also malic and succinic acid. The fruit contains minerals like calcium, phosphorus, iron, sodium and good amount of potassium (Table 18.1). Cranberries have rather high index of biological activity. Fruits contain ascorbic acid, colouring agent and phenolic substances (Khomych *et al.*, 2017, Table 18.2).

Anthocyanins predominate among flavonoids in berries, the content is 64 per cent of the content of phenolic compounds. Detected flavones and their derivatives were: myrecetin-3-o-arabinoside (6.69 mg/100g), querectin-3-*O*-glactoside (10.07 mg/100g), querecetin-3-o-arabinoside (1.84 mg/100g), quercetin-3-o-ramozide (3.86 mg/100g), campherol-3-o-glycoside (1.31 mg/100g), comphorl derivatives (0.67 mg/100g). The composition of phenolic compounds of cranberries contains oxycinnamic acid (19.46 mg/100g), caffeic acid (18.47 mg/100g) and vanillin acid (0.99 mg/100g). The share of cyanidin glycosides were 39 per cent of the total anthocyanin content, cyanidin-3-o-glatocide predominates, its contant was 8.12

mg/100g, cyanidin-3-o-arabinoside was 7.50 mg/100g and cyanidin-3-o-glucoside was 1.40 mg/100g (khomych *et al.*, 2017).

**Table 18.1: Nutritional Composition of Cranberry Fruits (per 100 g edible portion)***

| *Constituents* | *Amount* | *Constituents (mg)* | *Amount* |
|---|---|---|---|
| Water (per cent) | 87.90 | Niacin | 0.1 |
| Calories | 46.00 | Ascorbic acid | 11.0 |
| Protein (g) | 0.40 | Calcium | 14.0 |
| Fat (g) | 0.10 | Phosphorus | 10.0 |
| Carbohydrates (g) | 10.80 | Iron | 0.5 |
| Vitamin A (I.U.) | 40.00 | Sodium | 2.0 |
| Thiamine, B (mg) | 0.03 | Potassium | 82.0 |
| Riboflavin (mg) | 0.02 | | |

* Watt and Merrill (1963).

Paquette *et al.* (2017) reported that a 6-week dietary intervention with 333 mg of polyphenols from strawberry and cranberries improve insulin sensitivity in overweight and obese non-diabetic, insulin-resistant human subjects but was not effective in improving other cardiometabolic risk factors. Recent observations and clinical studies have raised interest in the potential health effects of cranberry consumption, an association that appears to be due to the phytochemical content of this fruit. The profile of cranberry bioactives is distinct from that of other berry fruit, being rich in A-type proanthocyanidins (PACs) in contrast to the B-type PACs present in most other fruit. Basic research has suggested number of potential mechanisms of action of cranberry bioactives, although further molecular studies are necessary. Human studies on the health effects of cranberry products have focused principally on urinary tract and cardiovascular health, with some attention also directed to oral health and gastrointestinal epithelia (Blumberg *et al.*, 2013).

The fruits are sold for both fresh and processed use. The pure and blended juices are becoming more popular than jelly and whole berry sauce (Dana, 1989). The medicinal values of cranberries are associated with the presence of benzoic acid (Igolight *et al.*, 1973).

## Origin and Distribution

Camp (1942) postulates that the genus *Vaccinium* is very old and its segments were highly differentiated in the Cretaceous period of the Mesozoic era. *V. microcarpon* is endemic to eastern North America, from Newfoundland to Minnesota and in the-south to Tenessee and North Carolina (Vanderkloet, 1988).

## Production

USA produced 375,000 tons of carnberries in 2015 which was 57 per cent of the total world's production. The other major producers were Canada (186,000 tons), Chile (85,000 tons), Belarus (7000 tons) and Azerbaijan (3000 tons). However, the

**Table 18.2: Composition of Organic Acids, Sugars and Biological Active Substances in Cranberries**

| *Organic acids (mg/100g)* | | | *Sugars (mg/100g)* | | | | | *Phenolic Substances (mg/100g)* | *Biological Activity cond. unt. act.* |
|---|---|---|---|---|---|---|---|---|---|
| *Citric Acid* | *Malic Acid* | *Succinic Acid* | *Glucose* | *Fructose* | *Sorbitol* | *L Ascorbic Acid (mg/100g)* | *Colouring Agent (mg/100g)* | | |
| 1662 | 965 | 312 | 1683 | 1819 | 382 | 16.60 | 48.70 | 90.00 | 2,840,00 |

(n=3, $P \leq 0.05$).

Compiled from Khomych *et al.* (2017).

annual growth rate between 2007 and 2015 was recorded maximum in Chile (18.3 per cent) followed by Canada (13.7 per cent) and USA (3 per cent).

## Taxonomy

On the basis of the same chromosome number of cranberries and blueberries, hybridization of some *Vaccinium* sp. with cranberries and the existence of intermediate species between cranberries and blueberries, Shultz (1944) suggested placement of cranberries in *Vaccinium* genus instead of *Oxycoccus* as was done by several authors. Recent taxomonic treatment of the section *Oxycoccus* made by VanderKloet (1998a,b) suggested only two species: (a) large-fruited, exclusively diploid American cranberry, *V. macrocarpon* Ait., which is endemic to North America, and (b) *V. oxyoccus* L., a northern hemisphere circumboreal, polyploidy complex existing as diploids (2n= 24), tetraploids (2n= 48) and hexaploids (2n= 72). The description of cranberry species is given below (Galletta, 1975):

### 1. *V. macrocarpon*

An endemic of eastern North America, as well as circumpolar diploid, tetraploid, and hexaploid relatives. All are woody, trailing, evergreen vines, typically found in sphagnum bogs, wet shores, headlands, muskegs or tundra. The red berries are born on ascending axillary branches and vary greatly in size, shape and colour both within and among genotypes. They are buoyant, being dispersed by water, birds and animals in the spring when the seeds readily germinate (Vander Kloet, 1983a,b).

### 2. *V. microcarpon*

the stem is slender, filiform with reddish brown bark. The leaves are strongly revolute, ovate with pointed apex. The fruits are 5-7 mm in diameter, pale pink and insipid. The species is endemic to Scandinavia, central Apls, Iceland, Asia northwestern North America.

### 3. *V. oxycoccus* var. *oxycoccus*

The creeping stem is slender with brown or black bark. The leaves are flat, slightly elliptic having pointed apex. The fruits are 8-12 mm in diameter, pink to red and often mottled at young age. It is found in North America, Europe and Asia.

### 4. *V. oxycoccus* var. *microphyllus*

It is a diminutive form of *V. oxycoccus* which has unilaterally arranged leaves on filiform stem. The fruits are less than 10 mm in diameter and pink in colour. Denmark, Greenland, Labrador, Newfoundland and Gulf of St. Lawrence are its native places.

### 5. *V. oxycoccus* var. *ovalifolium*

Its stem is stouter than *V. microcarpon* with dark brown or black bark. Flat leaves are slightly revolute. The fruits are 10-12 mm in diameter with bloom.

### 6. *V. hagerupii*

The vigorous vine with vertical shoots resembles a bush. The leaves are slightly revolute, ovate to elliptic. It is found in Finland, Denmark, East Prussia, Asia and North America.

## Cultivars

More than 100 cultivars were selected from population in native bogs during the past one and a half century of cranberry cultivation. Among the selections, Early Black, Howes, McFarlin and Searles Jumbo cultivars account for over 90 per cent of the total planting (Shoemaker, 1978). Several new cultivars have been selected or bred in the last fifty years of which few are grown commercially in different countries.

**Beaver**

It originated as a seedling in Beaver River, Nova Scotia, Canada. In comparison with Early Black, the fruits of this cultivar matures about a week earlier and are larger, more attractive and keep well in cold storage. The vine is moderately productive (Brooks and Olmo, 1972).

**Beckwith (McFarlin x Early Black)**

The fruit is large, oval-oblong, deep red, very high in flavour and makes excellent sauce (Galletta, 1975). The productive vine is promising in New Jersey.

**Bergman (Early Black x Searles)**

It was introduced in 1961 by USDA. The fruit size is medium, pear-shaped, skin red and keeping quality very good being suitable for long storage (Dana, 1983). It is a mid-season cultivar giving higher yield than Early Black or Howes. The cultivar is resistant to leaf hopper which spreads the false blossom virus disease.

**BE4 (Willapa Red) (Putative Aviator x McFarlin)**

Developed from USDA-ARS breeding programme. Mid-season cranberry with high yield potential. Fruit small, good tolerance to wet soils.

**CNJ 97-105-4 (Mullica Queen®) (Lemunyon x No. 35)**

Developed from Rutgers University, USA. Mid-season cranberry with high production potential. Large fruit (> 2g).

**CNJ 95-20-20 (Scarlet Knight ®) (Stevens x NJS 98-37)**

Developed from Rutgers University, USA. Early season cultivar with very early colour development with exceptionally high fruit anthocyanin content. Fruit large (>2g), ovate to very widely elliptic to occasionally elliptical.

**Early Black**

Early Black is a predominant cultivar in New Jersey and Massachusetts, having originated around 1845. The vine is comparatively slender, producing more uprights than runners (Dana, 1983). This is an early ripening cultivar with very good fruit colour, firmness and quality. The fruits colour well in storage and keeping quality is good to very good (DeMoranville, 2004).

**Franklin (Early Black x Howes)**

It was released in 1961 by USDA. Fruits are medium in size, nearly round, skin red to very dark red, and have good keeping quality. The cultivar is more productive than Early Black or Howes (Brooks and Olmo, 1972). It is resistant to leaf hopper, and fruit rot is also relatively less.

**Grygleski #1**

Early mid-season maturing cranberry, fruit medium, high yield potential and winter hardy.

**Grygleski #2**

Early mid-season maturing cranberry with high yield potential. Medium to large fruit, round, suitable for fresh fruit milling.

**Grygleski #3**

Early mid-season maturing cranberry, fruit medium, high yield potential and winter hardy.

**Howes**

It is a standard late cranberry cultivar in Massachusetts and New Jersey which originated in 1843 (Galetta, 1975). The fruit is of medium size and colour but has excellent keeping and shipping quality. Productivity is moderate.

**Hyred**

Developed from University of Wisconsion-Madison. Fruit 1.6-1.8g, elliptical blocky with squared stem end, sensitive to water logging.

**McFarlin**

McFarlin is one of the few cranberries grown in both east and west. The fruit ripens late; large, round-oblong, flower end conical, lacking in uniformity of size and shape. Plants are very productive, frost resistant, and resistant to false blossom.

**NJS 98-23 (Crimson Queen®) (Stevens x Ben Lear)**

Developed from Rutgers University, USA. Early season, large-fruited, highly productive. Fruit large (> 1.9g), elliptic.

**NJS 98-35 (Demoranville®) (Franklin x Ben Lear)**

Developed from Rutgers University, USA. Early season, large-fruited (>2g), highly productive cranberry.

**Pilgrim (Prolific x McFarlin)**

Fruit is large, long-oval, purplish red, and has good keeping quality. It is a prolific bearer and fruit ripens late. The uprights are moderately vigorous and resistant to leaf-hopper (Dana, 1983).

**Cranberry cultivar Early Black**

**Cranberry plantation**

## Searles Jumbo

It is a principal cultivar of Wisconsin and is also popular in Canada. Fruit is large, olive-shaped, uniformly red with moderate keeping quality but only fair in other qualities. It is a mid-season cultivar of excellent productivity (Galletta, 1975).

**Ripe Rot Caused by *Coleophoma empetri***

**Cranberry cultivar Stevens**

**Cranberry vines**

### Stevens (McFarlin x Potter)

It was introduced in 1950 by USDA. The cultivar is late mid-season ripening, having round, dark red, glossy, firm fruits of good keeping quality. The vine is vigorous and very productive.

### Wilcox (Howes x Searles)

Wilcox was introduced along with Stevens in 1950. Fruit size is medium, shape oval, skin deep red, flesh firm and keeping quality good. The vine is very vigorous, productive and resistant to false blossom disease; ripens very early.

## Climate

Temperature during growing period and sunshine are major climatic factors influencing the cranberry culture. Shoemaker (1978) stated that 2500 cumulative hours below 7°C are required to break the rest period of cranberries. Insufficient chilling is a limiting factor in its cultivation in many warmer regions which prolongs bloom period and can cause 'umbrella bloom'. Fluctuating temperatures during the chilling period decrease the chilling requirement whereas prevailing high temperature increases it. Fruit bud grows only during long days. Consequently, more chilling is required for bud break under short-day conditions. Besides chilling, exposure of the buds to a temperature above 7°C is also needed for normal flower production. Gibberellic acid treatment induces bud break up to 80 per cent without chilling and causes faster bud break after adequate chilling (Rigby and Dana, 1972).

Cranberries are subject to frost damage at below freezing temperature especially in spring and fall (Workmaster and Palta, 2006). The risk of damage is more prevalent on clear calm nights when leaves radiate infrared radiation (Sage and Sage, 2002), resulting in significantly lower bog temperatures compared to those in the surrounding uplands. Frost protection is considered essential to cranberry yield by ensuring that both the reproductive and vegetative meristem with in the buds are protected from damage. Sprinkler irrigation is provided in the orchard to protect from frost injury (Issa, 2012).

During winter season, the plants may die due to dry wind if the soil around roots is frozen restricting the compensation of transpired water. This situation is prevented by winter flooding. In spring, the cranberry buds can survive up to –4°C but it is always desirable to protect the plants from frost injury by flooding or sprinkling. Sometimes spring frosts cause crop failure.

Degaetand and Shulman (1987) evaluated the relationship between cranberry yield and meteorological factors and concluded that temperature and sunshine were the most important factors governing the growth and production of the cranberry, whereas precipitation, water table and soil moisture are of little importance. Warm temperature during mid-May to late June and mid-October to mid-November of the year prior to harvest increased berry production. Cold temperature during early May to mid-June also favours higher yield. Conversely, warmer weather in April and May adversely affects the production (Shoemaker, 1978).

Correlations of temperature to fruit composition indicated that warmer temperatures early in the season (around bloom and fruit set) had the most positive impact on total anthocyanins and total flavonolos in fruit (Heuvel and Aution, 2008).

## Soil

Commercially, cranberries are grown on peat, muck and sandy soils. High water table is a pre-requisite for the establishment of plantings on sandy soil because of its low moisture holding capacity. Marshlands with high soil acidity (pH 4.0-5.0) are natural habitats of the cranberry plants. The plant tolerates pH as low as 3.5 but grows very well at pH 6.5 in nutrient solution (Dana, 1989). At higher pH (7.0), the plant exhibits iron chlorosis due to faulty metabolism rather than from the deficiency of iron in the tissue. Thus, the plant does not demand low pH but the inability of other vegetation to survive in such a situation provides conducive conditions for cranberry plant. A pH range of 4.5 to 5.0 is generally considered best. Szpunar (1985) obtained best plant growth on acidifying the swamp soil (pH 6.6) with 10 t sulphur/ha two weeks before planting followed by top dressing with 5 t sulphur/ha in four doses. Szwonek *et al.* (2016) stated that vegetative and reproductive growth of cranberry cultivars were influenced by soil conditions. Vigorous growth of plants was observed in sandy and very acidic soils, compared to sandy-loam soil, even after reducing the pH to slightly acidic.

Maintaining optimum soil moisture level and draining pond water remains a challenge for many cranberry growers in south eastern Massachusetts. Wet soil moisture conditions have been implicated in increased root and fruit rot diseases, inhibition of root development (Caruso, 2012), whereas relatively dry soil moisture conditions have been linked increased photosynthesis and improved fruit yield (Caron *et al.*, 2016). Subsurface tiles are used for rapid removal of excess soil moisture. Tile drainage can significantly lower the water table, thereby reducing capillary rise and moisture levels in soil (Gillham, 1984).

## Propagation

Cranberries can be propagated both by seeds and vegetative methods but are raised commercially by cuttings.

### Seed Propagation

The seeds are extracted from large mature berries. Cold storage of seeds and treatment with gibberellic acid increase percentage of seed germination. According to Gronskis and Snickovstis (1989), the best field germination (87 per cent) and subsequent growth of the sprouts are achieved by soaking the seeds in 10 per cent $Na_2CO_3$ for 24 hours before sowing. The optimum quantity of seeds for getting about 21 to 27 growing sprouts per $m^2$ is 27 to 32 kg/ha. The seedlings, however, take longer period (6-7 years) to come into bearing than cuttings. Morosov (2000) stated that for germination of hybrid seeds, sowing of seed obtained from fresh berries is one of the most important condition for production of seedlings. Using seeds after 6 months in storage was found ineffective. Stratification of seed for 5-9 days allowed germination.

## Vegetative Propagation

### Cutting

Because of easy rooting, the cuttings are directly planted in the place of their growth. The cuttings made from two-year-old woody runners should be about 15-20 cm long containing 3 to 5 nodes (Gronskis and Snickovstis, 1989). Plants from cuttings rooted vertically showed higher growth than plants from cuttings rooted horizontally (Welker and Vass, 1983). These are secured from healthy plantings of a desired cultivar in spring before the resumption of new growth. Chong (1999) reported better rooting by using peat-ammended compost.

### Micropropagation

Scorza *et al.* (1984) reported the effects of auxin and cytokinin, combinations on *in vitro* development of cranberry node explants. They found 5- to 7-fold shoot proliferation on Anderson's medium. Serres *et al.* (1990) were able to establish shoot cultures on woody plant medium (WPM) + 0.1 µM 2iP. Adventitious buds were induced on stem sections when cultured on WPM + 10/µM 2iP and lµM thidiazuron (BIM). Field tests of micropropagated cranberry was found highly successful with greater than 90 per cent field survival and rapid growth allowing for quick ground coverage.

The protocol for regeneration from somatic tissue of cranberry have been standardized (Filipenya and Doglov,2000). Half-strength Woody Plant Medium (WPM) supplemented with 1 mg 2ip +0.5 mg TDZ/l or 5 mg 2ip + 0.2 mg TDZ/l produced the highest number of adventitious shoots per explants. Shoots from explants on this medium gave the highest frequency of elongation when transferred to growth regulator-free medium. Nodal explants were reported as more productive than shoot tips and the best total shoot production was obtained when nodal segments were cultured in nutrient medium supplemented with 2.5-5.0mg 2ip/l (Debnath and McRae, 2001). Sedlak and Paprstein (2011) reported highest multiplication rate (2.7) of shoot tip of cv. Howes on Anderson's rhododendron medium (AN) with 1mg/l of zeatin. The effects of different auxins (IBA, IAA and NAA) on root induction in WPM were tested. The percentage of rooting was 100 per cent for NAA, 86 per cent for IBA and 81 per cent for IAA.

## Planting

Soon after sanding, the cuttings are inserted in the soil at 30 cm spacings both ways. In closer planting, the ground is covered quickly, less weeding is required and higher returns are obtained (Shoemaker, 1978). The cuttings will dry out if they remain exposed more than 3 cm above the sand. Inserting of cutting is a better method than scattering over the soil surface (Gronskis and Snickovstis, 1989). The bog should be flooded for a day or two immediately after planting.

## Pruning

Davenport and Vorsa (1999) stated that excessive vegetative growth reduced yield in cranberry, which caused shading and promotion of fungal disease and decrease light penetration. Anthocyanin concentration (red pigment) of fruit depends on adequate light penetration (Strik and Poole, 1995). One practice used to manage the cranberry plant canopy is sanding. Sanding is used to rejuvenate old bog or curtail overly vegetative bogs (Strik and Poole, 1995). Typically, sanding is done every 2-5 years with verying depth between 1.2 and 5 cm (DeMoranville and Sander, 2000).

Marueci (1987) suggested pruning of vegetative shoots to make the bogs more productive. Pruning will reduce yield in the year of pruning which however, increased production considerably in the following year. Stirk and Poole (1992) suggested light pruning every other year for maximum benefit.

## Nutrition

Because of unusual cultural requirements and nature of growth, the nutrition of cranberries is complicated (Cain and Eck, 1966). In many cases, plants do not respond to fertilizer application. A balance between vegetative growth and fruiting is required to be maintained.

## Role of Nutrients

### Nitrogen

Among the macronutrients, nitrogen plays the most important role on growth, yield and fruit quality. In sand and soil culture experiments, N from both inorganic and organic sources increased the number of uprights, runner growth and total dry matter (Somogyi *et al.*, 1964). Eck (1962) obtained a significant increase in berry weight in response to P and a highly significant increase in berry size in response to N and P. Excessive N, however, reduced the yield, increased fruit rot and decreased the pigmentation and keeping quality of fruits (Atwood and Zuckerman, 1961; Eck, 1971). Shawa (1982) in another trial with urea and ammonium sulphate, each at 1.12 to 5.6 kg N/ha applied at 5, 50 and 80 per cent bloom. The best yield increment (32 per cent) was obtained from plants sprayed with urea at 4.48 kg N/ha. Ammonium sulphate at 2.24-5.6 kg N/ha however, slightly decreased yield.

Greidanus (1972) demonstrated that NH4-N is essential for cranberry because it is incapable to utilize nitrate in absence of N-reducing enzyme. They suggested application of ammonium nitrogen throughout the growing season. But Shawa (1979) got 26 per cent higher yield with sulphur-coated urea (SCU) and isobutylidene diurea (IBDU) than with urea. In order to prevent trampling of plants by three or four urea applications, he recommended a single application of SCU or IBDU during April-May.

Ruban and Kurlovich (1999) suggested application of 22kg ammonium sulphate/ha for increasing yield in high-yielding plots and 44kg ammonium sulphate/ha for low-yielding plots. The virgin soil characteristically maintain high soil moisture content and higher percentage of organic carbon. Davenport and Patten (2002) stated that mineral N in cultivated soil was predominantly in the nitrate from whereas in the virgin soils ammonium N was dominant. Ammonium fertilizers when applied and irrigated (solublized) were taken up by the plant within one day following application. Soil and root temperature for N uptake by cranberry vine was reported as 18-24°C (Roper *et al.*, 2004). Davenport and DeMoranville (2004) however, observed that soil N release from native organic matter was fairly consistent until soil temperature exceeded 21°C, indicating that when temperatures exceeded 21°C, planned fertilizer application should be reduced, particularly in highly organic soils.

### Phosphorus and Potassium

Cranberries have low requirements for both phosphorus and potassium which are adequately supplied by the soil (Dana, 1989; Greidanus and Dana, 1972). P deficiency restricts growth and leaves become purplish and pinkish red.

Most growers apply fertilizers that are complete blend of N, P and K, basing the application rate on the amount of N required. Depending on the N:P ratio of the chosen fertilizer, P rates may be in excess of plant needs. The recommended rate for P in cranberries with tissue tests above the critical value (0.1 per cent P) should be around 22 kg/ha P per season (around 50 kg $P_2O_5$). Phosphorus has been identified as a potential environmental pollutant discharging from cranberry system to inland water bodies. Best management practices should incorporate reduced P fertilizers and minimum retention time of harvest floods (DeMoranville *et al.*, 2009).

Potassium rarely becomes a limiting factor with cranberries and its concentration is about three times more in fruits than in leaves (Shoemaker, 1978). It is believed that high K produces firm fruit of good keeping quality. Roper (2009) could not found any response on yield, fruit size or fruit colour of $K_2O$ when applied from 0 to 896 kg/ha $K_2O$. KCl and $K_2SO_4$ were equally effective and did not show toxicity.

## Micronutrients

The deficiency symptoms of micronutrients have neither been seen in the fields nor have been demonstrated by solution culture experiments. On the basis of soil analysis, Fisher (1951) observed a positive correlation between available Fe in the soil and production and a negative association between production and available Al and Mn contents in the soil. No trend was found between production and soil levels of Ca, Mg, and P. Plants grown under a well-balanced fertilizer regime had Fe contents four times higher than under NPK regime alone (Sidorovich *et al.*, 1987b). At different soil zinc levels, Zn accumulated mainly in fruits (Sidorovich *et al.*, 1987a). Sidorovich *et al.* (1988) observed that the plant receiving B had highest B content in both vegetative and reproductive organs and gave the highest yield. Peterson *et al.* (1986) noted significant growth reduction in the greenhouse in solution culture when molybdenum was omitted but in a second experiment showed no reduction in growth. Doughty (1984) stated that spraying EDTA chelate of Zn or Fe to vines, showing leaf chlorosis similar to Fe deficiency, caused more growth from terminal buds. Demoranville and Deubert (1987), in a 3-year field experiment in USA, hand sprayed commercial foliar fertilizers containing Ca and B or Mn and Zn on cvs. Howes and Early Black during a period of low foliar levels of Ca and Mn, which was found to occur during the time of strong vegetative growth and floral development. The yield (3-year average) for areas treated with Ca, B was found to increase in 9 or 10 split application treatment (timed between late May and mid-July) for Early Black and in all 10 split application treatment for Howes, due to increased fruit set. The most effective treatment dates were the first week of June and the first week of July (full bloom) and the treatments increased the yield by 45 per cent and 68 per cent in cultivars Early Black and Howes, respectively, as compared with the untreated plants.

## Tissue Analysis

Adequate information is not available on the leaf analysis of cranberries. The nutrient levels in tissues do not show a significant relationship with plant vigour or yield. Furthermore, the ranges of nutrient levels are rather narrow. The desirable levels of various nutrient elements are given in Table 18.3. The lower or higher levels may show deficiency or excess of the nutrients.

**Table 18.3: Cranberry Tissue Nutrient Concentration Guidelines for Producing Beds***

| *Nutrient* | *Sufficient Range (per cent)*[1] | *Nutrient* | *Sufficient Range (ppm)*[1] |
|---|---|---|---|
| N | 0.90 to 1.10 | B | 15 to 60 |
| P | 0.10 to 0.20 | Cu | 4 to 10 |
| K | 0.40 to 0.75 | Mn[2] | Above 10 |
| Ca | 0.30 to 0.80 | Fe[2] | Above 20 |
| Mg | 0.15 to 0.25 | Zn | 15 to 30 |
| S | 0.08 to 0.25 | | |

1 Based on samples taken between August 15 and September 15.

2 No sufficient range has been identified for Mn and Fe.

* Hart *et al.* (2015).

Tissue N concentration increased linearly with increase N application but was greater in vegetative uprights than in fruiting uprights. Current season growth on vegetative uprights was more responsive to tissue N than on vegetative uprights. Chlorophyll concentration and net carbon exchange increased soil N but upright starch concentration and often total non-structural carbohydrate concentration decreased with increase soil N at midfruit development and preharvest, especially in vegetative uprights (Heuvel and Davenport, 2006).

## Fertilizer Recommendations

Shoemaker (1978) recommended 56 to 168 kg N and 56 to 112 kg P per ha. P should be applied on the basis of soil analysis. In New Jersey, 8-8-8 fertilizer is used at 68 kg in May and 68 kg in August. Eck (1985) obtained optimum yield from the cultivar Early Black at 40kg $P_2O_5$/ha/year. When the new uprights are more than 5 cm long, application of N fertilizers should be discontinued. P and K application did not increase the yield of cranberries, hence these may be applied at 4 to 5 years' interval.

Early fertilizer application in May is more advantageous than subsequent application as the mid-June application encourages vine growth at the expense of fruit. Late applications in August and November may increase the number of fruit buds and flowering uprights in the following spring.

Cranberry growers of USA are looking for ways to reduced off-site movement of N and P (Kennedy *et al.*, 2016). Controlled release fertilizers (CRF) have been suggested as a potential method of limiting nutrient losses, with increased nutrient use efficiency (NUE), equivalent production results should be possible at reduced N and P rates applied as CRF compared to the standard grower practice with soluble fertilizer (DeMoranville and Kennedy, 2017). A low-P CRF formulation (19-6-12) showed similar yield (46.728 kg/ha) compared with grower practice (10-12-26) where the reported yield was 50.808 kg/ha (DeMeMoranville and Kennedy, 2017).

## Irrigation

Cranberry evapotranspiration (ET) can be estimated using the Priestley-Taylor equation and a crop coefficient of 0.83 (Vanderleest and Bland, 2017). When ET exceeds of water demand, irrigation is applied to raise the soil water potential (SWP). The methods of irrigation in cranberry include sprinkler irrigation, subirrigation, or a combination of both methods (Caron *et al.*, 2017).

In cranberry production, efficient drainage systems are essential for the development of precision irrigation methods. Most cranberry fields are equipped with subsurface drainage systems used for water table control and excess water removal (Bulot *et al.*, 2017). A significant yield reduction (25 per cent) in one-half of a cranberry bed was noted where the drainage outlet was at a higher elevation than the drain (Pelletier *et al.*, 2017).

The flooding is done to save plants from severe winter cold, from frost in spring and autumn, at times of harvesting and to get rid of insects. New Jersey cranberry growers have historically used flood irrigation to protect their crops from freezing, but are now using sprinkler system to replace the practice of flooding for frost protection. There are a number of advantages of sprinkler irrigation over flooding including water conservation and improved climate control. Yield increase of up to 40 per cent has been reported (Gronwald and Haines. 1982).

As soon as winter sets in and temperature goes down to freeze the water, the bog is flooded in successive layers. A new water layer is applied only after the freezing of the previous layer. The ice blanket of frozen water prevents drying of vines and reduces fluctuations in temperature. The water is drained in early spring depending on the thawing of ice but in case of insect problem, this may be delayed for several weeks. Short duration floodings in spring and autumn save plants from frost damage and serve irrigation purpose during summer drought.

Sprinkler irrigation has been developed as a supplement to flooding in order to economize the use of water. Sprinkler frost protection is more water efficient, provides precise and nearly instantaneous protection, eliminates the possibility of crop injury from oxygen deficiency and reduces the spread of disease spores and weed seeds (Dana, 1989; Dana and Klingbeil, 1966). Sprinkling is also used to prevent scalding of berries and vines by high temperatures. Sprinkler irrigation is provided almost the year round in cranberry bogs (Shoemaker, 1978).

## Weed Control

Weed control is particularly necessary during the initial planting years until the entire ground is covered by the vines. The control measures include hand weeding, mowing and chemical treatments, the last one being more effective. The common weedicides used in the past are amitrole, dichlobenil and norflurazon. Dana (1989), however, reported major breakthrough in cranberry weed control by glyphosate as its use has increased cranberry production. Shawa (1983) stated that application of glyphosate controlled *Carex vesicaria, Aster subspicatus* and *Spiraea douglasii* and terbacil controlled *C. vesicaria,* while napropamide controlled, *A. subspicatus* and *Lotus corniculatus*. Application of sulphur (1.5 t/ha of poro-sul), in lowering pH by 1 unit, prevented growth of *Maianthemum dilatatum* and *Trifolium fimbriatum.* Postharvest application of norflurazon and 2, 4-D controlled *Rumex acetosella* and *Carex obnupta,* respectively. Repeat annual application of dichlobenil on commercial cranberry beds may be considered as part of a viable integrated weed management with no adverse effects on crop growth or yield (Sandler *et al.,* 2004).

Devlin and Deubert (1980) have reported a yield reduction of 80 per cent to 100 per cent in cranberries by swamp dodder weed (*Cascuta gronovii*). Pronamide at 0.6, 1.1 and 2.2 kg/ha controlled swamp dodder effectively as a pre-emergence herbicide without significantly reducing the cranberry yield (Bewick *et al.,* 1989). The microbial control of swamp dodder has been achieved by the spores of *Fusarium tricinctum* and *Alternaria* sp. (Bewick *et al.,* 1987).

Sandler (2009) reported that application of meotrione and imazamox, prior to flowering inhibits seed formation of dodder. Dewberry (*Rubus* spp.) and sawbrier (*Simlax glauca*) are woody cranberry weeds and are difficult to control, spread quickly and can cause significant loss of crop (Else *et al.,* 1995). Hand pulling young weeds, use of glyphosate, or spot renovation of heavily infested areas are the practices for controlling dewberry and sawbrier. Propane flame cultivation (FC) torches is a method of weed control where target plants are exposed to brief periods of high temperature to damage or kill plant tissue. The non-fatal response of cranberry to FC indicates that FC will cause less damage than glyphosate to cranberry plants that are incidentally exposed during spot treatment of weeds and thus could be integrated into weed control in certain situations, including organic farming (Ghantous *et al.,* 2013).

## Sanding

Shoemaker (1978) has recommended sanding and resanding in view of the many advantages, *i.e.,* levelling of the land, reduction of weeds in early years, stimulation of rooting, prevention of frost damage, control of certain insects and provision of best drainage and aeration. Apart from initial sanding, resanding can be done every third year by putting 6 to 25 mm sand at a time with hand or by dump trucks on the ice during winter. Resanding of bogs with ample water supply and heavy vines is not necessary.

# Flowering and Pollination

## Flowering

Bud dormancy breaks in mid-to late spring after receiving chilling (>1000 hours) and floral inflorescence are initiated in late summer or early fall. Three to seven flowers borne singly in axils of reduced leaves along the rachis and terminating in a leafy shoot flowering in early summer. Commercial production in cranberries begins after 4 or- 5 years from planting of cuttings. The fruits are borne on uprights which arise either from runners or from other uprights. The former are known as new uprights and the latter old uprights. Only old uprights flower but not all of them (Bergman, 1954). For optimum growth and yield, a population of uprights between 215 and 322 $dm^{-2}$ is desirable (Roberts and Struckmeyer, 1942). A further high density of uprights reduces yield due to both shading and poor pollination. The optimum upright length for fruit production is between 4 and 9 cm. Mixed buds containing flowers, leaves and a growing point develop terminally on the uprights and their differentiation occurs during late summer or fall prior to the dormant season. Eck (1986) reported July 4 as the date of flower induction for Bergman cultivar in British Columbia. The buds remain dormant during the winter, resuming growth in the spring.

Cranberry has a long flowering duration. Commencing in mid-June, it continues up to mid-July (Eck, 1986). Individual flowers are attached directly to the main axis of upright with a pedicel. A flower opens by separation of the petals which curl back exposing the stamens. The pollen is mature at the time of petal separation but stigma becomes receptive sometimes later. In the greenhouse, floral buds open throughout the day, the interval from petal separation to anthesis varying from 2 to 12 hours (Shoemaker, 1978). The pollen is not a single grain but is a tetrad which germinates into four functional tubes aiding in fruit set and a fruit with only few developing seeds matures (Roberts and Struckmeyer, 1942).

## Pollination

Pollination is a pre-requisite to fertilization, seed formation and fruit development in cranberry because seedless fruits are of no commercial value (Eck, 1986). Although cranberry is self-fruitful and floral morphology is helpful to self-pollination, yet greater fruit set, larger berries, and high seed counts occur following cross-pollination (Marucci and Filmer, 1964). Fruit abortion in cranberries has been explained partially due to inadequate number of pollen tubes and lack of subsequent fertilization (Birrenkott and Stang, 1989). Since wind and mechanical jarring of blossoms are ineffective in promoting pollination, the need for insect pollination, either indirectly as flower agitators or directly as pollinators has been advocated (Furror and Bain, 1946; Filmer, 1949). Filmer and Doehlert (1959) could harvest only 16 berries $dm^{-2}$ in caged plants receiving agitation as compared with 97 to 163 berries $dm^{-2}$ on insect-pollinated plants. Two to five honey-bee colonies, depending on the pollination conditions, have been recommended per hectare for cranberry pollination (Stewart and Marucci, 1970). Kevan *et al.* (1983) stated that the large cranberry is dependent on insects for pollination and fruit set.

Bumble bees (*Bombus* spp.) are the most common buzz pollinators of cranberry. Bumble bees can forage at low temperature, under wet conditions, early and late in the days, early in the season and late into fall (Goulson, 2003). Research was conducted on two commercial cranberry farms in western Newfoundland, Canada, to determine which bee species are most important for pollination of cranberry in this area, and whether importing the commercial bumble bee, *Bombus impatiens*, for supplemental pollination is desirable (Hicks and Sircom, 2016). The bumble bee, *Bombus ternarius* and helictid bee, *Lasioglossum* (*Dialictus*) spp. were the most common and abundant bees collected and are important for pollinating cranberry. The best time to place hives is when the cranberries are about 10 per cent bloom. A few species of *Osmia* and *Megachile* bees have shown to work cranberries in some areas of the Canada. The syrphid *Toxomerus* sp. was also thought to have important role in pollination. Birrenkott and Stang (1989) found 8 per cent increase in fruit set by supplementing insect pollination with hand pollination. They further reported that flowers receiving low amount of pollen (> 10 tetrads), or pollen that fails to germinate, are more likely to abort. More than 50 per cent fruit set has been observed in vines containing mixed cultivars (Eck, 1986).

## Fruit Set

Detailed yield component analyses have shown fruit set to be a limiting factor in cranberry production (Eaton and Kyte, 1978; Eaton *et al.*, 1983). A range of 30 per cent to 50 per cent fruit set is considered normal in Wisconsin cranberry fields (Birrenkott and Stang, 1989).

Baumann and Eaton (1986) concluded that competition occurs between the berries on an upright, the lower flowers being flower positions on a cranberry upright suppress fruit development by the remaining upper flowers, resulting in a temporal decline in fruit set (Birrenkott and Stang, 1989). Birrenkott and Stang (1990) made selective flower removal to evaluate interplant competition or inhibition within flowering uprights of 'Searles' cranberry. With one or both of the two earliest, *i.e.*, lowest, flowers developing into fruit, 25 per cent of the remaining flowers matured into fruit. Removal of earliest two flowers at preblossom or late bloom resulted ≈ 46 per cent fruit set for the remaining flowers. Slightly fewer upper flowers set (36 per cent) when the earliest flowers and fruits were removed at early fruit development. Hand-pollinated (upper) flowers set 58 per cent when the lowest two flowers were removed, compared to 17 per cent for the unthinned control. The failure of cranberry flowers to set fruits is termed as 'blossom blast' and may be caused by both internal and external factors. If all flowers set fruits, the plant is not in a position to carry the entire crop load. Hence, blossom blast is the result of natural attrition of the blossom (Marucci, 1966). Birrenkott and Stang (1986) stated that the flowering shoot (upright) of cranberry produces one to seven flowers. In the field normal fruit set ranges from 35-45 per cent. They investigated field pollination, pollen tube growth and upright density (intraplant shading) with 'Searles' cranberry grown under controlled environmental condition. Optimum environmental conditions with hand pollination resulted in 83 per cent fruit set. In the field, hand pollination in addition to insect pollination did not affect fruit set. Shading during bloom with hand pollination or natural pollination was found to reduce fruit set to 25-40 per cent. They also observed developed pollen tubes in the styles of undeveloped pistils which suggests that factors other than pollination are limiting cranberry fruit set.

Among the external factors influencing the fruit set, weather condition is the most important. Poor fruit set is associated with severe cold before flooding in the fall, severe frost in the spring before the expansion of floral buds, high temperatures in the spring coupled with low moisture supply, oxygen deficiency during flooding, nutrition imbalance, tipworm injury, false blossom disease and fungus attack (Eck, 1986; Wilcox, 1940). Adverse weather conditions also inhibit insect activities. The pollen germination is drastically reduced by fungicidal spray during bloom (Briston and Shawa, 1981; Shawa *et al.*, 1966).

The fruit set and yield can be increased by gibberellic acid treatment although fruits were small and parthenocarpic. (Devlin and Demoranville, 1967; Mainland and Eck. 1968; Stang and Birrenkott, 1989). The auxins, p-chlorophenoxyacetic acid and 2, 4, 5-trichlorophenoxyacetic acid have also been reported to enhance fruit set and yield (Doughty, 1962). Application of calcium and boron formulation has been shown to increase cranberry fruit production, possibly through enhanced pollen germination (DeMoranville and Deubert, 1987). Spraying of putrescine ($10^{-2}$M) on open pollinated 'Searles' flowers at 30 per cent and again at 60 per cent open bloom stage was found to increase fruit set by 3-4 per cent and final berry weight by 7 per cent, (Stang and Birrenkott, 1986).

## Fruit Growth, Development and Ripening

Chandler (1952) has characterized the cranberry fruit growth curve as single sigmoid and both large-fruited McFarlin and small-fruited Early Black and Howes cultivars followed the same pattern. The fruit development begins after fruit set, continues rapidly throughout the growing season and slows down near maturity. Demoranville (1960) demonstrated 12 per cent increase in average weight of 'Early Black' cranberries from the date of first harvest to the date of optimum harvest in both early and late-drawn bogs. Cranberries cv. Howes shows average 6 per cent increase in size for early-drawn bogs and a 4 per cent increase for late-drawn bogs. He further observed that the high 'Early Black' cranberry production coincided with the occurrence of the heaviest rainfall, whereas for 'Howes', the heaviest berries were produced in drier years. When the temperature during growing season is below normal, the berries become small.

Seeds are not only important for the development of berries but they also determine the fruit size. A strong relationship exists between the number of seeds contained in a berry and its ultimate size (Hall and Aalders, 1965; Rigby and Dana, 1971). Hall and Aalders (1965) calculated that each additional seed increased the weight of the 'Howes' berry by 36 mg and diameter by 0.09 mm or 2.1 per cent. Filmer *et al.* (1958) reported that berries averaging less than one seed per carpel were non-uniform in shape and small in size.

The sugar concentration increased from about 2.2 to 3.2 per cent as fruit ripens from white to red with glucose comprising 82 to 74 per cent of the total sugars, respectively. During the same time acid content decreased by only 22 per cent. Citric acid comprised over 50 per cent of the acid in white cranberry fruit, but declined to about 30 per cent in ripe fruit, while quinic and malic acids increased to 30 and 40 per cent, respectively. Total anthocyanins increased as colour developed, and total phenolics and antioxidant capacity remained relatively constant. Red cranberry fruit were firmer than white or turning fruit (Forney *et al.*, 2012).

Both respiration and ethylene production are highest during the early green stage of fruit development and thereafter decrease rapidly as the fruits grow and become steady at maturity. Earlier, a respiratory climacteric in cranberries was reported between 11 and 13 weeks after full bloom (Doughty *et al.*, 1961; Forsyth and Hall, 1967). But Abdallah and Palta (1989) report that cranberry is a non-climacteric fruit because there is no climacteric rise in respiration at the time of a rise in ethylene production near maturity. With the onset of red colour development, the ehtylene content in fruit increases reaching to a secondary maximum when the fruits are fully coloured.

Vvedenskaya and Vorsa (2004) quantified proanthocyanidins flavonols and anthocyanins in fruit during fruit growth and ripening in cranberries cvs. Ben Lear and Stevens. Concentrations of proanthocyanidins and total flavonols were highest in flower ovaries and early fruit development stages. Proanthocyanidins declined rapidly during ovary development stages and growth and then increased slightly during fruit ripening. Flavonol exhibited a less pronounced decline from levels in flowers and early fruit set stages, and increased slightly during ripening. Anthocyanin synthesis was initiated after fruit growth ceased and was greatest during fruit ripening. The flavonoid biosynthetic pathway channelled resources largely towards anthocyanins in 'Ben Lear' and towards both anthocyanins and proanthocyanidins in 'Stevens' during fruit ripening. Patterns of change over the fruit development period for levels of the major flavonol glycosides, quercetin-3-glactoside and quercetin-3-arabinofuranoside, were similar for the two cultivars. The cultivars exhibited differential rates of accumulation of quercetin-3(6″-benzoyl)-beta-glactoside, quercetin-3-arabinopyranoside, quercetin-3-rhamnoside, methoxyquercetin pentoside, and quercetin-3-(6″coumaroyl)-beta glactoside. Both cultivars and phonological stage influence the partitioning of the precursor pool in the biosynthetic pathway of flavanoids in cranberry fruit. YiFei *et al.* (2017) working with eight cranberry cultivars reported that anthocyanin biosynthesis initiated at early fruit development and reached in mature fruit, with significant differences between cultivars. Levels of various proanthocyanidin oligomers/polymers with different degree of polymerization were highly correlated within a cultivar during fruit development.

## Fruit Colour

The commercial value of cranberries depends upon the attractive red colour which is imparted by anthocyanin pigments present in the epidermal and subepidermal layers of the berry. The red anthocyanin, yellow flavanol and carotene pigments become dominant as the harvest period approaches and red colour can be enhanced considerably by a short delay in harvest but it reduces the keeping quality. The foliar spray of an insecticide, malathion, few days before harvest stimulates red pigment formation (Eaton *et al.*, 1970). Ethephon treatment also enhances the colour development but the results were inconsistent (Rigby *et al.*, 1972; Shawa, 1979). Farag *et al.* (1989) have shown that the fruit cuticle is of impermeable nature and this problem can be overcome by using tergitol (0.3 per cent v/v) and by increasing the surface binding and penetrability of ethephon by using compounds such as ethanol, ursolic acid, ascorbic acid and glycerine (Table 18.4).

**Table 18.4: The Effect of Field Applications (Large scale experiment) of some Ethrel Formulations on Anthocyanin Content and Fruit Size of Cranberry Fruit cv. Searles* (The anthocyanin content of the control (water) treatment was 28.2 mg/100 g fruit fresh weight.**

**This value is the mean of 64 separate measurements. Concentrations used were: ethrel 1000ppm, tergitol 0.3 per cent v/v, ethanol 2.5 per cent and glycerol 1.5 per cent).**

| *Treatments* | *Relative Anthocyanin Content (per cent)* | *Fruit Size (berries/100 g)*** |
|---|---|---|
| Water (control) | 100.0 | 86.7± 1.54 |
| Tergitol + ethanol + glycerol | 98.9 | 82.3± 2.1 |
| Ethrel + tergitol | 107. 1 | 87.8± 1.96 |
| Ethrel + tergitol + ethanol | 134.0 | 85.2± 1.54 |
| Ethrel + tergitol + ethanol + glycerol | 139.7 | 81.4± 1.98 |

** Mean of 32 separate samples. Duplicate samples obtained at four locations with each replication. Four replications were used for each treatment.

* Farag *et al.* (1989).

## Harvesting

Berry size and colour are used as maturity indices in cranberries. The fruit size varies a great deal, hence the fruit colour is the most important maturity index. Slight green colour is an indication of immaturity. Depending upon the cultivar, when 75 per cent of the surface of individual berries turn pink or red, the berries are considered fairly well-coloured. For fresh market, the berries should be plump, bright, firm, uniform in colour (light red to reddish black) and free from all kinds of blemishes. The optimum harvest dates of the cultivar Early Black for early-drawn and late-drawn bogs are September 13 to 19 and September 23 to 29, respectively. The optimum harvest dates for 'Howes' are October 3 to 9 for early-drawn bogs and October 9 for late-drawn bogs (Demoranville, 1960). A single picking is enough because of uniform ripening of cranberries.

### Manual Harvesting

Hand-picking is no longer used in cranberry culture but manual harvest is accomplished with scoop which is commonly made of wood. The Wisconsin scoop with larger hoop-like handles allows the picker to stand and pass the rake through vines to collect the berries (Dana, 1989). Scooping every year in the same direction decreases the damage and speeds the harvesting. An adult can harvest 163 to 544 kg berries per day with a scoop (Shoemaker, 1978).

## Mechanical Harvesting

Mechanical harvesting has almost replaced manual harvesting. Both dry-picking and water-picking are common but the later is increasing in popularity. The dry-picked berries have better keeping quality and are more suitable for fresh market sale. Water-picking results in higher yield and is used when berries are needed for processing. Machine-harvested cranberries are more damaged (35.4 per cent) than hand-harvested berries (2.9 per cent). With the Furford Picker-Pruner the major fruit injury was found when the first elevator paddle hits the fruit.

The Case and Getsinger self-propelled picking machines are used for mechanical harvest. In dry-picking, the fixed teeth of the Case picker pulls the berries from the vines and the chain-driven paddles transfer the free berries on the conveyor belt. In the Getsinger Retracto-tooth picker, a steel drum rotates and the offset teeth comb through the vines, pulling the berries free and dropping them on an inclined elevator belt.

Water raking not only speeds the harvest but also saves all the berries broken from the vines. The section ready to harvest is flooded with 15-25 cm deep water. Two systems of water-picking are common. In the first system, the berries are removed by a series of teeth and deposited in scrows drawn behind the machines. In the second, the machine (egg beater) knocks the berries from the vines with a rotating reel. The loose berries are then floated to a corner of the flooded section where a portable elevator lifts them into trucks (Shoemaker, 1978). The success of 'egg beater' resulted in the manufacture of larger machines to do total harvest for processing (Dana, 1989).

## Yield

Five yield components of cranberries identified are density of uprights, percentage of uprights flowering, number of flowers per upright, fruit set percentage and berry size (Eck, 1976; Shawa *et al.*, 1981). Shawa *et al.* (1981) have reported that the most important yield components are the number of flowering uprights per unit area and per cent fruit set, but the number of uprights, flowers per flowering upright and the size of individual berry are less important. They have attributed irregular cropping in cranberry to the variation in uprights flowering and fruit set. Doehlert (1953) demonstrated that the high production was a consequence of high percentage of uprights flowering because only 22 per cent of the uprights flowered in low yielding areas as compared with 32 per cent in high yielding areas. He further concluded that high yield was the result of an increase in the number of blooming uprights rather than an increase in the percentage of flowers setting fruits.

In Wisconsin, an average yield of 17 tonnes per hectare was reported in 1986 (Birrenkott and Stang, 1989).

## Storage

The harvested cranberries continue to develop red pigment when remain exposed to light. Likewise, ethylene treatment of the harvested berries enhances anthocyanin development in the presence of light but not in the dark (Craker, 1971).

The best colour development occurs at 7.2°C but they could be held best at 1.7°C. The berries are dried mechanically in heated air (29.4° to 32.2°C) before storing.

There is an increased interest in extending the market life of fresh cranberries as a result of the growing demand for fresh, healthful fruits. To supply of this growing demand, fruit quality must be maintained and postharvest losses reduced during extended storage and marketing periods (Forney, 2008).

Recommended storage temperature range from 2 to 7°C (Kader, 1997; Prange, 2004) and the relative humidity (RH) from 80 to 90 per cent. Postharvest loss of cranberry fruits is primarily the result of physiological breakdown and decay (Forney, 2003). Physiological breakdown is associated with overmature fruit, bruising, chilling injury, freezing, extended water immersion, anoxia and for other reasons. Physiological breakdown is characterized by a dull appearance, rubbery texture, and diffusion of red pigments throughout the fruit flesh (Bristow and Patten, 1995). Forney (2008) conducted experiment to determine the relationship of temperature and humidity on the storage life of fresh cranberry for three seasons. Each year, cranberries harvested from four commercial bogs were stored at temperatures ranging from 0 to 10°C in combination with RH ranging from 75 per cent to 98 per cent. Fruits were stored under these conditions for up to 6 months and after an additional week at 20°C, several parameters were studied. The percentage of marketable fruit declined substantially over time in all storage conditions with 41 per cent to 57 per cent becoming unmarketable after 2 months as a result of both decay and physiological breakdown. Relative humidity had a greater effect on fruit storage life than temperature and after 5 months, the amount of marketable fruit stored in high (98 per cent) and medium (88 per cent) RH was 71 per cent and 31 per cent less than that stored in low (75 per cent to 82 per cent) RH. Rates of fresh weight loss increased as RH in storage decreased and was 0.41 per cent, 0.81 per cent and 0.86 per cent per month in fruit stored in high, medium, and low RH, respectively. Physiological breakdown was highest in fruit stored at 10°C. Contrary to previous reports, no evidence of chilling injury was found in fruit stored at 0°C. Forney (2008) suggested that cranberry fruit should be stored at 0 to 7°C and 75 per cent to 82 per cent RH to retain marketable fruit.

Wang and Wang (2009) evaluated nine cultivars of cranberry during storage at 0-15°C for 3 or 4 months. Antioxidants activity (ORAC), anthocyanins and total phenolics content increased during storage. The highest increase occurred at 15°C. Susceptibility to chilling injury and decay varies with different cultivars. The cvs. Ben Lear, Cropper, Early Black and Stevens showed severe chilling injury (CI) and decay at the end of 3 or 4 months storage at 0°C. The cvs. Crowley, Howes and Pilgrim were relatively resistant to CI and decay while cvs. Franklin and Wilcox were moderately susceptible. The susceptibility was found to be related to the fatty acid composition and ratio of unsaturated to saturated fatty acid in phospholipids and glycolipids, but was not associated with ORAC, anthocyanins, phenolics or individual flavanoids. They suggested optimum storage temperature of 5°C with the least chilling injury symptoms and decay for all cranberry cultivars studied.

Cranberry stored at 22°C for 3 weeks in atmosphere of 5 or 10 kPa$CO_2$ with 3 kPa$O_2$ showed the same levels of rot as stored fruit, while fruit held in 100 kPa

$N_2$ become dull and water soaked in appearance and had a fermented odor (Stark *et al.*, 1969). However, Gunes *et al.* (2002) reported that fruits of cvs. Pilgrim and Stevens when stored at atmosphere of 30 kPa$CO_2$ and 21 kPa$O_2$ for 2 months at 3°C reduced decay and breakdown.

Hot water treatment at 50°C for up to 180 seconds of 'Stevens' cranberries reduced decay and physiological breakdown of fruit at storage temperature of 3 or 7°C and 86 per cent RH. Anderson and Smith (1971) reported that hot water treatments of 49°C for 150 or 300 seconds and 52°C for 150 seconds had some benefits in prolonging storage life. Coating with carnauba wax resulted lower decay incidence during storage at 13°C. Treatment with BioSave 110 (containing *Pseudomonas syringae*) with carnauba wax, reduced the total decay by 60, 42 and 33 per cent after 4, 8 and 16 weeks, respectively, (XiuHua *et al.*, 1999). Ozgen and Palta (2003) studied the effect of lysophosphatidylethanolamine (LPE) which is commercially derived from egg and soy lecithin, on anthocyanin accumulation and storage quality of cranberry cv. Stevens. Plants were sprayed with LPE at 200 mg/l, ethanol (5 per cent v/v) and a non-ionic surfactant Sylgard (0.05 per cent v/v) at 4 weeks before commercial harvest. Application of LPE resulted in 13-28 per cent increased in fruit anthocyanins compared with control. LPE treatment also resulted in 6-12 per cent increase in marketable fruit in cold storage. This influence of LPE on fruit quality was more apparent after one month of storage. The results suggested that a preharvest application of LPE may have the potential to enhance colour and prolong shelf life of cranberry fruit.

## Pests

### Black Headed Fireworm (*Rhopobota naevana*)

It over-winters as a small egg, on the underside of the leaf near the base of an upright. On hatching, the larvae crawl up to the tip of uprights and damage the leaves which turn brown and entire planting may look as if fireburnt.

Early detection and timely application of treatments are necessary to control black headed fireworm. Pheromone traps (3 traps for 4 hectares) attracts male moths. Parasitic wasp, *Trichogramma sibericum* parasitize fireworm eggs, and the larvae consume the contents of the host eggs. Mating can be disrupted by pheromone components released from spiral polyvinyl choloride (PVC) dispensers. The main pheromone component (Cz)-11-tetradecen-1-ol.acetate (Z11-14:AC) or a blend similar to natural pheromone (Z11-14:AC, Z11-14:OH and Z9-12:AC) released from PVC dispensers showed free flying males in treatment plot (0.8ha) located fewer than 5 per cent of individually caged virgin females, whereas mate location in most control plots ranged from 14 to 75 per cent (Fitzpatrick *et al.*, 2004).

### Cranberry Fruitworm (*Acrobasis vaccinii*)

Fruitworm feeds only on berries. The larvae are found in green fruit before harvest. In cool, late seasons, sometimes larvae remain in berries at harvest.

Mating disruption with a sprayable microencapsulated formulation of E11-14:AC at 25-187.5 g (AI)/ha is a promising strategy to manage fruitworm in cranberries (Polavarapu *et al.*, 2001). The cranberry fruitworm has a number of natural enemies. A parasitiod, *Trichogramma deion*, attacks the eggs of this pest.

### Sparganothis Fruitworm (*Sparaganothis sulfureana*)

The young larvae feed on new leaves and flower buds. Second generation larvae also feeds immature berries. Natural parasites Trichogrammatidae, Tachinidae, Braconidae and Ichneumonidae are known to control the Sparganothis fruit worm.

### Cranberry Weevil (*Anthonomus musculus*)

Cause damage by laying eggs in developing flower buds, larvae begin to feed on the internal parts of the flowers and flower buds did not produce fruit. Azinphosmethyl and chlorpyrifos are recommended if weevils are at threshold numbers. Spraying should be made before the onset of the blossom period (Mechaber and Chow, 1991).

### Cranberry Tipworm (*Dasyneura vaccinii*)

The uprights damaged by the first generation may form fruit bud. The second generation damages the upright tips which fail to flower altogether.

## Diseases

### Fruit Rots

Fruit rot occurring either in the field or storage are the main diseases of cranberries (Dana, 1989). Early rot (*Phyllosticta vaccinii*), bitter rot (*Glomerella cingulata*), end rot (*Fusicoccum putrefaciens*), cotton ball (*Monilinia oxycocci*), viscid rot (*Phomopsis vaccinii*), yellow rot (*Botrytis cinerea*) and others may cause damage individually or collectively. Storage rotting fungi including *Allantophomopsis cytisporea* and *A. lycopodina* were isolated at low frequencies, but were widely distributed within the growing regions of New Jersey, USA (Stiles and Oudemans, 1999). Fungicides are used to control fruit rots and the treatment should begin at the end of bloom to protect berries during the early stage of development.

### Twig Blight (*Lophodermium oxycocci, L. hypophyllum*)

The fungus infect one-year-old wood of uprights. Newly emerged leaves turn burn then turn brown and tan. Premature leaf fall and stunted growth of uprights are the symptoms of twig blight. Spraying of fungicides mancozeb and chlorothalonil are effective to control and spraying should be at the time when new leaves are emerging.

### Cottonball (*Monilinia oxycocci*)

The fungus affects young shoots, flowers and berries, causing tip blight of young shoots and fruit rot. McManus *et al.* (1999) in greenhouse tests recorded cottonball incidence of 25, 28, 31 and 38 per cent for cultivars Searles, Pilgrim, Ben Lear and Stevens, respectively. Spraying of fungicides at bud break and during bloom are recommended to control cottonball infection.

### Dieback

Cranberry fields in the lower mainland of British Columbia (B.C.) have been affected by a severe vine decline and death, referred to as cranberry dieback disorder (CDD), resulting in substantial yield losses. Symptoms of CDD appear as patches of dying plants with copper to burgundy colour leaves on drying uprights in spring and small to large blackened areas of dead vines in late summer. The most noticeable symptoms can be seen on diseased runners, where blackening of principal tissues and brown to dark-brown discolouration of pith tissue were distinguishable from healthy vines. *Phytopthora cinnamomi* and *Phytophthora* spp. Were detected from the soils in B.C. Several potential fungal pathogens were also isolated from tissue samples, *Coniothyrium sporulosum, Cylindrocarpon destructans* and a *Phomopsis* sp., suggesting that these pathogens may be responsible for CDD, either acting solely or synergistically (Sabaratnam *et al.,* 2009).

### Tobacco Streak Virus (TSV)

TSV virus affected plants caused low fruit set in uprights, berries were shrivelled and deformed. Once recovered, it behaves like normal uprights. A breakthrough in detecting of viruses in *Vaccinium* species has come from the development of a dsRNA extraction method that works reliably with blueberry and cranberry (Martin *et al.,* 2009).

## Breeding and Improvement

Like most of the other fruits, the improvement in cranberry began with the single seedling selections and more than 90 per cent cranberry production is obtained from such cultivars (Galletta, 1975). However, the cooperative cranberry breeding programme was initiated in 1929, in New Jersey and Massachusettes Agricultural Experiment Stations, the Wisconsin Department of Agriculture and Markets and the Wisconsin Cranberry Sales Company participated in it (Chandler *et al.,* 1947). The aim of the programme was to evolve high yielding, high quality cultivars with resistance to the false blossom disease. As a result, six cultivars, *viz.,* Beckwith, Bergman, Franklin. Pilgrim, Stevens and Wilcox were released for cultivation. Thereafter, another cultivar, Crowley, was evolved by the Washington Station which is early, more productive with large berries of better colour than the standard McFarlin (Doughty and Garren, 1970).

### Breeding Objectives

A large number of desirable characters have been listed as objectives of modern breeding programmes by various research workers (Chandler and Demoranville, 1958; Krust, 1965; Galletta, 1975).

### Vine Characteristics

1. The vine should be of normal vigour, *i.e.,* medium to coarse vine with medium runnering. Excessive runnering reduces the production of uprights.

2. A high density of uprights with high flowering percentage per unit area is essential for better production. The uprights should bear fruits at a uniform level above the bog.
3. Rapid rooting of cuttings is required so as to suppress the weeds by quickly covering the entire area.
4. Precocious vines will give commercial production in three years which otherwise may take five years. The fruit should have shorter growth-period enabling early maturity. Early maturing fruits escapes frost hazards and fetches higher price in the market.
5. The vines should have a high degree of ecological adaptability. These should tolerate severe cold in winter, spring and autumn frosts and water submersion. High photosynthesis capability is desirable under oxygen-deficient conditions (flooding).
6. Resistance of vines to insects, fungi and disorders is required to avoid low production or crop failures.

## Fruit Characteristics

1. The various traits of fruit quality which should be taken into consideration while assessing the hybrid seedlings are-aroma, firmness, size, flavour, organic acids, glossiness and attractiveness.
2. High anthocyanin pigment in cranberries is a commercial necessity. Sapers *et al.* (1986) suggested that the breeding should be done to increase the anthocyanin content of the berries rather than to breed small berries or synchronous ripening. The berries of Ben Lear, Crowley, Early Black and Franklin cultivars had as much as twice anthocyanin than some other cultivars. These cultivars can serve as good donor parents in breeding for anthocyanin pigment.
3. Round or oval berry-shape is more desirable because pointed ends will cause bruising in grading the fruit by screening.
4. A number of attributes contribute in extending the shelf life. The fruit is to be firmer, more acidic and resistant to cranberry fruit worm and fruit rotting fungi for high keeping quality.
5. The suitability of the berries for juice, sauce, jelly, preserves and wine manufacture must be ascertained before releasing a selection.

## Breeding Systems

Intraspecific hybridization and selection will provide quickest breeding advances in cranberries (Galletta, 1975). Since most of the vigorous, large-fruited vines are the selections from interclonal hybrids, it is speculated that inbreeding will result in depression (Kust, 1965). Tetraploid seedlings were vigorous but less productive (Galletta, 1975). Hegerup (1940) demonstrated chromosomal sterility in hexaploids. Possibility of interspecific breeding should also be explored which may provide interesting cranberry-blueberry hybrids. Inheritance pattern of various characters should be worked out in order to put cranberry breeding on sound footing.

Little is known of the response of various cranberry cultivars and selections to diseases and insects except for resistance to false bloom mycoplasma or its vector the blunt-nosed leafhopper (*Scleroracus vaccinii*) which were priorities in the USDA cooperative breeding programme (Chandler *et al.*, 1947). The cvs. Wilcox, Pilgrim and Shaw showed resistance to leafhopper and Early Black, Pilgrim, Franklin and Selection No 4 showed resistance to virus (Dana, 1983).

## Molecular Markers

In cranberry, randomly amplified polymorphic DNA (RAPD) markers were developed for genotyping and cultivar identification (Novy and Vorsa, 1996; Debnath, 2005). Polashock and Vorsa (2002) developed sequence-characterized amplified region (SCARs) markers for identification of gemplasm using nine primary sets designed from RAPD-identified polymorphoic markers. These primer sets generated 38 markers across a cranberry (cvs. Ben Lear, Early Black, Franklin, Howes, LeMunyon, McFarlin, Pilgrium, Stevens and Wilcox) germplasm collected. They stated that SCAR and RAPD markers can be employed for identifying closely related genotypes. However, the inferences of more distant genetic relationship are less certain. SCAR marker reactions provided more polymorphic markers on a per reaction basis than RAPD marker reactions and as such more readily separated closely related progeny.

Schlautman *et al.* (2017) designed and validated 16-multiplexing panels containing 61 evenly distributed simple sequence (SSR) markers, with non-overlapping allele ranges, throughout the 12 linkage groups. Several important cranberry cultivars and selections (n= 18) and a diploid accession of *V. oxycoccos* were genotyped with the multiplexing panels and separated through principal component analysis (PCA) to demonstrate their effectiveness for DNA fingerprinting and genetic diversity analysis. A subset of there multiplexing panels containing 12 SSR markers was used to genotype 174 seedlings from fruits collected in a commercial cranberry bed of the cultivar Stevens, and identification of intra cultivar heterogeneity was investigated in the bed to validate the use of the markers in such future applications. Furthermore, determining the likelihood that each seedling was self-or-cross-pollinated provided the first quantitative evidence ($P < 0.0001$) that the majority of seeds within commercial cranberry beds are self-pollinated. These markers can be used to assess the genetic homogeneity. Dong.An *et al.* (2015) screened 102 wild cranberry clones collected from four Canadian provinces and five cultivars with inter simple sequence repeat (IISR), expressed sequence tag-simple sequence repeat (EST-SSR) and EST-polymerase chain reaction (PCR) markers to validate the genetic diversity and relationship among them. EST-PCRs (0.54) and EST-SSRs (0.35) generated higher frequency of major alleles than IISRs (0.08), but ISSRs presented a higher level of polymorphism and greater polymorphic information content and expected heterozygosity than EST-SSRs and EST-PCRs. Combined cluster analysis by the unweighted pair-group method with arithmetic averages (UPGMA) separated the wild clones and cultivars into four main clusters, which was in agreement with the principle coordinate (PCo) analysis. Analysis of molecular variations detected sufficient variations among genotypes with in communities and among communities within provinces with ISSR (66 and 36 per cent, respectively), ECT-PCR (72 and 34 per cent respectively) and EST-SSR (72 and 34 per cent respectively) markers. Combined use of three types of molecular markers, for the first time in *Vaccinium* species, detected a sufficient degree of variations among cranberry genotypes, allowing for differentiation and rendering these technologies valuable for genotype identification in a diverse cranberry germplasm.

## Genetic Transformation

Serres *et al.* (1992) were the first to successfully transform cranberry using electrical discharge particle bombardement. Transforming using *Agrobacterium tumefaciens*, however, showed little success in obtaining stable translocates (Polashock and Vorsa, 2002). American cranberry was genetically transformed with the bar gene, conferring tolerance to the phosphinothricin-based herbicide glufosinate. Plants of one 'Pilgrim' transclone grown under greenhouse conditions were significantly injured by foliar treatments of 100 mg/glufosinate, although the injury was less severe compared to untransformed plants. However, the same transclone grown outdoors in coldframes survived foliar sprays of 500 mg/l glufosinate and higher, while untransformed plants were killed at 300 mg/l (Zeldin *et al.*, 2002).

## Genetic Map

The first genetic map of cranberry has been constructed (Georgi *et al.*, 2013), comprising 14 linkage groups totally 879.9 cM with an estimate coverage of 82.2 per cent. This map, based on four mapping population segregating for field fruit-rot resistance, contain 136 distinct loci. Mapped markers include blueberry-derived simple sequence repeat (SSR) and cranberry-derived sequence-characterized amplified region markers. In addition, SSR markers were developed near cranberry sequences resembling genes involved in flavonoid biosynthesis or defence against necrotrophic pathogens, or conserved orthologous set (COS) sequences. The cranberry SSRs were developed from next-generation cranberry genomic sequence assemblies, thus, the positions of these SSRs on the genomic map provide information about the genomic location of the sequence scaffold from which they were derived. The use of SSR markers near COS and other functional sequences, plus 33 SSR markers from blueberry, facilitates comparisons of this map of other plant species. Regions of the cranberry map were identified that showed conservation of synteny with *Vitis vinifera* and *Arabidopsis thaliana*. Positioned on this map are quantitative trait loci (QTL) for field fruit-rot resistance (FFRR), fruit weight, titratable acidity, and sound fruit yield (SFY). The SFYQTL is adjacent to one of the fruit weight QTL and may reflect pleiotropy. Two of the FFRR QTL are in regions of conserved synteny with grape and span defense gene markers, and the third FFRR QTL spans a flavonoid biosynthetic gene.

New genetic resources could aecelerate genetic gain in cranberry through characterization of its genomic structure and by enabling moleculer-assisted breeding strategies. To increase the availability of cranberry genomic resources, genotyping-by-sequencing (GBS) was used to discover and genotype and thousands of single nucleotide polymorphisms (SNPs) within three interrelated cranberry full-sib populations. Additional simple sequence repeat (SSR) loci were added to the SNP data sets and used to construct bin maps for the parents of the populations, which were then merged to create the first high-density cranberry composite map containing 6073 markers (5437 SNPs and 636 SSRs) on 12 linkege groups (LGs) spanning 1124 cM. Interestingly, higher rates of recombination were observed in maternal than paternal gamets. The large number of markers in common (mean of 57.3) and the high degree of observed collinerity (Mean Pair-wise Spearman rank correlations > 0.99) between the LGs of the parental maps demonstrates the utility of GBS in cranberry for identifying polymorphic SNP loci that are transferable between pedigrees and populations in future traits-association studies. Furthermore, the high-density of markers anchored within the component maps allowed identification of segregation distortion regions, placement of centromeres on each of the 12 LGs, and anchoring of genomic scaffolds. Collectively, the result represent an important contribution to the current understanding of cranberry genomic structure (Schleutman *et al.*, 2017). Genotyping by sequencing (GBs) data was used to rapidly saturate the cranberry genome with markers in a pseudo-testcross population. Collinearity between the saturated genetic map and cranberry SSR maps suggested that the SNP locations represented accurate marker order and chromosome structure of the cranberry genome. SNPs greatly improved current marker genome coverage, which allowed for genome-wide structure investigations such as segregation distortion, recombination, linkage disequilibrium, and synteny analysis. In the future, GBs can be used to accelerate cranberry molecular breeding through QTL mapping and genome-wide association studies (GWAS) (Covarrubias-Pazaran *et al.*, 2016).

## REFERENCES

Abdatlah, A.Y. and Palta, J. (1989). Acta Hort., 241:360-65.

Anderson, R.E. and Smith, Jr. W.L. (1971). USDA Market Res. Rept., 928, p. 8.

Anjou, K. and van Sydow, E. (1968). Ark. Kemi., 30: 9-14.

Atwood, W.M. and Zuckerman, B.M. (1961). Proc. Amer. Soc. Hort. Sci., 77: 359-66.

Bergman, H.C. (1954). Proc. 84th Ann. Meet. Amer. Cranberry Growers Assoc., pp. 11-26.

Bewick, T.A., Binning L.K. and Dana, M.N. (1989). HortSci., 24: 850.

Bewick, T.A., Binning, L.K., Stevenson, W.R. and Stewart, J. (1987). Marburg German Federal Republic. pp. 93-104.

Birrenkott, B.A. and Stang, E. (1989). Jr. Amer. Soc. Hort. Sci., 114: 733-37.

Birrenkott, B.A. and Stang, E.J. (1986). Abst. 22nd Int. Hort. Congress, California, USA, No. 1077.

Birrenkott, B.A. and Stang, E.J. (1989). Jr. Amer. Soc. Hort. Sci., 114: 723-37.

Blumberg, J.B., Camesano, T.A., Cassidy, A., Kris-Etherton, P., Howell, A., Manach, C., Ostertag, L.M., Sies, H., Shulas-Ray, A. and Vita, J. (2013) Adv. Nutr., 4: 618-32.

Briston, P.R. and Shawa, A.Y. (1981). Jr. Amer. Soc. Hort. Sci., 106: 290-92.

Bristow, P.R. and Patten, K.D. (1995). In: Compendium of blueberry and Cranberry diseases (eds. F.L. Caruso and D.C. Ramsdell). APS Press, os Paul, MN.

Brooks, R.M. and Olmo, H.P. (1972). Register of New Fruit and Nut Varieties, 2nd ed. Univ. of Calif. Press. Berkeley, pp. 213-14.

Bulot, D., Gumiere, S.J., Periard, Y., Lafond, J.A., Gallichand, J., Armaly-St. Gelaio, M.H. and Caron, J. (2017). Can. J. Soil. Sci., 97: 46-55.

Cain, J.C. and Eck, P. (1966). In : Fruit Nutrition, (ed. N.F. Childers) Horticultural Pubs., Rutgers State University, New Brunswick, New Jersey, pp. 114-23.

Camp, W.H. (1942). Brittonia, 4:189-204.

Caron, J., Bonin, S., Pepin, S., Kummer, L., Vanderleest, C. and Bland, W.L. (2016). Can. J. Soil. Sci., 96: 37-50.

Caron, J., Pelletier, V., Kennedy, C.D., Gallichand, J., Gumiere, S., Bonin, S., Bland, W.L. and Pepin, S. (2017). Can. J. Soil Sci., 97: 82-91.

Caruso, F. (2012). Diseases in Cranberry, (eds. M. Sylvia and D. Gauvin) Massachusets Extension Cir., p. 1-6.

Chandler, F.B. (1952). Cranberries, Natl. Maz., 17: 6-7.

Chandler, F.B. and Demoranville, I. (1958). Mass. Agr. Expt. Sta. Bull. p. 513.

Chandler, F.B., Wilcox, R.B., Bain, H.C., Berman, H.C. and Dermen, H. (1947). Cranberries, Natt. Mag., 12: 6-9.

Chong, C. (1999). Compost Sci. Utilization, 7(4): 6-14.

Coppola, E.D., Conrad, E.C. and Cotter, R. (1978). Jr. Assoc. Off. Anal. Chem., 61: 1490-92.

Covarrubias-Pazaran, G., Diaz-Garcia, L., Schlautman, B., Deutsch, J., Salazert, W., Hernandez-Ochoa, M., Grygleski, E., Steffan, S., Iorizzo, M., Polashock, J., Vorsa, N. and Zalapa, J. (2016). BMC Genomics, 17: 451. doi: 10.1186/s12864-016-2802-3.

Craker, L.E. (1971). HortSci., 6: 137-39.

Croteau, R. (1977). Jr. Food Biochem., 1: 317-26.

Dana, M.N. (1983). Fruit Var. J., 37: 88-95.

Dana, M.N. (1989). Acta Hort., 241: 287-94.

Dana, M.N. and Klingbeil, G.C. (1966). Wisc. Agr. Exp. Ext. Cir., p. 654.

Davenport, J.R. and DeMoranville, C. (2004). HortSci., 39: 80-83.

Davenport, J.R. and Patten, K.D. (2002). Acta Hort., 574: 341-46.

Davis, D.C. and Shawa, A.Y. (1983). Jr. Amer. Soc. Hort. Sci., 108: 444-47.

Debnath, S.C. (2005). Jr. Pl. Biochem. Biotech., 14(2): 173-77.

Debnath, S.C. and McRae, K.B. (2001). *In vitro* cellular and Development Biol. Plant, 37: 243-49.

Degaetand, A.T. and Shulman, M.D. (1987). Agric. and Forest Meteo., 40: 323-42.

DeMoranville, C., Howes, B., Schlezinger, D. and White, D. (2009). Acta Hort., 810: 633-40.

DeMoranville, C.J. (2004). Jr. Amer. Pom. Soc., 58(3): 126-28.

Demoranville, C.J. and Deubert, K.H. (1987). Jr. Hort. Sci., 62: 163-69.

DeMoranville, C.J. and Kennedy, C.D. (2017). Acta Hort., 1180: 313-19.

DeMoranville, C.J. and Sandler, H. (2000). Univ. Mass. Cranberry Station, Guide: Sanding.

Demoranville, *I.E.* (1960). Cranberries, Natl. Mag., 24: 10-11.

Devlin, R.M. and Demoranville, *I.E.* (1967). Physiol. Plantarum, 20: 587-92.

Devlin, R.M. and Deubert, K.H. (1980). Proc. Northeast Weed Sci. Soc., 24: 399-405.

Dewvenport, J.R. and Vorsa, N. (1999). Jr. Amer. Soc. Hort. Sci., 124: 90-93.

Doehlert, C.A. (1953). Proc.84th Ann. Conv. American Cranberry Growers Assoc., August, pp. 19-23.

Dong.An, ByKova, N.V. and Debnath, S.C. (2015). Can. J. Pl. Sci., 95: 1155-65.

Doughty, C.C. (1962). Proc. Amer. Soc. Hort. Sci., 86: 340-49.

Doughty, C.C. (1984). Canad. J. Pl. Sci., 64: 339-48.

Doughty, C.C. and Garren, R.Jr. (1970). Fruit Var. Hort. Dig., 24: 88-89.

Doughty, C.C., Patterson, M.E. and Shawa, A.Y. (1967). Proc. Amer. Soc. Hort. Sci., 91: 192-204.

Eaton, G.W. and Kyte, T.R. (1978). Jr. Amer. Soc. Hort. Sci., 103: 578-83.

Eaton, G.W., Shawa, A.Y. and Bowen, P.R. (1983). Scientia Hort., 20:179-84.

Eaton, G.W., Zuckerman,B.N., Shawa, A.Y., Eck, P., Dana M.N., Garren, R. and Lokhart, C.L. (1970). Jr. Amer. Soc. Hort. Sci., 97: 326-28.

Eck, P. (1962). Proc. Amer. Cranberry Growers Assoc., pp. 27-33.

Eck, P. (1971). HortSci., 6: 38-39.

Eck, P. (1976). Jr. Amer. Soc. Hort. Sci., 101: 375-77.

Eck, P. (1985). Acta Hort., 165: 299-301.

Eck, P. (1986). In: Handbook of Fruitset and Development, (ed. S.P. Monselise)., CRC Press, Boca Raton, Florida, pp. 109-117.

Else, M.J., Sandler, H.A. and Schluter, S. (1995). HortTech., 5: 302-305.

Farag, K.M., Palta, J.P.and Stang, E.J. (1989). Acta Hort., 241: 373-77.

Farror, C.L. and Bain, H.F. (1946). Amer. Bee J., 86: 503-504.

Filipenya, V.L. and Dolgov, S.V. (2000). Sadininkyste ir Darzininkyste, 19(3): 384-89.

Filmer, R,S. (1949). Proc. 80th Ann. Summer Meet. Amer. Cranberry Growers Assoc., pp. 14-22.

Filmer, R.S. and Doehlert, C.A. (1959). N.J. Agric. Exp. Stn. Circ., No. 588. p. 4.

Filmer, R.S. Marucci, P.E. and Moulter, H. (1958). Proc. 88th Ann. Meet. Amer. Cranberry Growers Assoc., pp. 22-30.

Filzpatrick, S.M., Troubridge, J.T. and Maurice, C. (2004). Canad. Entomol., 136: 91-108.

Fisher, R.A. (1951). Cranberries, Natl. Mag., 16: 8, 10.

Forney, C.F. (2003). HortTech., 13: 267-72.

Forney, C.F. (2008). HortSci., 43: 439-46.

Forney, C.F., Kolt, W., Jordan, J.A., Vinqvist-Tymchuk, M.R. and Fillmore, S.A.E. (2012). Acta Hort., 926: 331-37.

Forsyth, F.R. and Hall, I.V. (1967). Canad. J. Plant Sci., 47: 153-56.

Fuleki, T. and Francis, F.J. (1967). Phytochem., 6: 1705-08.

Galletta, C.J. (1975). In: Advances in Fruit Breeding, (eds. J. Janick and J.N. Moore). Purdue University Press, West Lafayette, lndiana, pp. 185-96.

Georgi, L., Jonson, C.J., Honing, J., Das, S.P., Rajah, V.D., Bhattacharya, D., Bassil, N., Rowland, L.J., Polashock, J. and Vorsa, N. (2013). TAG Theor. and Appl. Sci., 126: 673-92.

Ghantous, K.M., Sandler, H.A., Autio, W.R. and Jeranyama, P. (2013). HortSci., 48: 870-74.

Gillhan, R.W. (1984). J. Hydrol. (Amst.), 67: 307-24.

Goulson, D. (2003). Ann. Review of Ecol., Evolution and Systematic., 34: 1-26.

Greidanus, T. (1972). Jr. Amer. Soc. Hort. Sci., 97:272-77.

Greidanus, T. and Dana, M.N. (1972). Jr. Amer. Soc. Hort. Sci., 97: 326-28.

Gronskis, L. and Snickovstis, A. (1989). Acta Hort., 241:77-80.

Gronwald, R.F. and Haines, W.Sr. (1982). Proc. Conf. Environmentally Sound Water and Soil Management. USA pp. 284-89.

Gunes, G., Liu, R.H. and Watking, C.B. (2002). J. Agric. Food. Chem., 50: 5932-38.

Hagerup, O. (1940). Hereditas, 26 : 399-410.

Hall, I.V. and Aalders, L.E. (1965). Canad. J. Pl. Sci., 45:292.

Hart, J.M., Bernadine, C.S., DeMoranvill, C., Davenport, J.R. and Roper, T. (2015). Cranberries (South Soastal Region). Nutrient Management Guide. Oregon State Univ. Extn. Service, EM 8672.

Heuvel, J.E. Vanden and Autio, W.R. (2008). HortSci., 43: 1737-41.

Heuvel, J.E. Vanden and Devenport, J.R. (2006). HortSci., 41: 1552-58.

Hicks, B.J. and Sircon, J. (2016). Jr. Amer. Soc. Hort. Sci., 111: 869-72.

Hiirsalmi, H.M. (1989). Acta Hort., 241: 175-84.

Igolight, M.S., Gursel, E. and Zinsser, H.H. (1973). Urology, l: 1-10.

Issa, R. (2012). Thermal Sci., 16: 31-42.

Kader, A.A. (1977). Proc. 7th. Int. Contr. Atmos. Conf. Davis, CA., 3: 1-34.

Kennedy, C.D., Kleinman, P.J. and DeMoranville, C.J. (2016). J. Environ. Qual., 45: 285-93.

Kevan, P.G., Cadawski, R.M., Kevan, S.D. and Gadawski, S.E. (1983). Proc. Entomological Soc. Ontario, 114: 45-53.

Khomych, G., Matsuk, Y., Nakonechnaya, Y., Oliynyk and Medved, L. (2017) Tech. and Equip. of Fd. Prod., 6/11(90): 58-65.

Kust, T. (1965). Univ. Wis. Dept., Hort. Mimeo, p. 617.

Mainland, C.M. and Eck, P. (1968). Proc. Amer. Soc. Hort. Sci., 92:296-300.

Martin, R.R., Tzanetakis, I.E., Caruso, F.L. and Polashock, J.J. (2009). Acta Hort., 810: 299-304.

Marucci, P.E. (1966). Cranberries Natl. Mag., 30: 11-13.

Marucci, P.E. and Filmer, R.J, (1964). Proc. Amer. Cranberry Growers Assoc., pp 48-51.

Mechaber, W.L. and Chow, F.S. (1991). Cranberries, February, pp. 5-8.

Morosov, O.V. (2000). Vesti Natsyyanal-nai Akad. Navuk Belarusi Seryya Biyalagichny kh Navuk, 1: 5-8.

Novy, R.G. and Vorsa, N. (1996). Theor. and Appl. Genet., 92: 840-49.

Ozgen, M. and Palta, J.P. (2003). Acta Hort., 628: 141-46.

Paquette, M., Larque, A.S.M., Weisnagel, S.J., Desjardins, Y., Marois, J., Pilon, G., Dudonne, S., Marette, A., and Jacques, H. (2017) British J. Nutr., 117: 519-31.

Pelletier, V., Gallichand, J., Gumiere, S. and Caron, J. (2017). Canad. J. Soil, Sci., 97: 1-4.

Peterson, L.A., Krueger, A. and Wheeler, R.M. (1986). Abst. 22nd. Int. Hort. Congress, California, USA, No. 1085.

Polashock, J. and Vorsa, N. (2002a). Jr. Amer. Soc. Hort. Sci., 121: 210-15.

Polashock, J.J. and Vorsa, N. (2002b). Jr. Amer. Soc. Hort. Sci., 127: 677-84.

Polavarapu, S., Lonergan, G., Peng, H. and Neilsen, K. (2001). Jr. Economic Entomol., 94: 658-65.

Prange, R.K. (2004). U.S. Dept. of Agr. Handbook, No. 66.

Rigby, B. and Dana, M.N. (1971). HortSci., 6 : 495-96.

Rigby, B. and Dana, M.N. (1972). Jr. Amer. Soc. Hort. Sci., 97: 145-48.

Rigby, B., Dana, M.N. and Klingbell, G.C. (1972). HortSci., 7: 82-83.

Ripa, A. (1989). Acta Hort., 241: 284-86.

Roberts, R.H. and Struckmeyer, B.E. (1942). Plant Physiol., 18: 524-36.

Roper, T.R. (2009). Acta Hort., 810: 613-25.

Roper, T.R., Krueger, A.R., DeMoranville, C.J., Vorsa, N., Hart, J. and Poole, A.P. (2004). HortSci., 39: 588-90.

Ruban, N.N. and Kurlovich, T.V. (1999). Navuk. Belarusi Seryya Biyalagichnykh Navuk, 1: 5-10.

Sabaratnam, S., Dicarlo, A., Fitzpatric, S. and Forge, T. (2009). Acta Hort., 810: 417-24.

Sage, R. and Sage, T.L. (2002). Oecologia, 132: 501-508.

Sakamura, S. and Francis, F.J. (1961). Jr. Food. Sci., 26: 318-21.

Sandler, H.A., Mason, J., Autio, W.R. and Bewick, T.A. (2004). Weed Technol., 18: 648-57.

Sapers, C.M., Jones, S.B., Kelley, M.J., Phillips, J.G. and Stone, E.C. (1986). Jr. Amer. Soc. Hort. Sci., 111: 618-22.

Schlautmam, B., Bolivar-Medina, J., Hodapp, S. and Zalapa, J. (2017). Scientia Hort., 219: 280-86.

Schlautman, B., Covarrubias-Pazaran, G., Diaz-Garcia, L., Lorizzo, M., Polashock, J., Grygleski, E., Vorsa, N. and Zalapa, J. (2017). G3: Genes, Genomes, Genetics, 7: 1177-89.

Scorza, R., Welker, W.V. and Dunn, L.J. (1984). HortSci., 19:66-69.

Sedlak, J. and Paprstein, F. (2011). HortSci., 38: 159-62.

Seres, R., Stang, E., McCabe, D., Russell, D., Mahr, D. and McCown, B. (1992). Jr. Amer. Soc. Hort. Sci., 117: 174-180.

Serres, R.A., Mccown, B.H., Zeldin, E., Mccabe, D.E. and Stang, E.J. (1990). Abst. 23rd Int. Hort. Congress, Frienze, Italy, No. 1273.

Shawa, A.Y. (1933). In 33rd Ann. Weed Conf. USA. Washington Weed Assoc., pp.57-58.

Shawa, A.Y. (1979). Bull. College of Agri. Res. Centre, Wash. St. Univ., No. 0880, p. 5

Shawa, A.Y. (1982). HortSci., 17: 949-50.

Shawa, A.Y., Doughty, C.C. and Johson, F. (1965). Proc. Amer. Soc. Hort, Sci., 89: 255-58.

Shawa, A.Y., Eaton, G.W. and Bowan, P.A. (1981). Jr. Amer. Soc. Hort. Sci., 106: 474-77.

Shoemaker, J.S. (1978). Small Fruit Culture. The AVI Publ. Co. Westport, Connecticut, pp. 316-50.

Shultz, J.H. (1944). Ph.D. dissertation, Washington St. University.

Sidorovich, E.A., Rupasova, Zh.A., Zubkova, G.P., Ignatenko, V.A. and Rudakovskaya, R.N. (1987a). Sel' skokhozyaistvennaya Biologiya, 4: 28-33.

Sidorovich, E.A., Rupasowa, Zh.A., Zubkova, G.P. and Ignatenko, V.A. (1987b). Khimiya v Sel'skom Khozyaistve, 25: 41-44.

Sidorovich, E.A., Rupasova, Zh.A., Zubkova, G.P., Iznatenko, V.A. and Afanaskina. J.P. (1988). Rastitel'nye Resursy, 24: 54-60.

Somogyi, L.F., Childers, N.F. and Eck, P. (1964). Proc. Amer. Soc. Hort. Sci., 84: 280-88.

Stang, E.J. and Birrenkott, B.A. (1986). Abst. 22td. Int. Hort. Congress, California, USA, No. 1072.

Stark, R., Hall, I.V., Forsyth, F.R. and Dean, P.R. (1969). Cranberries, 37(6): 14,16.

Stewart, J.D. and Marucci, P.E. (1970). N.J. Agri. Exp. Stn. Circ., 588-A, 4.

Stiles, C.M. and Oudemans, P.V. (1999). Phytopathol., 89: 1127-30.

Strik, B.C. and Poole, A. (1992). HortSci., 26: 1327.

Strik, B.C. and Poole, A. (1995). HortSci., 30:47-49.

Szpunar, J.W. (1985). Acta Hort., 165: 333-36.

Szwonek, E., Kozinski, B., Smolarz, K., Bryk, H., Sas-Paszt, L., Derkowaska, E. and Easterbrooks, E. (2016). Acta Hort., 1117: 157-61.

Vander Kloet, S.P. (1983a). The Taxonomy of Vaccinium and Oxycoccus. Rhodora, 85: 1-43.

Vander Kloet, S.P. (1983b). Canad. J. Bot., 61: 256-66.

Vander Kloet, S.P. (1988). Res. Branch Agric. Can. Publ. 1828.

Vanderleest, C.P., Caron, J. and Bland, W. (2017). Canad. J. Soil. Sci., 97: 11-19.

Vvedenskaya, I.O. and Vorsa, N. (2004). Pl. Sci., 167: 1043-54.

Wang, C.Y. and Wang, S.Y. (2009). Acta Hort., 810: 853-61.

Wang, P.L., Du, C.T. and Francis, F.J. (1973). Jr. Fd. Sci., 43 1402-04.

Watt, B.K. and Merrill, A.L. (1963). Composition of foods. USDA, Agr. Handbook, 8.

Welker, W.W. and Vass, G.D. (1983). HortSci., 18 : 722-23.

Wilcox, R.B. (1940). Proc.70th Ann. Winter Meet. Cranberry Growers Assoc., pp. 14-22.

Workmaster, B.A. and Palta, J.P. (2006). Jr. Amer. Soc. Hort. Sci., 131: 327-37.

XiuHua, C., Grant, L.A. and Caruso, F. (1999). Proc. Fla. Sta. Hort. Soc., 112: 116-17.

YiFei, W., Johnson,-Cicalese, J., Singh, A.P. and Vorsa, N. (2017). Pl. Sci., 262: 91-102.

Zeldin, E.L., Jury, T.P., Serres, R.A. and McCown, B.H. (2002). Jr. Amer. Soc, Hort, Sci., 127: 502-507.

# Index

## C

## D



## E



## F



## G

## *H*



## *I*



## *J*



## *K*



## *L*



## *M*

## Q



## R



## S

## X



## Y

Almond Cultivar Ne Plus Ultra p. 563

Almond Cultivar Nonpareil p. 563

Verticillium Wilt p. 563

Almond Flower p. 564

Almond tree in bearing p. 564

Chestnut Blight p.611

Chinese Chestnut p.611

Chestnut Orchard p.611

Male Flower Stock p. 612

Bacterial Blight of Hazelnut p. 649

Catkins p. 649

Hazelnut Orchard p. 649

Hazelnut Cultivar Barcelona p. 650

Pecan Cultivar Desirable p.699

Downy Spot Symptom p.699

Pecan Cultivar Western p.699

Pecan Orchard p.699

Gummosis in Pistachio p. 761

Male Inflorescence of Pistachio p. 761

Pistachio Tree p.762

Pistachio cultivar Kerman p.762

Walnuit Cultivar Hartley p. 805

Anthracnose Disease p.805

Walnut in Bearing p.806

Male Flower Inflorescence p.806

Strawberry Cultivar Cambridge Vigour p. 867

Gray Mould p. 867

Strawberry Cultivar Honeoye p. 867

Verticillium Wilt p. 867

Black Currant Bush p. 967

Strawberry Cultivar Camarosa p. 867

Matted Row System p. 867

Strawberry with Mulching p. 967

Red Currant Bunch p. 967

Currant Plantation p. 967

Black Currant Cultivar Ben Nevis p. 968

Red Currant Bush p. 968

Gooseberry cultivar Invicta p.1017

Gooseberry cultivar Krusbar p.1017

European gooseberry p. 1018

Powdery Mildew symptom in gooseberry p. 1018

Raspberry Plantations p. 1047

Yellow Rust disease of raspberry p. 1048

Anthracnose disease symptom p. 1111

Blueberry cultivar Tifblue (Rabbiteye) p. 1111

Raspberry cultivar Malling Admiral p.1047

Raspberry cultivar Meeker p. 1048

Blueberry plantation p. 1111

Blueberry cultivar Jersey p. 1112

Blueberry plant showing symtoms of Shock Virus p.1112

Mummy Berry Symptoms p. 1112

Cranberry cultivar Early Black p. 1173

Cranberry plantation p.1173

Ripe Rot Caused by *Coleophoma empetri p. 1174*

Cranberry cultivar Stevens *p. 1174*

www.ingramcontent.com/pod-product-compliance
Lightning Source LLC
Chambersburg PA
CBHW071232230926
51849CB00001B/80
* 9 7 8 9 3 9 0 3 7 1 2 2 8 *